MW00453208

Engineering Computations and Modeling in MATLAB®/Simulink®

Engineering Computations and Modeling in MATLAB®/Simulink®

Oleg A. Yakimenko

Naval Postgraduate School
Monterey, California

EDUCATION SERIES

Joseph A. Schetz
Editor-in-Chief
Virginia Polytechnic Institute and State University
Blacksburg, Virginia

Published by the
AMERICAN INSTITUTE OF AERONAUTICS AND ASTRONAUTICS, INC.
1801 ALEXANDER BELL DRIVE, RESTON, VIRGINIA 20191-4344

American Institute of Aeronautics and Astronautics, Inc., Reston, Virginia

1 2 3 4 5

Library of Congress Cataloging-in-Publication Data
Yakimenko, Oleg A.
    Engineering computations and modeling in MATLAB/Simulink / Oleg A.
Yakimenko.
        p. cm.   (AIAA education Series)
        ISBN 978-1-60086-781-1
    1. Engineering mathematics—Data processing.   2. Computer
programming.   3. MATLAB.   4. SIMULINK.   I. Title.
        TA345.Y35 2011
        620.00285—dc22

                                                                    2010052540

MATLAB® and Simulink® are registered trademarks of the MathWorks, Inc., 3 Apple
Hill Drive, Natick, MA 01760-2098, USA, www.mathworks.com.

To Ella and Anatoliy

TABLE OF CONTENTS

LIST OF COLOR PLATES

PREFACE

This book is a resource for teaching an introductory course in programming and digital computations as a part of aeronautical and astronautical engineering curricula, and is based on the course currently taught at the Naval Postgraduate School in Monterey, CA. This course has been taught in different formats (one quarter, one semester, or even two semesters) to undergraduate and graduate students for over 20 years. The two major parts of the course are the basics of a high-level programming language, and using this language in engineering computations, that is, writing efficient code for applied engineering problems including an introduction to numerical methods. The purpose of including both parts is that if you concentrate only on the basics of the language, students will lack experience in using the language in practical applications, and therefore will forget it quickly. Hence, when they are introduced to numerical methods later, they will need to review programming techniques once again. It is the same concept as in learning a foreign language—you have to use it, that is, read, write, and most importantly speak. Otherwise, it will fade away in no time.

In older days, the programming portion of the course was based on FORTRAN, Pascal and BASIC, but for the last 10 years it has been based on MathWorks' MATLAB and Simulink. MATLAB is a unique programming language and development environment that features many advantages over other programming languages. It is very easy to use, constantly enhances its plotting capabilities, utilizes the latest trends in computer architecture, allows for working together with other languages (C/C++, Java), and enables real-time implementation for data acquisition, image processing, and control of autonomous vehicles and robots, that is, applications that many engineering students need for their graduate work. These factors have made MATLAB/Simulink quite popular in engineering universities worldwide. In the case of MATLAB, the aforementioned reason for introducing students to a high-level programming language and numerical methods simultaneously is even stronger because MATLAB represents a natural engineering environment that has a variety of the best numerical methods programmed effectively in its functions and toolboxes (collection of functions). Above all, the open-source paradigm allows you to look at the source code of a majority of these functions and, knowing the method coded within, to explore the code from the standpoint of programming constructions.

The goal of this book is to teach students how to use the MathWorks' products, MATLAB and Simulink, in applied engineering computations and effective modeling, by progressing from the basics of programming to writing complex scripts fully utilizing specific features of MATLAB. Therefore, the intended audience of this book is not only undergraduate and graduate students, who are using or intend to use MATLAB for the first time and have little or no experience in computer programming but also working professionals, who have some experience in using other programming languages and wish to learn quickly about the major features of the MATLAB/Simulink development environment. This book can also serve as a reference for professionals who use MATLAB regularly.

The content of the book is subdivided into four Parts as shown in Fig. 0.1. Basics of programming in MATLAB are considered in Part I. This Part introduces MATLAB/Simulink as a technical computing language, describes its development environment and basic operations, and emphasizes its uniqueness in handling arrays and structures. The Part then proceeds with teaching how to write programs in MATLAB, by progressing from single scripts and logical constructions to different types of functions and writing foolproof professional scripts. Special attention is dedicated to effective usage of MATLAB's plotting capabilities. This includes not only numerous 2-D and 3-D plotting functions but also the so-called handle graphics that allows a user to alter any attribute of a plot programmatically, which in turn enables creating animations.

Part II introduces one of the MATLAB's toolboxes, the Symbolic Math Toolbox. The reason for introducing this toolbox so early is that its functions can be used for developing and easy plotting analytical formulas for numerical methods considered in the following parts. A single chapter of this Part starts from basic symbolic computations, progresses through the introduction of a relatively new core engine of symbolic computations, MuPAD, and ends with showing how to use symbolic expressions in numerical computations.

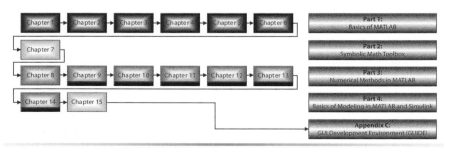

Content of the book subdivided into parts.

Part III is devoted to an overview of basic numerical methods coded in major MATLAB functions. It starts by discussing the accuracy of digital computations and then follows the same pattern of introducing the basic principles behind a numerical method, using the Symbolic Math Toolbox to derive analytical formulas, showing advantages and pitfalls of different approaches, and finally discussing the MATLAB functions realizing these methods (represented by the dark horizontal bar at the bottom of Chapters 8–14 in Fig. 0.1). The types of numerical computations covered in this manner include numerical linear algebra and eigenvalue problems, root finding and optimization, curve fitting, numerical differentiation and integration.

Numerical modeling in MATLAB and Simulink is discussed in Part IV. First, the initial-value problem is discussed in detail in Chapter 14, with different methods introduced to address it. It follows the same pattern as in Part III, utilizing Symbolic Math Toolbox to derive mathematical expressions for different methods. A variety of MATLAB solvers is introduced and thoroughly discussed. These solvers are also utilized in Simulink, the specialized modeling environment introduced in Chapter 15. Using several simple examples, this chapter covers the basics of modeling continuous and discrete systems in Simulink, as well as interacting with the MATLAB development environment, leaving further exploration to readers.

Each chapter of each Part in this book ends with the three simple problem sets. Depending on the specific application, each instructor can come up with a variety of additional problems for his or her students, but these simple examples are intended as part of the educational process. Thus, Appendix A includes the solutions. The hope is that by working through these solutions, readers will solidify their programming skills.

Appendix B complements the collection of MATLAB's Optimization Toolbox functions, considered in Chapter 10, with one more, very effective function, written and widely used by the author. This function realizes the Hooke-Jeeves method that searches for the minimum of a nonlinear function of multiple arguments without calculating its derivatives, and is given here as an example of professionally looking code, which mimics inputs and outputs of any other MATLAB optimization function.

Another wonderful MATLAB tool, which is included in the basic version of MATLAB—the Graphical User Interface Development environment (GUIDE)—is introduced in Appendix C. Readers who are interested in further improving their programming skills will find this part of the book very interesting. It takes a little bit more effort, but allows interested users to enhance their software with good-looking GUIs, simplifying input/output operations and making it easier to share code with others who may not necessarily want to understand how the code works, but rather intend to use it with no hassle.

Speaking of sharing, it is important to note that MATLAB allows the conversion of MATLAB software to standalone applications or shared libraries using MATLAB Compiler. Although MATLAB Compiler is not a part of the basic version of MATLAB, but rather one of its toolboxes that needs to be acquired separately, Appendix D introduces it and presents a simple example of converting code, developed in Chapter 2 and enhanced with the GUI in Appendix C to a standalone MATLAB-independent application (executable file).

Beside basic MATLAB and Simulink, MathWorks offers a variety of supplementary products addressing the specific needs of many industries. Surely, this includes the Aerospace industry. For completeness, the tools supporting technical computations and simulations of aerospace systems are presented in Appendix E.

The reference section includes publications that were used in the preparation of this book and a list of other textbooks that may be used to further enhance programming skills or to dive deeper into numerical methods. Of course, the MATLAB Help system was a major source of information about different MATLAB functions and numerical methods programmed in them.

Depending on the duration of a course, an instructor may choose to cover only the material in Chapters 1–6 (basic programming techniques). If time permits, he or she may also present examples of MATLAB functions utilizing different numerical methods without talking about actual algorithms, that is, covering the very last section of each of Chapters 8–14. For a one-semester or two-semester course (depending on students' background), it may be worth dedicating more time to explain the advantages and disadvantages of different numerical methods, that is, working students through the material of Chapters 8–14. The GUIDE tool is usually introduced at the end of the course, if at all, by showing a simple example and leaving further exploration to the students.

To cover the Symbolic Math Toolbox, discussed in Chapter 7, I usually dedicate two hours of lab time, rather than a lecture time, so that the students can follow the material of this chapter and familiarize themselves with this toolbox at their own pace. The same applies to Chapter 15, which introduces Simulink. I conclude a three-hour Simulink lab with a one-hour lecture summarizing basic concepts and approaches of the Simulink modeling environment and providing a guide for further study.

No doubt, it would be impossible to fully describe all the functions of MATLAB in one book. And with two MATLAB upgrades per year, such a book would become obsolete fairly quickly. Being a proficient applied programmer means knowing the basics and a variety of available resources so that, if needed, knowledge in certain areas can be built up quickly, or new topics can be quickly incorporated. In this regard, I hope that this book presents a variety of resources offered by MathWorks, and therefore accomplishes its goal.

To conclude, I would like to thank many of my students for all the why's and how's that allowed me to shape this book. I would also like to personally thank 2010 class students Sean Doherty and Ryan Decker for reviewing the final manuscript and the 2007 class student Melissa Corley for helping me with the last GUI of Appendix C. I am also grateful to my sponsors, who challenge me with complex, real-world problems that require me to constantly perfect my programming skills. Finally, I would like to thank my family—mom and dad, wife Tatiana, and kids, Aljona, Marina, Irina and Eugene—for encouraging me in writing this book and supporting me throughout the entire process. In fact, I promised my kids to write this book a long time ago while they were studying at university. Sorry that I did not deliver on time. Even though you became engineers and proficient programmers using other textbooks, I am sure I can still teach you a trick or two.

Oleg A. Yakimenko
July 2011

CONVENTIONS

To make it easier for you to follow the content, this book utilizes a set of the following conventions.

Variables in symbolic mathematical expressions appear in *italics*, for example, $y = \pi x - 1$. However, if the symbol represents a vector or matrix, rather than a scalar variable, **boldface type** is used, for example, $\mathbf{Ax} = \mathbf{b}$. Sometimes, such an expression may use subscripts to explicitly define the size of variables (number of rows and columns), for example, $\mathbf{A}_{n \times m} \mathbf{x}_{m \times 1} = \mathbf{b}_{n \times 1}$.

The newly introduced terms will appear in *italics*, whereas **boldface type** is also used to represent the name of a screen menu or an item that appears in such a menu (for example, **File** menu or **Save** option). The names of the keys on a computer keyboard are enclosed in less than and greater than signs, for example, <Enter>.

The `typewriter font` represents MATLAB commands, that is, any text you type in the Command Window, and any MATLAB response that appears on the screen. Moreover, to distinguish the actual command you enter from a MATLAB response, the command appears in **`typewriter bold font`**. For example, the following represents the command you enter (starting with a screen prompt >>) and the response you get:

```
>> z=6*5/2+1
z = 16
```

(It is assumed that you press the <Enter> key after you type a command, so this action is not shown with a separate symbol.) The larger fragments of MATLAB programs, scripts and functions, appear in `typewriter font` as well. Usually, they result in a plot and/or the output is suppressed anyway, so there is no need to use a bold font to distinguish them. For example,

```
x1=0:5; x2=(b-A(1)*x1)/A(2);    % defines the equation
plot(x1,x2), hold
xs=A\b;                         % defines a particular solution
plot(xs(1),xs(2),'rp')
xlabel('x_1'), ylabel('x_2'), axis equal
s=rref([A b]); xh(2)=1; xh(1)=-s(2)*xh(2); xp(2)=0; xp(1)=s(3);
% Displaying the set of LAEs, general and particular solutions
```

```
syms r, x=xp'+r*xh'; H=latex(x);
text(3.5,1,['$$' H '$$'],'interpreter','latex','fontsize',14)
text(xs(1)+0.3,xs(2),'Particular solution provided by A\b')
```

To highlight different constructions of code, MATLAB uses different colors. By default, green is used for the comments (text following the percent sign), and magenta—for the text strings (text embraced by apostrophes). In order to maintain this pattern even with a limited capability, this textbook employs gray and light blue colors for these two constructions (as seen in the above fragment).

Finally, the names of the files and directories appear in *italic Courier*, for example, *myfile.m* or ...*\MATLAB\R2011a\rtw*).

LIST OF ACRONYMS

2-D	two-dimensional
3-D	three-dimensional
ANSI	American National Standards Institute
BDF	backward differentiation formula
BVP	boundary-value problem
CAN	controller-area network
CG	center of gravity
COM	Microsoft component object model
CPU	central processing unit
CTF	component technology file
DATCOM	U.S. Air Force Digital Data Compendium
DDE	differential equation with delay
DFT	discrete Fourier transform
DoF	degree of freedom
FFT	fast Fourier transform
FPU	floating point unit
FSAL	first step as last step (principle)
GARCH	generalized autoregressive conditional heteroskedasticity
GNC	guidance, navigation and control
GPS	Global Positioning System
GUI	graphical user interface
GUIDE	graphical user interface development environment
HDL	hardware description language
HOT	the higher-order terms
HSB	hue-saturation-brightness
HSV	hue-saturation-value
HXML	hypertext markup language
IC	initial condition
IEEE	Institute of Electrical and Electronics Engineers
INS	inertial navigation system
IVP	initial-value problem
LAE	linear algebraic equation
LMI	linear matrix inequality
LSB	the least significant bit

MATLAB	matrix laboratory
MCR	MATLAB Compiler Runtime
MEX	MATLAB external interface
MSB	the most significant bit
NDF	numerical differentiation formulas
ODE	ordinary differential equation
OPC	object linking and embedding for process control
PC	personal computer
PDE	partial differential equation
PID	proportional-integral-derivative (controller)
RF	radio-frequency
RGB	red-green-blue
RK	Runge–Kutta (method)
SVD	singular value decomposition
uicontrol	user interface control
VPA	variable-precision arithmetic
VRML	virtual reality modeling language
WGS84	1984 World Geodetic System
XML	extensible markup language

Chapter 1

MATLAB/Simulink as a Technical Computing Language

- History of MATLAB Development
- Overview of MATLAB Family
- Overview of Simulink Family
- Student (Basic) Version of MATLAB and Simulink

MATLAB Toolboxes,
Simulink Blocksets

1.1 Introduction

Originally developed as a MATrix LABoratory almost three decades ago, today MATLAB integrates computation, visualization, and programming in a powerful full-featured, self-contained, easy-to-use, technical computing environment where problems and solutions are expressed in familiar mathematical notation. Typical uses for MATLAB include math and computation; algorithm development; modeling, simulation, and prototyping; data analysis, exploration, and visualization; scientific and engineering graphics; and graphical user interface (GUI) building and application development. The continuous enhancement of MATLAB makes it an ideal vehicle for exploring problems in all disciplines that manipulate mathematical content. This ability is multiplied by application-specific solutions using toolboxes, application development tools, and other products. The aim of this chapter is to provide an overview and interrelation of MathWorks' products and, by exploring the history of MATLAB development and its enhanced capabilities, present the MATLAB/Simulink product as a perfect tool for both introductory and enhanced programming involving engineering computations.

1.2 History of MATLAB Creation and Development

Officially, The MathWorks, Inc. was founded on 7 December 1984 in California by Jack Little, Cleve Moler, and Steve Bangert, but the basic concepts behind MATLAB were invented much earlier.

According to Moler, who maintains the Cleve's Corner on the company's Web site, www.mathworks.com (Fig. 1.1), which was one of the first 75 registered commercial Web sites, MATLAB's history started in 1960s when he wrote one of his first codes for solving the system of linear algebraic equations. The code was written in FORTRAN and used punched cards. When working on finite difference methods in 1967, Forsythe and Moler published a textbook containing working code in ALGOL, FORTRAN, and PL/I, which became an iconic book for many programmers. Later on, these algorithms, including those for matrix eigenvalue computation, were translated into FORTRAN to produce EISPACK, a collection of FORTRAN subroutines that compute the eigenvalues and eigenvectors for nine classes of matrices: complex general, complex Hermitian, real general, real symmetric, real symmetric banded, real symmetric tridiagonal, special real tridiagonal, generalized real, and generalized real symmetric matrices. In addition, it included two routines that used a singular value decomposition to solve certain least-squares problems. This was followed by LINPACK, a package of FORTRAN subroutines that analyzed and solved linear

Fig. 1.1 Home page of MathWorks' Web site.

Table 1.1   Functions Available in the First Version of FORTRAN MATLAB

ABS	ANS	ATAN	BASE	CHAR	CHOL	CHOP	CLEA	COND	CONJ	COS
DET	DIAG	DIAR	DISP	EDIT	EIG	ELSE	END	EPS	EXEC	EXIT
EXP	EYE	FILE	FLOP	FLPS	FOR	FUN	HESS	HILB	IF	IMAG
INV	KRON	LINE	LOAD	LOG	LONG	LU	MACR	MAGI	NORM	ONES
ORTH	PINV	PLOT	POLY	PRIN	PROD	QR	RAND	RANK	RCON	RAT
REAL	RETU	RREF	ROOT	ROUN	SAVE	SCHU	SHOR	SEMI	SIN	SIZE
SQRT	STOP	SUM	SVD	TRIL	TRIU	USER	WHAT	WHIL	WHO	WHY

equations and linear least-squares problems. This package was specifically designed for supercomputers used in the 1970s and early 1980s. Later on, it was superseded by LAPACK, which was written in FORTRAN 90 and designed to run efficiently on shared-memory computers.

During late 1970s, following the desire to allow users to use two packages, EISPACK and LINPACK, without writing FORTRAN programs, Moler used portions of these packages to develop the first version of MATLAB. The only data type was "matrix," and the HELP command listed all of the available functions with their names abbreviated (Table 1.1). There were no M-files or toolboxes, and the plots were very primitive, made by printing asterisks on the teletypes and typewriters that served as terminals. If you wanted to add a function, you had to modify FORTRAN source code and recompile the entire program. Even so, all the functionalities of future MATLAB versions were present.

It is indicative that Stanford University students, who used this initial version, split into two groups. Those from math and computer science departments were not impressed, because it was based on FORTRAN, not a particular powerful programming language, and it did not represent the current research work in numerical analysis (C and C++ on the one hand and Lisp and PROLOG on the another hand seemed to be much more innovative). However, engineering students liked the concept of MATLAB for its ability to handle matrices. Upon graduation, these students extended FORTRAN MATLAB to have more capability in control analysis and signal processing and, in the early 1980s, offered the resulting software as a commercial product.

At about the same time, in August 1981, IBM announced their first PC. Jack Little, the principal developer of the commercial FORTRAN MATLAB, suggested making MATLAB a matrix-based programming language, to which one could easily add new functions organized into toolboxes. So he and his colleague, Steve Bangert, reprogrammed

MATLAB in C, wrote the parser/interpreter, added M-files, the first few toolboxes, and more powerful graphics. It constituted the initial version of MATLAB, and some of the original code, proven by time, is still used today.

This version, MATLAB 1, implemented in C for MS-DOS PCs, was released by the newly established company, MathWorks, and presented at the 23rd IEEE Conference on Decision and Control in Las Vegas, Nevada, 12–14 December 1984.

Since then, MATLAB has been developed in parallel to PC. MATLAB 2, released in 1986, was enhanced with UNIX support. In 1990, the third version of MATLAB was enriched with Simulink, software for modeling, simulating, and analyzing dynamic systems. In 1992, 2-D and 3-D color graphics were added to MATLAB 4. The MATLAB Student Edition was also released at that time. Since 1993, MATLAB runs in Microsoft Windows; since 1995, it supports Linux. MATLAB 5 included a compiler, allowing conversion of MATLAB programs to C source code, featured data types, advanced visualization, a debugger, a profiler, and a GUI builder. MATLAB 6 featured a new desktop and provided Macintosh support. MATLAB 7 enabled single-precision and integer math, nested and anonymous functions, a plot editor, and interactive algorithm development. It also featured the Distributed Computing Toolbox (later renamed as Parallel Computing Toolbox).

Since 2006, MathWorks has released two updates a year: in winter (release A) and in fall (release B) adjusting the development environment, improving effectiveness and robustness of existing code, piloting new concepts, and fixing bugs, including those reported by the users of MATLAB and Simulink. You may learn about the new features and changes introduced in the latest versions of MATLAB in the following page of MathWorks' Web site: www.mathworks.com/help/doc-archives. html.

This textbook is based on the R2011a release of MATLAB/Simulink, but a vast majority of the functions and code samples presented in this book will also run in the previous versions as well.

1.3 Capabilities and Resources

As discussed in the previous section, MATLAB was designed for engineers and is very easy to use and interpret. Along with the basic core of functions, MATLAB features a variety of special-use libraries and toolboxes enabling the most efficient usage of modern numerical methods, something that, say, FORTRAN or C++ lacks. In general, by design, MATLAB suffers few disadvantages of other programming languages. Its code is compact, yet it is learning intuitive. It avoids the standard separate compilation and

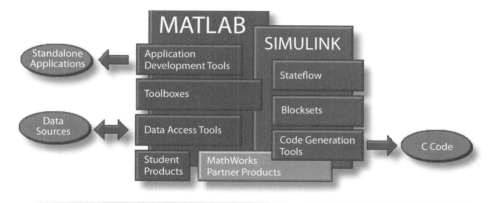

Fig. 1.2   MathWorks' MATLAB and Simulink family.

link sequence by using an interpreter. To make problems work it requires only a few lines of code compared with traditional languages. Because of interaction with general purpose code instead of specific computation, MATLAB is slower in terms of computational speed, but the usage of modern, fast computers mitigates this issue. The open source paradigm, unlike programming from the ground up, allows standard MATLAB pieces of code to be threaded together to translate into a new algorithm. The interactive features and built-in support of MATLAB are used to translate algorithms into functioning code in a fraction of the time needed in other languages. Another important feature is code portability: the code written for the PC version of MATLAB does not have to be changed for Macintosh or UNIX systems (of course, this does not hold if code relies on system-dependent resources).

With MATLAB as the foundation, MathWorks offers a variety of other products building on it. Figure 1.2 presents a general overview of these products with a more detailed discussion to follow in the following section. It is important that MATLAB allows you to convert its programs to C code. Reciprocally, it allows C, C++, and FORTRAN users to use their programs in MATLAB by compiling the so-called MATLAB external interface (MEX)-functions from the source code written in these languages. Also, MATLAB and Simulink allow you to get data from other software or external devices and compile standalone executable code.

To summarize, the family of MathWorks' MATLAB/Simulink products

- Presents a natural language for scientist and engineers all over the world
- Supports an open source paradigm
- Features many toolboxes with the best numerical algorithms
- Has powerful visualization features to present the results of calculations and simulations

- Is a dynamic and constantly growing software
- Is well documented and supported [including a very good support via the Web (Fig. 1.1)]
- Allows you to use programs written in other languages
- Converts easily to C code, executable files, and standalone applications
- Supports all operating systems (32-bit and 64-bit versions) and the latest advances in computer hardware (parallel and cluster computations)

For these reasons, MATLAB has become a standard tool for introductory and advanced courses in mathematics, engineering, and science in many universities around the world. In industry, it is a tool of choice for research, development, and analysis. Presenting a piece of code in MATLAB is academically equivalent to presenting a formula or algorithm; for example, the two functions presented in Fig. 1.3 (taken from www.wikipedia.org) represent a complete algorithm of converting a latitude–longitude–altitude triad to the local tangent plane coordinates.

Since the release of R2010b, more than 1200 books on or based on MATLAB and Simulink are available. Among them are several books on

From GPS measurements to ENU measurements: sample code

This code was written in *MATLAB*

Step 1: Convert GPS to ECEF

```matlab
function [X,Y,Z] = llh2xyzTest(lat,long, h)
    % Convert lat, long, height in WGS84 to ECEF X,Y,Z
    %lat and long given in decimal degrees.
    lat = lat/180*pi; %converting to radians
    long = long/180*pi; %converting to radians
    a = 6378137.0; % earth semimajor axis in meters
    f = 1/298.257223563; % reciprocal flattening
    e2 = 2*f -f^2; % eccentricity squared

    chi = sqrt(1-e2*(sin(lat)).^2);
    X = (a./chi +h).*cos(lat).*cos(long);
    Y = (a./chi +h).*cos(lat).*sin(long);
    Z = (a*(1-e2)./chi + h).*sin(lat);
```

Step 2: Convert ECEF to ENU

```matlab
function [e,n,u] = xyz2enuTest(Xr, Yr, Zr, X, Y, Z)
    % convert ECEF coordinates to local east, north, up

    phiP = atan2(Zr,sqrt(Xr^2 + Yr^2));
    lambda = atan2(Yr,Xr);

    e = -sin(lambda).*(X-Xr) + cos(lambda).*(Y-Yr);
    n = -sin(phiP).*cos(lambda).*(X-Xr) - sin(phiP).*sin(lambda).*(Y-Yr) + cos(phiP).*(Z-Zr);
    u = cos(phiP).*cos(lambda).*(X-Xr) + cos(phiP).*sin(lambda).*(Y-Yr) + sin(phiP).*(Z-Zr);
```

Fig. 1.3 MATLAB M-scripts.

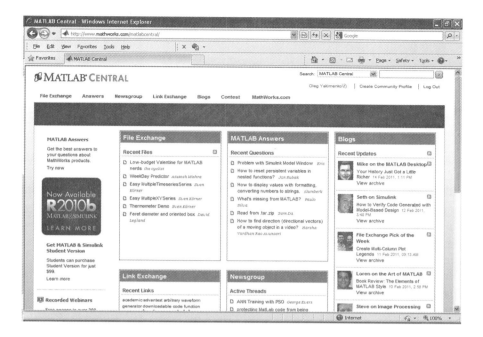

Fig. 1.4 MATLAB user community.

numerical methods that cover the best techniques and algorithms using MATLAB code and application-specific manuscripts concentrating on a specific area, such as classical feedback control, image processing, and financial analysis. The MathWorks' Web site features a user community page, where users all over the world post and discuss their algorithms (Fig. 1.4). It is safe to say that, before writing a function, you may want to check this particular page to see if such a function has been developed already. If so, you may simply download it (thanks to the open source paradigm) and modify to serve your own needs.

1.4 Aerospace Application Tools

The variety of toolboxes and blocksets available to extend MATLAB and Simulink (Fig. 1.2) will be addressed in the following section. Many of these tools can be and are widely used in aerospace industry. However, MATLAB features one toolbox and one blockset that were specifically designed to be used by aerospace scientists and engineers. The tools are mentioned briefly here to emphasize the attractiveness of MathWorks product for specialists of this industry (more details can be found in Appendix E).

The Aerospace Toolbox extends the MATLAB computing environment by providing standards-based environmental models. For instance, it features the 1984 World Geodetic System (WGS84) representation of Earth's gravity, the EGM96 Geopotential Model of Earth, a world magnetic model, the International Standard Atmosphere model, climatic data, and spherical harmonic gravity models of Moon and Mars. You can also use it to import aerodynamic and control coefficients directly from the U.S. Air Force Digital Data Compendium (DATCOM) to carry out preliminary control design and vehicle performance. The toolbox also provides an interface to the FlightGear Simulator (www.flightgear.org) to visualize flight data and vehicle dynamics in a 3-D environment and reconstruct behavioral anomalies in flight-test results. To ensure design consistency, the Aerospace Toolbox software provides utilities for unit conversions, coordinate transformations (such as the ones presented in Fig. 1.3), and quaternion math. Finally, it allows you to compute a variety of gas dynamics functions, including isentropic flow, normal shock, Rayleigh flow, Fanno flow, and Prandtl–Meyer flow.

The Aerospace Blockset expands Simulink with blocks for modeling and simulating three- and six-degree-of-freedom dynamics of manned and unmanned aircraft, spacecraft, and rocket-craft, including simulation of their propulsion systems, control systems, mass properties, and actuators. It also includes standards-based environmental models for atmosphere, gravity, wind, geoid height, and magnetic field based on the corresponding functions of the Aerospace Toolbox. In a similar manner, the Aerospace Blockset allows you to use Aerospace Toolbox utilities for converting units, transforming coordinate systems and spatial representations, and performing common aerospace math operations. Once you developed the model of your specific vehicle and tuned its guidance, navigation, and control algorithms, you can automatically generate code for production deployment and real-time execution in rapid prototyping and hardware-in-the-loop systems using the Real-Time Workshop and xPC Target tools briefly mentioned in Sec. 1.5.2.

1.5 Overview of MathWorks Products

MathWorks' Web site provides a nice interactive overview of its major products at www.mathworks.com/products/pfo (Fig. 1.5). Essentially, it represents the same hierarchy as in Fig. 1.2, but supplies a more detailed list of toolboxes and blocksets available for MATLAB and Simulink product families, respectively, and also lists the major application areas for these toolboxes and/or blocksets. The Aerospace Toolbox and Aerospace Blockset, introduced in the previous section, belong to the Control System Design and Analysis group of products portrayed in the top-right corner of

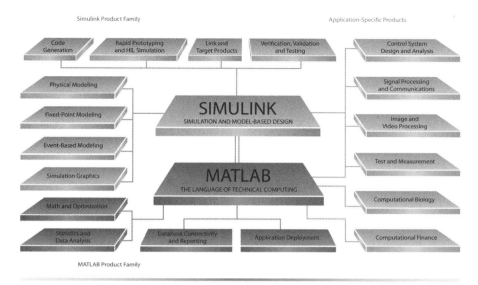

Fig. 1.5   Product families of MathWorks.

Fig. 1.5. By the way, the Control System Toolbox, also belonging to the same group, was one of the first toolboxes featured by the very first version of MATLAB.

Although this book basically describes the two major products of MathWorks, MATLAB and Simulink, for the sake of completeness, the following section briefly overviews other products available in the release R2010b. In release 2011a some of these products are combined into three new System Toolboxes for Design in MATLAB and Simulink:

• DSP System Toolbox (combining features of Signal Processing Blockset and Filter Design Toolbox)
• Communications System Toolbox (combining features of Communications Toolbox and Communications Blockset)
• Computer Vision System Toolbox (incorporating the functionality of Video and Image Processing Blockset and adding new computer vision algorithms)

Three new Code Generation products are also included:

• MATLAB Coder (allowing you to generate portable C/C++ code directly from MATLAB)
• Simulink Coder (combining the functionality of Real-Time Workshop and Stateflow Coder)
• Embedded Coder (combining the functionality of Real-Time Workshop Embedded Coder, Embedded IDE Link, and Target Support Package)

1.5.1 MATLAB Product Family

Being the kernel of all product families of MathWorks, MATLAB is an intuitive programming language and a user-friendly technical computing environment. MATLAB is easy to use and learn, and it provides core mathematics and advanced graphical tools for data analysis, visualization, and algorithm and application development. With almost 1000 mathematical, statistical, and engineering functions, satisfying a majority of engineers' and scientists' basic necessities, there is no need to go beyond the MATLAB environment. *Toolboxes*, collections of the highly optimized, application-specific functions, extend MATLAB functionality even farther. MATLAB toolboxes support applications involving signal and image processing, control system design, optimization, financial engineering, symbolic math, neural networks, and others, as shown in Fig. 1.5. In addition to the basis functions of MATLAB, the toolbox functions are built in the MATLAB language and can be easily viewed and modified, which means that you can access code, the so-called M-scripts, of any function. Tables 1.2–1.10 list available toolboxes by categories, as presented in Fig. 1.5.

Table 1.2   Math and Optimization Toolboxes

Toolbox	Brief Description
Global Optimization Toolbox	To find global solutions to optimization problems that contain multiple maxima or minima using global search, multistart, pattern search, genetic algorithm, and simulated annealing solvers (before R2010a, available as Genetic Algorithm and Direct Search Toolbox)
Optimization Toolbox	To solve standard and large-scale optimization problems
Partial Differential Equation Toolbox	To solve and analyze partial differential equations
Symbolic Math Toolbox	To perform computations using symbolic mathematics and variable-precision arithmetic (after R2008b includes functionality of the former Extended Symbolic Math Toolbox and is based on MuPAD engine)

Table 1.3   Statistics and Data Analysis Toolboxes

Toolbox	Brief Description
Curve Fitting Toolbox	To perform model fitting and analysis
Model-Based Calibration Toolbox	To calibrate complex powertrain systems
Neural Network Toolbox	To design and simulate neural networks
Spline Toolbox	To create and manipulate spline approximation models of data
Statistics Toolbox	To apply statistical algorithms and probability models

Table 1.4   Control System Design and Analysis Toolboxes

Toolbox	Brief Description
Aerospace Toolbox	To provide aerospace reference standards, environmental models, and allow aerodynamic coefficients importing
Control System Toolbox	To design and analyze controllers for dynamic closed-loop systems
Fuzzy Logic Toolbox	To design and simulate fuzzy logic systems
Model Predictive Control Toolbox	To control large, multivariable processes in the presence of constraints
Robust Control Toolbox	To design robust multivariable feedback control systems (includes linear matrix inequatlity (LMI) Control Toolbox and µ-Analysis and Synthesis Toolbox featured in the earlier versions)
System Identification Toolbox	To create linear dynamic models from measured input-output data

Table 1.5   Signal Processing and Communications Toolboxes

Toolbox	Brief Description
Communications Toolbox	To design and analyze communications algorithms
Fixed-Point Toolbox	To design and analyze fixed-point, adaptive, and multirate filters
Fixed-Point Toolbox	to design and execute fixed-point algorithms and analyze fixed-point data
Filter Design Hardware Description Language (HDL) Coder	To generate HDL code for fixed-point filters
Radio-Frequency (RF) Toolbox	To design, model, and analyze networks of RF components
Signal Processing Toolbox	To perform signal processing, analysis, and algorithm development
Wavelet Toolbox	To analyze and synthesize signals and images using wavelet techniques

Table 1.6   Image and Video Processing Toolboxes

Toolbox	Brief Description
Image Processing Toolbox	To perform image processing, analysis, and algorithm development
Image Acquisition Toolbox	To acquire images and video from industry-standard hardware
Mapping Toolbox	To analyze and visualize geographic information

Table 1.7 Test and Measurement Toolboxes

Toolbox	Brief Description
Data Acquisition Toolbox	To acquire and send out data from plug-in data acquisition boards
Instrument Control Toolbox	To control and communicate with test and measurement instruments
Object Linking and Embedding for Process Control (OPC) Toolbox	To read, write, and log data from OPC servers
System Test	To manage tests and analyze results for system verification and validation
Vehicle Network Toolbox	To send and receive controller-area network (CAN) packets directly from MATLAB or Simulink, allowing you to encode, decode, and filter them

Table 1.8 Computational Biology Toolboxes

Toolbox	Brief Description
Bioinformatics Toolbox	To read, analyze, and visualize genomic, proteomic, and microarray data
SimBiology	To model, simulate, and analyze biochemical pathways

Table 1.9 Computational Finance Toolboxes

Toolbox	Brief Description
Datafeed Toolbox	To acquire financial data from data service providers
Econometrics Toolbox	To select, calibrate, and analyze economic models for use in simulation and forecasting [before R2008b, known as generalized autoregressive conditional heteroskedasticity (GARCH) Toolbox]
Financial Derivatives Toolbox	To model and analyze equity and fixed-income derivatives
Financial Toolbox	To analyze financial data and develop financial algorithms
Fixed-Income Toolbox	To model and analyze fixed-income securities

Table 1.10 Database Connectivity and Reporting Toolboxes

Toolbox	Brief Description
Database Toolbox	To exchange data with relational databases
MATLAB Report Generator	To generate documentation for MATLAB applications and data

A collection of different data acquisition and access toolboxes (Tables 1.6-1.10) provide straightforward access to data from external devices and other software packages. These toolboxes allow you to easily get your data into MATLAB for analysis and usage in the models.

MathWorks' *application deployment tools* (Table 1.11) allow you to build standalone executables and software components from MATLAB code and share these standalone versions of your MATLAB applications with colleagues who use other environments. Specifically, MATLAB Compiler lets you to share your MATLAB application as an executable or a shared library (see Appendix D). Another useful tool, Spreadsheet Link EX, allows exchange of data between MATLAB and Excel, and thereby taking advantage of the familiar Excel environment while accessing the computational speed and visualization capabilities of MATLAB. Finally, there is an option of deploying MATLAB applications via the Web. Before release of 2006b, it was the MATLAB Web Server that would allow deploying. Lately, the MATLAB Web Server was substituted with two other products, MATLAB Builder NE to create a server-side .NET component and MATLAB Builder JA to create a server-side Java component.

To address architecture variety of the current computer systems, MathWorks provides two additional toolboxes, as shown in Table 1.12.

Table 1.11   Application Deployment Toolboxes

Toolbox	Brief Description
MATLAB Compiler	To build standalone executables and software components from MATLAB code
Spreadsheet Link EX	To allow using MATLAB from Microsoft Excel (also known as Excel Link in the earlier versions, before R2008a)
MATLAB Builder EX	To deploy MATLAB code as Microsoft Excel add-ins
MATLAB Builder NE	To deploy MATLAB code as .NET and Microsoft component object model (COM) components for Microsoft .NET Framework
MATLAB Builder JA	To deploy MATLAB code as Java classes for Java language

Table 1.12   Parallel and Distributed Computing Toolboxes

Toolbox	Brief Description
Parallel Computing Toolbox	Allows performing parallel computations on multicore computers and computer clusters (also known as Distributed Computing Toolbox in the earlier versions, before R2008a)
MATLAB Distributed Computing Server	Enables MATLAB and Simulink computations on computer clusters and server farms

1.5.2 Simulink Product Family

MathWorks' *Simulink*, introduced in the third version of MATLAB, is a simulation and prototyping environment providing a block diagram interface that is built on the core MATLAB numeric, graphics, and programming functionality. Using a drag-and-drop paradigm and an extensive block library, you can build a graphical block diagram of any dynamic system. Once the system is diagrammed, you can simulate and analyze its behavior.

Blocksets are collections of application-specific blocks that support multiple design areas, including electrical power-system modeling, digital signal processing, fixed-point algorithm development, and some others, as shown in Fig. 1.5 and Tables 1.13–1.18. These blocks can be incorporated directly into your Simulink models.

Table 1.13 Physical Modeling Blocksets

Blockset	Brief Description
Simscape	To model and simulate multidomain physical systems
SimMechanics	To model and simulate mechanical systems
SimPowerSystems	To model and simulate electrical power systems
SimDriveline	To model and simulate mechanical driveline systems
SimHydraulics	To model and simulate hydraulic systems
SimElectronics	To model and simulate electronic and electromechanical systems

Table 1.14 Simulation Graphics Blocksets

Blockset	Brief Description
Simulink 3-D Animation	To visualize and animate Simulink models in 3-D virtual environment using Virtual Reality Modeling Language (VRML) (before R2009a, also known as Virtual Reality Toolbox)
Gauges Blockset	To monitor signals with graphical instruments

Table 1.15 Control System Design and Analysis Blocksets

Blockset	Brief Description
Aerospace Blockset	To model and simulate aircraft, spacecraft, and propulsion systems
Simulink Control Design	To design and analyze control systems in Simulink

Blockset	Brief Description
Simulink Design Optimization	To improve designs by estimating and tuning model parameters using numerical optimization (available since R2009a and combines functionality of the former Simulink Parameter Estimation and Simulink Response Optimization blocksets)

Table 1.16 Signal Processing and Communications Blocksets

Blockset	Brief Description
Communications Blockset	To design and simulate the physical layer of communication systems and components
RF Blockset	To design and simulate the behavior of RF systems and components in a wireless system
Signal Processing Blockset[a]	To design and simulate signal processing systems and devices
Video and Image Processing Blockset[a]	To design and simulate video and image processing systems

[a]The latest versions of MATLAB allow you to use this blockset even outside Simulink.

Table 1.17 Verification, Validation, and Testing Blocksets

Blockset	Brief Description
EDA Simulator Link	To provide a bidirectional link between MATLAB/Simulink and HDL simulators using cosimulation interface
Embedded IDE Link	To generate, build, test, and optimize embedded code by connecting MATLAB/Simulink with embedded software development environment
Simulink Design Verifier	To generate tests and prove model properties using formal methods
Simulink Verification and Validation	To trace requirements, enforce modeling standards, and measure model coverage
SystemTest	To manage tests and analyze results for system verification and validation

Table 1.18 Fixed-Point and Report-Generating Tools

Blockset	Brief Description
Simulink Fixed-Point	To design and simulate fixed-point systems
Simulink Report Generator	To generate documentation for Simulink and Stateflow models

In addition to aforementioned blocksets, the Simulink product family also features *Stateflow* products (Table 1.19), a graphical simulation environment for modeling and designing event-driven systems. Stateflow provides Simulink users with an elegant solution for designing the control or protocol logic found in embedded systems.

Table 1.19 Stateflow Products

Tools	Brief Description
SimEvents	To model and simulate discrete-event systems
Stateflow	To design and simulate state machines and control logic

Table 1.20 Code-Generation Tools

Tool	Brief Description
Real-Time Workshop	To generate C code from Simulink models and MATLAB code
Real-Time Workshop Embedded Coder	To generate C and C++ code optimized for embedded systems
Simulink HDL Coder	To generate HDL code from Simulink models and MATLAB code
Simulink PLC Coder	To generates hardware-independent structured text in PLCopen extensible markup language (XML) and other file formats
Stateflow Coder	To generate C code from Stateflow charts
Target Support Package	To deploy embedded code onto a specific microcontroller
DO Qualification Kit and IEC Certification Kit	To assist to qualify Simulink and Simulink-derived software to meet certain software standards

Table 1.21 Rapid Prototyping and HIL Simulation Tools

Tool	Brief Description
Real-Time Windows Target	To run Simulink models on a PC in real time
xPC Target	To perform real-time rapid prototyping and hardware-in-the-loop simulation using PC hardware
xPC Target Embedded Option	To deploy real-time applications on PC hardware

The Simulink family also offers a variety of tools for code generation (Tables 1.20 and 1.21). Real-Time Workshop and Stateflow Coder generate customizable C-code directly from your Simulink and Stateflow diagrams for rapid prototyping, hardware-in-the-loop simulations, and desktop rapid simulation. Extension products generate efficient C-code for embedded systems applications.

In addition to aforementioned toolboxes and blocksets, more than 300 third-party hardware and software products are compatible with MATLAB/Simulink. These partner offerings include toolboxes, blocksets, and real-time workshop targets. MathWorks' Web site keeps track of them and allows you to find a specific product or application by industry, function, type, product, or company name (www.mathworks.com/products/connections).

1.5.3 MATLAB and Simulink Student Version

The last, but not least, powerful feature of MathWorks' product families gives special attention to beginner programmers and college and university students, by providing them with the student version of MATLAB and Simulink. This version includes full-featured versions of MATLAB and Simulink (the student version of Simulink enables you to create models that include up to 300 blocks) along with the key functions from

- Control System Toolbox
- Image Processing Toolbox
- Optimization Toolbox
- Signal Processing Blockset
- Signal Processing Toolbox
- Statistics Toolbox
- Symbolic Math Toolbox

It runs on Windows, Mac, and Linux and comes with a complete set of electronic documentation on DVD and two printed manuals, "Getting Started with MATLAB" and "Getting Started with Simulink." The low price of this product encourages students to install it on their own laptops, so that they can enjoy playing with MATLAB and using it for a variety of their engineering assignments. No wonder, then, that MATLAB and Simulink software has become a premier technical computing resource in thousands of universities worldwide.

Chapter 2

Development Environment and Basic Operations

- Desktop Overview, Help System, and Basic Procedures
- Basic Functions and Utilities
- Managing Work Session
- Problem Solving Methodology

plus, minus, mtimes, mpower, sin, cos, tan, atan, exp, log, sqrt, abs, round, fix, pi, i, Inf, NaN, format, disp, clear, close, clc, and many others

2.1 Introduction

This chapter provides a "quick start" introduction to MATLAB as an interactive calculator. It shows how to start MATLAB, make some basic calculations, and quit. The chapter starts with presenting an overall description of the MATLAB development environment, considering Desktop, multiple windows, basics of entering commands in the Control Window, and Editor. It then introduces the extended Help system available in MATLAB/Simulink, followed by presenting a variety of available functions and discussing the order of precedence. It further presents more details on managing the work session and addresses the way to change some of the development environment preferences. The chapter concludes with some commonly accepted methodologies for proficient programming and problem solving.

2.2 MATLAB Development Environment

To start MATLAB on a Windows platform, double-click on the MATLAB icon. You will then see the MATLAB *Desktop* (Fig. 2.1). The Desktop manages the *Command Window*

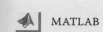 MATLAB

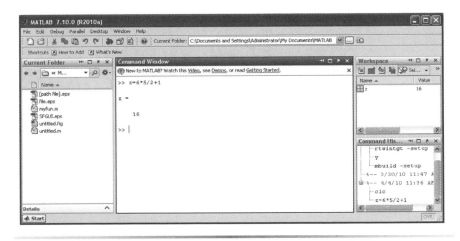

Fig. 2.1 Default desktop of MATLAB.

and *Help Browser*, as well as other tools. Two alternative ways to start MATLAB are

1. Clicking on the **Start** button, selecting **Programs**, choosing MATLAB folder, and selecting (clicking on) the MATLAB XX entry (XX denotes a specific version installed on your computer, e.g., 6.5, 7.0, or R2009b).
2. Using Windows Explorer, opening your top-level MATLAB installation directory (by default, in the *Program Files* folder) and double-clicking on *MATLAB XX.lnk*, a shortcut to the MATLAB executable, residing in the *bin* folder.

2.2.1 Desktop Overview

The default appearance of the MATLAB Desktop is slightly different for different versions of MATLAB, but most likely you will see something similar to Fig. 2.1, which represents the desktop for Releases 2010a and higher.

By default, the MATLAB Desktop (hereafter referred to as Desktop) is composed of four windows. They are the *Current Folder* window (on the left), Command Window (in the middle), *Workspace*, and *Command History* windows (one above another on the right). (In the earlier versions, you may see only three windows: the Command Window on the right, Command History window at the bottom left, and Current Folder and Workspace windows sharing the same space on the upper left.) Across the top of the Desktop you should be able to see the row of drop-down *menus*, the row of icons called the *Toolbar*, and another row, the so-called *Shortcuts Toolbar*, beneath it.

When you click on any of the four windows, its borders become blue, meaning that this window becomes an *active* window. When you type a command you will see a MATLAB response in the Command Window, but to do so, you must first click on this window to activate it.

At the right top corner of each of the four windows, you will see four icons you can click on. You may use these buttons to minimize the active window, maximize it, undock (from Desktop), and close it. When you minimize the window, it minimizes to appear as a tag by the left or right edge of

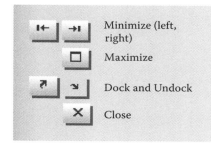

Desktop. If you undock the window, it becomes a regular Windows' window, which you can move around, minimize (to the bottom), maximize, and close. For undocked windows the Undock button changes to the Dock button, allowing you to *dock* this window back to the MATLAB Desktop.

You can resize MATLAB Desktop windows and move them around within Desktop even without undocking them, by simply pulling their edges or clicking on their upper portion and dragging them to the desired location. Therefore, MATLAB allows you to adjust the development environment to meet your personal preferences. If you want to return to the default layout, you should select the **Desktop** menu of MATLAB Desktop, choose the **Desktop Layout** option in the drop-down menu, and then click on the **Default** option in another menu that will extend to the right as shown in Fig. 2.2. You can also use the **Desktop** menu to minimize, maximize, and do all other actions on the active menu; check some other windows you want to appear within the Desktop along with the Command Window and three others. Along with the default desktop layout, you may choose some other predetermined layouts, including **All Tabbed** layout shown in Fig. 2.3.

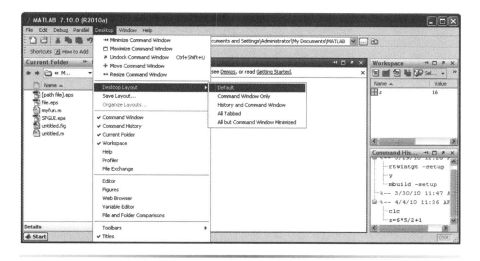

Fig. 2.2   **Desktop** menu of MATLAB Desktop.

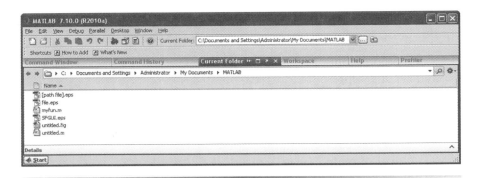

Fig. 2.3 All tabbed Desktop layout.

Next to the **Desktop** menu of the MATLAB Desktop is the **Window** menu. This menu allows you to navigate between all open windows. You can also do it by holding the <Ctrl> key on your keyboard and pressing the <0> key (for the Command window), <1> (for the Command History window), <2> (for the Current Folder window), and <3> (for the Workspace window). Alternatively, you can type in the following commands, `commandwindow`, `commandhistory`, `filebrowser`, and `workspace`, to open the corresponding Desktop window when it is closed or to make it active when it is open.

As you can see, you can achieve the same result either by manipulating Desktop tools (navigating through menus and pushing the icons) or programmatically (by issuing the commands in the Command Window). This is a very powerful and useful feature of the MATLAB development environment that we will use further, especially in Chapter 6.

2.2.2 Entering Expressions in the Command Window

As you can imagine, most of the time, you will communicate with the MATLAB program by typing instructions of various types in the Command Window. As seen in Fig. 2.1, MATLAB displays the prompt (>>) to indicate that it is ready to receive instructions. MATLAB is command-line oriented, (that is, it is an interpretative language), meaning that no compiling is required. Programming errors are flagged by error messages.

To see how simple it is to use MATLAB, try entering a few commands on your computer. Make sure the cursor is at the prompt in the Command Window and type in, for example, the following mathematical expression:

```
>> 3 * pi / 3.141592 \ 3 - 1 + 2 ^ 2
```

Press <Enter> to receive the MATLAB response

```
ans = 4.0000
```

This expressions contains only *scalars*, that is, real numbers that have an infinite decimal representation, including both rational numbers, such as 2 and −1/3, and irrational numbers, such as π and $\sqrt{2}$. However, one of the advantages of MATLAB is that it would handle arrays, groups of numbers, in exactly the same way (to be considered in the next chapter), making code readable and very compact.

If you make a typing mistake, just press the <Enter> key, wait until you get the prompt, and then retype the line. Because MATLAB retains your previous keystrokes in a command file, you can use the up-arrow key <↑> to scroll back through the commands. Press the key once to see the previous entry, twice to see the entry before that, and so on. Use the down-arrow key <↓> to scroll forward through the commands. When you find the line you want, you can edit it using the left- and right-arrow keys (<←> and <→>), <Back Space> key, and <Delete> key. Scrolling back and forth enables you to correct typing mistakes quickly. You can also see all previous keystrokes in the Command History window so that you can drag and drop any line from this window to the Command Window to execute it again. Press the <Enter> key to execute the corrected command or commands you dragged from the Command Window.

Note that to improve the readability of your code, you may use the spaces between operands (as we did it in the abovementioned example). When executing the commands, MATLAB simply ignores them. In the previous example, we used several spaces before and after the plus sign for an educational purpose. It would be a mistake, however, to break constant `pi` or number `3.141592`. In this case, MATLAB would return an error message, specifically pointing at its source, for example,

```
>> 3 * p i / 3.141592 \ 3 − 1 + 2 ^ 2
??? 3 * p i / 3.141592 \ 3 − 1 + 2 ^ 2

Error: Unexpected MATLAB expression.
```

The expression we dealt with contains all arithmetic operations available in MATLAB. They are listed in Table 2.1. Note that, MATLAB has two division operators. A good way to remember the difference between them is to note that the slash slants toward the denominator. For scalar operations, they are basically the same; hence, the right division operator is usually used. The left division operator is useful for solving sets of linear algebraic equations, as will be discussed in Chapter 9.

Now let us type

```
>> pi*1000
```

Table 2.1   Scalar Arithmetic Operations

Symbol	Operation	MATLAB Form	Result
^	Exponentiation	a^b	a^b
*	Multiplication	a*b	ab
/	Right division	a/b	a/b
\	Left division	a\b	b/a
+	Addition	a+b	$a+b$
−	Subtraction	a−b	$a-b$

which after hitting <Enter> yields

```
ans = 3.1416e+003
```

Note that as in the previous example, the result has only four decimal places. Does it mean that it is the precision MATLAB can deliver? Of course not. As will be discussed in detail in Chapter 8, MATLAB uses a fairly high *double precision* for its computations (meaning two four-byte words, eight bytes in total, are allocated for every variable). However, by default, it displays the results using the *short format*, rounding to just four decimal places. If it bothers you, do not worry, there are several ways to change it and they will be considered later in this chapter (Sec. 2.6) as well as in Sec. 4.9. To fit the results into the short format, MATLAB uses exponentiation to a power of 10, using the notation e, thus the result should be treated as 3.1415×10^3.

In both the examples, we did not specify a *variable* name to which the result should have been assigned to. Hence, MATLAB used the default name, ans (abbreviation for answer). You can use the variable ans for further calculations, but remember that it is a temporary variable, containing the most recent answer, which is overwritten every time you do not specify a variable name explicitly. This is illustrated by the following example, when ans changes its value from 30 to 20:

```
>> 6*5
ans = 30
>> 50-ans
ans = 20
```

To avoid this situation, you should use the specific variable names to contain a result of your expression. The valid variable name may contain 1–63 characters (defined by the namelengthmax function of MATLAB). It must start with a letter, but may use letters, digits, and underscore after the first letter (the only expectation is that the variable name cannot be keyword). Try to give your variables the meaningful names, so that anyone who looks at your code understands what it does. Also, be aware of the fact that MATLAB is case-sensitive; therefore, Rate, rate, and RATE represent three different variables.

When typing in expressions with the assignment operator =, MATLAB assigns the value generated on the right to the variable on the left. For example,

```
>> x = 3
```

assigns the value of 3 to x. If followed by

```
>> x = x + 1
```

it further assigns the value of 4 (3+1) to x.

The main advantage of using variables as opposed to their values is that mathematical expressions look better and you can use the same symbolic formula by simply changing the values of the variables it consists of. For instance, to compute a circumference of a circle with a radius of 10 units, it is wise to issue a meaningful statement like

```
>> R = 10; circumference = 2 * pi * R
```

as opposed to

```
>> x = 2 * pi * 10
```

Note that the first command consists of two statements—we are allowed to have several statements on the same line. A semicolon after the first statement simply suppresses the output (see Sec. 2.6). If you want to see the results of both statements, substitute the semicolon with a comma.

2.2.3 Command History Window and Editor

The Command History window (see examples in Figs. 2.1 and 2.2) shows all the previous keystrokes you entered in the Command Window. It is useful for keeping track of what you typed. Moreover, the Command History window keeps track of all sessions you opened MATLAB sorted by date/time, so that it is very easy to find what you are looking for and to collapse the branches you do not need at this time. You may also reexecute any previous command stored in the Command History window by double-clicking on it or by dragging it and dropping at the Command Window as was mentioned earlier. You can delete, cut, copy, evaluate, and save any portion of the previously issued commands. All you have to do is select the commands you need (use the <Ctrl> key to select the nonsequential commands) and click the right mouse button. As a result, a pop-up menu appears (Fig. 2.4a). As far as saving, you have two options: either to create (and save) the selected commands as a separate source code file (called M-file in MATLAB) or create a shortcut.

In the example, illustrated in Fig. 2.4a, we choose to create a shortcut, which brought up the Shortcut Editor window, as shown in Fig. 2.4b. (Dragging the selected commands to the Shortcuts Toolbar opens the Shortcut Editor window too.) The three specific commands you see in there allow

Fig. 2.4   a) Undocked Command History window with options available on a mouse right-click, b) Shortcut Editor window, and c) newly created **Prepare** shortcut on the shortcuts toolbar.

you to close all figure windows (discussed in Chapter 6), clear (delete) all variables from workspace, and return the cursor to the upper-left corner of the Command Window (see Sec. 2.6). You will repeat these three operations quite often so as to avoid typing them over and over again; we can create a shortcut button to appear on the Desktop Shortcuts Toolbar. For instance, if we label the chosen commands "Prepare" (as shown in Fig. 2.4b) and hit the **Save** button, the **Prepare** shortcut appears after **What's New** shortcut, as shown in Fig. 2.4c.

As shown in Fig. 2.4a, another option for working with the selected set of commands is to create a full-fledged M-file. Choosing this option brings up the *Editor* window, as shown in Fig. 2.5.

The MATLAB Editor acts as a basic word processor to write your script files or user-defined functions; that is, it provides graphical user interface

(GUI) for text editing. An alternative way of creating a new or editing an existing M-file is to use the **File** menu on the MATLAB Desktop, and choose the **New** or **Open** option, respectively. Finally, you can open the Editor programmatically by typing in `edit` in the Command Window. Then, you can either start typing in your code to create a new file or open an existing file. You can also copy the selected statements from the Command History, paste them into the opened M-file. Finally, with the statements selected, you can right-click, select the **Create M-File** option from the pop-up menu.

You can also do the same using the two first icons of the MATLAB Desktop

With the Editor window opened, you can use its menus and toolbar icons to access additional resources. The first nine icons in the first of the two toolbars, the Editor Toolbar, allows you to create a new file, open an existing file, save the current file, cut, copy, paste, undo, redo, print, are pretty much standard Microsoft Office icons. The next icon allows you to convert your script to an hypertext markup language (HTML) document, ready to be published on the web (and/or printed). MATLAB also allows publishing in other formats—Word, PowerPoint, and LaTeX. The **Find text** icon, allows searching your program for a specific text, whereas **Back** and **Forward** icons allow scrolling it back and forth.

Typically, you would want to create an M-file and save it, so that you can run it later in the MATLAB Command Window by simply typing its name.

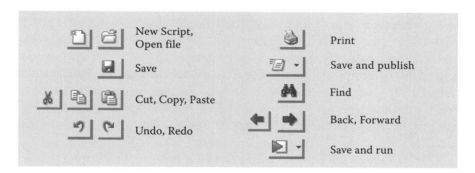

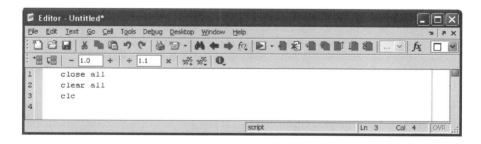

Fig. 2.5 The Editor window.

However, if your newly developed M-file contains some errors, instead of expected results you will see the warning message. Therefore, you will need to go back to the Editor, find and fix the errors, save the corrected version of your M-file, and then run it from the Command Window again. To avoid this back-and-forth procedure, the MATLAB Editor allows you to run and debug the program directly from the Editor window. You can do it by clicking the **Save and run** icon. If you want to estimate (run) only a portion of your program or even the entire script without saving it, you may select the portion you want to run and hit the <F9> key.

The remaining menus and icons, as well as the second toolbar of the Editor (the Cell Toolbar), help you debug your program (see Sec. 5.10.2).

2.2.4 Workspace Window and Variable Editor

By now, you must have created several variables, so that your workspace does not look empty containing only one variable, $z = 16$, as shown in Figs. 2.1 and 2.2. If not, try to type in the following commands to fill it up:

```
>> a=zeros(5); b=num2cell(ones(3,2)); c=rand(5,6);
>> h=single(ones(2,9)); t='aircraft';
>> s=struct(t,{'B787','A380'},'developer', {'US','EU'});
>> y=int8(45); z=pi/2; 6>3;
```

These commands create some variables (scalars and arrays) of different classes (to be discussed in the next two chapters), showing up in the Workspace window as presented in Fig. 2.6a. This Workspace window is a GUI that allows you to view and manage the contents of the MATLAB workspace (the allocation of memory for all variables). The Workspace browser shows the name of each variable (by the icon distinguishing its class), its value, array size, size in bytes, and class. These are the default columns of the Workspace window, but you can choose more or less of them by right-clicking on the name of the columns that opens a pop-up menu (shown in Fig. 2.6a) to choose what columns to display.

You can move the entire column to a new location by clicking on and dragging its header. You can resize the width of any column by clicking on and moving the divider between this and the next column. You can sort the content of the Workspace window by any column—simply click on this column header and observe the reverse order of any column (the second click reverses the sort order back to the original). The only exception is the Value column, which you cannot sort. To choose the column to be displayed and sort the content of the Workspace window, you can also use the **View** menu of Desktop, featuring two options: **Choose Columns** and **Sort By**. A third alternative is to right-click anywhere in the browser (except the Name and Value columns) and choose the **Sort By** option in the pop-up menu. You can also use this menu to refresh the

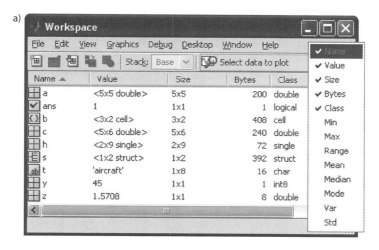

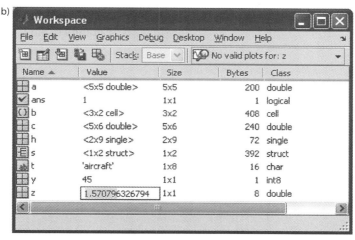

Fig. 2.6 a) Workspace window and b) value editing capability.

content of the browser (the **Refresh** option) and create a new variable (**New**). You can drag any variable to the Command Window so that you can use it in any expression.

A right-click on the name of the variable or its value brings up another menu, the full version of which is shown in Fig. 2.7a. A bottom portion of this menu (available only for numeric arrays) allows plotting all columns of data using the different plotting options. You can use this menu to bring up the Variable Editor (**Open Selection**), save, copy, duplicate, delete, rename, and even edit the value of the chosen variable right in the Workspace browser, as shown in Fig. 2.6b. For char-class variables and scalars, you can do this by selecting the variable (that is, click on it) and then clicking on the value again. (You can use this approach to rename any variable as well.) Note, when you

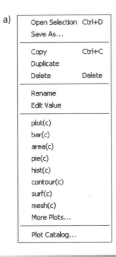
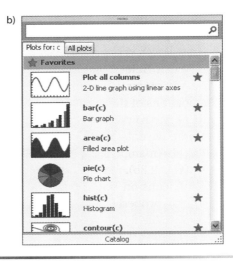

Fig. 2.7 a) One of the right-click pop-up menus and
b) a plot selection menu.

select a scalar for editing, its exact value shows up as opposed to a short-format version of it (which has gray background as seen in Fig. 2.6b).

The Workspace browser also allows you to preview the content of numeric arrays graphically. You can plot all columns of a selected numerical array by choosing the corresponding option of the pop-up menu shown in Fig. 2.7a or use the Plot Selection icon from the Workspace Toolbar (see Fig. 2.6), which brings a pop-down menu shown in Fig. 2.7b. The first five icons of this toolbar provide an alternative way of creating a new variable, opening selected variables in the Variable Editor (to be discussed next), loading data from outside (including directly from Excel spreadsheets), saving and deleting variables.

In addition to the previously discussed options, a simple double-click on the name or value of any variable opens the Variable Editor. You can also do it programmatically by using the `openvar` function. For example,

```
>> openvar c
```

brings up the window shown in Fig. 2.8a. The Variable Editor allows a variety of the options on editing and viewing variables in the workspace. In some sense, working in the Variable Editor is similar to working within a spreadsheet. Here, you can change array size, content, and format of the elements; cut, copy, paste, and delete them; exchange data with the

Command Window and with Excel; and create graphs and new variables from the current selection. In the latest versions of MATLAB, a new tool, the **Brush/Select Data** icon, was added to allow you to interactively mark, delete, modify, and save portions of the array (as an example, Fig. 2.8a features two rows marked using this tool, that is, the interactive data brushing mode).

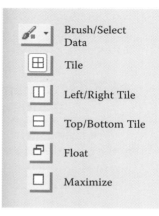

You can also manipulate several variables simultaneously (Fig. 2.8b). To this end, the rightmost group of the icons on the Variable Editor Toolbar allows you to properly arrange several windows within the Variable Editor. These icons allow you to arrange the open variable windows in an $n \times m$, left/right, top/bottom tile, and float pattern, as well as simply maximize the selected variable window to occupy all space of the Variable Editor (as in Fig. 2.8a). Among other things,

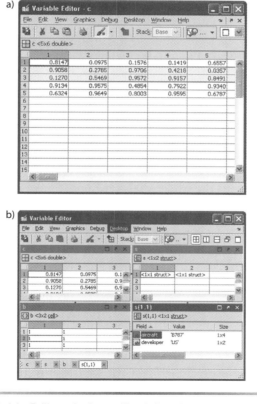

Fig. 2.8 Variable Editor window with a) the single variable and b) multiple simultaneously opened variables.

you can use the Variable Editor to browse and change the content of the cell arrays and structures (to be discussed in Chapter 4). As an example, Fig. 2.8b features one of the elements of the structure s.

2.2.5 Current Folder Window and Search Path Setting

The main destination of the Current Folder window (Fig. 2.1) is to navigate to the directory where your files are stored and show its content in a similar way to Microsoft Windows. Figure 2.9a shows the default view of the undocked Current Folder window. It also shows that beyond the name of the file, you can also add some additional information, including its size, date of its last modification, type, and description (the latter option adds the first line of each M-file in the current folder after its name).

Compared to the previous versions, starting from R2009a, the content of the Current Folder Toolbar has changed to include only a few icons. The first two icons allow you to go back and forward, switching between the current and the previous directory you looked at. Next is a typical Windows menu, showing the current directory and enabling a drop-down menu listing the previous directories you have used. The next icon allows to search for the files with a specific extension in the current folder and subfolders.

All other actions, which in prior versions were available as individual toolbar icons, are now hidden behind the **Actions** icon, which brings up a drop-down menu shown in Fig. 2.9b. This menu allows you to manage the content of the current folder as well as create several reports. For instance, the pop-up windows for the Code Analyzer Report (also known as M-Lint Code Check in previous versions) and the Dependency Report are presented in Fig. 2.10. Both windows feature a new icon, a refresh button.

Beginner students can simply ignore these reports, but later on when writing professional scripts and functions, you may want to explore what these reports have to offer in debugging your code. For every file in the current directory, the Code Analyzer Report displays a message suggesting the corrections and improvements you may want to consider. For example, a common Code Analyzer message is that a variable is defined but never used or that you should end a line with a semicolon to suppress the output (as seen in Fig. 2.10a).

When you open MATLAB, it sets your current folder to be at some default location. Therefore, unless you change the current directory every time you save something, it saves in this default folder. Using the default folder is perfectly fine, but our suggestion is that you create a new directory in the known location and navigate to it every time you open MATLAB, so that you know where your files are saved.

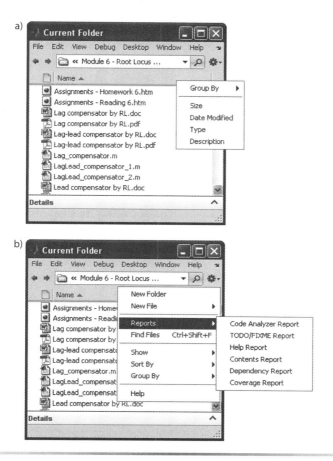

Fig. 2.9 a) Current Folder window and b) Actions Icon menu of the Current Folder toolbar.

The display box showing the current folder, where MATLAB looks for and saves files to, appears in the Current Folder window browser (Fig. 2.9) and also to the right from the **Help** icon on the Desktop toolbar (Fig. 2.1). The two icons which can alter this display box are **Browse for folder** and **Go up one level**. These buttons provide you with the quickest way to change the current directory. Of

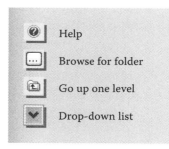

course, if you used this folder before, you can take advantage of it by simply choosing this directory from the drop-down list, invoked by clicking on the corresponding icon. (The Current Folder window browser allows you to do the same.)

When you create your first M-file using the MATLAB Editor and try to run it, it will suggest to save it first (if it was not saved already). Then, you

a)
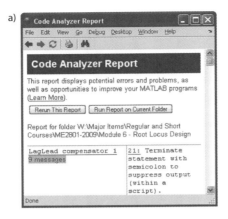

b)
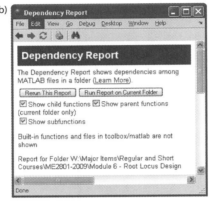

Fig. 2.10 a) Code Analyzer Report and b) Dependency Report windows of the Reports Option of the Actions Icon.

will have a chance to choose the directory to save it. If this directory happens to be different from the current folder, then the Editor will suggest you to make this directory a current directory (see example in Fig. 2.11).

This is important because you can only run the files from the current directory. When you are in the Command Window, you can run the files that are located either in the current folder or in a set of other directories MATLAB knows about, so that it can search through them looking for the file you are trying to execute. This set of "other directories" is referred to as a *search path*.

Fig. 2.11 Attempting to run a script from a not-current folder.

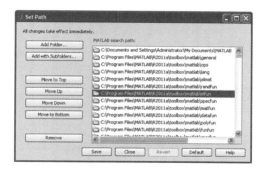

Fig. 2.12 Set Path dialog box.

To see which directories are on the search path or to change the search path, you should select the Desktop **File** button and then choose the **Set Path** option, which results in the dialog window shown in Fig. 2.12. Alternatively, you can open this window programmatically using the `pathtool` command.

The default search path shown in Fig. 2.12 allows you to use all the functions available in MATLAB and toolboxes that are installed on your computer from any directory. For instance, the `pi` function, *pi.m*, which we used in the previous sections, resides in the ...*MATLAB\R2011a\toolboxmatlab\elmat* directory (highlighted in Fig. 2.12). Nevertheless, we were able to use it from any other directory. Using this window, you can add your own directories, so that the files stored in are visible from any directory.

2.2.6 Desktop Menus, Toolbar Icons, and Start Button

Although most of your interaction will be in the Command Window, you can also use a row of standard Microsoft-type menus and a row of icons beneath it located at the top of the Desktop (Fig. 2.1). To conclude an introduction to the Desktop, let us formally mention them here.

The seven menu buttons of the Desktop are **File**, **Edit**, **Debug**, **Parallel**, **Desktop**, **Window**, and **Help**. (Most likely, your computer does not have the **Parallel** menu, because it shows up when the Parallel Computing Toolbox is installed.) The same buttons are also available in all four major windows even when they are undocked (except **Parallel**). Figure 2.13 shows the content of these menus. In addition to that, the Current Folder and Workspace windows have one and two more buttons, as shown in Fig. 2.14.

Any option in these menus can be selected either by clicking on it or using the shortcut key listed to the right of the item if available (for instance, <Ctrl>+<O>). The items followed by three dots (...) open a submenu or another window containing a dialog box. Clicking on a triangle (▶) opens another menu as well.

Most of the options in these menus are self-explanatory and some of them have been considered in the previous sections. For the sake of completeness, the following explains what the items are for from the two most often used menus, **File** and **Edit**. The options of the **File** menu cause the following actions:

New	Opens a dialog box that allows you to create a new program file, called an M-file, using a text editor called the Editor/Debugger, or a new Figure or Model file
Open...	Opens a dialog box that allows you to select a file for editing
Close Command Window	Closes the Command Window
Import Data...	Starts the Import Wizard which enables you to import data easily
Save Workspace As...	Opens a dialog box that enables you to save a file
Set Path...	Opens a dialog box that enables you to set the MATLAB search path
Preferences...	Opens a dialog box that enables you to set preferences for such items as fonts, colors, tab spacing, and so forth
Page Setup...	Opens a dialog box that enables setting printing options
Print...	Opens a dialog box that allows you to print all of the Command Window
Print Selection...	Opens a dialog box that enables you to print selected portions of the Command Window
File List	Contains a list of previously used files, in the order of the most recently used
Exit MATLAB	Closes MATLAB

The options of the **Edit** menu do the following:

Undo	Reverses the previous editing operation
Redo	Reverses the previous Undo operation
Cut	Removes the selected text and stores it for pasting later
Copy	Copies the selected text for pasting later, without removing it
Paste	Inserts any text on the clipboard at the current location of the cursor
Paste Special...	Inserts the contents of the clipboard into the workspace as one or more variables

File menu

New	
Open...	Ctrl+O
Close Command Window	Ctrl+W
Import Data...	
Save Workspace As...	Ctrl+S
Set Path...	
Preferences...	
Page Setup...	
Print...	Ctrl+P
Print Selection...	
1 H:\...n\Lag_compensator.m	
2 ...yakime$}Desktop}Drag.m	
3 H:\...Code\SMTrajectory.m	
4 H:\...B Code\SMGuidance.m	
Exit MATLAB	Ctrl+Q

Parallel menu

Select Configuration
Manage Configurations...

Edit menu

Undo	Ctrl+Z
Redo	Ctrl+Y
Cut	Ctrl+X
Copy	Ctrl+C
Paste	Ctrl+V
Paste to Workspace...	
Select All	Ctrl+A
Delete	Delete
Find...	Ctrl+F
Find Files...	Ctrl+Shift+F
Clear Command Window	
Clear Command History	
Clear Workspace	

Window menu

Close All Documents	
Next Tool	Ctrl+Tab
Previous Tool	Ctrl+Shift+Tab
Next Tab	Ctrl+Page Down
Previous Tab	Ctrl+Page Up
0 Command Window	Ctrl+0
1 Command History	Ctrl+1
2 Current Folder	Ctrl+2
3 Workspace	Ctrl+3

Debug menu

Open Files when Debugging	
Step	F10
Step In	F11
Step Out	Shift+F11
Continue	F5
Clear Breakpoints in All Files	
Stop if Errors/Warnings...	
Exit Debug Mode	Shift+F5

Help menu

Product Help	
Function Browser	Shift+F1
Using the Desktop	
Using the Command Window	
Web Resources	
Get Product Trials	
Check for Updates	
Licensing	
Demos	
Terms of Use	
Patents	
About MATLAB	

Desktop menu

Minimize Command Window	
Maximize Command Window	
Undock Command Window	Ctrl+Shift+U
Move Command Window	
Resize Command Window	
Desktop Layout	
Save Layout...	
Organize Layouts...	
Command Window	
Command History	
Current Folder	
Workspace	
Help	
Profiler	
File Exchange	
Editor	
Figures	
Web Browser	
Variable Editor	
File and Folder Comparisons	
Toolbars	
Titles	

Fig. 2.13 Content of the Command Windows menus.

Fig. 2.14 Additional Items for the a) Current Folder, and b) Workspace Windows.

Select All	Highlights all text in the Command Window
Delete or **Delete Selection**	Clears the variable (highlighted) in the Workspace Browser
Find or **Find Files**	Opens a dialog box that allows searching for files or for specified text within files
Clear Command Window	Removes all text from the Command Window
Clear Command History	Removes all text from the Command History window
Clear Workspace	Removes the values of all variables from the workspace

The toolbar below the menu bar provides icons as shortcuts to some commonly used options of Desktop menus. Clicking on the icon is equivalent to clicking on the menu and then clicking on the menu item; thus the icon eliminates one click of the mouse. The first seven icons from the left correspond to **New M-File**, **Open File**, **Cut**, **Copy**, **Paste**, **Undo**, and **Redo** menu items. The eighth icon activates **Simulink** (see Chapter 15). The ninth icon brings up the **GUIDE** tool (the MATLAB GUI design environment) discussed in Appendix C. The tenth icon opens the **Profiler**, which allows you to improve the performance of your code by finding the pieces "eating" most of the time during execution. The **Profiler** (see Sec. 5.10.3). As mentioned earlier, the question mark icon opens the **Help** browser window (discussed in the next section).

Finally, let us introduce the MATLAB **Start** button, located in the bottom left corner of the MATLAB Desktop. The **Start** button pop-up menu and its derivative menus (Fig. 2.15) provide an easy

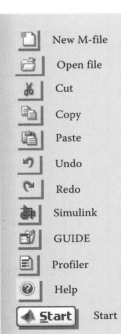

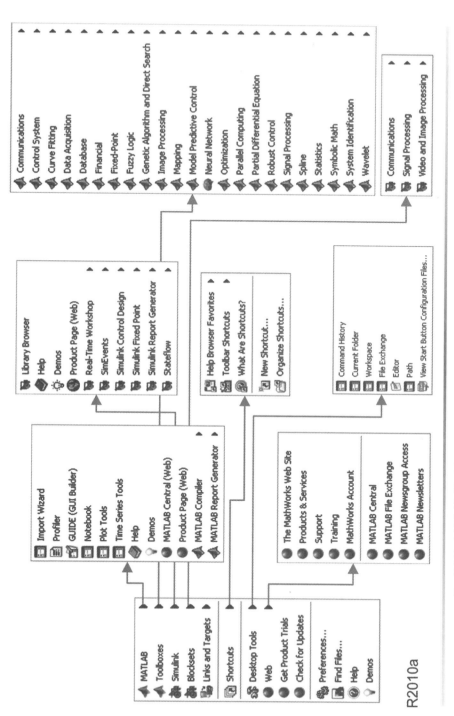

Fig. 2.15   MATLAB Start button menu (on the left) and some of the derivative menus.

access to tools, demos, shortcuts, and documentation in a manner similar to that of the **Start** button of Microsoft Windows.

2.3 Help System

Even a decade ago, the MATLAB documentation occupied several volumes. These days, every new release, taking place twice a year, introduces some changes, making the documentation on the previous release somewhat obsolete. For this reason, an extensive embedded MATLAB Help system becomes a major source of information. As seen in Fig. 2.1 (the first line in the Command Window), after MATLAB has been just installed, this help system immediately start to work offering you to watch a video, view demos, or read Getting Started notes.

MATLAB has several options to provide you with the relevant help when needed:

Help browser	This GUI allows you to find the most compete information and view online documentation for all MathWorks's products
Help functions	The functions `help`, `lookfor`, and `doc` can be used to display syntax information for a specified function
Other resources	Additional online help can be obtained by running various demos, contacting technical support, searching documentation, viewing a list of available books, and participating in a newsgroup

To obtain help on the MATLAB functions without even leaving the Command Window, you can use one of the following functions:

`help function_name`	Displays a description of the specific function `function_name` (displays the comment lines of the so-called *help text* the corresponding M-script starts with)
`lookfor topic`	Allows you to search for functions based on a keyword `topic` (lists all functions, the first comment line of the help text of which contains the word `topic`); if the suffix `-all` is added, the `lookfor` function searches the entire help text of all functions, not just the first line of it
`doc function_name`	Opens documentation on a `function_name` in the Help Browser (`doc` displays the start page of the Help Browser)

As discussed in the previous section, selecting the **Help** option from the **Help** menu of Desktop or clicking on the **Help** icon (or typing in the `doc` or `helpbrowser` command), opens the MATLAB Help browser (Fig. 2.16). As seen, it has two window panes: the Help Navigator pane on the left and the Display pane on the right.

The Help Navigator allows you to search for a specific topic or a function name in all installed toolboxes and blocksets and contains two tabs (as opposed to four in the earlier versions of MATLAB)

Contents   A contents listing tab
Search Results   A search tab having a find function and full text search
features

Using these tabs along with the **Search for** tool above them, you can get the most complete help on any topic you may be interested in.

The Display pane is used to view the documentation. Figure 2.17 provides you with an example of the viewing area in response for a `bin2dec` function search. With the list of the related topics appearing on the left, the Display pane shows an extended help on a specific topic with the keywords (`bin2dec`) highlighted everywhere in the text for your convenience. If you want to dedicate more space for a viewing area, drag the separator bar between the Display pan and Help Navigator pan. Figure 2.17 also shows the drop-down menu in response to a click on the **Actions** icon. Among other actions, you can print the article or even execute the selected MATLAB commands right from the Help browser. To see the results, we suggest to dock the Help browser to the Desktop first, using the Dock icon (otherwise, you will need to go to another window, MATLAB Desktop, to see them).

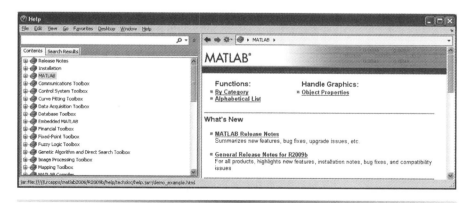

Fig. 2.16 Default view of the MATLAB Help browser.

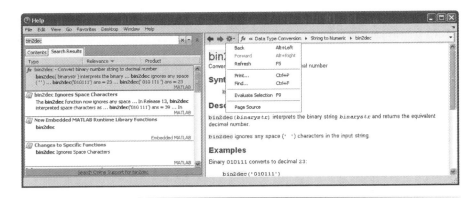

Fig. 2.17   Viewing search results in the Help browser.

The MATLAB help system provides you with demos on almost every major topic, including the Desktop and its tools. These demos are quite helpful, especially for self-studying and self-perfecting your programming skills. There are three basic kinds of demos:

• Video tutorials, which are movies that highlight a specific feature
• Published M-file examples, HTML files created from M-file scripts that show a step-by-step development
• M-file GUIs, the standalone tools that let you explore a specific topic

To find a specific demo, you need to expand the listing for a specific product, say MATLAB (see Fig. 2.16) and navigate to the **Demos** folder. When you expand the latter one, you will see all available demos grouped by categories (see example in Fig. 2.18). You may choose any category you need, say Getting Started, and select a specific demo from the list appearing in both of two panes (Fig. 2.18).

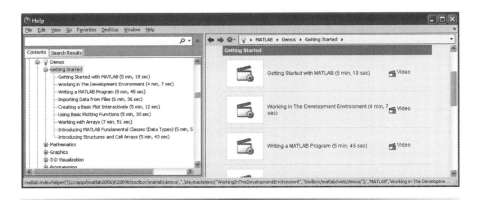

Fig. 2.18   Online demos available for the MATLAB getting started category.

To access demos, you can also click on the **Help** button of the Help browser and choose the **Demos** option. Alternatively, you can open the Demos tab programmatically from the Command Window. To do that, simply type in *demo* or go directly to the demos for a specific product or category. For example,

```
>> demo matlab desktop
```

which brings up the list of demos available for the Desktop Tools and Development Environment category.

The latest versions of MATLAB feature one more option to get quick help, while working in the Command Window—the use of the **Function Browser** icon. This new feature is explained in Sec. 2.6.

| fx | Browse for functions |

2.4 Basic Functions and Utilities

MATLAB provides you with a large number of standard elementary mathematical functions along with some specific functions to create and handle matrices, specialized functions, etc. Each toolbox enhances this list with more specialized functions (they can also be referred to as *utilities*). Many of them are available in the student version of MATLAB and Simulink.

The generalized syntax of MATLAB functions is

```
[outputs] = function_name(inputs)
```

Here, `inputs` and `outputs` may have multiple entries, but typically any function permits a simple straightforward syntax

```
x = function_name(y)
```

where `inputs` and `outputs` are represented by a single, most important, entry each. For the vast majority of the basic functions presented in this section, this simple format is the only format, for example,

```
>> x = sin(pi/2)
```

which returns

```
x = 1
```

Note that to enclose the function's argument(s) operated on by the function, MATLAB uses a pair of parentheses. If more than one output is expected, like for the `meshgrid` function, a pair of brackets must be used. If you want to explore more options (if available) on how to use any specific function, you can now use the MATLAB Help system introduced in the previous section.

Some of the functions, like `sqrt` and `sin`, are built in. They are part of the MATLAB core, and so they are very efficient, but the computational details are not readily accessible. The majority of functions though, like `dec2hex`, are implemented as M-files, allowing you to see the source code and even modify it if you need to (this will be addressed further in Sec. 5.9.1).

2.4.1 Elementary Functions

MATLAB allows you to overview the entire list of available elementary mathematical functions by issuing the following instruction:

```
>> help elfun
```

The slightly reordered result of this command shows available functions in groups as follows:

Trigonometric Functions (Operating with Radians)			
sin	Sine	sec	Secant
asin	Inverse sine	asec	Inverse secant
cos	Cosine	csc	Cosecant
acos	Inverse cosine	acsc	Inverse cosecant
tan	Tangent	cot	Cotangent
atan	Inverse tangent	acot	Inverse cotangent
atan2	Four quadrant inverse tangent	hypot	Square root of sum of squares
Trigonometric Functions (Operating with Degrees)			
sind	Sine	secd	Secant
asind	Inverse sine	asecd	Inverse secant
cosd	Cosine	cscd	Cosecant
cosd	Inverse cosine	acscd	Inverse cosecant
atand	Tangent	cotd	Cotangent
atan2d	Inverse tangent	acotd	Inverse cotangent
Hyperbolic Trigonometric Functions			
sinh	Hyperbolic sine	sech	Hyperbolic secant
asinh	Inverse hyperbolic sine	asech	Inverse hyperbolic secant
cosh	Hyperbolic cosine	csch	Hyperbolic cosecant
acosh	Inverse hyperbolic cosine	acsch	Inverse hyperbolic cosecant
tanh	Hyperbolic tangent	coth	Hyperbolic cotangent
atanh	Inverse hyperbolic tangent	acoth	Inverse hyperbolic cotangent
Exponential Functions			
exp	Exponential		
expm1	Accurate value of e^x-1		
log	Natural logarithm ($\ln(x)$)		
log1p	Accurate value of $\log(1+x)$		
log10	Common (base–10) logarithm ($\log_{10}(x)$)		

log2	Base–2 logarithm and dissect floating point number ($\log_2(x)$)
pow2	Base–2 power and scale floating point number
realpow	Power that will error out on complex result
reallog	Natural logarithm of real number
realsqrt	Square root of number greater than or equal to zero
sqrt	Square root
nthroot	Real nth root of real numbers
nextpow2	Next higher power of 2
Complex Functions	
abs	Absolute value
angle	Phase angle
complex	Construct complex data from real and imaginary parts
conj	Complex conjugate
imag	Complex imaginary part
real	Complex real part
unwrap	Unwrap phase angle
isreal	True for real array
cplxpair	Sort numbers into complex conjugate pairs
Rounding and Remainder	
fix	Round toward zero
floor	Round toward negative infinity
ceil	Round toward positive infinity
round	Round toward nearest integer
mod	Modulus (signed remainder after division)
rem	Remainder after division
sign	Signum (returns 1 if the element is greater than 0, 0 – if it equals 0, and −1 if it is less than 0)

For example, to compute the square root of 9, you type `sqrt(9)`. Note that taking the square root or logarithm of a negative number is not an error—the appropriate complex result will be produced automatically. Most of the aforementioned functions also accept complex numbers as arguments. Another comment is that the latest versions of MATLAB feature three special functions, `hypot`, `expm1`, and `log1p`, allowing you to compute certain quantities more accurately to avoid underflow and overflow and compensate for the roundoff errors (to be discussed in Chapter 8).

If you look in the subdirectories of the *toolbox\matlab* directory, you can find the M-file sources to many of the functions supplied with MATLAB beyond the "elfun" group. You can locate your *toolbox\matlab* directory by typing

```
>> dir([matlabroot '\toolbox\matlab\'])
```

which returns

```
.              datatypes  general   helptools  matfun     scribe    timefun

..             demos      graph2d   icons      mcc.enc    sparfun   timeseries

audiovideo  elfun         graph3d   imagesci   ops        specfun   uitools

codetools   elmat         guide     iofun      plottools  specgraph verctrl

datafun     funfun        hds       ang        polyfun    strfun    winfun
```

When one of these M-file functions is called, MATLAB parses and executes each line of code in the M-file. It saves the parsed version of the function in memory, eliminating parsing time on any further calls to this function.

2.4.2 Matrix Manipulating and Specialized Functions

To learn about functions, dedicated to creating elementary matrices and matrix manipulations, you can type

```
>> help elmat
```

To see the list of available specialized math functions, use the following command:

```
>> help specfun
```

For completeness, the returns of the above two commands are as follows:

Elementary Matrices			
zeros	Zeros array	linspace	Linearly spaced vector
ones	Ones array	logspace	Logarithmically spaced vector
eye	Identity matrix	freqspace	Spacing for frequency response
repmat	Replicate and tile array	meshgrid	X and Y arrays for 3-D plots
accumarray	Array with accumulation		
:	Regularly spaced vector and index into matrix		
Basic Array Information			
size	Size of array	numel	Number of elements
length	Length of vector	disp	Display matrix or text
ndims	Number of dimensions	isempty	True for empty array
isequal	True if arrays are numerically equal		
isequalwithequalnans	True if arrays are numerically equal, treating NaNs as equals		

Matrix Manipulation

cat	Concatenate arrays	flipdim	Flip matrix along dimension
reshape	Reshape array	rot90	Rotate matrix 90 deg
diag	Diagonal matrices	find	Find indices of nonzero elements
blkdiag	Block diagonal concatenation	end	Last index
tril	Extract lower triangular part	sub2ind	Linear index from subscripts
triu	Extract upper triangular part	ind2sub	Subscripts from linear index
fliplr	Flip matrix in left/right direction	bsxfun	Binary singleton expansion
flipud	Flip matrix in up/down direction		

Multidimensional Array Functions

ndgrid	Generate arrays for *N-D* functions and interpolation
permute	Permute array dimensions
ipermute	Inverse permute array dimensions
shiftdim	Shift dimensions
circshift	Shift array circularly
squeeze	Remove singleton dimensions

Array Utility Functions

isscalar	True for scalar	isvector	True for vector

Special Variables and Constants

eps	Relative accuracy	inf	Infinity
realmax	Largest positive number	nan	Not-a-number
realmin	Smallest positive number	isnan	True for not-a-number
intmax	Largest positive integer value	isinf	True for infinite elements
intmin	Smallest integer value	isfinite	True for finite elements
pi	Number π	j	Imaginary unit
i	Imaginary unit	why	Succinct answer

Specialized Matrices

compan	Companion matrix	gallery	Test matrices
hadamard	Hadamard matrix	magic	Magic square
hankel	Hankel matrix	pascal	Pascal matrix
hilb	Hilbert matrix	toeplitz	Toeplitz matrix
invhilb	Inverse Hilbert matrix	vander	Vandermonde matrix

(Continued)

Specialized Matrices	
`rosser`	Classic symmetric eigenvalue test problem
`wilkinson`	Wilkinson's eigenvalue test matrix

Specialized Math Functions	
`airy`	Airy functions
`besselj`	Bessel function of the first kind
`bessely`	Bessel function of the second kind
`besselh`	Bessel functions of the third kind (Hankel function)
`besseli`	Modified Bessel function of the first kind
`besselk`	Modified Bessel function of the second kind
`beta`	Beta function
`betainc`	Incomplete beta function
`betaincinv`	Inverse incomplete beta function
`betaln`	Logarithm of beta function
`ellipj`	Jacobi elliptic functions
`ellipke`	Complete elliptic integral
`erf`	Error function
`erfc`	Complementary error function
`erfcx`	Scaled complementary error function
`erfinv`	Inverse error function
`expint`	Exponential integral function
`gamma`	Gamma function
`gammainc`	Incomplete gamma function
`gammaincinv`	Inverse incomplete gamma function
`gammaln`	Logarithm of gamma function
`psi`	Psi (polygamma) function
`legendre`	Associated Legendre function
`cross`	Vector cross product
`dot`	Vector dot product

Number Theoretic Functions			
`factor`	Prime factors	`lcm`	Least common multiple
`isprime`	True for prime numbers	`rat`	Rational approximation
`primes`	Lists the prime numbers	`rats`	Rational output
`gcd`	Greatest common divisor	`perms`	All possible permutations
`nchoosek`	All combinations of *N* elements taken *K* at a time		
`factorial`	Factorial function		

Coordinate Transforms	
`cart2sph`	Transform Cartesian to spherical coordinates
`cart2pol`	Transform Cartesian to polar coordinates
`pol2cart`	Transform polar to Cartesian coordinates
`sph2cart`	Transform spherical to Cartesian coordinates
`hsv2rgb`	Convert hue–saturation–value colors to red–green–blue
`rgb2hsv`	Convert red–green–blue colors to hue–saturation–value

Some of these functions will be considered later in this section, some in the following chapters, when appropriate. The remaining ones are presented here just for your reference. Some other useful functions dealing with converting numbers between different scales of notation are

`bin2dec`	Converts binary string to decimal integer
`dec2bin`	Converts decimal integer to a binary string
`dec2hex`	Converts decimal integer to hexadecimal string
`hex2dec`	Converts hexadecimal string to decimal integer
`num2hex`	Converts singles and doubles to hexadecimal strings
`hex2num`	Converts hexadecimal number string to double-precision number
`base2dec`	Converts base-*B* string to decimal integer
`dec2base`	Converts decimal integer to base-*B* string

Some other useful functions to be used in this textbook are

`rand`	Generates uniformly distributed random numbers
`randn`	Generates normally distributed random numbers
`randi`	Generates uniformly distributed pseudorandom integers
`int2str`	Converts integer to string
`num2str`	Converts number to string
`str2num`	Converts string to number

2.4.3 Constants

Among the functions introduced in the previous section, there are some producing some useful constants. They are as follows:

`pi`	Generates the numerical equivalent to $\pi = 3.14159265$, the ratio of a circle's circumference to its diameter
`i` and `j`	Represents imaginary unit, $\sqrt{-1}$ (for complex numbers you may use an asterisk as in $2+3*i$ or have it implied as in $2+3i$)

realmax   Returns the largest floating-point number representable on your computer (anything larger overflows); for double precision realmax returns the number which is one bit less than 2^{1024} or about $1.7977e^{+308}$

realmin   Returns the smallest positive normalized floating-point number, for double precision realmin returns 2^{-1022}

The floating-point arithmetic and the latter two numbers will be discussed in detail in Chapter 8, but the point we would like to make here is that there are certain limitations, implied by a digital computer, that affect the range and precision of the numbers we can manipulate. Specifically, the floating-point numbers are distributed nonevenly, so that the distance between two closest numbers depends on the magnitude of the number itself. To this end, the MATLAB function eps(x) returns the positive distance from abs(x) to the next larger in magnitude floating-point number of the same precision as x. For example,

```
>> eps(realmax)
```

returns

```
ans = 1.9958e+292
```

while

```
>> eps(realmin)
```

yields

```
ans = 4.9407e-324
```

Obviously, the smaller numbers have better precision. Without an input argument, the function *eps* returns the value of eps(1), which happens to be 2.2204e-016 (or 2^{-52}). Again, these issues will be discussed later in Chapter 8.

The two remaining constants are

Inf   Returning infinity, ∞, when the result of operations like division by zero and overflow exceeds realmax

NaN   Returning an undefined numerical result, "Not-a-Number," produced by expressions like 0/0, Inf/Inf or Inf-Inf, as well as from arithmetic operations involving a NaN

The good thing about NaN is that it is still treated as a numerical result, and so if something in your code goes wrong yielding NaN, the program does not crash but simply produces NaN in all subsequent computations.

Let us conclude this section with a warning to be creative but careful in choosing the names of your variables so that none of them coincide with one of the MATLAB functions or constants. Otherwise, you may overwrite any of them with a new variable and get yourself in trouble. For example, if, by mistake, you introduce some variable `pi` equal to `2`, the following command will never give you an area of a 3-unit radius circle:

```
>> Area=pi*3^2
```

Similarly, if unintentionally you created a variable `cos=3`, then an attempt to use the `cos(pi)` function will result in the following error message:

```
??? Index exceeds matrix dimensions.
```

meaning that from now on MATLAB treats `cos` as a scalar, not a function. You will learn more about this in Sec. 5.8.6, but for now let us show how to restore the original constant value or function programmatically. If you get yourself in trouble, simply clear (delete) wrong-name variables using the `clear` function:

```
>> clear pi cos
```

This function will be formally introduced in Sec. 2.6.

2.4.4 Utilities

As mentioned earlier, each MATLAB toolbox has its own utilities. For example, among others, the following Mapping Toolbox units conversion functions (if available in your version of MATLAB) may be of a great use in a variety of engineering applications:

Angle Conversions	
`angl2str`	Formats angle strings
`angledim`	Converts angles units or encodings
`deg2dm`, `deg2dms`, `deg2rad`	Convert angles from degrees to deg:min encoding, deg:min:sec encoding and to radians, respectively
`dms2deg`, `dms2dm`, `dms2rad`, `dms2mat`	Convert angles from deg:min:sec encoding to degrees, deg:min encoding, radians and to `[deg min sec]` matrix, respectively
`mat2dms`	Converts `[deg min sec]` matrix to deg:min:sec encoding
`rad2deg`, `rad2dm`, `rad2dms`	Convert angles from radians to degrees, deg:min and deg:min:sec encoding, respectively
`str2angle`	Converts strings to angles in degrees
Distance Conversions	
`deg2km`, `deg2nm`, `deg2sm`	Convert distance from degrees to kilometers, nautical miles and statute miles, respectively
`dist2str`	Formats distance strings
`distdim`	Converts distance units

(Continued)

Distance Conversions	
`km2deg, km2nm, km2rad, km2sm`	Convert distance from kilometers to degrees, nautical miles, radians, and statute miles, respectively
`nm2deg, nm2km, nm2rad, nm2sm`	Convert distance from nautical miles to degrees, kilometers, radians, and statute miles, respectively
`rad2km, rad2nm, rad2sm`	Convert distance from radians to kilometers, nautical miles and statute miles, respectively
`sm2deg, sm2km, sm2nm, sm2rad`	Convert distance from statute miles to degrees, kilometers, nautical miles, and radians, respectively
Time Conversions	
`hms2hm, hms2hr, hms2sec, hms2mat`	Convert time from hrs:min:sec to hrs:min, hours, seconds, and to [hrs min sec] matrix, respectively
`hr2hm, hr2hms, hr2sec`	Convert time from hours to hrs:min, hrs:min:sec, and seconds, respectively
`mat2hms`	Converts [hrs min sec] matrix to hrs:min:sec
`ec2hm, ec2hms, ec2hr`	Convert time from seconds to hrs:min, hrs:min:sec, and hours, respectively
`time2str`	Formats time strings
`timedim`	Converts time units or encodings

2.5 Order of Precedence

You can build expressions that use any combination of arithmetic, relational, and logical operators (the latter two will be discussed further in Secs. 5.2 and 5.3, respectively). As in any programming language, the mathematical operations represented by the symbols +, −, *, /, \, and ^ follow a set of rules called *precedence*. Precedence levels determine the order in which MATLAB evaluates an expression. Mathematical expressions are evaluated starting from the left, with the exponentiation operation having the highest order of precedence, followed by multiplication and division with equal precedence, followed by addition and subtraction with equal precedence.

To alter the order of precedence you may use parentheses, ( ). Remember not to confuse them with the square, [ ], or curly brackets, { }, that are not allowed in MATLAB mathematical expressions because they have a special usage. Evaluation of mathematical expression begins with the innermost MATLAB pair of parentheses and proceeds outward.

To summarize, the precedence rules for MATLAB operators, ordered from highest precedence level to lowest precedence level, are shown below:

1. Parentheses ()
2. Power ^
3. Multiplication *, right division /, left division \
4. Addition +, subtraction −

This list will be updated in Sec. 5.3 (Table 5.1) to include precedence of relational and logical operators.

To better understand these rules, let us consider a simple example. Suppose, we are dealing with the following mathematical expression:

$$\frac{(\sin(0.3\pi - \frac{1}{32}\pi)+3)^2}{2\pi^3} - \frac{3}{4} \tag{2.1}$$

In MATLAB, it should appear like

```
(sin(0.3*pi-pi/32)+3)^2/2/pi^3-3/4
```

Of course, to avoid mistakes, you should feel free to insert parentheses wherever you are unsure of the effect precedence will have on the calculation. Besides, the usage of parentheses may also improve the readability of your MATLAB expressions. For example, Eq. (2.1) can also be represented as

```
(sin(0.3*pi-(1/32)*pi)+3)^2/(2*pi^3)-(3/4)
```

2.6 Managing Work Session

Now that you have some practice in typing your commands in the Command Window, let us explore some additional options on expedient management of your work session in MATLAB. We start by considering different display formats, followed by some useful hints and tricks for fast typing, and end by introducing special symbols and some useful functions.

As mentioned earlier, although all computations in MATLAB are done in double precision, by default, the results are displayed in the short format with only five digits (four decimal digits). For your convenience, MATLAB allows you to change the output display format by using the `format` command as follows:

`format`	Default format with four decimal digits, same as `short`
`format short`	Five-digit scaled fixed-point format (with four decimal digits)
`format long`	15-digit scaled fixed-point format (with 14 decimal digits)
`format short e`	Five-digit scaled floating-point format (with four decimal digits)
`format long e`	15-digit scaled floating-point format (with 14 decimal digits)
`format short g`	Best of five-digit fixed or floating-point format

format long g		Best of 15-digit fixed or floating-point format
format hex		Hexadecimal format
format +		The symbols +, −, and blank are printed for positive, negative, and zero elements, whereas imaginary parts are ignored
format bank		Fixed format for dollars and cents (two decimal digits)
format rat		Approximation by ratio of small integers

For example, the following set of commands

```
format      short,     disp(pi)
format      long,      disp(pi)
format      short e,   disp(pi)
format      long e,    disp(pi)
format      short g,   disp(pi)
format      long g,    disp(pi)
format      hex,       disp(pi)
format      +,         disp(pi)
format      bank,      disp(pi)
format      rat,       disp(pi)
format,                disp(pi)
```

returns

```
3.1416
3.141592653589793
3.1416e+000
3.141592653589793e+000
3.1416
3.14159265358979
400921fb54442d18
+
3.14
355/113
3.1416
```

The function disp used in the above example (and to be formally introduced later in this section) displays a value of a variable or in this particular case constant, without printing its name.

Using the format command, you may also change the output display spacing.

format compact	Suppresses excess line feeds to show more output in a simple screen

format loose   Adds extra line feeds to make output more
                    readable

You may check the current format setup by typing

get(0,'Format')

Now, let us consider a few typing hints and tricks. They are related to the usage of the keyboard arrow buttons, <Ctrl> and <Tab> keys to recall, edit, and reuse functions and variables you typed earlier. For example, suppose you mistakenly entered the line

```
>> GoldenRatio = 0.5 * (sqr(5) - 1)
```

MATLAB responds with the error message

```
Undefined function or variable 'sqr'
```

because you misspelled the name of the sqrt function. Instead of retyping the entire line, press the up-arrow key <↑> once to display the previously typed line. Press the left-arrow key <←> several times to move the cursor and add the missing t. You can also use the *smart recall* feature to recall a previously typed function or variable, whose first few characters you specify. For example, after you have entered the line starting with GoldenRatio, typing Gol and pressing the up-arrow key <↑> once recalls the last-typed line that starts with the function or variable, whose name begins with Gol. (This feature is case-sensitive.)

You can use the *tab completion* feature to reduce the amount of typing too. MATLAB automatically completes the name of a function, variable, or file if you type the first few letters of the name and press the <Tab> key. If the name is unique, it is automatically completed. For example, if you have a variable GoldenRatio and you type Gol and press <Tab>, MATLAB completes the name and displays GoldenRatio. Then, you can either press <Enter> to display the value of the variable or continue editing to create a new executable line that uses the variable GoldenRatio. If a few letters you typed in are not unique (there are other functions, filenames, variables, etc. on the search path starting with the letters you typed), upon pressing the <Tab> key, a pop-up menu shows up suggesting you a variety of options you can choose from (see example in Fig. 2.19a with the pop-up window in response to just two symbols, si, typed in).

You can also enjoy having a quick constantly available help provided in the latest versions of MATLAB—a **Browse for function** icon. This icon can be made always readily available to the left of the prompt as shown in Fig. 2.19b. Clicking on this icon (which is equivalent to pushing <Shift>+<F1>) invokes a miniaturized version of the Help browser, making it much easier to find any function. The function you are looking for much

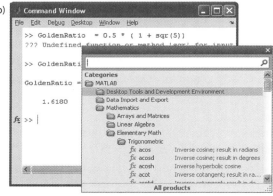

Fig. 2.19 Using a) the \<Tab\> Key and b) the Function Browser icon.

easier. To have this icon in the Command Window, you should enable it by right-clicking anywhere within the Command Window and checking the **Show Browser Function Icon** option.

Along with using the keyboard arrow keys by themselves, you can also use them in combination with the \<Ctrl\> key. To move through one word at a time as opposed to one character, press and hold \<Ctrl\> and then press \<→\> to move one word to the right (\<←\> to move one word to the left). Press \<Home\> to move to the beginning of a line and press \<End\> to move to the end of a line. Press \<Del\> to delete the character at the cursor and press \<Backspace\> to delete the character before the cursor. Press \<Esc\> to clear the entire line; press and hold \<Ctrl\>; and then press \<k\> to delete (kill) to the end of the line.

Next, Table 2.2 summarizes special symbols that may be useful when typing your commands.

Table 2.2 Use of Some Special Symbols

Symbol	Effect
, (comma)	Separates expressions on one line
; (semicolon)	Suppresses screen printing
... (ellipsis)	Continues a line

Once again, if a line does not end with a semicolon, MATLAB displays the results in the Command Window. Even if you suppress displaying the output of a command with the semicolon, MATLAB still computes and retains the variable's value in workspace. You can put several commands on the same line only if you separate them with a comma (if you want to see the result of each command) or semicolon (if you want to suppress displaying the results). For example,

```
>> x=2; y=6+x, x=y+7
y = 8
x = 15
```

(Note that the first value of x was not displayed. Also, note that the value of x has changed from 2 to 15.)

When a line does not end with a semicolon, the `display` function is being called internally to display the result (that is, 5+2 and display(5+2) produce the same result creating and displaying the default variable `ans`). If, however, you want to compute the result of some operation and just display it without creating any variable in workspace, as we did it in the beginning of this section, you might want to use the `disp` function. For instance, the command

```
>> disp(2*pi-1)
```

returns

```
5.2832
```

You can also use the `disp` command to display a hyperlink in the Command Window. For example, try

```
>> disp('-<a href = "http://www.wikipedia.org">Wikipedia</a>')
```

If you need to type a long line, you can use an ellipsis, by typing three periods, to delay execution. For example,

```
>> Number_of_AC_Damaged = 2;
>> Number_of_AC_Maintanance = 15;
>> Number_of_Aircraft_Grounded = Number_of_AC_Damaged ...
+ Number_of_AC_Maintanance
Number_of_Aircraft_Grounded = 17
```

Earlier in Sec. 2.2.3, we introduced three functions, `clear`, `close`, and `clc`, claiming that they are quite useful. Now, it is time to formally introduce them along with some other functions used to manage the work session. Although Table 2.3 presents just a list of these functions, brief descriptions of some of them are given later. Again, these functions simply allow executing some useful operations (that otherwise could be accomplished via buttons and menus) programmatically.

MATLAB retains the last value of a variable until you quit MATLAB or clear its value. Overlooking this fact commonly causes errors in MATLAB. For example, you may choose to use a variable `x` in a number of different calculations. If you forget to enter the correct value for `x`, MATLAB uses the last value, and you get an incorrect result. You can use the `clear` function to remove the values of all variables from memory (`clear all`), or you can use the form specifying what variables to clear specifically, for example,

```
>> clear var1 var2
```

clears two variables named `var1` and `var2`.

The effect of the `clc` command is different. It clears the Command Window of everything in the window display, but the values of the variables remain in the memory (workspace). After using `clc`, you cannot use the

Table 2.3   Commands for Managing the Work Session

Command	Effect
home	Moves the cursor to the upper-left corner of the Command Window
clc	Clears the Command Window and homes the cursor
exist('name')	Determines if a file or variable exists having the name name
who	Lists the variables currently in memory
whos	Lists the current variables and sizes and indicates if they have imaginary parts
clear var1 var2	Removes the variables var1 and var2 from memory
clear	Removes all variables from memory (workspace)
close('name')	Closes the named window
close all	Closes all the open figure windows
pwd	Displays the current working directory
cd('directory')	Sets the current working directory to directory
dir	Lists all files in the current working directory
what	Lists MATLAB files in the current directory
quit	Stops MATLAB

scroll bar to see the history of functions, but you can still use the up arrow (↑) to recall statements from the Command History window. In addition to that, the `home` command also moves the cursor to the upper-left corner of the Command Window and clears the visible portion of the window. However, as opposed to the `clc` command, you can still use the scroll bar to see what was on the screen before.

We know that typing the name of a variable and pressing <Enter> results in displaying its current value. If, however, the variable does not have a value, that is, if it does not exist, you will see an error message. To avoid getting an error of this sort, when writing a code it sometimes makes sense to check whether the variable you are trying to manipulate exists. The general syntax is to type `exist('x')` to see if the variable x is in use. If a 1 is returned, the variable exists; a 0 indicates that it does not exist.

The `who` function lists the names of all the variables in memory, but does not give their values. The form `who var1 var2` restricts the display to only those two variables specified. The wildcard character * can be used to display variables that match a pattern. For instance, `who  N*` finds all variables in the current workspace that start with N. For instance, if the variables from the previous example are still in workspace, then the `who N*` command will return

```
Your variables are:
Number_of_AC_Damaged        Number_of_Aircraft_Grounded
Number_of_AC_Maintanance
```

The `whos` function lists the variable names in the currently active workspace along with their size and class. Other syntaxes allowing you to specify what information you would like to display are also available.

You can quit MATLAB by typing `quit`. Alternative ways are clicking on the **File** menu and choosing the **Exit MATLAB** option, holding the <Ctrl> key and pressing the <Q> key, or simply click on the standard Window's Close icon in the upper right-hand corner of the Desktop.

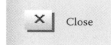

2.7 Changing Preferences

Quite often throughout this chapter, we have talked about the default setting or default values. However, MATLAB allows you to change any of these settings to match your personal preferences. In the MATLAB **File** menu, there is an option called **Preferences** that gives you a total control over settings. To familiarize yourself with this tool, we present some screens available within the **Preferences** option. As usual, there is an alternative way to open the **Preferences** window, programmatically by calling the corresponding `preferences` command.

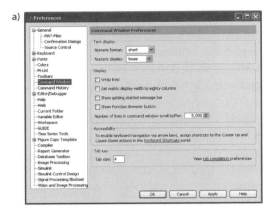

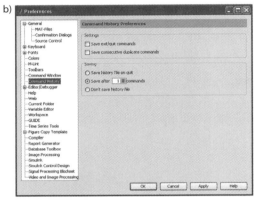

Fig. 2.20 a) Command Window Preferences and b) Command History Preference.

Depending on which window is active when you call this option, you may see something like what is shown in Fig. 2.20. Shown in this figure are the default preferences for the Command Window and Command History window, which gives you an idea what and where you can change using the **Preferences** option. For instance, in the Command Window Preferences (Fig. 2.20a), you can change the numeric format and display the same way the `format` function would do it programmatically, as described in the previous section. From Fig. 2.20b, you can see why the Command History window stores each command (the corresponding radiobutton is checked).

Figure 2.21a explores the Toolbars Preferences, showing why your Desktop may look differently than that presented throughout this chapter. You can populate any toolbar of the MATLAB development environment with the icons you want. Figure 2.21b shows Colors Preferences, explaining the rules for applying the highlighting colors.

Figure 2.22 shows some preferences from the expandable Editor/ Debugger preferences list. As seen from Fig. 2.22a, you can, for example, change right-hand text limit (although we would not recommend going beyond 82 symbols to avoid problems with printing your scripts on a Letter format paper). Figure 2.22b shows a screen, where you might want to make one change. When you open the Editor, MATLAB creates and automatically saves (every 5 min) an *autosave file*, to avoid loss of your data while typing in your code. Pretty soon, you will find your working directory filled with *asv*-files. Hence, we recommend checking the "Automatically delete autosave files" radiobutton, which will cause deletion of these files upon exiting the Editor. Of course, we assume that you will want to save your files before exiting if you really need them or that you have no intentions of saving these files anyway if you choose not to save them.

a)

b)
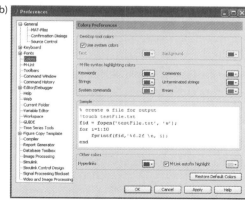

Fig. 2.21 a) Toolbars Preferences and b) Colors Preferences.

a)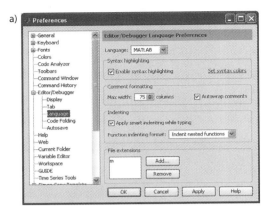

b)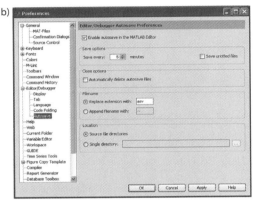

Fig. 2.22 a) Editor/Debugger Display and b) Autosave Preferences.

Another thing you may want to check and change, if appropriate, is that when copying the figures (to be discussed in Chapter 6), the Clipboard format is "Bitmap" and the Figure background color is "Force white background", as shown in Fig. 2.23a. Figure 2.23b shows us how to change the default array format in the Variable Editor. As seen, this format may be different from that of the Command Window (Fig. 2.20a).

2.8 Problem-Solving Methodology

To conclude this chapter, let us introduce a collective list of common steps involved in solving a problem on a computer. It might be more appropriate later in this textbook, when you have enough knowledge to write sophisticated programs, but introducing it as early as possible may eliminate a lot of troubles when writing even the simplest pieces of code, making programming enjoyable and rewarding. The engineering problem-solving methodology assumes that you:

1. Think about the main goal you want to achieve.
2. Define required inputs and desired outputs.
3. Collect all necessary information.
4. Start with a simplified version of the problem and clearly state your assumptions.
5. Draw a sketch of the problem and think of the names of your variables.
6. Check the dimensions and units of your variables.
7. Think about how you are going to proceed toward a solution and write down the general steps, maybe even labeling them (which will further be converted to the comments line in your code).

a)
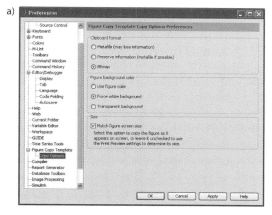

b)
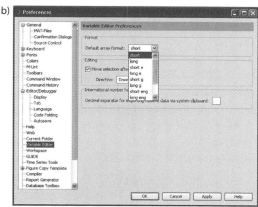

Fig. 2.23 a) Figure Copy Options Preferences and b) Variable Editor Preferences.

8. Make sure your program is robust enough to handle a variety of the expected inputs as well as interface properly with any applications you use.
9. Start writing code, checking its pieces if appropriate by printing (displaying) intermediate results as needed.
10. Devote some time to develop a concise, yet informative output structure, which later on will include producing informative plots.
11. Check the outputs of your program on some specific cases with known (expected) results.
12. Perform a "reality check" of the outputs your program produces in a general case.
13. If your algorithm involves some iterative computations (such as, optimization), try to limit the number of cycles first to be sure your program works properly.
14. Do not start batch computing unless you are absolutely sure your program runs properly for one dataset (having lots of pretty output does not necessarily mean it is not garbage).
15. After you debugged you program, make sure to suppress printing all unnecessary intermediate results, so that your program produces only those outputs you actually need.

Follow these rules closely at least once or twice to develop healthy programming habits. These habits will help you become a proficient applied programmer and make you an efficient scientist or engineer.

Problems

2.1   Calculate what speed it would take your aircraft, equipped with a wing having, say, $S = 4$ m^2, area, to get enough lift to take off. Assume that the total weight of $W = 100$ kg, air density $\rho = 1$ kg/m^3, gravitation due to gravity $g = 9.8$ m/s^2, and lift coefficient $C_L = 2$, and use the following formula:

$$V = \sqrt{\frac{2mg}{C_L \rho S}}$$

2.2   When cruising at a fixed altitude, a fixed angle of attack and a constant specific fuel consumption C_T (a ratio of consumed fuel per unit of thrust per unit of time), the range of an aircraft can be estimated using the following formula:

$$R = \frac{2}{c_T} \sqrt{\frac{2C_L g}{\rho S C_D^2}} \left(\sqrt{W_0} - \sqrt{W_f} \right)$$

where, in addition to notations of Problem 2.1, C_D is the drag coefficient and W_0 and W_f the aircraft's initial and final weights, respectively. In practice, flying at fixed angle of attack and constant Mach number M causes the aircraft to climb. In this case, the range can be estimated using Breguet range equation:

$$R = \frac{1}{c_T} \sqrt{\frac{2W_0 C_L g}{\rho S C_D^2}} \, \ell n \frac{W_0}{W_f}$$

where a is an average speed of sound at a given altitude (ambient temperature). Use the values, $C_L = 0.3$, $C_D = 0.03$, $c_T = 0.003/s$, $S = 20$ m², $W_0 = 2300$ kg, $W_f = 2100$ kg, and $\rho = 0.6$ kg/m³ to compute the range estimates using both formulas. Compare the results.

2.3 Use MATLAB Editor to develop and run your first M-file allowing you to define the weekday based on its date. Use the following formula:

$$WD = \left(D + \left[\frac{1}{5}(13m-1) \right] + Y + \left[\frac{Y}{4} \right] + \left[\frac{c}{4} \right] - 2c + 777 \right) \mathrm{mod} 7$$

(square brackets in this expression denote the integer part and so use the `fix` function). In this formula, d is the day of month (1 through 31), m the month's number in the Ancient Rome system ($m = 1$ for March, $m = 2$ for April, etc. starting with March as 1 and ending with February as 12, that is, January and February of any year are considered to be in the previous year), Y the number of the year in a century (0–99), and c the number of centuries. For example, for 22 January 1963, the inputs would be: $d = 22$, $m = 11$, $Y = 63$, and $c = 19$. The value WD defined using the above formula is treated as follows: "0" means Sunday, "1" means Monday, "2" means Tuesday, etc.

To make your script more useful, develop it as an interactive program. The following provides an example of all but one line (which you need to develop yourself):

```
% This is my first M-script
clc
clear all
    Year =input('\nEnter the year (YYYY): ');
    Month=input('Enter the month: ');
    Day =input('Enter the day: ');
    fprintf('nYou entered %3.0f/%2.0f/%4.0f',Month,Day,Year)
d=Day;
    m=Month-2;
    if m <= 0
```

```
      m=m+12; Year=Year-1; end % month's number in Ancient Rome
Y=mod(Year,100);              % year's number in the century
c=(Year-Y)/100;              % number of centuries
    WeekDay=_____;       % here goes your formula
Week={'Sunday' 'Monday' 'Tuesday' 'Wednesday' 'Thursday'...
       'Friday' 'Saturday'};
fprintf('\n\nYour weekday is: %s\n', Week{WeekDay+1}); % result
```

Check what was the week day when Sputnik 1 was launched (4 October 1957), when Yuri Gagarin became the first human in outer space (12 April 1961), and when Neil Armstrong set his boot on the surface of the Moon (21 July 1969).

Chapter 3 | Arrays and Array Operations

- Classification of Arrays and their Indexing
- Standard and Non-Standard Matrix and Array Operations
- Handling Polynomials as the Vectors
- Handling Text Strings as Arrays

```
linspace, logspace, cat,
diag, zeros, ones, rand,
eye, magic, pascal, length,
size, numel, ndims,
reshape, spy, cross, dot,
sort, min, max, median,
sum, find, poly, roots,
residue, char, num2str,
str2num, int2str, conv,
and others
```

3.1 Introduction

To acquaint you with the MATLAB development environment, Chapter 2 used simple arithmetic involving some basic operations with single variables (scalars). Now it is time to explore the MATLAB commands in more depth. As mentioned in Chapter 1, one of the MATLAB's strengths is its capability to handle collections of numbers, called arrays, as if they were a single variable. For example, to add two same-size arrays, A and B, you can issue a single command, C = A + B, and MATLAB will add up all corresponding elements in A and B to produce C. In most other programming languages, this operation would require more than one line of code. Even if, up to this point, Microsoft Excel remains your favorite tool for data analysis, upon completion of this chapter you may change your mind—you will find MATLAB to be a much easier and a more powerful instrument for such work, making it a natural choice for addressing a variety of engineering problems. With array being the basic building block in MATLAB, this chapter explains how to create, address, and edit different types of arrays, and how to use array operations, including addition, subtraction, multiplication, division, and exponentiation, to solve practical problems. It also introduces nonstandard array operations and some specific commonly used array-handling functions that can be used in addition to those listed in Sec. 2.3.1. Finally, this chapter explains how MATLAB array formalism is used to handle polynomials and text strings.

3.2 Types of Arrays and Indexing Their Elements

By definition, an *array* is a group of elements (values or variables) that are related and can be referred by one or more *indices*. Such a collection is usually called an array variable or simply an array. For example, the airport weather stations provide hourly weather data, so that hourly temperature readings at any specific day can be stored as a one-dimensional (1-D) array, composed of 24 values of temperature. Such an array is also referred as a *vector*. If, along with a temperature, you would like to store other parameters, like air density, dew point, visibility, etc., totaling n variables per hour, you will need to add another dimension, that is, you will end up dealing with a 2-D array, holding $24 \times n$ variables.

Now, suppose you need to store these data for the first day of each month. It adds another dimension, so that with a 3-D array you will be able to store your $24 \times n \times 12$ pieces of data. So far, we talked about only one airport, but what if we are gathering the data from 100 airports? To effectively handle $24 \times n \times 12 \times 100$ values, you will now need a 4-D array. For storing this type of data for 10 consecutive years, you may need a 5-D array, $24 \times n \times 12 \times 100 \times 10$ and so on.

By analogy with the mathematical concepts of vector and matrix, an array type with one or two dimensions is often called a *vector* type or *matrix* type, respectively. The MATLAB classification of arrays featuring two broad classes of arrays—2-D arrays or matrices, and multidimensional (n-D) arrays—is shown in Fig. 3.1. In this case, the vectors (traditionally,

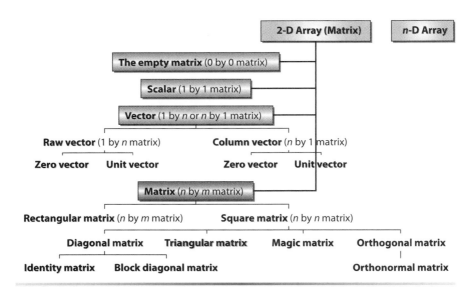

Fig. 3.1 MATLAB classification of arrays.

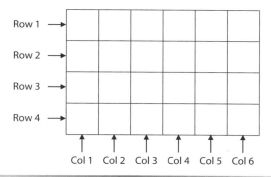

Fig. 3.2 Example of a 4 by 6 matrix.

1-D arrays) and even scalars are considered as subclasses of 2-D arrays. (Figure 3.1 also shows some other subclasses of a matrix which will be discussed later in this chapter.)

3.2.1 Matrices and Multidimensional Arrays

The first broad class of arrays in MATLAB is the 2-D arrays or matrices exemplified in Fig. 3.2.

A matrix composed of n rows and m columns is called an n by m matrix, so that these numbers, n and m, define a *size* of the matrix along each *dimension*. For example, the matrix shown in Fig. 3.2 is the 4 by 6 matrix and its symbolic form can be written as

$$A = [a_{ij}] = \begin{bmatrix} a_{11} & a_{12} & a_{13} & a_{14} & a_{15} & a_{16} \\ a_{21} & a_{22} & a_{23} & a_{24} & a_{25} & a_{26} \\ a_{31} & a_{32} & a_{33} & a_{34} & a_{35} & a_{36} \\ a_{41} & a_{42} & a_{43} & a_{44} & a_{45} & a_{46} \end{bmatrix} \tag{3.1}$$

The first subscript, i, changes in the vertical direction (refers to a *row* number) and the second subscript, j, changes in the horizontal direction (refers to a *column* numbers).

Therefore, a vector can be regarded as a special case of a matrix, where one of the dimensions is equal to 1. A row vector, with its elements stored "horizontally," is considered to be a 1 by n matrix, whereas a column vector with one element following another in a column format is considered to be an n by 1 matrix, as shown in the examples of Fig. 3.3.

From Fig. 3.1, it also follows that MATLAB scalars are treated as 1 by 1 matrices. For the sake of consistency, the empty (0 by 0) matrix is introduced.

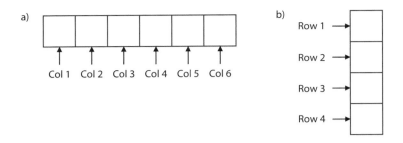

Fig. 3.3 Examples of a) row and b) column vectors.

To create a 0 by 0 matrix, use the square bracket operators with no value specified:

```
>> A = [];
```

This 0 by 0 matrix concept happens to be very helpful when the entire row(s) and/or column(s) are needed to be deleted and it will be discussed later.

Figure 3.4 presents an example of a 3-D array, where elements are organized as pages (slices or layers). Geometrical interpretation for arrays with a higher number of dimensions is not possible (we live in three dimensions ourselves), but you may consider a 4-D array to be a collection of several 3-D arrays, a 5-D array to be a collection of 4-D arrays, and so on.

3.2.2 Addressing Array Elements

Storing ordered data in arrays has a huge advantage because you can easily access any specific piece of it. Addressing the elements, and groups

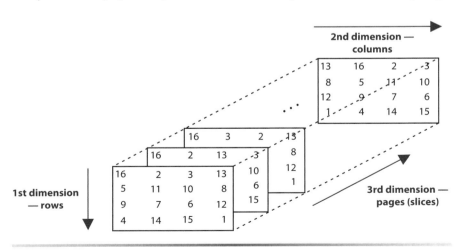

Fig. 3.4 Example of a 3-D array.

of elements, in the arrays is quite evident. For instance, given the 1-D array x with 24 elements

x(:)	Addresses the entire row (or column) of vector x, that is, all 24 elements
x(2:6)	Addresses four elements, with the indices 2–6, that is, x(2), x(3), x(4), x(5), and x(6)
x(20:end-2)	Addresses three elements, x(20), x(21), and x(22)
x(2:5:20)	Addresses four elements, x(2), x(7), x(12), and x(17)
x(6:-2:2)	Addresses three elements, x(6), x(4), and x(2)
x([1:3,5,22])	Addresses five elements, x(1), x(2), x(3), x(5), and x(22)

These six examples cover all possible ways of constructing an *indexing expression* to address array elements by their indices $ind_1, \ldots, ind_{last}$ (which must be positive numbers starting from 1). The general format to compute these indices is

```
first : increment : last
```

with the index of the first element to address, ind_1, being first; the next one, ind_2, first+increment; followed by ind_3, being first+2* increment; etc., all way up (or down) until $ind_{last} \leq$ last (in the case of positive increment) or $ind_{last} \geq$ last (in the case of negative increment). The first and last entries must be positive integers (1, 2, 3, ...), increment—a nonzero integer. The default value of increment (if it is omitted) is 1. The first and last entries may be omitted as well and in this case the list of indices includes all of the available range. The function end serves as the last available index and can be used by itself or in algebraic expression. Obviously, the following inequalities should hold: first ≤ last (in case increment > 0) and first ≥ last (in case increment < 0). That being said, any of the following addressing

```
x(4:4)
x(4:2:4)
x(4:-1:4)
```

is totally acceptable and returns just a single element, x(4). The next three

```
x(6:2:5)
x(8:0:10)
x(20:-1:22)
```

are acceptable (although making no sense), returning an empty matrix 1 by 0. However, the following two:

```
x(-2)
x(25)
```

return an error notifying that the index exceeds matrix dimensions.

The syntax employing brackets allows you to address a group of irregularly spaced elements. As shown in the previous example, a regular indexing expression can be a part of such addressing as well. Expressions in the brackets can be separated by space, comma, or semicolon.

The same rules apply to arrays with more than just one dimension as well. The only difference is that you need to provide an indexing expression for each dimension, separated by comma(s). For example,

`A(:,4)`	Addresses all the elements in the fourth column of a matrix `A`
`A([2 5],1:4)`	Addresses the elements of the first through fourth column in the second and fifth rows of a matrix `A`
`A(end,end-1)`	Addresses the last but one element in the last row of a matrix `A` (note that, the `end` function computes the last available index for each dimension separately)
`B(2,5,:)`	Addresses the fifth element in the second row on all pages of a 3-D array `B`

Keeping in mind that a vector can be regarded as a special case of a matrix, where one of the dimensions is equal to 1 (Fig. 3.1), the addressing `x(5)`, the so-called *linear indexing*, is equivalent to `x(1,5)` for the row vector, and addressing `x(3)` is equivalent to `x(3,1)` for the column vector (Fig. 3.5). This latter type of indexing is called *row–column subscripts* (row–column–page for the 3-D arrays, etc.).

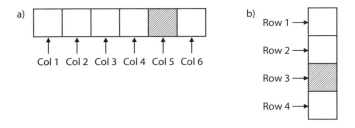

Fig. 3.5 Addressing elements of a) row and b) column vectors.

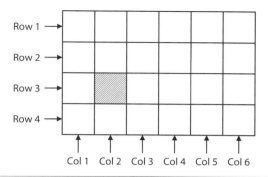

Fig. 3.6 Addressing elements of matrix A.

In computer's memory, arrays are stored, not in the shape they appear when displayed in the MATLAB Command window, but as a single column of elements, one element following another, with the indices of the leftmost dimension cycling the fastest. It means that in the case of a matrix, the elements are stored column by column (the columns index cycles the first), each appended to the last. In the case of a 3-D array, they are stored page by page with each page stored by columns and so on. Hence, MATLAB permits linear indexing even for multidimensional arrays. For instance, the easiest way to address the element located in the ith row and jth column of an n by m matrix A, is obviously A(i,j). However, if needed you may also use its linear indexing equivalent A(n*(j-1)+i). For an example shown in Fig. 3.6, addressing A(3,2) is equivalent to that of A(7).

The two useful MATLAB functions, sub2ind and ind2sub, help to convert subscripts to linear index and vice versa, for example, for the case shown in Fig. 3.6 the command

```
>> i=sub2ind(size(A),3,2)
```

returns i=7, and the command

```
>> [r,c]=ind2sub(size(A),6)
```

returns r = 2 and c = 2. These functions are applicable to multidimensional arrays as well (you simply increase the number of input/output arguments as appropriate).

Figure 3.7 summarizes different ways of addressing array elements and presents some more self-explanatory examples.

3.2.3 Creating Arrays

Usually, you only create and fill in simple vector arrays, while the larger arrays are obtained as a result of computations. So let us proceed with a

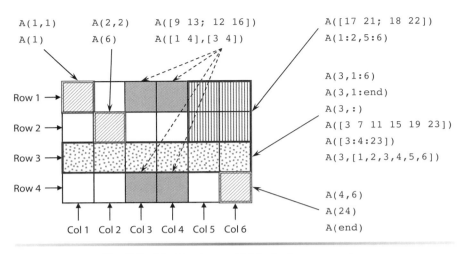

Fig. 3.7 Matrix element(s) indexing examples.

couple of simple examples of creating arrays manually and then introduce several useful functions simplifying this process. Let us start from the vectors. For example, the command a=[1,2,3] or a=[1 2 3], where comma, or space(s) are used to separate the elements, creates the same 1 by 3 row vector a:

```
>> a = [1,2,3]
a = 1 2 3
```

The command b = [4;5;6] (where a semicolon is used to separate elements) or b = [4,5,6]' (where the apostrophe is used to transpose a row vector) creates the same 3 by 1 column vector b

```
>> b=[4;5;6]
a = 4
    5
    6
```

The following two commands create examples of the *unit vectors* (with a vector magnitude or, to be more precise, Euclidian norm to be equal to 1):

```
>> c = [0, 1, 0]
c = 0   1   0
>> d = [sqrt(2)^-1; sqrt(2)^-1]'
d = 0.7071   0.7071
```

In the previous examples, we used comma or space to separate the elements in a row, and a semicolon to separate them in a column. On top of that we used transpose operation to transpose a row vector to a column vector and vice versa. Now we can use the same syntax to create matrices. For example, the same 3 by 2 matrix C can be created by issuing any of the following commands: `C = [1,4;2,5;3,6]`, `C = [1,2,3;4,5,6]'`, `C = [1:3;4:6]'`, `C = [a',b]`, and `C = [a;b']'` (in the two latter cases, we are using the vectors `a` and `b` created earlier)

```
>> C = [1:3;4:6]'
C = 1   4
    2   5
    3   6
```

If it is more convenient, you may also enter the elements of this matrix row-by-row—MATLAB will recognize what you are trying to do

```
>> C = [1   4
        2   5
        3   6];
```

(you will need to end each line hitting the <Enter> key).

Similarly, to create the first two pages of the 3-D array shown in Fig. 3.4 each page must be assigned individually, for example,

```
>> A(:,:,1) = [16,2,3,13;5,11,10,8;9,7,6,12;4,14,15,1]
>> A(:,:,2) = [16,2,13,3;5,11,8,10;9,7,12,6;4,14,1,15]
```

yields

```
A(:,:,1) = 16    2    3    13
            5   11   10     8
            9    7    6    12
            4   14   15     1
A(:,:,2) = 16    2   13     3
            5   11    8    10
            9    7   12     6
            4   14    1    15
```

Then, you can duplicate these pages as follows:

```
>> A(:,:,3:4) = A(:,:,1:2);
```

In addition to inputting elements' values manually as it was shown earlier, there are several other methods that can be used. Implicitly, we have introduced one of them when discussing indexing already. The simplest way to create regularly spaced vectors is to use a colon operator. Consider the following examples:

```
>> x = 1:10
x = 1    2    3    4    5    6    7    8    9    10
>> y = -1:0.1:1
y = Columns 1 through 8
    -1.0000  -0.9000  -0.8000  -0.7000  -0.6000  -0.5000  -0.4000  -0.3000
    Columns 9 through 16
    ...
```

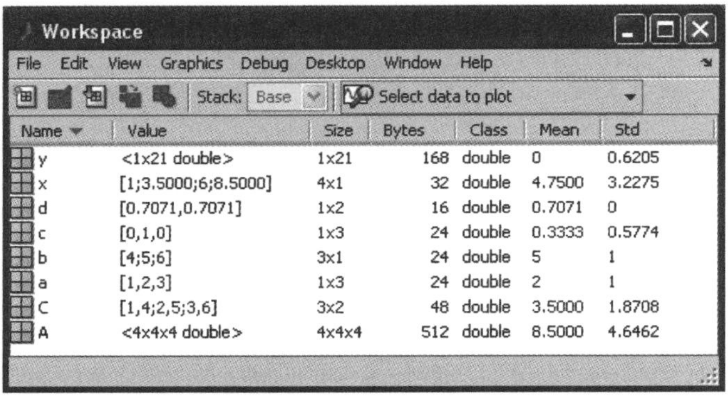

Fig. 3.8 Workspace with a variety of arrays.

Note that you can also use syntax with brackets, like x = [1 : 10], which is especially useful if you want to create a column vector:

```
>> x = [1 : 2.5 : 10.2]'
x = 1.0000
    3.5000
    6.0000
    8.5000
```

Obviously, all rules introduced in Sec. 3.1.2 apply, and the only difference is that all three values, the beginning value, increment, and the last value, can be given as the real numbers, not necessarily integer numbers as was the case for indexing.

Figure 3.8 shows the Workspace window with all arrays created so far in this section. As seen, to store each element of an array MATLAB dedicates eight bytes of memory (by default). For instance, to store $4 \times 4 \times 4 = 64$ elements of the array A 512 bytes are required. Also, Fig. 3.8 features some additional columns of the Workspace window allowing you to compute a mean value and standard deviation for each array (compare it with two of those given in Fig. 2.6).

An alternative way of doing essentially the same as a pair of brackets is to use the linspace function, linspace(x1,x2,n), which returns n numbers equally spaced between x1 and x2. For example,

```
>> xlin = linspace(1,100,30);
```

creates a row vector with 30 elements spread from $1(10^0)$ to 100 (10^2). The number of points n can be omitted and the default assumption value of 100 is used. The increment can be computed using a simple formula

$$\text{incr} = \frac{x_2 - x_1}{n-1} \tag{3.2}$$

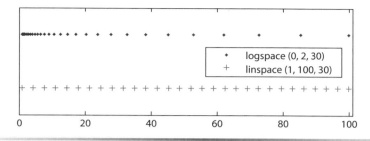

Fig. 3.9 Distribution of the elements in the vectors generated using the `linspace` and `logspace` functions.

Another command to create regularly spaced vectors, but in the logarithmic domain, is `logspace`. Syntax `logspace(a,b,n)` creates n logarithmically spaced points between `10^a` and `10^b`. For example,

```
>> xlog = linspace(0,2,30);
```

creates a row vector with 30 elements spread from 1 (10^0) to 100 (10^2). The number of points n can be omitted and the default assumption value of 50 is used. For better understanding, the distribution of elements in both vectors, `xlin` and `xlog`, is shown side-by-side in Fig. 3.9.

Once you have some vectors and matrices created already, you may use a variety of the functions to modify them. For instance, (block) diagonal matrices and diagonals of matrix can be created using the following functions (see illustrations in Fig. 3.10):

`diag(v)`	Puts elements of the vector v on the main diagonal of a *square* matrix, that is, the matrix that has the same number of rows and columns
`diag(v,k)`	Puts n elements of the vector v on the kth diagonal of a square matrix of the order n+abs(k) (for the main diagonal *k* = 0)
`diag(X)`	Returns a column vector formed from the elements of the main diagonal of a matrix X
`diag(X,k)`	Returns a column vector formed from the elements of the kth diagonal of a matrix X
`blkdiag(A,B,C,…)`	Outputs a block diagonal matrix, constructed from matrices A, B, C, …

When creating the matrix C in the beginning of this section, we introduced a couple of simple examples of how you could concatenate two vectors into a matrix: `C= [a',b]` and `C= [a;b']'`. Note that we had to use a transpose operation to bring the vectors to the same form (column vector or row vector). Moreover, it worked only because they had the same number of elements.

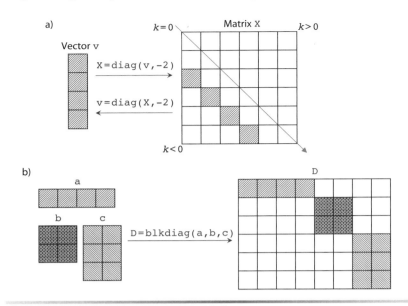

Fig. 3.10 Illustration of a) `diag` and b) `blkdiag` functions.

In the general case, to append one matrix to another MATLAB offers several special concatenation functions: `cat`, `vertcat`, and `horzcat`.

By definition, `cat(n,A,B,C,...)` creates a new array by concatenating the arrays A, B, and C along a dimension n (their length in the corresponding dimension should match). The `vertcat` and `horzcat` functions are subsets of the `cat` function and simplify things in the case of matrices as follows (see graphical illustration in Fig. 3.11):

`vertcat(A,B)`   Same as `cat(1,A,B)` and equivalent to `[A;B]`
`horzcat(A,B)`   Same as `cat(2,A,B)` and equivalent to `[A,B]`

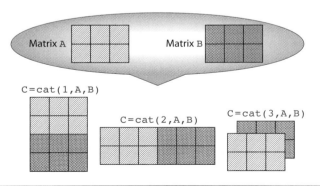

Fig. 3.11 Illustration of the `cat` function options.

MATLAB also features several useful functions to create special arrays (see examples of using these functions in Tables 3.1 and 3.2)

`ones(m,n,p,…)`	Creates an $m \times n \times p \times …$ array of ones (`ones(n)` creates a square $n \times n$ matrix of ones)
`zeros(m,n,p,…)`	Creates an $m \times n \times p \times …$ array of zeros (`zeros(n)` creates a square $n \times n$ of zeros, called a *null matrix* and denoted by **0**)
`rand(m,n,p,…)`	Creates an $m \times n \times p \times …$ array with uniformly distributed (on the interval (0.0, 1.0)) pseudorandom entries (`rand(n)` creates a square $n \times n$ matrix with random entries)
`randn(m,n,p,…)`	Creates an $m \times n \times p \times …$ array with normally distributed (chosen from a normal distribution with a mean value of 0, variance of 1, and standard deviation of 1) random entries (`randn(n)` creates a square $n \times n$ matrix with random entries)
`randi(imax,m,n,p,…)`	Creates an $m \times n \times p \times …$ array containing pseudorandom integer values drawn from the discrete uniform distribution on `1:imax` (`randi(imax)` generates a single number, whereas `randi(imax,n)` creates a square `n×n matrix`)
`eye(m,n)`	Creates an $m \times n$ matrix with one's on the diagonal and zeros elsewhere (`eye(n)` creates a square $n \times n$ matrix, called an *identity matrix* and denoted by **I**, where all of the elements are zeros except the diagonal elements, which are ones)
`magic(n)`	Creates a "magic" square constructed from the integers 1 through n^2 with equal row, column, and diagonal sums (produces valid magic squares for all $n > 0$ except $n = 2$)

Table 3.1 Examples of Natural-Number Arrays

Command	`ones(3,4)`	`zeros(3)`	`eye(3,4)`	`magic(3)`	`pascal(3)`
Output	1 1 1 1 1 1 1 1 1 1 1 1	0 0 0 0 0 0 0 0 0	1 0 0 0 0 1 0 0 0 0 1 0	8 1 6 3 5 7 4 9 2	1 1 1 1 2 3 1 3 6

Table 3.2 Examples of the `rand` and `randn` Functions Outputs

Command	rand(3)			randn(3)		
Output	0.8147	0.9134	0.2785	−0.4326	0.2877	1.1892
	0.9058	0.6324	0.5469	−1.6656	−1.1465	−0.0376
	0.1270	0.0975	0.9575	0.1253	1.1909	0.3273

`pascal(n)`   Creates the Pascal matrix of the order `n` (a symmetric positive definite matrix with integer entries), where any element is computed by adding the value of the element above it to the value of the element to its left, that is, $a_{ij} = a_{i-1;j} + a_{i;j-1}$, with the very first element $a_{11} = 1$

Note that the functions `eye`, `magic`, and `pascal` are only applicable to create matrices, whereas the remaining functions can also be used to produce multidimensional arrays (applying an obvious syntax change).

Two more functions help in creating lower and upper triangular matrices (Fig. 3.12)

`triu(B,k)`   Returns the upper triangular part of matrix `B` (on and above the `k`th diagonal of `B`)

`tril(B,k)`   Returns the lower triangular part of matrix `B` (on and below the `k`th diagonal of `B`)

Finally, the `repmat` function allows you to replicate and tile arrays. Specifically, the statement `repmat(A, [m,n])` creates a matrix consisting of

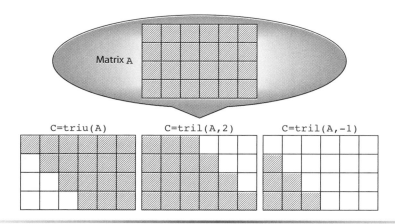

Fig. 3.12 Illustration of creation of lower and upper triangular matrices.

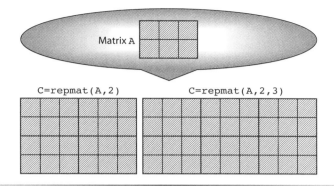

Fig. 3.13 Illustration of the `repmat` function.

an *m* by *n* tiling of copies of A, and the statement `repmat(A,n)` creates an *n* by *n* tiling (see graphical illustration in Fig. 3.13).

For instance, in the beginning of this section, we used the following command to replicate two layers of the 4 by 4 by 2 array A:

```
>> A(:,:,3:4) = A(:,:,1:2);
```

Instead, we could use

```
>> A=repmat(A,[1,1,2])
```

More functions, creating other specific arrays and matrices, are also available in MATLAB and its toolboxes. Hence, you are welcome to use MATLAB Help browser, to learn more.

3.2.4 Reshaping, Editing, and Displaying Arrays

Once you have an array you may reshape it to another size or dimensions using the `reshape(A,m,n,p,...)` function. Here A is the name of an array, and m, n, p, ... are the number of elements along each dimension in a proposed new (multidimensional) array. No need to mention that the number of elements in the new array, m*n*p*..., should be the same as the number of elements in A.

Consider an example. Suppose you have a 3 by 4 matrix you want to reshape. To better understand the underlying concept, we have to recall that arrays are stored in the computer's memory as a 1-D sequence, column vector, which is illustrated in Fig. 3.14a. Therefore, the `reshape(A,m,n)` command applied to matrix A will return an m by n matrix, whose elements are taken one by one from that column vector shown on the right-hand side of Fig. 3.14a. Of course, it will do it only if m*n yields 12, otherwise an error occurs indicating that A does not have m*n elements. Some possibilities of reshaping our specific array A are presented in Fig. 3.14b. On three occasions, we only changed the size of the

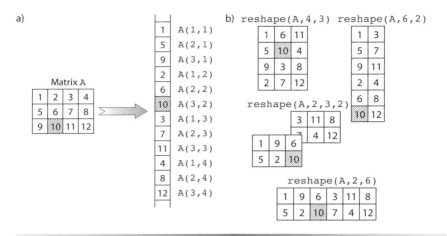

Fig. 3.14 Example of a) storing a 3 by 4 matrix in the computer's memory and b) reshaping it.

original matrix (to that of 4 by 3, 6 by 2, and 2 by 6) and once we reshaped our matrix to a 3-D array (2 by 3 by 2).

One more option to change the size (and content) of an existing array is to use the empty matrix formalism introduced in Fig. 3.1. For instance, creating the matrix Ah as

```
>> Ah = [0.7680    0.4387    0.3200    0.7446    0.6833
         0.9708    0.4983    0.9601    0.2679    0.2126
         0.9901    0.2140    0.7266    0.4399    0.8392
         0.7889    0.6435    0.4120    0.9334    0.6288];
```

and issuing the following commands

```
>> Ah(:,[2,4])=[]; Ah(3,:)=[]
```

eliminates the second and fourth columns as well as the third row from the matrix Ah and returns

```
Ah = 0.7680    0.3200    0.6833
     0.9708    0.9601    0.2126
     0.7889    0.4120    0.6288
```

So far, we have discussed changing arrays programmatically, but of course you can always employ the **Variable Editor** of the MATLAB development environment. Double-clicking a variable in the Workspace browser, or using `openvar variablename` command, brings up the **Variable Editor** window as shown in Fig. 3.15. You may use the **Variable Editor** to view and edit any variable (array) available in workspace as well as load data from somewhere else. It also allows you to visualize the content of numerical arrays using a variety of plotting functions (see drop-down window in Fig. 3.15).

MATLAB also has other means for viewing matrices. One of them, the `spy(A)` function, is especially useful for visualizing sparse matrices that have a lot of zero elements. Consider the following set of commands to create a sparse matrix z:

```
>> z=ones(100,200);
>> z(10:50,150:170)=0;
>> z(70:80,20:35)=0;
>> z(4:5:100,50:130)=0;
>> z(3:5:100,50:130)=0;
>> z(2:5:100,50:130)=0;
>> z=z-eye(100,200);
```

Then the command

```
>> spy(z)
```

brings up the window shown in Fig. 3.16. As seen, this command produces a template view of the sparse structure, where each point on the graph represents the location of a nonzero element. Below the graph, the number of nonzero elements (nz) is shown explicitly.

You may also use another MATLAB function, `imagesc(A)`, which finds the spread of elements in a matrix A and matches it to the full range of the current `colormap` (to be considered in Sec. 6.8), presenting the matrix as a color picture, where each color corresponds to a certain magnitude. For example, Fig. 3.17a shows how the matrix z, developed earlier, would look using the set of the following two commands:

```
>> imagesc(z)
>> colorbar
```

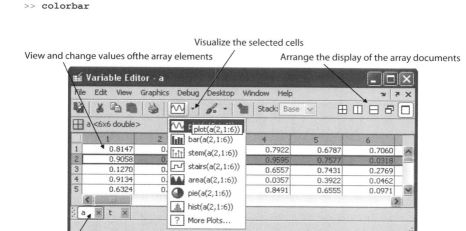

Fig. 3.15 Variable Editor window.

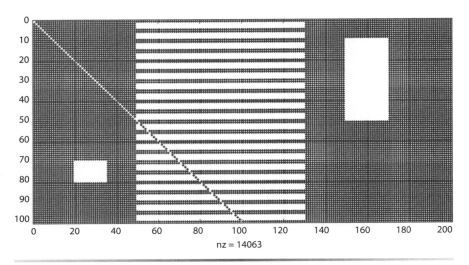

nz = 14063

Fig. 3.16 Visualization of a matrix sparsity structure using the `spy` function.

(the `colorbar` function resizes the current axes to display the current `colormap` to the right of the plot).

Figure 3.17b shows another example, visualizing the matrix Ah we introduced earlier, using another color map (to be discussed in Chapter 6) and slightly modified syntax of the `imagesc` function:

```
>> imagesc(Ah,[0.2 1])
>> colormap gray
>> colorbar
```

where the vector `[0.2 1]` specifies the range of the color bar explicitly. As will be discussed in Chapter 6, some other functions, typically used for 3-D plotting, can be utilized to visualize arrays as well.

Finally, we should warn you that when dealing with arrays and using the simple nonunique names for them, you have to be careful with what you have in workspace already. For example, suppose you have some 4 by 7 array x (say you created it by issuing the `x = zeros(4,7)` command) and later on you decided to use the same name x to create another array, completely forgetting that you have another one already. Well, if your new array has the same size, that is, 4 by 7, and you will be defining all elements of this array, you will have no trouble because the new values will override the old ones. However, if your new x has a smaller size you may run into a problem. For instance, typing

```
>> x(1:3,1:3)=ones(3)
```

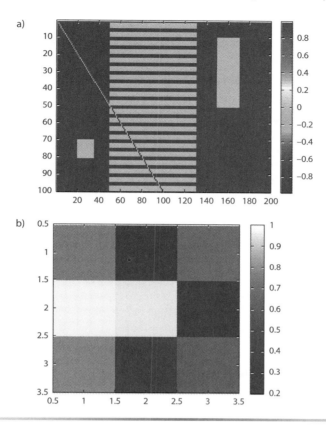

Fig. 3.17 Visualization of matrices a) `z` and b) `Ah` using the `imagesc` function.

"unexpectedly" returns

```
x = 1    1    1    0    0    0    0
    1    1    1    0    0    0    0
    1    1    1    0    0    0    0
    0    0    0    0    0    0    0
```

To avoid possible problems, either use the unique names or clear variables (arrays) you do not need anymore. In the latter case use

```
>> clear variablename
```

to delete just one variable `variablename` or

```
>> clear all
```

to clear the entire workspace. The wildcard character "`*`" can also be used to clear variables that match a pattern. For example,

```
>> clear x*
```

clears all variables (arrays) starting from x. (Now you understand why in the beginning of Sec. 2.1.3 we created a shortcut button **Prepare**, which includes the `clear all` command—it is not a bad idea to push it once in a while to start with the bland print.)

3.3 Array Operations

Applied to arrays, MATLAB allows you to use arithmetic operators +, -, *, /, and ^ in two different ways. The following considers both of them, starting from the conventional way of treating arrays resulting in standard matrix operations, and then proceeding with nonstandard operations.

3.3.1 Standard Array Operations

The three basic operations that can be applied to modify matrices are *matrix addition, scalar multiplication,* and *transposition.*

Let us start from transposition, which we have used throughout this chapter. The transposition

```
>> D=A';
```

interchanges rows and columns of the matrix A, so that

$$D = [d_{ij}] = [a_{ji}] \tag{3.3}$$

In fact, in MATLAB this operation is called a *complex conjugate transpose* because in the case the matrix A has complex elements, this operation not only interchanges rows and columns but also changes any complex number to its complex conjugate, for example, if $A = [a_{ij} + jb_{ij}]$, then $D = [a_{ji} + jb_{ji}]$. MATLAB also has an alternative that does not involve conjugation. The way to apply this type of transposition is to use the following syntax:

```
>> D=A.';
```

(It is the period character (.) that distinguishes between two transposition operations.) In this case, $D = [a_{ij} + jb_{ji}]$.

Given the same-size matrices A and B, matrix addition uses as simple syntax as

```
>> C = A + B;
```

The resulting, same-size matrix C is composed of the sum of the corresponding elements

$$C = [c_{ij}] = A + B = [a_{ij} + b_{ij}] = \begin{bmatrix} a_{11} + b_{11} & a_{12} + b_{21} & \cdots & a_{1n} + b_{1n} \\ a_{21} + b_{21} & a_{22} + b_{22} & \cdots & a_{2n} + b_{2n} \\ \vdots & \vdots & \ddots & \vdots \\ a_{m1} + b_{m1} & a_{m2} + b_{m2} & \cdots & a_{mn} + b_{mn} \end{bmatrix} \qquad (3.4)$$

The same-size requirement is very important. Hence, you cannot add the row and column vectors together. Scalar addition is the only exception to size rule. In this case, the scalar is added to each element in the array. For instance, if in the previous example B is a scalar, then

$$C = [c_{ij}] = A + B = [a_{ij} + B] = \begin{bmatrix} a_{11} + B & a_{12} + B & \cdots & a_{1n} + B \\ a_{21} + B & a_{22} + B & \cdots & a_{2n} + B \\ \vdots & \vdots & \ddots & \vdots \\ a_{m1} + B & a_{m2} + B & \cdots & a_{mn} + B \end{bmatrix} \qquad (3.5)$$

MATLAB also assures *commutative* property, that is,

A+B returns exactly the same result as B+A

However, the *associative* property

(A+B) +C should return the same result as A+ (B+C)

generally does not hold. Chapter 8 discusses how this is related to the way the numbers are represented in the computer's memory.

All of the above holds for the *matrix subtraction* as well (you just need to replace a addition sign with a subtraction sign).

The *scalar multiplication* k*A of a matrix A and a scalar k is given by multiplying every entry of A by k:

$$k*A = [ka_{ij}] = \begin{bmatrix} ka_{11} & ka_{12} & \cdots & ka_{1n} \\ ka_{21} & ka_{22} & \cdots & ka_{2n} \\ \vdots & \vdots & \ddots & \\ ka_{m1} & ka_{m2} & \cdots & ka_{mn} \end{bmatrix} \qquad (3.6)$$

Table 3.3   The Function Form of Standard Arithmetic Operators

Function	Description	Character
plus	Plus	+
uplus	Unary plus	+
mtimes	Matrix multiply	*
mpower	Matrix power	^
minus	Minus	−
uminus	Unary minus	−
mrdivide	Slash or right matrix divide	/
mldivide	Backslash or left matrix divide	\

The commutative property holds here as well, that is, k*A returns the same result as A*k. The same applies to the *scalar division*. In this case, A/k returns the same result as k\A.

Both transpose operations are compatible with addition and scalar multiplication, that is, (k*A)' returns the same result as k*A' and (A+B)' is equivalent to A'+B' (the same holds for .').

Two of the three aforementioned basic operations, addition (subtraction) and scalar multiplication (division) are also applicable to multidimensional arrays. An attempt to apply transposition, however, will return an error. For example,

```
>> ones(2,2,2)'
```

results in

```
??? Error using ==> ctranspose
Transpose on ND array is not defined.
```

Note that ctranspose is a *functional form* (as opposed to the *command form*) of transposition operator, ', so that you could use

```
>> ctranspose(ones(2,2,2))
```

instead. (The function form of the complex conjugate transpose is transform.) Other arithmetic operations have their function form as well and are shown in Table 3.3.

As opposed to the scalar multiplication, the matrix multiplication of two matrices A and B,

```
>> E=A*B;
```

can only occur when their inner dimensions agree in size (that is, conformable). If A is an $m \times n$ matrix and B is an $n \times p$ matrix, then the product A*B exists since n is the same inner dimension of both A and B (in the sense that the $m \times n$ matrix is multiplied by the $n \times p$ matrix, that is, $m \times n * n \times p$), that is, equal to each other. However, in the case when $m \neq p$, the product B*A does not exist (the inner dimension of B is p and the inner dimension of A is m). By definition, the ijth element of E is computed as a sum of the products of elements of ith row of A and jth column of B

$$E = [e_{ij}] = A*B = \left[\sum_{k=1}^{n} a_{ik} b_{kj} \right] \tag{3.7}$$

Therefore, even if both A and B are square matrices, so that both products A*B and B*A exist, the product A*B is not equal to B*A, that is, the commutative property does not hold. There are two exceptions:

1. If one of the matrices is a null matrix **0**, then A***0** yields the result as **0***A.
2. If one of the matrices is an identity matrix **I**, then A***I** provides the same result as **I***A.

You may expect that at least two other properties, associative and distributive, hold as it is the case in symbolic computations. In general, due to round off errors (to be discussed in Chapter 8), neither of them holds, that is, A*(B+C) is not exactly the same as A*B+A*C, and (A*B)*C may return a slightly different result compared to that of A*(B*C). Of course, we are talking about negligibly small errors, but they are present and may cause significant damage to the overall accuracy of your numerical computations. Chapter 8 discusses practical ways of mitigating these types of errors.

In the case when A is a row vector and B is a column vector with the same number of elements, the result of E=A*B is a scalar. This specific case is referred to as a *scalar* or *dot product* and has its own function form, dot, which can be used along with mtimes. For example,

```
>> E=dot([1 2 3 4],[4 3 2 1])
E=20
```

Now, if A is a column vector and B is a row vector, and they are composed of three elements each, then according to Eq. (3.7) their *matrix product* returns a 3 by 3 matrix as follows:

$$F = [f_{ij}] = A*B = \begin{bmatrix} a_1 b_1 & a_1 b_2 & a_1 b_3 \\ a_2 b_1 & a_2 b_2 & a_2 b_3 \\ a_3 b_1 & a_3 b_2 & a_3 b_3 \end{bmatrix} \tag{3.8}$$

However, there is another standard mathematical operation we may have between two three-element vectors, which is known as a *cross product*. In such a case

$$\mathrm{K} = \mathrm{A} \times \mathrm{B} = [a_2 b_3 - a_3 b_2 \quad a_3 b_1 - a_1 b_3 \quad a_1 b_2 - a_2 b_1] \tag{3.9}$$

($\times$ denotes a cross product operation). To compute a `cross` product, MATLAB has the `cross` function. The following set of commands produces the plot presented in Fig. 3.18a

```
A=[1.5,0,0]; B=[0,0.8,0];
K=cross(A,B);
        quiver3(0,0,0,A(1),A(2),A(3),'b','Linewidth',3), hold
        quiver3(0,0,0,B(1),B(2),B(3),'g--','Linewidth',3)
        quiver3(0,0,0,K(1),K(2),K(3),'r-.','Linewidth',3), axis equal
xlabel('axis x'), ylabel('axis y'), zlabel('axis z')
legend('Vector A','Vector B','cross(A,B)',2)
view([-140 20])
```

Specifically, the first line introduces two noncollinear vectors, and the second line utilizes the `cross` function, creating the vector, which is perpendicular to those two. The remaining lines utilize the plotting functions to be discussed in Chapter 6.

Using this example, let us introduce the remaining matrices presented in Fig. 3.1. An *orthogonal matrix* is a square matrix that has *orthogonal columns*. To define orthogonal columns, we consider each column as a separate column vector. Mathematically, if a matrix $\mathbf{V}$ is composed of n column vectors $[\mathbf{v}_1, \mathbf{v}_2, \ldots, \mathbf{v}_n]$, such that $\mathbf{v}_i^T \mathbf{v}_j = 0$, when $i \neq j$, then $\mathbf{v}_1, \mathbf{v}_2, \ldots, \mathbf{v}_n$ are said to be an *orthogonal set of vectors*.

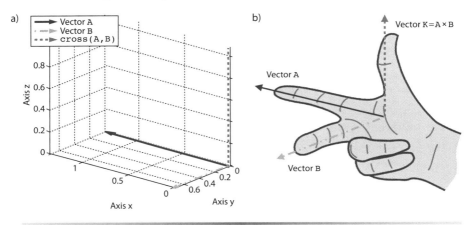

Fig. 3.18 Illustration of the `cross` function output producing a right-hand system according to the right-hand rule.

In our case, vectors A, B, and K form an orthogonal set of vectors, because the vectors A and B were chosen to be orthogonal. Formally, it can be easily proven by checking their pair-wise orthogonality

```
>> [A*B', A*K', B*K']
ans = 0    0    0
```

Therefore, the matrix

```
>> OrM = [A' B' K']
OrM = 1.5000           0           0
           0      0.8000           0
           0           0      1.2000
```

(we are using transposition because our vectors were defined as the row vectors) represents an example of orthogonal matrix. Moreover, the `cross` function complements its two input arguments, the 3-D vectors, in such a way that geometrically the three vectors, A, B, and K, form a right-hand system according to the right-hand rule (Fig. 3.18b).

In addition, if these orthogonal vectors are all unit vectors, they form an *orthonormal matrix*. A unique property of orthonormal matrices is that, $\mathbf{Q}^T\mathbf{Q} = \mathbf{I}, \mathbf{Q}\mathbf{Q}^T = \mathbf{I}$ (and $\mathbf{Q}^T = \mathbf{Q}^{-1}$). Let us now check if our orthogonal matrix is orthonormal

```
>> OrM * OrM'- eye(3)
ans = 1.2500           0           0
           0     -0.3600           0
           0           0      0.4400
```

Since the return is not a null matrix, the `OrM` matrix is not orthonormal, or the column vectors are not the unit vectors. Can we make it orthonormal? Yes. All we have to do is to normalize our column vectors. To do this, the `norm` function calculates several different types of matrix norms including the Euclidean norm of a vector

$$\| \mathbf{X} \| = \sqrt{\sum_{k=1}^{n} x_k^2} \tag{3.10}$$

For example,

```
>> a = [1,2,3,4]; norm(a)
```

returns

```
ans = 5.4772
```

Normalization implies dividing a vector by its norm, so that a normalized vector's norm is 1

```
>> a_norm = a/norm(a)
a_norm = 0.1826    0.3651    0.5477    0.7303
```

```
>> norm(a_norm)
ans = 1.0000
```

Equipped with this knowledge, we may now construct an example of an orthonormal matrix as follows:

```
>> OnM = [A'/norm(A) B'/norm(B) K'/norm(K)]
OnM = 1     0     0
      0     1     0
      0     0     1
```

Before we proceed with other operators of Table 3.3, let us finish with the `cross` function. This function is also applicable to matrices or even (multidimensional) arrays, which have at least one dimension of length 3. In such a case, the `cross` returns the cross product of A and B along the first dimension of length 3.

The *matrix power* operation (`mpower` or `^`) is applicable only to *square* matrices, which have the same number of rows and columns, in two ways. First, you may use it with a matrix base and a scalar exponent, A^n. If n is a positive integer, the power is computed by repeated squaring. For example,

```
>> magic(3)^3;
```

produces the same result as

```
>> magic(3)*magic(3)*magic(3)
ans = 1197    1029    1149
      1077    1125    1173
      1101    1221    1053
```

(If n is a negative integer, A is inverted first.) For other values of n, the calculation involves eigenvalues and eigenvectors, to be discussed in Chapter 9.

Second, you may have a scalar base with a matrix exponent n^A. In this case, the resultant n by n matrix is evaluated as the power series

$$n^A = \sum_{k=0}^{\infty} \frac{\ell n(n)^k A^k}{k!} \tag{3.11}$$

An expression involving matrix base with matrix exponent is not valid and returns an error

```
??? Error using ==> mpower
At least one operand must be scalar.
```

To perform a matrix division, A/B or A\B, using either one of the two remaining functions of Table 3.3 (`mrdivide` and `mldivide`), both matrices, A and B, must have the same number of rows. The only exception is when one of two matrices is, in fact, a scalar; specifically, when B happens to be a scalar in A/B or A being a scalar in $A\backslash B$ then the problem reduces to a scalar division (all elements of the resultant matrix are divided by a scalar). In general, this operation has to do with solving systems of linear algebraic equations and will be covered in detail later in Chapter 8.

3.3.2 Nonstandard Array Operations

Along with standard array (matrix) operations discussed in the previous section, MATLAB allows you to treat arrays as groups of numbers, not common-sense arrays and apply nonstandard operations to them. The list of these nonstandard operations can be found in Table 3.4 and differs from those shown in Table 3.3 by a period character (.), as was the case with transpose operation in the previous section. The intention of these non-standards operations is to utilize element-by-element operations as opposed to matrix-by-matrix operations.

Element-by-element multiplication is only applicable to the same-size arrays, and results in multiplying the corresponding elements in two matrices by each other, as shown for the 2-D case

$$C = [c_{ij}] = A.*B = [a_{ij}b_{ij}] = \begin{bmatrix} a_{11}b_{11} & a_{12}b_{21} & \cdots & a_{1n}b_{1n} \\ a_{21}b_{21} & a_{22}b_{22} & \cdots & a_{2n}b_{2n} \\ \vdots & \vdots & \ddots & \vdots \\ a_{m1}b_{m1} & a_{m2}b_{m2} & \cdots & a_{mn}b_{mn} \end{bmatrix} \quad (3.12)$$

For example,

```
>> A = [1 2 3 4; 4 3 2 1], B = [2 2 1 1; 1 2 1 2]
A = 1    2    3    4
    4    3    2    1
B = 2    2    1    1
    1    2    1    2
>> C = A.*B
C = 2    4    3    4
    4    6    2    2
```

Table 3.4 Function Form of Nonstandard Arithmetic Operators

Function	Description	Corresponding Character
times	Array multiply	.*
power	Array power	.^
rdivide	Right array divide	./
ldivide	Left array divide	.\

Similarly, *element-by-element division* on the same-size arrays results in

$$
C = [c_{ij}] = A./B = \left[\frac{a_{ij}}{b_{ij}}\right] =
\begin{bmatrix}
\dfrac{a_{11}}{b_{11}} & \dfrac{a_{12}}{b_{21}} & \cdots & \dfrac{a_{1n}}{b_{1n}} \\[2mm]
\dfrac{a_{21}}{b_{21}} & \dfrac{a_{22}}{b_{22}} & \cdots & \dfrac{a_{2n}}{b_{2n}} \\[2mm]
\vdots & \vdots & \ddots & \vdots \\[2mm]
\dfrac{a_{m1}}{b_{m1}} & \dfrac{a_{m2}}{b_{m2}} & \cdots & \dfrac{a_{mn}}{b_{mn}}
\end{bmatrix}
\tag{3.13}
$$

$$
C = [c_{ij}] = A.\backslash B = \left[\frac{b_{ij}}{a_{ij}}\right] =
\begin{bmatrix}
\dfrac{b_{11}}{a_{11}} & \dfrac{b_{12}}{a_{21}} & \cdots & \dfrac{b_{1n}}{a_{1n}} \\[2mm]
\dfrac{b_{21}}{a_{21}} & \dfrac{b_{22}}{a_{22}} & \cdots & \dfrac{b_{2n}}{a_{2n}} \\[2mm]
\vdots & \vdots & \ddots & \vdots \\[2mm]
\dfrac{b_{m1}}{a_{m1}} & \dfrac{b_{m2}}{a_{m2}} & \cdots & \dfrac{b_{mn}}{a_{mn}}
\end{bmatrix}
\tag{3.14}
$$

(both operations are also applicable to multidimensional arrays). Using the same matrices as the previous one, we obtain

```
>> C = A./B
C =  0.5000    1.0000    3.0000    4.0000
     4.0000    1.5000    2.0000    0.5000
>> C = A.B
C =  2.0000    1.0000    0.3333    0.2500
     0.2500    0.6667    0.5000    2.0000
```

Following the same paradigm, the *element-wise array power* operation, $A.\!\!\char94 B$, applied to same-sized arrays returns the same-sized array composed of the elements of the array A in the individual powers presented by the elements of the array B. For example,

$$
C = [c_{ij}] = A.\!\!\char94 B = [a_{ij}{}^{b_{ij}}] =
\begin{bmatrix}
a_{11}{}^{b_{11}} & a_{12}{}^{b_{21}} & \cdots & a_{1n}{}^{b_{1n}} \\[2mm]
a_{21}{}^{b_{21}} & a_{22}{}^{b_{22}} & \cdots & a_{2n}{}^{b_{2n}} \\[2mm]
\vdots & \vdots & \ddots & \vdots \\[2mm]
a_{m1}{}^{b_{m1}} & a_{m2}{}^{b_{m2}} & \cdots & a_{mn}{}^{b_{mn}}
\end{bmatrix}
\tag{3.15}
$$

Using the same matrices A and B we get

```
>> C = A.^B
ans =  1     4     3     4
       4     9     2     1
```

If B is a scalar, then A.^B raises each element of the array A to the power of the scalar B. For example,

```
>> C = A.^3
```

returns

```
C =  1      8     27     64
     64    27      8      1
```

If A is a scalar, then A.^B returns a new array of the same size as B with the scalar A raised to the power of the elements of the array B. Hence,

```
>> C = 5.^A
```

returns

```
C =   5      25    125    625
     625    125     25      5
```

3.4 Array Functions

To start with, it should be noted that the elementary functions introduced in Sec. 2.3.1 are applicable not only to scalars but also to arrays. Any of those functions applied to an array results in an element-wise operation, that is, this function will be applied to each element of this array returning the same-sized array composed of the results of individual operations. For instance, the command

```
>> sin([0 pi/4 pi/2 3*pi/4; pi -3*pi/4 -pi/2 -pi/4])
```

returns

```
ans =   0      0.7071     1.0000      0.7071
     0.0000    -0.7071    -1.0000     -0.7071
```

Similarly,

```
>> floor(randn(3,4,2))
```

returns

```
ans(:,:,1) = -1      0     -1     -2
              2      1     -1      0
             -1      0      0      1
ans(:,:,2) = -1     -2     -1      0
              0     -2      0      1
              1      0      0      0
```

Table 3.5 Descriptive Array Functions

Function	Brief Description
ndims(A)	Returns the number of dimensions in an array A
numel(A)	Returns the number of elements in an array or subscripted array expression
length(A)	Returns the size of the longest dimension of A (the number of elements in A, if A is a vector and the largest value of either rows or columns if A is a matrix)
size(A)	Returns the sizes of each dimension of the array A in a row vector with ndims(A) elements (if A is a scalar, which MATLAB regards as a 1 x 1 array, size(A) returns [1 1])
[m,n]=size(A)	Returns the number of rows in m and the number of columns in n if A is a matrix

showing the result of application of the `floor` function to each element of the 3-D array created using the `randn` function.

In addition to those elementary functions, MATLAB offers a variety of other array-handling functions applicable to both matrices (including vectors and scalars) and multidimensional arrays (including matrices and scalars). Depending on the number of dimensions used, the syntax of these functions may be slightly different, but the general idea is that these functions have the same formalism for a scalar and a multidimensional array. Tables 3.5–3.7 present the general overview of the major functions grouped into three categories (of course you may learn more about any specific function of your interest by checking the MATLAB Help system).

The first group of functions, presented in Table 3.5 allows you to analyze the array. For example, assume we have a 3-D array created using the following command:

```
>> A = rand(2,4,5);
```

Then, here is what we may learn about it by applying the functions of Table 3.5

```
>> ndims(A)          >> numel(A)          >> length(A)          >> size(A)

ans = 3              ans = 40             ans = 5              ans = 2 4 5
```

Although the `size` function permits another syntax

```
>> [l,m,n] = size(A)
l = 2
m = 4
n = 5
```

it seems to be redundant, since we got the same information using simple syntax, `size(A)`. If the number of outputs is less than the number of dimensions in an array, the very last output returns the product of missing sizes

```
>> [l,m] = size(A)

l = 2

m = 20
```

Additional syntax allows you to find a size along the specified dimension, for example,

```
>> size(A,2)

ans = 4
```

Also, there is an alternative way to find the number of dimensions without using the `ndims` function

```
>> length(size(A))

ans = 3
```

Using the functions of the second group, we can treat the elements of arrays as a sequence of numbers and perform some statistical analysis. Consider the following matrix:

```
>> B = reshape(1:12,3,4)

B = 1        4        7        10
    2        5        8        11
    3        6        9        12
```

Table 3.6 Statistical Array Functions

Function	Brief Description
`sort(A)`	Sorts the elements of A along the first nonsingleton dimension of an array (`sort(A,dim)` sorts along the dimension `dim`, `sort(…,'descend')` sorts the elements in the descending order)
`sortrows(A)`	Sorts the rows of a matrix or a column vector A in an ascending order
`mean(A)`	Returns the mean values of the elements, $\bar{x} = (1/n)\sum_{i=1}^{n} x_i$, along different dimensions of an array. Specifically, if A is a vector, `mean` returns the mean value of its elements, if A is a matrix `mean` produces a row vector containing the mean value of each column, and if A is an array, `mean` treats the values along the first non-singleton dimension as the vectors, returning an array of the mean values (the direction can be changed to `dim` by using mean(A, `dim`) command)

(Continued)

Table 3.6 Statistical Array Functions (Continued)

Function	Brief Description
`median(…)`	Returns the median values of the elements along different dimensions of an array with the syntax similar to that of `mean`
`sum(A)`	Sums the elements of each column of `A` and returns a row vector with these sums (sum(`A,dim`) sums along the dimension of `A` specified by scalar `dim`); if `A` is a multidimensional array, `sum` (`A`) treats the values along the first nonsingleton dimension as the vectors, returning an array of the row vectors
`std(A,[],dim)`	Returns the standard deviation for the vectors, produces a row vector containing the standard deviation along the dimension `dim` column for arrays (`dim` is optional, if omitted the first nonsingleton dimension is used by default); the place holder `[]` may be either 0 (default value) or 1 corresponding to the square root of an unbiased estimate of the variance , $s = \sqrt{1/(n-1)\sum_{i=1}^{n}(x_i - \bar{x})^2}$, and the second moment of the set of values about their mean, $s = \sqrt{(1/n)\sum_{i=1}^{n}(x_i - \bar{x})^2}$, respectively
`min(A)`	Returns the smallest elements along different dimensions of an array `A`. Specifically, if `A` is a vector `min(A)` returns the minimum value of `A`, if `A` is a matrix—a row vector containing the minimum value of each column of `A`; if `A` is a multidimensional array `min(A)` operates along the first nonsingleton dimension
`min(A,[],dim)`	Operates along the specific dimension `dim`
`[x,k]= min(…)`	Returns the minimum values in a row vector `x` along with indices in a row vector `k`
`min(A,B)`	Returns an array of the same size as `A` and `B` with the smallest elements taken from `A` or `B` (unless one of `A` and `B` is a scalar, their dimensions must match)
`max(…)`	finds the maximum values but it is similar to `min(…)`,

Now analyze the following self-explanatory examples. Let us start with the `sort` and `sortrows` functions with their major syntax being as follows:

```
>> sort(B,'descend')        >> sortrows(ans)        >> sort(B,2,'descend')
ans = 3   6   9   12        ans = 1   4   7   10    ans = 10   7   4   1
       2   5   8   11               2   5   8   11           11   8   5   2
       1   4   7   10               3   6   9   12           12   9   6   3
```

The `sum`, `mean`, `median`, `min`, `max`, and `std` functions have pretty much the same syntax:

>> sum(B)	>> mean(B)	>> median(B)
ans = 6 15 24 33	ans = 2 5 8 11	ans = 2 5 8 11

>> sum(B,2)	>> mean(B,2)	>> median(B,2)
ans = 22	ans = 5.5000	ans = 5.5000
26	6.5000	6.5000
30	7.5000	7.5000

>> min(B)	>> max (B)
ans = 1 4 7 10	ans = 3 6 9 12

>> min(B,[],2)	>> max(B,[],2)
ans = 1	ans = 10
2	11
3	12

>> std(B)	>> std(B,[],2)	>> std(B,1,2)
ans = 1 1 1 1	ans = 3.8730	ans = 3.3541
	3.8730	3.3541
	3.8730	3.3541

Now that we learned about the max function, we may construct an alternative to the length function as follows:

```
>> max(size(A))
ans = 5
```

In addition, the min and max functions assume two other syntax variations. First, along with finding the minimum and maximum values you can also locate their indices, for example,

```
>> a = ceil(randn(3)*10)
a = -15   -3    8
    -14    7   13
      6    9    7
>> [m,i] = max(a)
m = 6 9 13
i = 3 3   2
```

(note that if you try to replicate the first command that employs the random number generator randn, you will likely end up with a different matrix).

Second, these two functions allow you to compare two arrays, as shown in the following example:

```
>> b = ceil(randn(3)*10)
b = 12   -1 -10
   -12 -16   15
     0    3   -8
>> min(a,b)
```

```
ans = -15    -3   -10
       -14   -16    13
         0     3    -8
```

One more function to mention here is the commonly used `find` function (Table 3.7). Let us create a matrix with some zero elements using the `rand` and `round` functions that we already considered

```
>> c = round(rand(3,4)*2)

c = 2    0    1    1
    0    2    1    0
    1    1    0    0
```

Now, let us apply the `find` function, which finds the indices of nonzero elements

```
>> find(c)'
ans = 1    3    5    6    7    8    10
```

(we transposed the result for compactness). Note that `find (A)` also handles logical expressions. If A is a *logical expression* (to be considered in Chapter 5), then the output contains the nonzero elements of the logical array obtained by evaluating the expression A. We will use this feature later, in Chapter 5, but for consistency let us present a couple of examples here. For instance, using logical NOT, ~ (to be covered in Chapter 5), we may easily reverse the problem and look for zero elements rather than nonzero elements

```
>> find(~c)'
ans = 2    4    9    11    12
```

Table 3.7 Syntax of the `find` Function

Syntax	Brief Description
`find(A)`	Locates all nonzero elements of an array A, and returns their linear indices in a column vector
`[r,c,v]=find(...)`	Returns a vector v of the nonzero entries in the matrix A, as well as row and column indices, r and c, of these elements (to see the results in the most readable way, we recommend issuing one of the following two instructions: `[r c v]` or `r', c', v'`)
`find(A, k, ind)`	Returns at most the first (if `ind ='first'` or omitted) or the last (`ind ='last'`) k indices corresponding to the nonzero entries of an array A

Another example shows how to compare the elements of two matrices, a and b, introduced previously. For better understanding, let us first observe the result of comparison of two same-size matrices

```
>> a > b
ans = 0    0    1
      0    1    0
      1    1    1
```

The resulting matrix is composed of a *true* value, 1, and a *false* value, 0. For example, the first element of matrix a is −15 and matrix b is 12, hence an element-by-element comparison returns 0 (since −15 < 12). The same happens when comparing the second elements. However, a comparison of the third elements, 6 and 0, returns 1 (since 6 > 0), and so on. Now, let us explore another syntax of the find function (only applicable to matrices), which features three output variables

```
>> [i,j,x] = find(a>b);
>> [i j x]

ans = 3    1    1
      2    2    1
      3    2    1
      1    3    1
      3    3    1
```

Comparing the three output vectors, i, j, and k (we displayed them column-by-column in the form of a matrix), with the result of the previous command (a > b), we can clearly see that they represent row and column indices (subscripts) of the none-zero elements along with the values of these nonzero elements, respectively.

Let us go a little bit further and show how to use this information. Suppose you are required to show only those elements of the matrix a that satisfy an inequality a > b and zero for the rest of them. Here is how you could do this. We start from converting subscripts to a single index using the sub2ind function

```
>> t = sub2ind(size(a),i,j)'
ans = 3    5    6    7    9
```

Then, we pick the elements we need from the original matrix a

```
>> f(t) = a(t)
f = 0    0    6    0    7    9    8    0    7
```

And finally, we reshape the resulting matrix to match the size of the matrix `a`

```
>> g = reshape(f,size(a))
g = 0     0     8
    0     7     0
    6     9     7
```

The third syntax of the `find` function of Table 3.7 allows you to find not all but a few first or last nonzero elements, which may economize computations in the case of large matrices if you only need to know if it has nonzero elements at all.

In addition to the functions introduced in Tables 3.5–3.7 and in general applicable to multidimensional arrays (including matrices), some other useful MATLAB functions are defined for 2-D arrays (matrices) alone. Among them

det(A)	Calculates the *determinant* of a square matrix A
inv(A)	Computes the *inverse* of the square matrix A (a warning message is printed if A is badly scaled or nearly singular)
trace(A)	Summs the diagonal elements of A, which happens to be the sum of the *eigenvalues* of A

For instance, for the `g` matrix calculated earlier we will have

```
>> det(g)        >> inv(g)                                    >> trace(g)
ans = -336       ans = -0.1458   -0.2143   0.1667             ans = 14

                            0     0.1429        0

                       0.1250          0        0
```

Now that we know so much about basic programming in MATLAB we may check these results implicitly, for example,

```
>> 6*(-7*8)      >> g*inv(g)                                  >> sum(diag(g))
ans = -336       ans = 1.0000        0        0               ans = 14

                            0   1.0000        0

                       0.0000        0   1.0000
```

3.5 Using MATLAB Matrix Formalism to Handle Polynomials

Consider a (row) vector

```
>> c1 = [1 0 -2 5 -4]
c1 =   1     0    -2     5     4
```

Now let us write down the following forth-order polynomial:

$$P_1(x) = x^4 - 2x^2 + 5x - 4 \tag{3.16}$$

Do you see any resemblance? Well, our vector c1 looks like a collection of coefficients of a polynomial $P_1(x)$, written in a slightly different form of

$$P_1(x) = 1x^4 + 0x^3 - 2x^2 + 5x - 4 \tag{3.17}$$

Let us introduce another vector

```
>> c2 = [1 1 0]
c2 = 1      1      0
```

We can treat this vector as a collection of coefficients of another polynomial

$$P_2(x) = x^2 + x + 0 = x^2 + x \tag{3.18}$$

In fact, we can rewrite both polynomials in the following form to have a perfect match with the vectors of coefficients c1 and c2

$$P_1(x) = c1 \begin{bmatrix} x^4 \\ x^3 \\ x^2 \\ x \\ 1 \end{bmatrix} \qquad P_2(x) = c2 \begin{bmatrix} x^2 \\ x \\ 1 \end{bmatrix} \tag{3.19}$$

Now let us use the vectors c1 and c2 for the following two operations:

```
>> cp = c1 + [0 0 c2]
cp = 1      0     -1      6     -4
```

and

```
>> cm = c1 - [0 0 c2]
cm = 1      0     -3      4     -4
```

(Note that, we had to concatenate leading zeros to the vector c2 to get the same length as the vector c1, otherwise MATLAB would return an error). No surprise, but we can treat these two operations as if we were adding polynomials $P_1(x)$ and $P_2(x)$ together and subtracting one from another

$$P_+(x) = P_1(x) + P_2(x), \;\; P_-(x) = P_1(x) - P_2(x) \tag{3.20}$$

To automatize adding leading zeros you can use the following combination of the functions you already know:

```
horzcat(zeros(1,length(c1)-length(c2)),c2)
```

Multiplying two polynomials and performing polynomial division cannot be expressed in one line and requires some programming. Luckily, MATLAB provides us with these functions. To be more specific,

`c = conv(a,b)`   (Abbreviated from *convolution*) Performs multiplication of two vectors, `a` and `b`, in a polynomial sense, so that the order of the vector `c` is equal to the sum of the orders of `a` and `b`

`[q,r] = deconv(a,b)`   (*Deconvolution*) Divides `a` by `b` (in a polynomial sense), so that `q` is the quotient and `r` is the remainder (note that, to maintain the equality `a=conv(b,q)+r`, the vector `r` appears with the leading zeros to match the length of the vector `a`—this way it can be added to the result of `conv(b,q)` explicitly)

Let us explore these two functions

```
>> c3 = conv(c1,c2)
c3 = 1     1    -2     3     1    -4     0
>> [c4,c5] = deconv(c1,c2)
c4 = 1    -1    -1
c5 = 0     0     0     6    -4
```

Indeed, the resulting vectors of coefficients correspond to the correct solutions:

$$P_1(x)P_2(x) = x^6 + x^5 - 2x^4 + 3x^3 + x^2 - 4x, \quad \frac{P_1(x)}{P_2(x)} = x^2 - x - 1 + \frac{6x-4}{P_2(x)} \quad (3.21)$$

This is how MATLAB allows you to handle polynomials. You represent a nth order polynomial by a vector of the length $n+1$ arranging its coefficients so that the first element represents the coefficient by the highest-order monomial, x^n, and the last coefficient—the lowest order, x^0, that is, in descending powers.

In addition to the aforementioned two functions, MATLAB provides a nice visualization tool to better understand the results of polynomial operations, the `poly2sym` function. For example,

```
>> s = poly2sym (c3)
```

yields

```
s = x^6+x^5-2*x^4+3*x^3+x^2-4*x
```

The return s belongs to a *symbolic* class of variables (`class (s)` returns `sym`), which will be addressed in Chapter 7), but what is important here is that for this class of variables another function, `pretty`, yields even better representation

```
>> pretty(s)
```

$$x^6 + x^5 - 2x^4 + 3x^3 + x^2 - 4x$$

So far, we considered three out of several functions enabling polynomials handling. The complete list of these functions is presented in Table 3.8. Let us introduce a couple of more functions from this list here, whereas the remaining ones will be considered in the corresponding chapters later on.

The `poly(r)` function computes coefficients of the polynomial, whose roots are given by a vector r, for example,

```
>> h = poly([1; -4; 5])
```

returns

```
h = 1     -2     -19     20
```

Table 3.8 Polynomial-Handling Functions Summary

Function	Description
`conv (a,b)`	Multiplies two polynomials
`deconv(a,b)`	Divides two polynomials
`poly2sym(a)`	Converts a vector of coefficients a into symbolic polynomial representation
`poly(r)`	Creates a polynomial with the specified roots r
`polyder(a)`	Computes coefficients of a derivative polynomial defined by coefficients a (Chapter 12)
`polyfit(x,y,n)`	Performs polynomial curve fitting, that is, finds the coefficients of a polynomial of degree n that fits (x,y) data supplied in the vectors x and y the best (Chapter 11)
`polyval(a,x)`	Evaluates a polynomial a at a given set of arguments specified in x
`polyvalm(a,M)`	Evaluates a polynomial with coefficients a in a matrix sense
`residue(a,b)`	Performs partial-fraction expansion (computes residues)
`roots(a)`	Finds the roots of polynomial a (Chapter 10)
`sym2poly(s)`	Converts symbolic representation of a polynomial s to a vector of its coefficients

which means

$$H(x) = (x-1)(x+4)(x-5) = x^3 - 2x^2 - 19x + 20 \qquad (3.22)$$

You can check the result by applying the reciprocal function, `roots`, which computes the roots of a polynomial

```
>> roots(h)'
ans = 5.0000    -4.0000    1.0000
```

(we used transposition for compactness).

If you apply the `poly` function to an $n \times n$ matrix **A**, it will compute coefficients of the characteristic equation $|x\mathbf{I} - \mathbf{A}|$, where **I** is the $n \times n$ identity matrix.

The following self-explanatory examples show how to use the `polyval` and `polyvalm` functions evaluating it at a given set of points in a polynomial and matrix sense:

```
>> polyval([5; -4; 1],0:5)
ans = 1     2    13    34    65    106
>> polyval([5; -4; 1],eye(2))
ans = 2     1
      1     2
>> polyvalm([5; -4; 1],eye(2))
ans = 2     0
      0     2
```

By the way, note that in the three examples we defined a polynomial using a column vector, which is absolutely legitimate as well. These three commands are equivalent to

```
>> 5*[0 1 2 3 4 5].^2  -4*[0 1 2 3 4 5] + 1
>> 5*[1 0;     0 1].^2 -4*[1 0; 0 1]     + 1
>> 5*[1 0;     0 1]^2   -4*[1 0; 0 1]^1   + [1 0; 0 1]^0
```

respectively.

Finally, the `residue` function converts a quotient of polynomials to pole-residue representation, and back again, for example,

```
>> [r,p,k] = residue([2 -3 -9 16 -18],[1 -2 -5 6])
r = 3.0000
   -2.0000
    2.0000
p = 3.0000
   -2.0000
    1.0000
k = 2     1
```

which corresponds to

$$\frac{2x^4 - 3x^3 - 9x^2 + 16x - 18}{x^3 - 2x^2 - 5x + 6} = \frac{3}{x-3} - \frac{2}{x+2} + \frac{2}{x-1} + 2x + 1 \qquad (3.23)$$

You can check the result by applying this function "backwards"

```
>> [a,b] = residue([3 -2 2],[3 -2 1],[2 1])
a = 2      -3     -9     16     -18
b = 1      -2     -5      6
```

3.6 Handling Text Strings as Character Arrays

A *text string* is a variable that contains *characters*. To create a string, you should enclose the characters in single quotes. For example,

```
>> object = 'aircraft'
object = aircraft

>> number = '123.16'
number = 123.16
```

Both variables belong to the *char* class (type `class(object)` to get `char`). Note that the variable `number` is a string too, that is, not a decimal variable. It means that you cannot perform any arithmetic operations with it. The alternative way to create a string variable out of decimal number is to use `num2str` function. For example,

```
>> number = num2str(123.16)
```

produces the same result as the second command in the previous input. To convert integer numbers, there is another useful command, `int2num`. For example,

```
>> number = int2str(123.16)
```

returns

```
number = 123
```

Let us check the size of two character variables we just created

```
>> size(object)
ans = 1      8
>> size(number)
ans = 1      3
```

As you can see MATLAB does treat the test strings as character arrays dedicating one element for each character (in terms of memory, each character requires two bytes of memory). Therefore, a lot of noncomputational operations apply to character arrays as well. For example, you can concatenate strings, using brackets

```
>> B = ['Mean of 1:10 is ' num2str(mean(1:10)) '.']
```

which returns

```
B = Mean of 1:10 is 5.5.
```

or using a special function `strcat`

```
>> strcat('Mean of 1:10 is ',num2str(mean(1:10)),'.')
```

(we will deal with vertical concatenation a little bit later).

You can also address a specific character of a group of characters in the string the same way you would do it for the numeric array. For example,

```
>> object(1)
ans = a
>> object(4:end)
ans = craft
```

The `findstr` function may be useful for finding the locations of certain character(s) in a string

```
>> findstr(object,'a')
ans = 1    6
```

If you want to convert a string, which is an American Standard Code for Information Interchange (ASCII) character representation of a numeric value, back to numeric representation you need to use the `str2num` function. As a matter of fact this function can do much more, for instance,

```
>> str2num(['123' '+' '17'])
```

returns the sum of two numbers

```
ans = 140
```

Applying `str2num` to the variable `object`, which is not a representation of a numeric value, results in an empty matrix `[]`.

Now let us talk about vertical concatenation of the text strings. The text strings can be concatenated vertically as well. The only limitation is that they have to have the same length, that is, the same number of characters. For example, the command

```
>> carrierH = ['spacecraft' 'rotorcraft']
```

returns another string concatenated horizontally (1×20 character array)

```
carrierH = spacecraftrotorcraft
```

and the command

```
>> carrierV = ['spacecraft'; 'rotorcraft']
```

results in vertical concatenation and produces a 2×10 character array as follows:

```
carrierV = spacecraft
           rotorcraft
```

The command

```
>> vehicle = ['spacecraft';'aircraft';'ship';'submarine']
```

however, evokes an error

```
    All  rows  in  the  bracketed  expression  must  have  the  same  number  of
columns
```

That is why to concatenate strings vertically you better use a special function strvcat, which appends spaces to each string as necessary to form a valid matrix

```
>> vehicle = strvcat('spacecraft','aircraft','ship','submarine')

vehicle = spacecraft
          aircraft
          ship
          submarine
```

To assure that you do have spaces in you character array, you may run the findstr command looking for them, for example,

```
>> findstr(vehicle(3,:),' ')
```

returns

```
ans = 5   6   7   8   9   10
```

meaning that the length of ship is seven characters shorter than that of the longest row, spacecraft, and therefore for spaces were added to equalize them.

The find function happens to be very useful in handling character arrays as well. For example, the following command finds indices of the rows that start from s:

```
>> i = find(vehicle(:,1)=='s')
i = 1
    3
    4
```

followed by another command that retrieves the entire rows

```
>> vehs = vehicle(i,:)

ans = spacecraft
      ship
      submarine
```

Transposition is also a legitimate command for character arrays

```
>> h1 = vehs'

ans = sss
      phu
      aib
      cpm
      e a
      c r
      r i
      a n
      f e
      t
```

The `char` function, allows you to do the same thing as the `strvcat` function—it automatically pads each string with the blanks to form a valid matrix

```
>> vehicle = char('spacecraft','aircraft','ship','submarine')

vehicle = spacecraft
          aircraft
          ship
          submarine
```

Another syntax of the `char` function, `char (a)`, converts an array `a` that contains nonnegative integers representing character codes into a MATLAB character array. For example, the following command prints a 3 by 32 display of the printable ASCII characters in the range 32–127:

```
>> ascii = char(reshape(33:254,37,6)')

ascii =    !"#$%&'()*+,-./0123456789:;<=>?@ABCDE
           FGHIJKLMNOPQRSTUVWXYZ[\]^_`abcdefghij
           klmnopqrstuvwxyz{|} □□□□□□□□□□□□□□□□□□
           □□□□□□□□□□□□□□□□□□ ¡¢£¤¥¦§¨©ª«¬-®¯°±²³´
           µ¶·¸¹º»¼½¾¿ÀÁÂÃÄÅÆÇÈÉÊËÌÍÎÏÐÑÒÓÔÕÖ×ØÙ
           ÚÛÜÝÞßàáâãäåæçèéêëìíîïðñòóôõö÷øùúûüýþ
```

Compare this table to the one that shows in Microsoft Office when you are trying to insert a symbol (Fig. 3.18). The actual characters displayed depend on the character set encoding for a given font. MATLAB uses a font that Microsoft Office does not have (see Preferences as discussed in Sec. 2.6), but at least the first 127 characters are the same.

Let us show that we are talking about the same ASCII code by finding the character code for the letter Z

```
>> findstr(char(1:127),'Z')
ans = 90
```

(note that the `findstr` function only applies to the one row strings).

Now we can finally try to apply basic operations to character arrays and understand the results. For instance,

```
>> 'Z'-'X'
```

returns

```
ans = 2
```

and simply means how far the letter Z is behind the letter X.

Similarly, adding `'aircraft'` to `'ship'`

```
>> vehicle(2,:)+vehicle(3,:)
```

does not mean that we are creating a carrier but rather presents a vector, which contains a sum of character codes of the corresponding elements in both strings

```
ans = 212   209   219   211   146   129   134   148   64   64
```

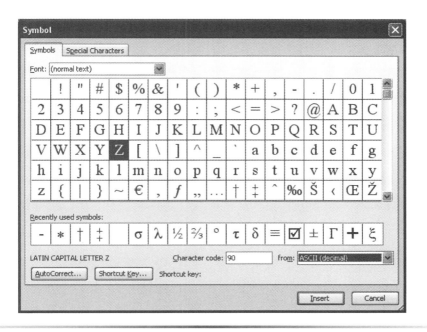

Fig. 3.18 Insert Symbol window in Microsoft Office.

Table 3.9 Strings Operations Functions

Function	Brief Description
blanks	Creates a string of blanks
char	Converts an input to character array (string)
deblank	Removes trailing blanks
eval	Executes string with MATLAB expression
findstr	Finds one string within another
lower	Converts string to lowercase
regexp	Matches regular expression
regexpi	Matches regular expression, ignoring case
regexprep	Replaces string using regular expression
strcat	Concatenates strings
strcmp	Compares strings
strcmpi	Compares strings, ignoring case
strfind	Finds one string within another
strjust	Justifies string
strmatch	Finds matches for string
strncmp	Compares first *n* characters of strings
strncmpi	Compares first *n* characters, ignoring case
strread	Reads formatted data from a string
strrep	Replaces a string within another
strtok	Finds token in string
strvcat	Concatenates strings vertically
upper	Converts string to uppercase

You can try to apply the `char` function to it

```
>> char(ans)
```

to see something like this

```
ans = ÔÑÛÓ□□□□@@
```

Let us consider one more example of using `findstr` and some other functions introduced earlier in this chapter to provide you with a piece of the code you may find useful. Suppose you want to permute letters in the word "space." First, you have to randomly pick the first letter. You can do it using the same procedure we used before, say `ceil` `(rand(1)*N)` generates a random number between `0` and `N`. Alternatively, you could simply use a `randi(N+1)-1` call. Then, you can throw this letter away, that is, reduce the length of the original string by one and proceed with picking the next letter, and so on. Hence, running the following commands:

```
>> f = 'space';
>> l1 = f(ceil(rand(1)*5)); f(findstr(f,l1))=[];
>> l2 = f(ceil(rand(1)*4)); f(findstr(f,l2))=[];
>> l3 = f(ceil(rand(1)*3)); f(findstr(f,l3))=[];
>> l4 = f(ceil(rand(1)*2)); f(findstr(f,l4))=[];
>> strcat(l1,l2,l3,l4,f)
```

may result in something like

```
ans = asecp
```

so that you may ask your friends to guess what meaningful word you can get out of it. We used this example to practice using several functions and show you a couple of tricks, but in practice, the same result could be achieved by issuing a single command

```
>> f = 'space';
>> f(randperm(length(f)))
```

where `randperm(n)` returns a random permutation of the integers `1:n`.

Table 3.9 lists the string operations functions that are available in MATLAB including those used in this section.

To conclude this section, let us use one more function of Table 3.9 to show how string concatenation allows you to create names of variables programmatically (you may want to use this feature when you become a proficient MATLAB user). Assume that we are computing some quantity that differs from one set of input data to another; say we are varying the number of data points. Suppose we have some vector `Speed` that was computed using 10 points and equals to `150`. Using two new MATLAB functions `genvarname` and `eval` as shown later, enables to reassign data from the vector `Speed` to a programmatically created vector `SpeedFor10Points` that directly indicates what dataset was used:

```
>> v = genvarname(['SpeedFor' num2str(10) 'Points']);
>> eval([v '=' 'Speed' ';']);
```

As shown in Fig. 3.19 upon execution `v='SpeedFor10Points'` and `SpeedFor10Points=150`

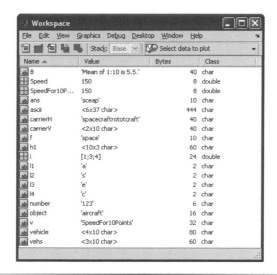

Fig. 3.19 Snapshot of Workspace with all variables introduced in this section.

Problems

3.1 Use vector notation for creating vectors with constant spacing, the `linspace` and concatenation functions to create a matrix

$$
B1 = \begin{pmatrix}
1 & 4 & 7 & 10 & 13 & 16 & 19 \\
72 & 66 & 60 & 54 & 48 & 42 & 36 \\
0 & 0.125 & 0.25 & 0.375 & 0.5 & 0.625 & 0.75 \\
0.3 & 0.4 & 0.5 & 0.6 & 1.2 & 1.4 & 1.6 \\
9 & 8 & 7 & 6 & 5 & 4 & 3
\end{pmatrix}
$$

Then, address the corresponding elements and replace them with the `zeros(1,4)`, `ones(1,4)`, `zeros(2,3)`, and `magic(3)` matrices to obtain

$$
B2 = \begin{pmatrix}
1 & 4 & 7 & 10 & 8 & 1 & 6 \\
0 & 66 & 0 & 54 & 0 & 42 & 0 \\
1 & 1 & 1 & 1 & 3 & 5 & 7 \\
0 & 0 & 0 & 0.6 & 4 & 9 & 2 \\
0 & 0 & 0 & 6 & 5 & 4 & 3
\end{pmatrix}
$$

Employ the spy function to visualize nonzero elements of the matrix B2 and the find function to find row and column indices of non-zero elements.

3.2 An air delivery system release altitude is h_0 = 1000m. However something went wrong and the system falls ballistically. Using the formula $h = h_0 - 0.5gt^2$, where t is the time in seconds and g = 9.8m/s^2 is the acceleration due to gravity, determine the altitude of the system at t = 0, 1, 2, 3, 4, 5, 6, 7, 8, 9, and 10 s. (Create a vector t and determine h using element-by-element calculation.)

3.3 Use arrays to handle polynomials and text strings. For the both cases considered below, employ either ceil(M*rand(1,N)) (for older versions of MATLAB) or randi(M,1,N) command to generate N random integers within the range [1 ; M]. Do the following:

   a. Create the seven-element and three-element vectors with M = 6 and M = 10, respectively. Treating these vectors as coefficients of polynomials, use the deconv function to find a quotient. Display a symbolic polynomial representation of this quotient (using the poly2sym and pretty functions).

   b. Use M = 26 and generate a sequence of N = 26 random integers. Using the char function convert this sequence to the text string containing letters A-Z (add 64 to the integers sequence, because the character code for letters starts from 65). Check whether your text string contains a letter A (use the findstr function). Reshape your text vector to a 2 by 13 matrix. Check whether this matrix has columns containing the same letter (use the find function applied to the difference between the two rows).

Chapter 4

Data Structures, Types of Files, Managing Data Input and Output

- Cell and Structure Arrays
- Importing and Exporting Data
- Displaying Formatted Data
- Interactive Input and Output

```
cell, struct, [], num2cell,
save, load, textread,
textscan, xlsread,
importdata, dlmread,
dlmwrite, fprintf,
sprintf, input, menu,
inputdlg, uigetfile,
pause, and others
```

4.1 Introduction

Before writing sophisticated MATLAB programs, this chapter concludes the introduction describing MATLAB as an effective technical environment and engineering tool. It also addresses some additional topics allowing you to become a proficient MATLAB user. It begins with a review on MATLAB data types and introduce two more types of data, namely, cell arrays and structure arrays, making storage/handling of heterogeneous data easier. Then, the two basic types of MATLAB files, MAT-files and M-files, are introduced. It is followed by presenting a few more MATLAB functions that allow you to manage workspace variables (`save` and `load` functions) and store all commands issued and all responses received during the current session (`diary`). It further proceeds with the unformatted import/export of data from different data files (`dlmread`, `dlmwrite`, `xlsread`, `textread`, and other functions) and writing formatted data to a file/string or displaying it on the screen (`fprintf` and `sprintf`). The chapter ends with presenting several user-friendly functions enabling interactive input of data (including `input` and `menu` functions) and managing the program flow using predefined (readily available) dialog boxes (`inputdlg`, `uigetfile`, `waitbar`, and other functions).

4.2 Data Types

MATLAB supports 15 fundamental and 2 additional data types. You may observe the names of these data types in the Class column of the Workspace browser as shown in Figs. 2.6, 3.8, and 3.19. Each of these data types can be in the form of a matrix or array. This matrix or array is a minimum of 0 by 0 in size (empty array) and can grow up to an *n*-dimensional array of any size. The hierarchy of the data types is shown in Fig. 4.1. Table 4.1 describes these data types in more detail, providing some self-explanatory examples. Figure 4.2 shows the content of MATLAB workspace after introducing the variables of Table 4.1, while clearly indicating their class.

Among the data types, 10 data types (shown in three boxes one below the other on the left of Fig. 4.1) are numerical data types enabling numerical computations. The major difference between them is the memory allocation to store one element of such a data type. By default, MATLAB assigns double-precision type, `double`, to all numerical data meaning that it allocates eight bytes of memory to store one numerical value. In Chapter 3, you have already performed some mathematical operations with this type of data. Chapter 7 will provide you more details on the ways numerical data are stored in computer's memory.

In Chapter 3, you had a chance to familiarize yourself with another data type, namely, `char`, assigned to data, which is represented by characters and text strings. The `logical` and `function handles` data types will be explained in Chapter 5. Symbolic variables, `sym`, not shown in Fig. 4.1, will be discussed in detail in Chapter 7.

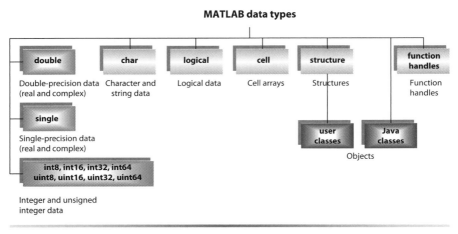

Fig. 4.1 Hierarchy of the basic MATLAB data types.

Table 4.1 Fundamental Data Types MATLAB

Data Type	Examples	Description

double	`>> w=2.5*10^300,`
	`   h=2-i`
	`w = 2.5000e+300`
	`h = 2.0000 - 1.0000i`
		An array of the double-precision numbers. It is the default numeric type in MATLAB

single	`>> single(3*10^300)` `ans = Inf`	An array of the single-precision numbers. Requires less storage space than `double` but has less precision and a smaller range

int8, int16, int32, int64
 	
uint8, uint16, uint32, uint64
	`>> a=int8(500)`
	`a = 127`
	
	`>> b=uint8(500)`
	`b = 255`
		An array of signed and unsigned integers. Requires less storage space than `single` or `double`. All integer types except for `int64` and `uint64` can be used in mathematical operations

logical	`>> f=rand(2)>0.5`
	`f = 1 1`
	`    0 1`
		An array of the logical values of `1`'s and `0`'s to represent the `true` and `false` values, respectively

char	`>> c='Summer'` `c = Summer`	An array of characters. The text strings are represented as the vectors of characters

cell	`>> d{1,1}=12;`
	`>> d{1,2}='Red';`
	`>> d{1,3}=magic(4);`
	`>> d`
	`d = [12] 'Red' [4x4`
	`    double]`
		An array of the indexed cells, each capable of storing an array of a different dimension and data type

structure	`>> u.day=12;`
	`>> u.color='Red';`
	`>> u.mat=magic(3);`
	`>> u`
	`u = day: 12`
	`    color: 'Red'`
	`      mat: [3x3`
	`           double]`
		An array of the C-like structures composed of the named fields capable of storing an array of different dimension and data type

function handle	`>> p=@sin` `p = @sin`	A pointer to a function (handle) that can be passed as an input argument to a function

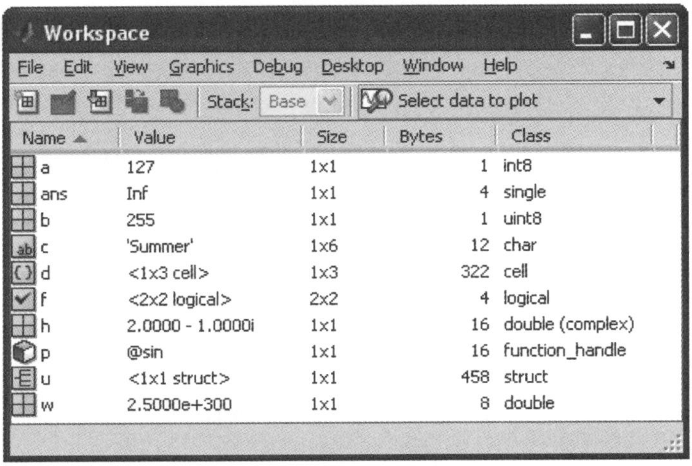

Fig. 4.2 MATLAB workspace with the variables defined in Table 4.1.

Finally, both user-defined object-oriented `user classes` and `Java classes` are beyond the scope of this book and will not be discussed. For the sake of completeness, however, it can be noted that all MATLAB data types are implemented as object-oriented classes. So you can add data types of your own to your MATLAB environment by creating additional classes. These user-defined classes define the structure of your new data type, and the M-file functions or methods that you write for each class define the behavior of that data type. And, a Java class is just another MATLAB data type providing an interface to the Java programming language, enabling creation of the objects from Java classes and calling Java methods on these objects.

We have briefly mentioned about all data types except two data types that are presented in Fig. 4.1. These two remaining fundamental data types, `cell` and `structure`, happen to be very useful in MATLAB programming and, therefore, deserve a special attention. They are explained in the following two sections.

4.3 Cell Arrays

We concluded Chapter 3 by introducing `char` variables and text strings, and we ran into the problem of not being able to store multiple different-length strings within a single array without padding each string with the blanks (using `char` function) to form a valid matrix. Now, let us revisit this example and accomplish the mission using *cell arrays*. And, the following command:

```
>> vehicle={'spacecraft';'aircraft';'ship';'submarine'}
```

creates a valid array containing four different-length text strings:

```
vehicle = 'spacecraft'
          'aircraft'
          'ship'
          'submarine'
```

As you can see, the only difference is that the values of the cells are assigned using the braces { } as opposed to parentheses (as we do it for ordinary arrays). After issuing this command, the workspace will be replaced with a 4 by 1 cell array. Each cell by itself occupies 60 bytes, so the total number of bytes required to store this array is 302 (60*4 bytes to maintain a cell plus 2 bytes each for 31 symbols).

Another detail is that you can address either the individual cell of a cell array (which is called *cell indexing*) or its content (*content indexing*). For instance,

```
>> a = vehicle(3)
```

returns

```
a = 'ship'
```

with a being a 1 by 1 cell, and

```
>> b = vehicle{3}
```

yields

```
b = ship
```

that is, the content of the third cell, which means you can use it in a variety of operations applicable to the variables of the type char, such as,

```
>> sort(b)
ans = hips
```

You can also mix both type of indexing. For example,

```
>> vehicle{2}(4:end)
```

where we use the braces to access the content of a specific cell and parentheses to address a certain part of it, returns

```
ans = craft
```

Note that, in the latter case, you can only address a content of a single cell, that is, the command

```
>> vehicle{2:3}(1:3)
```

returns an error

```
??? Bad cell reference operation
```

A cell array can hold not only text strings but also anything you want: a scalar, numerical array, text string, another cell, etc. Hence, a cell array allows us to effectively handle heterogeneous data. For instance, consider a 2 by 2 cell array c containing a matrix, a vector, and two text strings (its graphical interpretation is given in Fig. 4.3):

```
c{1,1} = '2-by-2';
c{1,2} = 'eigenvalues of eye(2)';
c{2,1} = eye(2);
c{2,2} = eig(eye(2));
```

MATLAB has a good function, cellplot, to visualize the structure and distinguish the contents of cell arrays. For instance, typing cellplot(vehicle) and cellplot(c) for the two examples stated earlier produce a graphical display of the cell array's contents shown in Fig. 4.4a and 4.4b, respectively.

Let us consider one more example. The following set of commands creates a cell array that contains some generalized data on five generations of fighter aircraft:

```
Gen{1,1,1}='Ramjet';          % type of engine
Gen{2,1,1}=[0.3 1; 0.8 6];    % thrust to weight ratio, CL_max, M_max, nz_max
Gen{1,2,1}=[12000 900];       % ceiling (m), range (km)
Gen{2,2,1}=[0.1; 4000];       % price (million dollars), number manufactured
Gen(1,1,2)={'Turbojet'};
Gen(2,1,2)={[0.4 0.6; 0.9 7]};
Gen(1,2,2)={[14000 1100]};
Gen(2,2,2)={[0.2; 34000]};
Gen(1,1,3:5)={'Turbojet w/AB' 'Turbojet w/AB' 'LBRTurbofan'};
Gen(2,1,3:5)={[0.6 0.8; 2.2 8] [0.73 1.6; 2 9] [0.6 1.1; 1.6 7]};
Gen(1,2,3:5)={[16000 1700]; [17000 2200]; [16000 2000]};
Gen(2,2,3:5)={[1.4; 43000], [30; 1000], [150; 1000]};
```

(The text appearing after the symbol % is a comment, which is ignored when you execute a command.) This set of commands features some

Cell 1,1	Cell 1,2
'2 by 2'	**'eigenvalues of eye(2)'**
Cell 2,1	Cell 2,2
$\begin{bmatrix} 1 & 0 \\ 0 & 1 \end{bmatrix}$	$\begin{bmatrix} 1 \\ 1 \end{bmatrix}$

Fig. 4.3 Graphical interpretation of the cell array c.

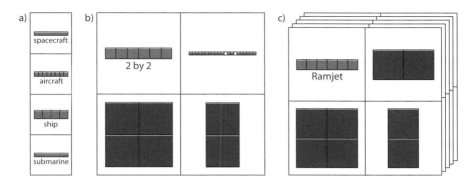

Fig. 4.4 Content of the cell array a) `vehicle`, b) `c`, and c) `Gen`.

alternatives you may want to exploit when assigning values to a cell array. The first four lines define data for the first generation (the first page or layer in Fig. 4.4c). The next four lines define the second generation. Here, we present another option of using the braces `{ }` on the right-hand side, that is, converting different objects to cell elements while performing an operation on the right as opposed to doing it when assigning the result of this operation to the cell element on the left as earlier. Finally, the last four lines use the bulk assignment (in this case, the braces may be used only on the right-hand side of expressions). Note that you may use cell arrays with space, comma, and semicolon separated lists just as you would do it with ordinary arrays.

One more comment about `cellplot(Gen)` command, which was used to generate Fig. 4.4c, is that it only shows what is presented, that is, you cannot see what lies behind the first layer (page). In the general case, if you want to display a content of some multidimensional cell array you have to do it layer-by-layer. For example, if you want to see the third page you should type `cellplot(Gen(:,:,3))`.

As mentioned earlier, you may address any specific portion of the content of a single cell by mixing two types of indexing, like `Gen{2,1,2}` `(1,2)`. However, you cannot mix indexing to access several cells simultaneously. Is there any way to overcome this inconvenience? Well, the following presents an example of how you could do this:

```
>> format bank
>> Data=[Gen{2,1,:}]
>> [a,b]=max([Data(3:4:end)])
>> format
```

The trick is to assign the content of multiple cells to some variable, in this case `Data`, and then treat this new variable in a usual way. The above commands yield the following output:

```
Data = Columns 1 through 5
                0.30      1.00      0.40      0.60      0.60
                0.80      6.00      0.90      7.00      2.20
        Columns 6 through 10
                0.80      0.73      1.60      0.60      1.10
                8.00      2.00      9.00      1.60      7.00
a = 1.60
b = 4.00
```

This means that the fourth generation aircraft has the largest lift coefficient C_{Lmax} (CL_max) of 1.6. Note that, we used the first `format` command (introduced in Sec. 2.5) to make the output more compact and then returned it back to default by issuing another `format` command.

To conclude this section, let us present three MATLAB functions that can be useful in creating cell arrays:

`cell(n,m,…)`	Creates an `n` by `m` by … cell array of empty matrices
`num2cell(b)`	Creates a cell array by placing each element of a numeric array `b` into a separate cell
`cellstr(a)`	Creates a cell array of strings from a character array `a`

Also, you may find useful nested cell arrays that have cells that contain cell arrays, which may also have cells that contain cell arrays, and so on. To create nested cell arrays, use nested braces `{ { } }`.

4.4 Structure Arrays

Like cell arrays, *structure arrays* may be composed of dissimilar ordinary arrays. The main difference is that structure arrays are accessed by

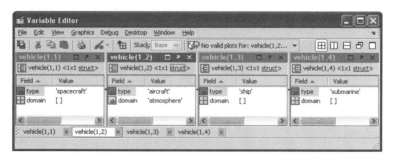

Fig. 4.5 Content of the `vehicle` structure.

Table 4.2 Manned Moon Landing Missions

Lander/Mission	Landing Date	Time on Surface	Crew	EVA Time
Eagle/Apollo 11	20 Jul. 1969	21:31	Neil Armstrong (2), Edwin Aldrin (2)	2:31
Intrepid/Apollo 12	19 Nov. 1969	31:31	Charles Conrad (4), Alan Bean (2)	7:45
Antares/Apollo 14	5 Feb. 1971	33:30	Alan Shepard (2), Edgar Mitchell (1)	9:21
Falcon/Apollo 15	30 Jul. 1971	66:55	David Scott (3), James Irwin (1)	18:33
Orion/Apollo 16	21 Apr. 1972	71:02	John Young (6), Charles Duke (1)	20:14
Challenger/ Apollo 17	11 Dec. 1972	74:59	Eugene Cernan (3), Harrison Schmitt (1)	22:04

named fields, not by standard indexing. In some sense, such indexing is similar to spreadsheet/column/row indexing in a Microsoft Excel spreadsheet. You may assign the values directly to the structure array using the following syntax:

```
variable.field_name=value
```

For example, the set of instructions

```
>> vehicle={'spacecraft';'aircraft';'ship';'submarine'}
>> vehicle.type ='spacecraft';
>> vehicle(2).type ='aircraft';
>> vehicle(3).type ='ship';
>> vehicle(4).type ='submarine'
```

creates a 1 by 4 structure as follows:

```
vehicle = 1x4 struct array with fields:
            type
```

It takes 124 bytes to create the very first element of a structure (even without content), 60 bytes for each additional element, and same 2 bytes per symbol. Hence, the `vehicle` structure occupies 366 bytes of memory.

You may add the additional fields by just typing the name of the new field after the period, like

```
>> vehicle(2).domain ='atmosphere';
```

In this case, this new field is added to every element of the structure. Of course, they hold an empty matrix until other values are assigned to them (Fig. 4.5).

Let us consider a more sophisticated (practical) example. Suppose we want to use a structure to store heterogeneous data shown in Table 4.2

(numbers in the parentheses indicate the total number of missions flown by each astronaut).

The following set of commands does the job (Fig. 4.6):

```
Lander.Name               = 'Eagle/Apollo 11';

Lander.LandingDate        = '20-Jul-69';

Lander.TimeonSurface      = '21:31';

Lander.Crew               = {'Neil Armstrong';'Edwin Aldrin'};

Lander.EVATime            = '02:31';

Lander.Flights            = [2 2];

Lander(2).Name            = 'Intrepid/Apollo 12';

Lander(2).LandingDate     = '19-Nov-69';

Lander(2).TimeonSurface   = '31:31';

Lander(2).Crew            = {'Charles Conrad';'Alan Bean'};

Lander(2).EVATime         = '07:45';

Lander(2).Flights         = [4 2];

Lander(3) = struct('Name','Antares/Apollo 14','LandingDate','05-Feb-71',...
     'TimeonSurface','33:30','Crew',{{'Alan Shepard';'Edgar Mitchell'}},...
     'EVATime','09:21','Flights',[2 1]);

Lander(4:6)= struct('Name',{'Falcon/Apollo 15','Orion/Apollo 16',...
     'Challenger/Apollo 17'},'LandingDate',{'30-Jul-71','21-Apr-72',...
     '11-Dec-72'},'TimeonSurface',{'66:55','71:02','74:59'},'Crew',...
     {{'David Scott';'James Irwin'},{'John Young';'Charles Duke'},...
     {'Eugene Cernan';'Harrison Schmitt'}},'EVATime',...
     {'18:33','20:14','22:04'},'Flights',{[3 1],[6 1],[3 1]});
```

Specifically, the first six lines take care of the first line in Table 4.2 (note how a cell array is used to keep multiple names in the field `Crew`). Similarly, the next six lines fill data for the second element of the structure `Lander`. To fill the remaining lines of Table 4.2, the bulk form is used. The

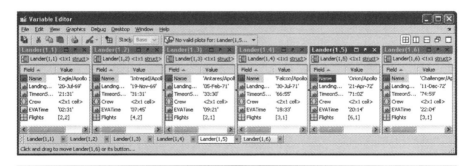

Fig. 4.6 Content of the `Lander` structure.

third element is created using the MATLAB function `struct`, which creates a structure array with the specified fields and values. The remaining elements are created using the same function `struct`, but employing cell arrays to assign multiple values to the same field of different elements. Note that in the two latter cases (when using the `struct` function), the field name should be always followed by its value, otherwise it would not work. If for any reason you do not want to assign the value right now but rather want to skip it and assign later, you must use a placeholder, an empty matrix `[ ]`.

Now, accessing the elements of structure arrays is as easy as follows. To access the entire element of a structure, type its name and number. For instance,

```
>> Lander(3)
```

returns

```
ans =    Name: 'Antares/Apollo 14'
   LandingDate: '05-Feb-71'
TimeonSurface: '33:30'
          Crew: {2x1 cell}
       EVATime: '09:21'
       Flights: [2 1]
```

To access a specific field, type a period after the structure array name followed by the field name. For example,

```
>> Lander(3).EVATime
```

returns a text string

```
ans = 09:21
```

To access elements within a field, add the index of this element. For example,

```
>> Lander(5).EVATime(1:2)
```

yields

```
ans = 20
```

As in the case with cell arrays, the best thing about keeping heterogeneous data together is that you can easily manipulate it using the common MATLAB operators and functions. However, be aware that you cannot address the same field of several elements directly. For instance, an attempt

```
>> Lander(:).LandingDate
```

returns

```
ans = 20-Jul-69
ans = 19-Nov-69
ans = 05-Feb-71
ans = 30-Jul-71
ans = 21-Apr-72
ans = 11-Dec-72
```

Hence, we need to use a couple of tricks. Here is one example of how you can still operate on the same field of multiple elements. The following call

```
>> Dates=reshape([Lander(:).LandingDate]',9,6)'
```

collects all multiple returns into one vector. By reshaping it, the call converts all data of the field `LandingDate` into a single array `Data` that you can use for further operations

```
Dates = 20-Jul-69
        19-Nov-69
        05-Feb-71
        30-Jul-71
        21-Apr-72
        11-Dec-72
```

Similarly, the following commands

```
>> y=[Lander(:).EVATime]';
>> y=reshape([y]',5,6)';
>> EVAMinutes = str2num(y(:,1:2))*60 + str2num(y(:,4:5))
```

first collect `EVATime` data to a single array, then reshape it into a column vector (containing the same-length text strings), and finally operate on it to convert the text strings representing time in hours and minutes to a numerical array showing the same value in minutes

```
EVAMinutes = 151
             465
             561
            1113
            1214
            1324
```

Of course, if the field contains a numerical value to begin with, the code can be more compact. For example, the following command

```
>> mean([Lander.Flights])
```

finds an average number of flights flown by 12 astronauts who landed on the Moon

```
ans=2.3333
```

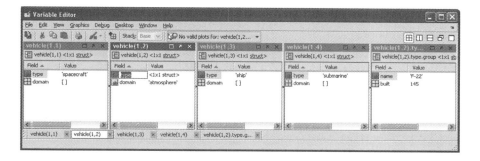

Fig. 4.7 Content of the modified structure `vehicle`.

As is the case with cell arrays, it should be noted that a structure element may be a structure by itself. For example, we may further extend the structure `vehicles`, introduced in the beginning of this section, to include more data (like Fig. 4.7).

```
>> vehicle(2).type.group ='Fighter';
>> vehicle(2).type.group.name ='F-22';
>> vehicle(2).type.group.built = 145;
```

More examples of using the cell arrays and structures will be presented in Chapter 6, and now let us move on to another topic.

4.5 Types of MATLAB Files

Before we proceed any further, let us formally introduce two basic types of files that MATLAB relies on. They are

1. **MAT-files** (with the default extension *.mat)*, which are double-precision binary MATLAB format files created by the `save` command and readable by the `load` command (described in the next section). These files can be created on one computer and later read using MATLAB on another computer with a different floating-point format, retaining as much accuracy and range as the disparate formats allow. They can also be manipulated by other programs, external to MATLAB.
2. **M-files** (with the default extension *.m)*, which are files that contain the multiple lines of code in the MATLAB language. You can create M-files using any text editor (including MATLAB's M-file Editor, introduced in Sec. 2.2.3 and described in more detail in Sec. 5.9.2), and use them as you would use any other MATLAB function or command. The two kinds of M-files are
   * *scripts*, which do not accept input arguments or return output arguments (they usually operate on data in the workspace) and

- *functions*, which can accept input arguments and return output arguments.

Note that, if you duplicate M-files names, MATLAB executes the one that occurs first in the search path (Fig. 2.12 and Sec. 5.7.6).

Many of the functions provided with MATLAB are implemented as M-files just like the M-files that you will create with MATLAB. Other MATLAB functions are precompiled executable programs called *built-ins* that run much more efficiently. Unlike M-file functions, you cannot see the source code for built-ins. Although most built-in functions do have an M-file associated with them, this file is mainly there to supply the help documentation for the function. Examples of built-in functions are `pi`, `sin`, `find`, `format`, `fprintf`, and `reshape` (Sec. 5.7.6).

Both types of files can be transported directly between machines: M-files because they are platform independent and MAT-files because they contain a computer signature in the file header. MATLAB checks the signature when it loads a file. If a signature indicates that a file is foreign, then MATLAB performs the necessary conversion.

The MATLAB environment allows a user to use data (matrices, text, etc.) stored by some other application. MATLAB easily "understands" the **American Standard Code for Information Interchange** (ASCII) format as well as some other formats that will be presented later in Sec. 4.6. In the latter case while loading data, it converts it into the proper format.

One more commonly used type of file is a **FIG-file** (with the default extension *.fig*) that contains figures and GUIs created in MATLAB. Simulink models are stored as **MDL-files** (having extension *.mdl*). You may also encounter **MEX-files**, built by using the `mex` function of MATLAB. These files are computer-dependent shared libraries with compiled and linked C, C++, or Fortran source files, which you can run in MATLAB exactly like MATLAB M-files and built-in functions (see Appendix D). Another MATLAB command, `pcode`, creates **P-files**, content-obscured executable files. Once created, a P-file takes precedence over the corresponding M-file for execution, even if the M-file is subsequently changed (Sec. 5.7.6). **DLL-files** are dynamic-link libraries. Different MATLAB toolboxes may exploit the files with other extensions.

4.6 Recording MATLAB Sessions and Storing Workspace Variables

Although the MATLAB Command History window presents a log of the statements most recently run in the Command window, there is another option allowing you to keep track of what commands were issued during the current session as well as what responses were obtained. It is the `diary` command that allows you to save interactive MATLAB section in the following manner:

`diary`	Toggles the switch to record the session on and off
`diary on`	Turns the diary switch on
`diary off`	Turns the diary switch off
`diary filename`	Records the session in the file with the specified name `filename`

You may also use the functional form of `diary`, for example, `diary('on')`. If the diary name (`filename`) is not specified, MATLAB creates a file named *diary* in the current directory.

The diary file features the following:

1. The diary file is created when the session is turned on, but not available for editing until the session is turned off.
2. The diary stores not only the commands themselves (as Command History window does) but also the response of the system, so it can be really useful in analyzing the session post facto.
3. The output of a diary is an ASCII file, suitable for printing or for inclusion in reports and other documents.
4. It can be edited using any word processor.
5. If the diary file already exists, the output is appended (!) to the end of the file.

You may check the status of the Diary command (`on/off`) by typing `get(0,'Diary')` or `get(0,'DiaryFile')` (be sure to use a zero 0, not a capital "Oh", O).

Along with recording the current session, you may also be interested in saving (and further retrieving) some specific workspace variables. That is where `save` and `load` commands become very useful. The `save` command may use one of the following common syntax:

`save`	Saves all workspace variables in the default binary file called *matlab.mat*
`save filename`	Saves all workspace variables in a binary file called *filename.mat*
`save filename var1 var2`	Saves only the variables `var1` and `var2`
`save filename -ascii`	Saves the supported data type in the eight-digit ASCII format
`save filename -double`	Saves the supported data types in the ASCII double-precision format (16 digits)
`save filename -tabs`	Saves supported data type in tab delimited format

By the term "supported data type," we mean numerical data and text strings (cell arrays, structures, handles, etc., can be saved in binary format only). Any of the above calls has its function form of the syntax, for example, `save('filename','var1','var2',…)`.

Once you stored your data using the `save` function, you can always load it back to the MATLAB development environment (workspace) using the `load` function in one of the reciprocal syntax

`load`	Retrieves variables in the file *matlab.mat* and loads them into workspace
`load filename`	Retrieves variables in the file specified and loads them into workspace
`load filename.ext`	Loads the ASCII file *filename. ext* into the matrix `filename` (the ASCII file must be space delimited)
`load filename var1 var2`	Retrieves two specified variables, `var1` and `var2`, from the MAT-file *filename*
`load –ascii filename`	Retrieves and loads variables from a space delimited ASCII file
`S=load('filename')`	Returns the content of *filename* in the variable `S`. If *filename* is a MAT-file, `S` is a structure containing fields matching the variables retrieved. If *filename* is an ASCII file, `S` is a double-precision array

You may also use the functional form of `load`, such as `load('filename')`, which may be the only option if the file name is stored in a string, when an output argument is requested (as in the last syntax), or if `filename` contains spaces.

4.7 Importing and Exporting Data

MATLAB allows you to import data (digital and textual) from a variety of different sources, including any ASCII file or even a Microsoft Excel spreadsheet. You can do it either programmatically or using the embedded MATLAB tool, Import Wizard.

4.7.1 Importing and Exporting Data Programmatically

When importing data programmatically, you must know some a priori information about the data files you are trying, specifically

1. How many data items are in each row?
2. Whether these items are numeric, text strings, or both?
3. Does each row or column have a descriptive header?
4. What character is used as a delimiter, that is, the character used to separate the data items from each other, or the column separator?

Depending on this information, you have a choice of using several MATLAB functions to import data into the MATLAB development environment most effectively. A complete list of all MATLAB functions assisting to import different types of data is given in Table 4.3.

We will only discuss a few of them, but if needed you are welcome to use the MATLAB Help system to learn about others. We start from reading numerical (textual) data, remembering that normally it is formatted in one of the following categories:

1. Space delimited
2. Tab delimited
3. Comma or semicolon delimited
4. Mixed text and numeric
5. Text headers only

The most common function to import numerical data, which is the `dlmread` function, has the following format:

```
M=dlmread('filename',delimiter)
```

This function allows you to read an ASCII delimited file `filename` into a matrix M, using the specified `delimiter`. Note that

1. The `filename` must contain only ASCII numeric data.
2. You may use `'\t'` to specify tab delimited data and `';'` to specify semicolon delimited data (if `delimiter` is not specified, a comma `','` is assumed to be the default delimiter).

Moreover, the `dlmread` function allows you to read only a portion of data. But, two additional formats are available for this

```
M=dlmread('filename',delimiter,R,C)
```

and

```
M=dlmread('filename',delimiter,range)
```

Table 4.3 Data Importing/Exporting Functions of MATLAB

File Format	File Content	Extension	Functions
MATLAB formatted	Saved MATLAB workspace	*.mat*	`load, save`
	Text	*any*	`textscan`
	Text	*any*	`textread`
Text/numeric data	Delimited numbers/text	*any*	`dlmread, dlmwrite`
	Comma-separated numbers/text	*.csv*	`csvread, csvwrite`
Extended Markup Language	XML-formatted text	*.xml*	`xmlread, xmlwrite`
Audio	NeXT/SUN sound	*.au*	`auread, auwrite`
	Microsoft WAVE sound	*.wav*	`wavread, wavwrite`
Movie	Audio/video	*.avi*	`aviread`
Scientific data	Data in common data format	*.cdf*	`cdfread, cdfwrite`
	Flexible image transport system data	*.fits*	`fitsread`
	Data in hierarchical data format	*.hdf*	`hdfread`
Spreadsheet	Microsoft Excel worksheet	*.xls*	`xlsread, xlswrite`
	Lotus 123 worksheet	*.wk1*	`wk1read, wk1write`
Graphics	TIFF image	*.tiff*	`imread, imwrite`
	PNG image	*.png*	
	HDF image	*.hdf*	
	BMP image	*.bmp*	
	JPEG image	*.jpeg*	
	GIF image	*.gif*	
	PCX image	*.pcx*	
	XWD image	*.xwd*	
	Cursor image	*.cur*	
	Icon image	*.ico*	

(note that in this case you must specify a delimiter even if it is `','`). In the first format, the values `R` and `C` specify the upper-left corner (row `R` and column `C`) of data that needs to be read (zero values, `R=0` and `C=0`, specify the first value in the file, which is the upper-left corner). The second format allows you to read a range specified by `range=[R1 C1 R2 C2]`, where (`R1,C1`) is the upper-left corner of data to be read and (`R2,C2`) is the lower-right corner. Range can also be specified using spreadsheet notation as in range=`'A1...B7'`. For instance, given the file *data.txt* that contains the comma-separated values

```
82,    10,    16,    15,    66,    76
13,    55,      ,    92,    85,    40
92,    96,    49,    80,    94,    66
64,    97,    81,    96,    68,    18
```

The following two commands produce the corresponding results shown beneath them:

```
>> dlmread('data.txt',',',2,4)
ans = 94 66
      68 18
>> dlmread('data.txt',',',[1,1,2,3])
ans = 55  0 92
      96 49 80
```

Note that the `dlmread` function fills empty delimited fields with zero.

In a similar manner, an array M can be exported to the ASCII data files by using the reciprocal function `dlmwrite`

```
dlmwrite('filename',M,delimiter)
dlmwrite('filename',M,delimiter,R,C)
```

The first command writes M into an ASCII-format file, using `delimiter` to separate the matrix elements. Data are written to the upper left-most cell of the spreadsheet `filename`. In the second command, data are written to the spreadsheet `filename`, starting at spreadsheet cells R and C, where R is the row offset and C is the column offset (again, R=0, C=0 specifies the first value in the file, which is the upper-left corner).

Along with the more general `dlmread` and `dlmwrite` functions, MATLAB has another pair of functions allowing you to read and write a comma-separated numeric data. They are `csvread` and `csvwrite`, respectively. Their syntax can be easily understood by exploring the following commands:

```
M=csvread('filename')
M=csvread('filename',R,C)
M=csvread('filename',R,C,range)
csvwrite('filename',M)
csvwrite('filename',M,R,C)
```

(R and C should match `range(1:2)`). For instance, for the same file *data.txt* the command

```
>> B=csvread('data.txt',2,0,[2,0,3,3])
```

creates

```
B = 92   96   49   80
    64   97   81   96
```

If your file contains mixed numeric–text data, you may only rely on the `textread` (`textscan`) function

```
[A,B,C,...]=textread('filename','format',N)
```

This function reads data from the file `filename` into the variables A, B, C, etc., using the specified `format` string N times (if $N < 0$ or it is not present at all, the entire file is read). Although the function `textread` is well suitable for reading mixed numeric–text data with fixed or free format, you still have to know this format, that is, organization of the file. The `textread` function matches and converts groups of characters from the input. Each input field is defined as a string of nonwhite-space characters (by default the white-space character is one of the following: `' '` – space, `'\b'` – backspace, or `'\t'` – horizontal tab) that extends to the next white-space or delimiter character, or to the maximum field width. For example, if the first line of the *mydata.dat* file is

```
John        LT 3.34 25 BS
```

then reading this line using the free format `%` (to be discussed in more detail in the next section of this chapter)

```
>> [name,rank,GPA,age,degree]=textread('mydata.dat','%s %s %f %d %s',1)
```

returns

```
name = 'John'
rank = 'LT'
GPA = 3.34000000000000
age = 15
degree = 'BS'
```

Note that, while the (repeated) white-space characters in the format string are ignored, the repeated delimiter characters are significant. The `format` string determines the number and types of return arguments. The number of return arguments is the number of items in the `format` string. The `format` string supports a subset of the conversion specifiers and conventions of the C language `fscanf` routine. The values for the `format` string are listed in Table 4.4.

More complex syntax of the `textread` function

```
[...] = textread(...,'param','value',...)
```

provides even more capabilities that may be very useful for reading data from the files created by other programs and containing a lot of redundant information. For instance, parameter `headerlines` allows you to skip the specified number of lines at the beginning of the file.

Table 4.4 Format Specifiers

Format Code	Action	Output
Literals (ordinary characters)	Ignore the matching characters. For example, in a file that has `MAE` followed by a number (for course number), to skip the `MAE` and read only the number, use `'MAE'` in the `format` string	None
`%d`	Read a signed integer value	Double array
`%u`	Read an integer value	Double array
`%f`	Read a floating-point value	Double array
`%s`	Read a white-space or delimiter-separated string	Cell array of strings
`%q`	Read a string, which could be in double quotes	Cell array of strings (does not include the double quotes)
`%c`	Read characters, including any white-space character	Character array
`%[ ]`	Read the longest string containing characters specified in the brackets	Cell array of strings
`%[^ ]`	Read the longest nonempty string containing characters that are not specified in the brackets	Cell array of strings
`%*` (instead of `%`)	Ignore the matching characters specified by `*`	No output
`%w ...` (instead of `%`)	Read field width specified by `w`. The `%f` format supports `%w.pf`, where `w` is the field width and `p` is the precision	

Consider the following practical example. Suppose the *Data.log* file contains some useful data (100 lines starting with symbol `c`) advanced by 10 lines of secondary information as follows:

```
v ATLAS
# Created by ATLAS on Fri Nov 14 09:48:20 2008.
#SSF 2.0 generic
######################################################################
#              Copyright © 1984-2008                          #
#              Silvaco Data Systems, Inc. All rights reserved #
# KEY=SVC86hn7562asdx                 PROD=345                #
######################################################################
j 4 1 2 94 95
```

```
k 2 2 0.342109921 0.670000017
c 1 -250 -0.87 0
c 2 -250 -0.866666667 0
c 3 -250 -0.863333333 0
...
```

Then the command

```
>> [a,b,c,d]=textread('Data.log','%*c %f %f %f %f',100,'headerlines',10)
```

skips the first 10 lines and reads 100 elements of data into the four column vectors a, b, c, and d.

Consider another example. Suppose the file *winds.txt* is organized as follows:

```
37 2
2008-12-16_00Z,DZID00 i,j=( 15.9, 15.6),lat,long=(54.832,83.104)
SFALT, SFPRES
381 968.3
LINE AGL(m) T(C) RH(%) WSPD(m/s) WDD P(mb)
1 2.0 11.76 50.93 9.33 182.93 968.04
2 15.2 11.64 51.16 10.97 182.91 966.53
3 56.2 11.26 52.30 13.47 182.37 961.80
...
```

The goal is to read not only the numerical data starting from line 6 (and we do not know how many lines of data are in there) but also some data from the header lines as well. The following set of commands does the job:

```
% Reading header information
[Y,M,D,H,f1,f2,f3,lat,lon]=...
textread('winds.txt',...
'%4c-%2c-%2c_%2c %s i,j=( %f , %f ),lat,long=( %f , %f',1,'headerlines',1);
clear f1 f2 f3
% Finding the length of the file
file=textread('winds.txt','%s','delimiter','\n'); N=length(file);
% Reading numerical data
[q,w,e,r,t,y,u]=textread('winds.txt','%f %f %f %f %f %f %f',...
                N-5,'headerlines',5);
```

It first reads the year, month, day, hour, latitude, and longitude information from the second line skipping all the text (and clearing auxiliary variables f1 ,..., f3). Then, it reads all lines of the file to find its length (number of lines) N. Finally, it reads N−5 lines of data into the matrix W. In the future release, the textread function will be substituted with the textscan function, which has a similar format. The only major difference is that textscan reads data into a cell array as opposed to the variables A, B, C, etc.

Another function that may be quite useful is the xlsread function, enabling reading numeric data and text directly from the Microsoft Excel spreadsheet files (with an extension *.xls*). Three possible formats are

```
A=xlsread('filename')
```

```
[A,B]=xlsread('filename')
[A,B]=xlsread('filename','sheetname')
```

The first format returns only numeric data in array A from the first sheet in the Microsoft Excel spreadsheet file named *filename*. With this syntax, the xlsread function automatically ignores leading rows or columns of text. However, if it encounters any cell which is located not in a leading row or column and is empty or contains text instead of numeric data, xlsread puts a NaN in its place.

The second format (syntax) of the xlsread function returns numeric data in array A and text data in cell array B. If the spreadsheet contains leading rows or columns of text, xlsread returns only those cells in B. If the spreadsheet contains text that is not in a row or column header, xlsread returns a cell array of the same size as the original spreadsheet with the text strings in the cells that correspond to text in the original spreadsheet. All cells that correspond to numeric data are empty. The third format (syntax) reads the sheet specified in sheetname.

For example, for the Microsoft Excel file shown in Fig. 4.8, the command

```
>> [numTS,txtTS]=xlsread('TS.xls')
```

returns

```
numTS = 22.0000              0          NaN        NaN
         8.0000         0.6000          NaN        NaN
        58.0000              0       7.0000          0

txtTS = Columns 1 through 5
     ' '          'Stage 1'           ' '        'Stage 2'            ' '
     'ID'     'Descent rate'  'Glide Ratio'  'Descent rate'  'Glide Ratio'
  'ADS 1'          ' '                ' '          ' '              ' '
  'ADS 2'          ' '                ' '          ' '              ' '
  'ADS 3'          ' '                ' '          ' '              ' '
```

Another function to explore is importdata('filename',delimiter), which loads data from filename using delimiter as the column separator (if text). The function importdata looks at the file extension to determine which helper function to use. If it can recognize the file extension, importdata calls the appropriate helper function, specifying the maximum number of output arguments. If it cannot recognize the file extension, importdata calls finfo to determine which helper function to use. If no helper function is defined for this file extension, importdata treats the file as delimited text. For instance, for the aforementioned file *Data.log,* the command

```
>> importdata('Data.log','\t')
```

returns a cell array putting each string into an individual cell

```
ans = 'v ATLAS'
      #[1x47 char]
      '#SSF 2.0 generic'
      #[1x70 char]
```

	A	B	C	D	E
1		Stage 1		Stage 2	
2	ID	Descent rate	Glide Ratio	Descent rate	Glide Ratio
3	ADS 1	22	0		
4	ADS 2	8	0.6		
5	ADS 3	58	0	7	0

Fig. 4.8   Content of the *TS.xls* file to be read into MATLAB workspace.

```
#[1x70 char]
#[1x70 char]
#[1x70 char]
#[1x70 char]
'j 4 1 2 94 95'
'k 2 2 0.342109921 0.670000017'
'c 1 -250 -0.87 0'
'c 2 -250 -0.866666667 0'
'c 3 -250 -0.863333333 0'
...
```

Most of the remaining functions of Table 4.3 are devoted to reading media files, for example, the `imread` function reads the true color, gray-scale, or the indexed image files, the `wavread` function reads the Microsoft wave sound files, the `aviread` function reads the movie files, etc. Some discussion on using the `aviread` function can be found at the end of Sec. 6.12, while an example of using the `imread` function is presented at the end of Sec. C.2. For more details, use MATLAB Help browser.

4.7.2 Importing Data Using the Import Wizard

Along with importing data into MATLAB programmatically as discussed previously, you can always do it using a special tool—the MATLAB Import Wizard. To activate the MATLAB's Import Wizard, you may select **Import Data** option from MATLAB Desktop **File** menu or click the **Import data** button of the Workspace window, or use the `uiimport` function. To use this tool, you do not even need to know

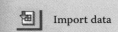 Import data

the format of data. You simply specify the file that contains data and the Import Wizard processes the file's content automatically. Figures 4.9 and 4.10 provide two examples of importing data from the ASCII text file and from a Microsoft Excel spreadsheet.

As seen, the Import Wizard automatically recognizes what type of data you are trying to import. You can always change the column separator of the data that should be read with in the case when the Import Wizard did not recognize it correctly (Fig. 4.9b). First, the Import Wizard separates the numeric data and creates a numeric matrix `data` from it. Then, it tries to store all headings in cell arrays. In the first example (Fig. 4.9a), when it is quite simple to separate the headings in a cell vector it does so and creates two essentially identical 1 by 2 cell arrays (row vectors), `textdata` and `colheaders`, both containing the text strings `'temp f'` and `'temp c'` (Fig. 4.9b only shows the tabs for these cell vectors). In the second example, when the numeric data are surrounded by text (Fig. 4.10a), the Import Wizard

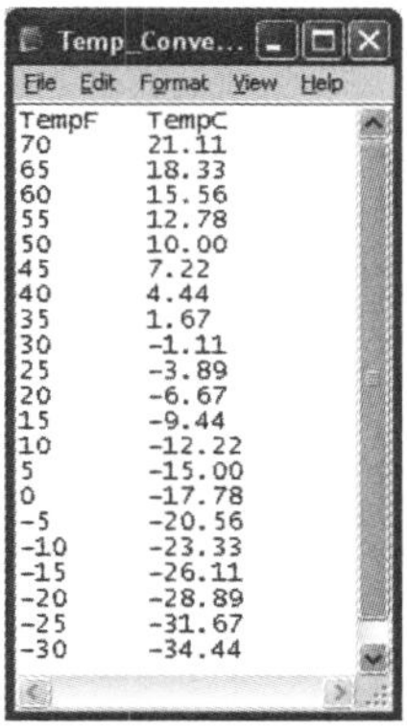

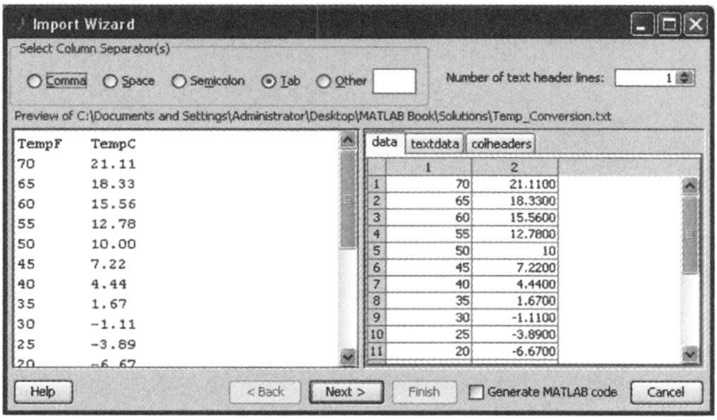

Fig. 4.9 Importing data from a) an ASCII text file using b) the Import Wizard.

a)
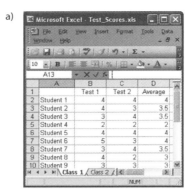

b)
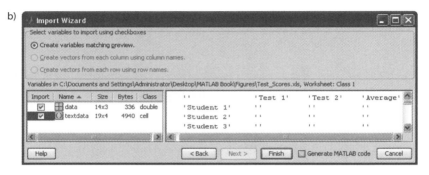

Fig. 4.10 Importing data from a) a Microsoft Excel spreadsheet using b) the Import Wizard.

gets confused, so it creates a `textdata` cell array shown in Fig. 4.10b (where all numerical data are substituted with the empty strings ` ' ' `). The Import Wizard represents the suggested imported data in the form exactly the same as it would (will) appear in the Workspace window (once accepted).

As shown in both Figs. 4.9b and 4.10b, you may want to use Import Wizard to automatically generate MATLAB code for you, so that you could use it if you need to import similar data ever again (just check off the corresponding radio button from the buttons of the Import Wizard window).

4.8 Writing and Displaying Formatted Data

So far, we primarily dealt with displaying unformatted data, as they are stored in computer's memory, having quite limited control on how we want it to be presented or saved. To be more specific, in Sec. 2.5 the `format` function was introduced to control the output format of numeric values displayed in the Command window. In the previous section (Table 4.4), specifiers were introduced to distinguish between the different types of data when reading from the files. This section elaborates on the specifiers' concept and shows how to apply a specific format to each variable to be presented on the screen or written on the file. Specifically, it is the `fprintf` function allowing you to do so. Its general format is

```
fprintf(fid,'format',A,…)
```

This function formats data in (the real part of) matrix `A` and any additional variables under control of the specified `format` string (`format` specifier) and writes it to the file associated with file identifier `fid`. An argument `fid` is an integer file identifier obtained from the function `fopen` (an example will be provided later in this section). This argument may also assume the value of `1` or be omitted—in both cases, `fprintf` will use a standard output onto a display screen. If you opened a file with `fopen`, you could close it later on with the reciprocal function `fclose`.

The format specifier `format` is a string containing C-language conversion specifications starting with the marker `%` and containing the following pattern (Fig. 4.11):

```
%[-][number1.number2]Y
```

(`[]` denotes optional fields). In this pattern

`%`	Starts the conversion specification (ending with the conversion character `Y`)
`-`	Specifies the alignment code
`number1`	Specifies the field width
`number2`	Specifies the number of digits to the right of the decimal point (precision)
`Y`	Specifies the notation of the output (format code)

The alignments codes (flags) are

`-`	Left-justifies the converted argument in its field
`+`	Prints always a sign character (+ or –)
`0`	Pads with zeros rather than spaces

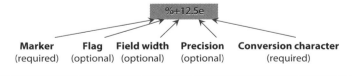

Marker	Flag	Field width	Precision	Conversion character
(required)	(optional)	(optional)	(optional)	(required)

Fig. 4.11 Example of the format specifier.

Field width and *precision* specifications are given by a digit string specifying the minimum number of digits to be printed (number1) or a digit string including a period (number1.number2) specifying the number of digits to be printed to the right of a decimal point.

The following *conversion characters* (including those already given in Table 4.4 for the textread function) specify the notation of the output:

%c	Single character
%d	Signed decimal format
%e	Exponential (scientific) format with lowercase e (as in 3.1415e+00)
%E	Exponential format with uppercase E (as in 3.1415E+00)
%f	Fixed-point decimal format
%g	%e or %f, whichever is shorter (insignificant zeros do not print)
%G	Same as %g, but using an uppercase G
%o	Unsigned octal format
%s	String of characters
%u	Unsigned decimal format
%x	Hexadecimal format (using lowercase letters a–f)
%X	Hexadecimal format (using uppercase letters A–F)

Again, for a better readability of the output within the format string you may place the following "escape" character sequences starting with symbol \ (allowing you to specify nonprinting characters):

\b	Backspace
\f	Form feed
\n	Start new line
\r	Carriage return
\t	Horizontal tab
\\	Backslash
' ' (two single quotes)	Apostrophe (single quotation mark)
%%	Percent character

Consider two examples allowing you to better understand how fprintf function actually works. The first example accommodates printing a simple scalar to the screen:

```
>> speed = 62.3;
>> fprintf('The speed is: %3.1f fpsn', speed)
The speed is: 62.3 fps
```

The second example allows you to print a matrix into the file *exp.txt*:

```
>> x = 0:.1:1;
>> y = [x; exp(x)];
>> fid = fopen('exp.txt','w');
>> fprintf(fid,'%6.2f %12.8fn',y);
>> fclose(fid)
```

(the `fopen` function defines where to write data, `'w'` specifies the permission to open file, or create new file for writing and discard existing contents if any). The newly created text file *exp.txt* now contains a short table of the exponential function

```
0.00    1.00000000
0.10    1.10517092
...     ...
1.00    2.71828183
```

If needed, data from this file can be read back by using the reciprocal function `fscanf` as

```
>> fid = fopen('exp.txt');
>> a = fscanf(fid,'%g %g',[2 inf]) % It has two rows now
>> a = a';
>> fclose(fid)
```

Here, the default permission in the `fopen` function (omitted second argument `'r'`) allows you to open the file *exp.txt* for reading alone.

Let us conclude this section with introducing another formatted output function

```
str = sprintf('format',A,...)
```

which is exactly the same as the `fprintf` function with the only difference that it writes formatted data to the string `str`. Similarly, the `sscanf` function reads formatted data from the string.

4.9 Interactive Input and Output

To conclude the introductory portion of describing MATLAB, let us present several functions enabling interactive input and output, and therefore, making the programming process more entertaining.

Let us start from two simplest options for friendly interactive input of data while running the script. First, you may enjoy using a simple straightforward function `input`. The command

```
>> x = input('text')
```

gives the user a prompt in the `text` string and then waits for an input from a keyboard. The input can be any MATLAB expression, which is evaluated using the variables in the current workspace and then returned in `x`. For instance, the command

```
>> Area = pi*input('Enter the radius: ')^2
```

prompts and expects you to enter a radius, after which it computes and returns the area

```
Enter the radius: 5
Area = 78.5398
```

To enter a text string a slightly modified format should be used

```
>> x=input('text','s')
```

This command gives the prompt in the `text` string and waits for a character string input. The typed input is not evaluated, so the characters are simply returned as a MATLAB string in `x`

```
>> Name = input('Enter the name:','s')
Enter the name: Yuri Gagarin
Name = Yuri Gagarin
```

The `text` string for the prompt may contain one or more '`\n`', allowing you to skip to the beginning of the next line. This enables the prompt string spanning several lines as in the following example:

```
>> N = input('Pick a number between\n 1 and 10\nYour choice: ')
Pick a number between 1 and 10
Your choice: 5
N = 5
```

More enhanced input capability is provided using the `menu` function. The command

```
k = menu('title','option1','option2',...)
```

displays the `title` string followed in sequence by the menu-item strings: `option1`, `option1`, ... It returns the number of the selected menu-item as a scalar `k` (if `option1` is chosen – k=1, if `option2` – k=2, etc.). For

Fig. 4.12   Example of the user's menu.

Table 4.5 Predefined Dialog Boxes in MATLAB

Function	Description
dialog	Creates dialog box
errordlg	Creates an error dialog box
helpdlg	Displays a help dialog box
inputdlg	Creates an input dialog box
listdlg	Creates a list selection dialog box
msgbox	Creates a message dialog box
pagesetupdlg	Displays a page setup dialog box
printdlg	Displays a print dialog box
questdlg	Creates question dialog box
uigetdir	Displays a dialog box to retrieve the name of a directory
uigetfile	Displays a dialog box to retrieve the name of a file for reading
uiputfile	Displays a dialog box to retrieve the name of a file for writing
uisetcolor	Sets ColorSpec using a dialog box
uisetfont	Sets a font using a dialog box
waitbar	Displays a wait bar
warndlg	Creates a warning dialog box

example, the following command:

```
>> k = menu('Choose a data marker','o','*','x');
```

can be used to interactively choose the type of the marker. In results in the menu shown in Fig. 4.12, you can make your choice by pushing any of the three buttons.

To complete this simple example, let us show how you can use the choice you made to plot say a piece of parabola. If marker 'o' was chosen, then the set of commands

```
>> x=1:10; y=x.^2;
>> type = ['o', '*', 'x'];
>> plot(x,y,type(k))
```

will generate the plot presented in Fig. 4.13. Note that k=menu('title', itemlist), where itemlist is a cell array containing a set of text strings, also represents a valid syntax. For instance, in the aforementioned example the cell array itemlist={'o' '*' 'x'} might be used.

In addition to the two simplest interactive input options we have just presented, MATLAB provides you with a variety of professionally looking dialog boxes you may use. Table 4.5 presents a complete list of these predefined dialog boxes (later on you will be able to develop your own menus using the GUIDE tools as discussed in Appendix C). Let us briefly introduce

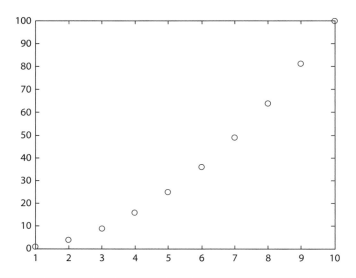

Fig. 4.13 Parabola with an interactively chosen marker.

several of them (you are welcome to use the Help system to get more details and to explore the others).

Let us start from the `questdlg` function. Its usage is as simple as in the following example. The command

```
>> k=questdlg('You want to continue?',...
'Premature Exit','Yes','No','Cancel','Yes');
```

opens the dialog box shown in Fig. 4.14a. Depending on the button you push, the variable k can became either `'Yes'` or `'No'` (char-type variable). The last `'Yes'` (must be `'Yes'`, `'No'`, or `'Cancel'`) specifies which push button is the default in the event that the <Esc> key is pressed.

The `waitbar` function is typically used inside a `for` loop (to be discussed in Chapter 5) that performs a lengthy computations. For example, the set of commands

```
>> h=waitbar(0,'Please wait...','Name','Status Bar');
>> Npoints=100;
>> for i=1:Npoints, % computation starts here
>> waitbar(i/Npoints,h,[num2str(i/Npoints*100,'%3.1f') '% completed'])
>> pause(0.1)
>> end
>> close(h)
```

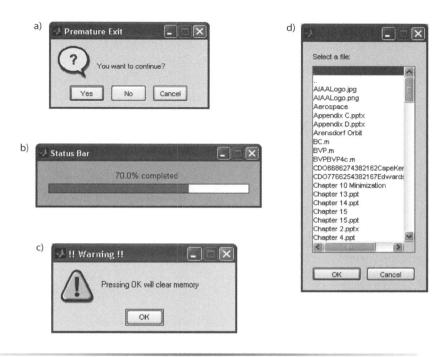

Fig. 4.14 Examples of predefined dialog boxes in MATLAB.

shows what percentage of calculation is complete. Although the computations progress, the horizontal bar in Fig. 4.14b fills up to the right and the number above it gradually increases from 0 to 100% (note how we used the formatted output '%3.1f' for this). Since there are no sophisticated computations involved in this example, the pause(0.1) line is used to slow it down, otherwise it is not needed.

The statement

```
>> warndlg('Pressing OK will clear memory','!! Warning !!')
```

uses the warndlg function and evokes the dialog box shown in Fig. 4.14c.

The next example presents a dialog box (Fig. 4.14d) that enables a user to select a file from the current directory

```
>> d = dir;
>> str = {d.name}
>> [s,v] = listdlg('PromptString','Select a file:','SelectionMode',...
'single','ListString',str)
```

The function listdlg returns two values, where s is the index to the selected file, and v is either 1 indicating that the selection has been made or

0 if no file was selected. For example, if the file *win32* was chosen, then the returns are as follows:

```
str = '.' '..' 'matlab.bat' 'registry' 'win32'
s = 5
v = 1
```

so that `str(5)='win32'`.

In this regard two other functions, `uigetfile` and `uiputfile`, seem to be more convenient in defining the name of the input and output file. For instance,

```
>> [FileName,PathName] = uigetfile('*.mat','Select the MAT-file');
```

allows you to bring up the window shown in Fig. 4.15a, navigate to the directory of your choice displaying only MAT-files (`'*.mat'` serves as a filer employing the * wildcard), and return the name and path of a chosen file in `char` variables `FileName` and `PathName`, respectively. Hence, the next command loading the selected file into the current workspace may look like as follows:

```
>> load([PathName FileName]);
```

a)

b)

Fig. 4.15 Dialog boxes for a) `uigetfile` and b) `uiputfile` functions.

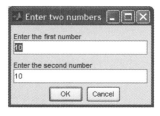

Fig. 4.16 Input dialog created using the `inputdlg` function.

Similarly, the `uiputfile` displays a modal dialog box used to select or specify a file for saving. For example, the following command

```
>> [FileName,PathName]=uiputfile('*.txt','Save the TXT-file','Trial');
```

brings up the window shown in Fig. 4.15b. (The optional default filename `Trial` appears in the **File name** field.) Now you can use output variables to save your workspace to the ASCII file:

```
>> save([PathName FileName],'-ascii');
```

One more function you might find very useful for entering the numerical data is `inputdlg`. The following fragment brings a scalable menu shown in Fig. 4.16:

```
>> prompt = {'Enter the first number','Enter the second number'};
>> dlg_title = 'Enter two numbers';
>> num_lines = 1;
>> def = {'10','10'};
>> answer = inputdlg(prompt,dlg_title,num_lines,def,'on');
>> a=str2num(answer{1});
>> b=str2num(answer{2});
```

(the optional `'on'` in the `inputdlg` call enables resizing the menu window). Note that the default values showing up in the corresponding fields are stored in the cell array as `char`-type variables. Similarly, the output `answer` is also a cell array and that is why we need the last two lines of code, accessing the content of the cells and converting it into the numerical format. Hence, after entering two numbers and hitting the **OK** button, these numbers will appear in the current workspace as decimal `a` and `b`, ready to be used in numeric computations.

Once again, you are welcome to learn more about these dialog boxes via the MATLAB Help system. As mentioned previously, in addition to these boxes, MATLAB provides a tool enabling development of your own GUI. This tool, GUIDE, presented in Appendix C, allows you to use panels, buttons, text fields, sliders, menus, and so on, and adjust an M-file that operates your GUI to meet your own needs.

Problems

4.1 The following table presents some of the speed records established by turbojet aircraft (it shows only those entries when the record went from one country to another):

Date	Country	Vehicle	Speed (mph)
7 Nov. 45	Great Britain	Gloster Meteor F Mk4	606
19 Jun. 47	USA	Lockheed XP-80R	624
7 Sep. 53	Great Britain	Hawker Hunter F Mk3	728
3 Oct. 53	USA	Douglas XF4D-1	753
10 Mar. 56	Great Britain	Fairey Delta Two	1132
12 Dec. 57	USA	McDonnell F-101A	1208
31 Oct. 59	USSR	Mikoyan E-66	1484
15 Dec. 59	USA	Convair F-106A	1526
7 Jul. 62	USSR	Mikoyan E-166	1666

Create and fill out a structure `Record`, composed of the following fields: `Date`, `Country`, `Vehicle`, and `Speed`. Manipulate the corresponding fields of this structure to: a) convert the records to m/s (1 mph = 0.44704 m/s) and b) find the country, which held its speed records the shortest and the longest. Hint: use the `datenum` function, for example, `datenum('7-Nov-45', 'dd-mmm-yyyy')`, to convert data to a serial date number.

4.2 Develop (and display) a 2 by 2 by 9 cell array containing the same information as in Problem 2.1. Manipulate the corresponding elements of this cell array to: a) convert the records to km/h (1 mph = 1.61 km/h) and b) compute the number of days each record held before being beaten by another country.

4.3 Issue the following set of commands that creates an ASCII file, containing temperatures in degrees of Fahrenheit and Celsius:

```
clear all, clc
T(:,1)=70:-10:-70;
T(:,2)=(T(:,1)-32)/1.8;
save('Temp.txt','-ascii')
clear
```

Now, write a script allowing you to upload data from the *Temp.txt* file into MATLAB environment (using `textread` or `dlmread` command), and then display it using formatted output in the following two-column left-adjusted format with a vertical separator (|):

```
|     Temp, F      |     Temp, C      |
|     70           |     21.1         |
|     60           |     15.6         |
|     50           |     10.0         |
|     40           |     4.4          |
|     ...          |     ...          |
```

Chapter 5 | Programming in MATLAB

- Relational Operators and Conditional Statements
- Loops, Switch Structure, Continue and Break Statements
- Developing the Professionally Looking Functions
- Editing, Debugging, and Profiling

lt, le, gt, ge, eq, ne, and, or, not, if-else, for, while, switch, continue, break, inline, function, which, isinf, iscell, isvector, ismatrix, all, any, global, nargin, nargout, fcnchk, feval, pause, echo, tic-toc, **and others**

5.1 Introduction

This chapter is devoted to writing effective MATLAB programs, scripts, and functions. It first addresses the relational and logical operators and functions, followed by several structures to alter straightforward stream of the program (if-else operators, for and while loops, continue and break statements, and switch structure). Next, it addresses writing of simplest MATLAB programs and scripts. The chapter continues with presenting different types of functions allow performing the same (group of) operations repeatedly. It also addresses the issue of local and global variables. It proceeds with a discussion on an open-source paradigm and the tricks to write the professional user-defined functions, including nargin, nargout, fcnchk, and feval. Some useful MATLAB functions for fast plotting, zero and minimum finding are introduced and analyzed later. The chapter ends with the tips on using the MATLAB tools including programmatic tools, such as the functions pause and echo, **M-File Editor**, **M-Lint** to debug your programs along with employing the MATLAB's **Profiler** to improve their performance.

5.2 Relational Operators

As in any other programming language, MATLAB has six traditional *relational operators* useful for making comparisons between expressions (arrays). They are

<	Less than (lt(A,B))
<=	Less than or equal to (le(A,B))
>	Greater than (gt(A,B))
>=	Greater than or equal to (ge(A,B))
==	Equal to (eq(A,B))
~=	Not equal to (ne(A,B))

When applying the relational operators, the following holds:

1. Arrays are compared on an element-by-element basis. Exception is when comparison is made to a scalar, then every element is compared with the scalar.
2. Result of comparison is
   0 if comparison is `false` and
   1 if comparison is `true`.
3. Precedence rules establish that
   • arithmetic operators have precedence over relational operators and
   • relational operators have equal precedence among themselves—they are evaluated in their order from left to right.

Next you will find several self-explanatory examples of using these relational operators

```
>> 5 > 2      >> le(0.1,-4)     >> 2 ~= 6-4      >> [0 4] == 0
ans = 1       ans = 0           ans = 0          ans = 1 0
```

Now that we know about relational operators let us discuss how useful they may be in addressing certain elements of arrays, or to be more specific, the element that obey certain conditions. One way of doing it is addressing these elements directly via placing the condition in parentheses right after the expression. For example, two commands

```
>> x = 10:-1:0;
>> z = x (x<3)
```

return

```
z = 2  1  0
```

A more sophisticated way, which enables more options, is addressing certain elements indirectly via the function `find(x)`, which was introduced in Sec. 3.4. Recall that when the input array x is a logical expression, the output of this function does not contain its nonzero entries. Instead, it contains the nonzero values returned after evaluating the logical expression. For instance, for the vector x we created in the previous example, the command

```
>> ind = find(x<3)
```

yields the indices of the elements that are less than 3:

```
ind = 9   10   11
```

To address the elements with these indices we could use

```
>> z = x(find(x<3))
z = 2   1   0
```

Alternately, the same result could be achieved without using logical expression, for example, by manipulating with the original vector with the help of the min function as follows:

```
>> z = x(find(min(x-3,0)))
z = 2   1   0
```

5.3 Logical (Boolean) Operators and Functions

MATLAB offers several logical operators and functions. For the same size arrays, it provides with three standard (Boolean) logical operators to perform element-by-element array operations. They are as follows:

    &   (Boolean AND), so that A&B or and(A,B) returns an array of the same dimension as A and B, which has ones where both A and B have nonzero elements and zeros where either A or B is zero

    | (Boolean OR), so that A|B or or(A,B) returns an array of the same dimension as A and B, which has ones where at least one element in A or B is nonzero and zeros where A and B are both zero

    ~   (Boolean NOT), so that ~A or not(A) returns an array of the same dimension as A, which has ones where A is zero and zeros where A is nonzero

One more logical operator, namely, xor(A,B) function (exclusive OR) is also available. This function is defined in terms of AND, OR, and NOT operators as

```
function X = xor(S,T)
X = (S|T) & ~(S&T);
```

Therefore, it returns zeros, where A and B are either both nonzero or both zero, and ones, where A or B is nonzero, but not both.

To better understand these operators, the following script provides with the so-called truth table, which defines the operations of the logical operators and the function `xor` (1 for `true` and 0 for `false`)

```
>> x = [1;1;0;0];
>> y = [1;0;1;0];
>> disp('    x      y      ~x      x|y     x&y    xor(x,y)')
>> disp([x,y,~x,x|y,x&y,xor(x,y)])
```

x	y	~x	x|y	x&y	xor(x,y)
1	1	0	1	1	0
1	0	0	1	0	1
0	1	1	1	0	1
0	0	1	0	0	0

MATLAB always gives the `&` operator precedence over the `|` operator. Although MATLAB typically evaluates expressions from left to right, the expression `a|b&c` is evaluated as `a|(b&c)`. Hence, it is a good idea to use parentheses to explicitly specify the intended precedence of statements containing combinations of `&` and `|`.

The short-circuit operators `&&` and `||` are also the logical AND and OR operators used in MATLAB to evaluate logical expressions

`&&` Returns 1 if both inputs evaluate to `true`, and 0 if they do not
`||` Returns 1 if either input, or both, evaluate to `true`, and 0 if they do not

These operators are used in the evaluation of compound expressions of the form

```
expression_1 && expression_2
```

where `expression_1` and `expression_2` each evaluate to a scalar, logical result. The `&&` and `||` operators support the so-called short-circuiting. This means that the second operand is evaluated only when the result is not fully determined by the first operand (sometimes using the element-wise operators `&` and `|` in this case may yield unexpected results). For instance, in the following statement

```
>> x = (b~=0) && (1/b>18.5)
```

it does not make sense to evaluate the relation on the right if the divisor, `b`, is zero. Therefore, the test on the left is put in to avoid generating a warning under these circumstances. By definition, if any operands of an `&&` expression are `false`, the entire expression must be `false`. So, if `(b ~= 0)` evaluates to `false`, MATLAB assumes the entire expression to be false and terminates its evaluation of the expression immediately, returning `x=0`.

As to the precedence, logical operations have lower precedence than arithmetic and relational operators with the exception of the `not` operator

Table 5.1 Operator Precedence

Order	Operator	Description
1	( )	Parentheses
2	.', .^, ', ^	Transpose, power, complex conjugate transpose, and matrix power
3	+, -, ~	Unary plus, unary minus, and logical negation
4	.*, ./, .\, *, /, \	Multiplication, right division, left division, matrix multiplication, matrix right division, and matrix left division
5	+, -	Addition and subtraction
6	:	Colon operator
7	<, <=, >, >=, ==, ~=	Less than, less than or equal to, greater than, greater than or equal to, equal to, and not equal to
8	&	Element-wise AND
9	|	Element-wise OR
10	&&	Short-circuit AND
11	||	Short-circuit OR

(~). For instance, command 0 | ~0 needs no parentheses. The and operator has always higher precedence than or (for MATLAB 6 and above). For example, because of this rule, the result of 1 | 0&0 is 1, not 0.

Now that we have considered all types of operations (algebraic, relational, and logical), the complete table of precedence looks as shown in Table 5.1.

Logical functions that are available in MATLAB allow you to write robust programs. Most of these functions are listed in Table 5.2. In addition to them, Table 5.3 lists cell, string, and structure tests functions.

Table 5.2 Functions Detecting the State of MATLAB Entities

Function	Brief Description
all(x) and all(A)	Returns a 1 if all the elements in the vector x or columns of matrix A are nonzero and a 0 otherwise
any(A)	Returns a row vector with same number of columns of A containing ones and zeros depending on whether or not the corresponding columns of A contain any nonzero elements
any(x)	Returns a scalar, that is, 1 if any of the elements in the vector x is nonzero and 0 otherwise

(Continued)

Table 5.2 Functions Detecting the State of MATLAB Entities (Continued)

Function	Brief Description
finite(A)	Returns an array of the same dimension as A with ones where the elements of A are finite
iscolumn(A)	Determines if input A is a column vector
isempty(A)	Returns a 1 if A is the empty matrix and 0 otherwise (checks whether A==[])
isfloat(A)	Determines whether input A is a floating-point array
isinf(A)	Returns an array of the same dimension as A with ones where A has inf and zeros elsewhere
isinteger(A)	Determines whether input A is an integer array
islogical(A)	Determines whether input A is a logical array
ismatrix(A)	Determines if input A is a matrix
isnan(A)	Returns an array of the same dimension as A with ones where A has NaN and zeros elsewhere
isnumeric(A)	Determines whether input A is a numeric array
isreal(A)	Returns a 1 if A has no elements with imaginary parts and 0 otherwise
isrow(A)	Determines if input A is a row vector
isscalar(A)	Determines whether input A is a scalar
issorted(A)	Determines if a set of elements A is in sorted order
isvector(A)	Determines if input A is a vector

5.4 Conditional Statements

MATLAB provides a standard programming means to construct conditional statements, which help to write programs that make decisions. To be able to do so, the program should contain one or more of the following statements: if, else and elseif. The end statement denotes the end of a conditional statement and is required to close the programming logic and cause the statements to be executed when the condition is true. The if statement has the following basic form (Fig. 5.1):

if *logical expression*
    statements
end

subject to regulations

1. A space is required between the if and the *logical expression*.
2. The *logical expression* may be a compound expression (multiple arithmetic, relational, and logical operators).

Table 5.3 Strings and Cells Operations Functions

Function	Brief Description
iscell(A)	Determines whether input A is a cell array
iscellstr(A)	Determines whether input A is a cell array of strings
ischar(A)	Determines whether input A is a character array
isdir('A')	Determines whether input A is a directory
iskeyword('A')	Determines whether input A is a MATLAB keyword
isletter(A)	Determines whether input A is composed of letters of the alphabet
isspace(A)	Determines whether input A is composed of whitespace characters
isstruct(A)	Determines whether input A is a structure array
isvarname('A')	Determines whether input A is a valid variable name

3. If the *logical expression* is performed on an array, then the conditional test is `true` only if all the elements of the *logical expression* are true (nonzero).
4. The *statements*, representing a series of MATLAB commands, are only executed if the *logical expression* is true.
5. For compactness the `if` structure with a short statement may be written on a single line: `if` *logical expression*, *statements*, `end`.

The `if` statements may be nested (Fig. 5.2). Note that in this case, each `if` statement has its corresponding `end` (for compactness you may type `end, end` on one line)

`if` *logical expression 1*
    statements 1
  `if` *logical expression 2*
      statements 2
  `end`
`end`

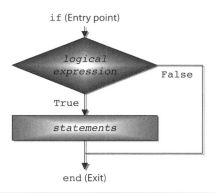

Fig. 5.1 Flowchart representing the `if` statement.

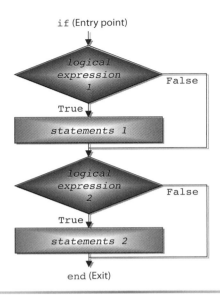

Fig. 5.2 Nested if statements.

If more than one action can occur as the result of a decision, the else and/or elseif statements should be used (depending on the number of actions). The basic form for the else command is represented as Fig. 5.3

if *logical expression*
    statements 1
else
    statements 2
end

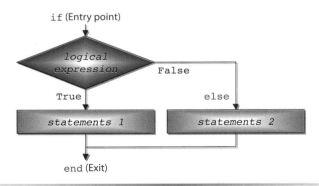

Fig. 5.3 Flowchart of the else structure.

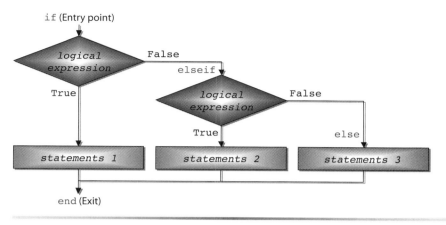

Fig. 5.4 Flowchart for the `elseif` structure.

The basic form for the `elseif` command may be represented as follows (Fig. 5.4):

```
if logical expression 1
     statements 1
elseif   logical expression 2
     statements 2
else
     statements 3
end
```

The general rule is that the `if` statement may have multiple `elseif` statements, but only one `else` statement. (Of course, all rules for the `if` portion are valid here too.)

When dealing with text variables (strings), you should be aware that two strings are equal *if and only if* every character is the same, including blank spaces (that is, uppercase and lowercase letters are not the same). MATLAB provides with the `strcmp(s1,s2)` function which compares two strings (`s1` and `s2`) to determine if they are equal.

5.5 Loops

A loop is a structure for repeating a calculation for a number of times. We call each repetition of the loop a pass. MATLAB supports two types of loops

1. *Implicit loops* (which)
2. *Explicit loops*
   - `for` loops
   - `while` loops

Implicit loops have been considered already on several occasions. First, when we studied the usage of a colon (for example, `x=0:5:100`); second, when array operations were introduced (for example, $y=\cos(x)$ or $z=\text{length}(x)$, where x is a vector); and finally, when matrix operations were considered (for example, multiplication of two matrices implies using the loops to sum the results of pair element multiplication). Hence, the following provides some details about using explicit loops.

5.5.1 for Loops

The for loops have the following basic form (Fig. 5.5):

```
for loop_variable = m:s:n
        statements
end
```

subject to a set of regulations

1. The expression m:s:n assigns loop variable an initial value m, and increments it by the value s until the result is lesser or equal to n. Looping breaks when the loop_variable exceeds the terminating value n.
2. Parameters m, s, and n may assume any real value or be real variables (no complex part) or expressions. Additionally
   - If the increment s is omitted, the default value is 1
   - If m=n, the loop will be executed only once
   - If the increment s is not an integer, round off errors may cause the loop to execute a different number of passes than expected
3. The statements are executed once during each pass using the current value of the loop_variable.

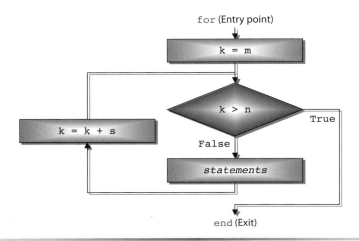

Fig. 5.5 Flowchart for the for loop (loop_variable is denoted with k here).

4. Do not attempt to alter the `loop_variable` inside the loop because MATLAB will not recognize this change (this is a sort of self-protection).
5. The `end` statement closes the programming logic and is mandatory.
6. Indenting is typically used to improve the readability, however the `for` loop may be written on a single line:

   ```
   for loop_variable=m:s:n, statements, end
   ```

7. Other loops and conditional statements may be *nested* within the `for` loop.

5.5.2 while Loops

The `while` loops are used to create a loop that will terminate when a certain condition is satisfied. They have the following basic form (Fig. 5.6):

while *logical expression*
        statements
end

subject to regulations

1. The *logical expression* should contain a *loop variable* that can be changed to make the expression false and end the loop.
2. The *loop variable* must have a value before the `while` statement is executed.

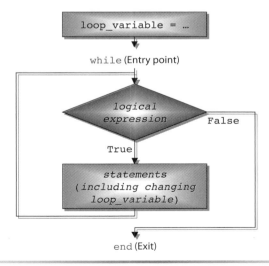

Fig. 5.6 Flowchart for the `while` loop.

3. Be cautious when using equality (==) or nonequality (~=) in the *logical expression*. If the compared quantities are not integers, the round off error may cause the loop to continue when it should have ended.

4. The *loop variable* must be changed somehow by the *statements* otherwise you may end up with an infinite loop (infinite loops are loops that never end, thst is, the *logical expression* is never false).

5. *Statements* are executed once during each pass, unless there are nested loops.

6. The mandatory end statement should close the programming logic.

5.5.3 continue and break Statements

There are two statements that can be used in conjunction with explicit loops to debug and/or improve the efficiency of the loops. They are continue and break.

The continue statement passes control to the next iteration of the for or while loop in which it appears, skipping any remaining statements in the body of the loop. For example,

```
s=0;  f=0;
for i=1:10
    s=s+1;
    continue
    f=f+1;
end
[s, f]
```

returns

```
ans = 10  0
```

because the line f=f+1; was never executed. In nested loops, continue passes control to the next iteration of the for or while loop enclosing it.

The break statement terminates the execution of a for or while loop. Statements in the loop that appear after the break statement are not executed. For example,

```
s=0;  f=0;
while s<=10
    s=s+1;
    if s==5
        break
    end
    f=f+1;
end
[s, f]
```

results in

```
ans = 5 4
```

In nested loops, `break` exits only from the loop in which it occurs. Control passes to the statement that follows the `end` of that loop. (Usually the use of the `break` statement in the final (debugged) program is discouraged. Instead, a properly constructed `while` loop should be used in its place.)

5.6 `switch` **Structure**

The `switch` structure provides an alternative to using `if`, `elseif`, and `else` commands, and for some applications `switch` structure is more readable than `if` structure. Its basic form is (Fig. 5.7)

`switch` *switch_expression* (scalar or text string)
      `case` *value 1*
             statements 1
      `case` *{value 2, value 3}*
             statements 2
  ...
      `otherwise`
             statements N

`end`

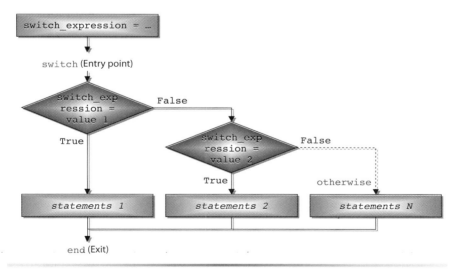

Fig. 5.7 Flowchart for the two-case `switch` structure with `otherwise` option.

When executing the `switch` structure, the *switch_expression* is compared to each *value*. If they match, the statements following that `case` statement are executed. The following rules apply:

1. Each `case value` statement must be on a single line.
2. The `switch` statement can handle multiple conditions in a single `case` statement by enclosing several *values* into a cell array.
3. Only the first matching `case` is executed.

The `otherwise` statement is optional. If it is not present and no `case` match occurs, then the next statement following the `end` is executed.

5.7 Script Files

A script file is an external file that contains a sequence of MATLAB statements. By typing the filename, subsequent MATLAB input is obtained from the file. Script files have a filename extension *.m* and are often called M-files.

Scripts are the simplest kind of M-file. They are useful for automating blocks of MATLAB commands, such as computations you have to perform repeatedly from the command line. Scripts can operate on existing data in the workspace, or they can create new data on which to operate. Although scripts do not return output arguments, any variables they create remain in the workspace, so you can use them in further computations. In addition, scripts can produce graphical output using commands like `plot`.

Scripts can contain any series of MATLAB statements. They require no declarations or begin/end delimiters. Like any M-file, scripts can contain *comments* (comments are strings or statements in an M-file that do not execute).

Any text following a percent sign, %, on a given line is a comment text. To distinguish it from the rest of the text, MATLAB makes its color green (the default color for comments as defined in the preferences shown in Fig. 2.21b). Comments can appear on lines by themselves, or you can append them to the end of any executable line, for instance,

```
% Earth's Parameters
e = 298.257223563; % WGS-84 Earth's ellipticity
f = 1 / e;         % Earth's flattening
eps2= 1-(1-f)^2;   % Earth's eccentricity (squared)
```

Another use for comments is when testing your files or looking for errors and would like to temporarily turn some lines of code into comments to see how the M-file runs without those lines.

You may also comment a contiguous group of lines without commenting each of them individually (this is useful when you want to identify the section of a file that is not working as expected). Simply type `%{` before the first line and `%}` after the last line you want to comment, as in the following example (for better understanding the comment block is indented):

```
g=9.81;
  %{
  g=-3.986004418e14/rp^3*(eye(3)+...
  1.5*1.0826267e-3*(re/rp)^2*(egm-5*sin(mu)*eye(3)))*p;
  %}
display(g)
```

This is referred to as a *block comment*. Note that, the lines that contain `%{` and `%}` cannot contain any other text. After typing the opening block comment symbol, `%{`, all subsequent lines assume the syntax highlighting color for comments until you type the closing block comment symbol, `%}`. At that point, only the lines between the block comment symbols have the syntax highlighting color for comments. You can easily extend a block comment without losing the original block comment, that is, create a nested block comment, as shown in the following example:

```
%{
g=9.81;
  %{
  g=-3.986004418e14/rp^3*(eye(3)+...
  1.5*1.0826267e-3*(re/rp)^2*(egm-5*sin(mu)*eye (3)))*p;
  %}
display(g)
%}
```

You can also comment out text within a multiline statement using the ellipsis (...). MATLAB ignores any text appearing after the ... on a line and continues processing on the next line. This effectively makes a comment out of anything on the current line that follows the .... For example, in

```
d=-omegab+R_becef'*...
    ... omegae*
  R_becef
```

the term appearing on the second line will be ignored.

Comments also determine what text will be displayed when you run the `help` function for your specific need. This option will be addressed in more detail in Secs. 5.7.2 and 5.8.1. Overall, using the extensive comments in an M-file to describe the goal of code, its flow, input and output arguments, or make special remarks is a good habit and is highly encouraged, especially if you intend to share this code with others or reuse it later yourself. In the latter case, it will help you to recall the details of your own code when you return to it after a while.

To display the contents of the specified file in the MATLAB Command window, you can use the command `type filename` or function `type` (`'filename'`). If you do not specify a `filename` extension and there is no `filename` file without an extension, the `type` function adds the *.m* extension by default. The `type` function checks the directories specified in the MATLAB search path, which makes it convenient for listing the contents of M-files on the screen.

It may be useful to use the `type` command with `more on` option to see the listing on one screen at a time. The command `more on` enables paging of the output in the MATLAB Command window. The command `more off` disables paging of the output, function `more(n)` displays n lines per page. By default, `more` is disabled. When enabled, `more` defaults to displaying 23 lines per page. You can use the keyboard to perform the following operations:

\<Return\>	Advances to the next line of output
\<Space\>	Advances to the next page of output
\<Q\>	Terminates display of the text

5.8 Functions

With two exceptions, a *function* is another type of M-file that contains several lines of code. Basically, it only differs from the usual script by the very first line containing the list of parameters the function works with and which are needed to be defined a priori. But before we consider this type of functions, let us introduce those two exceptions that do not need a full-scale M-file but can rather be represented by a single line of code.

5.8.1 Inline and Anonymous Functions

The *inline function* is one of the two types of MATLAB functions that do not require creating a separate file. The command

```
inline(expr)
```

constructs an inline function object from the MATLAB expression contained in the string `expr`. The input argument to the `inline` function is automatically determined by searching `expr` for an isolated lower case alphabetic character, other than i or j, that is not part of a word formed from several alphabetic characters. If no such character exists, x is used.

If the character is not unique, the one closest to x is used. If two characters are found, the one later in the alphabet is chosen. For example,

```
>> myfun = inline('3*sin(2*t^2)');
```

creates an inline function to represent the formula $3\sin(2\,t^2)$. It now may be used to perform computations such as

```
>> myfun(5)
ans = -0.7871
```

Three commands related to the `inline` function allow you to examine an inline function object and determine how it was created

`char(fun)`	Converts the inline function into a character array (this is identical to `formula(fun)`)
`argnames(fun)`	Returns the names of the input arguments of the inline object `fun` as a cell array of strings
`formula(fun)`	Returns the formula for the inline object `fun`

The fourth command,

`vectorize(fun)`	Inserts a . before any ^, *, or / in the formula for `fun` (the result is a vectorized version of the inline function)

If applied to the previous function, `myfun`, these four commands will return

```
3*sin(2*t^2)
't'
3*sin(2*t^2)
3.*sin(2.*t.^2)
```

respectively. Note that to be on the safer side, the `vectorize` function inserts a . everywhere, although in our particular example the vectorized version of `myfun` needs only the array power, `.^`. Vectorization of the function allows you to use it with the matrix inputs, for instance,

```
>> myfun([2:5])
```

yields

```
ans = 2.9681 -2.2530 1.6543 -0.7871
```

Consider another example. The following call to `inline` defines the function `f` to be dependent on three variables, `C`, `alpha`, and `x`, implicitly

```
>> f = inline('alpha*cos(C*x)')
f = Inline function:
    f(C,alpha,x) = alpha*cos(C*x)
```

The command

```
inline(expr,arg1,arg2,...)
```

allows you to construct an inline function whose input arguments are specified by the strings `arg1`, `arg2`, ..., explicitly. For the previous example, we could do it as follows:

```
>> f = inline('alpha*cos(C*x)','x','alpha','C')
f = Inline function:
    f(x,alpha,C) = alpha*cos(C*x)
```

(note that in this case you may define the order of arguments explicitly).

Be aware that one way or the another all variables entering the `expr` should be declared as the arguments, meaning that you cannot declare something like

```
>> f = inline('alpha*cos(C*x)','x')
```

You will not be able to use this function because you could not be able to specify `C` and `alpha` afterwards.

An *anonymous function* is another simple form of the MATLAB function that does not require an M-file. The syntax for creating an anonymous function from a mathematical expression is

```
fhandle = @(arglist) expression
```

The term `expression` represents the body of the function: the code that performs the main task that your function needs to accomplish. This consists of any single, valid MATLAB expression. The `arglist` is the comma-separated list of input arguments to be passed to the function (these two components are similar to the body and argument list components of any M-file function as addressed in the following section). Leading off the entire right side of this statement is an @ sign. The @ sign is the MATLAB operator that constructs a function *handle*, a MATLAB value that provides a means of calling a function indirectly. The syntax statement shown earlier constructs the anonymous function, returns a handle to this function, and stores the value of the handle in variable `fhandle`, which can be used to invoke the function. For instance, the following statement creates an anonymous function that finds the cubic root of a number

```
>> cubroot = @(x) power(x,1/3);
```

To execute the `cubroot` function defined previously, you simply type

```
>> cubroot(27)
ans = 3
```

For functions that do not take any input arguments, you can construct the anonymous function with an empty argument list as follows:

```
>> t = @() datestr(now);
```

When calling such a function, you should also use empty parentheses

```
>> t()
ans = 28-Oct-2006 12:39:38
```

Another example shows how to create and use a handle to the function of two variables

```
>> A = 3; B = 2;
>> sumAxBy = @(x,y) (A*x + B*y);
>> sumAxBy(5,6)
ans = 27
```

(Note that you have to define all parameters in the expression beyond its arguments beforehand.)

One more example demonstrates that the argument list may contain matrices as well (do not forget to use a vectorized version of the mathematical expression in this case)

```
>> myfun = @(x) sin(x(1)-2).*cos(x(2))
>> [x,y]=fminsearch(myfun,[2;-2])
```

Here `fminsearch` is the MATALB function that accepts mathematical expression and the initial guess on the varied parameters (x_1 and x_2 in this case). Although the `fminsearch` function is to be formally introduced in the following section, the important point to note here is that the function handle can be passed to another function as an argument. These two lines of code return the values of `myfun` arguments (in x), at which `myfun` reaches its minimum value y

```
x =  3.5708
    -3.1416
y = -1.0000
```

Speaking of handles, it should be noted that you can create a function handle to any MATLAB function, for example,

```
>> f_sin = @sin;
```

creates a handle to the built-in MATLAB function `sin`. You can now use this handle as if you were using the function itself

```
>> f_sin(1:3)
ans = 0.8415    0.9093    0.1411
```

The good thing about function handles is that you can store them in the data structures for later use. For instance, the following fragment creates a handle for an anonymous function, stores it along with other handles in a cell array, and then uses it as an element of this cell array to produce a plot:

```
>> sincos = @ (x) sin(x).*cos(x);
>> trigFun = {@sin, @cos, sincos, @tan};
>> plot(trigFun{3}(-pi:0.01:pi))
```

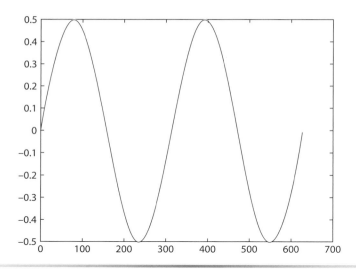

Fig. 5.8 Example of using a function handle as an input argument to another function.

The `plot` function (to be introduced in Chapter 6) creates a plot shown in Fig. 5.8.

There are several MATLAB functions that are specifically dedicated to work with the function handles

`functions`	Returns information describing a function handle
`func2str`	Constructs a function name string from a function handle
`str2func`	Constructs a function handle from a function name string
`save`	Saves a function handle from the current workspace to a MAT-file
`load`	Loads a function handle from a MAT-file into the current workspace
`isa`	Determines if a variable contains a function handle
`isequal`	Determines if two function handles are handles to the same function

You can to use the Help system to learn more about them and use them in your programs.

Be aware that MATLAB maps the handle to the function you specify and saves this mapping information in the handle. A function handle retains that same mapping even if its corresponding function goes out of scope. For example, if after creating the handle, you change the MATLAB path so that a different function of the same name now takes precedence, invoking the function handle still executes code to which the handle was originally mapped.

5.8.2 M-File Functions

As we saw in the previous section, only very simple expressions can be coded as inline and anonymous functions. More complex algorithms are programmed as an M-file functions. The first line of the M-file function defines the syntax, that is, the way this function can be called. Let us proceed with introducing four frequently used MATLAB functions with the goal of understanding how to call them and pass necessary parameters. Then we will formally introduce the function header, constituting its first line, and, in the following sections, address several issues of developing, storing, and using the user-defined functions.

The MATLAB `fplot` function plots the specified function within the certain limits. Its syntax is `fplot('function',limits)`, where `function` is either inline expression or the name of the (user-defined) function, `limits` is a vector of length 2 having the upper and lower limits of the plot. For example,

```
>> fplot('sin(x)*cos(x)', [0,pi])
```

plots the function `sin(x)*cos(x)` on the interval 0 to π.

Another example is the MATLAB `fzero` function, which finds the zero of a function of a single variable. Its syntax is

```
y = fzero('function', x0)
```

If `x0` is a scalar, it is the user-supplied initial guess of the zero (typically finds the closest zero to the initial guess). If `x0` is a vector of length 2, it is the interval (where the sign of the value of the function at `x0(1)` differs from the value of the function at `x0(2)`). For example,

```
>> z = fzero('cos',0)
```

finds the zero of *cosine* closest to `x=0`, whereas

```
>> z = fzero('cos(x)^2-0.5',[2 3])
```

returns the zero of $\cos^2(x) - 0.5$ found within the range $x = [2;3]$. (Note that, it is always a good idea to plot the function first using `fplot` command.)

The MATLAB `fminbnd` function finds the minimum (local or global) of a function of a single variable. Its syntax is

```
[x,y] = fminbnd('function',x1,x2)
```

It returns the value `x` that minimizes the function and the value `y` that is the minimum value of the function. Values `x1` and `x2` are the upper and lower limits in which to search for the minimum. Since there is no `fmaxbnd` function, to be able to find the maximum of the function, you have to redefine its output to be the negative of the desired function and use the `fminbnd` function.

Finally, the MATLAB `fminsearch` function used in the previous section finds the minimum of a function of more than one variable. Its general syntax is

```
[x,fval]=fminsearch('multi_var_function',x0)
```

Here `x0` is the user-supplied input vector containing the initial guesses of all the variables. This function returns `x`, a vector with the values of the variables that minimize the function, and `fval`, a scalar that is the minimum value of the function.

The theory (algorithms) behind `fzero`, `fminbnd`, and `fminsearch` functions will be considered in detail in Chapter 9, but here we can make several important points.

First, each function has its own workspace, where variables used by this function are stored. You cannot access this workspace from outside and vice versa (see more on this topic in Sec. 5.7.5). Hence, the list of input and output arguments is the only way of exchanging data (there are a few exceptions presented in the following sections though). Second, the professionally written functions allow some flexibility in defining input arguments. For example, the some of previous functions can accept the inline mathematical expression, name of the function, or its handle. Argument may assume a single scalar value or a vector, so that somewhere inside the function there is a logic distinguishing between different cases based on the type of the input argument and its dimension. Third, the M-file function may have different syntax accepting different number of arguments in the input. In fact, all of the aforementioned functions accept at least one extra argument defining some internal parameters used by these functions. The fact that these functions work fine even without this extra input argument means that these internal parameters assume some default values inside the functions and use them in the case when expected input argument is not present, rather than returning an error saying that there is not enough input arguments. The same applies to the number of output arguments, the very same function may spit out different number of arguments, depending on how we call it. All these features distinguish the professionally written function from the beginner-written functions. They are flexible and at the same time foolproof. We will use the open-source paradigm of MATLAB to look inside some of these functions in Sec. 5.8.1, but now let us proceed with writing our own, the initial function.

Although there is some flexibility, it is a good habit to have the name of a user-defined function, as defined in the first line of the M-file, the same as the name of the M-file without the *.m* extension. So assume we have the following function saved as the *stat.m* file in the current directory (or directory on the search path as addressed in Fig. 2.12)

```
function [mean,stdev] = stat(x)
% This function computes the mean value and
% standard deviation of a vector x
n = length(x);
mean = sum(x)/n;
stdev = sqrt(sum((x-mean).^2/n));
```

(Note that, there is no need to end the body of the function with any ending statement like `end` or `return`). This function accepts a single argument a vector and calculates its mean value and standard deviation. Now, you may address it in the way you would address any MATLAB function. For instance,

```
>> x = 0:100;
>> [mx,sx] = stat(x)
mx = 50
sx = 29.1548
```

You can also call it asking for just a single output argument

```
>> a = stat(0:10)
```

which returns only the mean value

```
a = 5
```

   Using this simple example of a complete M-file function, let us formally introduce the function definition line, header

```
function [output variables]=function_name(input variables)
```

This header defines everything, namely,

`function`	Indicates that this is an M-file function as opposed to an M-file script (it should use lowercase letters)
[output_variables]	Contains the optional list of output arguments, which may contain
Single argument	In this case, you may omit brackets
Multiple arguments	The list of the variables must be enclosed within square brackets
No output	Neither brackets nor an assignment operator is required

`function_name`	Uses up to 31 characters starting with a letter (while MATLAB is case sensitive, your operating system may not be, so you might be able to use capitalization)
(*input variables*)	Refers to the optional list of variables that are needed by the function to work properly

With respect to the help comment lines, they may be placed anywhere within the function, but any comments placed immediately after the function definition line will be displayed when the `help` command is used. For example, in the case of the function `stat` defined earlier

```
>> help stat
```

returns

```
This function computes the mean value and
standard deviation of a vector x
```

Sometimes your algorithm contains a portion that can be organized as a function, but to keep it short you prefer breaking it into several subfunctions, some of which may be very short and only used by another function, not by the main code. In this case, saving these pieces as separate M-files would create a mess in your current directory. The question is whether it is possible to save several functions in the same M-file. The following section addresses it.

5.8.3 Primary Function and Subfunctions

Each M-files can contain a code for more than one function. The function that appears first in the file is a *primary* function. Any function that follows the primary function in the same M-file is *subfunctions*. For instance, if the M-file *newst.m* contains the following:

```
function [av,sc] = newst(u)          % Primary function
% NEWST finds mean value of the input vector and also
%            scales it from 0 to 1
av = mean(u);
sc = scale(u);

function a = mean(v)                  % Subfunction
% Calculate average
a = sum(v)/length(v);

function m = scale(v)                 % Subfunction
% Scale from 0 to 1
m = (v-min(v))/(max(v)-min(v));
```

then the function `newst` presents a primary function, whereas functions `mean` and `scale` are subfunctions (Fig. 5.9). Note that, there is no need to

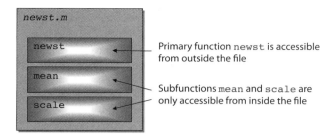

Primary function `newst` is accessible from outside the file

Subfunctions `mean` and `scale` are only accessible from inside the file

Fig. 5.9 Example of primary function and two subfunctions.

separate the functions by any statement, because the command `function` itself serves as an indicator that the new function has started and, therefore, the previous one has ended.

Primary functions have a wider scope than subfunctions. That is, primary functions can be invoked from outside of their M-file (from the MATLAB command line or from functions in other M-files), whereas subfunctions cannot be invoked. Subfunctions are visible only to the primary function and other subfunctions within their own M-file. When you call a function from within an M-file, MATLAB first checks the file to see if the function is a subfunction. Because of that, you can override existing M-files using subfunctions with the same name as other MATLAB functions (as in example the previous for the function `mean`).

You can write the help comments for subfunctions using the same rules that apply to primary functions. To display the help for a subfunction, precede the subfunction name with the name of the M-file that contains the subfunction (minus file extension) and a forward slash. For example, to get help on the function `scale`, we would need to type

```
help newst/scale
```

An interesting extension to the primary function/subfunction concept is the capability to keep the calling script and functions altogether in one file. For example, we can save the call to the function `stat` (from the previous section) and the function itself in the *allinone.m* file

```
function allinone
a = stat([1,2,3,4])
     function [mean,stdev] = stat(x)
     % This function computes the mean value and
     % standard deviation of a vector x
     n = length(x);
     mean = sum(x)/n;
     stdev = sqrt(sum((x-mean).^2/n));
```

and run it by issuing the command

```
>> allinone
```

Note that, the reason we were able to do this is that we organized the calling command as a function (with no inputs).

Next, we show that even more integration between the functions within one M-file can be achieved and exploited.

5.8.4 Nested Functions

The latest versions of MATLAB allow you defining one or more functions within another function. These inner functions are said to be *nested* within the function that contains them. You can also nest functions within other nested functions. To write a nested function, simply define one function within the body of another function in an M-file, for example,

```
function x = A(p1, p2)
...
  function y = B(p3)
  ...
  end
...
end
```

Note that, while primary functions and subfunction do not normally require a terminating end statement, this rule does not hold when you nest functions (Fig. 5.10). If an M-file contains one or more nested functions, you *must* terminate all functions (including subfunctions) in the M-file with end, whether or not they contain nested functions.

```
function A(x, y)              % Primary function
B=(x,y);
D(y);

    function B(x, y)         % Nested in A
    C(x);
    D(y);

        function C(x)        % Nested in B
        D(x);
        end
    end

    function D(x)            % Nested in A
    E(x);

        function E           % Nested in E
        ...
        end
    end
end
```

Fig. 5.10 Example of nested functions.

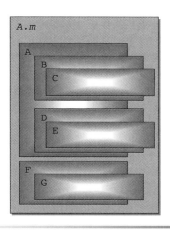

Function A can call subfunction F, nested functions B and D. It cannot, however, call next-level nested functions C and E, as well as nested function G

Function B can call the host function A, same-level nested function D, nested function C, and subfunction F

Function C can call the host function B, the next-level host function A, and subfunction F

Variables defined in the host function A are visible inside the nested functions B, C, D, E, and F

Fig. 5.11 Graphical illustration of nested functions accessibility.

Like other functions, a nested function has its own workspace where variables used by the function are stored. However, as opposed to the case when the functions are stored in one M-file as a primary function and subfunctions, the nested functions have access to the workspaces of all functions in which it is nested. For example, a variable that has a value assigned to it by the primary function can be read or overwritten by a function nested at any level within the primary. Similarly, a variable that is assigned in a nested function can be read or overwritten by any of the functions containing that function.

Consider a fragment of the M-file in Fig. 5.10 and its graphical interpretation in Fig. 5.11 (which also contains a subfunction F).

In this example, you can call a nested function

1. From the level immediately above it (function A can call B or D, but not C or E)
2. From a function nested at the same level within the same parent function (function B can call D, and D can call B)
3. From a function at any lower level (function C can call B or D, but not E)

You can also call a subfunction from any nested function in the same M-file (Fig. 5.11)

The following section continues the issue of "connecting" workspaces using global variables.

5.8.5 Passing Arguments as Global Variables

As you know already, MATLAB stores variables in a part of the memory called a workspace. The *base workspace* holds variables created during your interactive MATLAB session including those created as a result of running M-file scripts. Hence, as they belong to the same base workspace, variables created at the MATLAB command prompt can also be used by scripts. We refer to these variables as being *global*.

Functions, however, do not use the base workspace. Every function has its own *function workspace*. Each function workspace is kept separate from the base workspace and all other workspaces to protect the integrity of the data used by that function. Even subfunctions that are defined in the same M-file have a separate function workspace, that is, all variables within the body of the function are *local variables*, meaning that their values only available within the function.

As mentioned earlier, by design the nested functions are the only exception from this rule. Variables within nested functions are accessible to more than just their immediate function. As a general rule, the scope of a local variable is the largest containing function body in which the variable appears, and all functions nested within that function, that is, a variable that has a value assigned to it by the primary function can be read or overwritten by a function nested at any level within the primary (not vise versa). Similarly, a variable that is assigned in a nested function can be read or overwritten by any of the functions containing that function.

In addition to the aforementioned rules, the following also apply:

1. If a variable by the same name as one of your workspace variables is used by a function, any change made to that variable by the function is not passed back to the current workspace, unless it is an output variable.
2. All local variables are erased after the function finishes executing.
3. When calling the function you do not have to use the same name for input variables as those the function definition line calls the *input variables*. This makes functions "portable" to other applications, so that they do not need to be rewritten for the fear of using the same variable names in other calculations.

Therefore, the most secure way to extend the scope of a function variable is to pass it to other functions as an argument in the function call. Since MATLAB passes data only by value, you also need to add the variable to the return values of any function that modifies its value.

To minimize the number of variables needed to be passed through input variables, you may want to declare certain variables to be global (meaning that they are visible from inside/outside the function). This can be done by using the `global` function in the main M-script and within every function that needs access to its variables. The syntax for declaring variables global is

```
global var1 var2 var3
```

(once again, to make these variables "transparent" you have to place this statement in both the calling program and the function itself). If global variables are required, it is recommended using long distinctive names rather than the short ones frequently used, such as x, y, and z. To simplify handling global variables in MATLAB Editor, starting with R2011a, they are highlighted by a different color thus warning you that changing their values may span multiple workspaces.

While global variables seem to represent a good way to extend the variable scope, there is a pitfall. If you do declare some variables global, you need to make sure that no function with access to the variable overwrites its value unintentionally. Also, using global variables may restrict the universality of your functions, because you need to explain which variables have to be global and which should be passed through the list of input arguments. Although the beginners might enjoy using the global variables (which also saves the computational time), professionally developed programs utilize structures that may contain many heterogeneous data. This keeps the list of input variables as short as possible.

5.8.6 Search Order and Private Functions

As mentioned throughout the text several times already, intentionally or unintentionally you may end up having variables in your base workspace that have the same name as your M-file function, or you created and the M-file function has the name of one of the MATLAB functions. The question is how MATLAB manages to distinguish between them, if at all. The following explains it.

Suppose in the Command window you typed

```
>> test
```

In this case, the search order is as follows. First, MATLAB checks to see if `test` is a variable in your base workspace, and if so, its value is displayed in the Command window. If there is no such variable, MATLAB then looks in the current directory for a file named *test.m* and executes it if it is found. If not, MATLAB checks to see if `test` is one of its own built-in commands or functions and executes it if it is. If not, MATLAB then searches the directories in its search path in the defined order and in case the file *test.m* is found, it executes it. Finally, if nothing worked out, the message

```
??? Undefined function or variable 'test'.
```

appears in the Command window.

The `exist` function may help you sort everything out even without browsing the content of the current workspace and directory. The call

```
exist name
```

checks existence of variable, function, or directory var and returns one of the values shown in Table 5.4.

It can also be used as

```
exist name kind
```

Table 5.4   Order of Search and Values Returned by the `exist` Function

Order of Search	Return Value	Type of Entity
1	1	name is a variable in the workspace
2	5	name is a built-in MATLAB function
3	7	name is a directory
4	3	name is a MEX- or DLL-file on your MATLAB search path (Sec. 4.4)
5	4	name is a MDL-file on your MATLAB search path
6	6	name is a P-file on your MATLAB search path
7	2	name is a M-file on your MATLAB search path. It also returns 2 when name is the full pathname to a file or the name of an ordinary file on your MATLAB search path
8	8	name is a Java class (`exist` returns 0 if you start MATLAB with the `−nojvm` option)
	0	name does not exist

that is, specifying the `kind` of the entity: `builtin` (for built-in functions), `class` (for Java classes), `dir` (for directories), `file` (for files or directories), and `var` (for variables). You can also use the function form `exist('name','kind')`. An argument name may include an extension (that is, `exist('file.ext')`) to preclude conflicting with other similar filenames or a partial pathname (`exist('matlab/general/dir.m')`).

For example,

```
>> exist('pi')
```

returns

```
ans = 5
```

because `pi` is a built-in function, but

```
>> exist('pi.m')
```

returns

```
ans = 2
```

which is farther down in the order of the search, because the M-file *pi.m* also exists (Sec. 4.4). The call

```
>> pi=5; exist pi
```

returns

```
ans = 1
```

because you override whatever exists on the path with the variable in the current workspace.

One more place to look at can be defined using the *private functions*. Private functions are functions that reside in subdirectories with the special name *private*. These functions are called private because they are visible only to M-file functions and M-file scripts that meet the following two rules:

1. A function that calls a private function must be defined in an M-file that resides in the directory immediately above that private subdirectory.
2. A script that calls a private function must itself be called from an M-file function that has access to the private function according to rule number one.

For example, assume that you added your directory *mydir* to the search path, so that whatever is inside this directory is visible to the MATLAB calls and functions. Nevertheless, if inside this directory you create a directory with the name *private*, the functions which are in this directory can be only seen (called) from the *mydir* directory.

Because private functions are invisible outside the parent directory, they can use the same names as functions in other directories. This is useful if you want to create your own version of a particular function while retaining the original in another directory.

In the search sequence introduced in the beginning of this section, MATLAB looks for private functions before it turns to standard M-file functions.

5.9 Developing Professional Looking Functions

This section will show you that, with MATLAB, progressing from beginning programmer to the intermediate-level or professional programmer may not take too much of an effort compared with other languages. Primarily, this is due to the open-source paradigm. We will use it in the following section to peep at a couple of MATLAB M-file functions, followed by using what we learned in our own functions.

5.9.1 Open-source Paradigm

As mentioned previously, the distinctive feature of MATLAB is that it supports an open-source paradigm. Not only does it allow users to share their source code, for example, via MATLAB Central (Fig. 1.4), but it also enables learning about the numerical algorithms realized in MATLAB functions and learn about a trick or two from actual developers, by just looking at their code. For example, if we go to the MATLAB directory and search for the `fzero` function, we played with in Sec. 5.7.2, the return may look like the one shown in Fig. 5.12. Usually, you see multiple returns, residing in different subfolders, so navigate to the one that has the largest size—it will be the source code.

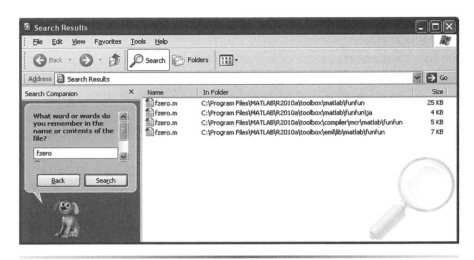

Fig. 5.12   Search results for the `fzero` file.

An alternative way to locate a source code for any MATLAB function is to use the `which` function. For example, the command

```
>> which fzero
```

returns

```
C:\...\MATLAB\R2010a\toolbox\matlab\funfun\fzero.m
```

By the way, many of the MATLAB functions are *overloaded*, meaning that they have multiple implementation to handle different data types of input arguments appropriately. When you use the `which` command with the `-all` option, MATLAB returns all occurrences of the file you are looking for. For example, the command

```
>> which -all sin
```

returns

```
built-in (C:\...\R2010a\toolbox\matlab\elfun\@double\sin)    % double method
built-in (C:\...\R2010a\toolbox\matlab\elfun\@single\sin)    % single method
C:\...\R2010a\toolbox\symbolic\@sym\sin.m                    % sym method
```

Let us open the `fzero` M-file function and have a look at the first portion of this specific code presented

```
function [b,fval,exitflag,output] = fzero(FunFcnIn,x,options,varargin)
%FZERO Single-variable nonlinear zero finding.
%   X = FZERO(FUN,X0) tries to find a zero of the function FUN near X0,
...
% Detect problem structure input
if nargin == 1
```

```
    if isa(FunFcnIn,'struct')
        [FunFcnIn,x,options] = separateOptimStruct(FunFcnIn);
    else % Single input and non-structure.
        error('MATLAB:fzero:InputArg',...
'The input should be either a structure or at least two arguments.');
    end
end
...
% Check for non-double inputs
if ~isa(x,'double')
    error('MATLAB:fzero:NonDoubleInput', ...
        'FZERO only accepts inputs of data type double.')
end
```

The first thing to note is that developers did a good job providing extensive comments. Again, the lines immediately following the function header represent the `help fzero` response. Also, the script has many foolproof checks preventing it from malfunctioning if wrong arguments were passed on into it. In this case, the function spits out the error messages telling exactly what went wrong, which may be a great help when debugging your code.

Second, after looking at the first few lines of code itself it becomes clear that although the header shows four input arguments, it will work fine with just one argument as well (the `nargin==1` checks whether the number of arguments is one). In this case, this input is expected to be a structure (the `isa` function checks whether the object is of a given class). If this is the case, the `separateOptimStruct` function breaks this structure into separate elements, one of which, `x`, represents the vector if initial guesses, which may alternatively be passed directly as a second input argument to the function. This is followed by checking whether each input belongs to the class it is supposed to belong (in our case, all elements of the vector `x` should be double-precision values).

Let us find another function, `fplot`, we used in Sec. 5.7.2 and look at its header and the following few lines of code

```
function [x0,y0] = fplot(varargin)
% FPLOT Plot function
% FPLOT(FUN,LIMS) plots the function FUN between the x-axis limits
...
% Examples:
%       fplot(@humps,[0 1])
%       fplot(@(x)[tan(x),sin(x),cos(x)], 2*pi*[-1 1 -1 1])
%       fplot(@(x) sin(1./x), [0.01 0.1], 1e-3)
%       f = @(x,n)abs(exp(-1j*x*(0:n-1))*ones(n,1));
%       fplot(@(x)f(x,10),[0 2*pi])
...
```

```
% Parse possible Axes input
[cax,args,nargs] = axescheck(varargin{:});
...
    fun = args{1};
    lims = args{2};
    args = args(3:end);
if (isvarname(fun))
    fun = ezfcnchk(fun);
else
    fun = fcnchk(fun);
end
...
% compute the first two points
y = feval(fun,x,args{4:end});
```

As shown in the Examples section of the Help comments, this function assumes multiple input arguments too, but by looking at the header we realize that these multiple inputs can be passed into the function as one object, the list of arguments. The `axescheck` function process this object and returns the processed argument list in `args` with `nargs` denoting the number of the elements in the input list (auxiliary output `cax` returns nothing, indicating that the first argument in the list is not Axes). For example,

```
>> [cax,args,nargs] = axescheck('sin',7,[4:6]);
>> [cax, args nargs]
```

returns

```
ans = 'sin' [7] [1x3 double] [3]
```

The important function happens after assigning the initial argument to the variable `fun`. The `fplot` function is supposed to plot whatever is stored in this variable and to be universal it has to distinguish what we actually passed on to it. This issue will be discussed in the following section, but let us mention here that it is the `fcnchk` function that takes care of converting a string containing parentheses, variables, and math into the inline function. If `fun` happens to be a valid MATLAB variable (determination made by the `isvarname` function), the `ezfcnchk` function attempts to distinguish between an input expression that is the identity function versus the name of an M-file or built-in function. Later on, the result of the either check (`fcnchk` or `ezfcnchk`) is evaluated using the `feval` function (see Sec. 5.9.2).

Sometimes you would want to use the open-source paradigm just to look at the parameters of a specific algorithm coded in MATLAB function.

For example, the following portion of the `ode23tb` M-file function specifies some parameters of the algorithm described in Sec. 14.5.2.

```
% Initialize method parameters
pow = 1/3;
alpha = 2 - sqrt(2);
d = alpha/2;
gg = sqrt(2)/4;
% Coefficients of the error estimate
c1 = (alpha - 1)/3;
c2 = 1/3;
c3 = -alpha/3;
% Coefficients for the predictors
p31 = 1.5 + sqrt(2);
p32 = 2.5 + 2*sqrt(2);
p33 = - (6 + 4.5*sqrt(2));
```

5.9.2 Passing a Function as an Argument

On several occasions (in Secs. 5.7.1 and 5.7.2), we showed that an inline expression, the name of the function, or the function handle can be passed in an argument list to another function, so that this latter function understands what you were trying to pass into it. In the previous section, we learned that it is possible because of special precautions taken in the beginning of MATLAB M-file functions. Let us explore this topic in more detail.

Let us formalize what input the MATLAB functions can possibly accept as a function input in general. In the function call

```
function(funIn,p1,p2,...)
```

where `function` is the name of an appropriate MATLAB function (`fzero`, `fminsearch`, etc.) and `p1, p2, ...` are corresponding parameters (if at all), the argument `funIn` must be one of the following:

1. The name of an M-file function, for example,
   ```
   >> z=fzero('cos',0)
   ```

2. A string with variable x, for example,
   ```
   >> fplot('sin(x)*x^2', [0,pi])
   ```

3. An inline function define beforehand, for example,
   ```
   >> myfun = inline('(x(1)-3)^2+(x(2)+2)^2');
   >> x = fminsearch(myfun,[1,1])
   ```

4. A function handle for an M-file function, for example,
   ```
   >> ezplot(@tan)
   ```

5. A function handle for an anonymous function, for example,

```
>> myfun = @(x,y) power(x,1/3).*sin(x).*cos(y)
>> [x,y] = meshgrid([1:0.1:5],[-3:0.1:3]);
>> surf(x,y,myfun(x,y))
```

Now let us see why any of these inputs is reliably recognized by the `function`. As discussed in the previous section, in the beginning of these functions (`fplot`, `fzero`, `fminsearch`, etc.) you may find the command

```
fun = fcnchk(fun);
```

which is responsible for this. What `fcnchk` does is it places `fun` into the inline function form. In fact, it is only needed for the case 2 in the aforementioned list of options, that is, when `fun` happens to be a mathematical expression. In all other cases, that is, when `fun` is a function handle (containing all information about the function itself), a name string (for example, `'sin'` or name of another M-script), or a MATLAB inline object, `fcnchk` simply returns `fun` without doing anything to it. You may also use another syntax

```
fun = fcnchk(fun,'vectorized')
```

that processes the text string representing a mathematical formula to produce a vectorized inline function `fun` (for example, replacing * with .*).

After `fun` is preprocessed with `fcnchk`, it is evaluated using the second important function, `feval`. Formally, the command

```
feval(fun,x1,...,xn)
```

evaluates the function `fun` at the given set of arguments, x1,...,xn.

The following example demonstrates all approaches you may use to pass the function/expression along with other parameters into another function. Suppose the actual function you are using to compute some expression is

```
function y = PassFun2(x,z)
global a b
y = a * sin(x) + b * sqrt(x) * cos(z);
```

(note that, some of the parameters are passed as a global variables, others as input variables). Then, you are using another function, `PassFun1` to pass some function and two arguments through

```
function x = PassFun1(fun,h,k)
fun = fcnchk(fun);
x = feval(fun, h, k);
```

(that is, where you use `fcnchk` and `feval` functions).

Finally, your main program may address your function `PassFun1` passing its first argument in three ways, using the handle for the M-file function `PassFun2`, name of the M-file function `PassFun2` itself, and mathematical expression (that again may invoke the M-file function `PassFun2`):

```
global a b
a = 1; b = -2;
g = PassFun1(@PassFun2,1,4)*PassFun1('PassFun2',5,7)+...
    PassFun1('3+sin(d)*PassFun2(d,r)',1,4)
```

(Any other function using two input variables, for example, function `sin-cos` of Sec. 5.7.1, could be used here instead of function `PassFun2`.) The above code produces a valid result

```
g = -4.4970
```

5.9.3 `nargin`, `nargout`, and `return` Functions

Another finding of Sec. 5.8.1 is that there is the MATLAB function `nargin` that determines the number of the elements in the list of input arguments. It allows your code to be flexible, so that the very same function could be used with a different number of input parameters without returning an error message. Based on the value of this parameter, you either compute a different portion of the code that does not rely on the parameters that were not provided, or use some default values for these missing parameters which must be defined in the beginning of your function.

Consider the function `testarg` that may accept one or two input arguments featuring different computations in each case

```
function [maxc] = testarg(a,pow)
    if (nargin == 1)
    c = a.^2;
    elseif (nargin == 2)
    c = a.^pow;
    end
plot(a,c)
    if nargout == 1
    maxc = max(c);
    end
```

As seen, the if-else-end structure takes care of two possible situations and uses only those input arguments that are available. This code also features another, reciprocal function `nargout`. Similar to `nargin`, the `nargout` function called inside the body of a user-defined function returns the number of output arguments that were used to call the function. In this specific case, the function `testarg` can be called with no arguments producing a plot $y=x^2$ or $y=x^{pow}$ or one output argument, which will be computed as $\max(x^2)$ or $\max(x^{pow})$, depending on the number of input arguments.

If being called outside the function, just at the command line, like `nargin(fun)` and `nargout(fun)`, these functions return the number of declared inputs and outputs for the M-file function `fun`, respectively.

The `return` command causes a normal return to the invoking function or to the `keyboard`. It also terminates the `keyboard` mode (discussed in the section). For instance, suppose we want to develop a function that computes a determinant of the input square function A. For some reason, if matrix A happens to be empty we want the output of our function to assume a value of 1 and quit executing the rest of the code in it. The following piece of code shows an example of how to program this using a `return` statement

```
function d = det(A)
%DET det(A) is the determinant of A.
if isempty(A)
        d = 1;
        return
else
    ...
end
```

Consider another typical example combining the output of `nargin` function, `isempty` function, and `return` command. The MATLAB function `median` has two syntaxes: `median(A)` and `median(A,dim)`. What you may find after the header of this function and the Help comments

```
function y = median(x,dim)
%MEDIAN Median value.
%...
```

is the following block of commands:

```
if nargin == 1
    dim = min(find(size(x)~=1));
    if isempty(dim), dim = 1; end
end
if isempty(x), y = []; return, end
```

If only one variable is supplied (no `dim` defined), the program determines the default value for `dim` by itself. If this function detects an empty input array, it immediately returns to the calling program with an empty array `y`.

5.10 Editing and Debugging

When you attempt to run a script file, MATLAB usually detects the more obvious syntax errors for you and displays a message describing the error and its location (Sec. 5.8.1 showed a couple of these error messages generated inside code). You may also encounter another type of errors, *runtime errors* or errors due to an incorrect mathematical procedure, such as divide by zero. Anyway, it happens quite rarely that you wrote a program and it runs as is, requiring corrections or adjustments. Most likely you will spend some time debugging your code. To this end, MATLAB offers several means that are discussed in the following sections.

5.10.1 Programmatic Tools

By default, the commands in the M-files do not display on the screen while code is running. Allowing the commands to be viewed as they execute, that is, command echoing, provides a very simple and yet effective tool for debugging or for demonstrations. The `echo` command allows you to control this process.

The `echo` command behaves in a slightly different manner for the script files and function files. For the script files, the use of `echo` is simple—echoing can be either `on` or `off`, in which case any script used is affected:

```
echo on    Turns on the echoing of commands in all script files
echo off   Turns off the echoing of commands in all script files
echo       Toggles the echo state
```

With the function files, the use of `echo` is more complicated. If `echo` is enabled (`echo on`) on a function file, the file is interpreted, rather than compiled. Each input line is displayed as it is executed which may substantially slow down your computations. You may `echo` some specific function only or all functions

```
echo fcnname on     Turns on echoing of the named function file
echo fcnname off    Turns off echoing of the named function file
echo fcnname        Toggles the echo state of the named function file
echo on all         Sets echoing on for all function files
echo off all        Sets echoing off for all function files
```

Another command useful for debugging purposes is the `pause` command. It waits for user's response as follows:

```
pause       Causes a procedure to stop and wait for the user to strike
            any key before continuing
```

`pause(n)` Pauses for n seconds before continuing, where n can also be a fraction (the resolution of the clock is platform specific but fractional pauses of 0.01 seconds should be supported on most platforms)

`pause off` Indicates that any subsequent `pause` or `pause(n)` commands should not actually pause (this allows normally interactive scripts to run unattended)

`pause on` Indicates that subsequent `pause` commands should pause

Finally, there is one more programmatic tool to use for debugging your M-files. It is the `keyboard` statement. When placed in an M-file, it stops execution of the file and gives control to the keyboard. The special status is indicated by a `K` appearing before the prompt

`K>>`

You can examine variables in the workspace created by this point and if necessary change them to see how your code executes the remaining computations (with these changes). To terminate the `keyboard` mode, type the `return` command and press <Return>.

5.10.2 Effective Usage of MATLAB Editor/Debugger

The overall description of the MATLAB Editor has already been provided in Sec. 2.1.3. Figure 5.13 shows an example of the M-file as it appears in the Editor window. Compared to Fig. 2.5, Fig. 5.13 also demonstrates a capability to split the screen into two portions, so that the different portions of the same file you are editing can be viewed in two different portions of the screen. (This is a very simple and yet very effective tool for debugging.) You split the window into two windows by double clicking on the dark gap right above the vertical scroll bar (Fig. 5.14a). You can click on and move the dividing horizontal bar between two windows up and down, and if you want to combine two windows back into a single window, you may double click on

Fig. 5.13 Split MATLAB Editor window.

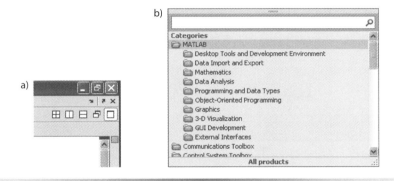

Fig. 5.14 a) Gap allowing you to split the MATLAB Editor window and b) help panel of the **Browse for functions** icon.

this bar. Note that, this option divides the Editor window into two windows to display the same file. If you have several M-files open, you can change the way they are displayed by using the group of buttons located at the far right end of the Editor Toolbar.

Most of the buttons on the Editor Toolbar were introduced in Sec. 2.1.3. Two more buttons are the **Show functions** button, which allows you to search for a function header (it is especially useful if you have multiple nested functions), and the **Browse for functions** button which brings a pop-up menu for a quick function reference as shown in Fig. 5.14b (as discussed in Sec. 2.5 this quick function reference is also available in the Command window). The remaining seven buttons are devoted specifically for the debugging purposes. Debugging with these buttons involves using the *breakpoints*.

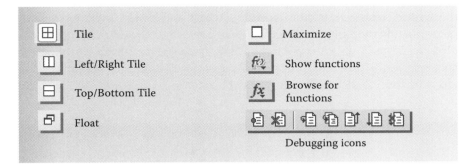

Before we proceed with the debugging mode, which uses these breakpoints, let us point your attention to the right vertical shoulder of the Editor window (Fig. 5.13). When you open an existing file, or right after you saved the new file you worked on, you may observe the colored box showing up on the top of this shoulder and a series of colored horizontal dashes (Fig. 5.15). If the box is red it means that you have fatal (syntax) errors that will prevent your program from running and returning the results as you would expect.

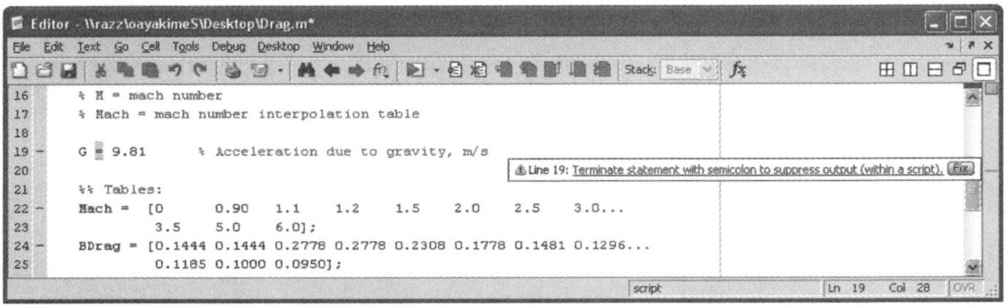

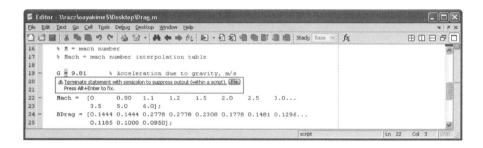

Fig. 5.15 Accessing warning messages using the warning messages bar.

The red horizontal dashes in this case correspond to the lines where these errors were found. You may navigate through them (either by clicking on the red square or on any specific horizontal dash) and correct your errors. The orange color for the warning box means that MATLAB has found some nonfatal warnings. For instance, Fig. 5.15 demonstrates a typical case when MATLAB suggests you to suppress the output (the warning line explaining what to do appears when you click on the corresponding horizontal dash). As seen the equality sign on the line in question has a dark background, also pointing your attention to this specific line. Placing a pointer over the equality sign brings the same warning menu (Fig. 5.16). When MATLAB sees no errors or has nothing to suggest the warning box turns green.

This convenient tool for quick and effective debugging is provided with the M-Lint Code Analyzer, introduced in Sec. 2.1.5. You can run this tool separately, as explained in Figs. 2.9b and 2.10, but now that it is built into the Editor/Debugger, it continuously checks your code for problems and recommends modifications to maximize performance and maintainability directly in the Code Analyzer rather than in a separate report. Moreover, it continually analyzes and updates messages as you work, so you can see the effect of your changes without having to save the M-file or rerun the Code

Fig. 5.16 Accessing warning messages placing a pointer over the equality sign.

Analyzer Report. If for some reason you would like to turn Code Analyzer off, you can select **File** => **Preferences** => **Code Analyzer** and clear the check box by **Enable integrated warning and error messages**. Some more details on using Code Analyzer will be provided in Sec. 5.10.3.

As you already know, to execute the entire script from the Editor window you may click on the Save and run button (in case the script is not saved, the pop-up window suggesting to save your 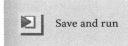 M-file will appear before file execution). You can also do it by pushing <F5>. You may also execute a single instruction or a block of commands. To do that you have to select it (them) and push <F9> (or choose the **Evaluate Selection** option of the **Text** menu of the toolbar). Alternatively, you may do the same by choosing **Evaluate Selection** option in the drop-down menu when you highlight and right-click on this instruction or block. You may apply this procedure even for entire script without saving your file first (you may use <Ctrl>+<A> to select the whole file).

After execution of the program from the MATLAB Editor, you can see the current values of the variables in the Workspace by placing and holding a pointer (cursor) over any variable. Its current value appears as a data tip (see example in Fig. 5.17).

Figure 5.17 also features a breakpoint, a red circle appearing to the right of the line number (22). Breakpoints are the points in the file where execution stops temporarily, so that you can examine the values of the variables up to that point and debug your program. The red icon (circle) shows up indicating that a valid breakpoint is set at this line (the file should be saved first). To set up a breakpoint, you should click on a horizontal dash to the right of the line number (Fig. 5.17 shows two other dashes, possible location of the breakpoints—one above and one below the breaking point). You can also use the **Set/ clear breakpoint** button to set up or clear a breakpoint at the current line (where the pointer is currently).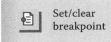

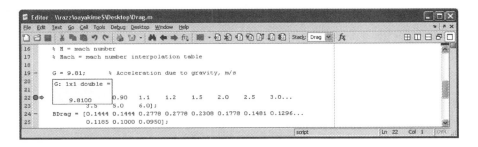

Fig. 5.17 Showing the values of variables and utilizing a breakpoint.

There are three basic types of breakpoints you can set in M-files:

1. A *standard breakpoint*, which stops at a specified line in an M-file (as shown in Fig. 5.17). The green arrow to the right of the breakpoint shows up after you run the file pointing to the line that was not executed still.
2. A *conditional breakpoint*, which stops at a specified line in your M-file only under specified conditions. To specify the condition, you may choose Set/Modify Condition in the context menu, which appears when you right-click on the breakpoint (Fig. 5.18a and 5.18b). The breakpoint then turns yellow.
3. An *error breakpoint*, which stops in any M-file when it produces the specified type of warning, error, or NaN or infinite value. To set error breakpoints, select **Debug** and then **Stop if Errors/Warnings**. In the resulting **Stop if Errors/Warnings for All Files** dialog box (Fig. 5.18c), specify error breakpoints on all appropriate tabs and click OK. To clear error breakpoints, select the **Never stop if...** option for all appropriate tabs and click OK.

When you set a breakpoint it may turn gray indicating that it is not valid (Fig. 5.19). This may occur in two cases—either the file was not saved since the changes were made (which is the case for Fig. 5.19, since the **Save** button is not faded as in Fig. 5.17), or there is a syntax error somewhere in the file. In the latter case when you

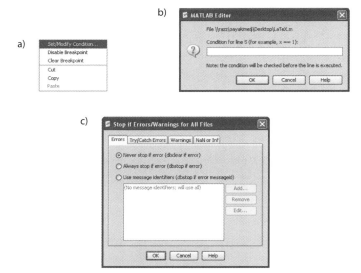

Fig. 5.18   a) Breakpoint context menu, b) menu for setting up a conditional breakpoint, and c) menu for setting up an error breakpoint.

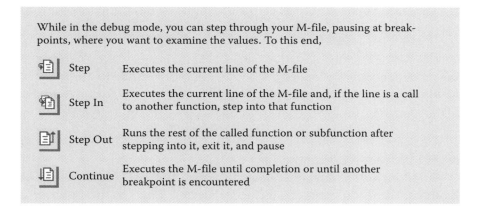

set a breakpoint, an error message appears indicating where the syntax error is.

Using the context menu (Fig. 5.18a), you can disable standard and conditional breakpoints so that MATLAB temporarily ignores them (in this case the breakpoint icon will appear with an cross over it), or you can clear (remove) them. Breakpoints are not maintained after you exit the MATLAB session.

Two remaining buttons are used to clear breakpoints and exit the debug mode. In the debug mode, the **Stack** field on the toolbar changes to reflect the current function (for example, in Fig. 5.17) and the corresponding workspace.

Let us make two final comments about the debug mode. First, you can use the aforementioned tools to work with both M-file scripts and functions. The only difference is that when debugging the functions all variables your function refers should be available in workspace. Second, when running the M-file from the MATLAB Editor in debug mode, as opposed to from the Command window by calling the name of the M-file, results in changing the prompt in the Command Window to K>>.

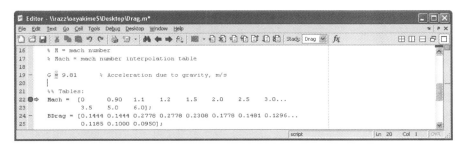

Fig. 5.19 MATLAB Editor window.

The latest versions of MATLAB also offer an additional option to conveniently debug M-files. When working with MATLAB, you often experiment with your code—modifying it, testing it, and updating it—until you have an M-file that does exactly what you want. That may include changing the numerical values of constants. Moreover, the overall structure of many M-file scripts seems to naturally consist of multiple sections. Especially, for the larger files, you often focus your efforts on a single section at a time, refining the code in just that section. To facilitate this process, MATLAB allows you to create and take advantage of the M-file *cells*, where a cell refers a defined section of code.

First, you have to enable a cell mode in the MATLAB Editor. Select the **Enable Cell Mode** option from the **Cell** menu. Items in the **Cell** menu become selectable and the Cell Toolbar appears as shown in Fig. 5.20. (MATLAB remembers the cell mode between sessions, so if cell mode was enabled when you quit MATLAB, it will be enabled the next time you start MATLAB.)

Next, you have to define the boundaries of the cells. Place the cursor just before the line you want to start the cell and then select **Cell => Insert Cell Divider** or click the cell divider button. MATLAB

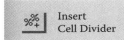

inserts a line after the cursor containing two percent signs (%%), which is the "start new cell" indicator. A cell consists of the line starting with %% and the lines that follow, up to the start of the next cell, which is identified by %% at the start of a line. You can also define a cell by entering two percent signs (%%) at the start of the line where you want to begin the new cell manually. Alternatively, you may select the lines of code to be in the cell and then select **Insert Cell Dividers Around Selection** option from the **Cell** menu. After the %%, type a space followed by a description of the cell. The Editor emphasizes the special meaning of the start of a cell by making any text following the percent signs and space boldface. The text on the %% line is called the cell title (like a section title). Including cell titles is optional, however, they improve the readability of the file and are very useful for publishing (to be described later in this section). Once you define the cells, you may use cell features to navigate quickly from cell to cell in your file, evaluating code in a cell, and viewing the results in the base workspace.

When the cursor is placed in any line within a cell, the Editor highlights the entire cell with a yellow background (Fig. 5.20). This identifies it as the current cell. The highlighted cell is the cell evaluated by the **Evaluate Current Cell** option on the **Cell** menu. You can also evaluate current cell by clicking the **Evaluate cell** button. A beep means there is an error, so you have to refer to the Command window to see the error message. Unlike file evaluation discussed earlier, the cell evaluation runs the current cell even if

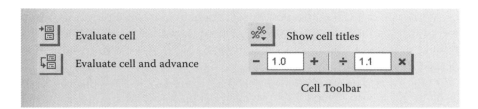

the file contains unsaved changes. Moreover, the file does not even have to be on the search path. The only requirement is that the values the current cell requires must already exist in the MATLAB workspace.

To run code in the current cell and move to the next cell, click on the **Cell** button of toolbar and then **Evaluate Current Cell and Advance** or click the **Evaluate and Advance** button. To run all code in the file, select **Cell => Evaluate Entire File**.

To move to the next cell without executing it, select the **Next Cell** option from the drop-down **Cell** menu. To move to the previous cell, select **Cell => Previous Cell**. To move to a specific cell, click the **Show cell titles** button and select the cell title you want to move to. Note that the cells without titles are not listed.

You can use the cell mode to operatively vary numbers in a cell. To modify a number in a cell, select the number (as shown for number 8 in Fig. 5.20) or place the cursor near it and use the value modification tool in the cell toolbar. Using this tool, you can specify a number in the number fields and press the appropriate math operator to add (increment), subtract (decrement), multiply, or divide the selected number. The cell then automatically reevaluates. You can use the numeric keypad operators instead of the operator buttons on the toolbar. Note that MATLAB does not automatically save changes you make to values using the cell toolbar, so if as a result of your trials you decided to change any number permanently, you have to save your M-file manually.

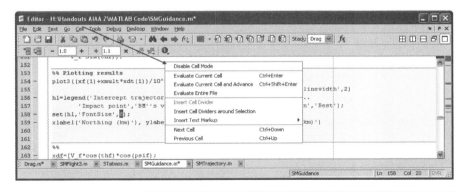

Fig. 5.20 Editor window with the Cell toolbar.

While you can still set breakpoints and debug a file containing cells, when you evaluate a file from the **Cell** menu, breakpoints are ignored. To run the file and stop at breakpoints, you should rather use the **Run/Continue** option of the **Debug** menu.

Cells are also useful if you want to share your results by publishing your work as an HTML, XML, LaTeX, Word, or PowerPoint document. When you publish the M-file, an output document consists of the M-file code itself, comments, and results. After organizing your code as a sequence of cells and adding text markup you may click on **File** menu, then select **Publish To** and choose an output format from those listed earlier. If the M-file contains unsaved changes, the menu item becomes **Save and Publish To**. You can also do it from the Command window using the `publish` function. Alternatively, you can publish to HTML by clicking the **Publish to HTML** button on the Editor toolbar. In the latter case, you have a nice option of representing equations symbolically. For example, the script appearing in the MATLAB Editor as shown in

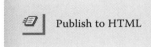 Publish to HTML

Fig. 5.21a will be represented in HTML Browser as presented in Fig. 5.21b.

To be able to take advantage of this option, you need to place the line with the equation that uses LaTeX language syntax somewhere in the

a)
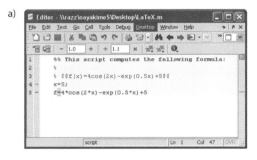

b)
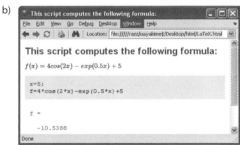

Fig. 5.21   a) Using LaTeX language in the comment section resulting in b) symbolic expressions in HTML document.

comments section after the cell header (the line starting with %%). As shown in Fig. 5.21, you should embrace your equation by $$ from both sides and have one empty line (starting from %) before and possibly after it (if your comments continue). Section 5.9.4 will present some more options for formatting MATLAB comments for publishing, whereas Sec. 6.4 will provide you with more details on the LaTeX syntax.

If needed, you can easily degrade a cell header to a simple comment line. All you have to do is to replace a double percent sign with the single percent sign. This changes the line from a cell to a standard comment and merges the cell with the preceding cell. You can also delete the entire line that contains the %%, which will have the same effect.

For the sake of completeness, it should be mentioned that the MATLAB Editor also provides a convenient means to simplify commenting text. You can comment the existing current line or a selection of existing lines from the **Text** menu by choosing the **Comment** option (<Ctrl>+<R>). Alternatively, you can do it if you right-click on the selection of lines and select the **Comment** option from the pop-up context menu. A comment symbol, %, will be added at the start of each selected line, and the color of the text becomes green (the color specified for comments by default). To uncomment the current line or a selected group of lines, select **Uncomment** from the **Text** menu, or right-click and select it from the context menu (that deletes the comment symbol, %).

You may use the **Display** option of the **Editor/Debugger** menu of the **Preferences** window (Fig. 2.22a) to automatically limit the comment lines to some maximum width while you type. By default, the maximum width is chosen to be 75 characters to ensure that the file prints without cropped text. As you reach the 75th column when typing a comment line, the comment automatically continues on the next line (the one-pixel-wide gray vertical line across the Editor window as shown in Figs. 5.13, 5.15–5.17, 5.19, and 5.20 defines the location of the last symbol for the comments). If you have them unwrapped

1. Select contiguous comment lines that you want to limit to the specified maximum width.
2. Select **Text** and then **Wrap Selected Comments**.

The selected comment lines will then be reformatted, so that no comment line in the selected area is longer than the maximum width. The lines that were shorter than the specified maximum will be merged to make longer lines if they are at the same level of indentation.

The latest versions of MATLAB also feature the File and Folders Comparisons tool to highlight the differences between two files. With a file opened in the Editor already, you can select **Tools** => **Compare Against**, and then browse to select the second file. The two files will appear aligned

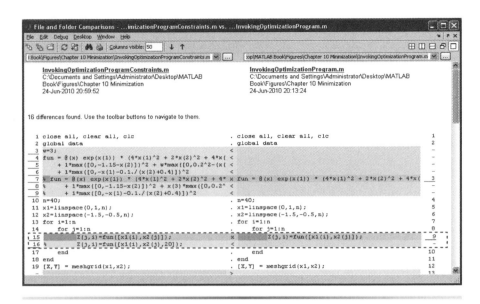

Fig. 5.22 File and Folder Comparison tool of the Editor.

in a window with highlights and indicators for any lines that differ, as shown in Fig. 5.22.

You can use the icons of the toolbar to swap positions of the two files, specify the number of columns shown (say 50, as in Fig. 5.22), and use down and up buttons to navigate between the blocks of code, where differences were detected (Fig. 5.22 features the second of such blocks enclosed with the dashed-line border box). You can also run the File and Folder Comparisons tool from the MATLAB Desktop by selecting **Desktop** => **File and Folder Comparison** (Fig. 2.13). It opens an empty window of this tool, which can be populated using three buttons: **New file comparison**, **New folder comparison,** and **Open**.

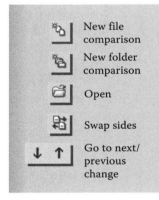

Let us conclude this section with several helpful bug finding tips

1. Always test your program with a simple version of the problem, whose answers can be checked by hand.
2. Display any intermediate calculations by removing semicolons at the end of the statements.
3. Try running your M-file function as a script by placing a comment sign (%) at the start of the function definition line (of course, you have to supply all variables this function may require).

5.10.3 Code Analyzer and Profiler

Once again, in the latest version of MATLAB, the Code Analyzer (M-Lint) is embedded into the Editor/Debugger assuring the best performance in creating syntax-error-free code. You can still use the Code Analyzer Report and other associated tools in separate windows if you choose to (Fig. 2.10). You can invoke this tool using the **Actions** icon, as explained in Sec. 2.1.5, or you can do it programmatically. A simple command

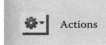

```
>> mlintrpt
```

scans all files with an .*m* file extension in the current folder for M-Lint messages and reports the results in a MATLAB Web browser. Other syntax include scanning a specific file or a specific directory

```
mlintrpt('filename','file')
mlintrpt('dirname','dir')
```

You can also apply the Code Analyzer preference settings to enable or suppress messages as indicated in the specified *settings.txt* file

```
mlintrpt('filename','file','settings.txt')
mlintrpt('dirname','dir','settings.txt')
```

These settings can be accessed by selecting **File** => **Preferences** => **Code Analyzer** (Fig. 5.23a).

Another function, `mlint`, allows you to look for the syntax errors in your file(s) even without opening the Code Analyzer browser. The results of any of the following commands will be displayed directly in the Command window:

```
mlint('filename')
mlint('filename','-config=settings.txt')
mlint('filename','-config=factory')
```

The typical output may include such messages as

```
L 2 (C 4-9): The value assigned here to variable 'Speed' might never be
used.
L 3 (C 5): Terminate statement with semicolon to suppress output (in
scripts).
L 7 (C 8): Use || instead of | as the OR operator in (scalar) conditional
statements.
L 15 (C 13): There may be a parenthesis imbalance around here.
L 21 (C 15-17): 'dim' might be growing inside the loop. Consider
preallocating for speed.
L 31 (C 20): Use of brackets [] is unnecessary. Use parenthesis to group,
if needed.
```

a)

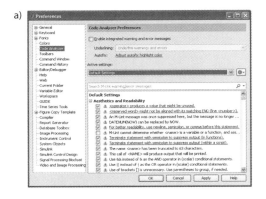

b)

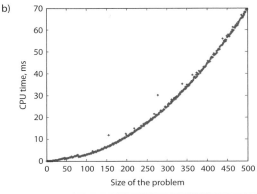

Fig. 5.23   a) Code analyzer preferences window and b) measuring required computational resources with the `tic-toc` pair.

Indicating the line and column of potential problem in code. Clicking on the line number opens the Editor and navigates directly to the candidate error, suggesting appropriate actions to fix it. If you wish not to show some specific warning, say about line termination with a semicolon, you may always uncheck the corresponding radio button in a Code Analyzer preferences browser (Fig. 5.23a).

Once you fixed all syntax errors and debugged your algorithm using any of the tools discussed in Secs. 5.9.1 and 5.9.2, the last thing you may want to consider (especially if we are talking about complex code that requires a lot of computational resources) is to possibly improve its effectiveness, that is, reduce central processing unit (CPU) time required to get the results.

The way to measure the total time required to run your code (or its particular portion) is to wrap it with two commands, `tic` and `toc`. This pair measures performance using a stopwatch timer. Specifically, `tic` starts a stopwatch timer, and `toc` prints the elapsed time since `tic` was used. You may assign this time to some variable for further use if needed. Consider the following self-explanatory example:

```
A=zeros(500,500); B=zeros(500,500); C=zeros(500,500);
for n = 1:500
  A = rand(n,n); B = rand(n,n);
    tic
    C = A.^B;
    C=sqrt(C);
    t(n) = toc;
end
plot(t*10^3,'o')
xlabel('Size of the problem'), ylabel('CPU time, ms')
```

The result is shown in Fig. 5.23b. As expected, the required computational resources (CPU time) grow with the size of the problem (not necessarily monotonically, but showing a general tendency).

Note that in principle, the first line in the aforementioned code is not required. However, repeatedly expanding the size of arrays A, B, and C in the `for` loop over time can adversely affect the performance of your program because

1. MATLAB has to spend time allocating more memory each time you add more elements to arrays.
2. This newly allocated memory is likely to be noncontiguous, thus slowing down any operations that MATLAB needs to perform on your arrays.

What we did in the previous example is called *preallocating memory* (or *preallocating arrays*), that is, we preallocated maximum possible amount of memory for all three arrays even before they were used for the first time. In this way, the program performs one memory allocation that reserves one contiguous block. Of course, in our example the advantages of memory preallocation may be negligible, but this is definitely one of the ways of improving performance of your code if it involves the loops with large variable arrays.

MATLAB features another more sophisticated tool that allows you to perform a detailed code profiling to find where a program spends most of the time (bottlenecks). That is where you should look at to optimize your code. The **Profiler** can be invoked by clicking the tenth icon on the MATLAB Desktop toolbar. You can enter the name of the file you are trying to profile in the **Run this code** window and then click **Start Profiling** button. The results for the piece of code we just considered are shown in Fig. 5.24a. Another alternative to bring the **Profiler** window up is to select **Desktop** => **Profiler** or use the `profile` function. For example, the commands

Profiler

```
profile on
ProfTest
profile viewer
```

will profile the aforementioned code (stored as the M-file *ProfTest*) and then display the results in the **Profiler** window.

The Profile Summary report presents statistics about the overall execution of the function and provides summary statistics for each function called. The report formats these values in four columns with

Function Name	Shows a list of all the functions and subfunctions called by the profiled function
Calls	Presents the number of times the function was called while profiling was on
Total Time	Indicates the total time spent in a function, including all child functions called, in seconds (the total time for the files, whose running time was inconsequential, shows as zero)
Self-Time	Shows the total time spent in a function, not including time for any child functions called (if it can determine it), in seconds

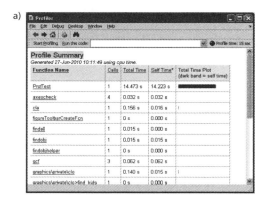

Fig. 5.24 a) Profiler summary and b) lines, where the most time was spent.

You can sort these results by clicking the title of the corresponding column. These results are also repeated visually by the two-color bars to the right of a function name.

Clicking the name of the file in the Profile summary brings up another window (Fig. 5.24b) providing specific details on each line of the code (when **Show busy lines** radio button is checked in). You can also access and review other information within this window by checking the corresponding radio button and then refreshing the results.

A general procedure for analyzing the results looks like follows. Once you identify which functions are consuming the most time, you can determine what they specifically do and whether there are ways to minimize their use and thus improve performance. Because programs may have several layers (one functions call other), your code might not explicitly call the most time-consuming functions. In this case, it is important to determine, which of your functions are responsible for such calls. Possible solutions include

1. Avoiding unnecessary computations (if possible)
2. Changing an algorithm to avoid costly computations (or functions)
3. Avoiding recomputation by storing intermediate results in variables and using them instead

One more example of using profiling will be presented in Sec. 14.6.

5.10.4 Formatting MATLAB Comments for Publishing

Sometimes, rather than trying to fix bugs by just reading code from the screen, it is better to print your work and have a closer look at it away from a computer. Maybe you would also want to include your debugged code to your report or even publish it on the Web. In this case, it is important how your printout looks like.

As mentioned in the previous section, MATLAB offers a nice capability to publish your work as an HTML document by simply clicking the **Publish to HTML** button on the Editor toolbar. If you click on the down arrow to the right of this button, you will be able to choose **Edit Publish Configuration for** `filename` option from the drop-down menu, which brings the Edit Configuration dialog box, shown in Fig. 5.25a. Alternatively, you can do the same by choosing **File** => **Publish** `filename` to publish with the default options or **File** => **Publish Configuration for** `filename` => **Edit Publish Configurations for** `filename` to publish with the customized options.

Figure 5.25b shows that HTML is not the only format available for publishing. Other formats are `"xml"` for Extensible Markup Language, `"latex"` for LaTeX, `"doc"` for Microsoft Word, `"ppt"` for Microsoft

a)
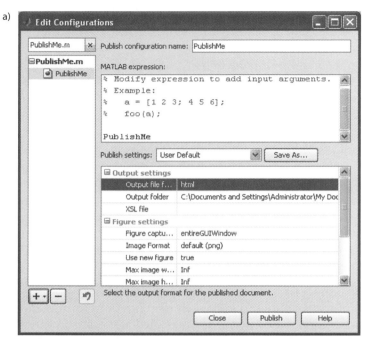

b)
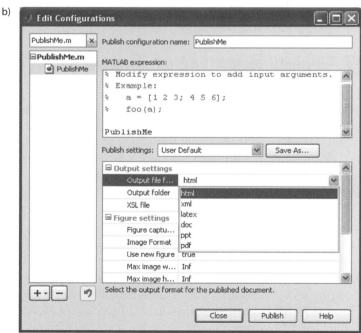

Fig. 5.25 a) Publishing Configurations window and b) available output formats.

a)
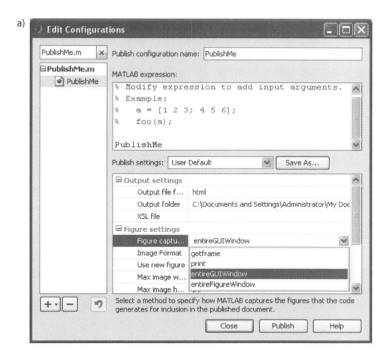

b)
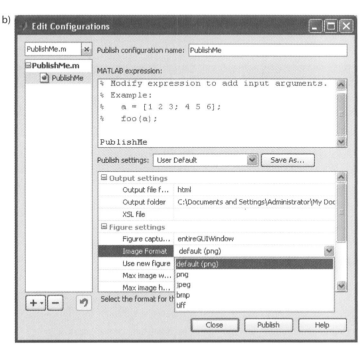

Fig. 5.26 a) Figure capturing options and b) choice of image format.

Power Point, and `"pdf"` for Portable Document Format. You may also change the way the figures will be incorporated into your published documents (Fig. 5.26) and tweak many other options.

Speaking of settings options shown in Fig. 5.26, Fig. 5.26a lists the options of how figure windows and GUI dialog boxes that your code creates appear in the published documents (background color and *window decorations*, that is, the title bar, toolbar, menu bar, and window border), and Fig. 5.26b shows the list of possible file type for images that MATLAB produces when publishing your program.

For better readability, MATLAB allows you to do a lot of different markups beyond cells and equations as shown in example in Fig. 5.21. Tables 5.5 and 5.6 summarize available options, whereas Figs. 5.27 and 5.28 show more examples of publishing a specially formatted script to an HTML document. Note that an image file you want to be inserted (*AIAALogo.jpg* in Fig. 5.28a) should reside in the same directory, where the HTML file is stored, that is, in the *html* directory. Also note that none of markups disturbs the flow of the program—they appear as comments. If you have significant portions of a formatted text you may save them as a separate M-file and then call it from the main file, where appropriate.

You do not need to memorize all these formatting options because you can always insert any of them by selecting the corresponding option from the pop-up list after clicking the **Cell** button and choosing **Insert Text Markup**

Table 5.5 Summary of Markup not Requiring a Formatted Block

Markup	Result in Published Document
`%% DOCUMENT TITLE` `% INTRODUCTORY TEXT`	Document title and introduction
`%% SECTION TITLE` `% DESCRIPTIVE TEXT`	Section title and description
`%%% SECTION TITLE` `% DESCRIPTIVE TEXT`	Section title without cell break
`%%`	Cell break without title or description
`% *BOLD TEXT*`	Bold text
`% _ITALIC TEXT_`	Italic text
`% |MONOSPACED TEXT|`	Monospaced text
`% <http://www.nps.edu/ADSC` `ADS Center>`	Hyperlinked text
`% TEXT(TM)`	Trademark symbol
`% TEXT(R)`	Registered trademark symbol
`snapnow;`	Code to force a snapshot of the current output

(Fig. 5.29). Note that, this list may be different for different versions of MATLAB.

When you publish M-file scripts, MATLAB runs the code and the published output includes the results (textual as in Fig. 5.21b or graphical as in Fig. 5.28b). To publish an M-file function, you should comment the function declaration statement. To show the results, you should supply all the input values you would normally pass over to this function via a list of input arguments.

Note that, to produce Fig. 5.28b we reduced the max image width in the options box of Fig. 5.25a to 200. You can save any changes you made to a publish settings file by clicking the **Save As** button.

You can also publish a MATLAB file with code cells, saving output to specified file, without opening the Editor by using the publish function

```
publish('file')

publish('file','format')

publish('file',options)

mydoc=publish('file',...)
```

The first syntax publishes *file.m* by running it in the base workspace, one cell at a time. It saves the results to an HTML output file, *file.html*, stored in an *html* subfolder of the folder containing *file.m*. Other syntax allows you to change an output file format (Fig. 5.25b) or any of the options shown in Figs. 5.25–5.26, and also name the output file (mydoc).

Table 5.6 Summary of Markup not Requiring a Formatted Block

Markup	Result in Published Document
`% <<FILENAME.JPG>> %`	Image
`% * ITEM 1` `% * ITEM 2`	Bulleted list
`% # ITEM 1` `% # ITEM 2`	Numbered list
`% $$ e^{\beta x} cos(5x) = 0$$`	TeX equation

`% <html>`
`% <table border=1><tr>`
`% <td>alpha</td>`
`% <td>beta</td></tr></table>`
`% </html>`
	HTML markup

`% % <latex>`
`% \begin{tabular}{|c|c|} \hline`
`% $n$ & $n!$ \\ \hline`
`% 1 & 1 \\`
`% 2 & 2 \\ \hline`
`% \end{tabular}`
`% </latex>`
	LaTeX markup (only works if you intend to publish it with the specified output file format LaTeX (Fig. 5.25b); if it is anything else, including HTML, then LaTeX markup will be excluded)

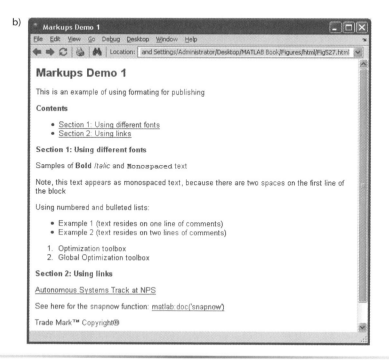

Fig. 5.27 a) Using markups to produce b) readable document.

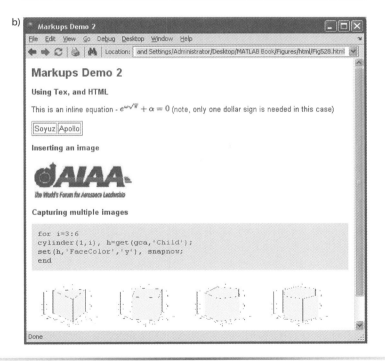

Fig. 5.28 a) Using markups to insert TeX and HTML code along with (multiple) images, resulting in b) complete HTML document.

Fig. 5.29 Markup options available in R2011a.

Another MATLAB function, `grabcode`, allows you to do exactly an opposite. The command

```
mycode=grabcode('name.html')
```

copies MATLAB code from the file *name.html* and pastes it into a document *mycode.m* in the Editor.

Problems

5.1   The rounded to the nearest integer average daily temperature (in °F) recorded by the Cape Canaveral Air Force Station in Florida and Edwards Air Force Base in California during the month of January of 2010 is given in the vectors below (to get data for any other period of time or place, go to the National Climatic Data Center web site, www7.ncdc.noaa.gov/CDO)

```
TCan = [ 62 50 46 44 43 41 43 45 41 36 41 42 48 51 60 68 71 61 ...
         56 55 66 72 65 70 69 59 56 54 59 67 61]
TEdw = [ 39 40 37 37 37 40 42 47 43 45 44 45 55 46 41 44 48 49 ...
         45 40 38 41 41 39 42 42 47 46 43 43 41]
```

Write a program (a script file) that utilizes the `find` function to answer the following questions:

- What was the average temperature for the month at each place?
- How many days was the temperature below its average at each location?

- How many days and which dates in the month was the temperature at the Edwards AFB higher than the temperature at Cape Canaveral?
- How many days and which dates in the month the difference between the temperatures at both places was within 3 degrees?
- Among those days found in the previous question, how many days the temperature at the Edwards AFB was higher than that of Cape Canaveral?

5.2 Write a function (inline or anonymous) computing the following expression $f(x)=\cos(x)\sin(10x)$. Use this function in conjunction with the MATLAB functions introduced in Sec. 5.7.2 to

- Plot the graph of this function within the range $x \in [0; 0.5]$
- Find a zero of this function within the range $x \in [0; \pi]$
- Find a minimum of this function within the range $x \in [0; \pi]$
- Find a maximum of this function within the range $x \in [0.5; 1.0]$

5.3 Write a function that returns parameters of the standard atmosphere at a certain altitude. You may use a tabular data for a certain altitudes, bottom edges of isothermal and nonisothermal layer or any approximations you could find, for examples at www.grc.nasa.gov/WWW/K-12/airplane/atmos.html.

Chapter 6

2-D and 3-D Plotting and Animation

- Anatomy of the Plot and Handle Graphics
- Creating Self-Contained Plots
- 2-D and 3-D Plotting Capabilities
- Interactive Plotting and Animation

`plot`, `plot3`, `line`, `figure`, `subplot`, `text`, `title`, `axes`, `legend`, `hold`, `grid`, `fplot`, `polar`, `stem`, `hist`, `mesh`, `surf`, `bar`, `bar3`, `cylinder`, `patch`, `get`, `set`, `gca`, `gcf`, `shading`, `gtext`, `ginput`, `movie`, `addframe`, **and many others**

6.1 Introduction

In engineering, the different kinds of plots are used to present and explain tabulated data produced by employing analytical formulas and/or a numerical experiment. This chapter addresses MATLAB's plotting capabilities that are the most advanced among all programming languages. The variety of the 2-D and 3-D plotting functions extends far beyond the simplest `plot` and `plot3` functions and include: `line`, `fplot`, `ezplot`, `ezcontour`, `ezcontourf`, `plotyy`, `loglog`, `semilogx`, `semilogy`, `polar`, `ezpolar`, `area`, `pie`, `bar`, `contour`, `contourf`, `ezcontour`, `comet`, `compass`, `feather`, `quiver`, `stem`, `stairs`, `hist`, `rose`, `fill`, `triplot`, `scatter`, `errorbar`, `ezplot3`, `mesh`, `meshc`, `meshz`, `ezmesh`, `ezmeshc`, `surf`, `surfc`, `ezsurf`, `ezsurfc`, `trisurf`, `bar3`, `pie3`, `contour3`, `quiver3`, `stem3`, `ribbon`, `waterfall`, `slice`, `comet3`, `cylinder`, `ellipsoid`, `sphere`, `fill3`, **and** `scatter3`. There are also many auxiliary functions such as `figure`, `subplot`, `title`, `axis`, `xlim`, `ylim`, `zlim`, `grid`, `xlabel`, `ylabel`, `zlabel`, `hold`, `legend`, `text`, `annotation`, `view`, `set`, `get`, `meshgrid`, `trimesh`, `colormap`, **and** `shading`. Some other functions enable interactive input (`gtext` and `ginput`) and creation and manipulation of animations (`getframe`, `frame2im`, `im2frame`, `movie`, `movie2avi`, **and** `addframe`). The chapter starts from the simplest 2-D plots addressing all environmental variables of a single plot, followed by discussion of line attributes; MATLAB's capabilities for enhancing the readability of the plots by formatting the text strings and including special symbols to them and the way to access the plot properties via a special function. Next, it considers an option of having more than just one plot within one Figure window and/or more than just one figure per single window. Special types of 2-D plots and different types of 3-D plots are addressed after this, followed by a brief discussion of how to manage a color palette and how to use the so-called easy-to-plot functions. The plot-editing options of MATLAB readily available within the Figure window are addressed next. For advanced users, this chapter presents capabilities of interactive plotting and movie video clips creating. The chapter concludes with a list of common-sense requirements that engineering plots should satisfy.

6.2 Anatomy of 2-D (x–y) Plots in MATLAB

To start, let us consider an anatomy of a 2-D plot. The typical plot consists of the following:

1. The horizontal axis, *abscissa*, and vertical axis, *ordinate*, both defined by the *scale*—the range and spacing of the numbers (*linear* or *logarithmic*) and *tick marks locations*.
2. The *curve* itself represented by a sequence of (connected) *data points* (*x*- and *y*-data). That can also include the *marker* used to identify data points.
3. The *text* inserts to show

Title	Identifies what the plot is about
Axes labels	Shows the name and units of the quantity plotted along each axis
Tick marks labels	Specifies numbers or letters that correspond to tick marks, and possibly
Legend	Differentiates between multiple sets of data on one plot and other supplemental information

The following basic *x*–*y* plotting functions allow you to create some of the afore-mentioned elements for a typical 2-D plot all at once (defining some of them explicitly and using some default values for the remaining ones):

`plot(y)`	Generates a plot of the array y vs their indices if y is a real vector, or plots the imaginary parts of y vs the real parts of y if y is complex
`plot(x,y)`	Generates a plot of the array y vs the array x on rectilinear axes

`plot(x,y,LineSpec,...)`	Plots all lines defined by the x–y–LineSpec triples, where LineSpec is a line specification that determines a line type, marker symbol, and color of the plotted lines (to be considered in Sec. 6.3)
`line(x,y)`	Adds a line defined in the vectors x and y to the current axes
`fplot('fun', limits)`	Generates a plot of the function described by fun within the specified limits
`ezplot('fun')`	Plots the function (or expression) described by fun within default domain $-2\pi < x < 2\pi$

Examples of using these functions are presented in Figs. 6.1 and 6.2. Specifically, the command

```
>> plot(sqrt([20:40]))
```

creates a plot of $\sqrt{x}$ vs the indices of the input vector, defining its axes and tick marks automatically (Fig. 6.1a). The line style and color for the curve itself are set to some default values. No other parts of the plot are defined. In contrast, the command

```
>> plot(20:40,sqrt([20:40]),'or-.')
```

(Fig. 6.1b) plots the same plot vs its argument explicitly defining the data point marker (circle), line style (dash-dotted line), and color (red). The command

```
>> fplot('cos',[0 pi])
```

(Fig. 6.2a) creates axes using the explicitly defined range for abscissa and calculating the corresponding range for ordinate based on the data points

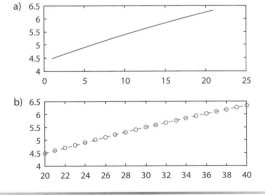

Fig. 6.1 Examples of using the `plot` function.

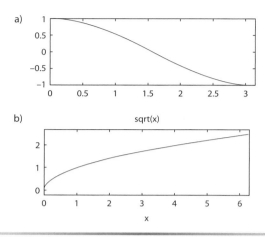

Fig. 6.2 Examples of using the a) `fplot` and b) `ezplot` functions.

computed automatically utilizing the function name (`cos`). Also, it automatically defines tick marks locations and adds the tick marks labels to them. The command

```
>> ezplot('sqrt(x)')
```

(Fig. 6.2b) creates a plot of the $\sqrt{x}$ function over the default domain $0 < x < 2\pi$ (the function is not defined for $x < 0$) adding the x-axis label (x) and title of the plot (`sqrt(x)`) automatically.

As mentioned previously, these functions allow other syntax, so some of the plot's properties can be changed within the function itself, but generally you need to employ other functions to do that. For example, the `axis` function allows you to manipulate with the first group of plot content (axes) as follows:

`axis([xmin xmax ymin ymax])`	Sets the minimum and maximum limits of the x- and y-axes. If you want to specify just one limit but want MATLAB to autoscale the other, use `Inf` or `-Inf` for autoscaled limits or use `xlim` and `ylim` functions
`axis square`	Selects the axes' limits so that the plot will be square
`axis equal`	Assures that the x- and y-axes have the same tick mark spacing
`axis tight`	Sets the limits to the range of the data
`axis manual`	Freezes the scaling at the current limits, so that if `hold` is turned on (Sec. 6.6), subsequent plots will use the same limits
`axis auto`	Returns the axis scaling to its default autoscaling mode

`axis vis3d`	Freezes aspect ratio properties to enable rotation of the 3-D objects and overrides stretch-to-fill
`v=axis`	Returns the current axis scaling in the vector v (so that you may use the same scaling for another plot by calling `axis(v)`)
`grid`	Displays gridlines at the tick marks corresponding to the tick labels. You may use `grid on` to turn gridlines on and `grid off` to turn them off. When used by itself, `grid` toggles the grid switch on or `off`

For example, the following set of commands results in the plot shown in Fig. 6.3a:

```
>> t=linspace(0,2*pi);
>> line(cos(t),sin(t)), grid
```

By default, axes are scaled automatically (in this particular case, the `axis tight` command would result in the same plot), and the `grid` function adds gridlines. However, if you substitute the second line with the following

```
>> line(cos(t),sin(t)), axis equal
```

you will have the same scale for both axes, and therefore a unit circle will appear as it is supposed to (Fig. 6.3b).

The third group of plot content (text) can be addressed using the following functions:

`title('text')`	Places `text` in a title at the top of a plot
`xlabel('text')`	Places a `text` label on the x-axis (abscissa)
`ylabel('text')`	Places a `text` label on the y-axis (ordinate)

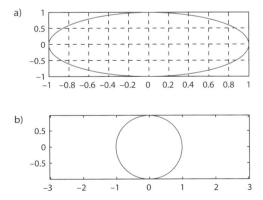

Fig. 6.3   Examples of changing axis properties.

`text(x,y,'text')`	Adds `text` in the quotes to a location (x,y) on the current axes, where (x,y) are expressed in the current plot units
`gtext('text')`	Places `text` in Figure window at a point specified by the mouse click
`annotation(type, parameters)`	Creates the specified annotation `type`, such as line, arrow, double arrow, text arrow, textbox, ellipse, and rectangle, and defines type-specific `parameters`

For example, the set of commands

```
>> title('My Plot'), xlabel('x-axis'), ylabel('y-axis')
>> text(0.5,0.5,'Text')
```

results in axes shown in Fig. 6.4a. The commands

```
>> annotation('arrow',[0.5 0.7],[0.7 0.8])
>> gtext('Chosen Location')
```

add an arrow, specifying *x*- and *y*-coordinates of its origin and end points. As you move the pointer into a Figure window, the pointer becomes cross-hairs to indicate that `gtext` is waiting for you to select a location of the text string as shown in Fig. 6.4b. Once you click a mouse button, the text `Chosen Location` will be added to your figure.

Note that by default, annotation objects use *normalized coordinates* to specify locations within the figure. In normalized coordinates, the point (0, 0) is always the lower-left corner and the point (1, 1) is always

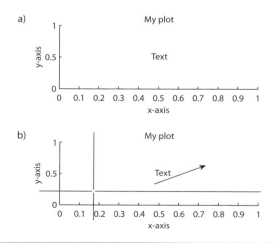

Fig. 6.4   Examples of adding the a) text and b) annotations to the plots.

the upper-right corner of the Figure window, regardless of the figure size and proportions. It is not convenient, so if you intend to use the `annotation` function, you might consider writing your own function to convert your axis coordinates into the normalized ones (Problem 6.1).

More options for changing the plot content and format will be discussed in Sec. 6.4. Specifically, it will be done using the *handles*. Section 6.11 will present an alternative method using the graphical user interface. Section 6.5 will show how to utilize different fonts and use TeX and LaTeX, document markup languages, to represent symbolic expressions. Before we proceed, let us briefly address how to change the properties of the line itself (corresponding to the second group of the plot content, curve).

6.3 **Line Specifications**

As shown in the previous section for the `plot` function, line specification `LineSpec` determines such properties as a line type, marker symbol, and color of the plotted lines. To this end, Table 6.1 presents all options MATLAB has to offer.

Note that circle and all following data markers are fillable markers (you may fill them with a color, not necessarily coinciding with the color of the edges).

Many plotting functions accept a `LineSpec` argument that defines three components used to specify the lines. You may specify the components as a quoted string after the data arguments in any order (as seen from

Table 6.1   `LineSpec` Options

Line Types		Data Markers		Colors	
Solid line	-	Dot (.)	.	Black	k
Dashed line	--	Asterisk (*)	*	Blue	b
Dash-dotted line	-.	Cross (x)	x	Cyan	c
Dotted line	:	Plus sign (+)	+	Green	g
		Circle (o)	o	Magenta	m
		Square (□)	s	Red	r
		Diamond (◇)	d	White	w
		Pentagram (five-point star)	p	Yellow	y
		Hexagram (six-point star)	h		
		Left triangle (◄)	<		
		Up triangle (▲)	^		
		Right triangle (►)	>		
		Down triangle (▼)	v		

Table 6.1 they are unique) or even missing some of them. For example,

```
>> plot([0:10].^2,'g>:')
```

plots a parabola using a dotted line (:), places right triangle markers (>) at the data points, and colors both the line and marker green (g) (Fig. 6.5a). Another command

```
>> fplot('tanh',[-pi, pi]/2,'md',0.1)
```

produces a plot of a hyperbolic tangent within $-\tfrac{1}{2}\pi \le x \le \tfrac{1}{2}\pi$ range, using (optional) 0.1 tolerance, with data points marked by magenta diamonds (Fig. 6.5b).

When using the `plot` function (as well as the `fplot` and some other functions), you can also specify other characteristics of lines using their graphics properties as additional (optional) arguments

LineWidth	Specifies the line's width (by default, it uses *points*, where 1 point is equal to 1/72 of an inch)
MarkerEdgeColor	Specifies the color of the marker or the edge color for filled markers (circle, square, diamond, pentagram, hexagram, and the four triangles)
MarkerFaceColor	Specifies the color of the face of filled markers
MarkerSize	Specifies the size of the marker in points

For example,

```
>> x = -pi:pi; y = sin(x);
>> plot(x,y,'Color','r','Marker','p','LineWidth',3,'MarkerSize',7)
```

plots y vs x using a three-point-wide red-color line with seven-point-size star-marker as shown in Fig. 6.6a. Note that Color and Marker properties were specified in a similar way rather than using the LineSpec string. On the contrary, the command

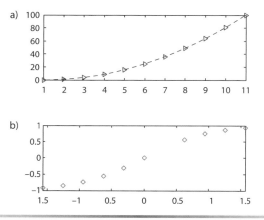

Fig. 6.5 Examples of using the LineSpec options.

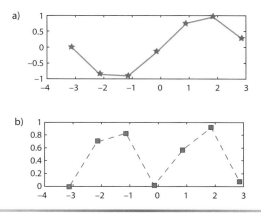

Fig. 6.6   Examples of varying line specification.

```
>> plot(x,y.^2,'--s','MarkerSize',10,'MarkerFaceColor','r')
```

blends both ways of specifying line properties together—via the `LineSpec` string, defining dashed line and a square marker, and via changing some of other properties explicitly (Fig. 6.6b).

By the way, the vector `x` in the previous example was created using a small number of elements on purpose. Obviously, that is the reason the curves in Fig. 6.6 look so ugly. The logical move is to increase the number of points; using at least 100 usually yields a good result (Fig. 6.7a)

```
>> x=linspace(0,pi); plot(x,cos(6*x).*exp(-x),'-.^')
```

However, marking all points obviously makes it too crowded.

There may be different reasons for adding markers (for example, distinguishing between the multiple curves), but usually you need just a few of

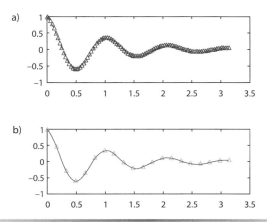

Fig. 6.7   a) Marking all data points and b) a preferred way of marking some data points.

them. The following shows how to achieve it using two separate sets of arguments (Fig. 6.7b)

```
>> x=linspace(0,pi); xm=linspace(0,pi,20);
>> plot(x,cos(6*x).*exp(-x),xm,cos(6*xm).*exp(-xm),'^')
```

Note that the `plot` function uses a syntax allowing you to plot multiple curves with just one command (to be discussed in Sec. 6.6).

6.4 Accessing Plot Properties via the `get` and `set` Functions

In the previous section, it was shown how some of the default values of the line properties can be accessed and changed. Promoting the open-source paradigm, MATLAB allows you to change any other property of any *graphics object*, including the Figure and its components (as shown in Fig. 6.8), to accommodate your own preferences.

There are three key things about the graphics objects you should know to be able to proficiently manipulate with them programmatically (another way is to use an interactive tool, **Plot Editor**, to be discussed in Sec. 6.11)

1. The figure, axis, and line objects obey a certain hierarchy, so that the Figure is the *parent* for the Axis, whereas the Line is its *children.*
2. Each object has its own handle or identifier.
3. You can access (change) all properties of any object using its handle.

To assign the handle to the line instead of issuing a simple `plot` (or `line`) command, we would rather do it using the following syntax

```
>> h_line=plot([10:-1:1]);
```

This command does exactly what the `plot([10:-1:1])` call would do, that is, creates a new Figure window and plots the line, but in addition to

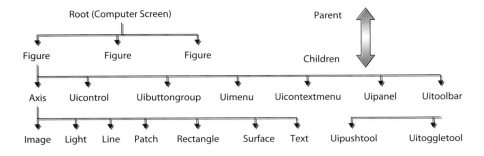

Fig. 6.8 Graphics objects hierarchy.

Table 6.2 Properties of the Line Object

```
>> get(h_line)
                  Color:    [0 0 1]
              EraseMode:    'normal'
              LineStyle:    '-'
              LineWidth:    0.5000
                 Marker:    'none'
             MarkerSize:    6
        MarkerEdgeColor:    'auto'
        MarkerFaceColor:    'none'
                  XData:    [1 2 3 4 5 6 7 8 9 10]
                  YData:    [10 9 8 7 6 5 4 3 2 1]
                  ZData:    [1x0 double]
          BeingDeleted:    'off'
         ButtonDownFcn:    []
              Children:    [0x1 double]
              Clipping:    'on'
             CreateFcn:    []
             DeleteFcn:    []
            BusyAction:    'queue'
      HandleVisibility:    'on'
               HitTest:    'on'
         Interruptible:    'on'
              Selected:    'off'
    SelectionHighlight:    'on'
                   Tag:    ''
                  Type:    'line'
         UIContextMenu:    []
              UserData:    []
               Visible:    'on'
                Parent:    157.0037
           DisplayName:    ''
             XDataMode:    'manual'
           XDataSource:    ''
           YDataSource:    ''
           ZDataSource:    ''
```

that it creates a handle to this line, h_line. This handle has some unique decimal number (identifier) assigned to it. Using this handle, we can access any property of this line. Type get(h_line) to see what these properties are (Table 6.2).

If you are only interested in some specific property you may address it directly, for instance, type get(h_line, 'Color') to obtain

```
ans = 0 0 1
```

which corresponds to blue in the red-green-blue (RGB) color representation (to be explained in Sec. 6.9).

As you might imagine, now we can change any property in a similar manner, that is, by addressing it directly using the set function. For example,

```
>> set(h_line,'Color','r','LineWidth',2.5)
```

changes the line color to red and increases the line width to 2.5 points.

Can we access the properties of axes? Certainly. We have to recall that the Axis object is a parent with respect to the Line object and, therefore, the command

```
>> h_axis=get(h_line,'parent')
```

retrieves the Axis handle for us (note that the property names are not case-sensitive). Now, all axis properties can be seen by issuing get (h_axis) command, the result of which is shown in Table 6.3. Obviously, the command

```
>> h_figure=get(h_axis,'parent')
```

retrieves the Figure handle. For your reference, the results of the get (h_figure) call are presented in Table 6.3 as well.

In the very same way we did for the line object, we now can set the values of specified properties for any other graphic object

Table 6.3 Properties of the Axis (Left Column) and Figure (Right Column) Objects

`>> get(h_axis)`	`>> get(h_figure)`
ActivePositionProperty = outerposition	Alphamap = [(1 by 64) double array]
ALim = [0 1]	BackingStore = on
ALimMode = auto	CloseRequestFcn = closereq
AmbientLightColor = [1 1 1]	Color = [0.8 0.8 0.8]
Box = on	Colormap = [(64 by 3) double array]
CameraPosition = [5.5 5.5 17.3205]	CurrentAxes = [157.004]
CameraPositionMode = auto	CurrentCharacter =
CameraTarget = [5.5 5.5 0]	CurrentObject = []
CameraTargetMode = auto	CurrentPoint = [0 0]
CameraUpVector = [0 1 0]	DockControls = on
CameraUpVectorMode = auto	DoubleBuffer = on
CameraViewAngle = [6.60861]	FileName =
CameraViewAngleMode = auto	FixedColors = [(10 by 3) double array]
CLim = [0 1]	IntegerHandle = on
CLimMode = auto	InvertHardcopy = on
Color = [1 1 1]	KeyPressFcn =
CurrentPoint = [(2 by 3) double array]	MenuBar = figure
ColorOrder = [(7 by 3) double array]	MinColormap = [64]
DataAspectRatio = [4.5 4.5 1]	Name =
DataAspectRatioMode = auto	NextPlot = add
DrawMode = normal	NumberTitle = on
FontAngle = normal	PaperUnits = inches
FontName = Helvetica	PaperOrientation = portrait
FontSize = [10]	PaperPosition = [0.25 2.5 8 6]
FontUnits = points	PaperPositionMode = manual
FontWeight = normal	PaperSize = [8.5 11]
GridLineStyle = :	PaperType = usletter
Layer = bottom	Pointer = arrow
LineStyleOrder = -	PointerShapeCData = [(16 by 16) double array]
LineWidth = [0.5]	
MinorGridLine Style = :	PointerShapeHotSpot = [1 1]
NextPlot = replace	Position = [232 246 560 420]

```
OuterPosition = [0 0 1 1]                    Renderer = painters
PlotBoxAspectRatio = [1 1 1]                 RendererMode = auto
PlotBoxAspectRatioMode = auto                Resize = on
Projection = orthographic                    ResizeFcn =
Position = [0.13 0.11 0.775 0.815]           SelectionType = normal
TickLength = [0.01 0.025]                     ShareColors = on
TickDir = in                                 ToolBar = auto
TickDirMode = auto                           Units = pixels
TightInset = [0.0321429 0.0404762            WindowButtonDownFcn =
  0.0142857 0.0190476]                       WindowButtonMotionFcn =
Title = [159.004]                            WindowButtonUpFcn =
Units = normalized                           WindowStyle = normal
View = [0 90]                                WVisual = 00 (RGB 32 GDI, Bitmap,
  XColor = [0 0 0]                             Window)
  XDir = normal                              WVisualMode = autoBeingDeleted = off
  XGrid = off                                ButtonDownFcn =
  XLabel = [160.004]                         Children = [157.004]
  XAxisLocation = bottom                     Clipping = on
  XLim = [1 10]                              CreateFcn =
  XLimMode = auto                            DeleteFcn =
  XMinorGrid = off                           BusyAction = queue
  XMinorTick = off                           HandleVisibility = on
  XScale = linear                            HitTest = on
  XTick = [(1 by 10) double array]           Interruptible = on
  XTickLabel = [1;2;3;4;5;6;7;8;9;10]        Parent = [0]
  XTickLabelMode = auto                      Selected = off
  XTickMode = auto                           SelectionHighlight = on
The above 14 lines are repeated for Y and Z axes    Tag =
BeingDeleted = off                           Type = figure
ButtonDownFcn =                              UIContextMenu = []
Children = [158.004]                         UserData = []
Clipping = on                               Visible = on
CreateFcn =
DeleteFcn =
BusyAction = queue
HandleVisibility = on
HitTest = on
Interruptible = on
Parent = [1]
Selected = off
SelectionHighlight = on
Tag =
Type = axes
UIContextMenu = []
UserData = []
Visible = on
```

(shown in Fig. 6.8) using its handle H. The general syntax of using the set function is

```
set(H,'PropertyName',PropertyValue)
```

The handle H can be a vector of handles, in which case set sets the properties' values for all the objects (unless you specifically indicate that you want to change the properties of just one element, say H(2).

Usually, MATLAB provides alternative ways of doing exactly the same thing. Getting the graphics objects handles is not an exception. Table 6.4

shows how you could get the handles differently from what was shown above (even if have not used `h_line=plot(...)` syntax to begin with).

Before Release R2010a of MATLAB, the best way to learn about a specific property of any graphics object (for instance, those shown in Tables 6.2 and 6.3) was to use the MATLAB Help Browser as shown in Fig. 6.9a. In the Contents folder (on the left), you would choose Handle Graphics Properties Browser that brings up a very convenient interactive Help window on the right. Using this browser, you could easily browse through the graphics objects hierarchical tree (Fig. 6.8) by clicking on the corresponding items. The list of properties for the selected item on the tree would appear in the middle of the screen. The complete information about some specific property you chose would be given to the right. Unfortunately, starting with Release R2010a this Browser is not available anymore, so you will need to look for each property by typing its name in the Search documentation and demos window above the Contents folder. For example, Fig. 6.9b shows the response of looking for the very first property of the figure object, `Alphamap`, as it appears in R2011a.

To conclude this section, let us consider one more example. The set of commands

```
axis([0,2*pi,-1,1])
set(gca,'XTick',[0:pi/2:2*pi])
set(gca,'XTickLabel',{'0';'0.5pi';'pi';'1.5pi';'2pi'})
xlabel('\phi, rad'), hyl=ylabel('cos(\phi)');
```

produces axes shown in Fig. 6.10. Specifically, the first line establishes ranges for both x- and y-axes, the second line sets the position of tick marks for the x-axis, the third line changes the tick labels from numbers to text strings, and finally the fourth line provides the names for the x- and y-axes.

Note that in the forth line we used TeX language to represent a Greek letter phi, ϕ (to be addressed in more detail in the next section). That is

Table 6.4 Alternative Ways of Retrieving Object Handles

Command	Description

`gcf` (stands for "get current figure")
`get (0,'CurrentFigure')`
`get(0,'Children')`
`get(gca,'Parent')`
	Returns the handle of the current figure

`gca` (stands for "get current axes")
`get(gcf,'Children')`
`get(gcf,'CurrentAxes')`
`get(get(0,'Children'),'Children')`
	Returns the handle to the current axes for the current figure

`get(gca,'children')`
`get(get(get(0,'Children'),`
`'Children'),'Children')`
	Returns the handle to the line(s)

a)
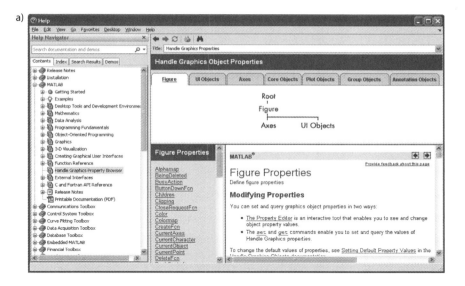

b)
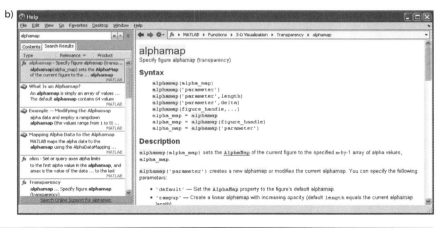

Fig. 6.9 Using MATLAB Help Browser to learn about graphics properties in a) previous versions of MATLAB and b) R2011a.

because axis labels allow you to use a TeX interpreter. Verify this by running the following command to retrieve the properties of the `ylabel` object (via handle `hyl` created earlier)

```
>> get(hyl)
```

Among other attributes you will see

...

```
Interpreter = tex
```

...

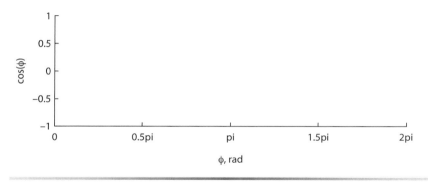

Fig. 6.10 Changing axes attributes.

Three available options are `tex`, `latex`, and `none`. By the way, it would be nice to have a symbolic π instead of `pi` in the *x*-axis tick labels as well. Unfortunately you do not have the `Interpreter` attribute among the properties of axes (left column of Table 6.3) and, therefore, cannot use TeX language for the tick labels.

6.5 Text Strings

In general, you can define a text that includes symbols and Greek letters using the `text` function. You can also include the character sequences in the string arguments of the `title`, `xlabel`, `ylabel`, and `zlabel` functions. What you are doing is that you are assigning the character sequence to the `String` property of any text object.

When the text `Interpreter` property is set to `Tex` (the default) or `Latex`, you can use a subset of TeX (LaTex) commands embedded in the string to produce special characters such as Greek letters and mathematical symbols as shown in Table 6.5. When `Interpreter` is set to `none`, no characters in the `String` are interpreted, and everything is displayed as typed.

You can also specify stream modifiers that control the font used

`\bf`	Forces using bold font
`\it`	Forces using italics font
`\sl`	Forces using oblique font (if available)
`\rm`	Forces using normal font
`\fontname{fontname}`	Specifies the name of the font family to use
`\fontsize{fontsize}`	Specifies the font size (in font units)

The first four modifiers are mutually exclusive. However, you can use `\fontname` in combination with one of other modifiers. Stream modifiers remain in effect until the end of the string, next font modifier or only within the context defined by braces `{ }`.

Table 6.5 Special Symbols

Character Sequence	Symbol	Character Sequence	Symbol	Character Sequence	Symbol
\alpha	α	\upsilon	υ	\sim	~
\beta	β	\phi	φ	\leq	≤
\gamma	γ	\chi	χ	\infty	∞
\delta	δ	\psi	ψ	\clubsuit	♣
\epsilon	ε	\omega	ω	\diamondsuit	♦
\zeta	ζ	\Gamma	Γ	\heartsuit	♥
\eta	η	\Delta	Δ	\spadesuit	♠
\theta	θ	\Theta	Θ	\leftrightarrow	↔
\vartheta	ϑ	\Lambda	Λ	\leftarrow	←
\iota	ι	\Xi	Ξ	\uparrow	↑
\kappa	κ	\Pi	Π	\rightarrow	→
\lambda	λ	\Sigma	Σ	\downarrow	↓
\mu	μ	\Upsilon	Υ	\circ	°
\nu	ν	\Phi	Φ	\pm	±
\xi	ξ	\Psi	ψ	\geq	≥
\pi	π	\Omega	Ω	\propto	∝
\rho	ρ	\forall	∀	\partial	∂
\sigma	σ	\exists	∃	\bullet	•
\varsigma	ς	\ni	∋	\div	÷
\tau	τ	\cong	≅	\neq	≠
\equiv	≡	\approx	≈	\aleph	ℵ
\Im	ℑ	\Re	ℜ	\wp	℘
\otimes	⊗	\oplus	⊕	\oslash	∅
\cap	∩	\cup	∪	\supseteq	⊇
\supset	⊃	\subseteq	⊆	\subset	⊂
\int	∫	\in	∈	\o	o
\rfloor	⌋	\lceil	⌈	\nabla	∇
\lfloor	⌊	\cdot	·	\ldots	...
\perp	⊥	\neg	¬	\prime	′
\wedge	∧	\times	×	\0	∅
\rceil	⌉	\surd	√	\mid		
\vee	∨	\varpi	ϖ	\copyright	©
\langle	⟨	\rangle	⟩		

The subscript character "_" and the superscript character "^" modify the character or substring defined in braces immediately following it. To print the special characters used to define the Tex strings, when `Interpreter` is `Tex`, prefix them with the backslash "\" character, for instance: \\, \ {, \ }, _, \^.

Let us summarize all possibilities in one illustrative example. The following set of commands:

```
fplot('0.25*exp(-.006*t)*sin(0.2*t)',[0 900])
a='\fontsize{12}{\itAe}^{-\alpha\itt}sin\beta{\itt}';
b='     \\  \alpha<<\beta';
title(strcat(a,b))
xlabel('\bf\itTime\rm, \musec')
ylabel('\fontname{Times}\fontsize{12}Amplitude')
```

(where concatenation of two strings is used with the only goal of avoiding a long line) produces the plot shown in Fig. 6.11a.

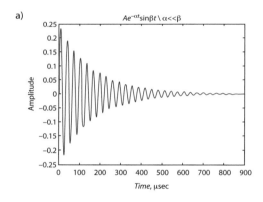

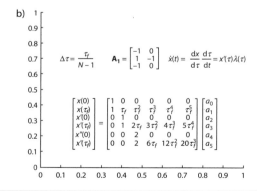

Fig. 6.11   Example of the enhanced text using a) TeX and b) LaTeX language.

Even more options are available using the LaTeX language. For instance, you can visualize fractions, matrices, derivatives as shown in Fig. 6.11b. The LaTeX sequence should be embraced with a double dollar sign, $\$\$$, and the LaTeX interpreter specified explicitly. The following commands used to produce the first line of expressions in Fig. 6.11 provide some self-explanatory examples (again, the concatenation is used to avoid long strings):

```
t1='\Delta\tau={\tau_f\over N-1}';
text(0.1,0.7,['$$' t1 '$$'],'Interpreter','latex')
t2='{\bf A_1}=\left[\matrix{-1&0\cr 1&-1\cr -1&0\cr}\right]';
text(0.35,0.7,['$$' t2 '$$'],'Interpreter','latex')
t3='\dot x(t)={dx\over d\tau}{d\tau\over dt}=x''(\tau)\lambda(\tau)';
text(0.65,0.7,['$$' t3 '$$'],'Interpreter','latex')
```

As mentioned in Sec. 5.12.2 LaTeX expressions can be used in the comment sections of your M-file scripts to improve readability when publishing them as HTML documents using the MATLAB Editor/Debugger. In Chapter 15 you will see that TeX commands can also be used in Simulink models (for exactly the same reason, that is, improving readability). Since using symbolic expressions does serve a good reason, MATLAB even offers the `latex` function that converts a symbolic expression to the LaTeX representation, to be further used elsewhere (to be covered in Sec. 7.11).

6.6 Overlays, Legends, Subplots, and Multiple Figures

So far (excluding Fig. 6.7b) we have only had a single line per figure. Of course, MATLAB allows you to do much more, having several plots on one graph, creating several plots within one Figure window and placing plots on separate (multiple) Figure windows. These options are discussed later.

Let us start with the *overlay* plots. An overlay plot is the one in which multiple lines or sets of data are shown together. The overlay plots can be created using several variants of employing the basic `plot` function as follows:

`plot(A)`	Plots the columns of A vs their indices generating n curves, one for each column (where A is an $m \times n$ matrix)
`plot(x,A)` or `plot(A,x)`	Generates m or n plots depending on the length of vector x (where A is an $m \times n$ matrix and x is a vector of length m or n)
`plot(A,B)`	Plots the columns of the matrix B the columns of the matrix A (where A and B are both $m \times n$ matrices)

`plot(x1,y1,x2,y2,`	Plots vector `y1` vs vector `x1`, `y2` vs
`LineSpec,x3,y3,…)`	`x2`, `y3` vs `x3`, etc., in one call (you can mix `xn–yn–LineSpec` triples with `xn–yn` pairs relying on default specifications)

(As you may recall we have used the latter option to create a plot shown in Fig. 6.7b. Let us have a look at one other option.) For example, the commands

```
>> a=1:10;
>> b=rand(10,3);
>> plot(a,b)
```

produce a plot shown in Fig. 6.12a. Note that in this case to distinguish between multiple lines, the `plot` function uses different colors, which is good but may not be enough. If you want to change a line style or add a marker, the only option you have is to access and change these lines properties via their handles. In this particular case, the command

```
>> h1=get(gca,'Children');
>> set(h1(2),'LineStyle','-.')
```

would access a 3×1 handle `h1`, referring to all three lines (even though we did not define this handle explicitly) and change the `LineStyle` attribute of the second line.

Overlay plots can also be created using the `line` function following the `plot` command (the `line` command does not create a separate plot, it just adds a line to the current plot). For example, the following two commands

```
>> plot(rand(1,6))
>> line(1:6,rand(1,6)+1,'LineStyle','-.')
```

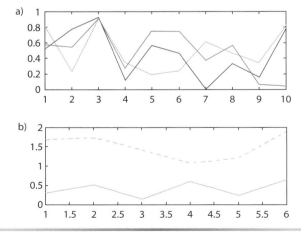

Fig. 6.12 Examples of overlay plots.

result in overlay plots shown in Fig. 6.12b. Similar to that of the `plot` func-tion if `x` and `y` are matrices of the same size, the call `line(x,y)` adds multiple lines, one line per column. Let us make two comments here. First, note that the `line` function does not support `x-y-LineSpec` triples, as the plot function does, so you have to specify line properties explicitly by "property name–property value" pairs. Second, both `x` and `y` values have to be specified explicitly and that constitutes another difference with the `plot` function, where you can plot a vector its indices.

As an alternative to using the `line` function, you may add multiple lines or sets of data to the current plot by issuing the multiple `plot` com-mands as well. However, in this case, you must hold the originally drawn axes by using the `hold` function. Specifically, use `hold  on` to turn the hold feature on and `hold  off` to turn it off. When used by itself, `hold` toggles the hold switch `on` or `off`. When `hold` feature is `on`, any `plot` command will add line(s) to the current plot. In this regard, the result of following two commands

```
>> plot(rand(1,6)), hold on
>> plot(rand(1,6)+1,'-.')
```

is identical to that of the previous two commands that used the `line` func-tion. Note that, you are only required to issue the hold command once, after plotting the very first plot. For example, in the latter case you may continue adding more lines using the `plot` function without issuing additional `hold` commands.

There is one more function allowing not only to plot two curves within one axis but also to use different *y*-axis for each of them. The `plotyy(x1,y1,x2,y2)` function generates a separate *y*-axes plot of `x1` `y1` with the `y1`-axis labeling on the left, and a plot of `x2` vs `y2` with the `y2`-axis labeling on the right. For example, the script

```
x=linspace(0,15);
y1=10*exp(-0.36*x).*sin(2.7*x);
y2=200*exp(-0.05*x).*sin(0.8*x);
[ax,h1,h2]=plotyy(x,y1,x,y2);
xlabel('Time, s')
set(get(ax(1),'Ylabel'),'String','Short Period Motion (alpha, ^o)')
set(get(ax(2),'Ylabel'),'String','Phugoid Motion (h, ft)')
set(h2,'LineStyle','.')
title('T-37 Longitudinal Dynamics')
```

produces a plot shown in Fig. 6.13. Note how tricky it is to access and change lines and axes properties (and that is the only way of doing this). Specifically, when issuing `plotyy` command we are getting the 2×1 han-dle to two axes `ax` and two scalar handles, `h1` and `h2`, to two lines. Then

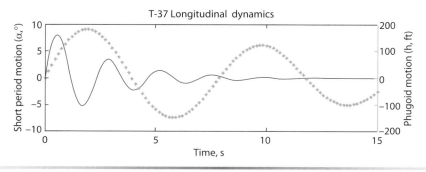

Fig. 6.13 Example of using the `plotyy` function.

we are using them to name both *y*-axes and change one of the lines line style (by default, they appear in blue and green, but using the same solid line style).

There is one more syntax of this function, allowing you to use other than the `plot` function to produce the graphic

```
plotyy(x1,y1,x2,y2,'name')
```

Here `name` is the name of any 2-D plotting function, which will be considered in the next section.

When having multiple lines within one figure, you would definitely want to add a legend. The `legend` function allows you to do this using the following syntax:

```
legend('string1','string2',...,Location)
```

In this call, `'string1'`, `'string2'`, ... are the labels for the data plotted, and `Location` places the legend in a specified location, namely,

    0   Stands for the automatic "best" placement which conflicts with data the least
    1   Stands for the upper right-hand corner (default)
    2   Stands for the upper left-hand corner
    3   Stands for the lower left-hand corner
    4   Stands for the lower right-hand corner
 −1   Stands for placing a legend to the right of the plot

These locations are presented in Fig. 6.14a graphically. Note that along with using the aforementioned numerical code to specify the legend position, the latest versions of MATLAB utilize an explicit attribute-value pair

```
legend('string1','string2',...,'Position','Specific Location')
```

Figure 6.14b names and shows these possible locations graphically. Two more options, `'Best'` and `'BestOutside'` are also available. The `Specific Location` string is not case-sensitive and can be abbreviated by sentinel letter (for example, S, se, and NwO). As usual, you can change any other property of the legend box, like background, edge color, orientation, font, etc.

MATLAB also allows you to divide the Figure window into an array of rectangular panes and use each one (of rectangular group of them) to produce a separate axes. The `subplot` function, which makes it possible, has the general syntax

```
subplot(m,n,p) or subplot(mnp) or subplot mnp
```

where m is the number of panes rows, n the number of columns of panes, and p the pane number counting sequentially across a top row, then to the next row, and so on.

The subplot function allows you to produce asymmetrical arrangements of subplots, by combining two and more rather than just one pane together. For example, the command

```
subplot(2,2,[2 4])
```

combines the second and fourth panes together, that is, allows you to use the right column of pans together. You can also specify the location of axes explicitly

```
subplot('Position',[left bottom width height])
```

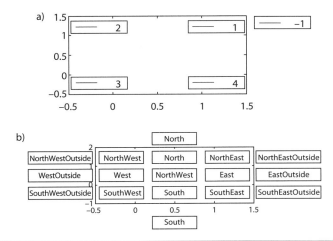

Fig. 6.14 Legend locations.

Here, a four-element vector provides a location (*x*- and *y*-coordinates) of the left-bottom corner, width and height of the axes object). As usual, by default these values are given in normalized coordinates in the range from 0 to 1. The following example explains the afore-mentioned options all together and produces a Figure window with the multiple panes as shown in Fig. 6.15a:

```
>> subplot(221), ezplot('sin')
>> subplot(2,2,[2,4]), ezplot('tan')
>> subplot('Position',[0.15,0.25,0.3,0.2]), ezplot('log')
```

By the way, now we can betray a small secret—in all preceding figures of this chapter, except Fig. 6.11, we used `subplot(221)` command to halve the height of the plot.

Be aware that when you use an explicit "position–four-element vector" pair and the location of a new subplot interferes with what you have plotted already, your new subplot eliminates all underlying subplots. For instance, if in the above case you use `[0.35,0.45,0.3,0.2]`, that is, move your last plot to an upper-right corner, it will eliminate both other plots. If you want to produce the overlapping plot you should rely on axes command like in the following example (Fig. 6.15b):

```
>> axes('Position',[0.1 0.6 0.4 0.3]), ezplot('tanh')
>> axes('Position',[0.3 0.35 0.4 0.3]), ezplot('cos')
>> axes('Position',[0.5 0.1 0.4 0.3]), ezplot('sqrt')
```

At last, if you have too much data to fit it on a single figure, you can dedicate several figures to display them. The `figure` function creates a new Figure graphics object. The simplest syntax,

```
>> figure
```

creates a new Figure using the default property values (the window for each Figure will be automatically numbered and by default titled as **Figure**

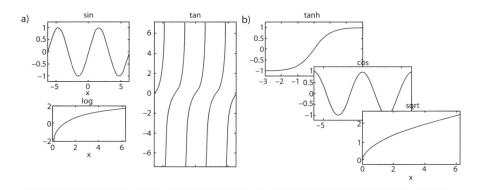

Fig. 6.15 Examples of using a) `subplot` and b) `axes` functions.

No.X, where **X** is the Figure's number). You may also specify the Figure number explicitly, like

```
figure(num)
```

(where num has to be an integer), which creates a new Figure window titled **Figure No.**num, or makes Figure num the current Figure (if it exists already), so that the subsequent plots are drawn in the current Figure. By the way, the num designator happens to be a handle of the Figure, so that the following set of commands is legitimate:

```
>> h=4
>> figure(h)
>> get(h)
```

Of course, you can change any property of the Figure by addressing it explicitly

```
figure('PropertyName',PropertyValue,...)
```

All other properties, not defined explicitly, maintain their default values. For example, the command

```
>> figure('NumberTitle','off','Name','New Window','Color','w')
```

suppresses numbering a Figure window, creating its name instead, and forces white background (as opposed to the default, grey) (Fig. 6.16). This latter feature maybe especially useful if you want to print your plots afterwards (it helps saving toner).

Recall, we tried to enforce a white background already by changing Figure Copy Options preferences (Fig. 2.23a). However, there is a big difference between these two approaches. If you want to copy your MATLAB figure (not Figure window) and paste it into a Microsoft Word document or Microsoft Power Point slide, you will need to choose the **Edit** drop-down menu from the Figure Toolbar and click **Copy Figure**. This copies the content of the figure with its background. So if you do not force it white, it will appear grey. That is where we can choose to change Figure Copy Options preferences (just once). Nevertheless, even if you changed this option as explained in Fig. 2.23a and try to publish your M-script that has the figure commands in it, you will end up having grey background. The above alternative of changing a background color while creating a figure, figure('Color','w'), relieves you from worrying about it once and forever.

To conclude this section, let us consider some more functions, you may find useful.

The close function deletes the current figure, or the specified figure (close(h)), or all figures (close all). The clf function (abbreviated from clear current figure) clear the current Figure window (deletes all

Fig. 6.16 Example of manipulating Figure's properties.

graphics objects) and also resets all of its properties such as `hold`, `axis`, etc. The `cla` function (clear current axes) deletes all the plots and text from the current axes (without resetting `axes`, properties).

The last function to be introduced in this section is

```
H=findobj('PropertyName',PropertyValue,...)
```

It returns the handles of all graphics objects having the property `PropertyName` set to the value `PropertyValue`. You can specify more than one property/value pair, in which case, `findobj` returns only those objects having all specified values. For example, the following set of commands:

```
>> x=linspace(-pi,pi,50);
>> plot(x,sin(x),x,cos(x),'rp',x,sin(x).*cos(x),'m+:')
```

produces three curves shown in Fig. 6.17a. Now, let us find and delete one of them

```
>> h=findobj('Marker','p');
>> delete(h)
```

The result is shown in Fig. 6.17b.

Using the logical operator, you can limit the search to the objects with the particular property values and along the specific branches of the hierarchy.

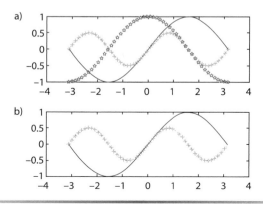

Fig. 6.17 Using the `findobj` function to find a Line with a certain property.

6.7 Special Types of 2-D Plots

By now, we covered all fundamentals of 2-D plotting, but we have considered just five basic plotting functions: `plot, line, fplot, ezplot,` and `plotyy`. MATLAB has much more to offer. This section lists most of available 2-D plotting function of MATLAB.

Figures 6.18–6.22 contain some illustrative examples of using these functions along with the corresponding sets of commands. The newly introduced functions appear in the boldface. We also use some more code around them to show you how to modify some of line properties. To learn about available syntaxes for any specific function, you should refer to the MATLAB Help system.

Figure 6.18 introduces some functions you can use to produce the plots with logarithmic base-10 or polar coordinates

`loglog(x,y)`	Plots data as logarithmic scales for both axes
`semilogx(x,y)`	Plots data with logarithmic *x*-axis and linear *y*-axis
`semilogy(x,y)`	Plots data with linear *x*-axis and logarithmic *y*-axis
`polar(theta,r)`	Generates a plot in polar coordinates (*θ,r*), where *θ* is the angular coordinate and *r* the radial coordinate
`compass(x,y)`	Plots arrows emanating from the origin
`rose(theta,r)`	Creates an angle histogram, which is a polar plot showing the distribution of values grouped according to their numeric range

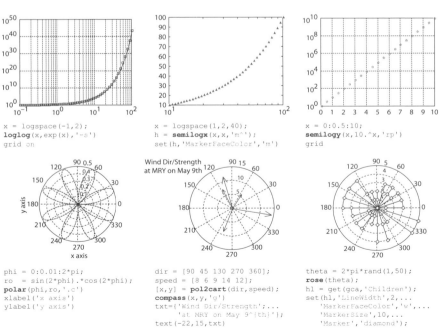

Fig. 6.18 Examples of logarithmic and polar plots.

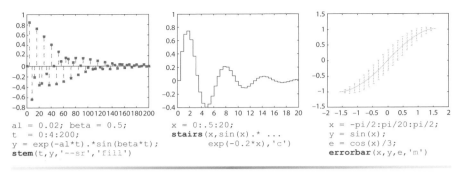

Fig. 6.19 Examples of discrete data and errorbar plots.

Note that as of R2010a, the polar plots are not very convenient to work with compared with say the `plot` function because the capabilities to access and change the polar plots attributes are somewhat limited (for example, you cannot change the tick marks labels or label axes properly).

Figure 6.19 provides examples of the discrete data and errorbar plots

`stem(x,y)`	Generates a stem plot of y vs x
`stairs(x,y)`	Generates a stairs plot of y vs x
`errorbar(x,y,e)`	Plots y vs x with the symmetric (±) errorbars `2*e(i)` long (arrays x, y, and e must be the same size)

Figure 6.20 presents a variety of contour plots

`contour(z)`	Displays isolines of matrix z (you may label the contour lines using `clabel` function)
`contourf(z)`	Displays isolines calculated from matrix z and fills the areas between the isolines using the constant colors (the color of the filled areas depends on the current Figure's colormap)

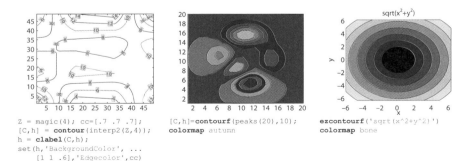

Fig. 6.20 Examples of contour plots.

ezcontour(f) and ezcontourf(f)	Plots the contour lines of $f(x, y)$, where f is a string that represents a mathematical function of two variables, such as x and y

The colormap function used in several examples changes a color matrix of the Figure's Colormap property (Table 6.3) and will be discussed in detail in Sec. 6.9.

Figure 6.21 illustrates MATLAB capabilities to produce the area plots, histograms, and direction-type plots

area(x)	Plots the values in each column of matrix x as a separate curve and fills the area between the curve and the x-axis
pie(x)	Displays the percentage that each element in a vector or matrix x contributes to the sum of all elements (when the sum of elements in the first input argument is equal to or greater than 1, pie normalizes the values, while the sum of the elements in the first input argument is less than 1, pie draws a partial pie)

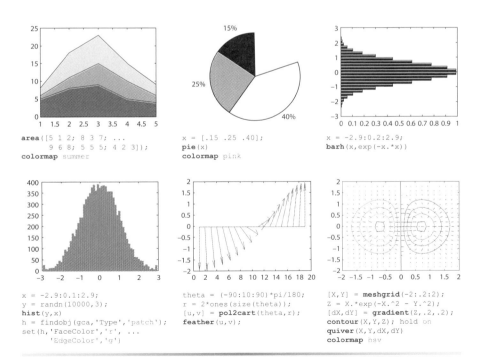

Fig. 6.21 Examples of area plots, histograms, and direction-type plots.

`bar(x)`	Displays the values in a vector or matrix `x` as the horizontal or vertical bars; `bar(...,width)` sets the relative bar width (0.8 is the default value) and controls the separation of bars within a group; `bar(...,'style')` specifies the style of the bars (`'style'` can be either `'grouped'`, which is a default mode, or `'stacked'`)
`barh(...)`	Creates the horizontal bars
`hist(y,x)`	Returns the distribution of `y` among the `length(x)` bins with the centers specified by `x` (where `x` is a vector)
`feather(u,v)`	Displays the vectors specified by `u` and `v`, containing the *x* and *y* components as the relative coordinates
`quiver(x,y,u,v)`	Displays the velocity vectors with components (`u`,`v`) as arrows originating at the points (`x`,`y`)

Figure 6.22 illustrates examples of the polygons and scatter plots

`fill(x,y,c)`	Creates filled polygons from data in `x` and `y` with a vertex color specified by `c` (`c` is a vector or matrix used as an index into the `colormap` function)
`triplot(tri,x,y,c)`	Creates a 2-D triangular plot, that is, the triangles defined in the *n*-by-3 matrix `tri` (a row of `tri` created with the `tri=delaunay(x,y)` command contains indices into the vectors `x` and `y` that define a single triangle), with a color specified by `c`
`scatter(x,y,s,c)`	Displays the colored circles at the locations specified by the same-size vectors `x` and `y` (`s` determines the area of each marker, `c` determines the colors of each marker)

To conclude this section, let us present one more example showing how you can combine different types of plots "within the same axes". The following set of commands produces Fig. 6.23:

```
generation=1:5;
number=[4.2 33.6 42.7 9.5 1];
price=[0.1 0.2 1 30 150];
bar(generation,number,'y')
xlabel('Generation')
ylabel('Number manufactored, thousands')
h1=gca;
```

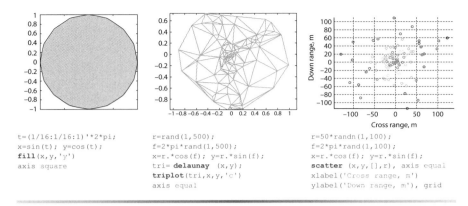

```
t=(1/16:1/16:1)'*2*pi;      r=rand(1,500);              r=50*randn(1,100);
x=sin(t); y=cos(t);         f=2*pi*rand(1,500);         f=2*pi*rand(1,100);
fill(x,y,'y')               x=r.*cos(f); y=r.*sin(f);   x=r.*cos(f); y=r.*sin(f);
axis square                 tri= delaunay (x,y);        scatter (x,y,[],r), axis equal
                            triplot(tri,x,y,'c')        xlabel('Cross range, m')
                            axis equal                  ylabel('Down range, m'), grid
```

Fig. 6.22 Examples of polygon and scatter plots.

```
% Setting the new axes atop the first ones
h2=axes('Position',get(h1,'Position'));
% Adding the second plot to the new axes
semilogy(generation,price,'LineWidth',3)
% Modifying second axes settings
set(h2,'Color','none','YAxisLocation','right','XLim',get(h1,'XLim'),...
       'XTickLabel',[],'TickLength',[0 0])
text(3.5,10,'Price','Rotation',58)
ylabel('Price, M$'), title('Jet Fighters')
```

It shows how to handle graphics to overlay data on a bar graph by creating another axes in the same position. It enables you to have an independent *y*-axis for the overlaid dataset, in contrast to the `hold`

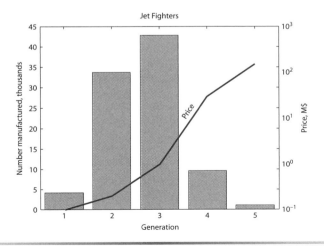

Fig. 6.23 Example of blending two different types of plots.

statement, which uses the same axes. To be more specific, using the handle to the second set of axes the preceding script modifies some of its properties. Most importantly, it makes it transparent and defines the location for the y-axis. Other modifications include setting the x-axis range to that of the first set of axes, eliminating x-axis labeling (because we already have one) and axes ticks (otherwise the log-scale ticks for the second set of axes would be superimposed on those of the first set).

6.8 3-D Plots

The basic MATLAB function enabling plotting the 3-D plots is plot3(x,y,z). It displays a 3-D plot of a set of data points defined by three coordinates, x, y, and z, stored in the same-size vectors x, y, and z. All we were talking about 2-D plots applies here as well (except the third coordinate now needs to be added when applicable). To place a label for the z-axis, the zlabel function should be used in the very same way as we did it for the 2-D plots. For instance, to plot a 3-D helix you may issue the following commands:

```
t=0:pi/50:10*pi;
plot3(sin(t),cos(t),t)
grid on, axis square
xlabel('x axis'), ylabel('y axis'), zlabel('z axis')
```

The result is shown in Fig. 6.24a. Another feature is the position the viewer looks at the 3-D plot. In MATLAB, the view point can be specified in terms of a point in 3-D space, or more often in terms of azimuth and elevation as shown in Fig. 6.25.

The view([az,el]) or simply view(az,el) command sets the viewing angle for a 3-D plot via azimuth and elevation. As shown in Fig. 6.25,

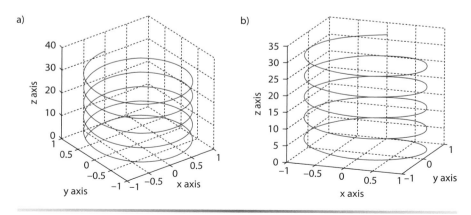

Fig. 6.24 Helix with the a) default and b) explicitly defined view setting.

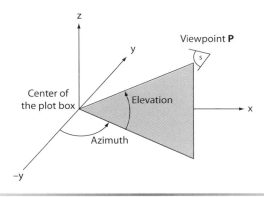

Fig. 6.25 Viewpoint definition in MATLAB.

the azimuth, `az`, is the horizontal rotation about the z-axis as measured in degrees from the negative y-axis (positive values indicate counterclockwise rotation of the viewpoint). The elevation, `el`, is the vertical elevation of the viewpoint in degrees (positive values of elevation correspond to moving above the object). By default, MATLAB uses `view(-37.5,30)`. For comparison, Fig. 6.24b shows another view of the helix depicted in Fig. 6.24a using `view(20,15)`.

The `view([x,y,z])` command sets the viewpoint along the ray originating at the origin of the Cartesian system of coordinates and point **P**(x,y,z) so that all what matters is the ratio between its coordinates. Therefore, this command is equivalent to that of `view(az,el)` with

```
el=180*atan2(z,sqrt(x^2+y^2))/pi; az=180*atan2(x,-y)/pi;
```

You can always return the view point to the default values by issuing the `view(3)` command. It sets `az=-37.5` and `el=30`. The `view(2)` command provides a bird-eye (2-D) view with `az=0` and `el=90`. (There is no `view(1)` call exists.)

The `plot(x,y,z)` function accepts not only the same-size vectors but also the same-size matrix arguments. In this case, instead of plotting a 3-D curve (spatial trajectory), it produces a surface with x and y matrices representing a rectangular mesh grid and matrix z elevation (function values) at the nodes of this grid. Hence, plotting the surface is a three-step process. First, you need to produce a grid. Next, you compute the function values at the nodes of this grid, $z = f(x, y)$. Finally, you pass three matrices to the plotting function to plot the surface.

The proper way to create a rectangular grid is to use the `meshgrid` function. Suppose the length of the vectors x and y is n and m, respectively. Then, the command

```
[X,Y]=meshgrid(x,y)
```

creates two $n \times m$ matrices X and Y ([X,Y]=meshgrid(x) is same as [X,Y]=meshgrid(x,x)). For example,

```
>> [X,Y]=meshgrid(1:10,1:3)
```

yields

X =	1	2	3	4	5	6	7	8	9	10
	1	2	3	4	5	6	7	8	9	10
	1	2	3	4	5	6	7	8	9	10
Y =	1	1	1	1	1	1	1	1	1	1
	2	2	2	2	2	2	2	2	2	2
	3	3	3	3	3	3	3	3	3	3

This function simply concatenates m rows of the vector x vertically for X, and n columns of the vector y horizontally for Y. Hence, the x-coordinate changes in the matrix X across the columns, whereas the y-coordinate changes in the matrix Y across the rows.

The next step is to compute the z values on this grid, for instance,

```
>> Z=sin(0.1*Y).*cos(0.1*X);
```

Finally, you may use the plot3 function

```
>> plot3(X,Y,Z)
```

which produces a surface shown in Fig. 6.26a. As seen, it simply plots n columns of matrix Z, shifting each new line along the x-axis. If you use another grid, say [X,Y]=meshgrid(1:40) (and repeat the following two commands), the result changes to that of Fig. 6.26b.

You may argue that it does not look like a surface. Well, MATLAB provides you with a variety of other functions producing a better result, but their syntax is essentially the same as that of the plot3 function. These functions are mesh, surf, and their derivatives

mesh(x,y,z)   Creates a 3-D mesh-surface plot
meshc(x,y,z)   Creates a 3-D mesh-surface plot with a contour plot
               underneath the surface

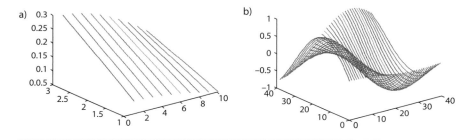

Fig. 6.26 Example of using the plot3 function to create a surface.

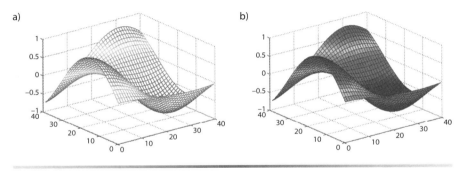

Fig. 6.27 Example of using a) `mesh(X,Y,Z)` and b) `surf(X,Y,Z)` functions to create a surface.

`meshz(x,y,z)`   Creates a 3-D mesh-surface plot with a series of vertical reference lines under the surface

`surf(x,y,z)`   Creates a shaded 3-D mesh-surface plot

`surfc(x,y,z)`   Creates a shaded 3-D mesh-surface plot with a contour plot underneath it

For example, using these two basic functions instead of `plot3` in the latter example (you still need to create a mesh and compute the *z* values) produces the surfaces shown in Fig. 6.27.

Figure 6.28 provides with some more examples resulted from the following set of commands (`peak` is the MATLAB function of two variables):

```
[x,y]=meshgrid(-3:0.125:3); z=peaks(x,y);
subplot(231), meshc(x,y,z)
axis([-3 3 -3 3 -10 5]); v=axis; title('meshc')
subplot(232), meshz(x,y,z), axis(v), title('meshz')
subplot(233), surfc(x,y,z), axis(v), title('surfc')
```

Note, how we captured axes settings for the first plot and used them for the remaining three.

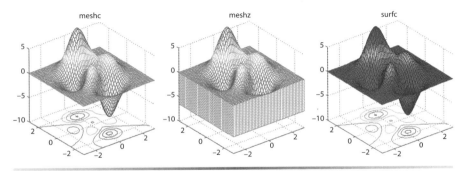

Fig. 6.28 Illustration of a variety of the surface plotting functions.

MATLAB also allows using the triangular rather than the rectangular grid. To create such a grid, the *Delaunay triangulation* connecting each point to its natural neighbors is used. Specifically, the function `delaunay(x,y)` returns a matrix, each row of which represents a triad of indices of the neighboring points, described by the same-size vectors x and y (the `delaunay` function was also used in the 2-D case of Fig. 6.22b). For example, the following code sets six arbitrary-chosen points as presented in Fig. 6.29a and returns the so-called face matrix for them:

```
>> N=6; x=N*rand(1,N); y=N*rand(1,N);
>> plot(x,y,'pr'), axis([0 N 0 N]), hold
>> text(x+0.1,y,cellstr(num2str([1:N]')),'FontWeight','bold')
>> xlabel('x axis'), ylabel('y axis')
>> tri = delaunay(x,y), K=length(tri);
tri =  5   4   1
       2   4   5
       5   6   2
       3   5   1
       6   5   3
```

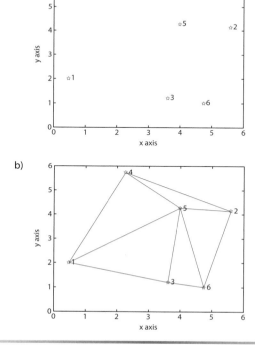

Fig. 6.29 Illustration of the Delaunay triangulation on six points.

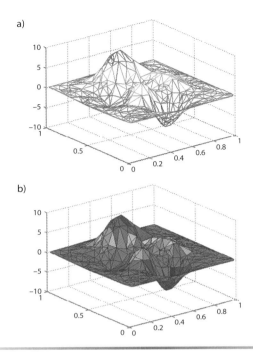

Fig. 6.30 a) Triangular mesh and b) surface plots.

This matrix specifies which points form the Delaunay triangles (row by row). A little bit more programming efforts:

```
>> for i=1:K
>> plot([x(tri(i,:)) x(tri(i,1))],[y(tri(i,:)) y(tri(i,1))])
>> end
```

allow you to visualize these triangles as presented in Fig. 6.29b.

The `trimesh(tri,x,y,z)` function uses these triangles as a mesh and produces a triangular-face surface. Similar to the previous example, the following set of commands first creates the data points (also called vertex vectors) x, y, z, and a face matrix `tri`, followed by creating a triangular mesh plot as shown in Fig. 6.30a:

```
>> x = rand(1,500);        y = rand(1,500);
>> tri = delaunay(x,y);    z = peaks(6*x-3,6*y-3);
>> trimesh(tri,x,y,z)
```

Utilizing the `trisurf` function would produce the result shown in Fig. 6.30b.

Before we proceed with other 3-D plotting functions, let us mention two kinds of unusual applications of the 3-D mesh-surface plots. The first one, which actually applies to the 2-D plots as well, is when for some reason you want to cut away a portion of a surface, so that it does not show up on

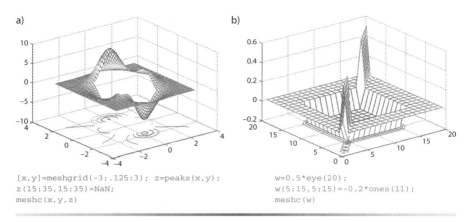

a)
```
[x,y]=meshgrid(-3:.125:3); z=peaks(x,y);
z(15:35,15:35)=NaN;
meshc(x,y,z)
```

b)
```
w=0.5*eye(20);
w(5:15,5:15)=-0.2*ones(11);
meshc(w)
```

Fig. 6.31 a) Examples of cropping the surface and b) using `meshc` to visualize a matrix.

the plot. It can be easily accomplished by assigning the NaN values to the *z* matrix at the corresponding grid nodes as shown in Fig. 6.31a.

The second unusual usage of the surface plots is when you use them to "visualize" matrices shown in Fig. 6.31b. A glimpse of such a plot may be much more effective compared to staring blankly at the bare numbers.

As in the 2-D case, MATLAB provides with a wide variety of different 3-D plotting functions as well. Most of them are listed later with examples shown in Figs. 6.32–6.35, 6.37–6.39 (again, the new graphic functions are marked out with the boldface, and we encourage you to use the MATLAB Help system to learn more about them if needed).

To start with, Fig. 6.32 presents the 3-D analogous of the 2-D `stem` function along with two more surface visualizing functions

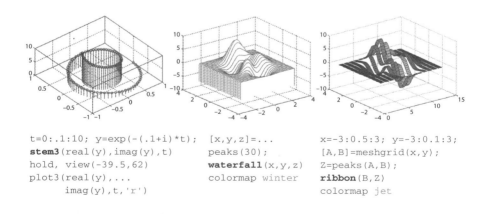

```
t=0:.1:10; y=exp(-(.1+i)*t);
stem3(real(y),imag(y),t)
hold, view(-39.5,62)
plot3(real(y),...
       imag(y),t,'r')
```

```
[x,y,z]=...
peaks(30);
waterfall(x,y,z)
colormap winter
```

```
x=-3:0.5:3; y=-3:0.1:3;
[A,B]=meshgrid(x,y);
Z=peaks(A,B);
ribbon(B,Z)
colormap jet
```

Fig. 6.32 3-D analogous of the `stem` function and special surface visualizing functions.

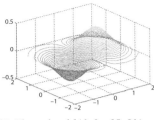

```
[X,Y]=meshgrid([-2:.25:2]);
Z=X.*exp(-X.^2-Y.^2);
contour3(X,Y,Z,50)
colormap spring
```

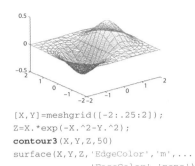

```
[X,Y]=meshgrid([-2:.25:2]);
Z=X.*exp(-X.^2-Y.^2);
contour3(X,Y,Z,50)
surface(X,Y,Z,'EdgeColor','m',...
                'FaceColor','none')
grid off, colormap gray
```

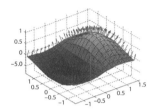

```
[X,Y]=meshgrid(-1.4:.2:1.4); Z=2*X.*exp(-X.^2-Y.^2);
[U,V,W] = surfnorm(X,Y,Z); quiver3(X,Y,Z,U,V,W,1)
hold, surf(X,Y,Z), colormap copper, axis equal
```

Fig. 6.33 Examples of the 3-D contour plots and surface with normals.

stem3(x,y)   Generates a stem plot of y vs x
waterfall(x,y,z) Draws a mesh similar to the meshz function, but
                      does not generate lines from the columns of
                      matrices (this produces a "waterfall" effect)
ribbon(x,y)   Plots x vs the columns of y as the 3-D strips (x
                      and y are the same-size vectors or matrices, or
                      alternatively, x can be a row or a column vector,
                      and y—a matrix with length(x) rows)

Figure 6.33 presents the 3-D contour plot function, an example of overlaying it with the surface net and provides an example of using the surfnorm function to compute surface normals, which are then visualized using the 3-D quiver plot

contour3(z)   Draws a contour plot of matrix z in a 3-D
                      view (z is interpreted as heights with
                      respect to the *x–y* plane). The number of
                      contour levels and the values of contour

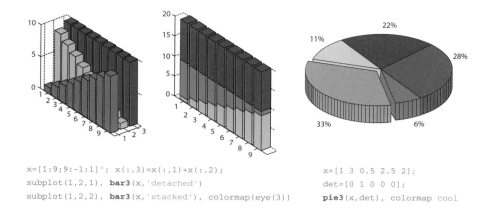

```
x=[1:9;9:-1:1]'; x(:,3)=x(:,1)+x(:,2);              x=[1 3 0.5 2.5 2];
subplot(1,2,1), bar3(x,'detached')                  det=[0 1 0 0 0];
subplot(1,2,2), bar3(x,'stacked'), colormap(eye(3)) pie3(x,det), colormap cool
```

Fig. 6.34 Examples of the 3-D bars and pie.

	levels are chosen automatically. The ranges of the *x*- and *y*-axes are `[1:n]` and `[1:m]`, where `[m,n]=size(z)`
`surfnorm(x,y,z)`	Computes and displays the unnormalized surface normals for the surface defined by the same-size matrices x, y, and z
`quiver3(x,y,z,u,v,w)`	Plots the vectors with components (*u, v, w*) originated at the points (*x, y, z*). Matrices x, y, z, u, v, and w must all be of the same size

Figure 6.34 shows examples of 3-D bar and pie functions

`bar3(x)` and `bar3h(x)`	Draws vertical and horizontal 3-D bar charts
`pie3(x,y)`	Draws a 3-D pie chart using the data in x with each element in x represented as a slice (optional same-size vector y containing zeros and ones defines whether this element is detached from the pie)

Figure 6.35 introduces different shapes available in MATLAB and shows how to manipulate with them (using handle graphics). Specifically,

`[x,y,z]=cylinder(r,n)`	Returns the *x*-, *y*-, and *z*-coordinates of a unit-height cylinder using (optional) r to define a profile curve and (optional) n to specify the number of equally spaced points around its circumference (20 equally spaced points by default)

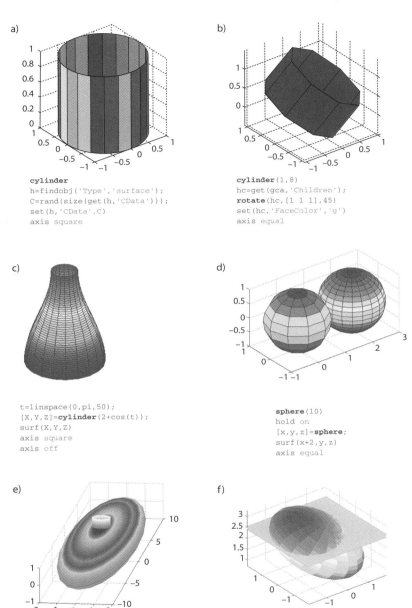

a)
```
cylinder
h=findobj('Type','surface');
C=rand(size(get(h,'CData')));
set(h,'CData',C)
axis square
```

b)
```
cylinder(1,8)
hc=get(gca,'Children');
rotate(hc,[1 1 1],45)
set(hc,'FaceColor','g')
axis equal
```

c)
```
t=linspace(0,pi,50);
[X,Y,Z]=cylinder(2+cos(t));
surf(X,Y,Z)
axis square
axis off
```

d)
```
sphere(10)
hold on
[x,y,z]=sphere;
surf(x+2,y,z)
axis equal
```

e)
```
ellipsoid(0,0,0,2,10,1)
hold, axis equal
[x,y,z]=cylinder(0.5,40);
surf(x,y,z+0.5)
shading interp
view([8,14])
colormap colorcube
```

f)
```
ellipsoid(0,0,0,2,2,1), hold
h=get(gca,'Children');
rotate(h,[3 3 3],34)
z=get(h,'Zdata'); shading flat
set(h,'Zdata',z+2), axis equal
hs=mesh([-2 2; -2 2],...
   [-2 -2; 2 2],2.5*ones(2));
set(hs,'FaceColor','c')
```

Fig. 6.35 Examples of visualizing the 3-D shapes.

`sphere`	Generates the x-, y-, and z-coordinates of a unit-radius sphere consisting of 20 by 20 faces for use with `surf` and `mesh` (`sphere(n)` generates a sphere consisting of n by n faces)
`ellipsoid(xc,yc,` `zc,xr,yr,zr)`	Graphs an ellipsoid with the center at (xc,yc,zc) and radii (xr, yr and zr) as a surface

Figure 6.35b, e, and f features two other functions, `rotate` and `shading`, that you can use. The `rotate` function (Fig. 6.35b and f) rotates a graphics object in 3-D space according to the right-hand rule (this graphics object must be a child of the same axes). Similarly to Fig. 6.25 the axis of rotation is defined by an origin and some point **P** (Fig. 6.36a). As in the case with the view function considered above, a point **P** can be defined via its Cartesian coordinates (x, y, z) `view` or spherical coordinates (ϑ, φ) (Fig. 6.36b).

The `shading` function controls the color shading of surface and patch graphics objects with three options available. They are the default `faceted` (flat shading with superimposed black mesh lines) as appears in Fig. 6.35d, `flat` (each mesh line segment and face has a constant color) as in Fig. 6.35f, and `interpolated` (the color varies across the line or face) as shown in Fig. 6.35e.

All surfaces in Fig. 6.35 are composed of the individual planes or faces. MATLAB offers two special functions allowing you to create and color individual polygons and patches

`fill3(x,y,z,c)`	Fills 3-D triangular polygons with the vertices defined by triplets (3 by m matrices) x, y, z, and c specifying their color
`patch(x,y,z,c)`	Creates a 3-D patch with the elements of x, y, and z specifying the vertices of a polygon and c specifying the color

Several examples are shown in Fig. 6.37.

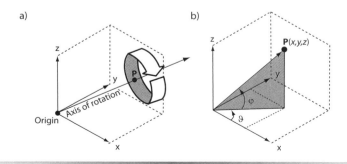

a)

b)

Fig. 6.36 a) The essence of the rotate function and b) defining the rotation axis.

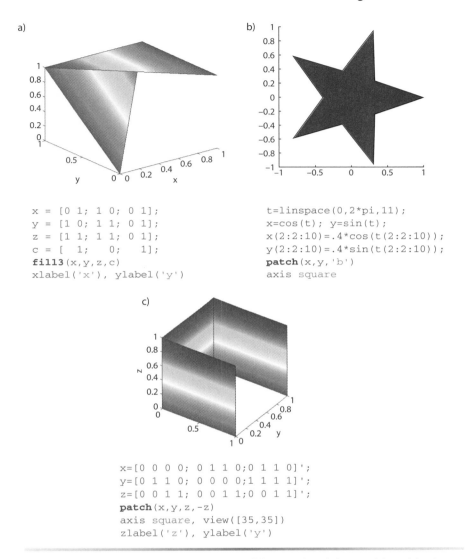

a)
```
x = [0 1; 1 0; 0 1];
y = [1 0; 1 1; 0 1];
z = [1 1; 1 1; 0 1];
c = [  1;    0;    1];
fill3(x,y,z,c)
xlabel('x'), ylabel('y')
```

b)
```
t=linspace(0,2*pi,11);
x=cos(t); y=sin(t);
x(2:2:10)=.4*cos(t(2:2:10));
y(2:2:10)=.4*sin(t(2:2:10));
patch(x,y,'b')
axis square
```

c)
```
x=[0 0 0 0; 0 1 1 0;0 1 1 0]';
y=[0 1 1 0; 0 0 0 0;1 1 1 1]';
z=[0 0 1 1; 0 0 1 1;0 0 1 1]';
patch(x,y,z,-z)
axis square, view([35,35])
zlabel('z'), ylabel('y')
```

Fig. 6.37 Examples of 3-D polygons filling and patch graphics objects.

The fill3 function in Fig. 3.37a produces two polygons defined by 3×2 matrices x, y, z—the slant one and the one parallel to the *Oxy* plane. It colors them defining the vertices with the hottest (1), and coldest (0) color. The matrix c may have the same size as other matrices, otherwise the color setup in a single 3 by 1 vector will be applied to all polygons. In our case, the second (horizontal) polygon, defined by three vertices (points), (1, 0, 1), (0, 1, 1), and (1, 1, 1), is colored so that the second vertex, (0, 1, 1), is the coldest one. As seen, the current (default) color map is scaled so that the color changes from being blue at the second vertex to red at two other vertices. To specify colors in Fig. 6.37c, the -z

a)   b)

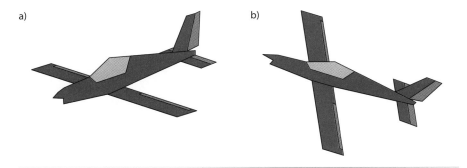

Fig. 6.38 Example of a) creating and b) working with a complex 3-D object using patch graphics.

matrix was used. This results in the hottest color for each patch appearing at the bottom.

Using patches, you can create any complex graphics objects for further use (for example, animation). For instance, the following script produces a 3-D model of an aircraft shown in Fig. 6.38a:

```
fx=[6.6 5.3 2.3 1.5 -2 -2.5 -10 -10 0 5.3 5.3]'; % fuselage
fy=zeros(11,1);
fz=[1.2 1.7 1.9 1.2 1.2 2.6 1.1 0.7 -0.2 0.3 0.8]';
    cx=[2.3 0 -2.5 -2 1.5]';                  % cockpit canopy
    cy=zeros(5,1);
    cz=[1.9 3.1 2.6 1.2 1.2]';
wingx=2.3*[1 -3 -3 -3+.6 -3+.6 -3 -3 -3+.6 -3+.6 -3 -3 1]'/4; % wing
wingy=[8.5 8.5 7.8 7.8 2.2 2.2 -2.2 -2.2 -7.8 -7.8 -8.5 -8.5]';
wingz=zeros(12,1);
    ax=2.3*[-3+.6 -3 -3 -3+.6]'/4;      % right aileron
    ay=[7.8 7.8 2.2 2.2]';
    az=zeros(4,1);
hsx=1.3*[0 -1 -1 0]'-8.1;               % horizontal stabilizer
hsy=5.1*[1 1 -1 -1]'/2;
hsz=0.7*ones(4,1);
    ex=0.6*[0 -1 -1 0]'-9.4;            % elevator
    ey=5.1*[1 1 -1 -1]'/2;
    ez=0.7*ones(4,1);
vsx=[0 -1.7 -2.7 -1.8]'-7.3;            % vertical stabilizer
vsy=zeros(4,1);
vsz=[1.65 4.5 4.5 1.25]';
    rx=[-2.7 -3.3 -3.1 -1.8]'-7.3;      % rudder
    ry=zeros(4,1);
    rz=[4.5 4.5 1.5 1.25]';
pf=patch(fx,fy,fz,'c'); pw=patch(wingx,wingy,wingz,'c');
pc=patch(cx,cy,cz,'b'); ps=patch([hsx vsx],[hsy vsy],[hsz vsz],'c');
pu=patch([ax ax ex rx],[ay -ay ey ry],[az az ez rz],'m');
axis equal, axis off, view(135,20)
```

You can gather individual patches together and have a single handle to the whole aircraft that can be used, for example, to rotate it (Fig. 6.38b)

```
haircraft=[pf pc pw ps pu];
rotate(haircraft,[25,35],-50)
```

The command `g=surf2patch(X,Y,Z)` allows you to convert the geometry (and color data) from the surface object defined by matrices `X`, `Y`, and `Z` (or by the handle) into the patch format and returns the face, vertex, and color data in the structure `g`. You can then pass this structure directly to the patch function as `patch(g)`.

Two more 3-D functions, presented in Fig. 6.39, enable creating the scatter plots (similar to their 2-D analogues shown in Fig. 6.22c) and visualizing volumetric data, slice by-slice

`scatter3(x,y,s,c)`   Displays colored circles at the locations specified by the same-size vectors `x`, `y`, and `z`, with `c` determining the colors of each marker

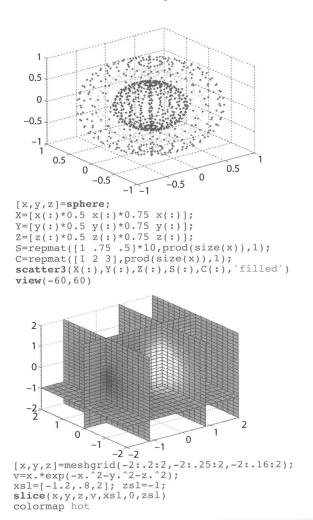

```
[x,y,z]=sphere;
X=[x(:)*0.5 x(:)*0.75 x(:)];
Y=[y(:)*0.5 y(:)*0.75 y(:)];
Z=[z(:)*0.5 z(:)*0.75 z(:)];
S=repmat([1 .75 .5]*10,prod(size(x)),1);
C=repmat([1 2 3],prod(size(x)),1);
scatter3(X(:),Y(:),Z(:),S(:),C(:),'filled')
view(-60,60)
```

```
[x,y,z]=meshgrid(-2:.2:2,-2:.25:2,-2:.16:2);
v=x.*exp(-x.^2-y.^2-z.^2);
xsl=[-1.2,.8,2]; zsl=-1;
slice(x,y,z,v,xsl,0,zsl)
colormap hot
```

Fig. 6.39 Examples of 3-D scatter plot and volume visualization.

slice(x,y,c)   Displays orthogonal slice planes through
                      volumetric data

For more details, especially on more than 20 other volume visualization functions, please refer to the MATLAB Help system.

6.9 Changing Color Palette

Table 6.1 presented eight preset colors you may use for lines, background, and text strings. For these applications, it is probably more than enough. However, when we came to visualizing the surfaces and shapes we used many more colors. If you click on any color changing menu, for instance in MATLAB Colors Preferences window, shown in Fig. 2.21b, you will see that in fact there are millions and millions of colors to choose from (256^3 to be more precise). Each color is defined by three numbers: either the RGB or hue-saturation-value (HSV) triplet. Sometime, the latter is also referred to as hue-saturation-brightness model.

Figure 6.40a shows the concept of the HSV model and Fig. 6.40b presents what options you have when you fix one of the HSV values and try to vary another two. In this model the first number, hue, ranges from 0 to 360 deg with each degree representing a distinct color starting from red (0 deg), up to the violet color (360 deg) (yellow is 120 deg, green −180 deg, and blue − 240 deg). The second number, saturation, represents the amount of color or, more exactly, its percentage, with its value ranging from 0 to 100, where 0 represents no color, whereas 100 represents the full color. Finally, the third number, value or brightness, allows you to enhance the color brightness by adding the white color, or reduce it by adding the black color. In this case, 0 represents the white color and 100 represents the black color.

The computer, however, "understands" RGB colors rather than HSV colors. Each pixel in your screen can be thought of as having three little

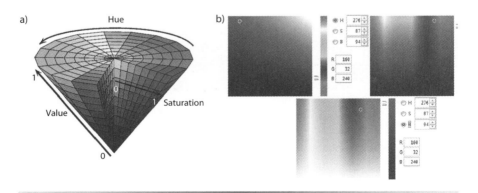

Fig. 6.40   HSV color model.

Table 6.6 RGB Values of Preset Colors

Short Name	Full Name	RGB Vector
'k'	'black'	[0 0 0]
'b'	'blue'	[0 0 1]
'c'	'cyan'	[0 1 1]
'g'	'green'	[0 1 0]
'm'	'magenta'	[1 0 1]
'r'	'red'	[1 0 0]
'w'	'white'	[1 1 1]
'y'	'yellow'	[1 1 0]

bulbs, which have RGB shades you can vary between 0 (bulb is off) and 255 (bulb is on). For more convenience, these numbers are scaled down to [0,1] range, so that the eight basic colors introduced in Table 6.1 have their RGB representation shown in Table 6.6.

Wherever you used the short names of these colors, they could be substituted with their full names or the corresponding RGB vector. Hence, to represent any other 16 million plus color we may now use an RGB triplet. In fact, we did it once already—see code corresponding to Fig. 6.20a.

MATLAB offers about a dozen functions for managing the colors, that is, performing color operations. Some of them are

`colorbar`	Turns on a color bar by the plot showing the color scale (Fig. 3.17)
`colormap`	Allows you to set and get the current color map
`colormapeditor`	Starts the color map editor allowing you to modify the existing and create new color maps
`hsv2rgb`	Converts an HSV color map to an RGB color map
`rgb2hsv`	Converts an RGB color map to an HSV color map
`rgbplot`	Plots a color map
`shading`	Sets color shading properties
`whitebg`	Changes the axes background color

Now, we will discuss the color maps used throughout this section. Let us use the `rgbplot` function from the above list and visualize a few of them (Fig. 6.41)

```
subplot 231, rgbplot(autumn),    axis tight, text(30,0.15,'autumn')
subplot 232, rgbplot(summer),    axis tight, text(30,0.15,'summer')
subplot 233, rgbplot(winter),    axis tight, text(30,0.15,'winter')
subplot 234, rgbplot(bone),      axis tight, text(30,0.15,'bone')
subplot 235, rgbplot(copper),    axis tight, text(30,0.15,'copper')
subplot 236, rgbplot(hsv), axis tight, text(30,0.15,'hsv')
```

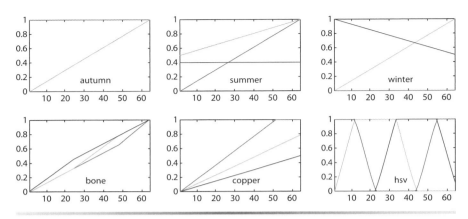

Fig. 6.41 Examples of preset color maps.

It turns out that these color maps are simply *m*-by-3 matrices of real numbers between 0.0 and 1.0, so that each row represents an RGB vector defining one color. All together, they specify some particular color palette. These color matrices show how RGB colors change their "weight" across a color scale. Hence, the `colormap` function simply allows you to change a color palette. The preset palettes (color maps) of MATLAB, most of which we have used in this and the previous sections already, are

`autumn`	Varies smoothly from red, through orange, to yellow (Fig. 6.20)
`bone`	Presents a grayscale color palette with a higher value for the blue component and is especially useful for adding an "electronic" look to grayscale images (Fig. 6.20)
`colorcube`	Contains as many regularly spaced colors in RGB color space as possible, while attempting to provide more steps of gray, pure red, pure green, and pure blue (Fig. 6.35)
`cool`	Consists of the colors that vary smoothly from cyan to magenta (Fig. 6.34)
`copper`	Varies smoothly from black to bright copper (Fig. 6.33)
`flag`	Consists of the colors red, white, blue, and black (this color map completely changes color with each index increment)
`gray`	Returns a linear grayscale color map (Fig. 6.33)
`hot`	Varies smoothly from black through shades of red, orange, and yellow, to white (Fig. 6.39)
`hsv`	Varies the hue component of the HSV color model (Fig. 6.41), so that this color map is particularly appropriate for displaying periodic functions (Fig. 6.21)

jet	Ranges from blue to red, and passes through the colors cyan, yellow, and orange (this color map is variation of the hsv map and was specifically designed for astrophysical fluid jet simulation) (Fig. 6.32)
pink	Contains pastel shades of pink and provides sepia tone colorization of grayscale photographs (Fig. 6.21)
prism	Repeats the six colors: red, orange, yellow, green, blue, and violet
spring	Consists of colors that are shades of magenta and yellow (Fig. 6.33)
summer	Consists of colors that are shades of green and yellow (Fig. 6.21)
white	Presents all white monochrome color map
winter	Consists of colors that are shades of blue and green (Fig. 6.32)

For all preset color maps, $m = 64$ (that is, there are 64 gradations). However, you can easily change it. For instance, colormap(gray(200)) creates a linear grayscale color map with 200 colors. Of course if needed, you can use the colormapeditor function to create a brand new colormap.

6.10 Easy-to-Use Function Plots

MATLAB provides with a nice tool allowing you to display a function without a burden of computing the arrays of points. It is a set of the so-called easy-to-use plotting functions. Two of them, fplot and ezplot, were introduced in the beginning of this chapter already. The rest of them have approximately the same syntax, that is,

ezfunction(expression, limits)

with limits being an optional parameter. If it is omitted, then the plotting function defines the range for plotting a graph automatically. These easy plotting functions are

ezcontour	Easy-to-use contour plotter
ezcontourf	Easy-to-use filled contour plotter
ezmesh	Easy-to-use 3-D mesh plotter
ezmeshc	Easy-to-use combination mesh/contour plotter
ezplot	Easy-to-use function plotter
ezplot3	Easy-to-use 3-D parametric curve plotter
ezpolar	Easy-to-use polar coordinate plotter
ezsurf	Easy-to-use 3-D colored surface plotter
ezsurfc	Easy-to-use combination surface/contour plotter

Some examples are shown in Fig. 6.42.

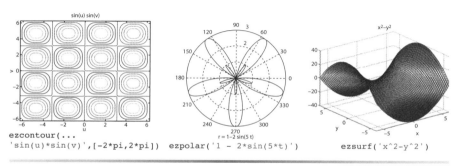

Fig. 6.42 Examples of plots produced with the easy-to-use functions.

6.11 Plot Editing

The Figure window in MATLAB allows you to do a lot of things with the plot that has already been created by running code. It includes editing, saving, copying, etc. As shown in Fig. 6.43, the Figure window has the following pretty much standard menus:

File Assists in opening, saving, exporting, and printing the figure. It also allows you to import data, save workspace and figure setup, and even generate a separate all-sufficient M-script for the current Figure window

Edit Helps cutting, copying, and pasting items such as legend or text titles, setting figure and axes properties, managing other MATLAB windows

View Allows you to select between **Figure toolbar** and/or **Camera toolbar** and/or **Plot Edit toolbar**. It also allows you to bring three additional panels to edit the figure

Insert Inserts labels, legends, titles, text, etc., rather than using the relevant commands in the Command window

Tools Operates on graphs with equipped sets of tools. These tools include calling the **Property Editor** (to be discussed next); zooming, panning, and 3-D rotating; aligning and distributing; data fitting (!) and data statistics (!) allowing the detailed analysis of a selected curve (Fig. 6.44, which will also be referred at Sec. 11.6)

Debug Allows you to debug the script the figure was created in

Fig. 6.43 Standard menus and icons of the Figure window.

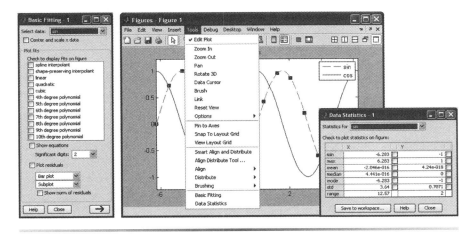

Fig. 6.44 Data fitting and data statistics menus available for each line.

Desktop	Allows you to dock the Figure window with the MATLAB Desktop window
Window	Allows you to arrange multiple figure windows on the screen and also switch between the opened MATLAB windows (Command Window, Command History, Current Directory, Workspace, etc.) and opened Figure window
Help	Provides assistance on different plotting-related issues

As it is the case for other MATLAB windows, three icons to the right of the menus bar allow you to dock/undock the current Figure window with/from the MATLAB Desktop and close it. (Sometimes, it is more convenient to have your Figure window docked to the MATLAB desktop so that you do not have to search for it among multiple other windows that might be opened too.)

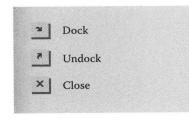

Specifically, the **Data Cursor** icon happens to be a very useful tool. It enables reading data directly from a graph by displaying the values of points you select on plotted lines, surfaces, images, and so on. When data cursor mode is enabled, you can

1. Click on any graphics object defined by data values and display the x, y, and z (if 3-D) values of the nearest data point.
2. Interpolate the values of points between the data points.
3. Display multiple data tips on the graphs.
4. Display the data values in a cursor window that you can locate anywhere in the Figure window or as a data tip (small text box) located next to the data point.

5. Export data values as workspace variables.
6. Print or export the graph with data tip or cursor window displayed for annotation purposes.

Figure 6.45 represents an example of using the data tips and also shows the content of the popping-up window that opens on the right click (with enabled **Data Cursor** icon).

As mentioned, MATLAB allows you to edit the figure without going back and changing the code, that is, after the Figure window was created. In the plot-editing mode, you can use a convenient GUI, called the **Property Editor**, to edit properties of objects in the graph. The **Property Editor** provides access to many properties of the `root`, `figure`, `axes`, `line`, `light`, `patch`, `image`, `surface`, `rectangle`, and `text` objects. For example, using the **Property Editor**, you can change the thickness of a line, add titles and axes labels, add lights, and perform many other plot-editing tasks.

You start the **Property Editor** by clicking the **Edit Plot** icon (or choosing the **Edit Plot** option from the **Tools** menu) and then by double-clicking an object in a graph, such as a line or axis. The **Property Editor** for this

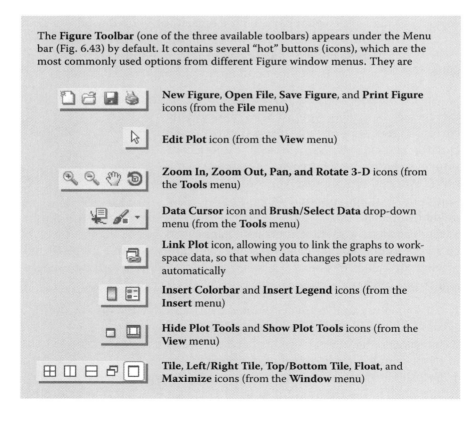

The **Figure Toolbar** (one of the three available toolbars) appears under the Menu bar (Fig. 6.43) by default. It contains several "hot" buttons (icons), which are the most commonly used options from different Figure window menus. They are

New Figure, **Open File**, **Save Figure**, and **Print Figure** icons (from the **File** menu)

Edit Plot icon (from the **View** menu)

Zoom In, Zoom Out, Pan, and Rotate 3-D icons (from the **Tools** menu)

Data Cursor icon and **Brush/Select Data** drop-down menu (from the **Tools** menu)

Link Plot icon, allowing you to link the graphs to workspace data, so that when data changes plots are redrawn automatically

Insert Colorbar and **Insert Legend** icons (from the **Insert** menu)

Hide Plot Tools and **Show Plot Tools** icons (from the **View** menu)

Tile, Left/Right Tile, Top/Bottom Tile, Float, and **Maximize** icons (from the **Window** menu)

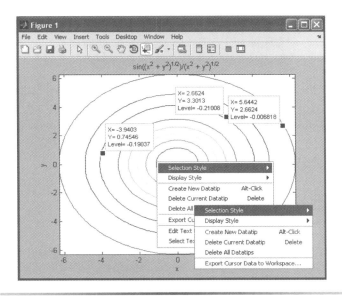

Fig. 6.45   Examples of using the Data Cursor icon.

object appears within the Figure window beneath the plot (Fig. 6.46). Once you started the **Property Editor**, keep it open throughout an editing session, since it provides access to all the objects in the graph. If you click another object in the graph, the **Property Editor** displays the set of panels associated with that type of objects (right click on the object brings a pop-up menu with pretty much the same options on object editing).

You can also start the **Plot Editor** by selecting the **Figure Properties**, **Axes Properties**, or the **Current Object Properties** from the **Edit** menu. One more way to start the **Property Editor** is to go to the **View** menu and choose the **Property Editor**

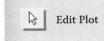

option. From the same menu, you can also bring up the **Figure Palette** and **Plot Browser** (Fig. 6.47). The **Figure Palette** enables you to perform the following tasks: add 2-D or 3-D axes to the figure or rearrange the existing plots (with the **New Subplots** panel), browse and plot workspace variables (with the **Variables** panel), add annotations to the graphs (with the **Annotations** panel). The **Plot Browser** allows you to select an object in the graph to edit among the list of all available objects. Any of these three options (**Figure Palette**, **Plot Browser,** or **Property Editor**) automatically enable plot-editing mode, if it is not already enabled.

To show the plot tools, that is, **Figure Palette**, **Plot Browser,** and **Property Editor**, you may also use the **Show Plot Tools** icon on the toolbar, or

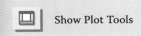

do it programmatically by issuing the `plottools` call in the **Command window**. If you want to bring any of plot tools separately, you should use `figurepalette`, `plotbrowser`, and `propertyeditor` functions, respectively).

Clicking the **More Properties** button in the **Property Editor** window, opens the **Inspector** window for the selected object (listed at the top of the **Inspector** window as seen in Fig. 6.47). It provides a complete list of all settable properties and displays their current value (these are the same properties shown in Tables 6.2 and 6.3). This allows you to look through and conveniently change any property of the selected object. Moreover, you can bring this window up even without opening the **Plot Editor** and disturbing the shape of your plot. The command

```
inspect(h)
```

where h is the handle of the object of your interest will do it for you.

It should be noted that having these Plotting Tools allow you to create plots without even knowing how to do it programmatically! You may load data into the current directory, say from Excel using the Import Wizard, then open Plotting Tools and develop a layout you want (as many different

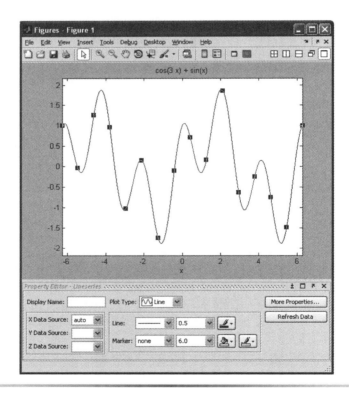

Fig. 6.46 Figure window supplemented with Property Editor window.

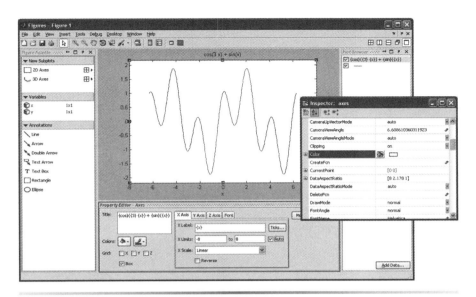

Fig. 6.47   Figure Palette, Plot Browser, Property Editor, and Inspector windows.

plots you want, any notations and shapes you would like to have, etc.). You may also align and distribute the objects within the figure space (using the **Tools** menu). Finally, you go to the **File** menu and choose a **Generate M-File** option. Believe it or not, MATLAB will generate an M-script, describing the complete layout for the figure you just created interactively, automatically! You may save this file and call it up later to plot another new set of data using the same layout.

Let us conclude this section with one practical tip. If you want to copy Figure to your Microsoft Word document or any other application without saving it on a disk, you should click **Edit** and choose the **Copy Figure** option. Now it resides in the buffer and is ready to be pasted. To get rid of gray background click **Edit**, **Copy Options** and check **Force white background** option rather than **Use Figure color** (default value), as was discussed in Sec. 2.7. You have to do this just once because MATLAB saves this change and uses white background every time you copy Figure into the buffer.

6.12   Interactive Plotting and Animation

For advanced users, MATLAB provides tools for interactive plotting, as well as creating and storing animations. This section presents several examples of how you can use them.

The `ginput` function enables you to use the mouse or the arrow keys to select points to plot. The `[x,y]=ginput` command returns the *x–y* coordinates of pointer's position when a mouse button or any key on the keyboard except a carriage return is pressed. It gathers an unlimited number of points until the <Return> key is pressed. The [x,y]=ginput(n) command gets exactly n points. The [x,y,button]=ginput(n) command along with vectors x and y returns a third vector, button, that contains integers specifying which mouse button was used (1, 2, 3 from the left) or ASCII numbers if a key on the keyboard was used.

The following example illustrates the use of ginput with the spline function to create a curve by parametric 2-D interpolation:

```
axis([0 10 0 10]), hold on
xy = []; n = 0; % Initially, the list of points is empty
%% Picking up multiple points by clicking left mouse button
% (Right mouse button means you are picking the last point)
but = 1;
while but == 1
   [xi,yi,but] = ginput(1);
   plot(xi,yi,'ro')
   n = n+1;
   xy(:,n) = [xi;yi];
end
%% Interpolating with a spline curve and finer spacing
t = 1:n; ts = 1:0.1:n;
xys = spline(t,xy,ts);
%% Plot the interpolated curve
plot(xys(1,:),xys(2,:),'b-'); hold off
```

First, a sequence of points, (x,y), is selected in the plane with ginput. Then two 1-D splines are passed through the points, evaluating them with a spacing 1/10 of the original spacing. The plot in Fig. 6.48 shows some typical output.

The two ways of generating moving, animated graphics in MATLAB are

1. *Erase Mode method*, which continually erases and then redraws the objects on the screen, making incremental changes with each redraw.
2. *Creating Movies approach*, which saves a number of different pictures and then plays them back as a movie.

The erase mode is one of the line specifications (Sec. 6.5) and is very useful in animation. This property controls the technique MATLAB uses to draw and erase the line objects. Alternative erase modes are useful for creating animated sequences, where control of the way individual objects are redrawn is necessary to improve performance and obtain the desired effect. The default EraseMode is normal allowing to redraw the affected

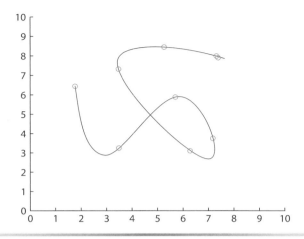

Fig. 6.48 Example of interactive plotting.

region of the display, performing the 3-D analysis necessary to ensure that all objects are rendered correctly. This mode produces the most accurate picture, but is the slowest. Two other modes (background and xor), described in more detail later in this section, are faster, but do not perform a complete redraw and, therefore, are less accurate. Finally, the none mode does not erase the line when it is moved or destroyed at all.

Consider the following set of commands that uses the Erase Mode method to slowly convert the sine wave into cosine wave:

```
x=0:0.2:2*pi; % Define the x-scale
y=sin(x);                % Compute sin(x)
z=cos(x);                % Compute cos(x)
plot(x,y)                % Plot sin(x) curve
set(gca,'xlim',[0 2*pi],'ylim',[-1 1]);
h_line=get(gca,'children'); % Get handle to the line
% Changing line properties
for i=1:1000
   pause(0.0005)
   w=i/1000;             % Set the weighting coefficient
   d=(1-w)*y+w*z;        % Blend sin(x) & cos(x) using w
% Change ydata for the line
   set(h_line,'ydata',d,'EraseMode','normal');
end
```

In the first four lines of code, we compute data for two dependences, $y = \sin(x)$ and $z = \cos(x)$ and plot the first dependence. The next two lines set the x- and y-axes limits and get a handle to the line (Sec. 6.5). Now, what we want to do by the remaining commands is to change the y-data for the line, keeping the remaining properties untouched. The *y*-data is computed as follows:

$$y_i = (1 - w_i)\sin(x) + w_i \cos(x), \quad \text{where } w_i = \frac{i}{1000}, \quad i = 0,1,2,\dots,1000 \qquad (6.1)$$

meaning that when $i = 0$, $w_i = 0$ and, therefore, $y_i = \sin(x)$. On the contrary, when $i \rightarrow 1000$, $w_i \rightarrow 1$, and $y_i \rightarrow \cos(x)$. Every time we change `Ydata`, the previous line is erased (the default value for the `EraseMode` property is set to `normal` anyway, so we just added this property-value pair here to highlight it). As a result, we will see a smooth conversion of a sinusoid to cosine wave.

Unfortunately, we cannot reproduce this transition here (you have to try it yourself), but Fig. 6.49 shows what you will eventually see on the figure if you run the above fragment with the `EraseMode` property set to `none`. Two other options would be

xor   Draws and erases the image by performing an exclusive OR (XOR) with the color of the screen beneath it (although this mode does not damage the color of the objects beneath the image, the image's color itself depends on the color of whatever is beneath it on the display)

background   Erases the image by drawing it in the axes background color (property `Color` of current axes) or the figure background color (property `Color` of the figure) if the axes color is set to none (as opposed to `xor` mode, this does damage objects that are behind the erased image, but images keep their color unchanged)

Note that while all line transitions are still visible on the screen, you cannot copy or print this figure, because MATLAB stores no information about its former location. The way we did it in Fig. 6.49 was taking a snap-shot of the current window with <Alt>+<PrtScrn> keyboard keys. You can also use the MATLAB `getframe` function (to be discussed next) or any other screen capture application to create an image of a figure containing non-`normal` mode objects.

To summarize, all you have to do to use the Erase Mode method is to create a 2-D or even 3-D scene composed of some objects, get handles to these objects, and then change their properties (*x-y-z* data, color, transparency, etc.). You will find this method widely used. For example, two MATLAB functions use the Erase Mode method to animate graphs. They are `comet` and `comet3` functions that can be used as a substitute for `plot` and `plot3`, respectively. A comet graphs are animated graphs in which a circle (the comet head) traces the data points on the screen. The comet body is a trailing segment that follows the head. The tail is a solid line that

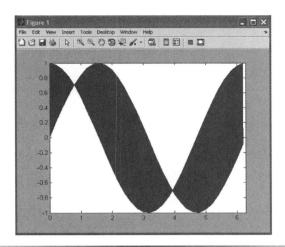

Fig. 6.49   Snap-shot of the current window after applying
`EraseMode none` mode.

traces the entire function. The following script provides an example of using these functions, with the final outputs shown in Fig. 6.50:

```
figure('Name','Fig.6.50a','Color','w')
t = 0:.0005:pi;
x = cos(2*t).*(cos(t).^2);
y = sin(2*t).*(sin(t).^2);
comet(x,y);
figure('Name','Fig.6.50b','Color','w')
t = -10*pi:pi/350:10*pi;
comet3((cos(2*t).^2).*sin(t),(sin(2*t).^2).*cos(t),t);
```

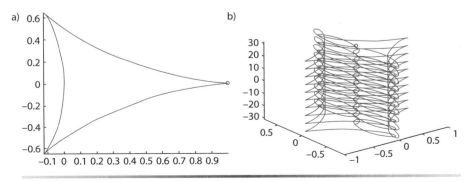

Fig. 6.50   Outputs of a) `comet` and b) `comet3` functions.

The only problem with the `comet` and `comet3` functions is that once again the trace left by the comet is created by using an `EraseMode` of `none`, which means you cannot print the graph (you get only the comet's head) and it disappears if you cause a redraw (for example, by resizing the window).

Consider one more example. Suppose you need to graphically represent the rotation of the body coordinate frame (tied to a satellite) with respect to some inertial frame. The following script takes advantage of the primary function (with no input parameters) and subfunction (which computes the rotation matrix), stored in one M-file, and results in animation, where one axis triad rotates with respect to another (Fig. 6.51 shows the tenth frame):

```
function rotation
figure('color','w')
quiver3(0,0,0,1.5,0,0,'LineWidth',2), hold on
quiver3(0,0,0,0,1.5,0,'LineWidth',2)
quiver3(0,0,0,0,0,1.5,'LineWidth',2)
    text(1.5,0,0,'n_1')
    text(0,1.5,0,'n_2')
    text(0,0,1.5,'n_3')
axis ([-1 1 -1 1 -1 1]), view(130,30)
xlabel('x_i'), ylabel('y_i'), zlabel('z_i');
R=eye(3);
h(1)=quiver3(0,0,0,R(1,1),R(1,2),R(1,3),'m','Linewidth',3);
h(2)=quiver3(0,0,0,R(2,1),R(2,2),R(2,3),'m','Linewidth',3);
h(3)=quiver3(0,0,0,R(3,1),R(3,2),R(3,3),'m','Linewidth',3);
    ht(1)=text(R(1,1),R(1,2),R(1,3),'\bf{b_1}');
    ht(2)=text(R(2,1),R(2,2),R(2,3),'\bf{b_2}');
    ht(3)=text(R(3,1),R(3,2),R(3,3),'\bf{b_3}');
ha(1)=text(-0.5,0,0,   ['\phi = 0 ^o']);
ha(2)=text(-0.5,0,0.2,['\theta = 0 ^o']);
ha(3)=text(-0.5,0,0.4,['\psi = 0 ^o']);
  for i = 1:100
  psi=4*pi*(i-1)/99; phi=2*pi*(i-1)/99; theta=pi*(i-1)/99;
  R=Euler2DCM(psi,theta,phi);
    for j=1:3
    set(h(j),'UData',R(j,1),'VData',R(j,2),'WData',R(j,3));
    set(ht(j),'Position',[R(j,:)]);
    end
phid=mod(phi*180/pi,360);     if phid>180,  phid=phid-360;     end
thetad=mod(theta*180/pi,360); if thetad>180, thetad=thetad-360; end
psid=mod(psi*180/pi,360);
    set(ha(1),'String',['\phi = '   int2str(phid)    ' ^o']);
    set(ha(2),'String',['\theta = ' int2str(thetad)  ' ^o']);
```

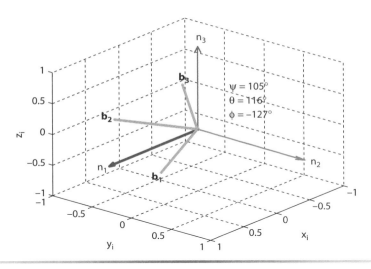

Fig. 6.51   Snap-shot of animated rotation.

```
    set(ha(3),'String',['\psi = '    int2str(psid)    ' ^o']);
 pause(0.001)
 end
function R=Euler2DCM(psi,theta,phi)
Rpsi = [cos(psi) sin(psi) 0; -sin(psi) cos(psi) 0; 0 0 1];
Rtheta = [cos(theta) 0 -sin(theta); 0 1 0; sin(theta) 0 cos(theta)];
Rphi = [1 0 0; 0 cos(phi) sin(phi); 0 -sin(phi) cos(phi)];
R = Rphi*Rtheta*Rpsi;
```

Another way of generating moving, animated graphics, the Creating Movies approach, is better suited to situations where each frame is fairly complex and cannot be redrawn rapidly. You create each movie frame in advance, so the original drawing time is not important during playback, which is just a matter of blitting the frame to the screen. A movie is not rendered in real time; it is simply a playback of previously rendered frames. Some useful functions are

`frame2im`	Converts a movie frame to indexed image
`getframe`	Captures a movie frame
`im2frame`	Converts an image to a movie frame
`movie`	Plays the recorded movie frames

The following self-explanatory script (based on that taken from the MATLAB Help system) provides with an example of using this second approach to create animations:

```
Z = peaks; surf(Z);
axis tight
```

```
set(gca,'nextplot','replacechildren');
% Record the movie
for j = 1:20
surf(sin(2*pi*j/20)*Z,Z)
F(j) = getframe;
end
k = questdlg('Ready to watch the recorded video?',...
             'Start the Movie','Yes','No','Yes');
if char(k(1))=='Y'
    movie(F,20)% Play the movie twenty times
end
```

This script animates the peaks function as the values of Z are scaled. The 'nextplot'-'replacechildren' pair sets current axes to keep their scale, that is, it removes all child objects, but do not resets axes properties while redrawing the surface. The questdlg function is used to create and display the question dialog box). Three of 20 generated frames are shown in Fig. 6.52 to give you an idea of what happens on the screen.

If you want to store the created frame-by-frame MATLAB movie into a standalone *avi*-file, playable anywhere outside MATLAB, you would need to use movie2avi function. The two self-explanatory syntax of this function are

```
movie2avi(mov,filename)
movie2avi(mov,filename,param,value,param,value,...)
```

Table 6.7 presents possible parameter settings.

For example, if you want to store the movie created in the previous fragment all you have to do is to add a line

```
movie2avi(F,'peaks.avi','compression','none','quality',100)
```

Be aware that the frame height and width will probably be padded to be a multiple of four as required by majority of codices. Another warning is that you should be careful when you create an *avi*-file on one computer to be

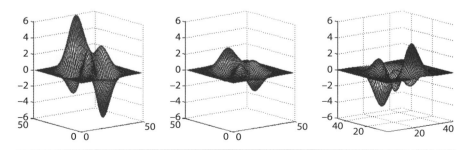

Fig. 6.52 Frames 1, 9, and 12 of the animated peaks function.

Table 6.7 Parameter Settings for the `movie2avi` Function

Parameter	Value		Default
`'colormap'` (see Sec. 6.8)	An *m*-by-3 matrix defining the `colormap` to be used for indexed *avi*-movies, where *m* must be no greater than 256 (236 if using `Indeo` compression). Note, you must set this parameter before calling `addframe`, unless you are using `addframe` with the MATLAB movie syntax.		There is no default `colormap`.
`'compression'`	A text string specifying the compression codec to use. The available codices are		
	on Windows platforms	On UNIX platforms	`'Indeo5'` on Windows.
	`'None'`, `'MSVC'`, `'RLE'`, `'Indeo3'`, `'Indeo5'`, `'Cinepak'`	`'None'`	`'None'` on UNIX.
	You may also use a custom compression codec, specifying the four-character code that identifies the codec (typically included in the codec documentation). The `addframe` function reports an error if it cannot find the specified custom compressor.		
`'fps'`	A scalar value specifying the speed of the *avi*-movie in frames per second (fps)		`15`
`'keyframe'`	For compressors that support temporal compression, this is the number of key frames per second.		`2`
`'quality'`	A number between `0` and `100` specifying the desired quality of output for compressed movies, so that the higher numbers result in higher video quality and larger file sizes.		`75`
`'videoname'`	A descriptive name for the video stream. This parameter must be no greater than 64 characters long.		`Filename`

played on another (that another computer may have no codec you created your movie with).

An alternative way of creating an *avi*-file is to use the `addframe` function. Its self-explanatory syntax

```
aviobj = addframe(aviobj,frame)
```

appends the data in `frame` to the *avi*-file identified by `aviobj`, which must be created beforehand by using the `avifile` function. To this end,

```
aviobj = avifile(filename)
```

creates an *avi*-file, giving it the name specified in `filename`, using default values for all *avi*-file object properties (Table 6.7). If `filename` does not include an extension, `avifile` appends *.avi* to it. The function `avifile` returns a handle to an *avi*-file object `aviobj`, which can be used in `addframe` function. Once the movie has been created, the `aviobj` object should be closed

```
aviobj = close(aviobj)
```

Let us modify the last program to accommodate these two commands:

```
Z = peaks; surf(Z);
   axis tight
   set(gca,'nextplot','replacechildren',...
          'Visible','off'); % Make axes invisible
   aviobj = avifile('peaks.avi','quality',100); % Create the avi-file
% Record the movie
for j = 1:20
      surf(sin(2*pi*j/20)*Z,Z)
      F = getframe;
      aviobj = addframe(aviobj,F);
end
% Close the avi-file
aviobj = close(aviobj);
```

You can always retrieve information about your *avi*-file using `aviinfo` function

```
aviinfo('filename')
```

and read the *avi*-movie `filename` back into the MATLAB movie structure `mov` using the `aviread` function

```
mov = aviread(filename)
```

After that, you can use the `movie` function again to view the movie `mov`. Note that as of MATLAB R2010b it is suggested to switch to three new functions, `VideoWriter`, `mmfileinfo`, and `VideoReader`, replacing `avifile`, `aviinfo`, and `aviread`, because in a future release the latter three functions will be removed. Also, since Microsoft Windows XP Service Pack 3 disables playback of Indeo 3 and Indeo 5 codecs in Windows Media Player and Internet Explorer, it is suggested to specify a compression value of `'None'`. The VideoWriter function supports files larger than 2 GB, and by default, creates files with Motion JPEG compression, which all platforms support.

6.13 Requirements to Engineering Plots

Now you know how to generate the plots, it will be quite appropriate to introduce a set of general engineering requirements applicable for any type of plot:

1. Choose the type of the plot and its scale wisely to have the best possible (that is, readable and understandable) representation of what you intend to show by this plot.
2. Use a title to identify what the plot is, especially when there are multiple plots of a similar type.
3. Label each axis and show the units used (place them in parentheses or after the comma right after the axis' title).
4. Assure that each axis has regularly spaced tick marks at convenient intervals to allow quick estimates of values.

If plotting more than one set of data or equation on the same plot:

1. Identify each line not only by different color (which MATLAB does for you automatically) but also with a different line style (and/or marker). While you may see the difference between colored lines while looking on the screen, it obviously disappears when the plot is printed out on a mono-color printer.
2. Place a legend distinguishing and explaining different curves.
3. Do not overload one plot by placing too much information on it, so if appropriate, use multiple x- and y-axes.

Finally, there are two more unwritten conventions:

- Measured data are usually represented by markers at each data point (different symbols then can be used for different sets of data). Connecting the data points may assume a relationship in the data that may not exist.
- On the contrary, equations are usually plotted as lines, not as individual data points.

Try to meet these requirements to produce informative, self-sufficient, and professional-looking plots.

Problems

6.1   As pointed out at the end of Sec. 6.2, the `annotation` function uses normalized coordinates to specify locations within the figure. Write a script (function) that uses the current axes handle and allows you to convert a pair of physical coordinates (x, y) into normalized coordinates

(x_n, y_n). Use this function to draw a box with its edges represented by arrows (using the annotation('arrow',...) commands).

6.2 A defect in a crystal lattice where a raw of atoms is missing is called an edge dislocation. The stress field around an edge dislocation is given by

$$\sigma_{xx} = \frac{-Gb}{2\pi(1-v)}\frac{y(3x^2+y^2)}{(x^2+y^2)^2}, \quad \sigma_{yy} = \frac{-Gb}{2\pi(1-v)}\frac{y(x^2+y^2)}{(x^2+y^2)^2},$$

$$\tau_{xy} = \frac{-Gb}{2\pi(1-v)}\frac{x(x^2+y^2)}{(x^2+y^2)^2}$$

where G is the shear modulus, b is the Burgers vector, and v the Poisson ratio. Plot the stress components (each in a separate figure) due to an edge dislocation in aluminum for which

$$G = 27.7{\cdot}10^9\text{Pa}, b = 0.286{\cdot}10^{-9}\text{m, and } v = 0.334$$

Plot the stresses in the domain -5×10^{-9}m$< $ x $ < 5\times10^{-9}$m and -5×10^{-9}m$<y<5\times10^{-9}$m. Plot the coordinates x and y in the horizontal plane, and the stresses in the vertical direction. Label each axis using Greek symbols.

6.3 Consider the following mode shape for a solid circular membrane clamped along its outer boundary ($r = 1$):

$$z(r, \phi) = J_1(3.8316r)\cos(\phi)$$

where $J_1(x)$ is the Bessel function of the first kind of the order of 1 (to address it properly, use besselj(1,x) command), whereas (r, ϕ) are the polar coordinates of any point on the membrane (the value 3.8316 corresponds to one of the natural frequencies for such a membrane). Do the following:

a. Develop a script allowing you to show the membrane's shape as a 3-D surface.
b. Modify the script of Sec. 6.12, animating the peaks function, to animate your membrane.

Chapter 7 / Symbolic Math Toolbox

- Symbolic Expressions in Linear Algebra and Calculus
- Solving Algebraic and Differential Equations
- Familiarization with MuPAD
- Using Symbolic Expressions in Numerical Calculations

```
sym,syms,diff,limit,int,
taylor,besselj,solve,
dsolve,collect,simple,
simplify,subs, pretty,
sym2poly,poly2sym,
ezplot,ezcontour,
matlabFunction,latex,
mupad,funtool,rsums,
taylortool, and others
```

7.1 Introduction

This chapter introduces one of the MATLAB toolboxes provided with basic MATLAB, the Symbolic Math Toolbox. This toolbox incorporates symbolic computation into the MATLAB numerical environment and can be very useful. In fact, the reason for introducing this toolbox so early is that we will rely on the functions of this toolbox to derive a lot of formulas in the following chapters. Not that we could not derive these formulas by hand or simply take them from some textbook on numerical methods, but among others, using the Symbolic Math Toolbox will allow you to get more hands-on experience in applied programming in the MATLAB development environment. As a result, you will become a quite proficient user not just in numerical computations, but in numerical and symbolic computations as well. The syntax for the Symbolic Math Toolbox is a natural extension of the normal MATLAB programming language, and therefore, there should be no difficulties following the material of this chapter after you have successfully completed the previous ones. The chapter starts from the very basics—explaining how to create symbolic objects and get help if needed. It then introduces all major functions of the Symbolic Math Toolbox by groups. The current engine of the Symbolic Math Toolbox features the MuPAD Notebook, optimized to handle and operate on symbolic math expressions. This capability is discussed later, followed by explaining how to incorporate the results of symbolic computations into numerical computations. Finally, three interactive graphical user interfaces (GUIs) that exploit the functions of the Symbolic Math Toolbox are presented.

7.2 Symbolic Math Toolbox Overview

Originally, the computational engine underlying the Symbolic Math Toolbox was the Maple kernel, developed at the University of Waterloo in Canada. Before R2007b, the *Basic* Symbolic Math Toolbox included in the Student Version was a collection of more than a hundred MATLAB functions allowing access to the functions in Maple's linear algebra package. The *Extended* Symbolic Math Toolbox augmented the basic toolbox by including access to nongraphics packages, programming features, and user-defined procedures in Maple.

As of R2007b+, the Symbolic Math Toolbox—now a single product— uses the MuPAD computational engine, originally developed by the MuPAD research group at the University of Paderborn, Germany. The functionality of the MuPAD language, together with included libraries, goes far beyond that of the previous Symbolic Math Toolbox software. The functionality of the Extended Symbolic Math Toolbox has been taken over by the MuPAD environment (MuPAD Notebook).

Most of the functions of the former Basic Symbolic Math Toolbox remained the same (the only difference is that a few functions allowing you to use the Maple functions were excluded). Therefore, for users of the former Basic Symbolic Math Toolbox, this change did not cause much of a problem, despite the fact that the new engine works slower and produces results that may be somewhat different from what MATLAB users were accustomed to. The major differences between the two engines are as follows:

1. Many computations return in a permuted order (such as `a+b` instead of `b+a`).
2. Trigonometric computations return in a different, mathematically equivalent form, which is probably the most annoying feature of the new engine (Sec. 7.6 provides an example).
3. The `solve` and `dsolve` functions may return a different number of solutions than before.

Fig. 7.1 Response to the `symengine` call in the case Maple software is not available.

Table 7.1 Areas Supported by the Symbolic Math Toolbox

Subject	Areas Supported
Linear algebra	Inverses, determinants, eigenvalues, singular value decomposition, and canonical forms of symbolic matrices
Calculus	Differentiation, integration, limits, summation, and Taylor series
Solving (systems of) equations	Symbolic and numerical solutions to algebraic and differential equations
Simplification and substitution	Methods of simplifying algebraic expressions
Integral transforms	Fourier, Laplace, z- transforms, and their corresponding inverse transforms
Conversions and VPA	Conversion and numerical evaluation of mathematical expressions to any specified accuracy
Special functions	Special functions in applied mathematics such as Airy functions, binomial coefficients, error functions, Bessel functions, Dirac delta function, Legendre polynomials, Jacobian, and Chebyshev functions
Plotting symbolic expressions	Easy (2-D) and (3-D) plotting of symbolic expressions

4. Most of Greek letters, such as `beta`, `zeta`, `theta`, and `psi` can only be used as symbolic variables and not in strings (because these symbols represent MuPAD reserved words).
5. Performance of numerical integration (`int`) is much slower than in the previous versions, and some integrals can no longer be evaluated, especially those involving the Bessel functions.
6. The `pretty` function output looks more bulkier than before.
7. Subexpressions no longer have partial subexpressions (previous syntax `%n`), and the `pretty` function no longer uses them.
8. The `simple` function returns different results.
9. The `simplify` function involving radicals and powers makes fewer assumptions on unknown symbols than in previous versions.
10. LaTeX output from the `latex` function looks different than before.

What is presented in this chapter shows the returns as they appear in R2010a version of MATLAB, but if you are using the previous versions, most likely you will see different returns. Also, if you own a compatible version of Maple software, you can choose to have Symbolic Math Toolbox software, use the Maple engine instead of a MuPAD engine, especially if you have existing Maple programs you would like to continue working with. In the latter case, it is the MATLAB `symengine` function that would allow you to make the choice. If Maple is not installed, you will get the response as shown in Fig. 7.1.

The areas supported by the Symbolic Math Toolbox (as they were supported by the former Basic Symbolic Math Toolbox) are listed in Table 7.1.

7.3 Getting Started

This section proceeds with explaining how to create symbolic objects, variables, and mathematical expressions, and how to access the Symbolic Math Toolbox help topics.

7.3.1 Creating Symbolic Variables

To construct a symbolic variable of class "sym," which is a new class compared to the MATLAB classes presented in Fig. 4.1, you should use the `sym` function as

```
>> sarg = sym('arg');
```

This command constructs a symbolic variable `sarg`. If the input `arg` is a string, the result is a symbolic number or variable. If the input argument is a numeric scalar or array, the result is a symbolic representation of the given numeric values. For instance, `x=sym('x')` creates the symbolic variable with the name `x` and stores the result in `x`, whereas `y=sym('1/10')` (or `y=sym(1/10)`) creates a symbolic number that avoids the floating-point approximations inherent in the value of 1/10 (to be discussed in Sec. 8.4.2). As usual, each symbolic variable must begin with a letter and can contain only alphanumeric characters.

If you have several variables to declare symbolic, you can use a shortcut as follows:

```
>> syms arg1 arg2 ...
```

which is a short-hand notation for

```
>> arg1 = sym('arg1');
>> arg2 = sym('arg2'); ...
```

Along with declaring symbolic variables, you can also assign some additional properties to them, for example,

```
>> x = sym('x','real');
>> y = sym('y','unreal');
```

In the first case, the command `conj(x)` will return x, whereas in the second case, `conj(y)` will result in `conj(y)`. By default, all symbolic variables are assigned to be `real`. Therefore, the second command can be used to clear symbolic variable y of its `real` property (if needed for any reason). The shortcut of this latter command would look like

```
>> syms y unreal
```

Table 7.2 Examples of Formatted Conversion to Symbolic Numbers

Command	Result
`>> sym(0.1,'r')`	`1/10`
`>> sym(0.1,'e')`	`eps/40+1/10`
`>> sym(0.1,'f')`	`3602879701896397/36028797018963968`
`>> sym(0.1,'d')`	`.10000000000000000555111512312578`
`>> digits(4), sym(0.1,'d')`	`0.1`
`>> sym(3*pi/4,'r')`	`(3*pi)/4`
`>> sym(3*pi/4,'e')`	`3*pi/4-(103*eps)/249`
`>> sym(3*pi/4,'f')`	`2652839157010665/1125899906842624`
`>> sym(3*pi/4,'d')`	`2.356`
`>> digits(25), sym(pi,'d')`	`3.141592653589793115997963`

In addition to the aforementioned syntax, the `sym` function has two more:

`sym('x','positive')` makes x a positive (real) variable

`sym('x','clear')` makes x a purely formal variable with no additional properties (that is, ensures that x is neither real nor positive)

Finally, the `sym` function can also be used to convert a floating-point numeric scalar or array A to a symbolic form using an optional `flag` as follows:

```
>> arg = sym(A,flag);
```

The optional second argument `flag` defines the technique for converting floating-point numbers. With some examples presented in Table 7.2, the `flag` is one of the following:

`'r'` (the default value)   Stands for "rational," meaning that floating-point numbers are obtained by evaluating expressions of the form `p/q`, `p*pi/q`, `sqrt(p)`, `2^q`, and `10^q` for modest-sized integers p and q

`'e'`   Stands for "estimate error," so that the `'r'` form is supplemented by a term involving the variable `eps` (Sec. 8.3.5), which estimates the difference between the theoretical rational expression and its actual floating-point value

`'f'`   Stands for "floating-point" (all values are represented in the form `N*2^e` or `-N*2^e`, where N and e are integers, N≥0

`'d'`   Stands for "decimal," so that the number of digits
                         `N` is taken from the current setting defined by a
                         preceding command `digits(N)` (fewer than 16
                         digits loses some accuracy, whereas more than 16
                         digits may not be warranted)

7.3.2 Constructing Symbolic Functions

Along with defining symbolic variables, as discussed in the previous section, you can also define a symbolic function. You can compose a symbolic function from symbolic variables you have already declared, like

```
>> syms a b c x
>> f=a*x^2+b*x+c;
```

Alternatively, you can create a symbolic function at once without even declaring any symbolic variables

```
>> f=sym('a*x^2+b*x+c');
```

Then, you can use this function in your manipulations. For example, the following command:

```
>> diff(f)
```

differentiates the function created using either approach shown earlier with respect to an independent variable x

```
ans = b + 2*a*x
```

This brings us to an important question. When manipulating the mathematical functions of more than one variable, the choice of independent variable is often clear from the context of the equation. However, how does MATLAB determine which variable is the independent one? Well, there is a simple rule. The default symbolic variable in a symbolic expression is the letter that is closest to x alphabetically. If there are two equally close variables, the letter that is later in the alphabet is chosen. For instance, the following set of commands

```
>> f=sym('var+zet^2');
>> diff(f)
```

will result in

```
ans = 2*zet
```

that is, the derivative will be taken with respect to the variable `zet`. Table 7.3 provides with some more examples on determining the default independent variable in symbolic mathematical expressions.

Table 7.3 Determining the Default Symbolic Variable

Expression	Default Variable
x^n	x
sin(a*t+b)	t
w*y+v*z	y
exp(i*theta)	theta
sqrt(pi*alpha)	alpha (pi has a predefined value in MATLAB)

If you are not sure, which variable is considered an independent variable, use the findsym function. Specifically,

findsym(S)   Returns all symbolic variables in S in lexicographical order
findsym(S,n)   Returns n variables alphabetically closest to x

Many functions allow you to specify the independent variable, which can alleviate some of the confusion. For example, for the function of two variables, var and zet, defined earlier, the command

```
>> diff(f,'var')
```

will produce

```
ans = 1
```

More complex symbolic functions can be constructed using the compose function. This function allows you to compose a symbolic function $f(g(y))$ from two other symbolic functions, $f=f(x)$ and $g=g(y)$, as in the following example:

```
>> f=sym('a*x^2+b*x+c');
>> g=sym('sin(y)');
>> h=compose(f,g)
h = a*sin(y)^2 + b*sin(y) + c
```

If for any reason you want to keep x as an independent variable, you should use the following variant:

```
>> syms x
>> h1=compose(f,g,x)
h1 = a*sin(x)^2+b*sin(x)+c
```

Note that, we have to explicitly declare a symbolic variable x before we use it.

Finally, here is another example of defining a symbolic function as simple as $f=f(x)$

```
>> f=sym('f(x)');
```

Now, based on this definition, you can construct the secondary functions. For instance, you can use the `subs` function (to be discussed further in Sec. 7.7.2) to have a symbolic representation of the forward-difference approximation for the first-order derivative

```
>> syms x h
>> dfdx=(subs(f,x,x+h)-f)/h
dfdx = (f(x+h)-f(x))/h
```

(Here, the function `subs` simply substitutes x with x+h.)

7.3.3 Getting Help

As in the case of MATLAB itself, there are several methods to get help on the Symbolic Math Toolbox functions as well. The first method is to type the following command in the MATLAB Command window:

```
>> help sym/function
```

The reason for this syntax is that there are many dual functions that use the same name but are applicable either for numeric or symbolic computations. For example, to get help for symbolic differentiation, you need to type

```
>> help sym/diff
```

which yields the following response:

```
DIFF Differentiate.
    DIFF(S) differentiates a symbolic expression S with respect to its
    free variable as determined by FINDSYM.
    DIFF(S,'v') or DIFF(S,sym('v')) differentiates S with respect to v.
    DIFF(S,n), for a positive integer n, differentiates S n times.
    DIFF(S,'v',n) and DIFF(S,n,'v') are also acceptable.
    Examples;
        x = sym('x');
        t = sym('t');
        diff(sin(x^2)) is 2*cos(x^2)*x
        diff(t^6,6) is 720.
    See also INT, JACOBIAN, FINDSYM.
```

If you type

```
>> help diff
```

you will get a different response corresponding to a completely different function

```
DIFF Difference and approximate derivative.
    DIFF(X), for a vector X, is [X(2)-X(1) X(3)-X(2) ... X(n)-X(n-1)].
    DIFF(X), for a matrix X, is the matrix of row differences,
        [X(2:n,:) - X(1:n-1,:)].
...
```

As discussed in Sec. 5.8.2, this method (`help sym/function`) simply displays the comment lines that are located at the beginning of the M-file function.

If you forgot the correct spelling of the function you are looking for, type several beginning letters in the Command window and then hit the <Tab> key to get a hint in a pop-up menu, or use a Function Browser button (if activated) as discussed in Sec. 2.8 (Fig. 2.9).

Another method of getting help is to click the Help button on the MATLAB toolbar and then click on the subgroup "Symbolic Math Toolbox" in the "Contents" folder. This opens 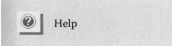 the Help window for the symbolic toolbox that will allow you to find the most complete information about any symbolic function. Alternatively, you may type in the name of the function you want to learn about in the search window of the Help browser. The third method is to go to the MathWorks' Web site and get missing information from there.

Now, because there are multiple functions bearing the same name, how does MATLAB know the difference between symbolic use and numerical use of the function? Well, it depends on the type of arguments that are passed on. If the type is symbolic, MATLAB "overloads" or provides symbol-specific implementation of the function.

7.4 Basic Operations

As mentioned in Sec. 7.2, all functions of the Symbolic Math Toolbox can be subdivided into different categories (Table 7.1). This section presents the list of basic operations that are supported by this toolbox. As you will see, many of the functions happen to be essentially the same as applied to the numerical data.

To start with, the Symbolic Math Toolbox supports all arithmetic operations, namely,

+	Addition
−	Subtraction
*	Multiplication
.*	Array multiplication
\	Left division
.\	Array left division
/	Right division
./	Array right division
^	Matrix or scalar raised to a power
.^	Array raised to a power
'	Complex conjugate transpose
.'	Real transpose

Beyond this, the Symbolic Math Toolbox also supports many standard elementary mathematical functions introduced earlier in Sec. 2.5 for numerical data. To be more specific, the so-called Basic Operations group of symbolic functions contains the following:

ccode	C code representation of a symbolic expression
ceil	Rounding a symbolic array toward positive infinity
conj	Symbolic complex conjugate
eq	Symbolic equality test
fix	Rounding a symbolic array toward zero
floor	Rounding a symbolic array toward negative infinity
fortran	FORTRAN representation of a symbolic expression
frac	Symbolic array elementwise fractional parts
imag	Imaginary parts of the elements of a symbolic array composed of the complex numbers
latex	LaTeX representation of a symbolic expression
log10	Logarithm base-10 of entries of a symbolic array
log2	Logarithm base-2 of entries of a symbolic array
mod	Symbolic array elementwise modulus
pretty	Pretty-print symbolic expressions
quorem	Symbolic matrix elementwise quotient and remainder
real	Real part of the elements of a symbolic array composed of the complex numbers
round	Symbolic array elementwise round
size	Determining symbolic array dimensions
sort	Sorting symbolic vectors or polynomials
sym	Defining the symbolic numbers, variables and objects
syms	Shortcut for constructing symbolic objects
symvar	Determining symbolic variables in expression

We have talked about two functions from this list, sym and syms, being the key functions in defining symbolic variables already, and the rest of them are self-explanatory and need no further explanation here. The only other function we would like to bring your attention to is the pretty function. Once you have your symbolic function created, the Symbolic Math Toolbox allows you to print a symbolic output in a format that resembles typeset mathematics (with the default linewidth of 79 symbols). The following example explains it all:

```
>> syms x y
>> y=(x^2+1/x)/sqrt((sin(x)+cos(x))/x^3.5);
>> pretty(y)
        1    2
        - + x
        x
```

```
---------------------
/ cos(x) + sin(x) \1/2
| -------------- |
|        7        |
|        -        |
|        2        |
\        x        /
```

(Note that, before R2007b+ the output of the `pretty` function were more compact.)

7.5 Linear Algebra and Calculus

7.5.1 Linear Algebra Functions

The Linear Algebra group of the Symbolic Math Toolbox contains the following functions:

colspace	Determines a basis for the column space
det	Calculates a symbolic matrix determinant
diag	Creates or extracts symbolic diagonals
eig	Finds symbolic eigenvalues and eigenvectors
expm	Calculates a symbolic matrix exponential
inv	Computes a symbolic matrix inverse
jordan	Determines the Jordan canonical form
null	Finds a basis for null space
poly	Creates characteristic polynomial of a matrix
rank	Calculates a symbolic matrix rank
rref	Finds a reduced row echelon form
svd	Performs a symbolic singular value decomposition
tril	Finds the lower triangular part of a symbolic matrix
triu	Finds the upper triangular part of a symbolic matrix

In essence, linear algebraic operations on the symbolic objects are the same as operations with numeric variables of double-precision accuracy, and will be considered further in Chapter 9. A couple of simple examples as applied to symbolic variables are given in Table 7.4.

7.5.2 Calculus Functions

The functions of the Calculus group of the Symbolic Math Toolbox are as follows:

diff	Differentiates a symbolic expression
int	Integrates a symbolic expression

Table 7.4 Examples of Handling Symbolic Matrices

Example 1	Example 2
	`>> H=hilb(3)`
`>> syms a b c d e f g h i`	H = 1.0000 0.5000 0.3333 0.5000 0.3333 0.2500 0.3333 0.2500 0.2000
`>> K = [a 0 b; 0 c d; e 0 1]`	`>> H=sym(H)`
K = [a, 0, b] [0, c, d] [e, 0, 1]	H = [1, 1/2, 1/3] [1/2, 1/3, 1/4] [1/3, 1/4, 1/5]
`>> detK=det(K)`	`>> detH=det(H)`
detK = a*c-b*c*e	detH = 1/2160
`>> K_inv=inv(K)`	`>> H_inv=inv(H)`
K_inv = [1/(a-b*e), 0, -b/(a-b*e)] [(d*e)/(c*(a-b*e)), 1/c, -(a*d)/ (c*(a-b*e))] [-e/(a-b*e), 0, a/(a-b*e)]	H_inv = [9, -36, 30] [-36, 192, -180] [30, -180, 180]

```
jacobian      Calculates the Jacobian matrix
limit         Finds a limit of a symbolic expression
symsum        Performs a symbolic summation of series
taylor        Creates a Taylor series expansion
```

We have introduced the first function of this group, the `diff` function, in Sec. 7.3.2 already. For completeness, let us give a couple of more examples here as well. Along with the main syntax, `diff(f)`, allowing you to differentiate a symbolic function `f` with respect to its default independent variable, this function also allows you to take a derivative with respect to some other variable `v`: `diff(f,'v')` (or `diff(f,v)`). Finally, you can obtain an nth order derivative by using the following syntax: `diff(f,n)`. The following examples are intended to summarize the usage of this function:

```
>> syms a x, f=sin(a*x^2);
>> dfdx=diff(f)
dfdx = 2*a*x*cos(a*x^2)
>> dfda=diff(f,a)
dfda = x^2*cos(a*x^2)
>> d2fdx2=diff(f,2)
d2fdx2 = 2*a*cos(a*x^2) - 4*a^2*x^2*sin(a*x^2)
```

Similarly, you can apply the `diff` function to a symbolic array, for example,

```
>> syms a x
   A=[cos(a*x),sin(a*x);-sin(a*x),cos(a*x)];
```

```
>> dAdx=diff(A)
dAdx =

    [ -a*sin(a*x),  a*cos(a*x)]
    [ -a*cos(a*x), -a*sin(a*x)]
```

Another function of this group, `jacobian(f,v)`, computes the Jacobian of a scalar or vector `f` with respect to `v`. The (i, j)th entry of the result is a partial derivative $(\partial f_i/\partial v_j)$. When `f` is a scalar, `jacobian(f,v)` returns the gradient of `f`, when `v` is a scalar, the result is the same as `diff(f,v)`. For instance, the following computes the Jacobian matrix **J** for the transformation from the Euclidean space (x, y, z) to spherical coordinates (r, λ, ϕ), where r is the distance from the origin, λ corresponds to the elevation or latitude, and ϕ denotes the azimuth or longitude

```
>> syms r lambda phi
>> x=r*cos(lambda)*cos(phi);
>> y=r*cos(lambda)*sin(phi);
>> z=r*sin(lambda);
>> J=jacobian([x;y;z],[r,lambda,phi])
J =

    [ cos(lambda)*cos(phi),  -r*cos(phi)*sin(lambda),  -r*cos(lambda)*sin(phi)]
    [ cos(lambda)*sin(phi),  -r*sin(lambda)*sin(phi),   r*cos(lambda)*cos(phi)]
    [          sin(lambda),           r*cos(lambda),                         0]
```

Symbolic integration using the `int` function is more difficult than symbolic differentiation, but in many cases, this function can still find a symbolic integral. A symbolic integral of an expression `f` is found by finding another symbolic expression `F` such that `diff(F)=f`. Similar to differentiation, you can specify the variable of integration if the default symbolic variable is not the variable of integration. If the `int` function cannot find the antiderivative, then the command line is returned unevaluated (as typed). Definite integrals may also be evaluated using the same function `int`, by specifying the limits of integration. Some examples are given as follows: first, let us compute $\int \sin(x)\mathrm{d}x$ and $\int_0^\pi \sin(x)\mathrm{d}x$

```
>> syms x, f=sin(x);
>> F=int(f)
F = -cos(x)
>> F02pi=int(f,0,pi)
F02pi = 2
```

Next, let us find another integral, $\int \sqrt{x}\log(x)\mathrm{d}x$, requesting to represent the result nicely using the `pretty` function, and then evaluating $\int_0^1 \sqrt{x}\log(x)\mathrm{d}x$ and converting the result into the fixed-point format

```
>> syms x, f=log(x)*sqrt(x);
>> pretty(int(f))
```

```
              3
              -
         2 /                 2 \
    2 x  | log(x)   -  -  |
         \                 3 /
         --------------------
                   3
```

```
>> F01=int(f,0,1)
F01 = -4/9
>> double(F01)
ans = -0.4444
```

The `limit` function of the Symbolic Math Toolbox allows computing the symbolic limits of a symbolic function. For example, let us use the fundamental theorem of calculus

$$f'(x) = \lim_{h \to 0} \frac{f(x+h) - f(x)}{h} \qquad (7.1)$$

If we let $f(x) = \cos(x)$, then the derivative should be $f'(x) = -\sin(x)$. The following demonstrates that rather than using the `diff` function, you could find the solution using the `limit` function as well:

```
>> syms x h, dydx=limit((cos(x+h)-cos(x))/h,h,0)
dydx = -sin(x)
```

As another example, let us determine the well-known limit $e^x = \lim_{n \to \infty} (1 + x/n)^n$. The following set of commands does the job:

```
>> syms x n
>> ex=limit((1+x/n)^n,n,inf)
ex = exp(x)
```

Equation (7.1) presents an example of a "two-sided" limit, that is, it does not matter from which side you approached the limit. If the function is discontinuous (has singularities), then the direction of approach may matter. For example, let us look at the following "one-sided" limits: $\lim_{x \to 0^+} x^{-1}$ and $\lim_{x \to 0^-} x^{-1}$. The `limit` function allows specifying what direction the limit is approached by adding the option `'right'` (to specify $x \to 0^+$) or `'left'` (to specify $x \to 0^-$) as follows:

```
>> syms x
>> lim_plus=limit(1/x,x,0,'right')
lim_plus = Inf
```

```
>> lim_minus=limit(1/x,x,0,'left')
lim_minus = -Inf
```

Note that if you use the default case, without specifying the direction of approach, you get

```
>> inv_x=limit(1/x,x,0)
inv_x = NaN
```

which means that the `limit` function could not find a consistent answer to the limit.

The indefinite or definite summation is done using the `symsum` function. For instance, to compute the indefinite sum

$$\sum_{i=0}^{n-1} \frac{a}{i} \tag{7.2}$$

you should issue the following two commands:

```
>> syms a n
>> symsum(a/n)
```

resulting in

```
ans = a*psi(n)
```

that is, $a\Psi(n)$ (the `psi` function will be introduced in Table 7.5). If only several terms are needed to be summed, say from $i = 6$ till $i = 15$, the command

```
>> symsum(a/n,6,15)
```

would produce the desired result

```
ans =74587/72072*a
```

Of course, you may determine an independent symbolic variable directly. Say for the example mentioned earlier, the command

```
>> symsum(a/n,a)
```

will compute the sum with respect to the variable `a`

```
ans = a^2/(2*n) - a/(2*n)
```

that is, $a(a - 1)/2/n$.

The limits may also include infinity. For example, the command

```
>> symsum(6/n^2,1,inf)
```

computes the infinite sum

```
ans = pi^2
```

and the command

```
>> symsum(a^n/sym('n!'),n,0,inf)
```

returns

```
ans = exp(a)
```

Finally, the `taylor` function of the Calculus group of the Symbolic Math Toolbox allows you to compute Taylor series approximation to the function $f(x)$ about a point $x = a$:

$$f(x) = \sum_{n=0}^{\infty} (x-a)^n \frac{f^{(n)}(a)}{n!} \tag{7.3}$$

(if the series is centered at zero, that is, $a = 0$, the series is also called a Maclaurin series). Specifically,

`taylor(f)`	Returns the fifth-order Maclaurin polynomial approximation to f
`taylor(f,n,v)`	Returns the (n-1)th-order Maclaurin polynomial approximation to f, where v specifies the independent variable in the expression
`taylor(f,n,v,a)`	Returns the Taylor series approximation to f about a

For instance, the following finds the seventh-order Maclaurin approximation of $f(x) = \sin(x)$:

```
>> syms x, f=sin(x);
>> M=taylor(f,8);
>> pretty(M)
        7     5    3
       x     x    x
   -  ---- + --- - -- + x
      5040   120   6
```

To find the Taylor series expansion of `sin(x)` about $\pi/2$, the following commands may be used:

```
>> T=taylor(f,8,pi/2);
>> pretty(T)
   / pi    \4   / pi    \2   / pi    \6
   | -- - x |   | -- - x |   | -- - x |
   \ 2    /    \ 2    /    \ 2    /
   ----------- - ----------- - ----------- + 1
       24            2            720
```

7.6 Solving Algebraic and Differential Equations

The Solution of Equations group of the Symbolic Math Toolbox includes four functions:

`compose`	Functional composition
`dsolve`	Symbolic solution of ordinary differential equations

finverse   Functional inverse

solve   Symbolic solution of algebraic equations

The `compose` function has been considered in Sec. 7.3.2 already, and the `finverse` function simply returns the functional inverse of the function $f(x)$. If exists, this inverse $g(x)$ satisfies the equation $g(f(x))=x$. For instance,

```
>> syms x
>> f=5*x^3;
>> pretty(finverse(f))

     2   1
     -   -
     3   3
     5   x
     -----
       5
```

```
>> syms x
>> f=tan(1/x);
>> pretty(finverse(f))

        1
     -------
     atan(x)
```

The remaining two functions, `solve` and `dsolve`, are dedicated to solve (systems of) algebraic and differential equations, respectively. To this end, one or more algebraic expressions (system of equations) can be solved using the `solve` function. For example, writing the quadratic equation and invoking the `solve` function as follows:

```
>> syms a b c x
>> f=a*x^2+b*x+c;
>> y=solve(f)
```

yields the two well-known roots

```
y = -(b + (b^2 - 4*a*c)^(1/2))/(2*a)
    -(b - (b^2 - 4*a*c)^(1/2))/(2*a)
```

Note that, the `solve` function can also accept a string rather than a symbolic function, so that

```
>> y=solve('a*x^2+b*x+c')
```

or

```
>> y=solve('a*x^2+b*x+c=0')
```

produces exactly the same result too. By default, it is assumed that you are trying to solve for the roots of $f(x) = 0$, but in the case of a string input, you may add an equal sign explicitly. As in the case of other symbolic functions, you may assign an independent variable explicitly. For example,

```
>> syms a b c x
>> f=a*x^2+b*x+c;
>> y=solve(f,b)
```

returns

```
y = -(a*x^2+c)/x
```

The `solve` function also allows you to solve multiple equations (system of equations). The general format is

```
solve(eq1,eq2,...,eqn)
```

or

```
solve(eq1,eq2,...,eqn,var1,var2,...,varn)
```

(if you need to specify variables explicitly, rather than rely on a `findsym` function findings). A self-explanatory example of solving several algebraic equations is given as follows:

```
>> syms a b c d x y t
>> [x,y]=solve(a*x+b*y-t,c*x+d*y)
x = (d*t)/(a*d - b*c)
y = -(c*t)/(a*d - b*c)
```

Note that, if the second command is changed to

```
>> g=solve(a*x+b*y-t,c*x+d*y)
```

it returns a structure

```
g = x: [1x1 sym]
    y: [1x1 sym]
```

Hence, to access solutions, you will need to refer to them explicitly, like

```
>> [g.x g.y]
```

which returns

```
ans = [ (d*t)/(a*d - b*c), -(c*t)/(a*d - b*c)]
```

The `dsolve` function finds symbolic solutions to ordinary differential equations (up to six). Differential equations should be specified by a text string containing the letter D to denote differentiation. The symbols D2, D3,..., and DN correspond to the second, third,..., and Nth-order derivative, respectively. The default independent variable is t, thus, say D2y would represent the Symbolic Math Toolbox equivalent of d^2y/dt^2. If the initial conditions are not specified, the solution contains constants of integration. For example,

```
>> x=dsolve('(Dx)^2+x^2=1')
```

returns

```
           1
          -1
cosh(C3 + t*i)
cosh(C7 - t*i)
```

By the way, the Maple-based Symbolic Math Toolbox used to return a different result

```
x =
    [            -1]
    [             1]
    [   sin(t-C1)]
    [  -sin(t-C1)]
```

with only one constant of integration `C1`.

The initial conditions can be specified by additional equations, for example,

```
>> y=dsolve('Dy=y^2+1','y(0)=1')
y = tan(pi/4 + t)
```

The following provides an example of a higher-order differential equation:

```
>> y=dsolve('D2y=cos(2*x)-y','y(0)=1','Dy(0)=0','x')
y =
(5*cos(x))/3 + sin(x)*(sin(3*x)/6 + sin(x)/2)  -  (2*cos(x)*(- 3*tan(x/2)^4
+ 6*tan(x/2)^2 + 1))/(3*(tan(x/2)^2 + 1)^3)
```

An attempt to simplify it (the `simplify` function is to be discussed in the following section) yields

```
>> simplify(y)
    1 - (8*(cos(x)/2 - 1/2)^2)/3
```

As mentioned in Sec. 7.2, the Maple-based Symbolic Math Toolbox used to return a trigonometrically equivalent but more compact result at once

```
y = 4/3*cos(x)-1/3*cos(2*x)
```

Finally, here is one more example related to solving more than one equation

```
>> [x,y]=dsolve('Dx=y','Dy=cos(t)','x(0)=1','y(0)=0')
x = 2 - cos(t)
y = sin(t)
```

7.7 Simplification and Substitution

The Simplification and Substitution group of the Symbolic Math Toolbox contains the following functions:

```
coeffs      Finds coefficients of a multivariate polynomial
collect     Collects coefficients
expand      Expands the polynomials and elementary functions
```

`factor`	Factorizes
`horner`	Finds the Horner nested polynomial representation
`numden`	Determines the numerator and denominator
`simple`	Searches for the simplest form of symbolic expression
`simplify`	Simplifies symbolic expressions
`subexpr`	Rewrites a symbolic expression in terms of common subexpressions
`subs`	Performs symbolic substitution

These functions, enabling the usage of the substitutes and making symbolic expressions more compact, are considered in the following section.

7.7.1 Simplification of Symbolic Expressions

Most of the functions of the Simplification group are self-explanatory. For instance, the `collect` function allows you to collects all the terms of like powers of the independent variable. For example,

```
>> syms x t
>> f=(1+x)*t+x*t;
>> f=collect(f)
```

returns

```
f = (2*t)*x + t
```

The `expand` function distributes products over the sums and applies other identities involving the functions of sums. For example,

```
>> syms x y
>> g=cos(x+y);
>> g=expand(g)
g = cos(x)*cos(y)-sin(x)*sin(y)
```

If `f` is a polynomial with rational coefficients, `factor(f)` expresses `f` as a product of polynomials of the lower degree with rational coefficients (if it cannot be done, `f` is returned unfactored). Two examples are presented as follows:

```
>> syms x                          >> syms x
>> f=x^3-6*x^2+11*x-6;             >> g=x^6+1;
>> f=factor(f)                     >> g=factor(g)
f = (x-3)*(x-1)*(x-2)             g = (x^2+1)*(x^4-x^2+1)
```

The `simplify` function of the Simplification group searches for the simplest form of a symbolic expression by applying several algebraic identities involving sums, integral powers, square roots, and other fractional powers, as well as some trigonometric, logarithmic, and exponential identities. For example, the commands

```
>> syms x
>> f=(1-x^2)/(1-x);
>> f=simplify(f)
```

return

```
f = x+1
```

The more general `simple` function goes beyond the `simplify` function, and in search for the simplest form of a symbolic expression that has the fewest number of characters applies all aforementioned functions (`collect`, `expand`, `factor`, and `simplify`) along with some other simplification functions. In the most general form, when `simple` is called without an output argument, all options `simple` looked through are displayed explicitly. For example,

```
>> syms a
>> f=(1/a^3+6/a^2+12/a+8)^(1/3);
>> simple(f)
simplify:
      ((2*a + 1)^3/a^3)^(1/3)
radsimp:
      (12/a + 6/a^2 + 1/a^3 + 8)^(1/3)
simplify(100):
      ((2*a + 1)^3/a^3)^(1/3)
combine(sincos):
      (12/a + 6/a^2 + 1/a^3 + 8)^(1/3)
combine(sinhcosh):
      (12/a + 6/a^2 + 1/a^3 + 8)^(1/3)
combine(ln):
      (12/a + 6/a^2 + 1/a^3 + 8)^(1/3)
factor:
      (12/a + 6/a^2 + 1/a^3 + 8)^(1/3)
expand:
      (12/a + 6/a^2 + 1/a^3 + 8)^(1/3)
combine:
      (12/a + 6/a^2 + 1/a^3 + 8)^(1/3)
rewrite(exp):
      (12/a + 6/a^2 + 1/a^3 + 8)^(1/3)
rewrite(sincos):
      (12/a + 6/a^2 + 1/a^3 + 8)^(1/3)
rewrite(sinhcosh):
      (12/a + 6/a^2 + 1/a^3 + 8)^(1/3)
```

```
rewrite(tan):
    (12/a + 6/a^2 + 1/a^3 + 8)^(1/3)
mwcos2sin:
    (12/a + 6/a^2 + 1/a^3 + 8)^(1/3)
collect(a):
    (12/a + 6/a^2 + 1/a^3 + 8)^(1/3)
ans = (2*a + 1)^3/a^3)^(1/3)
```

To suppress displaying all options, the `simple` function searched through, you should use the output argument as in the following example:

```
>> syms x
>> f=cos(x)^2+sin(x)^2;
>> f=simple(f)
f = 1
```

If you want to know how this result was achieved, you can use the following syntax:

```
>> [f,how]=simple(f)
f = 1
how = simplify
```

Two more functions, `coeffs` and `numden`, may be quite useful as well. The `coeffs(p,x)` call returns coefficients of polynomial `p` with respect to `x`. For instance,

```
>> syms x                          >> syms a b c x
>> f=x^3-2*x^2+3*x-4;               >> y=a+b*sin(x)+c*sin(2*x);
>> c=coeffs(f,x)                    >> c1=coeffs(y,sin(x))
c =[ -4, 3, -2, 1]                  >> c2=coeffs(expand(y),sin(x))
                                    c1 = [ a+c*sin(2*x), b]
                                    c2 = [ a, b+2*c*cos(x)]
```

(Note that the order of coefficients is opposite as compared to that described in Sec. 3.7 for numerical polynomials.)

The `numden` function of the Simplification group converts an input argument to a rational form, where the numerator and denominator are relatively prime polynomials with integer coefficients. For example,

```
>> syms x y
>> [n,d]=numden(x/y + y/x)
n = x^2+y^2
d = x*y
```

For another example considered in Sec. 7.4, application of the `numden` function

```
>> syms x y
>> y=(x^2+1/x)/sqrt((sin(x)+cos(x))/x^3.5);
>> [n,d]=numden(y)
```

would result in

```
n = x^3+1
d = x*((cos(x) + sin(x))/x^(7/2))^(1/2)
```

Finally, the `horner` function of the Simplification group produces a nested polynomial representation. For example,

```
>> syms x
>> f=x^3-6*x^2+11*x-6;
>> horner(f)
```

returns

```
c = x*(x*(x - 6) + 11) - 6
```

7.7.2 Substitutions

The Symbolic Math Toolbox has two functions that can be used for symbolic substitution: `subs` and `subexpr`. The very useful `subs` function can be invoked with one, two, or three input arguments. When invoked with only one input argument, `R=subs(S)`, it replaces all occurrences of variables in the symbolic expression `S` with the values obtained from the calling function or MATLAB workspace. For example,

```
>> f=sym('exp(a)*x^3+log(b)*x^2+c');
>> a=2; b=3; c=6;
>> r=subs(f)
r = exp(2)*x^3+log(3)*x^2+6
>> x=3;
>> r=subs(f)
r = 215.3920
```

When invoked with two input arguments, `R=subs(S,new)`, it replaces the default symbolic variable in `S` with `new`,

```
>> f=sym('exp(a)*x^3+log(b)*x^2+c');
>> r=subs(f,'sigma')
r = exp(a)*(sigma)^3+log(b)*(sigma)^2+c
```

Finally, the most general form involves invoking this function with three input arguments, `R=subs(S,old,new)`, which replaces `old` with `new` in the symbolic expression `S`. The variable `old` must be symbolic or a string representing a variable name, whereas the variable `new` may be symbolic or numeric. In the case that `old` and `new` represent several variables, as opposed to a single one, the list of the variables should be embraced with braces, `{ }`. For example,

```
>> syms a b alpha
>> f=cos(a)+sin(b);
```

```
>> g=subs(f,{a,b},{alpha,2})
g = sin(2) + cos(alpha)
```

7.8 Other Functions

Along with the major functions that have been considered in the previous sections, the Symbolic Math Toolbox contains a few more groups of functions. For completeness, let us briefly discuss about them as well.

7.8.1 Integral Transforms

The Integral Transforms group contains the functions enabling symbolic computation of some direct and inverse transforms

fourier   For Fourier integral transform, $F(\omega) = \int_{-\infty}^{\infty} f(t)e^{-i\omega t}dt$

ifourier   For inverse Fourier integral transform,

$$f(t) = \frac{1}{2\pi} \int_{-\infty}^{\infty} F(\omega)e^{i\omega t}d\omega$$

ztrans   For z-transform transform, $F(z) = \sum_{n=0}^{\infty} \frac{f(n)}{z^n}$

iztrans   For inverse z-transform,

$$f(n) = \frac{1}{2\pi i} \oint_{|z|=R} F(z)z^{n-1}dz, \ n=1,2,...$$

laplace   For Laplace transform, $F(s) = \int_{0}^{\infty} f(t)e^{-st}dt$

ilaplace   For inverse Laplace transform, $f(t) = \int_{c-i\infty}^{c+i\infty} F(s)e^{st}ds$

As seen, these transforms simply assume computing special integrals (or summation, in the case of the z-transform).

As representative examples, the following two sets of commands compute the Laplace transform of

$$f(t) = \frac{1}{t^2} \quad \text{and} \quad f(t) = \frac{1}{e^{at}}$$

```
>> syms t
>> laplace(t^2)
ans = 2/s^3
```

```
>> syms a t x
>> L=laplace(exp(-a*t),x)
L = 1/(a + x)
```

Note how the default output variable *s* in the latter expression was substituted with a symbol *x*. The inverse Laplace transform yields

```
>> syms s                    >> syms a t
>> f=ilaplace(2/s^3)         >> f=ilaplace(1/(t-a)^2)
f = t^2                      f = x*exp(a*x)
```

(Note that in the latter example, because we defined $F = F(t)$ as opposed to the default $F = F(s)$, `ilaplace` returns $f=f(x)$ instead of $f=f(t)$.)

▰ 7.8.2 Conversions and VPA

The Conversions group of the Symbolic Math Toolbox contains the following functions:

`double`	Converts a symbolic array to the MATLAB numeric form
`int8, int16, int32, int64`	Converts a symbolic array to signed integers
`poly2sym`	Brings a polynomial coefficient vector to a symbolic polynomial
`single`	Converts a symbolic array to a single precision
`sym2poly`	Converts symbolic-to-numeric polynomial
`uint8, uint16, uint32, uint64`	Converts a symbolic array to unsigned integers

Most of them are dedicated to convert symbolic numbers (see Table 7.2) to a numeric form with a variety of numeric formats of fixed-point numbers, to be explained in Chapter 8. Two of them, `poly2sym` and `sym2poly`, are designated to work with polynomials. They have been presented in Sec. 3.7 already, but for completeness, let us recall how to use them using a couple of simple examples. The `poly2sym` function takes a vector of polynomial coefficients and converts it to a complete symbolic representation of a polynomial like

```
>> coef = [1 0 1 2];
>> pol = poly2sym(coef)
pol = x^3+x+2
```

Obviously, the `sym2poly` function does the opposite, for example,

```
>> coef = sym2poly(pol)
coef = 1    0    1    2
```

The Variable Precision Arithmetic (VPA) group has only two functions:

`digits`	Sets variable precision accuracy
`vpa`	Uses VPA

The `digits` function specifies the number of significant decimal digits and is used to do VPA. The `digits(d)` command sets the current VPA accuracy to d digits. The default value is 32 digits. Typing `digits` without an input argument and pushing <Enter> returns the current VPA accuracy setting.

The `vpa` function uses VPA to compute the input argument to d decimal digits of accuracy, where d is the current setting of `digits`. Each element of the result is a symbolic expression. As an example, the following two calls return the symbolic representation of the *golden ratio* with the default (32 digits) and degraded accuracy of only five decimal digits:

```
>> GR1=vpa((1+sqrt(5))/2), GR2=vpa((1+sqrt(5))/2,5)
GR1 = 1.6180339887498949025257388711907
GR2 = 1.618
```

7.8.3 Special Functions

The Symbolic Math Toolbox also features several special functions. The Special functions group includes:

cosint   Computes the cosine integral (see Table 7.5)
dirac   Returns the Dirac delta function

$$\delta(x) = \begin{cases} \infty, & x=0 \\ 0, & x \neq 0 \end{cases} \quad \text{subject to} \quad \int\limits_{-\infty}^{\infty} \delta(x)\mathrm{d}x = 1$$

heaviside   Returns the Heaviside step function

$$H(x) = \int\limits_{-\infty}^{x} \delta(t)\mathrm{d}t$$

hypergeom   Constructs a generalized Barnes function (see Table 7.5)
lambertw   Evaluates the Lambert's *W* function (see Table 7.5)
sinint   Computes the sine (see Table 7.5)
zeta   Evaluates the Riemann zeta function (see Table 7.5)

These are special functions to symbolically represent and use certain mathematical functions. Besides, these several functions in the Symbolic Math Toolbox may return answers in terms of these functions.

Another way to access an even a bigger group of the special functions is to use the `mfunlist` call that lists special functions as presented in Tables 7.5 and 7.6. In these tables, n and m denote an integer argument; x, y, t, a, and b, the real argument; c, d, the vectors of real arguments (d ≥ **0**), and z the complex argument. Each function has its own specifics, and so for more detailed descriptions including any argument restrictions, you should refer to the MATLAB Help system.

These functions should not be run directly, but rather using the `mfun` function that has a general syntax

```
mfun('function',p1,p2,...,pk)
```

Table 7.5 Special Functions of MATLAB

Function	Description	Function	Expected Arguments Description
`bernoulli(n)` `bernoulli(n,t)`	Bernoulli numbers and polynomials: $\dfrac{te^{xt}}{e^t-1} = \sum_{n=0}^{\infty} B_n(x)\dfrac{t^n}{n!}$, $0 < \lvert t \rvert < 2\pi$	`euler(n)` `euler(n,z)`	Euler numbers and polynomials: $\dfrac{2e^{xt}}{e^t+1} = \sum_{n=0}^{\infty} E_n\dfrac{t^n}{n!}$, $\lvert t \rvert < 0.5\pi$
`besseli(y,x)` `besselj(y,x)` `besselk(y,x)` `bessely(y,x)`	Bessel function of the first kind Bessel function of the second kind	`fresnels(x)` `fresnelc(x)`	Fresnel sine and cosine integrals: $S(x) = \int_0^x \sin\left(\dfrac{\pi}{2}t^2\right)dt, \; C(x) = \int_0^x \cos\left(\dfrac{\pi}{2}t^2\right)dt$
`beta(x,y)`	Beta function: $B(x,y) = \dfrac{\Gamma(x)\Gamma(y)}{\Gamma(x+y)}$	`gamma(z)` `gamma(a,z)`	Gamma and incomplete gamma functions: $\Gamma(z) = \int_0^{\infty} e^{-t}t^{z-1}dt, \; \Gamma(a,z) = \int_z^{\infty} e^{-t}t^{a-1}dt$
`binomial(m,n)`	Binomial coefficients: $\left(\dfrac{m}{n}\right) = \dfrac{m!}{n!(m-n)!}$	`harmonic(n)`	Harmonic function: $h(n) = \sum_{k=1}^{n} \dfrac{1}{k!}$
`chi(z)`	Hyperbolic cosine integral: $Chi(z) = \gamma + \ln(z) + \int_0^z \dfrac{\cosh(t)-1}{t}dt$	`hypergeom(c,d,x)`	Generalized hypergeometric function
`ci(x)`	Cosine integral: $Ci(z) = \gamma + \ln(x) + \int_0^x \dfrac{\cos(t)-1}{t}dt$	`lngamma(z)`	Logarithm of the gamma function: $\ln(\Gamma(z))$
`dawson(x)`	Dawson integral: $F(x) = e^{-x^2}\int_0^x e^{t^2}dt$	`li(x)`	Logarithmic integral: $Li(x) = Ei(\ln(x))$

(Continued)

Table 7.5 Special Functions of MATLAB (Continued)

Function	Description	Function	Expected Arguments Description
`dilog(x)`	Dilogarithm integral: $f(x) = \int_1^x \frac{ln(t)}{1-t}\,dt$	`psi(z)` `psi(n,z)`	Digamma and polygamma functions: $\Psi(x) = \dfrac{\Gamma'(x)}{\Gamma(x)}, \; \Psi^{(n)}(z) = \dfrac{d^n}{dz^n}\Psi(z)$
`ei(x)` `ei(n,z)`	Exponential integrals: $Ei(x) = \int_{-\infty}^x e^t\, t^{-1}\,dt, \; Ei(n,z) = \int_1^\infty e^{-zt}\, t^{-n}\,dt$	`shi(z)`	Hyperbolic sine integral: $Shi(z) = \int_0^z \frac{\sinh(t)}{t}\,dt$

`elliptick(k)`
`elliptice(k)`
`ellipticpi(a,k)`
	Complete elliptic integrals of the first, second, and third kind	`si(x)`	Sine integral: $Si(x) = \int_0^x \frac{\sin(t)}{t}\,dt$

`ellipticck(k)`
`ellipticce(k)`
`ellipticpi(a,k)`
	Associated complete elliptic integrals of the first, second, and third kind	`ssi(z)`	Shifted sine integral: $Ssi(z) = Si(z) - 0.5\pi$

`ellipticF(x,k)`
`ellipticE(x,k)`
`ellipticPi(x,a,k)`
	Incomplete elliptic integrals of the first, second, and third kind	`w(z)`
		`w(n,z)`
			Lambert's W function $z = w(z)e^{w(z)}$ and the Kth branch of W function

`erf(z)` `erfc(z)`	Error function and complementary error function: $erf(z) = \frac{2}{\sqrt{\pi}}\int_0^z e^{-t^2}\,dt, \; erfc(z) = 1 - erf(z)$	`zeta(z)` `zeta(n,z)`	Riemann zeta function: $\zeta(z) = \sum_{n=1}^\infty \frac{1}{n^z}$

Table 7.6 Orthogonal Polynomials

Function	Description (Recursive Definition)
`T(n,x)` and `U(n,x)`	Chebyshev polynomials of the 1st and 2nd kind
	$T_0(x) = 1$, $T_1(x) = x$, $T_n(x) = 2xT_{n-1}(x) - T_{n-2}(x)$
	$U_0(x) = 1$, $U_1(x) = 2x$, $U_n(x) = 2xU_{n-1}(x) - U_{n-2}(x)$
`G(n,a,x)`	Gegenbauer polynomials
	$C_0^{(\alpha)}(x) = 1$, $C_1^{(\alpha)}(x) = 2\alpha x$, $C_n^{(\alpha)}(x) = 1/n(2x(n + \alpha - 1) C_{n-1}^{(\alpha)}(x) - (n + 2\alpha - 2) C_{n-2}^{(\alpha)}(x))$
`H(n,x)`	Hermite polynomials
	$H_0(x) = 1$, $H_1(x) = x$, $H_n(x) = xH_{n-1}(x) - (n-1)H_{n-2}(x)$
`P(n,a,b,x)`	Jacobi polynomials
	$P_0^{(\alpha,\beta)}(x) = 1$, $P_1^{(\alpha,\beta)}(x) = 1/2((2 + \alpha + \beta)x + (\alpha - \beta))$
	$P_n^{(\alpha,\beta)}(x) = \dfrac{(\rho-1)((\alpha^2 - \beta^2) + \rho(\rho-2)x)P_{n-1}^{(\alpha,\beta)} - 2(n-1+\alpha)(n-1+\beta)\rho P_{n-2}^{(\alpha,\beta)}}{2n(n+\alpha+\beta)(\rho-2)}$
	where $\rho = 2n + \alpha + \beta$
`L(n,x)` and `L(n,a,x)`	Simple and generalized Laguerre polynomials
	$L_0(x) = 1$, $L_1(x) = 1 - x$, $L_n(x) = 1/n((2n-1-x)L_{n-1}(x) - (n-1)L_{n-2}(x))$
	$L_0^{(\alpha)}(x) = 1$, $L_1^{(\alpha)}(x) = 1 + \alpha - x$, $L_n^{(\alpha)}(x) = 1/n((2n-1+\alpha-x)L_{n-1}^{(\alpha)}(x) - (n-1+\alpha)L_{n-2}^{(\alpha)}(x))$
`P(n,x)`	Legendre polynomials
	$P_0(x) = 1$, $P_1(x) = x$, $P_n(x) = 1/n((2n-1)xP_{n-1}(x) - (n-1)P_{n-2}(x))$

where `function` is the name of the function (Tables 7.5 and 7.6), and `p`'s are the numeric inputs to `function`. As a result, this call numerically evaluates one of the symbolic engine's special mathematical functions. Compared to the running time for standard MATLAB calculations, calculations involving special functions are typically much slower.

7.9 MuPAD Development Environment

As mentioned in Sec. 7.2, starting from R2007b+, the Symbolic Math Toolbox of MATLAB relies on the MuPAD engine. In addition to covering common mathematical tasks as described in the previous sections, the functions of MuPAD language cover specialized areas, such as number theory, combinatorics, and others. You can extend its built-in functionality even further by writing the custom symbolic functions and libraries. Although you can operate with the MuPAD language functions from the MATLAB development environment (to be addressed at the end of this section), let us briefly introduce the main program that came with a new engine, MuPAD Notebook, representing a separate development environment.

You may start MuPAD by launching the specialized MuPAD interface

```
>> mupadwelcome
```

which brings the window shown in Fig. 7.2, or by opening a new blank MuPAD notebook directly

```
>> mupad
```

resulting in a window shown in Fig. 7.3. In the first case, you can access MuPAD Help (by clicking one of the three options in the **First Steps** pane), open an existing (saved) notebook file (from the ones displayed in the **Open Recent File** pane on the right), launch a new notebook (**New Notebook** option) or editor (**New Editor** option), or navigate to another existing MuPAD notebook file (**Open File** option).

The MuPAD notebook toolbar (Fig. 7.3) features the Standard Toolbar and Format Toolbar on the top, Input Area, occupying most of the notebook window and Command Bar, a symbol palette for accessing the common MuPAD functions, on the right. While the first 12 and the last two buttons of the Standard Toolbar are pretty much standard Microsoft-type buttons, the remaining four allow you to add a text area (called Text Paragraph) or input area (called Calculation) into the Input Area, to start and stop symbolic evaluation. The standard (Microsoft Office look-like) instruments of the Format Toolbar allow you to change the font and attributes of the entered text. The commands for execution, evaluation, or plotting are entered in the input areas. The Command Bar (on the right) helps to enter these commands with the proper syntax.

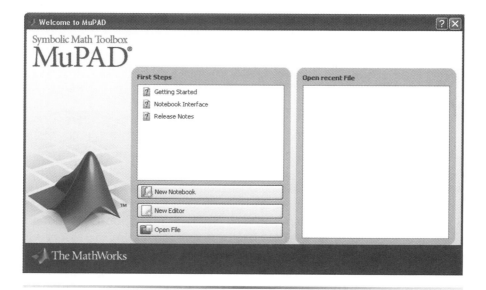

Fig. 7.2 MuPAD interface window.

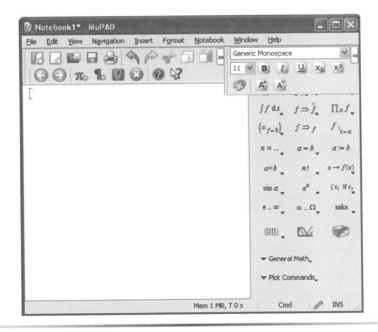

Fig. 7.3 MuPAD Notebook window.

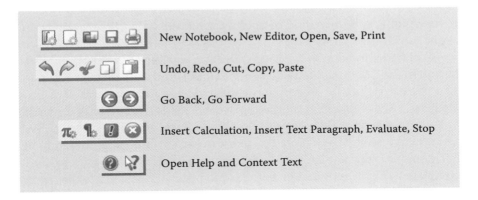

The MuPAD notebook interface differs from the MATLAB interface as follows. Similar to MATLAB, the commands typed in an input area are not evaluated until you press <Enter>. However, the MATLAB method of recalling a previous command by pushing the <↑> key does not have the same effect in a MuPAD notebook. MuPAD notebook uses the arrow keys for navigation, similar to the most word processors. You can edit the commands typed in any input area. For example, you can change a command, correct its syntax, or try different values of parameters simply by selecting the area you wish to change and typing over it. Pressing <Enter> forces the notebook to evaluate the result. Some other differences are shown in Tables 7.7 and 7.8.

Table 7.7 Major Differences between MATLAB and MuPAD Syntax

	MATLAB	MuPAD
Assignment	`=`	`=:`
List variables	`whos`	`anames(All,User)`
Numerical value of an expression	`double(expression)`	`float(expression)`
Suppress output	`;`	`:`
Enter a matrix	`[a,b,c; d,e,f]`	`matrix([a,b,c], [d,e,f])`
`{a,b,c}`	cell array	set
Linear algebra commands	nothing extra is needed	`linalg::` prefix or `use(linalg)`
Autocompletion	\<Tab\>	\<Ctrl\>+\<Space\>
Equality and inequality comparison	`==`, `~=`	`=`, `<>`

Let us consider a simple example. Suppose we want to compute an integral $\int_0^\infty \sin(x)x^{-1}dx$. We start from opening the blank MuPAD window (Fig. 7.3) and typing

```
f:=sin(x)/x
```

in the input area demarcated by a left bracket. Pressing \<Enter\> yields the result shown in Fig. 7.4a.

To place an integral in the correct syntax, you need to click the integral button in the Command Bar and select definite limits as shown in Fig. 7.4b. The correct syntax for integration appears in the input area as shown in Fig. 7.5a. Now we need to fill the replaceable fields marked with a symbol #. You can use the \<Tab\> key to select these fields and type `f`, `x`, `0`, and `infinity`, respectively. Note that you can use an autocompletion option. For example, you may type `infi` and then use \<Ctrl\>+\<Space\> to autocomplete it. Upon completion, you need to press \<Enter\> to see the result, shown in Fig. 7.5b.

You can supplement your code with comments by inserting the text paragraph. You may use the corresponding button or use shortcuts: \<Ctrl\>+\<T\> to insert a text paragraph after the current calculation area or \<Ctrl\>+\<Shift\>+\<T\> to insert a text paragraph above the current calculation area. Figure 7.6a presents an example. It also shows the next step of choosing the simplest plotting command to plot the integrand. More plotting options are available under the **Plot Commands** tab (you can also look for the `plotfunc2d` and `plotfunc3d` functions in the MuPAD notebook Help system, which is pretty similar to that of MATLAB). After filling the replaceable field and pushing the <Enter> key, the plot will appear as shown in Fig. 7.6b. You can save this Notebook for the further use (by default it saves it with an extension *.mn*) or export to *html* or *pdf* file (by choosing an appropriate type in the **Save** window of the **Export...** option of the **File** menu).

Now, you may scroll up and down between the input areas, but if you introduce any changes in one specific area and push the <Enter> key, only this selected area will be reevaluated. For example, if you slightly change the integrand as shown in Fig. 7.7a and reevaluate this function, only the integrand itself will change, not affecting the integral and the plot, that is, the changes will not automatically propagate throughout the notebook. To have the changes cascade to all parts of the notebook, from top to bottom, you should select the **Evaluate All** option from the **Notebook** menu (as also shown in Fig. 7.7a). The result is presented in Fig. 7.7b. Now your notebook is consistent.

The same applies when you open a saved MuPAD notebook file. Although you can see all the commands you saved, they are not synchronized with the MuPAD engine, until you select **Notebook** and hit the **Evaluate All`** option.

Table 7.8 Major Differences between the MATLAB and MuPAD Functions

MATLAB	MuPAD
Inf	infinity
pi	PI
i	I
NaN	undefined
fix	trunc
log	ln
asin, acos, atan, asinh, acosh, atanh, asc, asec, acot, acsch, asech, acoth	arcsin, arccos, arctan, arcsinh, arccosh, arctanh, arcsc, arcsec, arccot, arccsch, arcsech, arccoth
besselj, bessely, besseli, besselk	besselJ, besselY, besselI, besselK
lambertw	lambertW
sinint, cosint	Si, Ci
eilergamma, catalan	EULER, CATALAN
conj	conjugate
laplace, ilaplace	transform::laplace, transform::ilaplace
ztrans, iztrans	transform::ztrans, transform::iztrans

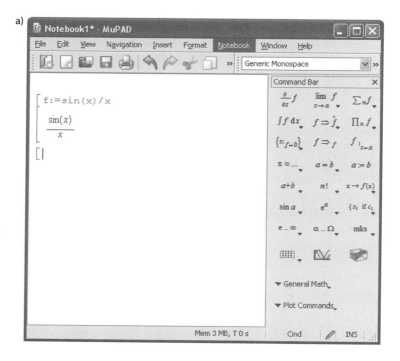

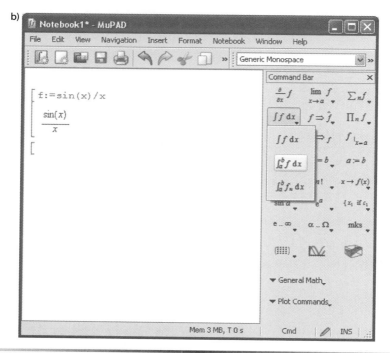

Fig. 7.4 Typing mathematical expressions in the MuPAD notebook.

a)
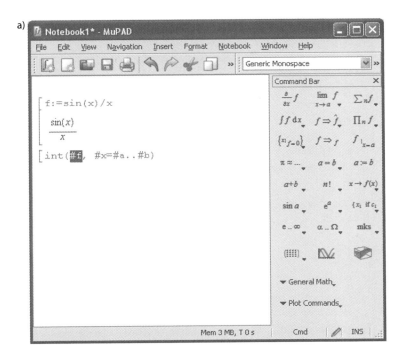

b)
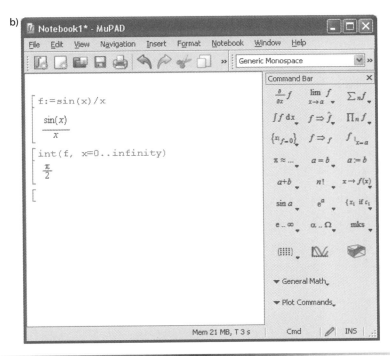

Fig. 7.5 Filling the replaceable fields in the MuPAD notebook.

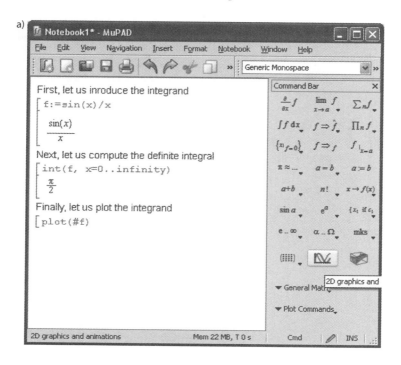

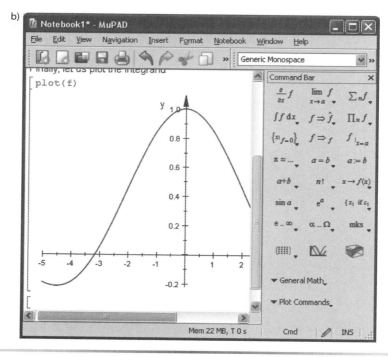

Fig. 7.6   Adding comments and plots in the MuPAD notebook.

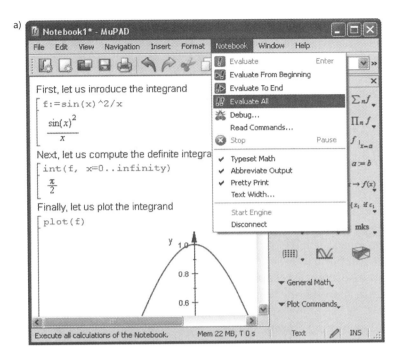

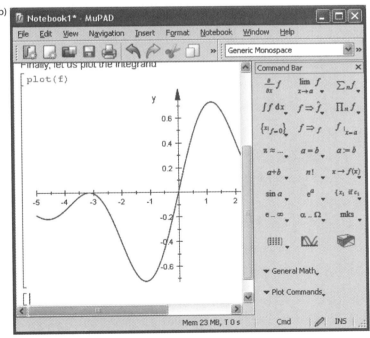

Fig. 7.7 Cascading calculations and synchronizing the notebook with the MuPAD engine.

To conclude the brief introduction of MuPAD notebook, we must say that you can also convert the results into MathML and TeX. You can embed graphics, animations, and descriptive text within your notebook. An editor, debugger, and other programming utilities of MuPAD provide tools for authoring custom symbolic functions and libraries in the MuPAD language. The MuPAD language supports multiple programming styles including imperative, functional, and object-oriented programming. The language treats variables as symbolic by default and is optimized for handling and operating on symbolic math expressions.

Let us also show how to copy variables and expressions between the MATLAB workspace and MuPAD notebooks. The simplest way of doing it is just copying it from one window (MuPAD notebook or the MATLAB Command Window) and pasting to another. Of course in this case, they will not necessarily be immediately usable, you may want to introduce some corrections to reflect the differences depicted in Tables 7.7 and 7.8. You can also copy a response as shown in Fig. 7.8 and paste it into the MATLAB Command window to obtain its symbolic representation ($sin(x)^2/x$ in our case). If you paste it into the word document, it will appear as a bitmap graphic object.

However, the proper way of integrating both environments, so that these variables and expressions could be used immediately, is to do it programmatically. You can copy a variable in a MuPAD notebook to a variable in the MATLAB workspace using a MATLAB command. Similarly, you can

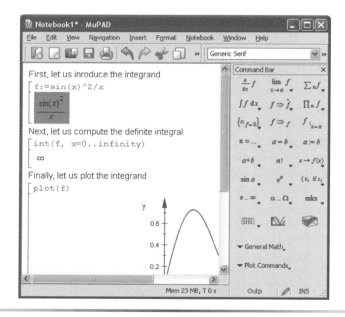

Fig. 7.8 Copying MuPAD expressions.

copy a variable or symbolic expression in the MATLAB workspace to a variable in a MuPAD notebook using another MATLAB command. In either case, you need to know the handle to the MuPAD notebook you want to address. Hence, from the very beginning, you would better use the following call:

```
>> h_ntb=mupad;
```

The only difference is that this call creates a handle to the notebook you intend to use.

Suppose that following this call, you introduced the function shown in Fig. 7.9a. Now, you can create a symbolic variable in the MATLAB workspace by using the getVar function

```
>> f = getVar(h_ntb,'f')
```

which returns

```
f = exp(y)*asin(x)
```

Let us introduce a new function g in MATLAB as follows:

```
>> g=sqrt(f)
g = (exp(y)*asin(x))^(1/2)
```

To be able to pass this new variable with its (symbolic) value to the MuPAD notebook, we can employ the setVar function

```
>> setVar(h_ntb,'w',g)
```

which assigns the symbolic expression g in the MATLAB workspace to the variable w in the MuPAD notebook n_ntb. Now in the MuPAD notebook, we can call this new variable created in MATLAB to get the result shown in Fig. 7.9b.

7.10 Plotting Symbolic Expressions

The last group of the Symbolic Math Toolbox functions, the Plotting functions, allows you to create graphical representations of the results of symbolic computations. In fact, these easy-to-create functions have been introduced in Chapter 6 (Sec. 6.10) already. They are

ezcontour	Contour plotter
ezcontourf	Filled contour plotter
ezmesh	3D mesh plotter
ezmeshc	Combined mesh and contour plotter
ezplot	2-D parametric curve plotter
ezplot3	3-D parametric curve plotter
ezpolar	Polar coordinate plotter

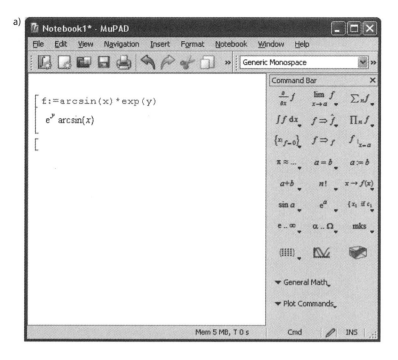

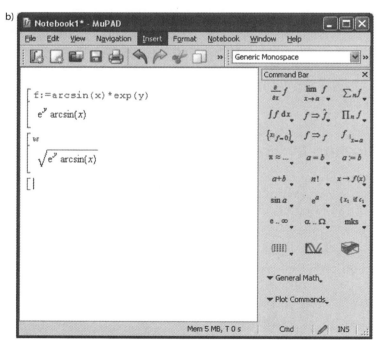

Fig. 7.9 Using the a) `getVar` and b) `setVar` functions.

```
ezsurf          3-D colored surface plotter
ezsurfc         Combined surface and contour plotter
```

Earlier, these plotting functions used the text strings as input arguments for defining mathematical expressions to plot. It turns out that these functions have much more to offer and can work with symbolic input arguments as well. For example, let us compute the Bessel function $J_1(z)$ by calling the corresponding Symbolic Math Toolbox function, and plot it using the `ezplot` function as follows:

```
>> syms z
>> g=besselj(1,z);
>> ezplot(g)
```

These commands produce the result shown in Fig. 7.10a. The default domain for `ezplot` is $-2\pi \le x \le 2\pi$, but you can alter it by including a vector with the lower and upper limits for the x axis (and y axis, if necessary), for example (Fig. 7.10b),

```
>> ezplot(g,[0 10*pi])
```

Of course, the beauty of using the functions from the Plotting Functions group of the Symbolic Math Toolbox is that there is no need to compute the vectors of data points before calling these functions, which results in extreme compactness of code.

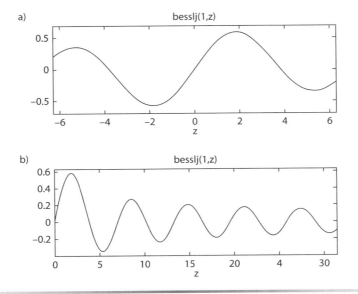

Fig. 7.10 Example of the `ezplot` function output with a) default and b) specified x domains.

a)

$x = t,\ y = -\cos(t)+2,\ z = \sin(t)$

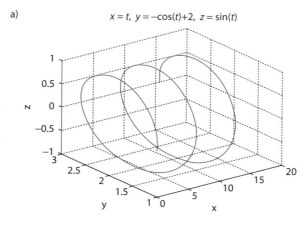

b)

$x = r\cos(\phi),\ y = r\sin(\phi),\ z = 1+\text{besselj}(1,3r)\cos(\phi)$

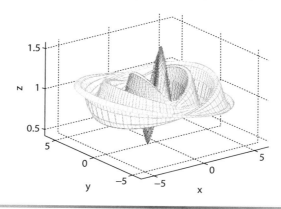

Fig. 7.11 Examples of exploiting a) ezplot3 and b) ezmesh functions.

Figure 7.11a shows another example, when the 3-D plotting function, ezplot3, was used to present the results of solving the system of two differential equations as discussed at the very end of Sec. 7.6:

```
>> syms t
>> [x,y]=dsolve('Dx=y','Dy=cos(t)','x(0)=1','y(0)=0');
>> ezplot3(t,x,y,[0 5*pi])
```

Finally, Fig. 7.11b shows an example of using the surface-plotting function, ezmesh. The commands that produce this nice membrane without creating a mesh grid (as you would need to do in the case of the mesh function) are as follows:

```
>> syms r phi
>> x=r*cos(phi); y=r*sin(phi);
>> z=1+besselj(1,3*r)*cos(phi);
>> ezmesh(x,y,z)
```

7.11 Using Symbolic Expressions in Numerical Operations

MATLAB offers several ways of using symbolic expressions in numerical operations. In what follows, we start from showing how to use the `fcnchk` and `inline` functions introduced earlier, in Secs. 5.11.2 and 5.8.1, respectively, and then proceed with the `matlabFunction` function, specifically designed for these purposes.

7.11.1 Employing the `fcnchk` and `Inline` Functions

As shown in Sec. 7.10, the easy-to-use plotting functions allow you to use symbolic expressions as input arguments. However, other functions of MATLAB do not necessarily permit it. For example, although the call

```
>> syms x, ezplot(sin(x))
```

does produce an expected sinusoid, the following commands, intended to find a minimum of a sinusoid $f(x) = \sin x$ near $x = 1$ using the MATLAB `fminsearch` function (to be addressed further in Chapter 10):

```
>> syms x, fminsearch(sin(x),1)
```

return an error

```
??? Error using ==> fcnchk at 108
If FUN is a MATLAB object, it must have an feval method.
Error in ==> fminsearch at 178
funfcn = fcnchk(funfcn,length(varargin));
```

As usual, the detailed explanation of the error provides a clue for resolving the issue. Indeed, the reason for the error is stated fairly clearly—the wrong type of an input argument provided to the `fcnchk` function, which appears in `fminsearch` to convert the first input argument into an inline object (as was discussed in Sec. 5.8.6). Hence, to use `fminsearch` and other MATLAB functions with the symbolic objects, you first need to convert them into the format understood by the `fcnchk` function. Specifically, the symbolic object has to be converted to a string. The MATLAB function that does it is `char`. Replacing `sin(x)` with `char(sin(x))` in the aforementioned example fixes the problem.

Of course, in the latter example, we could use a string as opposed to a symbolic object from the very beginning, like

```
>> fminsearch('sin(x)',1)
```

However, it would be a waste of time if we performed complex symbolic computations and were not able to use the result in numerical computations. Therefore, the recipe provided earlier is a key to effective use of the Symbolic Math Toolbox.

Once your symbolic mathematical expression is converted into a string, you may do some other useful things with it as well. For instance, you can convert it into an inline function. The two ways of doing it are to use either `fcnchk` or `inline` function. The only difference is that if you want your function to be vectorized (a . inserted before any ^, *, or /), you use a slightly different format as shown as follows:

```
f=fcnchk(f,'vectorized') vs f=inline(vectorize(f))
```

To conclude this discussion, let us consider the following general example. Suppose that dealing with the Symbolic Math Toolbox, you produced some symbolic mathematical expression (we will use one of those obtained in Sec. 7.6). You then wish to plot it (using the `ezplot` function, you can do it right away using just a symbolic object as an input argument). After that, you want to find one of the zeros of this mathematical expression. To do that, you will need to use the MATLAB `fzero` function, introduced in Sec. 5.8.2 (and to be considered in more detail in Chapter 10). As you know already, for this purpose, you will need to convert your symbolic mathematical expression into a string or inline function. Finally, you would like to add your initial guess and a zero to the plot produced earlier (using the `ezplot` function). The following set of commands provides a complete solution to the problem:

```
fsym=dsolve('Dy=y^2+1','y(0)=1');
ezplot(fsym), hold, grid
```

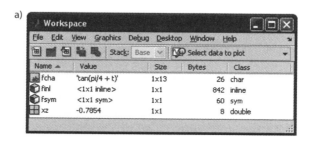

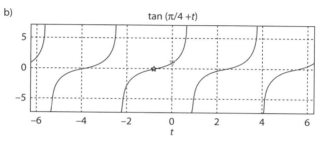

Fig. 7.12 Example of using symbolic expressions in numerical calculations.

```
fcha=char(fsym);
finl=fcnchk(fcha);
xz=fzero(finl,0);
plot(0,finl(0),'gV')
plot(xz,finl(xz),'rp')
```

The resulting Workspace and Figure windows are shown in Fig. 7.12a and 7.12b, respectively.

Note that instead of `fcnchk`, you could use the `inline` function as well. Also, in this particular example, the inline function was the only possible input argument to `fzero`. The reason is that the default output argument for the `dsolve` function is t, not x (as required by the `fzero` function). So, if you want to use a string, you would need to do the following adjustments to the first and fifth lines:

```
fsym=dsolve('Dy=y^2+1','y(0)=1','x');
. . .
xz=fzero(fcha,0);
```

which would also change t to x in Fig. 7.12b.

7.11.2 Using the `matlabFunction` Function

The latest versions of MATLAB feature another and even more powerful way allowing you to use symbolic expressions in numerical operations. Specifically, the `matlabFunction` from the former Extended Symbolic Math Toolbox allows you to convert the symbolic expression f into an anonymous function with the handle g. This function features several syntax

```
g = matlabFunction(f)
g = matlabFunction(f1,f2,…)
g = matlabFunction(f,option1,value1,…)
```

You can use handle g to calculate numerical values as if you were substituting numbers for variables in a symbolic expression. For example, the following set of commands creates a handle to the function that calculates the dynamic pressure using the static pressure ps and Mach number M inputs:

```
>> syms ps M
>> q = 0.7*ps*M^2;
>> hq = matlabFunction(q)
hq = @(M,ps)M.^2.*ps.*(7.0./1.0e1)
```

Now, the call

```
>> hq(0.5,101325)
```

returns the dynamic pressure (in pascals) for $M = 0.5$ and a standard (static) atmospheric pressure

```
ans = 1.7732e+004
```

Because `matlabFunction` creates a vectorized function, you can use the function just created with the matrix inputs as well. For example,

```
>> M = [0.2,0.2,0.6]; ps = [101325, 60000, 101325];
>> hq(M,ps)
```

yields

```
ans = 1.0e+004 *
      0.2837    0.1680    2.5534
```

Note that, `matlabFunction` generates input variables in alphabetical order from a symbolic expression, and that is why the function handle `hq` has `M` before `ps`. However, if needed, you can specify the order of input variables in the function handle using the `vars` option, by passing a cell array of strings or symbolic arrays, or a vector of symbolic variables. For example, the command to create an anonymous function with the reverse order of inputs can look like

```
>> hq = matlabFunction(q,'vars',{'ps', 'M'})
```

or

```
>> hq = matlabFunction(q,'vars',[ps M])
```

In either case, the return is

```
hq = @(ps,M)M.^2.*ps.*(7.0./1.0e1)
```

In addition to that, the `matlabFunction` allows you to create an M-file function as opposed to anonymous function. All you have to do is to specify the file name using the `file` option (if you do not specify the path to the file, `matlabFunction` creates this file in the current directory). The following example generates an M-file function that calculates a rotation matrix from the Earth-Centered, Earth-Fixed coordinate frame to the East-North-Up local tangent plane using two inputs, `lat` and `lon`

```
syms lat lon
Rlon = [cos(lon) sin(lon) 0; -sin(lon) cos(lon) 0; 0 0 1];
Rlat = [cos(lat) 0 sin(lat); 0 1 0; -sin(lat) 0 cos(lat)];
R = Rlat*Rlon;
F = matlabFunction(R,'file','RotationMatrix.m')
```

The file *RotationMatrix.m* contains the following code:

```
function R = RotationMatrix(lat,lon)
%ROTATIONMATRIX
%    R = ROTATIONMATRIX(LAT,LON)
```

```
% This function was generated by the Symbolic Math Toolbox version 5.4.
% 30-Apr-2010 15:22:36
t2 = cos(lat);
t3 = sin(lon);
t4 = cos(lon);
t5 = sin(lat);
R = reshape([t2.*t4,-t3,-t4.*t5,t2.*t3,t4,-t3.*t5,t5,0.0,t2],[3, 3]);
```

As seen, `matlabFunction` generates several intermediate variables, which constitutes the *optimized* code. Using intermediate variables results in more efficient code (because of reusing intermediate expressions) and makes it easier to read (by keeping expressions short). Once again, if you do not want the default alphabetical order of input variables, you may always use the `vars` option. Continuing the example mentioned earlier,

```
F=matlabFunction(R,'file','RotationMatrix.m','vars',[lon lat])
```

generates a file equivalent to the previous one, with a different order of inputs

```
function R = RotationMatrix (lon,lat)
...
```

The `matlabFunction` function also allows you to have multiple output variables with customization of their names by using the `output` option. For example, if in the previous example, you use a slightly modified call

```
F=matlabFunction(R,Rlat,Rlon,'file',...
            'RotationMatrix.m','outputs',{'R','Rlat','Rlon'})
```

it will result in generating the following function:

```
function [R,Rlat,Rlon] = RotationMatrix(lat,lon)
%ROTATIONMATRIX
% [R,RLAT,RLON] = ROTATIONMATRIX(LAT,LON)
% This function was generated by the Symbolic Math Toolbox version 5.4.
% 30-Apr-2010 15:40:02
t2 = cos(lat);
t3 = sin(lon);
t4 = cos(lon);
t5 = sin(lat);
R = reshape([t2.*t4,-t3,-t4.*t5,t2.*t3,t4,-t3.*t5,t5,0.0,t2],[3, 3]);
if nargout > 1
    Rlat = reshape([t2,0.0,-t5,0.0,1.0,0.0,t5,0.0,t2],[3, 3]);
end
if nargout > 2
    Rlon = reshape([t4,-t3,0.0,t3,t4,0.0,0.0,0.0,1.0],[3, 3]);
end
```

As seen, this latter function automatically incorporates possibilities to be called using one, two, or three output arguments (that is, uses the `nargout` function discussed in Sec. 5.11.3).

7.12 LaTeX Representation of Symbolic Expression

There is one more possible application of symbolic expressions beyond using them in numerical operations and plotting them with the help of the easy-to-create functions. The LaTeX expressions introduced in Sec. 6.5 may have a good use of them.

The `latex(S)` function, formally introduced in Sec. 7.4, returns the LaTeX representation of the symbolic expression S. For example, let us create several symbolic expressions using the following commands:

```
>> syms x y t
>> F1 = diff(atan(x))
>> F2 = sym(hilb(3))
>> F4 =(x^2+1/x)/sqrt((sin(x)+cos(x))/x^3.5)
>> F5 = taylor(sin(x),8,pi/2)
```

They return

```
F1 = 1/(x^2 + 1)
F2 = [    1, 1/2, 1/3]
     [ 1/2, 1/3, 1/4]
     [ 1/3, 1/4, 1/5]
F4 = (1/x + x^2)/((cos(x) + sin(x))/x^(7/2))^(1/2)
F5 = (pi/2 - x)^4/24 - (pi/2 - x)^2/2 - (pi/2 - x)^6/720
```

Now, let us use the `latex` function

```
>> H1=latex(F1)
>> H2=latex(F2)
>> H4=latex(F4)
>> H5=latex(F5)
```

which converts the aforementioned symbolic expressions to the LaTeX format

```
H1 = frac{1}{x^2 + 1}
H2 = \left(\begin{array}{ccc} 1 & \frac{1}{2} &
     \frac{1}{3}\\ \frac{1}{2} & \frac{1}{3} &
     \frac{1}{4}\\ \frac{1}{3} & \frac{1}{4} &
     \frac{1}{5} \end{array} \right)
H4 = \frac{\frac{1}{x} + x^2}{\sqrt{\frac{\cos\!
     \left(x\right) + \sin\!\left(x\right)}
     {x^{\frac{7}{2}}}}}
```

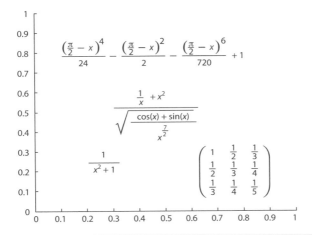

Fig. 7.13 Converting symbolic expressions to the LaTeX strings.

```
H5 = \frac{{\left(\frac{\pi}{2} - x\right)}^4}{24} - \frac{{\left(\
     pi}{2} - x\right)}^2}{2} -
     \frac{{\left(\frac{\pi}{2} - x\right)}^6}{720} + 1
```

You can use these LaTeX expressions for publishing M-scripts (see Fig. 5.23) or adding them to the plots, like

```
>> text(.2,.25,['$$' H1 '$$'],'interpreter','latex','fontsize',14)
>> text(.6,.2, ['$$' H2 '$$'],'interpreter','latex','fontsize',16)
>> text(.3,.5, ['$$' H4 '$$'],'interpreter','latex','fontsize',18)
>> text(.1,.8, ['$$' H5 '$$'],'interpreter','latex','fontsize',14)
```

which results in displaying them within the default axes as shown in Fig. 7.13.

7.13 Symbolic Calculators

As for any other MATLAB toolbox of Simulink blockset, the Symbolic Math Toolbox has several demos providing hints and examples on using a variety of functions. Among these demos, there are three interactive GUIs that exploit different functions of this toolbox and present the results graphically. These GUIs are as follows:

funtool	Function calculator
rsums	Interactive evaluation of the Riemann sums
taylortool	Taylor series calculator

For educational purposes, let us briefly mention these GUIs here, because once you learn how to use the Mathworks's GUI design

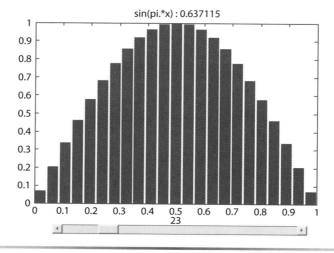

Fig. 7.14 Example of using the interactive Riemann sums calculator of the Symbolic Math Toolbox.

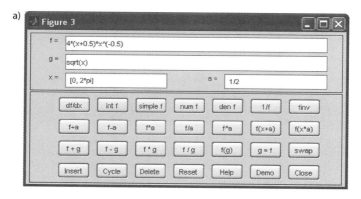

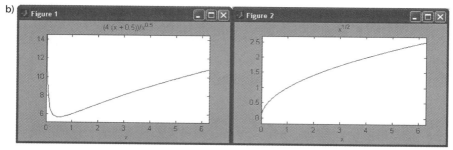

Fig. 7.15 Examples of using the function calculator of the Symbolic Math Toolbox.

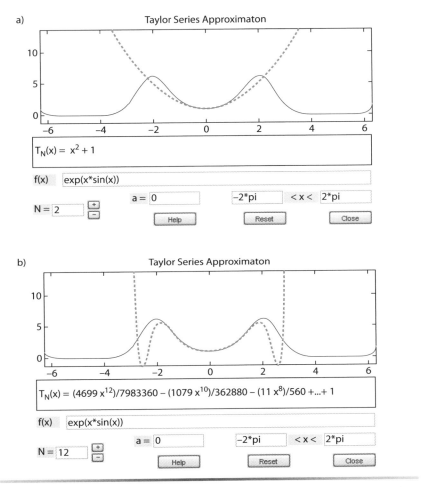

Fig. 7.16 Examples of using the Taylor tool of the Symbolic Math Toolbox for $f(x) = e^{x\sin(x)}$ with a) two and b) 12 terms.

environment, GUIDE, presented in Appendix B, you will be able to create such GUIs yourself.

We start from the `rsums` function, which can be used in one of the following formats: `rsums(f)`, `rsums(f,a,b)`, or `rsums(f,[a,b])`, where `f` is the function (a symbolic expression or string), whereas `a` and `b` are the limits (the default values are `0` and `1`, respectively). A Riemann sum is a method for numerical evaluation of a definite integral (approximating the total area underneath a curve on a graph). In fact, there are several approaches including Left Rectangles sum, Right Rectangles sum, and Trapezoidal rule (to be considered in Chapter 13 in detail). For example, the following call:

```
>> syms x, rsums(sin(pi*x))
```

displays a graph of $f(x) = \sin(\pi x)$ within $x \in [0;1]$ and allows you to adjust the number of terms taken in the Riemann sum by using a slider below the graph (Fig. 7.14 shows 23 terms). The number of terms available ranges from 2 to 128.

The `funtool` call invokes three windows of a visual function calculator (Fig. 7.15) that manipulates and displays two functions of one variable, $f(x)$ and $g(x)$. At the click of a button, this function calculator draws a graph representing the sum, product, difference, or ratio of two functions that you specify (the result shows up in the **Figure 1** window).

Finally, the Taylor series calculator that can be invoked by typing `taylortool` or `taylortool(f)` in the Command Window, brings up a display (Fig. 7.16) allowing you to graphically explore the quality of the Taylor series of a function of interest `f` with a specified number of terms [see Eq. (7.3)].

As shown in Fig. 7.16, increasing the number of terms in the Taylor series expansion allows you to have a better approximation of the function.

Problems

7.1 Use the `limit` function to compute $\lim_{x \to 0^+}(1-x)^{x^{-2}}$. Show that it is different from $\lim_{x \to 0^-}(1-x)^{x^{-2}}$. Prove your findings graphically (using the `ezplot` function).

7.2 The equation for a curve called the "lemniscate" (of Bernoulli) in polar coordinates (r,θ) is $r^2 = a^2 \cos(2\theta)$. Use the `subs` function to find the equation for this curve in Cartesian coordinates (x,y), where $x = r\cos\theta$ and $y = r\sin\theta$. Use the `ezplot` function to plot several lemniscates computed for different values of coefficient a in the same figure.

7.3 Use the `dsolve` function to solve two differential equations: $\dot{x} - y = 0$ and $\dot{y} + 2x + 5 = 0$ with $x(0) = 0$ and $y(0) = 1$. Plot the results in the range $t=[0;5]$s using the `fplot` function. Use the `fzero` function to compute the location of the first three zeros for both $x(t)$ and $y(t)$. Add them to the plot. Add the axis labels, and using the `text` function show mathematical expression of each solution by its curve.

Chapter 8 Accuracy of Digital Computations

- Understanding Computer Number Representation
- Understanding Round-Off Errors
- Understanding Truncation Errors
- Numerical Errors Control

```
bin2dec, dec2bin, dec2hex,
hex2dec, num2hex, hex2num,
base2dec, dec2base,
uint8…uint64, int8…int64,
intmax, intmin, eps,
realmax, realmin, single,
double, vpa
```

8.1 Introduction

Before proceeding with any numerical method, it is important to understand what to expect in terms of accuracy and precision when programming mathematical formulas on a computer. The primary objective of this chapter is to acquaint you with the major sources, quantification, and minimization of errors involved in numerical computations. The first source of errors is the computer itself, because it is limited in its ability to represent the magnitudes and precision of numbers. Second, this imperfection is aggravated further by inappropriate handling of these numbers. Lastly, the mathematical approximation or numerical method may introduce some additional errors. Specific objectives and topics covered in this chapter are divided into four sections. It starts from introducing concepts of accuracy and precision of calculations and measurements. Next, it discusses the alternative ways of representing numbers in computer's memory, followed by showing how a limited ability to represent the numbers results in round-off errors. Another source of numerical errors, truncation errors, associated with the fact that mathematical expressions, even as simple as π, are represented by approximations rather than exact formulas, is presented later. The chapter ends with some practical advices on how to control (mitigate) these errors.

8.2 Accuracy and Precision of Calculations and Measurements

The errors associated with both calculations and measurements can be characterized with regard to their *accuracy* and *precision*. Accuracy refers to how closely a computed or measured value agrees with the true value. Precision refers to how closely individual computed or measured values agree with each other. *Inaccuracy* (also called *bias*) is defined as systematic deviation from the truth, whereas imprecision (also called *uncertainty*), however refers to the magnitude of the scatter. These concepts are illustrated in Fig. 8.1 where you can assume that the true value for some 2-D quantity is located at the (0,0) point.

The general requirement to any numerical method to be discussed in the following chapters (Chapters 9–14) is that it should be sufficiently accurate or unbiased to meet the requirements of a particular problem, and also precise enough, so that you do not get the different results every time you get a numerical solution. We will use the collective term *error* to represent both the inaccuracy and imprecision of our estimates. For each numerical method, we explicitly address the issue of the upper bound of the error this method guarantees, thus making you understand the method's limitations.

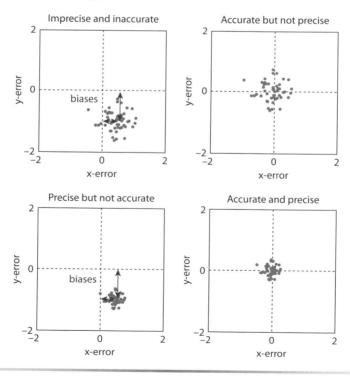

Fig. 8.1 Illustration of accuracy and precision concepts.

8.3 Computer Number Representation

Recall that by default, MATLAB allocates 8 bytes per each numeric variable, which means it deals with *double-precision* numbers. However, MATLAB also allows you to represent numbers using other formats or *data types*. The term data type refers to the way in which a computer represents the numbers in its memory. A data type determines the amount of storage allocated to a number, the method used to encode the number's value as a pattern of binary digits, and the operations available for manipulating the type. Most computers provide a choice of data types for representing numbers, each with specific advantages in the areas of precision, dynamic range, performance, and memory usage. To enable you to take advantage of different ways to represent numerical data to optimize (increase) the performance of MATLAB programs (and decrease the size of code generated from the Simulink models), MATLAB allows you to specify the data types of MATLAB variables. Simulink builds on this capability by allowing you to specify the data types of Simulink signals and block parameters, which is particularly useful in real-time control applications.

Numerical round-off errors are directly related to the manner in which numbers are stored in a computer. The fundamental unit whereby information is represented is called a *word*. This is an entity that consists of a string of *binary* digits, or *bits*. Numbers are typically stored in one or more words. To understand how this is accomplished, we must first review different numerical systems and a way they represent the numbers.

8.3.1 Numerical Systems

A number system is merely a convention for representing quantities. Because we have 10 fingers and 10 toes, the number system that we are most familiar with is the *decimal*, or *base-10*, number system. A base is the number used as the reference for constructing the system. The base-10 system uses 10 digits—0, 1, 2, 3, 4, 5, 6, 7, 8, and 9, traditionally called Arabic—to represent the numbers.

In fact, real Arabic digits, Hindi digits, represented by the values U + 0660 through U + 0669 in Unicode as shown in the following table, were adopted from writing in India and look quite different from Western or European digits as Unicode Character Standard refers them.

European Digits	0	1	2	3	4	5	6	7	8	9
Arabic Digits	٠	١	٢	٣	٤	٥	٦	٧	٨	٩

(By the way, although Arabic text is written right-to-left, numbers are written the same way as in left-to-right languages, with the most significant digit on the left.)

By themselves, these digits are satisfactory for counting from 0 to 9. For larger quantities, combinations of these basic digits are used, with the position or *place value* specifying the magnitude. The rightmost digit in a whole number represents a number from 0 to 9 or a multiple of 10 in the zeroth power. The second digit from the right represents a multiple of 10 (in the first power). The third digit from the right represents a multiple of 100 (10 in the second power) and so on. If we have a fractional number, the same paradigm applies—the first number to the right of a decimal point is a multiple of 0.1 (10 in the minus first power), etc. For example, if we have the number 3451.67, then we have three groups of 1000, four groups of 100, five groups of 10, one group of 1, six groups of 0.1, and seven groups of 0.01, or

$$3 \times 10^3 + 4 \times 10^2 + 5 \times 10^1 + 1 \times 10^0 + 6 \times 10^{-1} + 7 \times 10^{-2} = 3451.67$$

This type of representation is called *positional notation.*

Now, because the decimal system is so familiar, it is not commonly realized that there are alternatives. Ancient peoples knew *base-12* or *duodecimal* system (originated in ancient Mesopotamia and still used today to count time and as the English system of units), *base-20* or *vigesimal* system (used by the ancient Mayans and traced in many European languages), *base-60* or *sexagesimal* system (also used in ancient Mesopotamia by Sumerians and Babylonians), and others. Some systems, like *quinary* or *base-5* system (derived from having five fingers on either hand) are used subconsciously as a sub-base system. The main usage of a quinary system, for example, is as a *biquinary* system, that is, the decimal system using five as a sub-base. Another example of a sub-base system is a base-60 system mentioned earlier, which uses 10 as a sub-base. If human beings happened to have three fingers (like the Asgard race in the *Stargate* TV serial), we would probably use a *ternary*, or a *base-3* representation. Such a number system would use only three digits, 0, 1, and 2, but could still use positional notation. In this case, the ternary number 103.22_3 (we use a subscript to denote the base of numeration, *radix*, of a specific nondecimal numerical system) would be converted to the decimal system as follows:

$$110.22_3 = 1 \times 3^2 + 1 \times 3^1 + 0 \times 3^0 + 2 \times 3^{-1} + 2 \times 3^{-2} = 9 + 3 + 0 + 2/3 + 2/9$$
$$= 12 + 8/9 = 12.88(8)$$

In fact, related to computers, it has been proven (in the numbers theory) that it is the ternary system providing the most effective (including compact) computations. That is why in the beginning of the computer era there were several known attempts to create a computer that would use trinary arithmetic. However, these attempts failed because it is quite difficult to support this system electronically. On the contrary, having two levels of a signal, on/off electronic components, happened to be the easiest solution for digital computers, that is why all modern computers use a *binary*, or base-2 numerical system (for the record, the modern binary number system was first fully documented by Gottfried Leibniz in the 17th century in his article "Explication de l'Arithmétique Binaire.") and its derivatives, *octal*, or *base-8*, and *hexadecimal, base-16* systems. Just as with the decimal and ternary systems, numeric quantities in these three numerical systems can be represented using positional notation as shown in the following examples:

$$\mathbf{101.11}_2 = \mathbf{1} \times 2^2 + \mathbf{0} \times 2^1 + \mathbf{1} \times 2^0 + \mathbf{1} \times 2^{-1} + \mathbf{1} \times 2^{-2}$$
$$= 4 + 0 + 1 + 0.5 + 0.25 = 5.75$$

$$\mathbf{268.42}_8 = \mathbf{2} \times 8^2 + \mathbf{6} \times 8^1 + \mathbf{7} \times 8^0 + \mathbf{4} \times 8^{-1} + \mathbf{2} \times 8^{-2}$$
$$= 128 + 48 + 7 + 0.5 + 1/32 = 183.53125$$

$$\mathbf{1F0.B}_{16} = \mathbf{1} \times 16^2 + \mathbf{F} \times 16^1 + \mathbf{0} \times 16^0 + \mathbf{B} \times 16^{-1} + \mathbf{2} \times 16^{-2}$$
$$= 256 + 240 + 0 + 11/16 = 496.6875$$

(Note that, because we have only 10 digits—0, 1, 2, 3, 4, 5, 6, 7, 8, 9—and for the base-16 system we need six more, the letters **A** through **F** are used to represent them as follows: **A** – 10, **B** – 11, **C** – 12, **D** – 13, **E** – 14, **F** – 15.)

By the way, to play with these numerical systems you are welcome to use the Microsoft embedded Calculator, which can be accessed via Start/Programs/Accessories/Calculator. You are probably familiar with the standard (simplified) format already. You can change it to the advanced (scientific) format by choosing **Scientific** option from the **View** menu (Fig. 8.2).

As seen from Fig. 8.2b, by checking the corresponding radiobutton located right below the input window you may convert the number currently displayed in there into hexadecimal, decimal, octal, or binary representation. As you recall from Sec. 2.4.2, MATLAB offers several functions to convert between these representations (bin2dec, dec2bin, dec2hex, hex2dec, num2hex, hex2num, base2dec, dec2base). Figure 8.3a presents general ideas of how to convert numbers from one numerical system to another, whereas Fig. 8.3b provides with basics of binary arithmetic.

a)
b)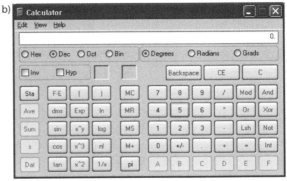

Fig. 8.2 Changing Microsoft Calculator from a) Standard to b) Scientific format.

a)

Decimal to binary	Decimal to ternary	Decimal to octal
136_{10}	136_{10}	136_{10}

Operation	Remainder
$136 \div 2 = 68$	0
$68 \div 2 = 34$	0
$34 \div 2 = 17$	0
$17 \div 2 = 8$	1
$8 \div 2 = 4$	0
$4 \div 2 = 2$	0
$2 \div 2 = 1$	0
$1 \div 2 = 0$	1

Operation	Remainder
$136 \div 3 = 45$	1
$45 \div 3 = 15$	0
$15 \div 3 = 5$	0
$5 \div 3 = 1$	2
$1 \div 3 = 0$	1

Operation	Remainder
$136 \div 8 = 17$	0
$17 \div 8 = 2$	1
$2 \div 8 = 0$	2

10001000_2 12001_3 210_8

b) **Addition**

$0 + 0 \rightarrow 0$
$0 + 1 \rightarrow 1$
$1 + 0 \rightarrow 1$
$1 + 1 \rightarrow 0$, **carry 1** (since $1 + 1 = 0 + 1 \times 10$ in binary)

```
 1 1 1 1 1  (carried digits)
 0 1 1 0 1
+  1 0 1 1 1
-------------
=1 0 0 1 0 0
```

Subtraction

$0 - 0 \rightarrow 0$
$0 - 1 \rightarrow 1$,
borrow 1
$1 - 0 \rightarrow 1$
$1 - 1 \rightarrow 0$

```
 * * * *  (columns borrowed from)
 1 1 0 1 1 1 0
 -   1 0 1 1 1
---------------
=1 0 1 0 1 1 1
```

Multiplication

```
        1 0 1 1
      × 1 0 1 0
      ---------
        0 0 0 0
    +   1 0 1 1
    +   0 0 0 0
    + 1 0 1 1
    -----------
    = 1 1 0 1 1 1 0
```

Division

```
1 1 0 1 1 1 | 1 0 1
-----------
-1 0 1        1 0 1 (quotient)
 0 1 1
 -0 0 0
 -------
   1 1 1
  -1 0 1
  -------
    1 0 (remainder)
```

Fig. 8.3 a) Illustration of an algorithm of converting from the decimal to other numerical systems and b) the basics of binary arithmetic.

Finally, the following example demonstrates that the octal and hexadecimal systems are indeed the derivatives from the binary system because you can easily convert any number represented in the binary system to the other two by appropriate grouping of the digits in the binary number (from right to left for the integer part and from left to right for the decimal part adding zeros if necessary) and computing decimal values for these three- and four-digit groups, respectively,

$$11001110001111.1_2 = 11\ 001\ 110\ 001\ 111\ .\ 100_2 = 31617.4_8$$

$$11001110001111.1_2 = 11\ 0011\ 1000\ 1111\ .\ 1000_2 = 338F.8_{16}$$

8.3.2 Handling Integer Numbers

As mentioned earlier, the binary system fits into the organization of a computer's memory storage the best. Each bite can be either 1 or 0. A sequence of 8 bits, *byte* (alteration and blend of "bit" and "bite"), is processed as a single unit of information. In digital hardware, numbers are stored in binary words [a fixed-length sequence of binary digits (ones and zeros)]. One computer word is usually made of 32 bits or 4 bytes (alternatively, it may be as small as 1 byte and all way up to 8 and more bytes). The way in which hardware components or software functions interpret this sequence of ones and zeros) is described by a *data type*.

Consider a simple example when we want to use just 1 byte (8 bits) to store an integer number. We start filling this byte with ones from the rightmost bit b_0, called the least significant bit (LSB), all the way to the leftmost bit b_7, called the most significant bit (MSB), as shown in Fig. 8.4a. As seen, 8 bits may accommodate 256 positive integer numbers from 0 to 255.

Now, what if we want to represent negative integers too? In mathematics, negative numbers in any base are represented in the usual way, by prefixing them with a minus sign. However, in a computer, there is no single way of representing a number's sign. The straightforward approach would be simply allocating the MSB to be a *sign bit* designated to store, say, 0 for positive numbers and 1 for negative numbers (Fig. 8.4b). For example, decimal −47 encoded in an 8-bit word this way is 10101111_2 (first "1" stands for a "−" sign). This approach, known as *sign-and-magnitude*, is directly comparable to the common way of showing a sign (placing a "+" or "−" next to the number's magnitude). However, it results in having two representation of zero, plus zero, and minus zero, as shown in Fig. 8.5a.

A better way to introduce negative numbers would be having an agreement to subtract some specified quantity, bias, from whatever number we get

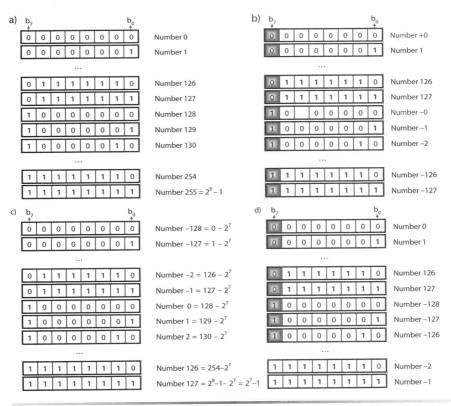

Fig. 8.4 Eight-bit word format for a) unsigned, b) sign-and-magnitude, c) excess-M, and d) two's complement integers.

Fig. 8.5 Two-zero problem when using a) sign-and-magnitude and b) one's complement method to represent negative numbers.

from a 1-byte representation of Fig. 8.4a. To make it more or less symmetric with respect to zero when we are dealing with an N-bit word, we may use

$$bias_N = 2^{N-1} \qquad (8.1)$$

For example, in case $N = 8$ (1-byte word), the bias computed according to Eq. (8.1) becomes $2^7 = 128$. This bias results in having different numbers for exactly the same binary words (cf. Fig. 8.4c with Fig. 8.4a). This method of representing signed numbers is known as *excess-M* method, where M is a biasing value, $M = bias_N$ (Fig. 8.6). For instance, the case of $M = 128$ Eq. (8.1) would be referred to as an excess-128 method. As seen from Fig. 8.6, after biasing, zero will be defined by a binary representation of the number M (10000000_2 for the excess-128 method applied to the 8-bit word) and $-M$ will be represented by the all-zeros bit pattern (Fig. 8.4c). The excess-M method is primarily used within the floating-point number format to be discussed further in Sec. 8.2.4.

Two more alternatives to represent negative numbers are the *one's* and *two's* complements. The one's complement form of a binary number is the bitwise not ($\sim$) applied to it (that is every *bit inversion*)—the complement of its positive counterpart. As an example, the one's complement form of 00101111_2 (47) becomes 11010000_2 (-47). The range of signed numbers using one's complement in the 8-bit word is -127 to $+128$. Like sign-and-magnitude representation, the one's complement still suffers from having two representations for zero: 00000000_2 ($+0$) and 11111111_2 (-0) (Fig. 8.5b).

The problem of multiple representations of 0 is circumvented by a system called two's complement. In two's complement, negative numbers

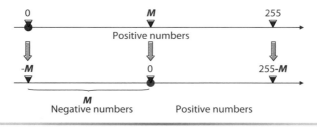

Fig. 8.6 Introducing a bias for representing negative numbers.

are represented by the bit pattern which is one greater (in an unsigned sense) than the one's complement of the positive value. Negating a number (converting positive to negative numbers or vice versa) is done by inverting all the bits and then adding 1 to that result. The MSB of a signed two's complement binary numeral still indicates a sign. If the leftmost bit is 0, the number is interpreted as a nonnegative binary number. If the MSB (leftmost) is 1, the bits contain a negative number in the two's complement form. For example, the first step for converting an 8-bit binary representation of the positive decimal value 5, 00000101_2 would be inverting all the bits to yield 11111010_2 (at this point, the numeral become the one's complement of the decimal value 5). Now adding 1 gives 11111011_2, a signed binary numeral representing the decimal value -5 in the two's complement form. The initial bit is 1, and so the numeral is interpreted as a negative value. This procedure works conversely too. For example, inverting the bits of -5 gives 00000100_2, and adding one gives the final value 00000101_2, that is, positive decimal 5 (see Fig. 8.4d for the rest of the numbers in the 8-bit binary representation). Note that the two's complement of zero, 00000000_2, is zero itself: inverting gives all ones and adding one changes the ones back to zeros (the overflow is ignored). The two's complement of the most negative number, 10000000_2, sometimes called the *weird number*, also ends up as itself.

The decimal value of a two's complement representation of a negative binary number can be easily obtained in the usual manner by not accounting the MSB and adding -2^{N-1} to the number. For instance, the decimal value for a signed binary numeral in the two's complement form 11111011_2 can be obtained as $-2^7+2^6+2^5+2^4+2^3+2^1+2^0 = -5$. In this sense, it is a sort of similar to biasing binary numbers starting from 1 in the MSB with the bias defined by Eq. (8.1), whereas disregarding this leftmost bit.

What is very important here is that when you apply any of the aforementioned methods to represent negative numbers you cannot discern whether the resulting binary numbers are signed or unsigned data types and what method was used merely by inspection. This information is not explicitly encoded within the word. Everything depends on the a priori agreement.

The latest versions of MATLAB support several formats of signed and unsigned integers, depending on how many bits are used to store a single number:

`uint8(X)`, `uint16(X)`, `uint32(X)`, `uint64(X)`	Convert a scalar (array) X to unsigned integer scalar (array) that requires 8, 16, 32, and 64 bits per number, respectively
`int8(X)`, `int16(X)`, `int32(X)`, `int64(X)`	Convert a scalar (array) X to signed integer scalar (array) that requires 8, 16, 32, and 64 bits per number, respectively

Table 8.1   Smallest and Largest Unsigned Integer Numbers

	uint8	uint16	uint32	uint64
Largest integer (intmax)	255 (2^8–1)	65535 (2^{16}–1)	4294967295 (2^{32}–1)	18446744073709551615 (2^{64}–1)
Smallest integer (intmin)	0	0	0	0

Tables 8.1 and 8.2 define the range of the integers that can be represented using the aforementioned types of unsigned and signed integer numbers (to find the range limits, the corresponding commands, such as intmin('int8') or intmax('int64'), can be used).

Now, we are ready to introduce an error that can occur using these types of data. For instance, the command

```
>> y=int8(300)
```

returns an erroneous response

```
y = 127
```

because 300 exceeds the maximum value that can be represented by int8 data type. Although this example is obvious, the following example might be easily overlooked even by the experienced programmer:

```
>> 3*int8(100)
```

```
y = 127
```

Although int8(100) returns the correct result, it also defines the output data type. Hence, we run into the same problem as in the previous example.

Now that you know how integer numbers are stored in a computer and what integer data types are available in MATLAB, let us explore what difference it makes to represent noninteger, that is, real numbers.

8.3.3 Fixed-Point Format

One way to represent noninteger numbers is to follow the same routine we just considered, but this time to additionally introduce a virtual binary point in the binary word as shown in Fig. 8.7a. As a matter of fact, we can assume that, while considering the integer numbers, this point was already present and located to the right of the LSB (Fig. 8.7b).

Fig. 8.7   a) Introducing a binary point and b) expanding it to the case of integer numbers.

| 1 | 0 | 1 | 1 | 0 | 1 | 1 | 1 | $1 \times 2^3 + 1 \times 2^1 + 1 \times 2^0 + 1 \times 2^{-2} + 1 \times 2^{-3} + 1 \times 2^{-4} = 11.4375$ |

| 1 | 0 | 1 | 1 | 0 | 1 | 1 | 1 | $1 \times 2^5 + 1 \times 2^3 + 1 \times 2^2 + 1 \times 2^0 + 1 \times 2^{-1} + 1 \times 2^{-2} = 45.75$ |

Fig. 8.8 Examples of representing the real numbers using fixed-point data type.

This type of representation is called a *fixed-point* data type, and the binary point is the means by which the fixed-point values are scaled. Depending on where we think this binary point is located, we may have different numbers represented by the same sequence of ones and zeros (Fig. 8.8).

Essentially, we can cast the placement of the binary point as a scaling problem. In the example of Fig. 8.8, moving the binary point two places to the right means multiplying by 2^2. Compared to Fig. 8.7b moving the binary point w places to the left means multiplying whatever integer number we had by 2^{-w}. Therefore, the main difference between different fixed-point data types is a location of their binary point. Accounting for the biasing to represent negative real numbers as well (as discussed in the previous section), the fixed-point real numbers employing the excess-M method can be decoded as

$$number = 2^{-w}(integer - bias_N) = 2^{-w}\sum_{i=0}^{N-1} b_i 2^i - 2^{-w}2^{N-1} \qquad (8.2)$$

that is, be defined by their word length in bits N, their fraction length w (slope 2^{-w}) and the bias (defining whether they are signed or unsigned). Again, whether a fixed-point value is signed or unsigned is usually not encoded within the binary word explicitly (that is, there is no sign bit). Instead, the sign information is implicitly defined within the computer architecture.

The limits of representation [Eq. (8.2)] (its range) with the bias defined by Eq. (8.1) are

$$range = 2^{-w} [-2^{N-1}; 2^{N-1} - 1] \qquad (8.3)$$

Table 8.2 Smallest and Largest Signed Integer Numbers

	int8	int16	int32	int64
Largest integer (intmax)	127 (2^7-1)	32767 ($2^{15}-1$)	2147483647 ($2^{31}-1$)	9223372036854775807 ($2^{63}-1$)
Smallest integer (intmin)	-128 (-2^7)	-32768 (-2^{15})	-2147483648 (-2^{31})	-9223372036854775808 (-2^{63})

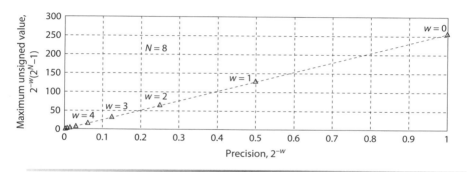

Fig. 8.9 Maximum unsigned value versus precision for the 8-bit fixed-point numbers.

Obviously, by letting $w = 0$, Eqs. (8.2) and (8.3) naturally extend to include the excess-$bias_N$ representation of integer numbers.

Along with the error associated with the limited range [Eq. (8.3)], now we may observe the error associated with the data type's *precision*, which is characterized by the distance between successive numbers in the representation. Consider the same example, where we use just 1 byte to represent the real numbers. Regardless the scaling and biasing this 1 byte can still accommodate only 256 different numbers. And the distance between any two successive numbers is

$$precision = 2^{-w} \qquad (8.4)$$

meaning that we cannot possibly have more precise solution while operating with the fixed-point data type numbers as the one predescribed by Eq. (8.4). For instance, if $w = 0$ the distance between two successive numbers is 1 (that is, no error for integer numbers), for $w = 1$ it is 0.5, for $w = 2$ it is 0.25, etc. If $w = N$, then we would have the best precision (2^{-8} in our case), and ... the worst (the smallest) range. The range-precision tradeoff, by the way, is always the case (Fig. 8.9). It should also be noted that for the fixed-point data type, the relative precision worsens regardless of the slope, although real number becomes smaller:

$$\frac{precision}{real - world \ number} = \frac{1}{\sum_{i=0}^{N-1} b_i 2^i - bias_N} \qquad (8.5)$$

Although the Simulink fixed-point blockset (specifically designed to handle fixed-point numbers and fixed-point arithmetic) does support the fixed-point format with the word sizes up to 128 bits, MATLAB itself uses another form of representing the real numbers as discussed later.

8.3.4 Floating-Point Format

An alternative way to represent real numbers is to use the *floating-point* data types. A floating-point number x can be represented by two numbers m and e, such that

$$x = mb^e \tag{8.6}$$

where b is the *base of numeration*, number e is called the *exponent*, and the *significand, m,* or, informally, *mantissa*, is a p-digit number of the form $\pm d.ddd \ldots ddd$, where each digit d is an integer between 0 and $b - 1$ inclusive. If the leading digit of m is nonzero, then the number is said to be *normalized*. A general representation of a normalized nonzero base-2 floating-point number

$$x = (-1)^s (1 + f) 2^e \tag{8.7}$$

includes the *fraction f*, which satisfies an inequality $0 \le f < 1$, and a separate sign bit s to represent positive and negative numbers, $s = 0$ and $s = 1$, respectively. Thus, a normalized binary number has a mantissa of the form $1.f$, where f has a fixed size for a given data type. Because the leftmost mantissa bit is always a 1, it is unnecessary to store this bit and therefore it is hidden (implicit). Thus, an N-bit fraction stores an $(N + 1)$-bit number.

Following this setup for the 8-bit word, we need to allocate the MSB to store the sign s, then devote some number of bits to represent exponent e and the rest of them will characterize the fraction f as shown, for instance, in Fig. 8.10.

To decode the number shown in Fig. 8.10, we may use the following formula:

$$x = (-1)^{MSB} \left(1 + \sum_{i=1}^{F} \frac{b_{F-i}}{2^i} \right) 2^{\sum_{i=0}^{N-F-2} b_{F+i}2^i - bias^{**}} \tag{8.8}$$

where F is the number of bits to represent the fraction, whereas the *bias*[**] is introduced into the exponent to be able to represent both negative and positive numbers. Therefore, one of the advantages of floating-point data type is that it allows a large range of magnitudes to be represented within a given number of bits, which is not possible in fixed-point notation. To have

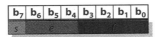

Fig. 8.10 Introducing a floating-point format.

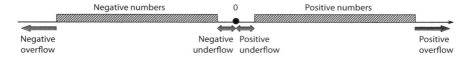

Fig. 8.11   Range of floating-point numbers.

a more or less symmetric range of the exponent, this bias can be computed using Eq. (8.1) as $bias^{**} = bias_{N-F-1} = 2^{N-F-2}$ (to exceed the number of negative exponents by one) or

$$bias^{**} = 2^{N-F-2} - 1 \qquad (8.9)$$

(to have an opposite). With Eq. (8.9), the range of the exponent is $[-2^{N-F-2}+1; 2^{N-F-2}]$. In practice, however, some values of the exponent are reserved for flagging Inf (infinity), NaN (not-a-number), and denormalized numbers (see further), and so the true range of values is as follows:

$$[e_{min}; e_{max}] = [-2^{N-F-2} + 2; 2^{N-F-2} - 1] \qquad (8.10)$$

Figure 8.11 provides a graphical illustration of the range of representable numbers using Eq. (8.8). The absolute value of the minimum number accounting for the exponent's range [Eq. (8.10)] is given by

$$V_{min} = 2^{-2^{N-F-2}+2} \qquad (8.11)$$

and the absolute value of the maximum number is given by

$$V_{max} = (2 - 2^{-F})2^{2^{N-F-2}-1} \qquad (8.12)$$

As seen from Fig. 8.11, there is always a gap around zero in floating-point arithmetic. To fill this gap and therefore provide a gradual underflow allowing the calculation to lose precision slowly, rather than all at once, *denormal* or subnormal numbers are used. These numbers, denormalizable to the form of Eq. (8.8), using mantissa of the form $0.f$, rather than $1.f$. Denormal numbers are encoded with a biased exponent $e = 0$, but are interpreted with the value of the smallest allowed exponent. Zero has a special bit pattern, where $e = 0$ and $f = 0$ (due to the sign-and-magnitude representation of floating-point numbers, there are two representations of zero, one positive and one negative).

To summarize, positive numbers greater than V_{max} and negative numbers less than $-V_{max}$ are overflows, so that they are mapped to +Inf and −Inf, respectively. Positive numbers less than V_{min} and negative numbers greater than $-V_{max}$ are either underflows or denormalized numbers, including zero.

Figure 8.12 provides an example of the signed real numbers that can be stored in 8 bits ($N = 8$) using the fixed-point and floating-point formats. For the fixed-point format, given by Eq. (8.2), we assume the binary point to reside between the third and fourth bit (counting from the left), so that $w = 3$. To have about the same range for the floating-point format, given by Eq.(8.8) with the bias, Eq. (8.9), we let $F = 4$.

According to Eq. (8.11), the underflow around zero, which is clearly seen in Fig. 8.12, ranges from −0.25 to 0.25, the range of the floating-point representation, determined by Eq. (8.12) in this case is from −15.5 to 15.5. Note that compared to that of the fixed-point numbers [Eq. (8.3)], the range for the floating-point numbers is fully symmetric. Also note that because we reserved two exponents to keep `Inf`, `NaN` and denormalized numbers, the 8 bits for the floating-point numbers in Fig. 8.12 now represent only 192 rather than 256 different numbers (2 signs × 6 different exponents × 16 different fractions).

As seen from Fig. 8.12 the floating-point numbers exhibit one more very important feature. As opposed to the fixed numbers they are not distributed evenly, that is, the precision of the floating-point numbers varies and goes from as small as

$$precision_{min} = 2^{e_{min}-F} \qquad (8.13)$$

for small numbers, to as large as

$$precision_{max} = 2^{e_{max}-F} \qquad (8.14)$$

for large numbers. For $N = 8$ and $F = 4$ (Fig. 8.10) it corresponds to 0.0156 and 0.5, respectively. To have a single unified merit, the distance from 1.0 to the next largest floating-point number referred to as a *roundoff constant* is used. It corresponds to exponent equal to zero and therefore is defined as

$$\varepsilon = 2^{-F} \qquad (8.15)$$

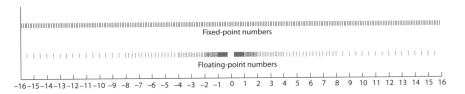

Fig. 8.12 Example of signed fixed- and floating-point numbers stored in the 8-bit word.

For the aforementioned example of $F = 4$, it yields $\varepsilon = 0.0625$. This constant also happens to be the upper bound on the relative precision. As opposed to that of the fixed-point numbers (Eq. (8.5), the relative precision of floating-point numbers does not depend on the magnitude of the real number and stays almost the same within all range of numbers. Increasing the exponent field (decreasing e_{min} and increasing e_{max}) improves the precision for the small numbers [Eq. (8.13)], but simultaneously leads to worsening the precision for the large numbers [Eq. (8.14)]. Increasing the fraction field leads to the better precision [Eqs. (8.13)–(8.15)].

In the previous discussion, we made several assumptions on what part of the N-bit word should be dedicated to keep the fraction f and how to introduce the biased exponent e. By choosing different N and F, we obviously obtain different precision and different range of the representable floating-point numbers. To be able to operate with these numbers we have to know their format up front, otherwise if some other computer uses different representation, the code will produce erroneous results (if at all).

Back in the 1960s to 1980s, the situation was even more complicated. Not only each computer had its own floating-point number system, but some of them were binary; some were decimal. Among the binary computers, some used 2 as the base; others used 8 or 16. Each of them had different precision. In 1985, the Institute of Electrical and Electronics Engineers (IEEE) Standards Board and the American National Standards Institute (ANSI) adopted the ANSI/IEEE Standard 754-1985 for binary floating-point arithmetic. This was the culmination of almost a decade of work by almost a hundred-person working group of mathematicians, computer scientists, and engineers from universities, computer manufacturers, and microprocessor companies.

Since 1985, all computers designed use ANSI/IEEE floating-point arithmetic. This does not mean that they all provide exactly the same results, because there is some flexibility within the standard. However, it does mean that we can rely on a *machine-independent* model of how floating-point arithmetic behaves. Among other things, ANSI/IEEE standard (recently renewed as IEEE 754-2008 standard) specifies four floating-point number formats of which single-precision format (using a 32-bit word) and double-precision format (using a 64-bit word) are the most widely used.

8.3.5 Standard Single-Precision and Double-Precision Formats

As discussed earlier to enable representing both small and large numbers, the exponent in the floating-point representations is biased. The ANSI/IEEE

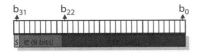

Fig. 8.13   Single-precision (32-bit word) format.

floating-point standard defined the exponent field of a single-precision number as an 8-bit excess-127 field (according to Eq. (8.9) $bias^{**} = 2^{8-1} - 1 = 127$). The double-precision exponent field uses an 11-bit excess-1023 field ($bias^{**} = 2^{11-1} - 1 = 1023$).

Therefore, the *single-precision floating-point format* is a 32-bit word divided into the 1-bit sign indicator s, 8-bit biased exponent e, and 23-bit fraction f as shown in Fig. 8.13.

Formally, the relationship between this format and its representation of real numbers is given by

$$x = (-1)^s (1 + f) 2^{e-127}, \text{ where } 0 < e < 255 \text{ and } f = \sum_{i=1}^{23} \frac{b_i}{2^i} \ (b_i = \{0,1\}) \ (8.16)$$

The *double-precision floating-point format* is based on a 64-bit word divided into the 1-bit sign indicator s, 11-bit biased exponent e, and 52-bit fraction f as shown in Fig. 8.14.

Formal relationship between this format and its representation of real numbers is given by

$$x = (-1)^s (1 + f) 2^{e-1023}, \text{ where } 0 < e < 2047 \text{ and } f = \sum_{i=1}^{52} \frac{b_i}{2^i} \ (b_i = \{0,1\}) \ (8.17)$$

Because some values of the exponent are reserved for flagging different numbers as discussed in the previous section, the true exponent values for single precision range from -126 to 127 (a pure excess-127 would provide -127 to 128 range) and for double precision range from -1022 to 1023 (a pure excess-1023 would provide -1023 to 1024 range).

MATLAB has traditionally used the IEEE double-precision format (which is the default format for all mathematical operations). The latest versions of MATLAB also support single-precision arithmetic. Although a single-precision format saves space, but it is not necessarily much faster.

Fig. 8.14   Double-precision (64-bit word) format.

Table 8.3 Floating-Point Relative Precision, the Smallest and Largest Floating-Point Number

	Single Precision Format	Double Precision Format
Round-off error	eps('single')=2^{-23}	eps=2^{-52}
Underflow	realmin('single')=2^{-126}	realmin=2^{-1022}
Overflow	realmax('single')= (2-eps('single'))2^{127}	realmax=(2-eps)2^{1023}

You may convert any array X to the single-precision format and back to the double-precision format by using the following functions:

single(X)   Converts X to single-precision
double(X)   Converts X to double-precision

Based on Eqs. (8.16) and (8.17) with account of a reduced exponent range, the roundoff constant (relative precision) for singles and doubles as well as min/max numbers they can represent are estimated as presented in Tables 8.3 and 8.4.

In Table 8.3, the floating-point relative precision eps returns the distance from 1.0 to the next largest number [Eq. (8.15)], realmin returns the smallest positive normalized floating-point number on your computer [Eq. (8.11)] (anything smaller underflows or is denormal), realmax returns the largest floating-point number representable on your computer [Eq. (8.12)] (anything larger overflows).

8.4 Round-off Errors

As we have learned already, due to the fact that digital computers have the range and precision limits on their ability to represent numbers, they cannot represent some quantities exactly causing *round-off errors*. These errors are important to engineering and scientific problem solving because they can make calculations unstable or lead to the erroneous results. The latter can be obvious or, if they lead to subtler discrepancies, quite difficult to detect.

Table 8.4 Decimal Representations of the Values of Table 8.3

	Single Precision Format	Double Precision Format
Round-off error	1.1921e – 007 = 1.1921×10^{-7}	2.2204e – 016 = 2.2204×10^{-16}
Underflow	1.1755e – 038 = 1.1755×10^{-38}	2.2251e – 308 = 2.2251×10^{-308}
Overflow	3.4028e + 038 = $3.4028 \times 10^{+38}$	1.7977e + 308 = $1.7977 \times 10^{+308}$

It turns out that the range and precision limits are not the only source for round-off errors involved in numerical calculations. Certain numerical manipulations can aggravate the precision due to round-off errors even farther. This can result from both mathematical considerations as well as from the way in which computers perform arithmetic operations. Before we proceed with considering these cases, let us review the fixed-point and floating-point formats from the standpoint of round-off errors.

8.4.1 Fixed-Point vs Floating-Point Numbers

As long as the numeric value uses only the number of digits specified after the decimal point, fixed-point values can exactly represent all values up to its maximum value (determined by the number of bits in its representation). This is in contrast to floating-point representations, which include an automatically managed exponent, but given the same number of bits in its representation cannot represent as many digits accurately.

That is why a common use for fixed-point numbers is for storing monetary values, where the inexact values of floating-point numbers are often a liability. Fixed-point representations are also sometimes used if either the executing processor does not have any floating-point unit (FPU) or if fixed-point arithmetic provides an improved performance necessary for an application. For instance, some audio codecs use fixed-point arithmetic because to save money, many audio decoding hardware devices do not have an FPU, and audio decoding requires enough performance that a software implementation of floating-point on low-speed devices would not produce a real-time output.

Very few computer languages, however, include built-in support for fixed-point values, because for most applications, floating-point representations are fast enough and accurate enough. Floating-point representations are more flexible than fixed-point representations, because they can handle a wider dynamic range. Floating-point representations are also slightly easier to use, because they do not require programmers to specify the number of digits after the decimal point.

Historically, fixed-point representations were the norm for decimal data types (for example, in PL/1 or Cobol). The Ada programming language includes built-in support for both fixed-point and floating-point. However, if they are needed, fixed-point numbers can be implemented even in programming languages like C and C++ that do not include such support built-in. MATLAB also relies on floating-point arithmetic, although it supports fixed-point numbers (in Simulink).

Floating-point numbers usually behave very similarly to the real numbers they are used to approximate. However, this can easily lead programmers into overconfidently ignoring the need for numerical analysis. There are many cases, where floating-point numbers do not model real numbers well.

For example, the decimal fraction 0.1 or 0.01 cannot be exactly represented in any binary floating-point format (to be discussed later).

In general, when fixed precision is required, fixed-point arithmetic is usually a better choice. However, a floating-point representation is more likely to be appropriate when relative (proportional) precision over a range of scales is needed. The common errors in floating-point computation are to be discussed next.

8.4.2 Range and Precision

Formally, the errors associated with range and precision are:

1. Overflow (which for the floating-point numbers yields an infinity)
2. Underflow (often defined as an inexact tiny result outside the range of the normal numbers for a format), which yields zero, a subnormal number, or the smallest normal number
3. Rounding and nonrepresentable numbers

As follows from Secs. 8.2.3 and 8.2.4, all types of data in MATLAB have their own limits. These limits are given in Tables 8.1–8.4. If you try to assign the number, which exceeds these limits, you will get an erroneous (undetermined) result. For integers, it will be the boundary number of a specific range, for floating-point numbers—infinity.

Another source of errors for floating-point numbers is their limited precision. To start with, some numbers cannot be represented exactly. For example, irrational numbers such as π, e, or $\sqrt{5}$ cannot be expressed by a finite number of significant figures. Therefore, they cannot be represented exactly by the computer. In general, for computer tools that use a 32-bit word (single precision), the mantissa can be expressed to about eight base-10 digits of precision. Thus, π can be found as

```
>> format long, single(pi)
ans = 3.1415927
```

Performing arithmetic operations, for example, 1/3 or 6/7, will also be affected by floating-point numbers precision. For example, for 1/3 the single-precision format produces 0.6666667 instead of 0.(6) and for 6/7— yields 0.8571429 instead of 0.(857142).

For tools using a 64-bit word, the precision increases to about 15 base-10 digits. For example, the number π is represented as

```
>> format long, double(pi)
ans = 3.14159265358979
```

In addition, because computers use a binary, or base-2, representation they cannot possibly precisely represent certain exact base-10

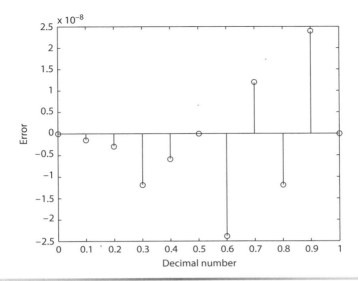

Fig. 8.15   Error in representing the floating-point numbers.

numbers. For example, the exact base-10 quantity 0.1 cannot be represented exactly in a base-2 system. It may be hard to see with MATLAB's
default 64-bit floating-point numbers, but if we intentionally degrade
the precision down to the `single` format you will be able to see that
there is a difference between two representations indicating that 1/10 is
not represented precisely

```
>> 0.1-single(0.1)
ans = -1.4901161e-009
```

(in the case the precise representation exist, the return would be zero).

If you think of 0.1 being a some kind of an exceptional number, run the
following script:

```
i=[0:0.1:1];
er=i-single(i);
stem(i,er), xlabel('Decimal number'), ylabel('Error')
```

The result it produces is shown in Fig. 8.15.

As seen, among those 11 fractions, only three: 0, $0.5 = 2^{-1}$ and $1 = 2^0$,
have the exact representation. The remaining ones have not. The MAT-
LAB's 64-bit word format does not make these errors disappear, but simply
makes them much smaller.

Certain numerical methods require extremely large numbers of arithmetic
manipulations to arrive at their final results. In addition, these computations are often interdependent. That is, the later calculations are dependent
on the results of the earlier ones. Consequently, even though an individual

round-off error could be small, the cumulative effect over the course of a large computation can be significant.

Another implication is that floating-point arithmetic is neither *associative* nor *distributive*. This means that in general for floating-point numbers x, y, and z

$$(x+y) + z \neq x + (y + z), \quad (xy) \, z \neq x \, (yz), \quad x \, (y + z) \neq xy + xz \qquad (8.18)$$

For instance, while

```
>> (1e100 - 1e100) + 1
```

produces

```
ans = 1
```

a slightly different command

```
>> 1e100 - (1e100 - 1)
```

yields an erroneous result

```
ans = 0
```

That is because the unity in the second example was absorbed by the much larger number. Specifically, the command

```
>> eps(1e100)
```

shows that the very next largest number is greater than `1e100` by

```
ans = 1.9427e+084
```

As was demonstrated in Fig. 8.12, the precision for the large numbers is much bigger than that of the small numbers, and so no wonder that `1` was absorbed. Even if you used say `1e83` instead of `1`, it would be absorbed anyway.

8.5 Truncation Errors

The results of numerical computations can also be affected by another type of errors caused by using approximations in place of exact mathematical expressions. This type of errors is known as truncation errors. The following formally introduces several definitions associated with truncation errors and proceeds with a simple example.

8.5.1 Error Definitions

Let us start from the basic relation establishing the relationship between the exact, or true, expression ζ and its approximation ζ^*

$$\zeta = \zeta^* + \delta_\zeta^t \qquad (8.19)$$

where δ_ζ^t is used to designate the exact value of the error (the subscript t is included to designate that this is the "true" error). By rearranging Eq. (8.19), we find the *true error* δ_ζ^t as

$$\delta_\zeta^t = \zeta - \zeta^*$$

(8.20)

A shortcoming of this definition is that it takes no account of the order of magnitude of the value under examination. For example, an error of a centimeter is much more significant if we are measuring a wing thickness that its span. One way to account for the magnitudes of the quantities being evaluated is to normalize the error to the true value

$$\overline{\delta}_\zeta^t = \frac{\delta_\zeta^t}{\zeta}$$

(8.21)

The sign of expressions in Eqs. (8.20) and (8.21) can be either positive or negative. However, when performing computations, we may be less concerned with the sign of the error, compared to its magnitude. Thus, in practice we usually deal with the nonnegative *absolute error*, $\left|\delta_\zeta^t\right|$, and *relative error*

$$\varepsilon^t = \left|\overline{\delta}_\zeta^t\right|$$

(8.22)

Note that the relative error is a unitless number expressed as a fraction. However, it can also be expressed as a *percent error*

$$\varepsilon^t\,[\%] = \varepsilon^t \cdot 100\%$$

(8.23)

For example, a relative error of 10^{-2} corresponds to a 1% percent error.

Equations (8.19)–(8.22) imply that we know the true value ζ. However, such information is rarely available. Hence, we need to use the estimates of these parameters, that is, the relative error of Eq. (8.22) becomes

$$\varepsilon^a = \left|\frac{\hat{\zeta} - \zeta^*}{\hat{\zeta}}\right| \quad (\varepsilon^a\,[\%] = \varepsilon^a \cdot 100\%)$$

(8.24)

Here the subscript a signifies that the error is normalized to an approximate value of ζ.

Computing ε^a in the absence of knowledge regarding the true value of ζ constitutes one of the challenges of numerical methods. Different numerical methods use the different approaches to address it. For instance, certain numerical methods rely on *iterations*. In this case, a previous value of ζ, ζ_{i-1}

is used to produce the next iteration ζ_i ($i = 1,2,...$). This process is performed repeatedly, or *iteratively*, to successively compute (hopefully) better and better approximations. For such cases, the error is often estimated based on the difference between previous and present approximations, so that Eq. (8.24) becomes

$$\varepsilon_i^a = \left| \frac{\zeta_i - \zeta_{i-1}}{\zeta_i} \right| \tag{8.25}$$

This error is estimated at each iteration and is compared with some prespecified percent *tolerance* ε^s. If the iterative procedure converges, ε_i^a become smaller and smaller and at some point satisfy an inequality

$$\varepsilon_i^a \leq \varepsilon^s \tag{8.26}$$

This relationship is referred to as a *stopping criterion*. If it is satisfied, the result is assumed to be within the prespecified acceptable level ε^s.

In practice, the true value ζ can be a very small number or zero. To avoid singularity in Eq. (8.25), the stopping criterion assumes a more robust form

$$\frac{|\zeta_i - \zeta_{i-1}|}{\max\left(|\zeta_i|, \dfrac{\varepsilon^{abs}}{\varepsilon^{rel}} \right)} \leq \varepsilon^{rel} \tag{8.27}$$

where ε^{abs} and ε^{rel} are absolute and relative tolerances, defined by the user, respectively. In MATLAB, the default value of ε^{abs} is usually 10^{-6}, and the default value of ε^{rel} is of the order of 10^{-3}, which corresponds to 0.1% accuracy (ε^{rel} 100% = ε^s). When $|\zeta_i| \rightarrow 0$, Eq. (8.27) reduces to

$$|\zeta_i - \zeta_{i-1}| \leq \varepsilon^{abs} \tag{8.28}$$

Because the ratio of tolerances is small (of the order of 10^{-3}), some algorithms use the sum of the two terms in denominator of Eq. (8.27) instead of finding a maximum value.

It is also convenient to relate these errors to the number of significant figures in the approximation. It can be shown that the result, which is correct to at least N significant figures ($N = 1,2,...$), assures the upper bound on the percent error

$$\varepsilon^{u.b.} = \frac{10^{2-N}}{2} \% \tag{8.29}$$

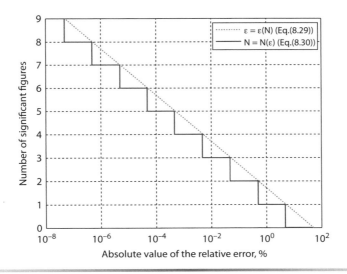

Fig. 8.16 Number of the correct significant figures vs percent error.

Equation (8.29) can be inverted to yield

$$N = [2 - \log_{10}(2\varepsilon)] \tag{8.30}$$

(where [] denotes the integer part). This dependence is presented graphically in Fig. 8.16. Given the number N, we may find the upper bound on the percent error, and vice versa, given the estimate of the percent error, this dependence yields the number of correct significant figures.

Consider a simple example. As well known, any function $f(x)$ can be represented as the *Taylor series expansion*, an infinite sum of terms calculated from the values of its derivatives at a single point $x = a$

$$f(x) = f(a) + f'(a)(x-a) + \frac{f''(a)}{2!}(x-a)^2 + \frac{f'''(a)}{3!}(x-a)^3 + \cdots$$
$$= \sum_{i=0}^{\infty} \frac{f^{(i)}(a)}{i!}(x-a)^i \tag{8.31}$$

which for $a = 0$ is also known as a *Maclaurin series expansion*

$$f(x) = f(0) + f'(a)x + \frac{f''(0)}{2!}x^2 + \frac{f'''(0)}{3!}x^3 + \cdots = \sum_{i=0}^{\infty} \frac{f^{(i)}(0)}{i!}x^i \tag{8.32}$$

Table 8.5 Accuracy of Taylor Series Expansion to Estimate an Exponential

Number of Terms	Approximation	$\varepsilon^t(\%)$	$\varepsilon^a(\%)$	$N(\varepsilon^t)$	$N(\varepsilon^a)$	Actual N	$\varepsilon^{u.b}(\%)$
1	1	63.21		0		0	50
2	2	26.424	50	0	0	1	5
3	2.5	8.0301	20	0	0	1	5
4	2.67	1.89885	6.25	1	0	1	5
5	2.708	0.36599	1.5385	2	1	2	0.5
6	2.7167	0.059418	0.30675	2	2	3	0.05
7	2.71806	0.008324	0.051099	3	2	4	0.005
8	2.718254	0.001025	0.007299	4	3	5	0.0005
9	2.7182788	0.000113	0.000912	5	4	5	0.0005
10	2.71828152	1.11e-05	0.000101	6	5	7	0.00005

Let use Eq. (8.32) to approximate the exponential function e^x

$$e^x = 1 + x + \frac{x^2}{2!} + \frac{x^3}{3!} + \cdots = \sum_{i=0}^{\infty} \frac{x^i}{i!} \qquad (8.33)$$

at $x = 1$. The exponential is an irregular number, and so the more the terms we use in Eq. (8.33), the better the accuracy we get. In this example, we will consider say a 16-digit representation

$$e = 2.718281828459046\ldots \qquad (8.34)$$

as a true value (in MATLAB you may use either `format long` or `vpa(exp(1),16)` command to obtain this value). Now, the question is how the number of elements in the series [Eq. (8.33)] affects the accuracy of approximation.

According to Eq. (8.25), the estimate of a percent error for approximation of Eq. (8.33) can be computed as

$$\varepsilon_n^a = \frac{\dfrac{x^n}{n!}}{\displaystyle\sum_{i=0}^{n} \dfrac{x^i}{i!}} 100\% \qquad (8.35)$$

In this particular case, because the true value is known [Eq. (8.34)], we can also compute a true value of the percent error as

$$\varepsilon_n^t = \frac{\left| \displaystyle\sum_{i=0}^{n} \dfrac{x^i}{i!} - 2.718281828459046 \right|}{2.718281828459046} 100\% \qquad (8.36)$$

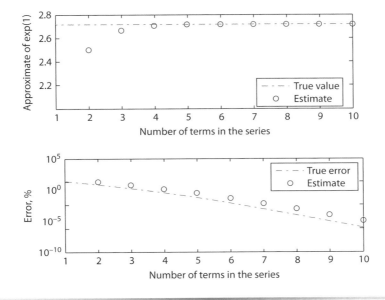

Fig. 8.17 Approximation of exponential with the Taylor series.

Finally, we can employ Eq. (8.30) to compute the number of significant figures based on the accuracy of our approximation. The results produced by Eqs. (8.33), (8.36), (8.35), and (8.30) with with ε equal to ε_n^t and ε_n^a, respectively, are presented in Table 8.5. They are also illustrated in Fig. 8.17. Obviously, compared to Eq. (8.36), Eq. (8.35) lags one step, but that is the best we can possibly do. Hence, approximation of the absolute relative error is always conservative.

The last two columns in Table 8.5 show the actual number of correct significant figures (for your convenience the correct digits in approximations appear in the boldface) and the upper bound on the error. For example, for the five terms the actual number of correct significant figures happens to be 2, which corresponds to the upper bound of 0.5%. The true error ε^t for this line is less than 0.5% and the predicted $N(\varepsilon^t)$ is equal to 2. The estimate ε^a exceeds 0.5% tolerance, and therefore $N(\varepsilon^a)$ is less than 2. Hence, Eq. (8.25) provides quite conservative estimates. The true error is always way less than that. Consequently, the estimate of the number of correct significant figures [Eq. (8.30)] based on ε^a is conservative too. Moreover, even the estimate of N based on true relative error ε^t provides a worse result that it is in reality. That is, both Eqs. (8.25) and (8.30) ensure that the result is at least as good as they specify.

8.5.2 Application of Error Formulas to Fundamental Arithmetic Operations

Let us address one more issue here. Suppose we have some arbitrary function of several variables written in the most general form as

$$F = f(u_1, u_2, ..., u_m) \tag{8.37}$$

Also, assume that independent variables u_1, u_2, ..., u_m are subject to errors Δu_1, Δu_2, ..., Δu_m, respectively. These errors in *us* will obviously cause an error in the function F

$$F + \Delta F = f(u_1 + \Delta u_1, u_2 + \Delta u_2, ..., u_m + \Delta u_m) \tag{8.38}$$

The question now is how ΔF is related to errors Δu_1, Δu_2, ..., Δu_m.

To find the expression for ΔF. we should expand the right-hand side of Eq. (8.38) using the Taylor's series expansion for a function of several variables

$$f(u_1 + \Delta u_1, u_2 + \Delta u_2, ..., u_m + \Delta u_m) = f(u_1, u_2, ..., u_m) + \sum_{i=1}^{m} \frac{\partial f}{\partial u_i} \Delta u_i + HOT$$
$$\tag{8.39}$$

where *HOT* stands for the higher-order terms (including squares, products and higher powers of Δu_i). Because errors in *us* are considered to be relatively small, we can neglect these *HOT*, and after substituting Eqs. (8.39) and (8.37) in Eq. (8.38), we obtain the general expression for computing the error of the function

$$\Delta F = \sum_{i=1}^{m} \frac{\partial f}{\partial u_i} \Delta u_i \tag{8.40}$$

or the relative error

$$\frac{\Delta F}{F} = \sum_{i=1}^{m} \frac{\partial f}{\partial u_i} \frac{\Delta u_i}{F} \tag{8.41}$$

These two expressions happen to be very useful. Particularly, they allow you to estimate the errors while dealing with basic operations in arithmetic. For addition and subtraction

$$F = u_1 \pm u_2 \tag{8.42}$$

and therefore

$$\Delta F = \Delta u_1 \pm \Delta u_2 \tag{8.43}$$

The errors Δu_1 and Δu_2 may be either positive or negative, and so the error cannot possibly be less than

$$\left.|\Delta F|\right._{\min} = \left|\,|\Delta u_1| - |\Delta u_2|\,\right| \tag{8.44}$$

and theoretically (if no other errors are involved) should be less than

$$\left.|\Delta F|\right._{\max} = |\Delta u_1| + |\Delta u_2| \tag{8.45}$$

The absolute error of the sum/difference of approximate numbers in the worse-case scenario is therefore bounded by the algebraic sum of their absolute errors. That implies the proper way to handle addition or subtraction of two numbers of different accuracies: to round them to the worst accuracy first. Say, we want to find a sum of three numbers, 2314.2, 3.2713, and 0.56723, that are known to be correct in their last figures, but not farther. Instead of adding them directly as

$$
\begin{array}{r}
2314.2 \\
+\quad 3.2713 \\
+\quad 0.56723 \\
\hline
= 2318.03853
\end{array}
$$

which implies that we actually add 2314.20000, 3.27130, 0.56723, that is, considering additional nonexisting significant figures in the first two numbers, we should first round them off to two decimals (one more than in the least accurate number), and then add them up

$$
\begin{array}{rl}
2314.2 & \\
+\quad 3.27 & \text{(rounding off the more accurate factor)} \\
+\quad 0.57 & \text{(rounding off the more accurate factor)} \\
\hline
= 2318.04 & \\
= 2318.0 & \text{(rounding off the result)}
\end{array}
$$

The last line represents rounding the result to one decimal as in the least accurate number. The reason for that is that by retaining two decimals in the more accurate numbers we eliminate the errors inherent in these numbers and thus reducing the error of the sum to that of the least accurate number [neglecting all but one term in Eq. (8.45)]. Of course, when using the computer, you would probably sum them up as it is, but it would be incorrect to present the final result as 2318.03853, although according to Eq. (8.45), the final result cannot have the better accuracy than that of the least accurate number. In our case, the correct representation of the result would be 2318.0.

Similarly, when one approximate number is to be subtracted from another, they must both be rounded off to the same place before subtracting (in the case of more numbers involved in subtraction one more decimal should be left as in example of summation of three numbers earlier). For example, subtracting 3.2713 from 2314.2 assuming that both numbers are only correct to their last figures, as

$$
\begin{array}{r}
2314.2 \\
- \quad 3.2713 \\
\hline
= 2310.9287
\end{array}
$$

would be absurd, as opposed to

$$
\begin{array}{r}
2314.2 \\
- \quad 3.3 \\
\hline
= 2310.9
\end{array}
\quad \text{(rounding off the more accurate factor)}
$$

Loss of significant figures by subtraction, *subtractive cancellation,* can also be cast in terms of Eq. (8.45). Suppose two numbers 64.567 and 64.564 are each correct to five figures. Then their difference, 64.567 − 64.564 = 0.003, is only correct to one figure.

For multiplication and division, it is convenient to consider the relative error [Eq. (8.41)] rather than the absolute error. Similarly to Eqs. (8.44) and (8.45) for both multiplication and division, we may compute the lower and upper bounds of the relative error (in the case where no other errors present) to be

$$
\left| \frac{\Delta F}{F} \right|_{\min} = \left| \left| \frac{\Delta u_1}{u_1} \right| - \left| \frac{\Delta u_2}{u_2} \right| \right| \tag{8.46}
$$

and

$$
\left| \frac{\Delta F}{F} \right|_{\max} = \left| \frac{\Delta u_1}{u_1} \right| + \left| \frac{\Delta u_2}{u_2} \right| \tag{8.47}
$$

Therefore, when it is required to find a product or division of two approximate numbers of different accuracies, to minimize the upper bound on the error [Eq. (8.47)], the more accurate number should be rounded off, in order to contain one more significant figure than the least accurate factor. By doing so, we eliminate the error due to the more accurate factor thus making the error of the product/division due solely to the error of the less accurate number [neglecting one term in Eq. (8.47)]. The final result should be given to as many significant figures as are contained in the least accurate factor.

For example, having numbers 3.45 and 2.6781 with all figures trustworthy, the appropriate way to multiply them would be

$$
\begin{array}{r}
3.45 \\
\times \, 2.678 \\
\hline
= 9.23910 \\
\hline
= 9.24
\end{array}
\quad
\begin{array}{l}
\text{(rounding off the more accurate factor)} \\
\\
\text{(rounding off the result)}
\end{array}
$$

rather than

$$
\begin{array}{r}
3.45 \\
\times\ 2.6781 \\
\hline
=\ 9.239445
\end{array}
$$

In conclusion, let us consider one last example using division. Suppose we need to find the quotient $\sqrt{3} = 1.732$ assuming that the numerator is correct to its last figure but no father. The correct way of handling it would be presenting a square root of 3 as $\sqrt{3} = 1.732$ so that

$$
\begin{array}{l}
41.2 \\
\underline{/\qquad 1.732} \qquad \text{(limiting } \sqrt{3} \text{ to four significant figures)} \\
= 23.7875288683603 \\
= 23.8 \ \text{(rounding off the result)}
\end{array}
$$

Let us prove it. If a number is correct to N significant figures, it is evident that its absolute error cannot be greater than half a unit in the Nth place. Computing the right-hand side terms in Eq. (8.46) yields

$$
\left|\frac{\Delta u_1}{u_1}\right| = \frac{0.05}{4.12} \approx 0.01 \quad \text{and} \quad \left|\frac{\Delta u_2}{u_2}\right| = \frac{0.005}{1.732} \approx 0.003
$$

thus producing the lower bound on the possible error of the result to be ≈ 0.007. Now, if we leave two decimals in the result for F, we will have

$$
\left|\frac{\Delta F}{F}\right| = \frac{0.05}{23.79} \approx 0.002
$$

which is obviously less than it could possibly be. On the contrary, with only one decimal the relative error will be

$$
\left|\frac{\Delta F}{F}\right| = \frac{0.5}{23.8} \approx 0.02
$$

which at least do not violate the lower bound. In this particular case because of the rounding off the result, we worsen the estimate Eq. (8.47). But that is the only choice (between incorrect and less accurate).

8.6 Control of Numerical Errors

The total numerical error is the summation of the truncation and round-off errors. In general, the only way to minimize round-off errors is to increase the number of significant figures of the computer. Further, we have noted that round-off error may increase due to subtractive cancellation or due to an increase in the number of computations in an analysis. Say, if we perform certain amount of digital computations in K points then, if we double K, the total round-off error is likely to (at least) be doubled too. In contrast, in many numerical methods, the truncation error is proportional to some power of $h \sim K^{-1}$ ($h \ll 1$) and therefore can be reduced by decreasing the step size h. Because a decrease in step size can lead to subtractive cancellation or to an increase in computations, the truncation errors are decreased as the round-off errors are increased.

Therefore, we are facing the following dilemma. The decrease of one component of the total error leads to an increase of the other component. If the total error is as shown in Fig. 8.18, the challenge is to identify the *point of diminishing returns*, where the round-off error begins to negate the benefits of step-size reduction. When using MATLAB, such situations are relatively uncommon because of its 15-digit precision. Nevertheless, they sometimes do occur and suggest a sort of "numerical uncertainty principle" that places an absolute limit on the accuracy that may be obtained using certain computerized numerical methods. On top of that, we might also want to account for the computational recourses. Decreasing the step size leads to increasing the computational time, t_{CPU}. Hence, you may cast this decrease as a penalty P for having too small step size, that is, replace the round-off error in Fig. 8.18 with the required CPU time

$$P = \varepsilon + wt_{CPU} \tag{8.48}$$

(ε accounts for the total error and w is the weighting coefficient).

In this case, even for the 15-digit precision of MATLAB, you will always end up having some step size after which a further decrease of truncation error will result in too much efforts drastically slowing the computation process.

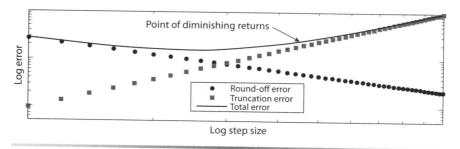

Fig. 8.18 Graphical depiction of the tradeoff between the round-off and truncation error.

For most practical cases, we do not know the exact error associated with numerical methods. The exception, of course, is when we know the exact solution, which makes our numerical approximations unnecessary. Therefore, for most engineering and scientific applications, we must settle for some estimate of the error in our calculations.

There are no systematic and general approaches to evaluating numerical errors for all problems. In many cases, the error estimates are based on the experience and judgment of the engineer or scientist. Although error analysis is, to a certain extent, an art; there are several practical programming guidelines we can suggest.

First and foremost, avoid subtracting two nearly equal numbers. Loss of significance almost always occurs when this is done. Sometimes you can rearrange or reformulate the problem to avoid subtractive cancellation. Furthermore, when adding and subtracting numbers, it is best to sort the numbers and work with the smallest numbers first, which may prevent losing significance.

Second, an attempt to predict total numerical errors using theoretical formulations should be made. In the flowing chapters, the Taylor series expansion, like the ones shown in Eqs. (8.33) and (8.39), will be our primary tool for doing it. Note that usually these estimates happen to be quite conservative. Sometimes there is an alternative way to estimate the accuracy of your results by seeing if the results satisfy some condition or by substituting the results back into the original equation.

Finally, you should be prepared to perform numerical experiments to increase awareness of computational errors and possible ill-conditioned problems. Such experiments may involve repeating the computations with a different step size or method and comparing the results. You may employ sensitivity analysis to see how your solution changes when you change model parameters or input values. You may want to try different numerical algorithms that have different theoretical foundations, are based on different computational strategies, or have different convergence properties and stability characteristics.

Problems

8.1 Write a script, which will convert any nonnegative integer into the octal, ternal, or binary number. Use the `dec2base` function to check the correctness of your results.

8.2 Perform analysis, similar to that presented in Table 8.5 (except the two last columns) and Fig. 8.17, on the Maclaurin series expansion (at $x = 0$) of $\sin(x)$ for $x = 1$. Considering 1 to 5 terms of the expansion, compute the value, $\varepsilon^t(\%)$, $\varepsilon^a(\%)$, $N(\varepsilon^t)$, $N(\varepsilon^a)$ and plot the results.

8.3 Try to reproduce Fig. 8.18 for the following problem. Consider evaluating the following integral numerically:

$$I = \int_0^{10} x^3 dx$$

Although its exact value is known, $I^t = 0.25 \cdot 10^4$, for the numerical estimate, let us break the entire interval $[0;10]$ onto n equal subintervals and use the approximate formula:

$$I_n^a = \frac{10}{n^4} \sum_{i=1}^{n} \left(i - \frac{1}{2} \right)^3$$

The step size h is inversely proportional to the number of the terms in the preceding sum, $h = 10n^{-1}$. For $n = 1, 2, \ldots, 7$, compute the true error $\varepsilon = |I^t - I_n^a|$. Using the tic and toc functions, record the CPU time required to obtain a solution for each number of intervals. Use Eq. (8.48) with $w = 0.01s^{-1}$ to estimate the overall penalty. Plot all three dependences, $\varepsilon(n)$, $t_{CPU}(n)$, and $P(n)$, vs the step size $h(n)$ and comment on the point of diminishing returns.

Chapter 9

Numerical Linear Algebra and Eigenvalue Problems

- Solving Systems of Linear Algebraic Equations
- Understanding Matrix Decompositions
- Using Linear Algebra to Solve Differential Equations
- Using the MATLAB Functions

```
mldivide, mrdivide,
rref, linsolve, lu, qr,
svd, eig, det, rank, inv,
null
```

9.1 Introduction

T his chapter considers the usage of basic MATLAB functions to solve applied engineering problems effectively by employing different numerical methods. Specifically, it deals with the problem of solving systems of linear algebraic equations (LAEs). MATLAB features a variety of functions to address these types of problems, mldivide, mrdivide, rref, and linsolve, each using a different method. The goal of this chapter is to help you to understand the essence of these methods and the difference between them, so that you could employ them appropriately. In addition to the aforementioned functions, this chapter presents several functions dedicated to matrix decompositions, such as LU (lu), QR (qr), SVD (svd), and eigen (eig) decomposition, used to compute a determinant of a matrix (det), its rank (rank), inverse (inv), and null space (null). The latter decomposition, eigendecomposition, turns to be very useful to solve the systems of linear ordinary differential equations (ODEs) and so despite the fact that this specific topic will be addressed later in Chapter 14, this chapter shows how to use this decomposition to solve differential equations analytically. In what follows, we first present the theoretical background behind all major methods devoted to addressing the linear algebra problems and then show examples of using the corresponding MATLAB commands.

9.2 Classification of LAEs

Consider a system of LAEs in the following form:

$$\mathbf{A}\mathbf{x} = \mathbf{b} \tag{9.1}$$

In this matrix, equation $\mathbf{x}$ is the vector of unknown variables, $\mathbf{A}$ the matrix of coefficients, and $\mathbf{b}$ the vector of inhomogeneous terms:

$$\mathbf{A} = [a_{ij}] = \begin{bmatrix} a_{11} & a_{12} & \cdots & a_{1n} \\ a_{21} & a_{22} & \cdots & a_{2n} \\ \vdots & \vdots & \ddots & \vdots \\ a_{m1} & a_{m2} & \cdots & a_{mn} \end{bmatrix}, \quad \mathbf{x} = \begin{bmatrix} x_1 \\ x_2 \\ \vdots \\ x_n \end{bmatrix}, \quad \mathbf{b} = \begin{bmatrix} b_1 \\ b_2 \\ \vdots \\ b_m \end{bmatrix} \tag{9.2}$$

Therefore, a set of LAEs is a collection of linear equations involving the same set of variables.

Depending on the matrix $\mathbf{A}$ and vector $\mathbf{b}$, this system of equations has zero, one, or infinitively many solutions as shown in Fig. 9.1. No other possibilities exist.

As used in this chapter, the most general case is when $\mathbf{A}$ is a square n by n matrix ($m = n$). As seen from Fig. 9.1, even then several possibilities exist. The deciding point on the type of a solution the set of LAEs yields is examining its matrix $\mathbf{A}$. Figure 9.1 suggests to either check whether matrix $\mathbf{A}$ is invertible (in this case, another n by n matrix $\mathbf{B}$ exists, such that

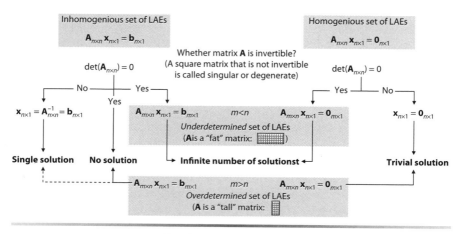

Fig. 9.1 Variety of LAEs.

AB = **I**) or whether its determinant is not equal to zero. In fact, some other equivalent checks may be applied:

1. **A** is invertible.
2. Determinant of **A** is not equal to zero.
3. Rank of **A** is equal to n.
4. The columns of **A** are linearly independent.
5. The transpose $\mathbf{A}^T$ is an invertible matrix.
6. The matrix times its transpose, $\mathbf{A}^T\mathbf{A}$, is an invertible matrix.
7. The number 0 is not an eigenvalue of **A** (**A** is a *nonsingular* matrix).

Some of these concepts along with all possibilities depicted in Fig. 9.1 will be addressed later in this chapter, but the major emphasis will be given to solving the sets of *inhomogeneous* equations, assuming that **b** ≠ **0** (**0** is a vector of zeros), that is, that vector **b** has at least one nonzero component (otherwise, Eq. (9.1) is referred to as a *homogeneous* set of equations).

9.3 Inhomogeneous Sets of LAEs

The case when a matrix **A** in Eq. (9.1) is square, has a full rank, and at least one element of the vector **b** is not equal to zero leads to a single (unique) solution as shown in Fig. 9.2. For the sake of briefness, Fig. 9.2 shows that the solution can be obtained by simply inverting the matrix **A**. In fact, none of MATLAB functions does this. Instead, several other, more computationally effective, methods are employed as discussed later.

9.3.1 Gauss Elimination

Gauss or *Gaussian elimination* is the most widely used method for solving a set of inhomogeneous LAEs known from ancient times and formally presented by Gauss in the early 1800s. It consists of the *forward elimination* process followed by the *backward substitution* process and makes the entire approach very suitable for computer applications. The MATLAB functions `mldivide` (\) and `mrdivide` (/) use exactly this method.

The forward elimination process of the Gauss elimination method starts from multiplying the first equation by a multiplier $l_{21} = a_{21} / a_{11}$ and subtracting it from the second equation to eliminate the first term of the second equation. In matrix notation, it means pre- (which means, left-) multiplying both sides of Eq. (9.1) by the *elementary matrix* $\mathbf{L}^*_{21}$

$$\mathbf{L}^*_{21}\mathbf{A}\mathbf{x} = \begin{bmatrix} 1 & 0 & \cdots & 0 \\ -l_{21} & 1 & \cdots & 0 \\ \vdots & \vdots & \ddots & \vdots \\ 0 & 0 & \cdots & 1 \end{bmatrix} \mathbf{A}\mathbf{x} = \begin{bmatrix} 1 & 0 & \cdots & 0 \\ -l_{21} & 1 & \cdots & 0 \\ \vdots & \vdots & \ddots & \vdots \\ 0 & 0 & \cdots & 1 \end{bmatrix} \mathbf{b} = \mathbf{L}^*_{21}\mathbf{b} \qquad (9.3)$$

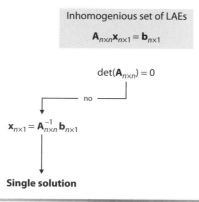

Fig. 9.2 General case of a single solution for a full-rank square matrix.

As seen, the elementary matrix $\mathbf{L}_{ij}^*$ is an identity matrix with one zero element at the intersection of the ith row and jth column substituted with the corresponding multiplier l_{ij}.

Likewise, multiplying the first equation of system [Eq. (9.1)] by $l_{i1} = a_{i1}/a_{11}$ and subtracting it from the ith equation eliminates the first term from the ith equation ($i = 3, 4, ..., n$). When applied for all but the last rows, system (9.1) transforms to

$$
\begin{aligned}
a_{11}x_1 + a_{12}x_2 + a_{13}x_3 + \cdots + a_{1n}x_n &= b_1 \\
a'_{22}x_2 + a'_{23}x_3 + \cdots + a'_{2n}x_n &= b'_2 \\
&\ \vdots \\
a'_{n2}x_2 + a'_{n3}x_3 + \cdots + a'_{nn}x_n &= b'_n
\end{aligned}
\tag{9.4}
$$

where

$$
a'_{ij} = a_{ij} - \frac{a_{i1}}{a_{11}}a_{1j} \quad \text{and} \quad b'_i = b_i - \frac{a_{i1}}{a_{11}}b_1, \quad i = 2, 3, ..., n, \quad j = 2, 3, ..., n
$$

(prime does not denote a partial derivative here, it is just a notation to distinguish the new coefficients from the previous ones).

The coefficient in the denominator of multipliers l_{i1}, a_{11}, is called the *pivot*, the first pivot. In matrix notation, it means premultiplying both sides of Eq. (9.1) by corresponding elementary matrices $\mathbf{L}_{i1}^*$ one after another, so that the system (9.4) corresponds to

$$
\mathbf{L}_{n1}^* \cdots \mathbf{L}_{31}^* \mathbf{L}_{21}^* \mathbf{A}\mathbf{x} = \left(\prod_{i=2}^{n} \mathbf{L}_{i1}^{*T} \right)^T \quad \mathbf{A}\mathbf{x} = \left(\prod_{i=2}^{n} \mathbf{L}_{i1}^{*T} \right)^T \mathbf{b}
\tag{9.5}
$$

(The transposition of matrices is used to write them in conjunction with the product symbol, which runs from 2 up to n as opposed to from n down to 2.)

Now that we managed to eliminate the first term from all but the first equations, it is time to eliminate the second term. For all equations starting from the third one and down, we eliminate the second term by multiplying the second equation in Eq. (9.4) by $l_{i2} = a'_{i1}/a'_{22}$ and subtracting it from the ith equation ($i = 3, 4, ..., n$). The coefficient a'_{22} becomes the second pivot. In matrix notation, it means continuously premultiplying both sides of Eq. (9.5) by the corresponding elementary matrices $\mathbf{L}^*_{i2}$

$$\mathbf{L}^*_{n2} \ldots \mathbf{L}^*_{42}\mathbf{L}^*_{32}\mathbf{L}^*_{n1} \ldots \mathbf{L}^*_{31}\mathbf{L}^*_{21}\mathbf{Ax} = \left(\prod_{i=3}^{n}\mathbf{L}^{*T}_{i2}\right)^T \left(\prod_{i=2}^{n}\mathbf{L}^{*T}_{i1}\right)^T \mathbf{Ax}$$

$$= \left(\prod_{i=3}^{n}\mathbf{L}^{*T}_{i2}\right)^T \left(\prod_{i=2}^{n}\mathbf{L}^{*T}_{i1}\right)^T \mathbf{b} \qquad (9.6)$$

This process is continued until the set of equations is converted into the *row echelon* form

$$\begin{aligned} a_{11}x_1 + a_{12}x_2 + a_{13}x_3 + \cdots + a_{1n}x_n &= b_1 \\ a'_{22}x_2 + a'_{23}x_3 + \cdots + a'_{2n}x_n &= b'_2 \\ ''a_{33}x_3 + \cdots +''a_{3n}x_n &= b'' \\ &\vdots \\ a_{nn}^{(n-1)}x_n &= b_n^{(n-1)} \end{aligned} \qquad (9.7)$$

with the leading terms in each equation being pivots.

In matrix notation, Eq. (9.7) is equivalent to

$$\mathbf{L}^*\mathbf{Ax} = \mathbf{L}^*\mathbf{b} \qquad (9.8)$$

or simply

$$\mathbf{Ux} = \mathbf{c} \qquad (9.9)$$

By construction, the matrix

$$\mathbf{U} = \mathbf{L}^*\mathbf{A} \qquad (9.10)$$

is the *upper triangular* matrix or sometimes called the *right matrix* (because all entries below the diagonal are zero), and

$$\mathbf{c} = \mathbf{L}^*\mathbf{b} \qquad (9.11)$$

is the new vector of inhomogeneous terms. Both **U** and **c** are obtained via premultiplication of the original matrix **A** and the original vector **b** by the compound matrix

$$\mathbf{L}^* = \mathbf{L}^*_{n,n-1} \left(\prod_{i=n-1}^{n} \mathbf{L}^{*T}_{i,n-2} \right)^T \cdots \left(\prod_{i=3}^{n} \mathbf{L}^{*T}_{i2} \right)^T \left(\prod_{i=2}^{n} \mathbf{L}^{*T}_{i1} \right)^T \tag{9.12}$$

which by construction happens to be a *lower triangular* matrix (*left matrix*). The inverse of matrix **L**** is also a lower triangular matrix with ones on the main diagonal and corresponding multipliers below it

$$\mathbf{L} = (\mathbf{L}^*)^{-1} = \begin{bmatrix} 1 & 0 & \cdots & 0 & 0 \\ l_{21} & 1 & \cdots & 0 & 0 \\ \vdots & \vdots & \ddots & \vdots & \vdots \\ l_{n-1,1} & l_{n-1,2} & \cdots & 1 & 0 \\ l_{n1} & l_{n2} & \cdots & l_{n,n-1} & 1 \end{bmatrix} \tag{9.13}$$

The effectiveness of any algorithm for solving systems of LAEs is judged upon the number of operations needed to complete it. Let us ignore operations on the right-hand side terms (which is negligibly small compared to those performed on the elements of the matrix **A**), and also count each multiplication–subtraction as a single operation. At the first step [Eq. (9.4)], all n^2 coefficients of the matrix **A** need to be changed except the n coefficients of the first row making the total of $n^2 - n$ operations required. Similarly, the second step [Eq. (9.6)] requires $(n-1)^2 - (n-1)$ operations, etc. Altogether, the total number of operations is

$$\sum_{i=1}^{n} (i^2 - i) = \sum_{i=1}^{n} i^2 - \sum_{i=1}^{n} i = \frac{n(n+1)(2n+1)}{6} - \frac{n(n+1)}{2} = \frac{n^3 - n}{3} \tag{9.14}$$

(the standard formulas for the sum of the first n numbers and for the sum of the first n squares are used here). For large n, a good estimate for the number of operations required for the elimination process is $(1/3)n^3$ (the more precise numbers are $n(n+1)/2$ divisions, $(2n^3 + 3n^2 - 5n)/6$ multiplications, and the same number of subtractions, totaling to $(2/3)n^3$).

Now that we reduced the original system of linear equations [Eq. (9.1)] to that of system (9.9), we may proceed with the *backward substitution* process. It starts from resolving the last equation [see Eq. (9.7)] resulting in

$$x_n = \frac{b_n^{(n-1)}}{a^{(n-1)}} \tag{9.15}$$

Substituting this value into the $(n-1)$th equation yields

$$x_{n-1} = \frac{b_{n-1}^{(n-2)} - a_{n-1,n}^{(n-1)} x_n}{a_{n-1,n-1}^{(n-2)}} \tag{9.16}$$

Using solutions (9.15) and (9.16) in the $(n-2)$th equation yields the solution for x_{n-2}, and so on until from the first equation of system [Eq. (9.7)] we get

$$x_1 = \frac{b_1 - \sum_{j=2}^{n} a_{1j} x_j}{a_{11}} \tag{9.17}$$

which completes the backward substitution. For your convenience, the entire process is presented graphically in Fig. 9.3.

Obviously, the backward substitution process involves fewer operations and therefore is performed faster than that of the forward elimination. Specifically, the last unknown is found in only one arithmetic operation, a division by the last pivot. The second to the last unknown requires two operations and so on. Hence, the total number of operations required for the backward substitution is

$$\sum_{i=1}^{n} i = \frac{n(n+1)}{2} \approx 0.5n^2 \tag{9.18}$$

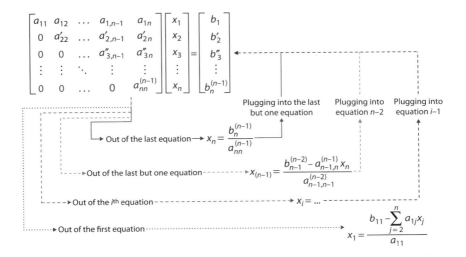

Fig. 9.3 Visualization of backward substitution.

The number of arithmetic operations required to find a solution is only one characteristic of a numerical procedure. Its accuracy is another. From this standpoint we know that for the floating-point arithmetic, the best precision is obtained in the small number areas as opposed to the large numbers (Fig. 8.12). That is why it is important what pivot values the original matrix A has. The extreme case is when one of the pivots happens to be a very small number close to zero. (Even if the original matrix may not have zeros, any diagonal coefficient may become zero in the process of forward elimination.) This would result in the corresponding multiplier being a huge number, so that in the consequent operations, we would be adding together very small and very large numbers. That is exactly where rightmost significant digits will be lost, as explained in examples of Sec. 8.5.2.

Consider a classical example

$$\begin{bmatrix} 0.00001 & 1 \\ 1 & 1 \end{bmatrix} \begin{bmatrix} x_1 \\ x_2 \end{bmatrix} = \begin{bmatrix} 1 \\ 2 \end{bmatrix} \tag{9.19}$$

After forward elimination with the multiplier equal to 100,000, Eq. (9.19) transforms to

$$\begin{bmatrix} 0.00001 & 1 \\ 0 & -99,999 \end{bmatrix} \begin{bmatrix} x_1 \\ x_2 \end{bmatrix} = \begin{bmatrix} 1 \\ -99,998 \end{bmatrix} \tag{9.20}$$

with two pivots of 0.00001 and −99,999, that is absolutely out of scale. The backward substitution yields $x_2 = 0.99999$ and $x_1 = 1$. On a hypothetical computer which keeps less than five significant digits, due to the round-off error, the solution for x_2 would be $x_2 = 1$, which is wrong only in the fifth place. Substituting this "erroneous" solution back into the first equation results in $x_1 = 0$.

If we swap the equations

$$\begin{bmatrix} 1 & 1 \\ 0.00001 & 1 \end{bmatrix} \begin{bmatrix} x_1 \\ x_2 \end{bmatrix} = \begin{bmatrix} 2 \\ 1 \end{bmatrix} \tag{9.21}$$

a forward elimination with the multiplier equal to 10^{-15} yields

$$\begin{bmatrix} 1 & 1 \\ 0 & 0.99999 \end{bmatrix} \begin{bmatrix} x_1 \\ x_2 \end{bmatrix} = \begin{bmatrix} 2 \\ 0.99998 \end{bmatrix} \tag{9.22}$$

with about the same-magnitude pivots of 1 and 0.99999, respectively. The backward substitution should yield $x_2 = 0.99999$ and $x_1 = 1.00001$. However, because of the round-off error on our hypothetical computer, $x_2 = 1$. Substituting this value into the first equation results in $x_1 = 1$. Hence, swapping equations assures that even in the case of round-off error, we still have an accurate solution.

That is why each step of the forward elimination process of the Gauss method also involves the *partial pivoting*, that is, changing the sequential order of equations, so that at each step the pivot has the largest absolute value among all other column elements. If at any step, another row looks more attractive (has a larger absolute value of the leading element), it is swapped with the current pivot row. If a zero diagonal element is unavoidable even with pivoting, it indicates that the problem has no unique solution, that is, that the matrix **A** happens to be a singular matrix.

If row swapping has occurred, it means that both sides of Eq. (9.1) were premultiplied by the *permutation matrix* **P** that produced a desirable row exchange, for example, the matrix

$$\mathbf{P}_1^* = \begin{bmatrix} 0 & 0 & 1 & 0 \\ 0 & 1 & 0 & 0 \\ 1 & 0 & 0 & 0 \\ 0 & 0 & 0 & 1 \end{bmatrix} \tag{9.23}$$

swaps the first and the third rows at the first step (the unities moves from their positions in the identity matrix as shown in Eq. (9.23) with the arrows).

Overall, although pivoting adds more operations to the computational cost of an algorithm, these additional operations may be necessary for the algorithm to work at all. In fact, these additional operations are worthwhile because they add numerical stability to the final result.

As opposed to the partial pivoting, the *complete* or *maximal pivoting* considers all entries in the whole matrix, interchanging both rows and columns. Complete pivoting is rarely used because of the many additional computations it introduces. Lastly, *scaled partial pivoting* is a variation of the partial pivoting, so that the pivoting element is chosen as the entry that is the largest relative to remaining elements in the same row.

9.3.2 Gauss–Jordan Elimination

Gauss–Jordan elimination is a variant of Gauss elimination in which in place of the backward substitution, the *backward elimination* process is implemented to eliminate all coefficients except the pivots, which become unity. Starting with Eq. (9.7), we ended up with after forward elimination, the last row is divided by a_{nn}^{n-1} yielding

$$
\left[
\begin{array}{cccccc|c}
a_{11} & a_{12} & \cdots & a_{1,n-2} & a_{1,n-1} & a_{1,n} & b_1 \\
0 & a_{22}' & \cdots & a_{2,n-2}' & a_{2,n-1}' & a_{2,n}' & b_2' \\
\vdots & \vdots & \ddots & \vdots & \vdots & \vdots & \vdots \\
0 & 0 & \cdots & 0 & a_{n-1,n-1}^{(n-2)} & a_{n-1,n}^{(n-2)} & b_{n-1}^{(n-2)} \\
0 & 0 & \cdots & 0 & 0 & 1 & \overline{b}_n
\end{array}
\right]
\tag{9.24}
$$

where $\overline{b}_n = b_n^{(n-1)}/a_{nn}^{(n-1)}$. Note that, in Eq. (9.24), we used the *augmented matrix* form, which for compactness combines the elements of a matrix **A** and vector **b** together.

Next, multiplying the nth row by $a_{in}^{(i-1)}$ and subtracting it from the ith row ($i = n - 1, n - 2, ..., 1$) eliminates the $a_{in}^{(i-1)}$ coefficients from the remaining $n - 1$ rows upwards

$$
\left[
\begin{array}{cccccc|c}
a_{11} & a_{12} & \cdots & a_{1,n-2} & a_{1,n-1} & 0 & \overline{b}_1 \\
0 & a_{22}' & \cdots & a_{2,n-2}' & a_{2,n-1}' & 0 & \overline{b}_2 \\
\vdots & \vdots & \ddots & \vdots & \vdots & \vdots & \vdots \\
0 & 0 & \cdots & 0 & a_{n-1,n-1}^{(n-2)} & 0 & \overline{b}_{n-1} \\
0 & 0 & \cdots & 0 & 0 & 1 & \overline{b}_n
\end{array}
\right]
\tag{9.25}
$$

Here, $\overline{b}_i = b_i^{(i-1)} - \overline{b}_n a_{in}^{(i-1)}$, $i = 1, 2, ..., n - 1$. Continuing this process of normalizing the $(n - 1)$th row and eliminating $a_{i,n-1}^{(i-1)}$ ($i=n-2,n-1,...,1$), coefficients from the remaining $n-2$ rows upwards yield

$$
\left[
\begin{array}{cccccc|c}
a_{11} & a_{12} & \cdots & a_{1,n-2} & 0 & 0 & \overline{b}_1 \\
0 & a_{22}' & \cdots & a_{2,n-2}' & 0 & 0 & \overline{b}_2' \\
\vdots & \vdots & \ddots & \vdots & \vdots & \vdots & \vdots \\
0 & 0 & \cdots & a_{n-2,n-2}^{(n-3)} & 0 & 0 & \overline{b}_{n-2}' \\
0 & 0 & \cdots & 0 & 1 & 0 & \overline{b}_{n-1}' \\
0 & 0 & \cdots & 0 & 0 & 1 & \overline{b}_n
\end{array}
\right]
\tag{9.26}
$$

with $\quad \overline{b}_{n-1}' = \dfrac{\overline{b}_{n-1}}{a_{n-1,n-1}^{(n-2)}} \quad$ and $\quad \overline{b}_i' = \overline{b}_i - \overline{b}_{n-1}' a_{i,n-1}^{(i-1)}, \quad i = 1, 2, ..., n - 2.$

Repeating the elimination process for the remaining rows results in the *reduced row echelon* form

$$
\begin{bmatrix}
1 & 0 & \cdots & 0 & 0 & \overline{b}_1^{(n-1)} \\
0 & 1 & \cdots & 0 & 0 & \overline{b}_2^{(n-2)} \\
\vdots & \vdots & \ddots & \vdots & \vdots & \vdots \\
0 & 0 & \cdots & 1 & 0 & \overline{b}'_{n-1} \\
0 & 0 & \cdots & 0 & 1 & \overline{b}_n
\end{bmatrix}
\tag{9.27}
$$

which ends the backward elimination process. The solution becomes

$$
x_i = \overline{b}_i^{(n-i)}, \quad i = 1, 2, \ldots, n
\tag{9.28}
$$

that is, the rightmost column of the reduced row echelon happens to be a column-vector containing the final solution. This algorithm is implemented in the MATLAB's `rref` function.

9.3.3 LU Decomposition

As shown in Sec. 9.3.2, the forward elimination process of the Gauss elimination method resulted in converting the original matrix **A** in Eq. (9.1) into the upper triangular matrix **U** in Eq. (9.9). Premultiplying both sides of Eq. (9.10) by the inverse of a matrix $\mathbf{L}^*$ and accounting for Eq. (9.13) results in

$$
\mathbf{A} = \mathbf{LU}
\tag{9.29}
$$

which is referred to as the *LU decomposition* of a matrix **A**.

Formally, the LU decomposition scheme is a transformation of a nonsingular (with a nonzero determinant) matrix **A** to a product of two matrices, one being a lower triangular matrix and another an upper triangular matrix. For example, for a 3×3 matrix, the LU decomposition yields

$$
\begin{bmatrix}
a_{11} & a_{12} & a_{13} \\
a_{21} & a_{22} & a_{23} \\
a_{31} & a_{32} & a_{33}
\end{bmatrix}
=
\begin{bmatrix}
1 & 0 & 0 \\
l_{21} & 1 & 0 \\
l_{31} & l_{32} & 1
\end{bmatrix}
\begin{bmatrix}
u_{11} & u_{12} & u_{13} \\
0 & u_{22} & u_{23} \\
0 & 0 & u_{33}
\end{bmatrix}
\tag{9.30}
$$

Following the forward elimination routine resulting in Eq. (9.7), a general scheme to perform the LU decomposition of a square n by n matrix is as follows:

The first row of **U** is obtained as	$u_{1j} = a_{1j}$	for $j = 1, 2, \ldots, n$
The first column of **L** is obtained as	$l_{i1} = \dfrac{a_{i1}}{u_{11}}$	for $i = 2, 3, \ldots, n$
The second row of **U** is obtained by	$u_{2j} = a_{2j} - l_{21}u_{1j}$	for $i = 2, 3, \ldots, n$
The second column of **L** is obtained by	$l_{i2} = \dfrac{a_{i2} - l_{i1}u_{12}}{u_{22}}$	for $i = 3, 4, \ldots, n$
...		
The *s*th row of **U** is obtained by	$u_{sj} = a_{sj} - \displaystyle\sum_{k=1}^{s-1} l_{sk}u_{kj}$	for $j = s, s+1, \ldots, n$
The *s*th column of **L** is obtained by	$l_{is} = \dfrac{a_{is} - \displaystyle\sum_{k=1}^{s-1} l_{ik}u_{ks}}{u_{ss}}$	for $i = s+1, s+2, \ldots, n$
...		

Note that in this process, the diagonal elements of **L** are not calculated because they are unity. Although the LU decomposition follows exactly the same steps as a regular forward elimination procedure, defining matrices **L** and **U** explicitly has its computational advantage. Especially, in the case when there is a need to solve multiple sets of LAEs featuring the same matrix **A**

$$\mathbf{Ax}_1 = \mathbf{b}_1, \quad \ldots, \quad \mathbf{Ax}_p = \mathbf{b}_p \qquad (9.31)$$

Recall that the forward elimination (LU decomposition of matrix **A**) requires $\sim n^3$ operations, whereas the backward substitution only requires $\sim n^2$ operations. If we were to solve all p systems of Eq. (9.31) "from scratch" that would require $\sim p(n^3 + n^2)$ operations. Instead, all we have to do is to perform the LU decomposition of matrix **A** just once and then use its results to resolve all p systems of Eq. (9.31).

Using Eq. (9.29), we may rewrite Eq. (9.1) as

$$\mathbf{Ax} = \mathbf{LUx} = \mathbf{b} \qquad (9.32)$$

or as a combination of two sets

$$\mathbf{Lc} = \mathbf{b} \quad \text{and} \quad \mathbf{Ux} = \mathbf{c} \tag{9.33}$$

for each $\mathbf{b}_k$, $k = 1, 2, ..., p$. (For simplicity, we are not showing the permutation matrix $\mathbf{P}$ assuming that it is incorporated in $\mathbf{A}$ and $\mathbf{b}$.) Because $\mathbf{L} = [l_{ij}]$ is a lower triangular matrix, we can easily solve the first equation in Eq. (9.33) for each $\mathbf{c}_k = [c_1, c_2, ..., c_n]^T$ (for different $\mathbf{b}_k = [b_1, b_2, ..., b_n]^T$) using *forward substitution*

$$c_1 = b_1, \quad c_i = b_i - \sum_{j=1}^{i-1} l_{ij} c_j, \quad \text{for } i = 2, 3, ..., n \tag{9.34}$$

requiring altogether $\sim pn^2$ operations. We then can use the solution for $\mathbf{c}_k$ [Eq. (9.34)] and employ the upper triangular matrix $\mathbf{U} = [u_{ij}]$ to solve for the corresponding $\mathbf{x}_k = [x_1, x_2, ..., x_n]^T$ using *backward substitution*

$$x_n = \frac{c_n}{u_{nn}}, \quad x_i = c_i - \sum_{j=i+1}^{n} u_{ij} x_j, \quad \text{for } i = n-1, n-2, ..., 1 \tag{9.35}$$

also requiring $\sim pn^2$ operations altogether. Therefore, instead of $\sim p(n^3 + n^2)$ operations we only have $\sim n^3 + 2pn^2$, that is of the order of p times less.

Additionally, because the upper part of $\mathbf{L}$ and the lower part of $\mathbf{U}$ contain only zeros, we can store both matrices in the same n by n matrix (to save memory) as

$$\begin{bmatrix} u_{11} & u_{12} & \cdots & u_{1,n-1} & u_{1n} \\ l_{21} & u_{22} & \cdots & u_{2,n-1} & u_{2n} \\ \vdots & \vdots & \ddots & \vdots & \vdots \\ l_{n-1,1} & l_{n-1,2} & \cdots & u_{n-1,n-1} & u_{n-1,n} \\ l_{n1} & l_{n2} & \cdots & l_{n,n-1} & u_{nn} \end{bmatrix} \tag{9.36}$$

The memory usage can be reduced even further by noticing that each element in $\mathbf{A}$ is used for calculation of l_{ij} or u_{ij} during the factorization process only once. Thus, once l_{ij} or u_{ij} is computed, it can be overwritten into a_{ij} making it unnecessary to create a separate matrix.

The `lu` function expresses a matrix $\mathbf{A}$ as the product of two essentially triangular matrices [Eq. (9.29)], with the only difference that a lower triangular matrix is returned permutated. As explained later in Sec. 9.7.1, there is a syntax, however, allowing you to obtain a lower triangular matrix **L** with a unit diagonal, and a permutation matrix **P** separately, so that

$$PA = LU \tag{9.37}$$

Another function to solve the linear system [Eq. (9.1)], `linsolve`, employs the LU factorization when **A** is square as well. Examples will be provided in Sec. 9.7.1.

9.3.4 Inverse and Determinant of a Square Matrix

By definition, the *inverse of a square matrix*, A^{-1} satisfies the relationship

$$AA^{-1} = I \tag{9.38}$$

You may treat Eq. (9.38) as the set of matrix equations

$$Ax_1 = \begin{bmatrix} 1 \\ 0 \\ \vdots \\ 0 \end{bmatrix}, \ Ax_2 = \begin{bmatrix} 0 \\ 1 \\ \vdots \\ 0 \end{bmatrix}, \ \dots, \ Ax_n = \begin{bmatrix} 0 \\ 0 \\ \vdots \\ 1 \end{bmatrix} \tag{9.39}$$

where solutions $x_1, x_2, \dots, x_n$ constitute the columns of coefficients of the inverse matrix $A^{-1} = [x_1 \ x_2 \ \cdots \ x_n]$. This is exactly what the LU decomposition is good for. You may find the matrix inverse by performing the LU decomposition of a matrix **A**, followed by solving Eq. (9.24) for each set of Eq. (9.39). In fact, in this specific case, you can solve all n systems simultaneously.

If you recall, in Sec. 9.3.2, we introduced the augmented matrix, which we used for the forward and backward elimination to produce a reduced row echelon form of Eq. (9.27). We can do essentially the same here augmenting the matrix **A** with the same-size identity matrix **I** to begin with

$$\left[\begin{array}{cccc|cccc} a_{11} & a_{12} & \cdots & a_{1n} & 1 & 0 & \cdots & 0 \\ a_{21} & a_{22} & \cdots & a_{2n} & 0 & 1 & \cdots & 0 \\ \vdots & \vdots & \ddots & \vdots & \vdots & \vdots & \ddots & \vdots \\ a_{n1} & a_{n2} & \cdots & a_{nn} & 0 & 0 & \cdots & 1 \end{array} \right] \Leftrightarrow [A \mid I] \tag{9.40}$$

Table 9.1   Budget of Numerical Operations for Solving n Sets of LAEs
with Two Different Approaches

Gauss Elimination Method	Number of Operations	LU Decomposition-Based Approach	Number of Operations
Gauss elimination	$\dfrac{n^3-n}{3} \underset{n\to\infty}{\sim} \dfrac{n^3}{3}$	Matrix decomposition	$\dfrac{n^3-n}{3} \underset{n\to\infty}{\sim} \dfrac{n^3}{3}$
Backward substitution	$\dfrac{n^2-n}{2} \underset{n\to\infty}{\sim} \dfrac{n^2}{2}$	Solving two systems (Eq. 9.33)	$n^2+n \underset{n\to\infty}{\sim} n^2$
Total for the n sets of LAEs	$n\left(\dfrac{n^3}{3}+\dfrac{n^2}{2}\right) - \dfrac{n^4}{3}$	Total for the n sets of LAEs	$\dfrac{n^3}{3}+n\cdot n^2 \sim \dfrac{4}{3}n^3$

Following the Gauss–Jordan elimination procedure in exactly the same way we did it in Sec. 9.3.2 solves Eq. (9.40) by bringing it to the form

$$
\begin{bmatrix}
1 & 0 & \cdots & 0 \\
0 & 1 & \cdots & 0 \\
\vdots & \vdots & \ddots & \vdots \\
0 & 0 & \cdots & 1
\end{bmatrix}
\left.
\begin{matrix}
\overline{a}_{11}^{(n-1)} & \overline{a}_{12}^{(n-1)} & \cdots & \overline{a}_{1n}^{(n-1)} \\
\overline{a}_{21}^{(n-2)} & \overline{a}_{22}^{(n-2)} & \cdots & \overline{a}_{2n}^{(n-2)} \\
\vdots & \vdots & \ddots & \vdots \\
\overline{a}_{n1}' & \overline{a}_{n2}' & \cdots & \overline{a}_{nn}'
\end{matrix}
\right]
\Leftrightarrow \left[\mathbf{I} \mid \mathbf{A}^{-1}\right] \qquad (9.41)
$$

If the Gauss–Jordan elimination becomes unsolvable even with pivoting, meaning the matrix $\mathbf{A}$ is singular, then the inverse does not exist.

The computational effectiveness of the approach based on the LU decomposition is compared with that of the Gaussian elimination in Table 9.1. While solving a single set of LAEs seems to favor the Gauss method (consider two first lines of the table), if it comes to solving the n sets of LAEs having the same matrix $\mathbf{A}$ the LU decomposition has an obvious advantage. In this case (see the last row of the table) you only have to perform the LU decomposition once and then solve two auxiliary systems of LAEs n times, while for the Gauss method you have to do both steps, forward elimination and backward substitution for each set of LAEs, that is, the sum of operations in the two first rows should be multiplied by n.

The MATLAB function `inv` uses the LU decomposition approach for $n > 2$, whereas for $n = 2$, it used the analytical formula. To obtain it, you may use

```
>> syms a b c d
>> A=[a b; c d]; inv(A)
```

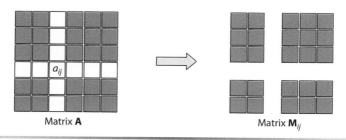

Matrix **A**   Matrix **M**$_{ij}$

Fig. 9.4   Constructing a matrix **M**$_{ij}$.

which returns

```
ans = [d/(a*d-b*c),  -b/(a*d-b*c)]
      [-c/(a*d-b*c),  a/(a*d-b*c)]
```

One of the ways to compute the *determinant* of a matrix **A**, det(**A**), a special number associated with any square matrix, is to use Laplace's formula expanding a determinant along a row or column. For example, using elements of the first column yields

$$\det(\mathbf{A}) = \begin{cases} a_{11}, & n=1 \\ \sum_{i=1}^{n} a_{i1}(-1)^{i+1}\det(\mathbf{M}_{i1}), & n>1 \end{cases} \qquad (9.42)$$

where $(-1)^{i+1}\det(\mathbf{M}_{i1})$ is called the *cofactor* of an element a_{i1}, and $\det(\mathbf{M}_{i1})$ is referred to as a *minor* of matrix **A**. The matrix **M**$_{ij}$ is obtained from **A** by removing the ith row and the jth column, that is. those that contain the coefficient a_{ij} (Fig. 9.4).

In turn, determinants of matrices **M**$_{i1}$, i = 1, 2, ..., n (which are one order less than the order of the original matrix **A**) can be found using the same Eq. (9.42), but written for matrices **M**$_{i1}$ now, and so on. The process stops when cofactors become scalars (order-one matrices).

In general, any row or column may be used to find the determinant of a matrix. For the sparse matrices (matrices containing zero blocks), the best choice would be the column or row containing maximum number of zero elements. The approach based on Eq. (9.42) can only be computationally effective for relatively small matrices.

Luckily, the determinant holds a very useful property

$$\det(\mathbf{AB}) = \det(\mathbf{A})\det(\mathbf{B}) \qquad (9.43)$$

Using this property, you may compute the determinate using LU decomposition in the form of Eq. (9.37)

$$\det(\mathbf{A}) = \det(\mathbf{L})\det(\mathbf{U}) / \det(\mathbf{P}) = \det(\mathbf{U}) / \det(\mathbf{P}) \qquad (9.44)$$

By construction,

$$\det(\mathbf{L}) = 1, \ \det(\mathbf{U}) = \prod_{i=1}^{n} u_{ii} \qquad (9.45)$$

and the determinant of permutation matrix $\mathbf{P}$ is either 1 for an even number of permutations or -1 for an uneven number of permutations. The built-in `det` function of MATLAB computes the determinant using LU decomposition as presented earlier.

9.3.5 Iterative Improvement of a Solution

Solving a system of LAEs with a large n involves a large number of mathematical operations. Because of round-off errors, the accuracy of a solution may degrade fairly fast. Also, you can easily loose a couple of significant figures and make your matrix $\mathbf{A}$ pretty close to a singular matrix, which you would not expect.

If it happens, there is a neat trick to restore the full machine precision called the *iterative improvement of a solution*. Suppose that a vector $\mathbf{x}$ is the exact solution of the linear set [Eq. (9.1)]. However, you do not know it. You only know the solution $\mathbf{x} + \delta\mathbf{x}$, which is slightly off by some unknown error $\delta\mathbf{x}$.

As a result, when multiplied by the matrix $\mathbf{A}$, this inaccurate solution yields a product, which is slightly discrepant from the desired right-hand side vector $\mathbf{b}$

$$\mathbf{A}(\mathbf{x} + \delta\mathbf{x}) = \mathbf{b} + \delta\mathbf{b} \qquad (9.46)$$

Subtracting Eq. (9.1) from Eq. (9.46) gives

$$\mathbf{A}\delta\mathbf{x} = \delta\mathbf{b} \qquad (9.47)$$

Also, from Eq. (9.46), you may express $\delta\mathbf{b}$ as

$$\delta\mathbf{b} = \mathbf{A}(\mathbf{x} + \delta\mathbf{x}) - \mathbf{b} \qquad (9.48)$$

and substitute it into Eq. (9.47), which results in

$$A\delta x = A(x + \delta x) - b \qquad (9.49)$$

In this last equation, the right-hand side is known, because $x + \delta x$ is the inaccurate solution you want to improve. Hence, you may solve it for the error δx and then subtract this error from your inaccurate solution to get an improved solution.

An important extra benefit occurs if you obtained the original solution via the LU decomposition. In this case, you have the LU decomposed form of A already, so that all we need to do to solve Eq. (9.49) is to compute the right-hand side and perform a backward substitution.

9.4 Matrix Decompositions

The LU decomposition introduced in Sec. 9.3.3 is not the only decomposition of a matrix A. Other decompositions that may be very useful for solving the sets of LAEs are also available.

The *QR decomposition* (the qr function of MATLAB), which is applicable to square and m by n matrices, yields

$$A = QR \qquad (9.50)$$

where Q is an m by m orthonormal matrix (introduced and discussed in Sec. 3.3.1) and R is an m by n upper triangular matrix. The fact that Q is orthonormal, meaning $Q^T Q = I$ and $Q^{-1} = Q^T$ allows you to rewrite Eq. (9.1) as

$$Rx = Q^T b \qquad (9.51)$$

The methods for obtaining matrices Q and R include Gram–Schmidt process, Householder transformations, and Given rotations. Obviously, Eq. (9.51) is easier to solve because R is triangular (basically, you can proceed with the backward elimination at once).

This decomposition is used in the mldivide (\) and linsolve functions of MATLAB for determining the effective rank of underdetermined and overdetermined cases (nonsquare matrix A), which will be discussed in Sec. 9.5. It is also used in the polyfit function (Sec. 11.6)

Eigen or *spectral decomposition* (the eig function of MATLAB), applicable to square matrices only, brings the original matrix A to

$$A = V\Lambda V^{-1} \qquad (9.52)$$

where Λ is a diagonal matrix

$$\Lambda = \begin{bmatrix} \lambda_1 & 0 & \cdots & 0 \\ 0 & \lambda_2 & \cdots & 0 \\ \vdots & \vdots & \ddots & \vdots \\ 0 & 0 & \cdots & \lambda_n \end{bmatrix} = I[\lambda_1, \lambda_2, \ldots, \lambda_n]^T = diag(\lambda_1, \lambda_2, \ldots, \lambda_n) \qquad (9.53)$$

formed by the *eigenvalues* λ_i, $i = 1, 2, \ldots, n$ of matrix A, and the columns v_i, $i = 1, 2, \ldots, n$ of the orthonormal matrix

$$V = \begin{bmatrix} v_1 & v_2 & \cdots & v_n \end{bmatrix} \qquad (9.54)$$

are the corresponding *eigenvectors* of A, satisfying equations

$$Av_i = \lambda_i v_i, \quad i = 1, 2, \ldots, n \qquad (9.55)$$

or in the matrix form

$$AV = V\Lambda \qquad (9.56)$$

This decomposition is used for solving a system of linear ODEs, as will be explained in Sec. 9.6.1.

The function `eig` is also used to produce a diagonal matrix Λ and a full matrix V, whose columns are the corresponding eigenvectors for the generalized case

$$AV = BV\Lambda \qquad (9.57)$$

In this case, it employs the *generalized Shur decomposition* (the `qz` function), which operates on two square matrices A and B and decomposes them to

$$A = QSZ^* \quad \text{and} \quad B = QTZ^* \qquad (9.58)$$

where Q and Z are orthonormal matrices (* denotes conjugate transpose), whereas S and T are upper triangular matrices. Then, the generalized eigenvalues in a matrix Λ of Eq. (9.57) are the ratios of the corresponding diagonal elements of S to those of T.

At last, the *singular value decomposition* (SVD), available for square and nonsquare matrices, yields

$$\mathbf{A} = \mathbf{U}\mathbf{\Sigma}\mathbf{V}^*$$

(9.59)

where $\mathbf{\Sigma}$ is a nonnegative diagonal matrix, whereas $\mathbf{U}$ and $\mathbf{V}$ are orthonormal matrices. The diagonal elements of $\mathbf{\Sigma}$ are called the *singular values* of $\mathbf{A}$. Like the eigendecomposition, the SVD involves finding basis directions along which matrix multiplication is equivalent to scalar multiplication [see Eq. (9.55)]. To be more specific, the SVD seeks for singular values σ, such that

$$\mathbf{A}\mathbf{v}_i = \sigma_i\mathbf{u}_i \quad \text{and} \quad \mathbf{A}^*\mathbf{u}_i = \sigma_i\mathbf{v}_i, \ i = 1, 2, \dots, n$$

(9.60)

The vectors $\mathbf{u}_i$ and $\mathbf{v}_i$, the columns of $\mathbf{U}$ and $\mathbf{V}$ in Eq. (9.59), are called the *left-singular* and *right-singular vectors* for σ_i. They are the eigenvectors of $\mathbf{A}\mathbf{A}^*$ and $\mathbf{A}^*\mathbf{A}$, respectively. The triad $\mathbf{U}$, $\mathbf{\Sigma}$, and $\mathbf{V}^*$, in Eq. (9.59) can be thought of as a set of orthonormal input basis vector directions ($\mathbf{V}$), output directions ($\mathbf{U}$) and gains multiplying each corresponding input to produce the corresponding output.

Compared to eigendecomposition, the SVD has a greater generality because the matrix under consideration need not be square. As a result, it has many applications. This includes computing the pseudoinverse of a nonsquare matrix (the `pinv` function for Moore–Penrose pseudoinverse of a matrix), solving homogeneous linear equations, evaluating range (`orth`), null space (`null`), and rank (`rank`) of a matrix $\mathbf{A}$.

Specifically, the rank of a matrix $\mathbf{A}$, mentioned in Sec. 9.2, defines the maximum number of linearly independent columns or rows (they are the same). The rank of an m by n matrix is at most $\min(m,n)$. A matrix that has a rank as large as possible is said to have *full rank*; otherwise, the matrix is *rank-deficient*. An inhomogeneous system [Eq. (9.1)] with a square matrix yields a single solution only if the rank of a matrix $\mathbf{A}$ is equal to n. In control theory, the rank of a matrix is used to determine whether a linear or linearized system describing vehicle or plant dynamics is controllable or observable. The `rank` function of MATLAB computes the rank as the number of nonzero (nonsingular) elements of a matrix $\mathbf{\Sigma}$. Obviously, the numerical determination requires a criterion for deciding when a value, such as a singular value from the SVD, should be treated as zero. Hence, the key lines in the `rank` M-file function look as simple as

```
s = svd(A);
tol = max(size(A))*eps(max(s));
r = sum(s > tol);
```

returning the rank in `k` related to the floating-point relative precision estimated at `max(s)`.

Most of the functions introduced in this section are the core functions of MATLAB. They are the built-in (embedded) functions relying on the algorithms programmed in LAPACK package, mentioned in Sec. 1.2.1. Some examples of using these functions will be provided in Sec. 9.7.1.

9.5 Special Cases of LAEs

As shown in Sec. 9.3, in the case when the determinant of matrix **A** is not equal to zero (or a matrix **A** has an inverse or it has a full rank, etc.), Eq. (9.1) with $\mathbf{b} \neq \mathbf{0}$ has a single unique solution. Examples of such a solution for the 2-D and 3-D cases are shown in Fig. 9.5.

In the 2-D case ($n = 2$), each equation describes a line and therefore the point that belongs to both LAEs is the point, where the two lines intersect (Fig. 9.5a). In the 3-D case, each equation describes a plane. Intersection of two planes produces a line, which being crossed by the third plane yields a single point belonging to all three planes (Fig. 9.5b). By the way, the script that produced three planes shown in Fig. 9.5b uses the `fill3` function and `FaceAlpha` property changing a transparency

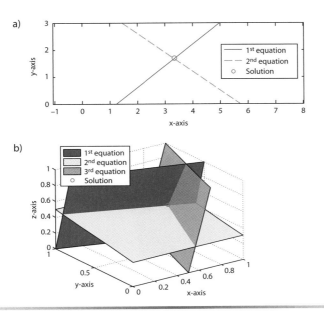

Fig. 9.5 Examples of the unique solution in a) 2-D and b) 3-D cases.

```
x=[0;1;1;0];          y=[1;0.7;0.5;0.8]; z=[0;0;1;1];
fill3(x,y,z,'b'), hold
x=[0;1;1;0];          y=[1;1;0;0];        z=[0.5;0.5;0.3;0.3];
fill3(x,y,z,'y')
x=[0.5;0.8;1;0.7]; y=[0;1;1;0];          z=[0;0;1;1];
h=fill3(x,y,z,'g'); set(h,'FaceAlpha',0.9)
```

Even in these simple examples, you can easily think of a situation where you may have no solutions or infinite number of solutions. On top of this, you may have a situation when you have more equations than unknowns ($m > n$) or vice versa. These cases are considered next.

9.5.1 Inconsistent Sets of LAEs

Let us change the pictorial representation of the 2-D and 3-D cases shown in Fig. 9.5 to those of Fig. 9.6. Obviously, in this case, no solution exists. In both cases, lines (Fig. 9.6a) and planes (Fig. 9.6b) happen to be parallel to each other.

In logic, a consistent theory is one that does not contain a contradiction. If we think of a solution (intersection of lines, planes, etc.) as a contradiction, then we can treat the cases shown in Fig. 9.6 as *inconsistent* sets of LAEs (the ones shown in Fig. 9.5 are *consistent* then). How does inconsistency transfer to properties of the matrix **A** and vector **b** in Eq. (9.1)?

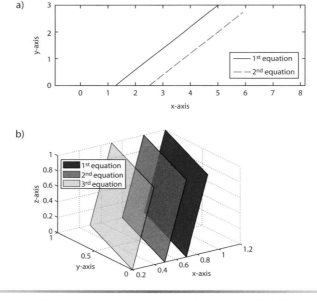

Fig. 9.6 Example of no solution in a) 2-D and b) 3-D cases.

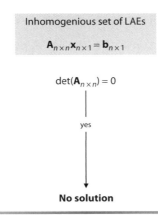

Fig. 9.7 Inconsistent systems of LAE.

Consider equations depicted in Fig. 9.6a

$$\mathbf{Ax} = \begin{bmatrix} -0.8 & 1 \\ 0.2 & -0.25 \end{bmatrix}\begin{bmatrix} x \\ y \end{bmatrix} = \begin{bmatrix} -1 \\ -0.5 \end{bmatrix} = \mathbf{b} \qquad (9.61)$$

The determinant of the matrix $\mathbf{A}$ in this case is equal to zero (the rank is equal to one, not two), which indicates that two rows are linearly dependent. Indeed it is true, because the second row can be obtained by multiplying the first row by -0.25. Hence, the first indicator of possible inconsistency of a set of LAEs is a rank-deficient matrix $\mathbf{A}$.

Let us write an augmented matrix $\mathbf{A}$ for the system (9.61)

$$\begin{bmatrix} -0.8 & 1 & \vline & -1 \\ 0.2 & -0.25 & \vline & -0.5 \end{bmatrix} \qquad (9.62)$$

The reduced row echelon form of this matrix is

$$\begin{bmatrix} 1 & -1.25 & \vline & 0 \\ 0 & 0 & \vline & 1 \end{bmatrix} \qquad (9.63)$$

where the last equation, $0 = 1$, constitutes a false statement [cf. Eq. (9.63) with Eq. (9.27)]. As a result, the system (9.61) cannot possibly have a solution.

To summarize, inconsistency occurs when the rows of matrix $\mathbf{A}$ are linearly dependent, but the constant terms on the right-hand side of LAEs do not satisfy the dependence relation. This case corresponds to a partition of Fig. 9.1 repeated in Fig. 9.7. Examining the reduced row echelon form of the augmented matrix is the best way to reveal this.

9.5.2 Underdetermined Sets of LAEs

What if in the example with the linearly dependent rows of matrix **A** considered earlier, the right-hand side of LAEs satisfies the dependence relation of the rows on the left-hand side as well? This case corresponds to

$$\mathbf{Ax} = \begin{bmatrix} -0.8 & 1 \\ 0.2 & -0.25 \end{bmatrix} \begin{bmatrix} x \\ y \end{bmatrix} = \begin{bmatrix} -1 \\ 0.25 \end{bmatrix} = \mathbf{b} \tag{9.64}$$

The determinant of the matrix **A** will still be zero (the rank equal to one), but the reduced row echelon will look differently

$$\begin{bmatrix} 1 & -1.25 & 1.25 \\ 0 & 0 & 0 \end{bmatrix} \tag{9.65}$$

There is no inconsistency in the last equation, however, the first equation describes the line, rather than a single point. Indeed, Eq. (9.64) has infinite number of solutions because two equations are essentially the same: the second one is the first one multiplied by −0.25. If plotted, these two equations show up as two coinciding lines (Fig. 9.8a).

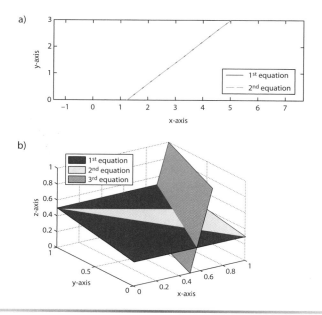

Fig. 9.8 Example of the infinite numbers of solution in the case when a) $n = 2$ and b) $n = 3$.

Hence, from the very beginning, we could have started from a single equation with two variables

$$\mathbf{A}\mathbf{x} = \begin{bmatrix} -0.8 & 1 \end{bmatrix} \begin{bmatrix} x \\ y \end{bmatrix} = -1 = \mathbf{b} \tag{9.66}$$

Figure 9.8b shows another, 3-D, example when two equations (planes) coincide (are linearly dependent). The first plane may coincide with the second (as it is the case in Fig.9.8b) or with the third one, alternatively the second plane may coincide with the third one. This results in the line of intersection of two remaining planes to be the solution of the set of LAEs. In fact, in the 3-D case, the situation can be even worse: if all three planes coincide, that all points belonging to this plane are the solutions of the degenerate system of LAEs.

The set of LAEs [Eq. (9.1)] such that $m < n$ is called *underdetermined*. You may have fewer equations than unknowns from the very beginning, or find yourself in this situation if the original square matrix $\mathbf{A}$ is degenerate (rank-deficient), as shown for the 2-D and 3-D cases earlier. In any case, such a set of LAEs results in an infinite number of solutions. It is true for both inhomogeneous equations and homogeneous ones as shown in Fig. 9.9. The only difference is that in the latter case all solutions will pass through the origin of the coordinate frame.

The way to write a solution for the set (9.64) [or Eq. (9.66)] is to present it parametrically. The solution space (formed by the infinite number of solutions) consists of a *particular solution* $\mathbf{x}_p$ (any point on the line) added to a solution of a homogeneous set of LAEs (when $\mathbf{b} = \mathbf{0}$), $\mathbf{x}_h$ (defining a vector belonging to this line)

$$\mathbf{x} = \mathbf{x}_p + \mathbf{x}_h \tag{9.67}$$

Specifically, for Eq. (9.64), the row echelon form [Eq. (9.65)] suggests

$$x = 1.25\,y + 1.25 \tag{9.68}$$

If $\mathbf{b} = \mathbf{0}$, we would have

$$x = 1.25y \tag{9.69}$$

Considering y to be a *free* (independent) parameter, meaning that it can assume any value r, the solution of a homogeneous set [Eq. (9.69)] can be presented as

$$\mathbf{x}_h = r \begin{bmatrix} 1.25 \\ 1 \end{bmatrix} \tag{9.70}$$

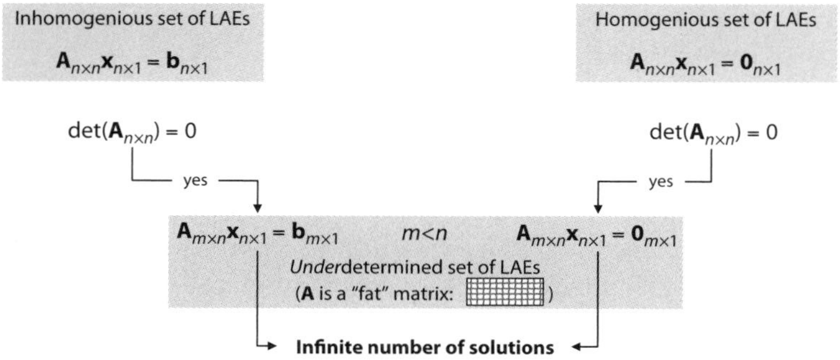

Fig. 9.9 Underdetermined systems of LAEs.

Now, choosing, for instance, $y = 0$ and substituting it to Eq. (9.68) yields a particular solution

$$\mathbf{x}_p = \begin{bmatrix} 1.25 \\ 0 \end{bmatrix} \tag{9.71}$$

Hence, the complete solution of this underdetermined (yet, consistent) set of LAEs (defining a line on the plot as shown in Fig. 9.8a) is a combination of Eqs. (9.70) and (9.71)

$$\mathbf{x} = \begin{bmatrix} 1.25 \\ 0 \end{bmatrix} + r \begin{bmatrix} 1.25 \\ 1 \end{bmatrix} \tag{9.72}$$

Consider the 3-D case, similar to that shown in Fig. 9.8b, when the set of LAEs involving three unknowns reduces to two equations

$$\mathbf{Ax} = \begin{bmatrix} 1 & 3 & 5 \\ -4 & 4 & 2 \end{bmatrix} \begin{bmatrix} x \\ y \\ z \end{bmatrix} = \begin{bmatrix} 0 \\ 4 \end{bmatrix} = \mathbf{b} \tag{9.73}$$

The reduced row echelon for this equation is

$$\begin{bmatrix} 1 & 0 & 0.875 & | & -0.75 \\ 0 & 1 & 1.375 & | & 0.25 \end{bmatrix} \tag{9.74}$$

Following the same routine as presented earlier and allowing z to be a free parameter, we can first write

$$\mathbf{x}_h = r \begin{bmatrix} -0.875 \\ -1.375 \\ 1 \end{bmatrix} \tag{9.75}$$

and then pick some particular solution. If we choose say $z = 0$, Eq. (9.74) yields $x = -0.75$ and $y = 0.25$, so that the complete parametrical solution of Eq. (9.73) can be presented as

$$\mathbf{x} = \begin{bmatrix} -0.75 \\ 0.25 \\ 0 \end{bmatrix} + r \begin{bmatrix} -0.875 \\ -1.375 \\ 1 \end{bmatrix} \tag{9.76}$$

In general, in the case when $m < n$ $(rank(\mathbf{A}) = m)$, a solution of Eq. (9.1) requires $n - m$ free parameters, and so it is represented as a combination of $n - m$ vectors, which are said to be in the *null space* of the matrix $\mathbf{A}$. For example, consider the case when all three planes in Fig. 9.8b coincide, so that the original set of LAEs can be reduced to a single equation, say

$$\mathbf{A}\mathbf{x} = \begin{bmatrix} -4 & 4 & 2 \end{bmatrix} \begin{bmatrix} x \\ y \\ z \end{bmatrix} = 4 = \mathbf{b} \tag{9.77}$$

In this case, the solution would require two free parameters, since $n - m = 3 - 1 = 2$. The reduced row echelon form yields

$$\begin{bmatrix} 1 & -1 & -0.5 \mid -1 \end{bmatrix} \tag{9.78}$$

Choosing y and z to be two independent free parameters and allowing $y = 1$, $z = 0$, and then $y = 0$, $z = 1$ results in

$$\mathbf{x}_h = r \begin{bmatrix} 1 \\ 1 \\ 0 \end{bmatrix} + s \begin{bmatrix} 0.5 \\ 0 \\ 1 \end{bmatrix} \tag{9.79}$$

The reason to choose these values for y and z is to assure that two vectors are not linearly dependent. Picking some particular solution results in

$$\mathbf{x} = \begin{bmatrix} -1 \\ 0 \\ 0 \end{bmatrix} + r \begin{bmatrix} 1 \\ 1 \\ 0 \end{bmatrix} + s \begin{bmatrix} 0.5 \\ 0 \\ 1 \end{bmatrix} \tag{9.80}$$

describing the original plane [Eq. (9.77)]. Obviously representation [Eq. (9.80)] is not unique and depends on the chosen basis [Eq. (9.81)]. We will elaborate on this example a little bit farther, employing `svd` and `null` functions, in Sec. 9.7.1.

9.5.3 Overdetermined Sets of LAEs

What if the number of equations exceeds the number of unknowns, which corresponds to the case $m > n$ in Eq. (9.1)?

Such a system is called *overdetermined* set of equations. Of course, we may get lucky, so that all equations still produce a single solution as

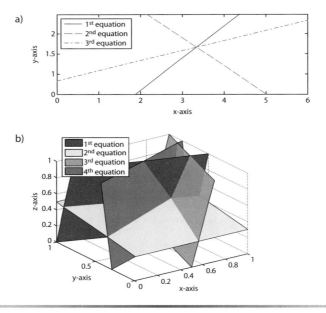

Fig. 9.10 Special case of linearly dependent equations in a) 2-D and b) 3-D.

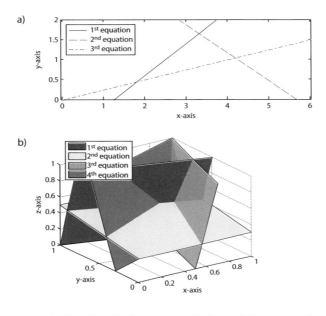

Fig. 9.11 General case of linearly independent equations in a) 2-D and b) 3-D.

shown in examples of Fig. 9.10. In fact, this is always the case for the overdetermined homogeneous set of LAEs resulting in a trivial solution $\mathbf{x} = \mathbf{0}$. For overdetermined inhomogeneous systems, this simply means that the equations are linearly dependent, so that the overdetermined system can be reduced to a general n by n system of inhomogeneous equations.

In the general case, though, such a system of inhomogeneous LAEs has no solution (examples are presented in Fig. 9.11), meaning that all equations do not intersect at one point. This situation is reflected in Fig. 9.12, which is the last remaining partition of Fig. 9.1 we need to discuss.

Although there is no strict solution to Eq. (9.1) in this case, Fig. 9.12 offers some kind of a compromise solution that comes closest to satisfying all equations simultaneously (depicted in Fig. 9.12 with dashed lines). This closeness can be defined in the *least-squares* sense (to be discussed in Chapter 10), meaning that we are trying to find a point which is equidistant from all hyperplanes (lines in the 2-D case and planes in the 3-D case). Mathematically, this least-squares solution $\mathbf{x}_*$ delivers a minimum value to the sum of squared differences between the left- and right-hand sides of Eq. (9.1)

$$S(\mathbf{x}) = \|\mathbf{Ax} - \mathbf{b}\|^2 = (\mathbf{Ax} - \mathbf{b})^T(\mathbf{Ax} - \mathbf{b}) \tag{9.81}$$

[see definition of the norm in Eq. (3.10)], which can be expressed as

$$\mathbf{x}_* = \arg\min_{\mathbf{x}} S(\mathbf{x}) \tag{9.82}$$

Finding a minimum means that the derivative of S with respect to $\mathbf{x}$ should be equal to zero. Simplifying Eq. (9.81)

$$\begin{aligned}S &= (\mathbf{Ax} - \mathbf{b})^T(\mathbf{Ax} - \mathbf{b}) = (\mathbf{x}^T\mathbf{A}^T - \mathbf{b}^T)(\mathbf{Ax} - \mathbf{b})\\ &= \mathbf{x}^T\mathbf{A}^T\mathbf{Ax} - \mathbf{b}^T\mathbf{Ax} - \mathbf{x}^T\mathbf{A}^T\mathbf{b} + \mathbf{b}^T\mathbf{b}\end{aligned} \tag{9.83}$$

and taking a derivative yields

$$\begin{aligned}\frac{dS}{d\mathbf{x}} &= \frac{d}{d\mathbf{x}}(\mathbf{x}^T\mathbf{A}^T\mathbf{Ax} - \mathbf{b}^T\mathbf{Ax} - \mathbf{x}^T\mathbf{A}^T\mathbf{b} + \mathbf{b}^T\mathbf{b})\\ &= 2\mathbf{A}^T\mathbf{Ax} - \mathbf{A}^T\mathbf{b} - \mathbf{A}^T\mathbf{b} + 0 = 0\end{aligned} \tag{9.84}$$

Equation (9.84) implies that for the column vectors, the following relations hold:

$$\frac{d}{d\mathbf{x}}\mathbf{Ax} = \mathbf{A}^T, \quad \frac{d}{d\mathbf{x}}\mathbf{x}^T\mathbf{A}^T\mathbf{Ax} = 2\mathbf{A}^T\mathbf{Ax}$$

Simplifying and rearranging the terms in Eq. (9.84) results in

$$(\mathbf{A}^T\mathbf{A})\mathbf{x} = \mathbf{A}^T\mathbf{b} \tag{9.85}$$

The MATLAB functions `mldivide` (\) and `mrdivide` (/) do exactly this—they reduce the set (9.1) with a nonsquare m by n matrix $\mathbf{A}$ to the

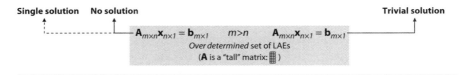

Fig. 9.12 Overdetermined systems of LAE.

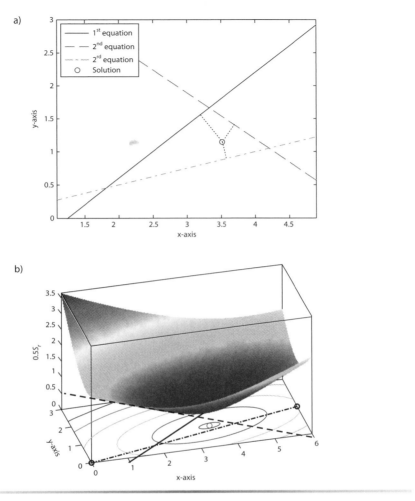

Fig. 9.13 a) Least-squares solution and b) $S(x)$ surface.

standard set of n equations with n unknowns by premultiplying both sides with the transpose of matrix $\mathbf{A}$, $\mathbf{A}^T$. Solving Eq. (9.85) determines a least-squares solution [Eq. (9.82)].

As an example, Fig. 9.13a presents the least-squares solution (marked with a circle) for the situation of Fig. 9.11a. It also shows the distances from this solution to all three lines, the sum of squares of which [Eq. (9.81)] was minimized. Figure 9.13b shows the surface $S(\mathbf{x})$ (with the contour plot and whatever was shown in Fig. 9.13a beneath it) clearly indicating that the solution we found delivers a minimum of $S(\mathbf{x})$ [Eq. (9.82)].

9.5.4 Homogeneous Sets of LAEs and Matrix Eigenvalue Problems

In Sec. 9.3, we dealt with a set of inhomogeneous equations [Eq. (9.1)] with a square matrix $\mathbf{A}$, which has a single unique solution when $\det(\mathbf{A}) \neq 0$. Similarly, when $\det(\mathbf{A}) \neq 0$ a set of n by n homogeneous LAEs

$$\mathbf{A}\mathbf{x} = \mathbf{0} \tag{9.86}$$

has a single unique solution too

$$\mathbf{x} = \mathbf{0} \tag{9.87}$$

This solution is referred to as a *trivial solution*. You may consider the 2-D and 3-D examples of Fig. 9.5 describing this trivial solution with the only difference that the intersection of lines (Fig. 9.5a) and planes (Fig. 9.5b) occurs at the origin of the coordinate frame.

When equations happen to be linearly dependent (homogeneous equations are always consistent), that is, when $\det(\mathbf{A}) = 0$, infinite number of solutions exist similar to the cases shown in examples of Fig. 9.8. Again, the only difference is that the intersection of hyperplanes (lines in the 2-D case and planes in the 3-D case) always involve the $\mathbf{x} = \mathbf{0}$ point. The solution then can be represented parametrically similar to those of Eqs. (9.72), (9.76), and (9.80) without a particular solution (which in the case of homogeneous equations is simply $\mathbf{x}_p = \mathbf{0}$).

Let us introduce the *eigenvalue equation*

$$\mathbf{A}\mathbf{x} = \lambda \mathbf{x} \tag{9.88}$$

where $\lambda = \text{const}$. This equation has a lot of applications. For example, if a matrix $\mathbf{A}$ represents a linear transformation, such as rotation, reflection, and stretching, Eq. (9.88) seeks for such a nonzero vector $\mathbf{x}$, called an *eigenvector*, that its direction is not changed by the transformation $\mathbf{A}$. The only affect this transformation has on the eigenvector is that it is scaled by a factor of λ, called the *eigenvalue* of $\mathbf{A}$ (corresponding to the eigenvector $\mathbf{x}$, because such a solution is not unique). We can rewrite Eq. (9.88) in the form of a standard set of homogeneous LAEs

$$\mathbf{A}\mathbf{x} - \lambda \mathbf{I}\mathbf{x} = (\mathbf{A} - \lambda \mathbf{I})\mathbf{x} = \mathbf{0} \tag{9.89}$$

where the identity matrix $\mathbf{I}$ is used to be able to combine λ with the matrix $\mathbf{A}$. In the scalar form, Eq. (9.89) looks as

$$
\begin{aligned}
(a_{11}-\lambda)x_1 + a_{12}x_2 + \cdots + a_{1n}x_n &= 0 \\
a_{21}x_1 + (a_{22}-\lambda)x_2 + \cdots + a_{2n}x_n &= 0 \\
&\vdots \\
a_{n1}x_1 + a_{n2}x_2 + \cdots + (a_{nn}-\lambda)x_n &= 0
\end{aligned}
\tag{9.90}
$$

By now, we know that Eq. (9.90) has a nontrivial solution only if the determinant of its matrix is zero

$$
\det(\mathbf{A}-\lambda\mathbf{I}) =
\begin{vmatrix}
a_{11}-\lambda & a_{12} & \cdots & a_{1n} \\
a_{21} & a_{22}-\lambda & \cdots & a_{2n} \\
\vdots & \vdots & \ddots & \vdots \\
a_{n1} & a_{n2} & \cdots & a_{nn}-\lambda
\end{vmatrix} = 0
\tag{9.91}
$$

This leads to the *characteristic equation*

$$
f(\lambda) \equiv \det(\mathbf{A}-\lambda\mathbf{I}) = 0
\tag{9.92}
$$

which is a polynomial of order λ^n. The eigenvalues are the roots of Eq. (9.92). Once these eigenvalues are determined, the nontrivial solution to the set of homogeneous equations [Eq. (9.89)] is found for each eigenvalue as was shown earlier in Eq. (9.55). Typically, all eigenvectors are written as one matrix $\mathbf{V}$ [Eq. (9.53)], called the *eigenvector matrix* or *modal matrix*. The collection of all eigenvectors of the matrix $\mathbf{A}$ form a *fundamental set of solutions*.

A unique feature of an eigenvalue problem is the *similarity transformation*, which follows from Eq. (9.56) and ties the original matrix $\mathbf{A}$, modal matrix $\mathbf{V}$, and eigenvalues matrix $\mathbf{\Lambda}$ [Eq. (9.53)] together, so that

$$
\mathbf{V}^{-1}\mathbf{A}\mathbf{V} = \mathbf{\Lambda}
\tag{9.93}
$$

(by construction, eigenvectors $\mathbf{v}_i$, $i = 1, 2,\ldots, n$ are linearly independent and therefore the inverse of modal matrix $\mathbf{V}$ exists).

9.6 Applying Linear Algebra to Solve Systems of ODEs

Many principles of physics, engineering, and other sciences are described in terms of ODEs, that is, equations involving variables and their derivatives. Different methods for solving ODEs numerically along with the corresponding MATLAB functions will be introduced in Chapter 14. The goal of this section is to show how the solution developed for the problem of Sec. 9.5.4 can be applied to solve ODEs analytically. It is the *Laplace transform* that makes it possible to handle linear ODEs as LAEs. The following section shows a similarity between solving a single linear ODE and a set of the first-order linear ODEs, and then extends it to the case of multiple ODEs via the state-space representation.

9.6.1 Treating a System of the First-Order ODEs as a Single Equation

Consider a single homogeneous first-order ODE

$$\dot{x} = ax \tag{9.94}$$

where a is a constant and $x = x(t)$ is an unknown function to be determined. Applied to ODEs, "homogeneous" means that there are no terms except those involving the function and its derivative. Equation (9.92) is ordinary because it involves only one independent variable and is referred to as being of first order because of the highest order of a derivative.

As well known, the solution of this equation can be expressed in the following form:

$$x(t) = e^{at}c \tag{9.95}$$

where c is a constant. This constant is defined by an *initial condition*

$$x(t_0) = x_0 \tag{9.96}$$

so that

$$c = e^{-at_0}x_0 \tag{9.97}$$

If $t_0 = 0$, Eq. (9.97) reduces to

$$c = x_0 \tag{9.98}$$

Because every solution of Eq. (9.94) must be of the form of Eq. (9.95), the latter one is called the *general solution*.

We can also obtain the solution (9.95) by applying the Laplace transform to Eq. (9.94) yielding

$$sX(s) = aX(s) \quad \text{or} \quad (a-s)X(s) = 0 \tag{9.99}$$

(At this point, simply assume that this transform replaces the function $x(t)$ with some function $X(s)$ and the derivative $\dot{x}(t)$ with $X(s)$ multiplied by a constant s.)

Equation (9.99) resembles Eq. (9.89), so that you may follow the routine of Sec. 9.5.4 and find the (only) eigenvalue $s = a$ and corresponding eigenvector $v = 1$ (or any constant). Therefore, we can express the general solution [Eq. (9.95)] using the eigenvalue—eigenvector pair as

$$x(t) = v e^{st} c \tag{9.100}$$

The constant c is defined by the initial condition at $t_0 = 0$

$$c = v^{-1} x_0 \tag{9.101}$$

In case of n first-order linear homogeneous ODEs with constant coefficients instead of Eqs. (9.94) and (9.96), we have

$$\dot{x} = Ax \tag{9.102}$$

and

$$x(t_0) = x_0 \tag{9.103}$$

where A is a square matrix of coefficients (let us call it a *state matrix*) and $x_i = \bar{b}_i^{(n-i)}$, $i = 1, 2, ..., n$, the unknown vector to be determined. Applying the Laplace transform to Eq. (9.102) yields

$$sX(s) = AX(s) \quad \text{or} \quad (A - sI)X(s) = 0 \tag{9.104}$$

which is a vector analog of Eq. (9.99). Because of that, the general solution of Eqs. (9.102) and (9.103) can be written using exactly the same pattern as in Eq. (9.100) using the corresponding matrices and vectors instead of scalars

$$x(t) = V e^{\Lambda t} c = v_1 e^{\lambda_1 t} c_1 + v_2 e^{\lambda_2 t} c_2 + \cdots + v_n e^{\lambda_n t} c_n \tag{9.105}$$

In this equation,

$$e^{\Lambda t} = I \left[e^{\lambda_1 t}, e^{\lambda_2 t}, ..., e^{\lambda_n t} \right]^T = diag(e^{\lambda_1 t}, e^{\lambda_2 t}, ..., e^{\lambda_n t})$$

and

$$c = \left[c_1, c_2, ..., c_n \right]^T$$

is a vector of constants determined by n initial conditions similarly to that of Eq. (9.101)

$$c = V^{-1} x_0 \tag{9.106}$$

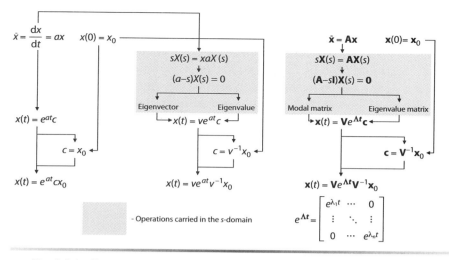

Fig. 9.14 Similarity between a single differential equation and a set of first-order differential equations.

To summarize, knowing eigenvalues and eigenvectors of the state matrix **A** defining the set of the first-order ODEs [Eq. (9.102)] allows you to find its general solution [Eq. (9.105)]. Given initial conditions [Eq. (9.103)], the unknown coefficients in [Eq. (9.105)] can be determined using Eq. (9.106). Figure 9.14 shows the logic of this solution side-by-side with that of a single ODE. Section 9.7.2 will show a couple of practical examples of employing the MATLAB functions to solve a system of ODEs numerically, but before we proceed let us show how higher-order differential equations can be brought to the form of Eq. (9.102).

9.6.2 State-Space Representation of ODEs

In practice, when deriving equations driving dynamics of some mechanical system, you usually end up with a set of the higher-order differential equations (involving higher-order derivatives) as opposed to a single first-order ODE or several ODEs written in the form of Eq. (9.102). The question then is how to reduce these higher-order ODEs to the compact form of Eq. (9.102) called the *state-space form* representation of ODEs?

As an example, let us derive a system of differential equations governing the oscillatory motion of the frictionless inertialess mass–spring system shown in Fig. 9.15.

Using the second Newton's law, we can write

$$
\begin{aligned}
m_1 \ddot{y}_1 &= -k_1 y_1 + k_2 (y_2 - y_1) \\
m_2 \ddot{y}_2 &= -k_2 (y_2 - y_1)
\end{aligned}
\tag{9.107}
$$

where $y_1 = y_1(t)$ and $y_2 = y_2(t)$ are the displacements of the masses m_1 and m_2, such that $y_1 = 0$ and $y_2 = 0$ correspond to the position of static equilibrium, and the terms on the right-hand side describe the forces produced by the compressed (expanded) springs with k_1 and k_2 denoting spring constants. These equations can be rewritten as

$$
\ddot{y}_1 = -\frac{k_1 + k_2}{m_1} y_1 + \frac{k_2}{m_1} y_2
$$

$$
\ddot{y}_2 = \frac{k_2}{m_2} y_1 - \frac{k_2}{m_2} y_2
$$

(9.108)

constituting a system of two second-order equations, which we would like to bring to the form of Eq. (9.102).

Let us introduce the *state vector*, **x**, having four states

$$
\mathbf{x} = \left[x_1, x_2, x_3, x_4 \right]^T = \left[y_1, \dot{y}_1, y_2, \dot{y}_2 \right]^T
$$

(9.109)

The number of states is defined by the total order of a system, which in the case of Eq. (9.108) totals to $2 \times 2 = 4$. Taking the derivative of the state vector with account of Eq. (9.108) yields

$$
\mathbf{x} = \begin{bmatrix} \dot{x}_1 \\ \dot{x}_2 \\ \dot{x}_3 \\ \dot{x}_4 \end{bmatrix} = \begin{bmatrix} \dot{y}_1 \\ \ddot{y}_1 \\ \dot{y}_2 \\ \ddot{y}_2 \end{bmatrix} = \begin{bmatrix} \dot{y}_1 \\ -\dfrac{k_1 + k_2}{m_1} y_1 + \dfrac{k_2}{m_1} y_2 \\ \dot{y}_2 \\ \dfrac{k_2}{m_2} y_1 - \dfrac{k_2}{m_2} y_2 \end{bmatrix} = \begin{bmatrix} x_2 \\ -\dfrac{k_1 + k_2}{m_1} x_1 + \dfrac{k_2}{m_1} x_3 \\ x_4 \\ \dfrac{k_2}{m_2} x_1 - \dfrac{k_2}{m_2} x_3 \end{bmatrix} = \mathbf{A}\mathbf{x}
$$

(9.110)

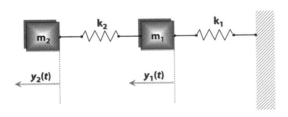

Fig. 9.15 Two-body mass–spring system.

where the state matrix $\mathbf{A}$ is

$$\mathbf{A} = \begin{bmatrix} 0 & 1 & 0 & 0 \\ -\dfrac{k_1 + k_2}{m_1} & 0 & \dfrac{k_2}{m_1} & 0 \\ 0 & 0 & 0 & 1 \\ \dfrac{k_2}{m_2} & 0 & -\dfrac{k_2}{m_2} & 0 \end{bmatrix} \tag{9.111}$$

Employing the findings of Sec. 9.6.1, we can write a solution of this equation at once

$$\mathbf{x}(t) = \mathbf{V}e^{\Lambda t}\mathbf{c} = \mathbf{v}_1 e^{\lambda_1 t}c_1 + \mathbf{v}_2 e^{\lambda_2 t}c_2 + \mathbf{v}_3 e^{\lambda_3 t}c_3 + \mathbf{v}_4 e^{\lambda_4 t}c_4 \tag{9.112}$$

The vector $\mathbf{c} = [c_1, c_2, c_3, c_4]^T$ is defined by the initial conditions, which in the case of the state vector [Eq. (9.109)] yields

$$\mathbf{c} = \mathbf{V}^{-1}\mathbf{x}_0 = \mathbf{V}^{-1}[y_1(0), \dot{y}_1(0), y_2(0), \dot{y}_2(0)]^T \tag{9.113}$$

9.7 Using the MATLAB Functions

9.7.1 Solving Systems of LAEs

MATLAB features several functions allowing you to address Eq. (9.1). They are

`x=mldivide(A,b)` or `x=A\b`	Uses Gauss elimination with apartial pivoting
`y=rref([A b]); x=y(:,end)`	Employs Gauss–Jordan elimination with a partial pivoting
`x=linsolve(A,b)`	Relies on LU factorization with a partial pivoting

The primary method to solve a set of LAEs is to use the left matrix division operator or backslash, \. The syntax of this built-in function is as simple as

`x=A\b`

If A is a full-rank matrix (which is determined using the QR factorization), and b is a vector (having the same number of rows as A), then the `mldivide` function (\) returns a single solution of an inhomogeneous set of

LAEs using the Gaussian elimination. The partial pivoting is used to reduce the possibility of singularities.

If A is an *m* by *n* matrix with $m \neq n$ and b is a column vector with *m* components, or a matrix with several such columns, then $x = A \backslash b$ returns a solution in the least-squares sense. It means that for the underdetermined or overdetermined system, instead of attempting to solve the original system of equations A*x=b with a nonsquare matrix A, MATLAB tries to solve the system A'*A*x=A'*b Eq. (9.85), where A' stands for the transpose of the matrix A, and therefore A'*A is a square matrix.

The QR factorization (performing the orthogonal–triangular decomposition of a matrix A) can be used for both square and rectangular matrices explicitly

```
[Q,R]=qr(A); y=Q'*b; x=R\y
```

The right matrix division operator, /, or mrdivide is a very short function, which simply calls the mldivide function. The correct syntax is

```
x=mrdivide(b', A')'      or      x=(b'/A')'
```

Gauss–Jordan elimination in MATLAB is accomplished with the rref function, which simplest syntax is

```
R=rref(Aa)
```

where Aa=cat(2,A,b) is an augmented matrix [A|b]. The aforementioned call returns the reduced row echelon form of Aa in R, with the last column being the solution of the equation A*x=b. Be aware that the round-off errors may cause problems in computing reduced row echelon of a matrix. Partially, the reason for this is that the rref function tests for negligible column elements using some internally computed tolerance of max(size(A))*eps*norm(A,inf). You may, however, use another syntax of the rref function and define the tolerance tol explicitly

```
R=rref(Aa,tol)
```

Once the reduced row echelon is found, the unique solution (in the case of nonsingular square matrix A) can be found as

```
x=y(:,end)
```

The inv function computes the inverse of a square matrix. Hence, this function may also be employed to solve Eq. (9.1) as

```
x=inv(A)*b
```

In the general case (for $n > 3$), the algorithm of inv relies on the LU decomposition as well as algorithm of linsolve function, which can be called as

```
x=linsolve(A,b)
```

The LU decomposition is performed with the `lu` function. The general syntax is

```
[L,U,P]=lu(A)
```

It returns the lower triangular matrix `L`, upper triangular matrix `U`, and permutation matrix `P` (an identity matrix with the rows shuffled), so that `P*A=L*U`. The alternate syntax of the `lu` function is

```
[Lp,U]=lu(A)
```

which creates an upper triangular matrix in `U` and a "pseudolower triangular matrix" in `Lp`, so that `A=Lp*U` (`Lp=inv(P)*L`). This pseudotriangular matrix `Lp` is a product of lower triangular and permutation matrices, and therefore may not look as a lower triangular.

Among other functions introduced in this chapter, the `det` function

```
d=det(A)
```

computes the determinant of a square matrix using the LU decomposition

```
d=sign(det(P))*prod(diag(U))
```

The `svd` function with a general syntax

```
[U,S,V] = svd(A)
```

performs an SVD. The singular values of a matrix **A** are returned as the diagonal elements of the matrix `S`. These values can be also obtained using a simpler syntax of `svd`, `s=svd(A)`, returning them in the vector `s`.

The SVD decomposition is used to determine the rank of a matrix via the `rank` function

```
r=rank(A)
```

This call simply returns the number of singular values of `A` that are larger than some internally defined tolerance (based on the maximum singular value of the matrix `A`). The SVD algorithm is the most time consuming, but also the most reliable. The `rank` function also allows you to specify the tolerance explicitly

```
r=rank(A,tol)
```

Finally, the `null` function

```
Z=null(A)
```

returns an orthonormal basis for the null space of `A` obtained from the SVD as well.

Consider several examples. Let us define the matrix `A` and vector `b`

```
>> A=[2 1 1; 4 -6 0; -2 7 2]; b=[5; -2; 9]; n=length(b)
```

To start with, the following script shows how a piece of code that realizes the forward elimination of the Gauss method may look like

```
Ab=[A b]                          % displays the original augm. matrix
i=1; j=1;
while i<=n                        % loops over the entire matrix
% possible swapping (partial pivoting)
   for k=i+1:n
   l=A(k,j)/A(i,j);               % computes multiplier
   A(k,j:n)=A(k,j:n) -l*A(i,j:n); % subtracts pivot row
   b(k)=b(k)-l*b(i);
   Ab=[A b]                       % displays augm. matrix at every step
   end
i=i+1; j=j+1;
end
```

This code is pretty straightforward, following derivations of Sec. 9.3.1, and for your convenience displays every step as an augmented matrix [A b], so that the sequential elimination of the three leading terms in the matrix A along with the changes in the vector b show up as follows:

```
 2   1   1   5  :  2   1   1    5  :  2   1   1    5  :  2   1   1    5
 4  -6   0  -2  :  0  -8  -2  -12  :  0  -8  -2  -12  :  0  -8  -2  -12
-2   7   2   9  : -2   7   2    9  :  0   8   3   14  :  0   0   1    2
```

By the way, the determinant of the original matrix and the upper triangular matrix happens to be the same

```
>> d=det(A)
d = -16
```

The aforementioned code does not involve pivoting, but you can easily add it by replacing the comment

```
% possible swapping (partial pivoting)
```

with the following:

```
% Finding value and index of the largest element in the column j
[p,m]=max(abs(A(i:n,j))); m=m+i-1;
A([i m],j:n)=A([m i],j:n);b([i m])=b([m i]); % swaps ith and mth rows
```

In this case, at the very first step, the first line will be swapped with the second one, so that the elimination process develops as follows:

```
 2   1   1   5  :  4  -6   0  -2  :  4  -6   0  -2  :  4  -6   0  -2
 4  -6   0  -2  :  0   4   1   6  :  0   4   1   6  :  0   4   1   6
-2   7   2   9  : -2   7   2   9  :  0   4   2   8  :  0   0   1   2
```

Because of this swapping, the determinant of a final matrix changes its sign

```
>> d=det(A)
d = -16
```

The backward substitution process realizing Eqs. (9.15) and (9.16) can then be coded as

```
x(n)=b(n)/A(n,n);                    % computes x(n)
 for k=n-1:-1:1
 b(1:k)=b(1:k)-A(1:k,k+1)*x(k+1);  % sustitutes x(k+1) to all rows above k
 x(k)=b(k)/A(k,k);                  % computes x(k)
 end
x'
```

Continuing the aforementioned example (that is, applying this latter code to the lower triangular matrix we ended up with after the forward elimination process), we get

```
x = 1
    1
    2
```

Let us now define the augmented matrix to be used to demonstrate the essence of the Gauss–Jordan elimination process described of Sec. 9.3.2. For instance, we are interested in finding the inverse of the matrix magic(3). Then, we define

```
>> A=[magic(3) eye(3)]; [m,n]=size(A);
```

The piece of code realizing the Gauss–Jordan elimination (for the general case of a nonsquare matrix) is essentially a combination of the aforementioned two pieces of the Gauss elimination

```
A                              % displays the original augmented matrix
i=1; j=1;
while i<=m && j<=n             % loops over the entire matrix
% Finding value and index of the largest element in the column j
[p,k]=max(abs(A(i:m,j))); k=k+i-1;
%  if p<=tol
%  A(i:m,j)=zeros(m-i+1,1);     % zeroing out the negligible column
%  j=j+1;
%  else
    A([i k],j:n) = A([k i],j:n); % swaps i-th and m-th rows
    A(i,j:n) = A(i,j:n)/A(i,j);   % divides the pivot row by the pivot element
for k=[1:i-1 i+1:m]
A(k,j:n)=A(k,j:n)-A(k,j)*A(i,j:n); % subtracts pivot row from all other rows
end
```

```
A                       % displays the results after finishing each column
  i= i+1; j=j+1;
% end
end
```

That is exactly what appears in the MALTAB `rref` function. The commented lines use internally defined or supplied tolerance `tol` to zero the negligibly small elements. Also note how k=i line is skipped in the `for` loop. Applied to our matrix A, this code returns

```
A =  8     1     6     1     0     0
     3     5     7     0     1     0
     4     9     2     0     0     1

A = 1.0000  0.1250   0.7500   0.1250        0         0
         0  4.6250   4.7500  -0.3750   1.0000         0
         0  8.5000  -1.0000  -0.5000        0    1.0000

A = 1.0000       0   0.7647   0.1324        0   -0.0147
         0  1.0000  -0.1176  -0.0588        0    0.1176
         0       0   5.2941  -0.1029   1.0000   -0.5441

A = 1.0000       0        0   0.1472  -0.1444    0.0639
         0  1.0000        0  -0.0611   0.0222    0.1056
         0       0   1.0000  -0.0194   0.1889   -0.1028
```

Next, let us show an example of employing the `lu` function. Consider some random matrix, scaled to the range of [0;10]

```
>> A=fix(10*rand(4))
A =  8     1     8     1
     9     2     2     2
     5     8     9     6
     1     2     3     4
```

In this case, the full syntax of the `lu` function

```
>> [L,U,P]=lu(A)
```

returns three matrices as explained in Sec. 9.3.3

```
L = 1.0000        0        0        0
    0.5556   1.0000        0        0
    0.8889  -0.1129   1.0000        0
    0.1111   0.2581   0.1043   1.0000

U = 9.0000   2.0000   2.0000   2.0000
         0   6.8889   7.8889   4.8889
         0        0   7.1129  -0.2258
         0        0        0    2.5397
```

```
P =    0        1        0        0
       0        0        1        0
       1        0        0        0
       0        0        0        1
```

Although the syntax

```
>> [Lp,U]=lu(A)
```

yields

```
Lp = 0.8889        -0.1129        1.0000              0
     1.0000              0              0              0
     0.5556         1.0000              0              0
     0.1111         0.2581         0.1043         1.0000

U =  9.0000         2.0000         2.0000         2.0000
          0         6.8889         7.8889         4.8889
          0              0         7.1129        -0.2258
          0              0              0         2.5397
```

with the `Lp` matrix being a combination of `L` and `P` matrices, so that `Lp=inv(P)*L`. Finally, the simplest syntax

```
>> lu(A)
```

returns the results in a single matrix

```
ans =

     9.0000         2.0000         2.0000         2.0000
     0.5556         6.8889         7.8889         4.8889
     0.8889        -0.1129         7.1129        -0.2258
     0.1111         0.2581         0.1043         2.5397
```

Let us now compare the performance of different functions allowing you to solve the set of LAEs. We define some random matrix `A` and vector `b`

```
>> A=rand(1000); b=rand(1000,1);
```

and apply the set of commands examining different options

```
tic,           x1=mldivide(A,b);                    toc
tic,           x2=mrdivide(b',A')';                 toc
tic,           y=rref([A b]); x3=y(:,end);          toc
tic,           [l,u]=lu(A); c=l\b; x4=u\c;          toc
tic,           [q,r]=qr(A); y=q'*b; x5=r\y;         toc
tic,           x6=linsolve(A,b);                    toc
tic,           x7=inv(A)*b;                         toc
[norm(A*x1-b); norm(A*x2-b); norm(A*x3-b); norm(A*x4-b); ...
 norm(A*x5-b); norm(A*x6-b); norm(A*x7-b)]
```

Specifically, we are interested in the CPU time required to solve the problem and the attained accuracy of a solution. The typical output is shown in Table 9.2.

As seen, the QR decomposition features the best accuracy and that is why it is used to estimate the rank of a matrix for the Gauss elimination. It takes a little bit more time though to compute the solution. On the contrary, the `inv` function takes about the same CPU time, but features the worst accuracy. The `rref` function seems to be the worst from the standpoint of required CPU resources. On the positive side, this function provides more information about the solution, especially in the cases of an inconsistent and underdetermined system. Let us demonstrate it.

Consider an inconsistent set of LAEs

```
>> A=[1 1; -1 -1]; b=[2; 3];
```

Obviously, the rows of the matrix A are linearly dependent. We can prove it, for example, by employing the `det` and `rank` commands

```
>> d=det(A), r=rank(A)
d =0
r =1
```

Any of the following calls

```
x=A\b
x=(b'/A')'
x=linsolve(A,b)
x=inv(A)*b
```

returns

```
Warning: Matrix is singular to working precision.
x = -Inf
      Inf
```

Table 9.2 Comparison of Performance of Different MATLAB Functions to Solve the System of LAEs

Method	Elapsed time, seconds	Accuracy ($\times 10^{-10}$)
/	0.181244	0.0658
\	0.193403	0.0658
rref	48.336460	0.0399
lu	0.191984	0.0671
qr	0.639089	0.0044
linsolve	0.176448	0.0658
inv	0.473522	0.1059

The rref call

```
>> x=rref([A b])
```

allows you to really understand what is going onv

```
x = 1    1    0
    0    0    1
```

By looking at the last row, we realize that the system is inconsistent (Sec. 9.5.1).

Now, consider an underdetermined system of LAEs

```
>> A=[1 1]; b=[0];
```

The calls

```
x=A\b
x=(b'/A')'
x=linsolve(A,b)
```

return a particular solution

```
x = 0
    0
```

whereas the call

```
s=rref([A b])
```

reveals

```
s = 1 1 0
```

allowing you to define a general solution parametrically as say $x = r[-1\ 1]^T$ (see Sec. 9.5.2). Let us modify the aforementioned example to the inhomogeneous system

```
>> A=[1 1]; b=[2];
```

The calls

```
x=A\b
x=(b'/A')'
x=linsolve(A,b)
```

return a particular solution

```
x = 2
    0
```

whereas the call

```
>> s=rref([A b])
```

returns

```
s = 1  1  2
```

which may be used to establish a general solution as $x = [2, 0]^T + r[-1, 1]^T$. The following script reproduces this solution as shown in Fig. 9.16:

```
x1=0:5; x2=(b-A(1)*x1)/A(2);          % defines the equation
plot(x1,x2), hold
xs=A\b;                               % defines a particular solution
plot(xs(1),xs(2),'rp')
xlabel('x_1'), ylabel('x_2'), axis equal
s=rref([A b]); xh(2)=1; xh(1)=-s(2)*xh(2); xp(2)=0; xp(1)=s(3);
% Displaying the set of LAEs, general and particular solutions
text(1,-1.5,'Set of LAEs:')
f='%+2.0f'; t1=' x_1 '; t2=' x_2 = ';
text(0.91,-1.9,[num2str(A(1)) t1 num2str(A(2),f) t2 num2str(b)],...
               'fontsize',10)
text(-xh(1)+0.3,xh(2),'General solution by rref:')
syms r, x=xp'+r*xh'; H=latex(x);
text(3.5,1,['$$' H '$$'],'interpreter','latex','fontsize',14)
text(xs(1)+0.3,xs(2),'Particular solution provided by Ab')
```

Obviously, for nonsquare matrices, the `inv` function fails to produce a solution

```
x=inv(A)*b
inv(a)*b
??? Error using ==> inv
Matrix must be square.
```

Consider another undetermined system

```
>> A=[1 2 3; 2 5 1]; b=[1; 2];
```

The calls

```
x=A\b
x=(b'/A')'
x=linsolve(A,b)
```

return a particular solution

```
x = 0
    0.3846
    0.0769
```

Employing

```
>> s=rref([A b])
s =    1    0     13     1
       0    1     -5     0
```

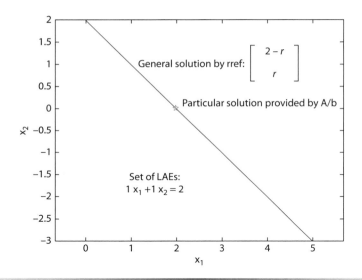

Fig. 9.16 Example of undetermined system of LAEs.

allows you to find a general solution

```
>> xh(3)=1; xh(1:2)=-s(1:2,3)*xh(3), xp(3)=0; xp(1:2)=s(1:2,4)
xh = -13       5       1
xp =    1      0       0
```

that is, $x = [1, 0, 0]^T + r[-13, 5, 1]^T$.

Let us normalize the aforementioned solution of the homogeneous system

```
>> xh=xh/norm(xh)
xh = -0.9309
       0.3581
       0.0716
```

and employ a SVD (Sec. 9.3)

```
>> [U,S,V]=svd(A)
U = -0.5145    -0.8575
    -0.8575     0.5145
S =  6.2450         0         0
          0    2.2361         0
V =
    -0.3570     0.0767    -0.9309
    -0.8513     0.3835     0.3581
    -0.3845    -0.9204     0.0716
```

As seen, the last column of matrix V corresponding to the zero column of matrix S contains the solution of homogeneous system that matches that of the normalized vector xh. To retrieve just this vector, we can also employ the null function

```
>> null(A)
```

resulting in

```
ans = -0.9309
        0.3581
        0.0716
```

Let us see how this works for another undetermined system, which has three linearly dependent rows. Take the one described by Eq. (9.77)

```
>> A = [-4 4 2]; b=[4];
```

The rank of the matrix A in this case is equal to one

```
>> r = rank(A)
r = 1
```

and the rref function returns what was shown in Eq. (9.78)

```
>> rref([A b])
ans = 1.0000    -1.0000    -0.5000    -1.0000
```

from which we can derive the solution as was shown in Eq. (9.80). The solution of the homogeneous set of LAEs, which in this case represents a plane, can be described by any two noncollinear vectors belonging to this plane. We choose these two vectors to be [1;1;0] and [05;0;1] [Eq. (9.79)]. Chosen quite arbitrarily, these vectors are not orthogonal

```
>> [1 1 0]*[0.5;0; 1]
ans =    0.5
```

Now let us examine the corresponding return of the svd function and null function

```
>> [U,S,V]=svd(A)
U = 1
S = 6    0    0
V = -0.6667     0.6667     0.3333
     0.6667     0.7333    -0.1333
     0.3333    -0.1333     0.9333

>> xh=null(A)
xh = 0.6667     0.3333
     0.7333    -0.1333
    -0.1333     0.9333
```

The two basis vectors suggested by these two functions belong to the same plane as those of Eq. (9.79), but on top of that happen to be orthonormal

```
>> xh(:,1)'*xh(:,2)
ans = 0
```

Being complemented with a particular solution

```
>> xp=A\b
xp = -1
        0
        0
```

they constitute a general solution in the following form:

$$
\mathbf{x} = \begin{bmatrix} -1 \\ 0 \\ 0 \end{bmatrix} + r \begin{bmatrix} \frac{2}{3} \\ \frac{11}{15} \\ \frac{-2}{15} \end{bmatrix} + s \begin{bmatrix} \frac{1}{3} \\ \frac{-2}{15} \\ \frac{14}{15} \end{bmatrix}
\tag{9.114}
$$

(we used the `format rat` call to come up with these ratios). The bottom line is that you can use the `rref` function to examine the underdetermined system, and then conveniently obtain a general solution using the backslash operator \ (for a particular solution) and `null` function (for the solution of the homogeneous set of LAEs).

In the case of the overdetermined system (Sec. 9.5.3), for instance,

```
>> A=[2 1; 2 5; 4 4]; b=[1; 2; 3];
```

the `rref` function returns no result

```
>> x=rref([A b])
x = 1   0   0
    0   1   0
    0   0   1
```

whereas any of other functions

```
x=A\b
x=(b'/A')'
x=linsolve(A,b)
[l,u]=lu(A); c=l\b; x4=u\c
[q,r]=qr(A); y=q'*b; x5=r\y
x=inv(A'*A)*A'*b
```

produce the least-squares solution

```
x = 0.2500
    0.2857
```

9.7.2 Solving Eigenvalue Problems and Differential Equations

The major function of MATLAB specifically dedicated to address the eigenvalue problem is `eig`. The simplest call

```
lambda=eig(A)
```

returns the vector `lambda` containing eigenvalues of matrix `A`. For example,

```
>> A=magic(3); eig(magic(3))
```

yields

```
ans = 15.0000
       4.8990
      -4.8990
```

An alternate syntax for the `eig` function is

```
[V,D]=eig(A)
```

which returns eigenvectors of matrix `A` in the modal matrix `V` and eigenvalues in the diagonal matrix `D`, for example,

```
>> [V,D]=eig(A)
V = -0.5774    -0.8131    -0.3416
    -0.5774     0.4714    -0.4714
    -0.5774     0.3416     0.8131
D = 15.0000        0          0
         0     4.8990         0
         0         0     -4.8990
```

The eigenvalues are normalized, but not necessarily orthonormal

```
>> [norm(V(:,1)) norm(V(:,2)) norm(V(:,3))]
ans = 1    1    1
>> [V(:,1)'*V(:,2) V(:,2)'*V(:,3) V(:,3)'*V(:,1)]
ans = -0.0000    0.3333         0
```

The similarity transformation of Eq. (9.93) does hold, but you should be careful examining it, because the straight comparison

```
>> A*V==V*D
```

yields a confusing result

```
ans = 0    0    0
      0    0    0
      0    0    0
```

The more robust approach

```
>> A*V-V*D
ans = 1.0e-014 *

        0.1776        0.4885        0.0222
        0.3553        0.0444        0.0444
       -0.1776        0.1332        0.1776
```

explains the results obtained with the straightforward approach—the equality holds to the degree of round-off errors.

In principle, you may obtain these results using other functions of MATLAB. For educational purposes, let us use an alternative (maybe clearer) approach and then compare it with that of the `eig` function. Let us start from employing the `poly` function to obtain coefficients of the characteristic polynomial $|A-\lambda I|$

```
>> p = poly(A)
p = 1.0000    -15.0000    -24.0000    360.0000
```

Then, you can find its roots (eigenvalues) using the `roots` function (to be discussed in Chapter 10)

```
>> lambda=roots(p)
lambda =  15.0000
         -4.8990
          4.8990
```

These eigenvalues happen to be the same as mentioned earlier, but in a slightly different order. Next, you may want to find eigenvectors by solving the corresponding system of LAEs. The best way would be employing the `rref` function

```
>> rr=rref([A-lambda(1)*eye(3)])
```

which returns the reduced row echelon form of matrix $A-\lambda I$

```
rr = 1     0    -1
     0     1    -1
     0     0     0
```

Following the routine outlined in Sec. 9.5.2, you may choose the third element to be a free parameter, which results in an eigenvector

```
>> v=-rr(:,3); v(3)=1
v = 1
    1
    1
```

If normalized

```
>> v=v/norm(v)
```

it coincides with that obtained using the aforementioned `eig` function (it points to the opposite direction, which is all right),

```
v = -0.5774
    -0.5774
    -0.5774
```

Doing the same operations for all three eigenvalues

```
for i=1:3
rr=rref([A-lambda(i)*eye(3)]);
v=-rr(:,3); v(3)=1;
V(:,i)=v/norm(v);
end
```

returns

```
V = 0.5774    -0.3416    -0.8131
    0.5774    -0.4714     0.4714
    0.5774     0.8131     0.3416
```

which is the same as the result returned by the `eig` function, but corresponding to a different order of eigenvalues. However, although the results are similar, the specially developed `eig` function does computations thousands of times faster than an alternative algorithm. To this end, the following script employs the `tic-toc` pair introduced in Sec. 5.10.3 to compare both approaches

```
A=magic(5); n=length(A);
tic, [V,D]=eig(A); toc          % performance of the eig function
tic
p=poly(A); lambda=roots(p);
for i=1:n
rr=-rref([A-lambda(i)*eye(n)]);
v=rr(:,n); v(n)=1; V(:,i)=v/norm(v); D=diag(lambda);
end
toc                             % performance of the alternative approach
```

This script returns something like

```
Elapsed time is 0.000079 seconds.
Elapsed time is 0.033833 seconds.
```

You may see some different results on your computer, but on the average, the `eig` function performs much faster.

Finally, let us use the `eig` function to solve the system of linear ODEs introduced in Eqs. (9.110) and (9.111), with some (dimensionless)

numerical values for parameters of the spring–mass system shown in Fig. 9.15

$$m_1 = 1, \quad m_2 = 1, \quad k_1 = 3, \quad \text{and} \quad k_2 = 2 \tag{9.115}$$

and the initial conditions

$$y_1(0) = 1, \; \dot{y}_1(0) = -2\sqrt{6}, \; y_2(0) = 2, \; \dot{y}_2(0) = \sqrt{6} \tag{9.116}$$

The following script performs eigendecomposition of the state matrix and presents the results in full accordance with Eq. (9.112)

```
% Defining parameters of the system
m1=1; m2=1; k1=3; k2=2;
% Defining the initial conditions
y0=[1, 2]'; ydot0=[-2*sqrt(6), sqrt(6)]';
% Forming the state matrix
A=[0 1 0 0; -(k1+k2)/m1 0 k2/m1 0; 0 0 0 1; k2/m2 0 -k2/m2 0];
% Solving ODEs using eigendecomposition of matrix A
[V,D]=eig(A);
c=inv(V)*[y0(1) ydot0(1) y0(2) ydot0(2)]';
% Developing symbolic (analytical) representation of a solution
syms t
syms('y','real')
y=V*diag(diag(exp(D*t)))*c;
subplot(2,2,1), ezplot(y(1),[0 10]), title('Position of Mass 1')
subplot(2,2,2), ezplot(y(2),[0 10]), title('Velocity of Mass 1')
subplot(2,2,3), ezplot(y(3),[0 10]), title('Position of Mass 2')
subplot(2,2,4), ezplot(y(4),[0 10]), title('Velocity of Mass 2')
```

Because the results are given analytically, the `ezplot` function is employed to show time histories of all the four states. They show up correctly (Fig. 9.17) and agree with the analytical solution defined as

$$\mathbf{x}(t) = \begin{bmatrix} x_1(t) \\ x_2(t) \\ x_3(t) \\ x_4(t) \end{bmatrix} = \begin{bmatrix} y_1(t) \\ \dot{y}_1(t) \\ y_2(t) \\ \dot{y}_2(t) \end{bmatrix} = \begin{bmatrix} \cos(t) - 2\sin(t\sqrt{6}) \\ -\sin(t) - 2\sqrt{6}\cos(t\sqrt{6}) \\ 2\cos(t) + \sin(t\sqrt{6}) \\ -2\sin(t) + \sqrt{6}\cos(t\sqrt{6}) \end{bmatrix} \tag{9.117}$$

Unfortunately, the MATLAB Symbolic Math toolbox cannot simplify symbolic expressions for components of the state vector **y** to the form of Eq. (9.117), and so plotting them is about all you can do.

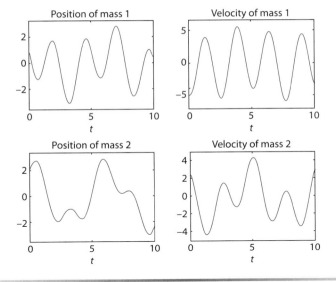

Fig. 9.17 Dynamics of the two–body mass–spring system of Fig. 9.15.

Problems

9.1 Consider the following set of LAEs:

$$3x_3 + 5x_4 = b_1$$
$$3x_1 + 7x_2 - 4x_3 + 5x_4 = b_2$$
$$5x_1 - 6x_2 + 7x_3 + 8x_4 = b_3$$
$$4x_2 + 5x_4 = b_4$$

where b_i, $i = 1,2,3,4$ are the uniformly distributed pseudorandom numbers (use `b=rand(4,1)` command). Solve it employing all appropriate functions introduced in this chapter, that is `\`, `inv`, `rref`, `linsolve`, `lu` and `qr`. Compare the results and check one of solutions (any of them) by substituting it back to the system of LAEs. Comment on the results.

9.2 One of the methods to solve inhomogeneous system of LAEs, we did not consider in this chapter, is the Cramer's rule. The solution is expressed explicitly in terms of determinants of the n by n (square) coefficient matrix `A` and matrices obtained from `A` by replacing one column by the vector of the right-hand-side parts of equations, `b`

$$x_i = \frac{\det(\tilde{\mathbf{A}}_i)}{\det(\mathbf{A})}, \quad i = 1,2,...,n$$

where

$$
\tilde{\mathbf{A}}_i = \begin{bmatrix}
a_{11} & \cdots & a_{1,i-1} & b_1 & a_{1,i+1} & \cdots & a_{1n} \\
a_{21} & \cdots & a_{2,i-1} & b_1 & a_{2,i+1} & \cdots & a_{2n} \\
\vdots & \vdots & \vdots & \vdots & \vdots & \vdots & \vdots \\
a_{n1} & \cdots & a_{n,i-1} & b_n & a_{n,i+1} & \cdots & a_{nn}
\end{bmatrix}
$$

Although the Cramer's rule is important theoretically, for large matrices, it has little practical value because the computation of large determinants is somewhat cumbersome and not precise. Develop the M-file function realizing this algorithm and compare its performance with that of the Gauss elimination method.

9.3 Consider the following simple scheme of transporting cargo between four airports:

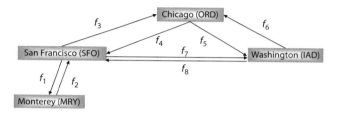

Suppose that you know that for some specific day, the total cargo handled in Monterey (incoming and outgoing) was 100 conditional units, in San Francisco, 600; in Chicago, 800; and Washington, DC, 600. In addition to this, you know that $f_1=f_2$, while the difference between incoming and outgoing cargo in Washington was 100 units. Finally, you know that 200 units of cargo were transported from Washington to Chicago. Using 8 unknown variables, f_1 trough f_8, develop the equations describing this traffic. Determine whether you have enough independent equations to find a unique solution. If not, add additional reasonable observation(s) to come up with the feasible results (so that $f_i>0$, $i=1,\ldots,8$). Vary your observation to see how sensitive to these changes your solution is.

Chapter 10

Root Finding and Introduction to Optimization

- Understanding Bracketing Methods
- Understanding Open Methods
- Similarities Between Root Finding and Optimization
- Using the MATLAB Functions

```
fzero, roots,
fminsearch, fminbnd,
fminunc, fmincon
```

10.1 Introduction

Suppose, we need to find zero of a scalar function of a single variable, $f(x)$, meaning that we are required to find the root of the equation. In other words, solve the following:

$$f(x) = 0 \qquad (10.1)$$

For some cases, the roots of Eq. (10.1) can be obtained analytically (for instance, one can easily solve a quadratic equation), for others an approximate solution technique is needed (for example, for complex algebraic functions, not analytically defined functions and transcendental or nonalgebraic equations). MATLAB offers two basic functions to address these types of problems. The fzero function finds one root of a continuous function of a single variable, and the roots function finds all roots of a polynomial. Specifically, the fzero function uses the so-called *zeroin algorithm*, exploiting several numerical techniques based on two different approaches. That is why this chapter starts with introducing them. The first approach, the so-called *bracketing methods*, requires two initial guesses obeying certain properties, and the second one, *open methods*, requires only one initial guess. The chapter proceeds with explaining the zeroin algorithm and addressing several root-finding-related problems, followed by introducing an algorithm behind the roots function.

It turns out that the function minimization (maximization) problem

$$x_* = \arg(\min_x g(x)) \tag{10.2}$$

is closely related to the root-finding problem (Eq. (10.1)). Indeed, solving Eq. (10.2) implies finding the solution of Eq. (10.1) with $f(x)=g'(x)$, where $g'(x)=dg/dx$ is the derivative of function $g(x)$. Therefore in general, you can still use the `fzero` function to address Eq. (10.2). However in addition to this, MATLAB offers a variety of other tools to deal with Eq. (10.2)-type problems, that is, solve standard and large-scale optimization problems. These tools are combined into the Optimization Toolbox and even more advanced (specialized) Global Optimization Toolbox (Table 1.2). Several basic functions from the first toolbox are available to users of the basic MATLAB core and, therefore, are also introduced in this chapter. They are `fminbnd` to minimize a function of one variable on a fixed interval, `fminsearch` to minimize a function of several unconstrained variables, `fminunc` to find a minimum of an unconstrained multivariable function, and `fmincon` to find a minimum of a constrained multivariable function. These functions developed to address the problem of Eq. (10.2) can be used to solve Eq. (10.1) as well, because the latter one is in general an equivalent to

$$x_* = \arg(\min_x f^2(x)) \tag{10.3}$$

For the sake of completeness this chapter presents a variety of methods for solving Eq. (10.1), followed by the discussion of several basic techniques to address Eq. (10.2).

10.2 Bracketing Methods

To find the solution of Eq. (10.1), two groups of methods can be employed. This and the following chapters introduce them one after another. The *bracketing methods*, introduced in this chapter, exploit the fact that the function $f(x)$ typically changes a sign in the vicinity of its root x_* (see example in Fig. 10.1). Bracketing means that two initial guesses are required. These guesses must "bracket," or be on either side of, the root (for example, $x=a$ and $x=c$ in Fig. 10.1 constitute a valid bracket). With each iteration, bracketing methods systematically reduce the width of the bracket and, hence, home in on the correct answer. Therefore, these methods guarantee unconditional convergence of numerical algorithm. This is the main advantage of bracketing methods. Moreover, given a required tolerance ε, the number of iterations necessary to find an approximate solution can be easily estimated. Methods considered in this chapter are the bisection, false position, and golden section.

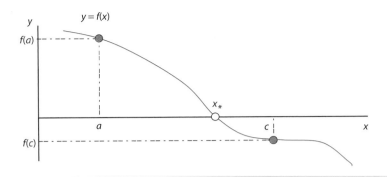

Fig 10.1 Typical geometry in the close vicinity of a single-variable function root.

10.2.1 Bisection Method

The *bisection method* is the simplest and most robust method for root finding (if the root is localized within the initially assumed interval). The bisection method relies on the principle that if an interval [a;c] has one root, then the sign of $f(x)$ at $x = a$ and $x = b$ are opposite to each other, or either $f(a)$ or $f(c)$ is equal to zero. Mathematically, this can be represented by the expression

$$f(a) f(c) \leq 0 \tag{10.4}$$

The numerical algorithm looks as follows (for better understanding several consecutive steps are shown graphically in Fig. 10.2):

1. Given $f(x)$, find an initial interval enclosing the root (like $x \in [a;c]$ in Fig. 10.2a).
2. Check inequality of Eq. (10.4).
3. Find a trial point (Figs. 10.2b, 10.2d, and 10.2f).

$$x = b = \frac{1}{2}(a+c) = \frac{1}{2}a + \frac{1}{2}c \tag{10.5}$$

4. If the size of new interval [a;b] is less or equal to the predefined tolerance ε

$$|a - b| \leq \varepsilon \tag{10.6}$$

stop iterations assuming the solution at $x_* = b$.
5. If Eq. (10.6) does not hold, continue iterations by choosing the correct interval among the two available ones ([a;b] or [b;c]).

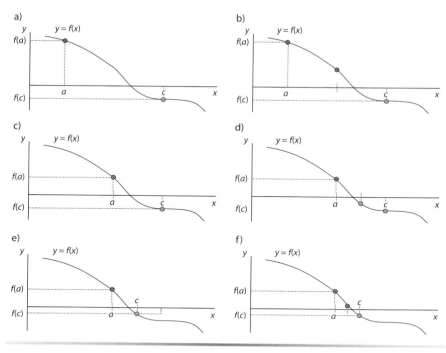

Fig. 10.2   Illustration of several consecutive steps using the bisection method.

- If inequality $f(a)$ at $f(b) \le 0$ holds, then assign $c=b$ (Fig. 10.2e)
- Otherwise, assign $a=b$ (Fig. 10.2c).

6. Go to Step 3 with a new value of c or a and repeat iteration with the halved interval.

Because you are halving the original interval $[a_0;c_0]$ at each iteration, exiting algorithm after n iterations means

$$\frac{|c_0 - a_0|}{2^n} \le \varepsilon \qquad (10.7)$$

This obvious relation allows you to have an exact number of iterations needed to satisfy the tolerance ε

$$n \ge \frac{\ln\dfrac{|c_0 - a_0|}{\varepsilon}}{\ln(2)} \qquad (10.8)$$

(The greater or equal sign in Eq. (10.8) means that to produce an integer number, the right-hand side expression should be rounded toward infinity.)

Given the bounds of initial root bracketing interval, x_0 and x_1, function $f(x)$ developed as an anonymous, inline function or an M-file, and relative tolerance ε^{rel}, [see Eq. (8.27)], the bisection method can be programmed as

```
a=x0; c=x1;
if f(a)*f(c)<=0
    while abs(c-a)>eps*max(abs(c),1e-3)    % meeting the relative tolerance
      b=(a+c)/2;                           % halving the interval
if f(a)*f(b)<=0
  c=b;
else
  a=b;
 end
end
disp([b,f(b)])                             % displaying the results

else
warndlg('The interval does not enclose the root','!! Warning !!')
end
```

Once again, the first step is very important. It helps establishing the correct initial interval and to bracket (isolate) the correct root you are trying to find. If you are not careful in choosing the initial interval, you may enclose a singularity instead of (or in addition to) the root, as in the example shown in Fig. 10.3. That is why plotting your function, even with the easiest `ezplot` function, should be your first check.

Suppose, you overlooked a singularity and picked an interval like the one shown in Fig. 10.3 that still satisfies Eq. (10.4). The bisection method will still work producing the result, in our case converging to a singularity.

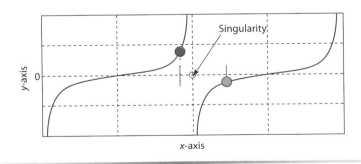

Fig. 10.3 Case of singularity.

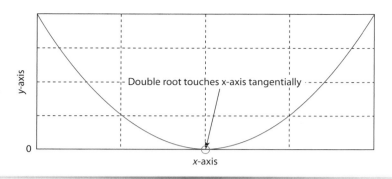

Fig. 10.4 Example of abscissa-touching function.

How can you possibly recognize this? Well, it is a good idea to do a second check examining whether

$$f(b) \to 0 \qquad (10.9)$$

That will allow you to tell if the method converged to a singularity, rather than a root.

Obviously, any bracketing method relying on Eq. (10.4) cannot find a root if the function $f(x)$ only touches the abscissa axis, not crossing it (an example is shown in Fig. 10.4). In this case, the root-finding problem can be recast as an extremum-finding problem [Eq. (10.2)] (even without squaring the function) and solved using of the minimum-finding algorithms (Sec. 10.7 discusses this issue farther).

Finally, bracketing methods may also not be applied to finding the complex roots because of the obvious reason that the criterion for defining a bracket (that is, a sign change) does not translate to the complex domain.

10.2.2 False-Position Method

Although the bisection method is a perfectly valid technique for determining the roots, its "brute-force" approach is relatively inefficient. A simple modification of Eq. (10.5) allows you to reduce the number of iterations necessary to find the root.

A major shortcoming of bisection method is that, in dividing the current interval into equal halves, no account is taken of the magnitudes of $f(a)$ and $f(c)$. The easiest way to take it into account and, therefore,

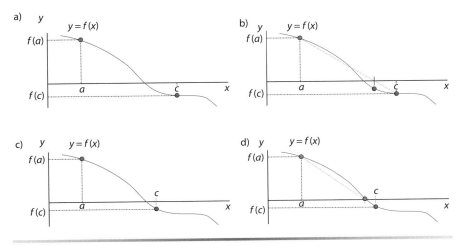

Fig. 10.5 Illustration of several consecutive steps using the false-position algorithm.

improve convergence performance is to join $f(a)$ and $f(c)$ by a straight line and consider the intersection of this line with the x-axis as the next trial point as shown in Fig. 10.5. Then, the new trial point b can be determined as

$$b = c - \frac{f(c)(a-c)}{f(a)-f(c)} = \frac{-f(c)}{f(a)-f(c)}a + \frac{f(a)}{f(a)-f(c)}c \qquad (10.10)$$

replacing Eq. (10.5). This is the *false-position formula*. According to the numerical routine outlined in the previous section, the new point b replaces whichever of the two initial guesses, a or c, and yields a function value with the same sign as $f(b)$. As expected, weighting coefficients $w_a = -f(c)/(f(a)-f(c))$ and $w_c = (f(a)/f(a)-f(c))$ ($w_a + w_c = 1$) are now proportional to the closeness of $f(a)$ and $f(c)$ to the abscissa axis (if $|f(a)| < |f(c)|$, then $w_a > w_c$ and otherwise).

Given the bounds of initial root bracketing interval, x_0 and x_1, function $f(x)$ developed as an anonymous, inline function or an M-file, and relative tolerance ε^{rel}, the false-position method could be programmed as

```
a=x0; c=x1;
if f(a)*f(c)<=0
while abs(c-a)>eps*max(abs(c),1e-3) % meeting the relative tolerance
wa=-f(c)/(f(a)-f(c)); wc=1-wa;      % computing weighting coefficients
b=wa*a+wc*c;                        % producing the next iteration
    if f(a)*f(b)<=0
    c=b;
```

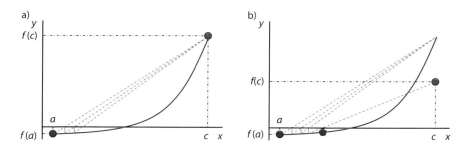

Fig. 10.6 Illustration of potential pitfall of a) false-position algorithm and b) one possible solution.

```
    else
        a=b;
    end
    end
disp([b,f(b)])                    % displaying the results
else
warndlg('The interval does not enclose the root','!! Warning !!')
end
```

Although the false-position method would seem to always be the bracketing method of preference, there are cases when it performs poorly. It happens, for instance, when the function $f(x)$ at one of the two initial guesses, a or c, is flat (that is, its derivative is close to zero) and is much smaller than at the other end of the interval. This means degradation of one of the weighting coefficients so that the information about another point in Eq. (10.10) will be neglected, resulting in the new point to be very close to the one of the ends of the interval as shown in Fig. 10.6a.

Because of this "one-sidedness" the false-position algorithm will only assure a very slow convergence. The remedy is to periodically check whether one of the bounds is stuck, and if this occurs, either apply one step of the bisection method or halve the function value at the bound opposite to the stagnant one (artificially increasing its weight), as shown in Fig. 10.6b. The latter procedure is known as the *modified false-position method*.

10.2.3 Golden-Section Method

As opposed to varying weighting coefficients depending on geometry as it was the case for the false-position method, the *golden-section method*, applied to root-finding problems, basically uses another pair of constant coefficients, different from those 0.5 and 0.5 of the bisection method [Eq. (10.5)]. It implies $w_a \neq w_c$. It is similar in sprit to other methods, but to make it work (to assure a fair choice between the two same-sized intervals), two rather than one intermediate point are required at the very first iteration (after bracketing) (Fig. 10.7).

Following the notations of Fig. 10.7, we want to start from the interval of size $(l_1 + l_2)$ ($[a;c]$), then reduce it to l_1 ($[b;c]$), then to l_2 ($[b;c]$), etc. For consistency, the following condition should hold:

$$\frac{l_1+l_2}{l_1}=\frac{l_1}{l_2} \tag{10.11}$$

Introducing a notation $R = l_2/l_1$ yields

$$1+R=\frac{1}{R} \tag{10.12}$$

meaning

$$R^2 + R - 1 = 0 \tag{10.13}$$

Resolving for the positive root of Eq. (10.13) gives

$$R^\otimes = \frac{\sqrt{5}-1}{2} \approx 0.61803 \tag{10.14}$$

This value of $R^\otimes$ has been known since antiquity and is called the *golden ratio*. It is related to an important mathematical series known as the *Fibonacci numbers*, which are 0, 1, 1, 2, 3, 5, 8, 13, ... The Fibonacci numbers

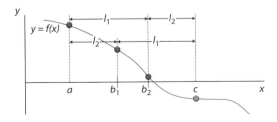

Fig. 10.7 Finding two intermediate points for the golden-section algorithm.

start from $n_1^F = 0$ and $n_2^F = 1$ and then the next number n_i^F is computed as the sum of the previous two

$$n_i^F = n_{i-1}^F + n_{i-2}^F, \quad i = 3, 4, \ldots, \infty \tag{10.15}$$

It happens that

$$\lim_{i \to \infty} n_i^F = \frac{n_i^F}{n_{i+1}^F} = R^\otimes \tag{10.16}$$

That is, the ratio of consecutive Fibonacci numbers approaches the golden ratio $R^\otimes$ (and very quickly we might add: 0/1=0, 1/1=1, 1/2=0.5, 2/3=0.667, 3/5=0.6, 5/8=0.625, 8/13=0.615, and so on).

The numerical algorithm for zero finding looks simlar to that of bisection and false-position methods

1. Given $f(x)$, determine the initial interval $[a;c]$.
2. Assure that inequality of Eq. (10.4) holds.
3. Find two trial points as

$$b_1 = c - R^\otimes(c - a) \text{ and } b_2 = a + R^\otimes(c - a) \tag{10.17}$$

4. If the size of new interval $[a=b_1]$ is less or equal to the predefined tolerance ε

$$|a - b_1| \le \varepsilon \tag{10.18}$$

   stop iterations assuming the solution at $x_* = b_1$.
5. If Eq. (10.18) does not hold, continue iterations by choosing the correct interval among the two possible ones ($[a;b_2]$ and $[b_1;c]$)
   - If inequality $f(a)f(b_2) \le 0$ holds, then assign $c = b_2$, $b_2 = b_1$, and find $b_2 = c - R^\otimes(c - a)$
   - Otherwise assign $a = b_1$, $b_1 = b_2$, and find $b_2 = a + R^\otimes(c - a)$
6. Go to Step 4 with a new interval.

The weighting coefficients in Eq. (10.17) are $R^\otimes$ and $1 - R^\otimes$.

Given the bounds of initial root bracketing interval, x_0 and x_1, function $f(x)$ developed as an anonymous, inline function or an M-file, and relative tolerance ε^{rel}, the golden-section method might be programmed as

```
a=x0; c=x1; R=(sqrt(5)-1)/2;
b1=c-R*(c-a); b2=a+R*(c-a);
if f(a)*f(c)<=0
    while abs(a-b1)>eps*max(abs(b1),1e-3)% meeting the relative tolerance
```

```
    if f(a)*f(b2)<=0
        c=b2;
        b2=b1;
        b1=c-R*(c-a);
    else
        a=b1;
        b1=b2;
        b2=a+R*(c-a);
    end
end
disp([b1,f(b1)])                    % displaying the results
else
warndlg('The interval does not enclose the root','!! Warning !!')
end
```

The few first iterations of this algorithm, as applied to the function shown in Fig. 10.8a, are shown in Figs. 10.8b–f. As seen, starting from the second iteration, we only have to find a single trial point at each step, because by construction we have another one from the previous iteration.

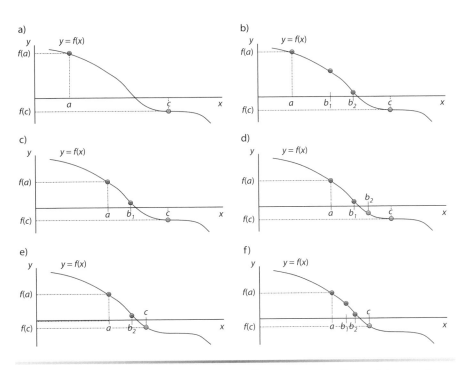

Fig. 10.8 Illustration of several consecutive steps using the golden-section algorithm.

Therefore, except the very first iteration, the number of computations remains the same as in any other bracketing method. However, compared to the bisection method with the termination condition defined by Eq. (10.6) resulting in Eqs. (10.7) and (10.8), in the golden-section method, the size of the interval $|a-c|$ shrinks slower (by the factor of $R^\circledR > 0.5$). For example, after say 10 iterations, the interval shrinks to 0.618^{10} or about 0.8% of its original length, whereas for the bisection algorithm, it reduces to 0.5^{10} or about 0.1%.

Because at each step there is an extra point to play with, the golden-section method belongs to the *ternary search algorithms*, a technique usually used for finding the minimum or maximum of functions (see Sec. 10.7.1), and in this capacity, it can be easily modified to handle situations like the one presented in Fig. 10.4.

10.3 Open Methods

As opposed to bracketing methods, open methods require only a single value of x to start iterations [some open methods may still require two or even more, but not necessarily bracketing the root as required by Eq. (10.4)]. As such, they generally do not guarantee convergence as computations progress unless the initial guess x_0 is chosen wisely, that is, very close to the root x_* (as shown in Fig. 10.9). However, even then, the numerical algorithm may still diverge. Nevertheless, the advantage of open methods is that if they converge, they demonstrate much better convergence rate compared with that of bracketing methods. The methods belonging to this groups and to be considered in this section are the *fixed-point iteration method, Newton's method, secant method* (requiring two points to start the iterative procedure), and *inverse quadratic interpolation method* (requiring three initial points).

10.3.1 Simple Fixed-Point Iteration

Let us start from the most simple method, which is known as *fixed-point iteration* or *one-point iteration* or *successive substitution* method. Before continuing, we need to slightly modify the original problem. Let us add x to both sides of Eq. (10.1) to get

$$x + f(x) = x \tag{10.19}$$

or

$$x = g(x) \tag{10.20}$$

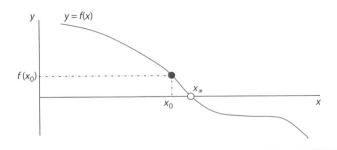

Fig. 10.9 Choosing the initial guess for open methods.

(we simply introduced a new notation, $g(x) \equiv x + f(x)$). Solving Eq. (10.1) automatically solves Eq. (10.19) and vice versa—if we manage to solve Eq. (10.20) for some x_*, the very same x_* is then the solution of Eq. (10.1).

Note that, to obtain the equation that looks like Eq. (10.20), we could add say Kx to both sides of Eq. (10.1) as well

$$Kx + f(x) = Kx \tag{10.21}$$

This would lead to

$$g(x) \equiv x + K^{-1}f(x) \tag{10.22}$$

Hence, the reduction of the original equation of the form (10.1) to that of Eq. (10.20) can be accomplished in different ways. Sometimes, you can even "extract" x from $f(x)$ analytically. Say, if $f(x) = x^3 + ax^2 + bx + c$, then you can simply rewrite it as Eq. (10.20) with $g(x) = -b^{-1}(x^3 + ax^2 + bx + c)$.

The utility of Eq. (10.20) is that it explicitly provides a formula to predict a new value of x as a function of an old value

$$x_{i+1} = g(x_i), \; i = 0,1,... \tag{10.23}$$

As a matter of fact, instead of finding where the function $f(x)$ crosses the abscissa axis, we are now trying to find the intersection of two curves, $y = x$ and $y = g(x)$ (Fig. 10.10a).

Let us now address the convergence issue. Figures 10.10b and 10.11a,b complement Fig. 10.10a and show three other possible situations of relative position of curve $y = g(x)$ with respect to $y = x$. As seen, among those three, only the case of Fig. 10.10b exhibits convergence, whereas the two other cases (depicted in Figs. 10.11a and b) diverge. What causes this divergence?

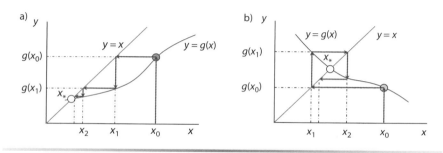

Fig. 10.10 Illustration of converging iterations in the simple fixed-point method.

To answer this question, we should look at the value of derivative $g'(x) = dg(x)/dx$ in the vicinity of the root. For example, in the case of Figs. 10.10a and b, $0 < g'(x_*) < 1$ and $-1 < g'(x_*) < 0$, respectively. In the case of Fig. 10.11a, $g'(x_*) > 1$ and in the case of Fig. 10.11b, $g'(x_*) < -1$. Therefore, the convergence of the algorithm can only be guaranteed when in the vicinity of the root $|g'(x)| < 1$. In addition to that, if the derivative $g'(x)$ is positive (Fig. 10.10a), the error exhibits a mono-tonic behavior (not changing its sign from one iteration to another), whereas if $g'(x)$ is negative (Fig. 10.10b), the error oscillates (changes its sign).

That is where having a capability to have a different function $g(x)$, when converting Eq. (10.1) to the form of Eq. (10.19), may really help. For instance, we need to find a root of $f(x)=e^{x-1}-2$ (where $x_* = 1.69314$). If we convert this equation to that of the form Eq. (10.20) by adding x to both sides of the original equation, we get $g(x) = e^{x-1} - 2$. The problem with this is that $|g'(x)|_{x=x_*} \approx 3 \geq 1$, that is, successive substitution method would diverge. However, if we add $-4x$ to the both sides of the original equation, we get $g(x) = x - 0.25\,(e^{x-1} - 2)$, with $g'(x)|_{x=x_*} = 0.5$, and so the successive substitution method would produce a solution.

Consider Eq. (10.22). The derivative of $g(x)$ in this general case is

$$g'(x) = 1 + K^{-1} f'(x) \tag{10.24}$$

For convergence it is required to have

$$-1 < 1 + K^{-1} f'(x) < 1 \tag{10.25}$$

or

$$-2 < K^{-1} f'(x) < 0 \tag{10.26}$$

Hence, assuming that the initial guess is fairly close to the solution, we can always choose K as

$$K < -0.5 f'(x_0), \text{ if } f'(x_0) > 0$$
$$K > -0.5 f'(x_0), \text{ if } f'(x_0) < 0 \qquad (10.27)$$

In the case of the function $f(x) = e^{x-1} - 2$ considered earlier, Eq. (10.27) estimated at $x=x_*$ would yield $K < -0.3$, but the actual value of K that would guarantee convergence may depend on the initial guess x_0 and function geometry in the vicinity of the solution. Hence, despite all precautions any open method must have a safeguard—a firm limit on the number of iterations.

Given the initial guess, function $f(x)$ and relative tolerance, the successive substitution algorithm may look as follows:

```
x=x0;
h=0.01*x; if h==0, h=0.01; end
K=-4*(f(x+h)-f(x))/h;            % choosing the gain K
for k=1:500                      % limiting the number of iterations
a=x;                             % keeping the current iteration
x=x+f(x)/K;                      % finding the next iteration
if abs(x-a)<=eps*max(abs(x),1e-3), break, end % checking the accuracy
end
disp([x,f(x),k])                 % displaying the results
```

(the second and third lines of this script are dedicated to estimating the first-order derivative numerically, which will be discussed in more detail in the following section). As seen, we actually combined Eq. (10.22) with an estimate of the gain [Eq. (10.27)], so that Eq. (10.23) takes the form

$$x_{i+1} = x_i - \frac{f(x_i)}{4\hat{f}(x_0)}, \quad i = 0, 1, 2, \ldots \qquad (10.28)$$

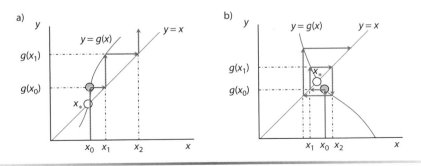

Fig. 10.11 Pitfalls of the successive substitution algorithm.

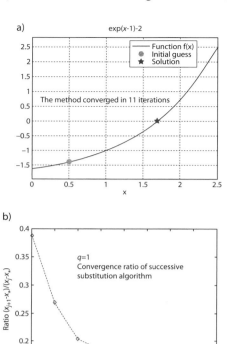

Fig. 10.12   a) Example of a solution using successive substitution algorithm and b) a proof of its linear convergence.

(we chose the coefficient 4 in denominator quite arbitrary with the only goal of satisfying Eq. (10.27) for sure, at least at the initial point).

Let us use the previous code and find the root of $f(x) = e^{x-1} - 2$ by issuing the following commands:

```
f=inline('exp(x-1)-2');
x0=0.5; eps=1e-6;
...                  % here comes the successive substitution algorithm code
ezplot(f, [0 2.5]), grid, hold
plot(x0,f(x0),'go','MarkerSize',10,'MarkerFaceColor','g')
plot(x,f(x),'rp','MarkerSize',12,'MarkerFaceColor','r')
legend('Function f(x)','Initial guess','Solution')
text(0.6,0.7,2,['The method converged in ' int2str(k) ' iterations'])
```

The results are presented graphically in Fig. 10.12a. With the given initial guess and tolerance, it took 11 iterations to converge. How good is this?

In dealing with a sequence of successive approximations for any iterative method, the *rate of convergence* $\mu \in [0;1]$ can be formally defined as

$$\mu = \lim_{i \to \infty} \frac{|x_{i+1} - x_*|}{|x_i - x_*|^q} \qquad (10.29)$$

If for $q=1$ the limit [Eq. (10.29)] converges to some number $\mu \in (0;1)$, the sequence x_i, $i = 0,1, \dots , \infty$ is said to converge *linearly* to x_*. If Eq. (10.29) holds with $\mu = 1$, then the sequence is said to converge *sublinearly*. One says that the sequence converges *superlinearly* if it converges to $\mu = 0$. In the latter case, you may consider the higher values of the order q. Convergence to $\mu \in (0;1)$ with $q = 2$ is called *quadratic convergence*, convergence with $q = 3$ is called *cubic*, etc.

The ratio $|x_{i+1} - x_*| / |x_i - x_*|$ for bracketing methods not necessarily converges, but most of the time stays within $\mu \in (0;1)$, therefore they assure a linear convergence. The behavior of this ratio in the example we just considered is presented in Fig. 10.12b. As seen, it converges to some number $\mu \in (0;1)$ (with $q = 1$) and therefore proves the linear convergence property of the simple fixed-point iterations method.

10.3.2 Newton's Method

Of all the root-finding methods, perhaps the most widely used is the *Newton's method* (also known as *Newton–Raphson method*). It relies on evaluating the tangential lines to $f(x)$ and is applicable to finding both real and complex roots.

Newton's method is derived from a Taylor series expansion about the point x_i

$$f(x_{i+1}) = f(x_i) + (x_{i+1} - x_i)f'(x_i) + O(\Delta x^2) \qquad (10.30)$$

where $f'(x_i)$ is the first-order derivative of function $f(x)$, $\Delta x = x - x_i$, and $O(\Delta x^2)$ is the truncation error. Neglecting the truncation error, the tangential line passing through the point $(x_i, f(x_i))$ is given by

$$f(x_i) + f'(x_i)(x_{i+1} - x_i) = 0 \qquad (10.31)$$

Solving for x_{i+1} yields the formula for Newton's method successive approximations

$$x_{i+1} = x_i - \frac{f(x_i)}{f'(x_i)} \qquad (10.32)$$

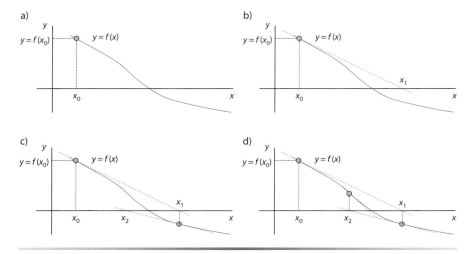

Fig. 10.13 Illustration of several consecutive iterations using the Newton's method.

For better understanding, several iterations are demonstrated in Fig. 10.13. To enhance robustness of this formula, Raphson suggested to use the following modification with $\alpha < 1$

$$x_{i+1} = x_i - \alpha \frac{f(x_i)}{f'(x_i)} \tag{10.33}$$

As seen, Eq. (10.33) has exactly the same form as Eq. (10.28). The only difference is that the first-order derivative in denominator is now updated at every iteration, ensuring a better robustness. To be more precise, the Newton's method guarantees quadratic convergence, meaning that the number of significant figures of accuracy approximately doubles with the each iteration [according to Eq. (10.29) the true percent relative error is roughly proportional to the square of the previous iteration error]. This can be easily proved using the same example we used in the previous section. The Newton's method can be programmed as follows:

```
x=x0;
fdot=inline(char(diff(char(f))));
for k=1:500                    % limiting the number of iterations
a=x;                           % keeping the current iteration
x=x-f(x)/fdot(x);              % finding the next iteration
    if abs(x-a)<=eps max(abs(x),1e-3), break, end % exit condition
end
disp([x,f(x),k])               % displaying the results
```

Note what we did in the second line to analytically compute the first-order derivative of the inline function f. If we run it for $f(x) = e^{x-1} - 2$ with the same settings as in the previous section, we will arrive to the solution much faster compared with the simple fixed-point iteration algorithm (cf. Fig. 10.14a with Fig. 10.12a). As seen from the top plot of Fig. 10.14b, the rate of convergence μ [Eq. (10.29)] with $q = 2$ is equal to 0.5. Moreover, although Eq. (10.32), adaptive to the first-order derivative, assures convergence in a fairly descent range of initial guesses, Eq. (10.28) diverges if we start from say $x_0 = 0$ (because the derivative estimated at x_0 is quite different from that estimated at x_*).

Now, let us address an issue of computing the first-order derivative. In the particular example considered earlier with $f(x)$ defined analytically, we know that $f'(x) = e^{x-1}$, and so we took advantage of it supplying (computing symbolically) an analytical formula for the derivative. In practice, however, the function $f(x)$ may not be defined analytically, meaning that its first-order

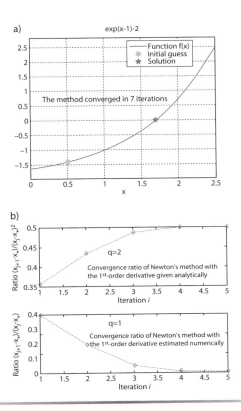

Fig. 10.14 a) Example of solution using the Newton's method and b) its rate of convergence.

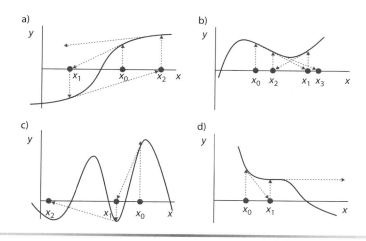

Fig. 10.15 Pitfalls of the Newton's method.

derivative has to be estimated numerically. Although this issue is covered in detail in Chapter 12, the easiest way to do this is to use a forward difference approximation

$$f'(x_i) \cong \frac{f(x_i + h) - f(x_i)}{h} \qquad (10.34)$$

where h has to be a sufficiently small value to assure a certain accuracy. We have used this approximation in the script following Eq. (10.27) already (we chose h to be 1% of x or 0.01 if x is zero). In that script it caused no problems because the derivative was computed only once and then multiplied by some arbitrary number (4). However, if we use this approximation instead of an exact analytical formula in the Newton's method, its convergence rate would degrade to superlinear (shown in the bottom plot of Fig. 10.14b).

Although the Newton–Raphson method is often very efficient, there are situations where it performs poorly. For instance, it cannot handle multiple roots. However, even when dealing with simple roots, difficulties with convergence can also arise. Some of them are presented in Fig. 10.15.

Specifically, Fig. 10.15a depicts the case where an *inflection point* (that is, $f''(x) = 0$) occurs in the vicinity of root. Notice that iterations beginning at x_0 progressively diverge from the root. A typical example of such a case is finding a root of the function $f(x) = \text{sign}(x - a)\sqrt{|x - a|}$. The Newton's method will iterate forever. Figure 10.15b illustrates the tendency of the Newton–Raphson technique to oscillate around a local maximum or minimum. Moreover, as shown in Fig. 10.15b, if a near-zero slope is reached, whereupon the solution may be sent far from the area of interest. (By the way, that is one of the cases where a step-limiting gain α in Eq. (10.33) may help.) In the particular situation of Fig. 10.15b, one may argue that the problem was caused by unfortunate choice of the initial guess, so that $f'(x_*)\,f'(x_0) < 0$, but even if we start from x_1 we would run into the same problem. Figure 10.15c shows how an initial guess that is close to one root can jump to another location several roots away. The tendency to move away from the area of interest is caused by the varying first-order derivative $(\mathrm{sign}(f'(x)|_{x\sim x_*}) = var)$, so that the near-zero slopes are encountered. Obviously, a zero slope $(f'(x)|_{x\sim x_*} \sim 0)$ is always a disaster because it causes division by zero in both Eq. (10.32) and Eq. (10.33). Graphically, it means that the solution shoots off horizontally and never hits the x-axis (Fig. 10.15d).

Thus, no general convergence criterion for the Newton–Raphson method exist. Its convergence depends on the nature of the function and on the accuracy of the initial guess. The only remedy is to have an initial guess that is "sufficiently" close to the root. For some functions, no guess will work at all! Good guesses are usually predicated on knowledge of the physical problem setting or on visualization of the function that provides insight into its behavior in the vicinity of root. The lack of a general convergence criterion also suggests a script realizing this algorithm (in fact, any open-method algorithm) be designed to recognize slow convergence or even divergence, making the corresponding adjustment iterations or simply quitting if nothing helps.

To summarize, a good programming paradigm as applied to realization of the Newton–Raphson method (and any other open method) implies

- Plotting the function to choose a good initial guess
- Alerting a user and taking into account the possibility that $f'(x)$ might become zero at any time during the computation
- Including an upper limit on the number of iterations to guard against oscillating, slowly convergent or divergent solutions that could persist interminably
- Always checking the solution (substituting it into the original function to compute whether the result is close to zero)

The last check partially guards against those cases where slow or oscillating convergence may lead to a small value of ε_a^i, although the solution is still far from a root.

10.3.3 Secant Method

As discussed in the previous section, evaluation of the first-order derivative (in the case of a not analytically given function) causes a certain problem. As suggested by Eq. (10.35), the derivative can be approximated by a forward difference with as small h as possible to make the estimate of the derivative more accurate. As a result, instead of just one evaluation of function at the each iteration for the analytically given functions, we end up with two evaluations for the not analytically given functions, which obviously degrades algorithm's performance.

An attractive alternative for estimating the derivative or at least the tendency of function change is to use the previous iteration as follows:

$$f'(x_i) \approx \frac{f(x_i) - f(x_{i-1})}{x_i - x_{i-1}} \tag{10.35}$$

This constitutes a very rough estimate of the derivative using a backward difference approximation (to be discussed in Chapter 12). Substituting Eq. (10.35) into Eq. (10.32) yields the formula known as the *secant method*

$$x_{i+1} = x_i - \frac{f(x_i)(x_i - x_{i-1})}{f(x_i) - f(x_{i-1})} \tag{10.36}$$

To initialize iterations, two points are required. As shown in Fig. 10.16, you start from x_0 and x_1 to produce x_2, then use x_1 and x_2 to produce x_3, and so on.

The code for this algorithm may look as follows:

```
a=x0;
h=0.01*a; if h==0, h=0.01; end
x1=a+h;
for k=1:500                        % limiting the number of iterations
    fdot=(f(x1)-f(a))/(x1-a);      % estimating the derivative
    a=x1;
    x=x1-f(x1)/fdot;               % finding the next iteration
    x1=x;
        if abs(x-a)<=eps*abs(x), break, end    % checking the accuracy
end
disp([x,f(x),k])                   % displaying the results
```

Obviously, there is a similarity between the secant method and the false-position method. Both use two initial estimates to compute an approximation of the slope of the function that is used to project to the

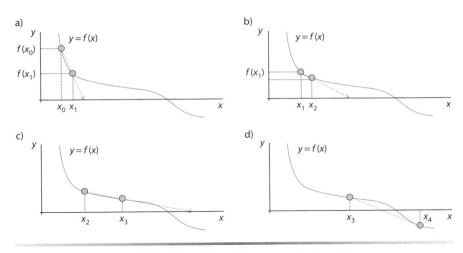

Fig. 10.16 Illustration of the secant algorithm.

x-axis for a new estimate of the root. However, a critical difference between the methods is how one of the initial values is replaced by a new estimate. In the false-position method, the latest estimate of the root replaces whichever of the original values yielding a function value with the same sign as the replaced one, therefore always bracketing the root. Consequently, the method always converges. In contrast, the secant method replaces the values in a strict sequence, with the new value replacing the previous one. As a result, the two values can lie on the same side of the root (Fig. 10. 16a–c), as opposed to the different sides of the root (Fig. 10.16d). As discussed earlier, this may lead to divergence.

10.3.4 Inverse Quadratic Interpolation Method

Basically, what we did in the previous section is that we used two trial points to obtain a local linear approximation of the function, and then, by finding the intersection of this approximation with the *x*-axis, we found the next trial point. This allowed us to obtain an algorithm assuring close to quadratic convergence (referred to as *superlinear convergence*). As it turns out, to increase the order of convergence, the same idea can be explored farther to involve three trial points for a local approximation of the function.

If three trial points are used (not necessarily located on one side of the root, as shown in Fig. 10.17a though), a quadratic approximation of the function $f(x), f^*(x)$, may be used (Fig. 10.17b). This approximation based on three points $(x_{i-2}, f(x_{i-2}))$, $(x_{i-1}, f(x_{i-1}))$, and $(x_i, f(x_i))$ is given by equation

$$f^*(x) = a(x - x_i)^2 + b(x - x_i) + c \tag{10.37}$$

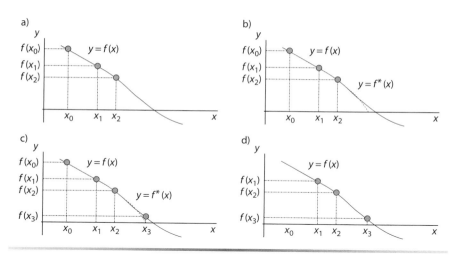

Fig. 10.17 Illustration of the quadratic interpolation algorithm.

where coefficients a, b, and c are computed as

$$a = \frac{f(x_{i-2})(x_{i-1} - x_i) + f(x_{i-1})(x_i - x_{i-2}) + f(x_i)(x_{i-2} - x_{i-1})}{(x_{i-2} - x_i)(x_i - x_{i-1})(x_{i-1} - x_{i-2})}$$

$$b = \frac{-f(x_{i-2})(x_{i-1} - x_i)^2 + f(x_{i-1})(x_i - x_{i-2})^2 - f(x_i)(x_{i-2} - x_{i-1})(x_{i-2} + x_{i-1} - 2x_i)}{(x_{i-2} - x_i)(x_i - x_{i-1})(x_{i-1} - x_{i-2})}$$

$$c = f(x_i) \tag{10.38}$$

To start this algorithm, you have to have two approximations, x_0 and x_1, to be obtained somehow. Among the two roots of $f^*(x) = 0$ defined by the well-known formula

$$x = x_i + \frac{-b \pm \sqrt{b^2 - 4ac}}{2a} \tag{10.39}$$

the nearest to the point x_i may be chosen to constitute the next guess x_{i+1} (Fig. 10.17c). Then, the entire procedure is repeated for the new triplet and so on.

The problem with this approach though is that for a relatively flat function $f(x)$, the parabola may not intersect the x-axis (Figs. 10.18a and 10.18b), so that a quadratic three-point approximation $f^*(x)$ does not necessarily have the real roots (formally, it happens when $b^2 - 4ac < 0$). Sometimes this can be regarded as an advantage (for example, some methods benefit from using the complex roots to find the complex zeros of a polynomial). However, for this particular application, we want to avoid complex arithmetic and rather have a single real root guaranteed at each iteration. Well, the trick is to swing the parabola (Figs. 10.18c and 10.18d).

Hence, instead of using three points, $(x_{i-2}, f(x_{i-2}))$, $(x_{i-1}, f(x_{i-1}))$, and $(x_i, f(x_i))$, to develop an approximation in x, we should be looking at approximation in $f(x)$

$$x = P(f(x)) = \tilde{a}f(x)^2 + \tilde{b}f(x) + \tilde{c} \tag{10.40}$$

where coefficients of a "sideways" parabola, $P(f(x))$, are determined by the interpolation conditions $x_{i-2} = P(f(x_{i-2}))$, $x_{i-1} = P(f(x_{i-1}))$ and $x_i = P(f(x_i))$. By construction, Eq. (10.40) always intersects the x-axis at $x = P(0) = \tilde{c}(f(x) = 0)$, which becomes the next, $(i+1)$th, iteration. Given the three points and function $f(x)$, the entire procedure referred to as the *inverse quadratic interpolation method* can be programmed as follows:

```
a=x0; b=x1; c=x2; k=0;
while abs(c-b)>eps*abs(c)
    x=polyfit([f(a),f(b),f(c)],[a,b,c],2); % producing a quadratic fit
    a=b;
    b=c;
    c=x(3);                        % finding the next iteration
    k=k+1;
    if k>100, break, end           % limiting the number of iterations
end
disp([c,f(c)])                     % displaying the results
```

Note that, to perform a polynomial curve fitting we used the `polyfit` function, which was introduced in Table 3.8 of Sec. 3.5 (and will be discussed in detail in Chapter 11). This single line can be replaced with the following fragment that does essentially the same math:

```
A = [f(a)^2 f(a) 1;
     f(b)^2 f(b) 1;
     f(c)^2 f(c) 1];
B = [a;b;c];
x = A\B;
```

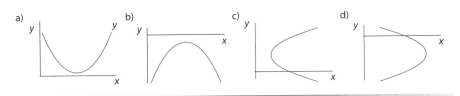

Fig. 10.18   a, b) Quadratic vs c, d) inverse quadratic interpolation.

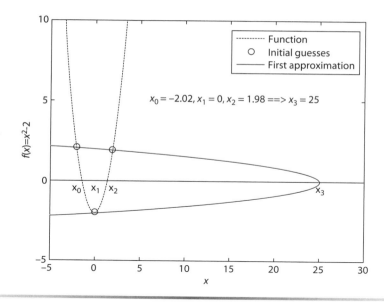

Fig. 10.19 Pitfall of the inverse quadratic interpolation method.

If it converges, the inverse quadratic interpolation method provides *superquadratic convergence*. The trouble with the "pure" inverse quadratic interpolation algorithm, however, is that the polynomial interpolation requires the values of $f(x_{i-2})$, $f(x_{i-1})$, and $f(xi)$ to be distinct. Yet, generally, this is not guaranteed. For example, if we try to compute $\sqrt{2}$ using $f(x)=x^2-2$ and start with $x_0=-2$, $x_1=0$, and $x_2=2$, we would end up with $f(x_0)=f(x_2)$, and so the first step is undefined. If we start somewhere nearby this singular situation, say with $x_0=-2.01$, $x_1=0$, and $x_2=1.99$, the very first iteration will bring us to $x_3=50$ (Fig. 10.19). Hence, the inverse quadratic interpolation method needs to be supervised to decide whether the iteration it provides is of any good.

10.4 Zeroin Algorithm

Figure 10.20 summarizes the findings of the two previous sections and sorts all methods by their convergence properties. As seen, although bracketing methods guarantee convergence, the speed of convergence is slower than that of the open methods. The latter ones, however, do not guarantee the convergence.

Therefore, the idea is to combine the best properties of both approaches and that is what is being realized in the zeroin algorithm (`fzero` function). This algorithm originated by van Wijngaarden, Dekker, and others at the Mathematical Center in Amsterdam in the 1960s combines the reliability of bisection method with the convergence speed of secant and inverse quadratic interpolation methods. It is implemented in MATLAB as Richard Brent's version, which looks as follows:

1. Start with a and b so that $f(a)$ and $f(b)$ have the opposite signs.
2. Use a secant step to produce c between a and b.
3. Check the exiting conditions, which are $|b-a| < \varepsilon|b|$ or $f(b) = 0$.
4. Arrange a, b, and c so that
   - $f(a)$ and $f(b)$ have the opposite signs meaning that zero is bracketed between them
   - $|f(b)| \leq |f(a)|$ meaning that b is closer to zero as compared to a
   - c represents the previous value of b
5. If $c \neq a$, meaning that a, b, and c are distinct points, perform an inverse quadratic interpolation step.
6. If $c = a$, perform a secant step.
7. If the inverse quadratic interpolation or secant step produces "reasonable" iteration (the definition of "reasonable" is rather technical, but essentially it means that the point is inside the current interval $[a;b]$ and not too close to the end points), accept it.
8. If the last step produces a point, which is not in the interval $[a;b]$, use a bisection iteration and go back to step 3.

By construction, this combination guarantees the decrease of interval length at each iteration (due to bracketing) and the fastest convergence when the function $f(x)$ is well behaved. As an example, Fig. 10.21 represents two sets of three sequential steps (shown in the top and bottom rows). These snapshots were obtained using the `fzerogui` function written by Cleve Moler. The function itself is shown as a dashed line, whereas a short vertical line in each plot represents a bisection point (halving the current interval $[a;b]$ or $[b;a]$).

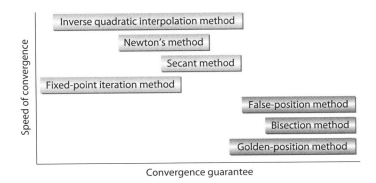

Fig. 10.20 Summary of bracketing and open methods.

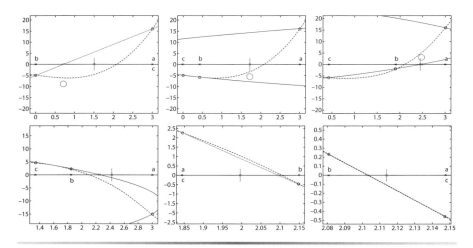

Fig. 10.21 Illustration of the zeroin algorithm (two sets of three sequential steps).

The first set of sequential steps (first row) shows that at the very first step (left top plot), the interval [b;a] enclosures zero (b and a are arranged so that $|f(b)| \le |f(a)|$ and point c coincides with a). Because there are no three distinct points available, a secant step is made to produce the third point (depicted with a circle). According to the procedure, the new interval [b;a] enclosures zero, and b and a are arranged so that $|f(b)| \le |f(a)|$. The point c now represents an old value of the point b (second plot in the upper row). Because three distinct points are now available, an inverse quadratic interpolation step is made to produce the next candidate iteration. Unfortunately, it turns out that this candidate point does not belong to the interval [b;a] (the intersection with the abscissa axis occurs far to the left). Because of this, the candidate point is rejected, and the bisection step point (depicted with a circle) is taken as the next iteration. This point becomes point b and its old value is assigned to c. The new three points are rearranged as shown in the last plot in the top row of plots. This time, the inverse quadratic interpolation step produces a reasonable solution (depicted as a circle with the abscissa slightly to the right of a bisection point).

The second sequence in Fig. 10.21 starts with a situation which is very similar to that we ended up with the previous sequence (the bottom left plot). Having three distinct points, an inverse quadratic interpolation step is made. Because the new point happens to be much closer to zero, it becomes a new point b, causing point a to accept an old value of b (the middle plot of the bottom-row sequence). By definition, c also accepts the old value of b. Hence, now that c=a, we have no other choice as to produce a secant step. As seen from the right bottom-row plot, the situation with the point *b* jumping over zero repeats, and we are left with only two points to produce the next iteration again.

A major advantage of the zeroin algorithm is that it is foolproof. It never loses track of the zero trapped in a shrinking interval. On the average, for well-behaved functions, this combination of root-finding methods saves about half function evaluations per zero (compared to the other methods), exploiting a rapidly converging method only when it is reliable. It uses slow, but reliable methods otherwise.

10.5 Multiple Roots and Value Finding

This section addresses several specific topics related to zero or value finding. It starts from the issue of bracketing the individual roots of a function. It then proceeds with the handling of multiple roots, that is, with a situation when the function $f(x)$ does not cross the abscissa axis, but rather touches it. At last, the section addresses the value-finding problem.

10.5.1 Multiple Roots Bracketing

As shown in Secs. 10.2 and 10.4, bracketing methods rely on root localization. If the original interval is found, any of these methods assure convergence. The question then is how to automatize finding such an interval, especially in the case when the function has multiple roots (that is, you need to find multiple intervals). Later you will find a self-explanatory M-file function `lroot` (abbreviated form for localize a root) that does the job (this function includes several professional elements, such as, `nargin`, `isempty`, `fcnchk`, and `feval` functions that were discussed in Sec. 5.9)

```
function Br=lroot(fun,xmin,xmax,nx,flag)
if nargin<5, flag=0; end     % denying the graphical output
if nargin<4 | isempty(nx), nx=20; end % the default number of brackets
% Creating vectors of potential brackets and corresponding f(x) values
x=linspace(xmin-.1,xmax+.1,nx);
f=fcnchk(fun); f=feval(f,x);
% Searching for the brackets
nb=0; Br=[];                  % Br is null unless brackets are found
for k=1:length(f)-1
    if sign(f(k))*sign(f(k+1))<=0      % true if sign(f(x)) changes
        nb=nb+1;
        Br(nb,1)=x(k); Br(nb,2)=x(k+1);   % saving the bracket bounds
    end
end
```

```
% Plotting function f(x) with the brackets
if flag~=0
% Preparing data to draw boxes that indicate brackets
ytop=max(f); ybot=min(f);              % y-coordinates of the box
ybox=0.05*[ybot ytop ytop ybot ybot]; % around a bracket
c=[0.7 0.6 0.7];                       % RGB color for bracket the boxes
hold on
for i=1:nb
    fill([Br(i,1) Br(i,1) Br(i,2) Br(i,2) Br(i,1)],ybox,c); % adding box
end
xp=linspace(xmin-.1,xmax+.1);      % vector of arguments for plotting
fp=fcnchk(fun); fp=feval(fp,xp);   % vector of f(x) values brackets
plot(xp,fp,'Linewidth',2), plot(get(gca,'xlim'),[0 0],'-.k');
xlabel('x'); ylabel(['f(x) defined in/as' char(fun)]);
hold off
end
    if isempty(Br)                    % warning message
    warning ('No brackets found.  Check [xmin, xmax] or increase nx');
    end
```

The inputs to this function are fun, the name of M-file function evaluating $f(x)$; xmin, xmax, endpoints of a larger interval the individual roots will be bracketed within; nx, the number of samples along x axis to be used to test for bracket endpoints; and flag, indicating whether to display the results graphically. The latter two inputs are optional. The only constraint is that nx should be large enough (greater than a number of potential roots within the defined interval [xmin; xmax]) to localize each root individually. The default value for nx, established in the second line of the script, is 20.

This function finds subintervals of function argument x that contain the sign changes of $f(x)$, so that each root may have its own brackets (to guaranteed though). The output of this function is a two-column matrix of bracket limits Br, so that Br(k, 1) is the left bound and Br(k, 2) is the right bound for the kth root. If no brackets were found, Br returns an empty matrix. In addition to this, the results can be presented graphically featuring the plot of $f(x)$ with the determined bracketing intervals overlaid over it.

You may call this function using either syntax

```
lroot(fun,xmin,xmax)
lroot(fun,xmin,xmax,nx)
lroot(fun,xmin,xmax,nx,flag)
lroot(fun,xmin,xmax,[],flag)
```

For example, the call

```
>> lroot(@cos,-2*pi,2*pi,[],1);
```

returns four intervals shown in Fig. 10.22a, and the call

```
>> b=lroot('4*cos(2*x)-exp(0.5*x)+4',-3,4,6,1)
```

returns five intervals

```
b = -3.1000    -1.6600
     -1.6600    -0.2200
     -0.2200     1.2200
      1.2200     2.6600
      2.6600     4.1000
```

shown in Fig. 10.22b. In this particular case, if we issue the same command with nx equal to 5, the lroot function would only localize one root, and so nx = 5 is the smallest number to localize all five roots.

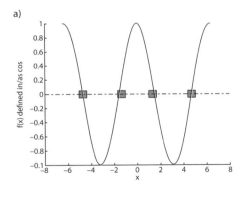

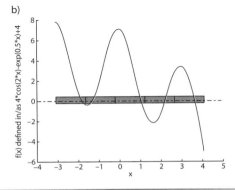

Fig. 10.22 Localizing individual roots.

The brackets found using `lroot` can be further passed over to the `fzero` function for root finding as shown later in Sec. 10.9.1.

10.5.2 Handling Root Multiplicity

The root x_* is called a zero of *multiplicity k* of function $f(x)$ if there exist a real number $l \neq 0$ such that

$$l = \lim_{x \to x_*} \frac{f(x)}{|x - x_*|^k} \tag{10.41}$$

Geometrically, zeros of nonunity multiplicity show up as a point where a function is tangent to the x-axis as shown in Fig. 10.23.

As mentioned already, neither bracketing method nor open methods can explicitly handle multiple roots. The fact that the function does not change a sign at the even multiple roots (*double, quadruple*, etc.) precludes the use of bracketing methods. The open methods may diverge.

The latter is especially true for the Newton–Raphson and secant methods, both of which contain the current derivative (or its estimate) in the denominator [Eqs. (10.30) and (10.34), respectively]. Because at the multiple roots, not only $f(x)$, but also $f'(x)$ goes to zero, both recurrent formulas encounter division by a very small number as they approach the root. That is why it is important to have a safeguard in root-finding software, as suggested at the end of Sec. 10.3.2, warning about $f'(x)$ approaching a zero. Even if the convergence for multiple roots is achieved, for both Newton–Raphson and secant methods, it degrades to linear convergence.

To restore a superb (quadratic) convergence, you may use a slightly modified version of the Newton's equation (10.32)

$$x_{i+1} = x_i - m\frac{f(x_i)}{f'(x_i)} \tag{10.42}$$

where m is a multiplicity of the root. However, it implies foreknowledge of root's multiplicity, which is not always available.

Another alternative, suggested by Ralston and Rabinowitz, is to replace $f(x)$ with a new function

$$u(x) = \frac{f(x)}{f'(x)} \tag{10.43}$$

which by construction has the same roots. Then, Eq. (10.31) for the multiple roots becomes

$$x_{i+1} = x_i - \frac{f(x_i)f'(x_i)}{(f'(x_i))^2 - f(x_i)f''(x_i)} \tag{10.44}$$

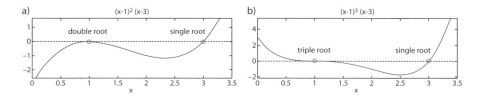

Fig. 10.23 Examples of roots of multiplicity a) two and b) three as compared to a single root.

As seen, having $f'(x) \to 0$ poses no problem anymore. Obviously, implementation of this formula requires more computational resources (to compute or estimate $f''(x)$ along with $f'(x)$ and $f(x)$) and therefore should be used with caution. To conserve some computational resources, you may also consider employing a modified formula for a secant method

$$x_{i+1} = x_i - \frac{u(x_i)(x_i - x_{i-1})}{u(x_i) - u(x_{i-1})} \qquad (10.45)$$

suited for the multiple roots [cf. with Eq. (10.37)]. If neither of the previous two options helps, you can always reformulate a multiple zero finding problem to that of finding function's minimum, and solve it using optimization tools of MATLAB (discussed in Sec. 10.7).

10.5.3 Value-Finding Problem

Up to this point, we were dealing with a zero-finding problem. However, algorithms introduced in the previous sections of this chapter are suitable for a value-finding problem as well. This problem can be formulated as follows: given a function $F(x)$ and a value η, find ξ so that $F(\xi) = \eta$.

Obviously, to handle this problem, we can use any zero finder on a translated function

$$f(x) = F(x) - \eta \qquad (10.46)$$

This gives you the desired ξ so that $F(\xi) = 0$, and hence $F(\xi) = \eta$.

10.6 Roots of Polynomials

In principle, any of the approaches discussed in Secs. 10.2 and 10.3 may be applied to find the roots of a polynomial equation

$$f_n(x) = a_n x^n + a_{n-1} x^{n-1} + \cdots + a_1 x + a_0 \qquad (10.47)$$

where n is the order of a polynomial and a_i, $i = 0, 1, ..., n$ are its real constant coefficients. To be more accurate, the applicability and efficacy of these approaches depend on whether the roots of Eq. (10.47) are complex. If only real roots exist, then any of the previously described methods can be employed. Yet, the problem of finding the good initial guesses would complicate both bracketing and open methods, whereas as usual, open methods can be susceptible to divergence.

Once solved for one root, the root location procedure may be continued to find the next root. However, if the initial guess is not accurate you may end up finding the same root again. Therefore, it makes sense to remove the found root before attempting to find another one. The removal process is referred to as *polynomial deflation*. Several computer algorithms are available for performing this operation. For instance, the following algorithm

```
n=length(a)
r=a(1);
a(1)=0;
for i=2:n
        s=a(i);
        a(i)=r;
        r=s+r*t;
end
```

uses a simple method, the so-called *synthetic vision approach* of dividing a polynomial of Eq. (10.47), defined via the vector of its coefficients a by a monomial $x-t$. If t is a root of polynomial, the reminder r will be zero, and coefficients of the quotient will be stored in the same vector a (with a(1), corresponding to x^n, equal to zero).

Because each calculated root is known only approximately, the deflation is sensitive to round-off errors. In some cases, they can grow to the point that the results can become meaningless. One way to reduce round-off errors is to consider each successive root estimate obtained during deflation as a good first guess. These can then be used as a starting guess, and the root determined again with the original nondeflated polynomial. This procedure is referred to as *root polishing*.

For finding the complex roots, only open methods can be employed. Among them, the conventional Newton–Raphson method provides a viable approach. In this case, the deflation algorithm should be capable of dividing the polynomial by a binomial rather than just a monomial, as it was the case for the real roots. As you might expect, the procedure is susceptible to convergence problems. For this reason, special methods to find the real and complex roots of polynomials (still related to the open approach) have been developed. The MathWorks's MATLAB instruments an algorithm involving computing eigenvalues of the *companion matrix*

$$
C = \begin{bmatrix}
-\dfrac{a_{n-1}}{a_n} & -\dfrac{a_{n-2}}{a_n} & \cdots & -\dfrac{a_1}{a_n} & -\dfrac{a_0}{a_n} \\
1 & 0 & 0 & 0 & 0 \\
0 & 1 & 0 & 0 & 0 \\
\cdots & \cdots & \cdots & \cdots & \cdots \\
0 & 0 & 0 & 1 & 0
\end{bmatrix}
\tag{10.48}
$$

Because the characteristic polynomial of this matrix yields the nth-order polynomial [Eq. (10.47)], these eigenvalues represent the roots of this polynomial. It is as simple as that! (This example again demonstrates the beauty of dealing with matrices, rather than with algebraic equations.)

The MATLAB function `roots` can essentially be reduced to the three lines of code that construct the companion matrix and compute its eigenvalues

```
C = diag(ones(n-1,1),-1);
C(1,:) = -a(2:n+1)./a(1);
eig(C)
```

Moreover, there is a special function to compute a companion matrix, `company`, that can simplify code even further to just a single line

```
eig(compan(a))
```

Some examples of using the `roots` function will be provided in Sec. 10.9.

10.7 From Root Finding to 1-D Unconstrained Optimization

Root location and optimization are related in the sense that both involve guessing and searching for an argument, where a function $f(x)$ satisfies a certain condition. The only difference between the two types of problems is that root location involves searching for zeros [Eq. (10.1)], whereas optimization involves searching for an extremum (minimum or maximum) value [Eq. (10.2)].

It should be pointed out that if you have an algorithm that allows you to find a (local) minimum value of the function $f(x)$, you can always use it to find a maximum as well, because

$$
\max_x(f(x)) = \min_x(-f(x))
\tag{10.49}
$$

As a result, MATLAB features several functions that allow you to minimize a function. These functions can be used for zero finding as well [Eq. (10.3)]. Similarly, the root-finding functions may be applied to find an extremum, because the first-order derivative, $f'(x)$, turns zero there. Such a point, x_*, may be either a local maximum, if $f''(x_*) < 0$, or local minimum if $f''(x_*) > 0$, or *saddle point* if $f''(x_*) = 0$ (to be sure, you should check the order of the first nonzero derivative, $f^{(n+1)}(x_*)$, as well).

In what follows, just for educational purposes, we will consider three simple algorithms derived directly from those of root-finding techniques, presented in Secs. 10.2 and 10.3, which are used in the `fminbnd` and `fminunc` functions of MATLAB. The `fminunc` function is then addressed in more detail in Sec. 10.8, which also formally introduces two more functions, `fminsearch` and `fmincon`.

10.7.1 Golden-Section Method

The ternary search golden-section method presented in Sec. 10.2.3 can be easily modified to search for a minimum value of a 1-D *unimodal* function as follows (a function $f(x)$ between two bounds is called unimodal if for some value m (local minimum) it is monotonically decreasing for $x<m$ and monotonically increasing for $x>m$):

1. Given $f(x)$, determine the initial interval $[a;c]$ containing a single maximum.
2. Find two trial points as prescribed by Eq. (10.17).
3. If the size of a new interval $[a;b_1]$ is less or equal to the predefined tolerance as in Eq. (10.18), stop iterations assuming the solution at $x_* = b_1$.
4. If Eq. (10.18) does not hold, continue iterations by choosing the correct interval among the two possible ones ($[a;b_2]$ and $[b_1;c]$)
   - If inequality $f(b_1) < f(b)_2$ holds, then assign $c = b_2$, $b_2 = b_1$, and find $b_1 = c - R^{\otimes}(c-a)$
   - Otherwise assign $a = b_1$, $b_1 = b_2$, and find $b_2 = a + R^{\otimes}(c-a)$
5. Go to step 3 with a new interval.

Figure 10.24 features a few consecutive steps of this algorithm. This is a bracketing method, and therefore if the initial interval is chosen correctly (that is, if it contains a single minimum), the algorithm guarantees the convergence. This algorithm is implemented in the `fminbnd` function of MATLAB, which solves for $\min_x f(x)$ subject to $x_{min} \leq x \leq x_{max}$.

As you already know, this algorithm does not provide the fastest convergence and that is why `fminbnd` is supplemented with another derivative-free algorithm, which is considered in the following section.

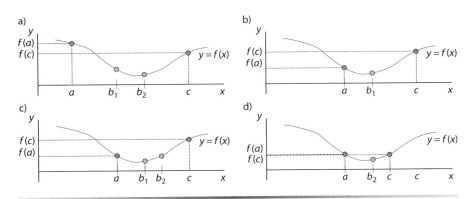

Fig. 10.24 Illustration of the golden-section algorithm to find a minimum of a function.

10.7.2 Quadratic Interpolation Method

Quadratic interpolation takes advantage of the fact that a second-order polynomial often provides a good approximation to the shape of the function $f(x)$ near an extremum as shown in Fig. 10.25.

As in the inverse quadratic interpolation method of Sec. 10.3.4, having three iterations, $(x_{i-2}, f(x_{i-2}))$, $(x_{i-1}, f(x_{i-1}))$, and $(x_i, f(x_i))$, we can first construct a quadratic function [Eq. (10.38)] with coefficients [Eq. (10.39)] that fits a parabola to these points. Then, we define the extremum value by differentiating this approximation and equating it to zero, which results in the next candidate iteration to be

$$x_{i+1} = x_i - \frac{b}{2a} \tag{10.50}$$

or

$$x_{i+1} = \frac{1}{2} \frac{f(x_{i-2})(x_{i-1}^2 - x_i^2) + f(x_{i-1})(x_i^2 - x_{i-2}^2) + f(x_i)(x_{i-2}^2 - x_{i-1}^2)}{f(x_{i-2})(x_{i-1} - x_i) + f(x_{i-1})(x_i - x_{i-2}) + f(x_i)(x_{i-2} - x_{i-1})} \tag{10.51}$$

If this candidate is reasonable (falls within $[x_{i-2}; x_i]$), it is accepted and a new quadratic approximation is made using a new triple, $(x_{i-1}, f(x_{i-1}))$, $(x_i, f(x_i))$, and $(x_{i+1}, f(x_{i+1}))$. The MATLAB script that computes Eq. (10.51) is given as

```
%% Defining symbolic variables
syms x a b c fa fb fc xim2 xim1 xi fim2 fim1 f_i
%% Putting together and solving a system of linear algebraic equations
A = [(a-c)^2 (a-c) 1;
     (b-c)^2 (b-c) 1;
     (c-c)^2 (c-c) 1];
```

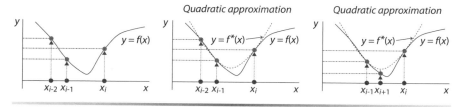

Fig. 10.25 Quadratic interpolation in the vicinity of function's extremum.

```
B = [fa; fb; fc];
coef=A\B;
%% Finding an argument corresponding to parabolas' maximum
m=c-coef(2)/coef(1)/2;
%% Making substitutions
m=subs(m,{a,b,c,fa,fb,fc},{xim2,xim1,xi,fim2,fim1,f_i});
%% Reorganizing the terms and displaying the result
m=simplify(m); m=collect(m,'fim2');
m=collect(m,'fim1'); m=collect(m,'f_i'); pretty(m)
```

It returns

```
         2      2                      2          2           2                2
((xim1 - xim2 ) f_i + fim2 xi  - fim1 xi  + fim1 xim2  - fim2 xim1 )/
((2 xim1 - 2 xim2) f_i + 2 fim2 xi - 2 fim1 xi + 2 fim1 xim2 - 2 fim2 xim1)
```

10.7.3 Newton's Method

Although the fminbnd function uses a combination of two derivative-free methods, the fminunc function relies on the Newton's method, which is adapted in the minimum-finding problems as

$$x_{i+1} = x_i - \frac{f'(x_i)}{f''(x_i)} \tag{10.52}$$

As seen, Eq. (10.52) uses the fact that at extremum, the first-order derivative is equal to zero. Being an open method, this algorithm shares all advantages and disadvantages of the original Newton's method [Eq. (10.32)] developed for root finding. If the first- and second-order derivatives are not readily available, a secant-like version of the Newton's method employing appropriate finite-difference approximations like in Eq. (10.36) can be used.

However, in general, methods to avoid time-consuming computation of the second-order derivative in the denominator of Eq. (10.52), which poses a serious issue, is briefly addressed in the following section.

10.8 Introduction to Multidimensional Optimization

Among four major MATLAB optimization functions, `fminbnd`, `fminsearch`, `fminunc`, and `fmincon` (the complete list of functions will be presented in Sec. 10.9.2), only the first one handles a function of a single variable, $f(x)$. Other functions deal with a more general case when the function depends on several variables $f(\mathbf{x}) = f(x_1, x_2, ..., x_n)$ (Fig. 10.26). Among the remaining three, `fminunc` and `fmincon`, one way or another use an analog of Eq. (10.52). In the case of multivariable function, this equation becomes

$$\mathbf{x}_{i+1} = \mathbf{x}_i - \frac{\nabla f(\mathbf{x}_i)}{|\mathbf{H}(\mathbf{x}_i)|} \tag{10.53}$$

where instead of the first- and second-order derivatives the *gradient* of f

$$\nabla f(\mathbf{x}) = \left[\frac{\partial f}{\partial x_1}, \frac{\partial f}{\partial x_2}, ..., \frac{\partial f}{\partial x_n} \right]^T = \left[f_{x_1}, f_{x_2}, ..., f_{x_n} \right]^T \tag{10.54}$$

and the *Hessian* or *Hessian* matrix

$$\mathbf{H}(\mathbf{x}) = \begin{bmatrix} \frac{\partial^2 f}{\partial x_1^2} & \frac{\partial^2 f}{\partial x_1 \partial x_2} & \cdots & \frac{\partial^2 f}{\partial x_1 \partial x_n} \\ \frac{\partial^2 f}{\partial x_2 \partial x_1} & \frac{\partial^2 f}{\partial x_2^2} & \cdots & \frac{\partial^2 f}{\partial x_2 \partial x_n} \\ \cdots & \cdots & \cdots & \cdots \\ \frac{\partial^2 f}{\partial x_n \partial x_1} & \frac{\partial^2 f}{\partial x_n \partial x_2} & \cdots & \frac{\partial^2 f}{\partial x_n^2} \end{bmatrix} = \begin{bmatrix} f_{x_1 x_1} & f_{x_1 x_2} & \cdots & f_{x_1 x_n} \\ f_{x_1 x_2} & f_{x_2 x_2} & \cdots & f_{x_2 x_n} \\ \cdots & \cdots & \cdots & \cdots \\ f_{x_1 x_n} & f_{x_2 x_n} & \cdots & f_{x_n x_n} \end{bmatrix} \tag{10.55}$$

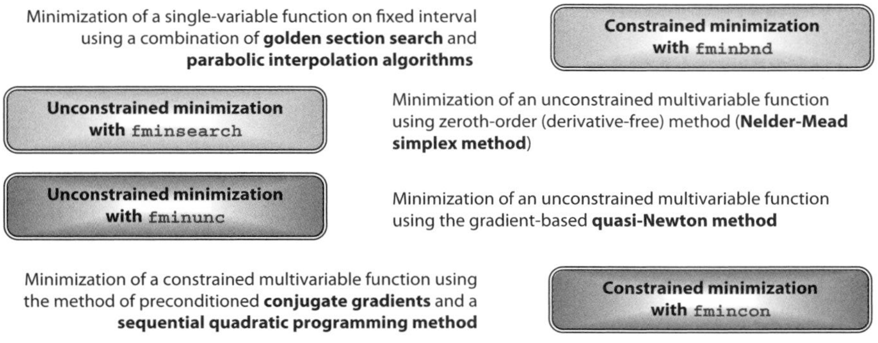

Minimization of a single-variable function on fixed interval using a combination of **golden section search** and **parabolic interpolation algorithms**

Constrained minimization with `fminbnd`

Unconstrained minimization with `fminsearch`

Minimization of an unconstrained multivariable function using zeroth-order (derivative-free) method (**Nelder-Mead simplex method**)

Unconstrained minimization with `fminunc`

Minimization of an unconstrained multivariable function using the gradient-based **quasi-Newton method**

Minimization of a constrained multivariable function using the method of preconditioned **conjugate gradients** and a **sequential quadratic programming method**

Constrained minimization with `fmincon`

Fig. 10.26 Major optimization functions of MATLAB.

are used. That is why these methods are called *gradient-based* methods (as opposed to *nongradient* methods that do not evaluate a gradient). As you can imagine, a numerical estimation of the *Hessian* matrix, especially for a large n causes a major problem.

One way to avoid computing a Hessian is to use the *gradient descent* method. To find a local minimum of $f(\mathbf{x})$, this method takes the steps that are proportional to the negative of the gradient (or approximate gradient) of the function at the current point

$$\mathbf{x}_{i+1} = \mathbf{x}_i - \alpha_i \nabla f(x_i) \tag{10.56}$$

The value of the step size $0 < \alpha_i << 1$ can be constant or variable (changing at every iteration). Geometrically, these steps are orthogonal to the multidimensional surface (in the 2-D case, they are orthogonal to the contour line going through that point).

Figure 10.27 presents a qualitative comparison of the gradient descent and Newton's methods for minimizing a function of the two variables with a small step size. The geometric interpretation of Newton's method in this case is that at each iteration i, Eq. (10.53) approximates $f(\mathbf{x})$ by a quadratic function around $\mathbf{x}_i$, and then takes a step toward the minimum of that quadratic function. As seen, because the Newton's method uses curvature information (residing in a Hessian), it results in a more direct route to the minimum; therefore, the capability to estimate a Hessian matrix is essential for the convergence performance.

In *quasi-Newton methods*, the Hessian matrix is approximated at each (or every other) step using the updates specified by gradient evaluations (or approximate gradient evaluations). Quasi-Newton methods are a generalization of the secant method to find the root of the first-order derivative for

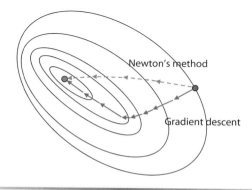

Fig. 10.27 Newton's method vs gradient descent.

multidimensional problems. In multidimensions, the secant equation does not specify a unique solution, and quasi-Newton methods differ in how they do that. The Broyden–Fletcher–Goldfarb–Shanno (BFGS) method, implemented in the MATLAB's fminunc function is one of the most popular members of this class. For completeness, its numerical procedure is provided as follows:

1. Given an initial guess $\mathbf{x}_0$ and approximate Hessian matrix $\hat{\mathbf{H}}_0$, which can be as simple as an identity matrix I meaning the gradient descent step [Eq. (10.56)], repeat the following steps until

$$\left\|\mathbf{x}_{i+1} - \mathbf{x}_i\right\| \le \varepsilon \tag{10.57}$$

2. Obtain a direction $\mathbf{p}_i$ by solving

$$\hat{\mathbf{H}}_i \mathbf{p}_i = -\nabla f(\mathbf{x}_i) \tag{10.58}$$

3. Perform a line search to find an acceptable step size α_i and compute the next iteration as

$$\mathbf{x}_{i+1} = \mathbf{x}_i + \alpha_i \mathbf{p}_i \tag{10.59}$$

4. Set $\mathbf{s}_i = \alpha_i \mathbf{p}_i$, evaluate $\mathbf{y}_i = \nabla f(\mathbf{x}_{i+1}) - \nabla f(\mathbf{x}_i)$ and update an estimate of the Hessian as

$$\hat{\mathbf{H}}_{i+1} = \hat{\mathbf{H}}_i + \frac{\mathbf{y}_i \mathbf{y}_i^T}{\mathbf{y}_i^T \mathbf{s}_i} - \frac{\hat{\mathbf{H}}_i \mathbf{s}_i (\hat{\mathbf{H}}_i \mathbf{s}_i)^T}{\mathbf{s}_i^T \hat{\mathbf{H}}_i \mathbf{s}_i} \tag{10.60}$$

The *line search* mentioned in Step 3, searches along the search direction $\mathbf{p}_i$ attempting to decrease the objective function $f(\mathbf{x})$. There may be different variants, but one of the basic ones is to start with the value of one and continually halve the step length (bisection) until a reduction is achieved. Then, a quadratic approximation (using the current and two other points along the direction of search) is used to determine the best step size.

The fminsearch function of MATLAB uses the nongradient Nelder–Mead simplex method, which is referred to be one of the most effective *direct search* methods. The term direct (or pattern) search was introduced by Hooke and Jeeves, the authors of another gradient-free method. The Hooke–Jeeves algorithm (realized in the fminhj function provided in Appendix B) simply samples the function around the current point, one argument at a time, by the steps of same magnitude, makes one or several steps in the direction of the largest decrease and then repeats the pattern search again (halving the step size when no further decrease is observed). The Nelder–Mead method bases its pattern on a *simplex*, which in *n*-dimensional space is characterized by the $n+1$ distinct vectors, simplex vertices. At each step of the search, a new point in or near the current simplex is generated. The function value at the new point is compared with the function's values at the vertices of the simplex and, usually, one of the vertices is replaced by the new point, giving a new simplex. This step is repeated until the diameter of the simplex is less than the specified tolerance. Both Hooke–Jeeves and Nelder–Mead methods are also referred to as the *random-walk*, *univariate* (changing one variable at a time), or *cyclic coordinate descent* methods.

Finally, the fmincon function deals with the constrained optimization. As opposed to addressing a simple problem

$$\mathbf{x}_* = \arg(\min_x f(\mathbf{x})) \qquad (10.61)$$

This function tries to find the solution to Eq. (10.61) subject to a variety of constraints. These constraints represent lower- and upper-bound constraints

$$\mathbf{x}_{lb} \le \mathbf{x} \le \mathbf{x}_{ub} \qquad (10.62)$$

linear inequality and equality constraints

$$\mathbf{A}(\mathbf{x}) \le \mathbf{b}, \ \mathbf{A}^{eq}(\mathbf{x}) = \mathbf{b}^{eq} \qquad (10.63)$$

nonlinear inequality and equality constraints

$$\mathbf{c}(\mathbf{x}) \le 0, \ \mathbf{c}^{eq}(\mathbf{x}) = 0 \qquad (10.64)$$

This is generally referred to as *constrained nonlinear optimization* or *nonlinear programming* problem. Depending on the scale of the problem, the fmincon function employs sequential quadratic programming, trust-region, or preconditioned conjugate gradient method. For the medium-scale algorithm, the function computes a quasi-Newton approximation to the Hessian of the *Lagrangian* that combines the objective function [Eq. (10.61)] together with constraints [Eqs. (10.62)–(10.64)] using Lagrange multipliers

$$L(\mathbf{x}, \boldsymbol{\mu}, \boldsymbol{\lambda}) = f(\mathbf{x}) + \boldsymbol{\mu}^T [\mathbf{x}_{lb} - \mathbf{x}; \ \mathbf{x} - \mathbf{x}_{ub}; \ \mathbf{A}(\mathbf{x}) - \mathbf{b}; \mathbf{c}(\mathbf{x})]$$
$$+ \boldsymbol{\lambda}^T [\mathbf{A}^{eq}(\mathbf{x}) - \mathbf{b}^{eq}; \mathbf{c}^{eq}(\mathbf{x})] \qquad (10.65)$$

An estimate of the Hessian is updated at each iteration using the BFGS formula [Eq. (10.60)].

Examples of using `fminbnd`, `fminsearch`, `fminunc`, and `fmincon` functions are provided in Sec. 10.9.2.

10.9 Using the MATLAB Functions

10.9.1 Root-Finding Functions

As mentioned previously, MATLAB features only two functions for the root finding. They are `fzero` applicable for any function of one variable and `roots` used to find the roots of polynomials. The `fzero` function using the zeroin algorithm discussed in Sec. 10.4 has the following general syntax:

```
x = fzero(fun,x0)
```

This function tries to find a zero of `fun` having x0 as an initial guess. If x0 is a scalar, the zero-finding routine proceeds with an open-method step in an attempt to find an initial bracketing interval. If x0 is a vector of length two, `fzero` continues with a bracketing iteration assuming x0 to be an interval bracketing a zero.

Alternatively, you may use one of the following syntax, enabling more options:

```
x = fzero(fun,x0,options)
[x,fval] = fzero(...)
[x,fval,exitflag] = fzero(...)
[x,fval,exitflag,output] = fzero(...)
```

In addition to the candidate solution, x, `fval` returns the value of objective function `fun` at the solution x (a safety check to assure it is zero), and `output` returns a structure that contains information about the iterative process

- Algorithm used
- Number of function evaluations
- Number of iterations taken to find a bracketing interval (in the case the search started from a single guess)
- Number of zero-finding iterations

The `exitflag` returns the exit condition of `fzero`, as MATLAB sees it (Table 10.1). The optional structure `options`, which can be modified using the `optimset` function, allows you to change some parameters of the algorithm to meet your preferences (Table 10.2).

Table 10.1   Exit Conditions for the `fzero` Function

exitflag Value	Description
1	Algorithm converged to a solution successfully based on termination tolerance `options.TolX`
-1	Algorithm was terminated by the output function (provided by a user)
-3	NaN or `Inf` function value was encountered during the search for an interval containing a sign change
-4	Complex function value was encountered during search for an interval containing a sign change
-5	There is a suspicion that `fzero` might have converged to a singular point
-6	Algorithm could not detect a change in sign of the function to establish the initial bracketing interval

The default values of these parameters can be found by issuing the following command:

```
>> optimset fzero
```

which returns a long list of parameters (because the `optimset` structure is also used by other functions of the Optimization Toolbox of MATLAB) with empty fields, but among those, three with the nonempty fields are

```
      Display: 'notify'
         TolX: 2.2204e-016
   FunValCheck: 'off'
```

To change any of these values, the field name followed by its new value should appear in the `optimset` function, for example,

```
>> options=optimset('Display','iter');
```

The most simple call may look like

```
>> fzero(@sin,3)
```

which returns the root of $f(x) = \sin(x)$ closest to $x_0 = 3$

```
ans = 3.1416
```

The `fzero` function also accepts inline function, for example,

```
>> x=fzero(inline('sin(5*x)+cos(x)'),1)
```

finds the root of $f(x) = \sin(5x) + \cos(x)$ closest to $x = 1$

```
x = 0.7854
```

A more sophisticated call

```
>> Opt=optimset('display','iter');
>> [X,fval,exitflag]=fzero('cos(x)+x',[-2 2],Opt)
```

allows you to obtain a more detailed information

```
Func-count        x                  f(x)              Procedure

2                    2                1.58385           initial

3                0.416147             1.3308            interpolation

4               -0.791927           -0.0894511         bisection

5               -0.715839            0.038704          interpolation

6               -0.738818            0.000446888       interpolation

7               -0.739085           -8.91982e-008      interpolation

8               -0.739085            5.26001e-012      interpolation

9               -0.739085                0             interpolation

Zero found in the interval [-2, 2]

X          =    -0.7391

fval       =     0

exitflag   =     1
```

As mentioned throughout this chapter, another way to check the validity of the obtained solution is to present it graphically. For example, the commands

```
>> ff='cos(x)+x';
>> X=fzero(ff,1,optimset('display','iter','TolX',10^-2))
>> ezplot(ff), hold, plot(X,0,'sm')
>> grid
```

return the results of finding a bracketing interval (where `fzero` spends most of the time) and several inverse quadratic interpolation iterations, along with the plot, shown in Fig. 10.28a.

Table 10.2 Modifiable Optimization Parameters for the `fzero` Function

Parameter	Description
Display	It changes the way the results are displayed: `'off'` displays no output; `'iter'` displays output at each iteration; `'final'` displays just the final output; and `'notify'` (default) displays output only if the function does not converge
FunValCheck	It checks whether the objective function values are valid: `'on'` displays an error when the objective function returns a value that is complex or NaN and `'off'` (default) displays no error
OutputFcn	It is an user-defined function that is called at each iteration and can interrupt the iteration procedure
PlotFcns	It is an user-defined plot function that is called at each iteration and may be used to visualize intermediate results
TolX	It terminates tolerance on x

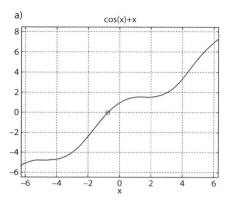

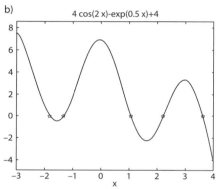

Fig. 10.28 a) Checking the results of root-finding graphically and b) finding multiple roots employing the `lroots` function.

```
Search for an interval around 1 containing a sign change:
Func-count      a              f(a)       b        f(b)         Procedure
    1                 1          1.5403     1        1.5403    initial interval

...
   21               0.36         1.2959    1.64      1.57085   search

...
   26            -0.810193     -0.120835   2.28      1.62877   search

Search for a zero in the interval [-0.810193, 2.28]:
Func-count      x              f(x)             Procedure
   26           -0.810193     -0.120835       initial
   27           -0.810193     -0.120835       interpolation
   28           -0.736766      0.00387855     interpolation
   29           -0.736766      0.00387855     interpolation
```

```
Zero found in the interval [-0.810193, 2.28]
X = -0.7368
```

Figure 10.28b presents the results of finding multiple roots using the `lroot` function introduced in Sec. 10.5.1. We use brackets found in Fig. 10.22b as the intervals supplied to the `fzero` function

```
f=inline('4*cos(2*x)-exp(0.5*x)+4');
b=lroot(f,-3,4);
n=length(b); x=zeros(n);
ezplot(f,[-3 4]), hold
for i=1:n
      x(i)=fzero(f,b(i,:));
      plot(x(i),0,'rp')
end
```

In the situation, when the function only touches the x-axis, but does not cross it, `fzero` returns no valid zeros and executes until `Inf`, `NaN`, or a complex value is detected

```
>> fzero('x^2',0.5)
```

```
Exiting fzero: aborting search for an interval containing a sign change
because NaN or Inf function value encountered during search.
(Function value at -1.7162e+154 is Inf.)
Check function or try again with a different starting value.
ans = NaN
```

You may request a more detailed report

```
>> [x,fval,exitflag,output]=fzero('x^2',0.5)
```

which yields

```
Exiting fzero: aborting search for an interval containing a sign change
because NaN or Inf function value encountered during search.
(Function value at -1.7162e+154 is Inf.)
Check function or try again with a different starting value.
x = NaN
fval = NaN
exitflag = -3
output = intervaliter: 1038
             iterations: 0
              funcCount: 2076
              algorithm: 'bisection, interpolation'
                message: [1x238 char]
```

As seen, after 1038 open-method iterations, this call fails to find an initial bracketing interval. Likewise, for the function $f(x) = (x-1)^4 (x-3)$, similar to that shown in Fig. 10.23a, although initiated very close to the quadruple root, the `fzero` function converges to the second, single root

```
>> fzero('(x-1)^4*(x-3)',1.000001)
ans = 3
```

The following script is intended to resolve this problem by introducing a new function as suggested by Eq. (10.43)

```
syms x
f=(x-1)^2*(x-3);
f_prime=diff(f);
u=f/f_prime;
x1=fzero(char(u),-2); x2=fzero(char(u),4);
ezplot(char(f)), hold
h=ezplot(char(u));
set(h,'Color',[0.8 0.1 0],'LineStyle','--')
plot(x1,0,'cs'), plot(x2,0,'mo')
legend('original function f(x)','new function u(x)',...
       'quadruple root','single root',2)
```

The results are shown in Fig. 10.29. This time, being initiated at $x_0 = -2$ and $x_0 = 4$, `fzero` successfully converges to $x_* = 1$ and $x_* = 3$, respectively.

The syntax for the `roots` function is

```
r=roots(p)
```

Here `p` is a vector containing $n+1$ coefficients of the nth-order polynomial in the following form

$$f_n(x) = a_1 x^n + a_2 x^{n-1} + \cdots + a_n x + a_{n+1} \tag{10.66}$$

For example, the command

```
>> r=roots([1 -10 35 -50 24])
```

which means that you are trying to find the roots of the fourth-order polynomial

$$f_4(x) = x^4 - 10x^3 + 35x^2 - 50x + 24 \tag{10.67}$$

returns

```
r = 4.0000
    3.0000
    2.0000
    1.0000
```

that is,

$$f_4(x) = x^4 - 10x^3 + 35x^2 - 50x + 24 = (x-1)(x-2)(x-3)(x-4) \tag{10.68}$$

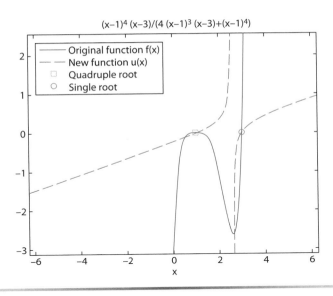

Fig. 10.29 Example of finding a quadruple root.

As mentioned, the `roots` function simply employs two other functions, `compan` and `eig`. These functions can be employed by themselves

```
>> A = compan([1 -10 35 -50 24])
>> r = eig(A)
```

to yield the companion matrix

```
A = 10    -35     50    -24
     1      0      0      0
     0      1      0      0
     0      0      1      0
```

and its eigenvalues

```
r =  4.0000
     3.0000
     2.0000
     1.0000
```

which are the roots of the polynomial.

Let us illustrate that the location of the roots can be very sensitive to perturbations in the coefficients of the polynomial. The following script produces the so-called Wilkinson's polynomial

$$f_{20}(x) = (x-1)(x-2)\ldots(x-20) \tag{10.69}$$

then slightly disturbs one of its coefficients and attempts to find the roots of this disturbed polynomial (Fig. 10.30a)

```
% Composing a 20th-order polynomial with roots x=[1:20]
a = poly(1:20);
or = roots(a);
% Plotting the original roots (marked by the red crosses)
plot(1:20,or,'xr')
xlabel('Root number'), ylabel('Root value'), hold
delta = 2^(-10);
a(3) = a(3)-delta; % disturbing one of coefficients
z=roots(a);         % finding the roots of disturbed polynomial
% Adding the 'disturbed' roots marked by blue circles
plot(1:20,z,'o')
legend('original roots','roots of disturbed polynomial')
```

Specifically, the coefficient $a_3 = 20{,}615$ is replaced with $a_3 = 20{,}614.999$ which may be seen as a negligible error. Yet, some of the roots of the disturbed polynomial happen to be greatly displaced. In fact, many of the roots become complex (Fig. 10.30a shows the real parts only).

The problem is that despite the fact that the original roots $r_i = i$, $i = 1,2,\ldots,20$ seem to be widely spaced (and therefore disturbing one of the coefficients should not do much of a damage), it is the ratio $r_i/(r_i - r_j)$ that defines the "closeness" of other roots in terms of stability. You may run the following script

```
r=20:-1:1;
d1=r(3)-r;   d1=d1(d1~=0); s1=r(3)^19/abs(prod(d1))
d2=r(20)-r;  d2=d2(d2~=0); s2=r(20)^19/abs(prod(d2))
```

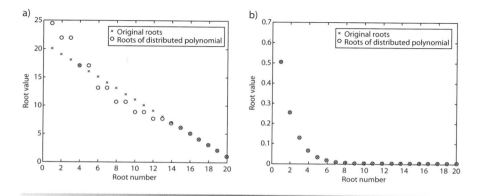

Fig. 10.30 a) Sensitive and b) nonsensitive Wikinson's polynomials.

to find out that for the third root ($r_3=18$), the product

$$\prod_{\substack{j=1 \\ j\neq 3}}^{20} \frac{|r_3|}{|r_3 - r_j|} \tag{10.70}$$

is of the order ~10^8, whereas for the last root ($r_{20}=1$), it is ~10^{-18}. No wonder the larger roots are so sensitive to even tiny disturbances—they have many close roots around in the sense that the distance $|r_i - r_j|$ between them is smaller than $|r_i|$.

Figure 10.30b shows the similar results for another Wilkinson's polynomial

$$f_{20}(x) = (x - 2^{-1})(x - 2^{-2}) \dots (x - 2^{-20}) \tag{10.71}$$

produced using the same script as earlier with only two lines changed as

```
a = poly(2.^-[1:20]);      % composing a new 20th-order polynomial
...
delta = a(3)*2^(-10)/20615; % same relative error as in the previous case
...
```

Computing the sensitivity coefficient [Eq. (10.70)] in this case

```
r=2.^-[20:-1:1];
d1=r(3)-r;   d1=d1 (d1~=0); s1=r(3)^19/abs(prod(d1))
d2=r(20)-r; d2=d2 (d2~=0); s2=r(20)^19/abs(prod(d2))
```

yields the orders of ~1 and 10^{-46}, respectively.

10.9.2 Basic Optimization Functions

Formally, all optimization functions of MATLAB belong to the Optimization Toolbox (the majority of them are still available in the basic MATLAB). This toolbox is composed of five distinctive groups

Minimization	For solving minimization problems
Equation Solving	For equation solving
Least Squares	For solving least squares (curve fitting) problems
Utilities	For getting and setting optimization options
Graphical User Interface (GUI)	For selecting a solver (and its options) and run problems in a user-friendly way

Specifically, the Minimization group consists of the following functions:

bintprog	Solves binary integer programming problems
fgoalattain	Solves multi-objective goal attainment problems

`fminbnd`	Finds minimum of single-variable function on fixed interval
`fmincon`	Finds minimum of constrained nonlinear multivariable function
`fminimax`	Solves minimax constraint problem
`fminsearch`	Finds minimum of unconstrained multivariable function using derivative-free method
`fminunc`	Finds minimum of unconstrained multivariable function
`fseminf`	Finds minimum of semi-infinitely constrained multivariable nonlinear function
`linprog`	Solves linear programming problems
`quadprog`	Solves quadratic programming problems
`ktrlink`	Provides interface to the (third party) nonlinear optimization package KNITRO

The Equation Solving group consists of

`\` and `/`	Solves linear equations of the form $\mathbf{Ax} = \mathbf{b}$
`fsolve`	Solves system of nonlinear equations
`fzero`	Finds root of continuous function of one variable

As seen, this group contains the `fzero` function dealt with in the previous section as well as `rdivide` and `ldivide` functions considered in Chapter 9. The Least Squares group consists of the following functions:

`\` and `/`	Solves linear equations of the form $\mathbf{Ax} = \mathbf{b}$ with the rank deficient or nonsquare matrix $\mathbf{A}$
`lsqcurvefit`	Solves nonlinear curve-fitting, (data-fitting) problems in least-squares sense
`lsqlin`	Solves constrained linear least-squares problems
`lsqnonlin`	Solves nonlinear least-squares (nonlinear data fitting) problems
`lsqnonneg`	Solves nonnegative least-squares constraint problem.

The Utilities group is formed by

`fzmult`	Performs multiplication with fundamental nullspace basis
`gangstr`	Zeros out "small" entries subject to structural rank
`optimget`	Getts optimization options values
`optimset`	Creates or edits optimization options structure

At last, the call

```
>> optimtool
```

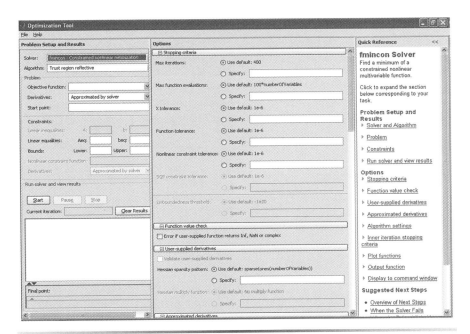

Fig. 10.31   Optimization Toolbox GUI.

brings up the Optimization Toolbox GUI (Fig. 10.31), allowing you to select solver, optimization options, and run problems using its friendly environment rather than programming everything in the Command window.

Let us present several examples of using four major minimization functions shown in Fig. 10.26. The algorithm implemented by the `fminbnd`
function is based on the golden-section search and parabolic interpolation.
The simplest call is

```
x = fminbnd(fun,x1,x2)
```

It returns a value `x` that is a local minimum of the function `fun` residing in
the interval `x1≤x≤x2`.

Similar to `fzero` function (see Sec. 10.9.1), you can use an additional
(optional) input `options` and up to four outputs, so that the most comprehensive call is

```
[x,fval,exitflag,output] = fzero(fun,x1,x2,options)
```

The `exitflag` values for `fminbnd` are given in Table 10.3, and the list
of modifiable `options` is pretty much the same as for `fzero` function,
and so Table 10.4 presents only two additional parameters beyond those
given in Table 10.2.
The default values of modifiable parameters can be obtained by calling

```
>> optimset fminbnd
```

Table 10.3 Exit Conditions for the `fminbnd` Function

exitflag Value	Description
1	Algorithm converged to a solution x based on `options.TolX`
0	Maximum number of function evaluations or iterations was reached
-1	Algorithm was terminated by the output function
-2	Bounds are inconsistent (x1 > x2)

which among others returns

```
        Display: 'notify'
    MaxFunEvals: 500
        MaxIter: 500
           TolX: 1.0000e-004
    FunValCheck: 'off'
```

As in the case of `fzero` function, `fval` returns the value of objective function `fun` calculated at the solution x, and `output` returns a structure with the following fields:

`output.algorithm`   Defines algorithm used
`output.funcCount`   Shows the number of function evaluations
`output.iterations`   Contains the number of iterations
`output.message`   Keeps an exit message

For instance, the call

```
>> s = fminbnd (@cos,3,4,optimset('TolX',1e-12,'Display','off'))
```

computes π being the minimum of $f(x) = \sin(x)$ within the interval $x \in [3;4]$ to about 12 decimal places, suppresses the output, and returns the function value at s.

The simplest call of the `fminsearch` function looks like that of `fminbnd`

```
x=fminsearch(fun,x0)
```

Iterations start at x0, and a value of a local minimum is returned in x. The most comprehensive call is

```
[x,fval,exitflag,output] = fminsearch(fun,x0,options)
```

Table 10.4 Additional (to those of Table 10.2) Modifiable Parameters for the `fminbnd` Function

Parameter	Description
MaxFunEvals	Maximum number of function evaluations allowed
MaxIter	Maximum number of iterations allowed

The default values of nonempty fields of the `options` structure can be obtained as usual via

```
>> optimset fminsearch
```

which returns

```
        Display: 'notify'
    MaxFunEvals: '200*numberofvariables'
        MaxIter: '200*numberofvariables'
         TolFun: 1.0000e-004
           TolX: 1.0000e-004
    FunValCheck: 'off'
```

As seen, the only new field (in addition to those of `fminbnd`) is `TolFun`, establishing termination tolerance on the function value. The `exitflag` values are the first three of Table 10.3 with the only difference for `flag=1`. In addition to satisfying the termination criteria `options.TolX` on **x**, $f(\mathbf{x})$ should satisfy the convergence criteria `options.TolFun` as well. The `output` has the same fields as that of the `fminbnd` function.

The simplest call for the `fminunc` function is similar to those of `fminbnd` and `fminsearch`

```
x=fminunc(fun,x0)
```

The most comprehensive call features two more output parameters along with the value of objective function at the solution `x (fval)`

```
[x,fval,exitflag,output,grad,hessian]=fminunc(fun,x0,options)
```

Because it is based on the quasi-Newton method, it also returns the values of the gradient (in `grad`) and Hessian (in `hessian`). The call

```
>> optimset fminunc
```

returns a lot of nonempty fields, but among those

```
        Display: 'final'
    MaxFunEvals: '100*numberofvariables'
        MaxIter: 400
         TolFun: 1.0000e-006
           TolX: 1.0000e-006
    FunValCheck: 'off'
        Hessian: 'off'
      HessUpdate: 'bfgs'
```

If the Hessian field is `'on'`, `fminunc` uses a user-defined Hessian, if it is `'off'`, `fminunc` approximates the Hessian using finite differences. As for the `HessUpdate` option, `fminunc` allows you to chose either Eq. (10.60) (`'bfgs'`) or Eq. (10.56) (`'steepdesc'`). The `exitflag` values are presented in Table 10.5.

Table 10.5 Exit Conditions for the `fminunc` Function

exitflag Value	Description
1	Magnitude of gradient became smaller than the specified tolerance `options.TolFun`
2	Change in x was smaller than the specified tolerance `options.TolX`
3	Change in the objective function value was less than the specified tolerance `options.TolFun`
0	Number of iterations exceeded `options.MaxIter` or number of function evaluations exceeded `options.MaxFunEvals`
-1	Algorithm was terminated by the output function
-2	Line search cannot find an acceptable point along the current search direction

Finally, the simplest call for the `fmincon` function is

```
x=fmincon(fun,x0,A,b)
```

The most comprehensive ones include up to 10 inputs and 7 outputs

```
x=fmincon(fun,x0,A,b,Aeq,beq,lb,ub,nonlcon,options)
[x,fval,exitflag,output,lambda,grad,hessian]=fmincon(...)
```

Additional inputs are to provide different type of constraints [Eqs. (10.62)–(10.64)] and one extra output, `lambda`, returns a structure whose fields contain the Lagrange multipliers at the solution x [vectors μ and λ in Eq. (10.65)]. The order of the input arguments is important, and so if some of the constraints are not present, an empty array, [ ], should be used. Among others, the call

```
>> optimset fmincon
```

returns

```
         Display: 'final'
     MaxFunEvals: '100*numberofvariables'
         MaxIter: 400
          TolFun: 1.0000e-006
            TolX: 1.0000e-006
     FunValCheck: 'off'
         Hessian: 'off'
          TolCon: 1.0000e-006
```

The last field defines the termination tolerance on the constraint violation. The `exitflag` values are given in Table 10.6.

Now, let us apply all the aforementioned functions to solve the same problem of minimizing the function of two variables

$$f(\mathbf{x}) = e^{x_1}(4x_1^2 + 2x_2^2 + 4x_1x_2 + 2x_2 + 1) \qquad (10.72)$$

Table 10.6 Exit Conditions for the `fmincon` Function

exitflag Value	Description
1	Gradient at the solution became less than `options.TolFun` and maximum constraint violation was less than `options.TolCon`
2	Change in `x` was smaller than the specified tolerance `options.TolX`
3	Change in the objective function value was less than the specified tolerance `options.TolFun`
4	Magnitude of the search direction was less than `2*options.TolX` and constraint violation was less than `options.TolCon`
5	Magnitude of directional derivative in search direction was less than `2*options.TolFun` and maximum constraint violation was less than `options.TolCon`
0	Number of iterations exceeded `options.MaxIter` or number of function evaluations exceeded `options.MaxFunEvals`
-1	Algorithm was terminated by the output function
-2	No feasible point was found

First, we program Eq. (10.72) as an anonymous function

```
>> fun = @(x) exp(x(1))*(4*x(1)^2+2*x(2)^2+4*x(1)*x(2)+2*x(2)+1);
```

and define the initial guess

```
>> x0 = [0.95;-0.55];
```

Also, let us degrade the default termination tolerance to achieve the results faster

```
>> options=optimset('TolFun',1e-3,'TolX',1e-3);
```

You may employ the `fminsearch` function as

```
>> [x,fval,exitflag,output]=fminsearch(fun,x0,options)
```

The output of this set of commands appears as follows:

```
x = 0.5003
   -1.0002
fval = 2.5164e-007
exitflag = 1
output = iterations: 32
         funcCount: 59
         algorithm: 'Nelder-Mead simplex direct search'
           message: [1x196 char]
```

Because we know the exact solution ($\mathbf{x}_* = [0.5; -1]$), we can additionally estimate the accuracy of the obtained solution

```
>> dx = norm([x(1)-0.5;x(2)+1])
```

which returns

```
dx = 3.3557e-004
```

It took 32 iterations (plus additional 27 function evaluations needed to create a simplex where appropriate) to get from the initial ("wild") guess to the solution. The required tolerance of `1e-3` was met with the errors for $\mathbf{x}_*$ and $f(\mathbf{x}_*)$ of the order of `1e-4` and `1e-7`, respectively). The results of iterations are shown in Fig. 10.32. Figure 10.32a features a 3-D plot, also showing the objective function's topology in the vicinity of the minimum value and in Fig. 10.32b, iterations overlie over a contour plot. It takes 18 iterations to arrive at the vicinity of the solution and then additional 14 iterations (which are happening in the close vicinity of the solution) to satisfy the tolerance on $\mathbf{x}_*$.

The advantages of using the `fminsearch` function is that because it uses a gradient-free algorithm it allows you to solve even nondifferentiable problems and can often handle discontinuities, particularly if it does not occur near the solution. For the same reason, the convergence of `fminsearch` is the slowest.

Next, let us minimize the same function [Eq. (10.72)] using the `fminunc` function. The command

```
>> [x,fval,exitflag,output]=fminunc(fun,x0,options)
```

returns

```
        x = 0.5000
            -0.9997
        fval = 1.9208e-007
        exitflag = 1
        output = iterations: 7
                    funcCount: 27
                    stepsize: 1
                firstorderopt: 0.0016
                    algorithm: 'medium-scale: Quasi-Newton line search'
                    message: [1x440 char]
```

The command

```
>> dx = norm([x(1)-0.5;x(2)+1])
```

returns

```
dx = 2.6151e-004
```

As opposed to the gradient-free algorithm it only took seven iterations (plus additional 20 function evaluations needed to estimate a direction and step size at each iteration) to get from the initial guess to the solution with an accuracy of 2.6e-4 for $\mathbf{x}_*$ and 1.9e-7 for $f(\mathbf{x}_*)$ (Fig. 10.33a). Obviously, fminunc performs much better than fminsearch.

However, although the fminunc function allows you to solve differentiable (continuous) problems like the one we just considered very fast, it is not applicable for solving nondifferentiable problems and cannot handle discontinuity as fminsearch does. Also, for solving problems that are the sums of squares, another gradient-based function, lsqnonlin, might be even a better choice.

For the sake of comparison, Fig. 10.33b also presents the results of using the fminhj function realizing another gradient-free method (mentioned

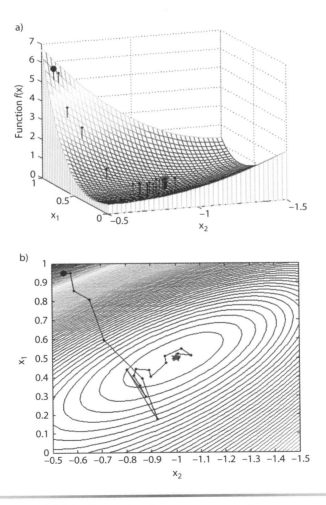

Fig. 10.32 Example of fminsearch performance.

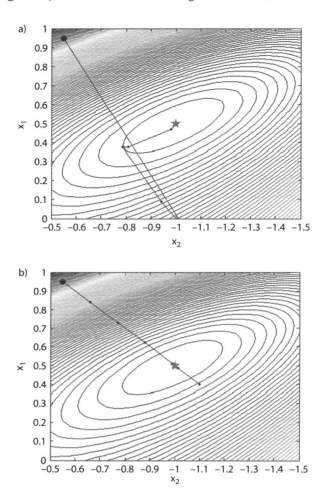

Fig. 10.33 Example of the performance of a) `fminunc` and b) Hooke–Jeeves algorithm.

in Sec. 10.8). To match the tolerances achieved for `fminsearch` and `fminunc` solutions, this function was initialized with

```
>> options=optimset('TolFun',1e-6,'TolX',1e-4);
```

and returned the error of $1.4e-4$ for $\mathbf{x}_*$ and $1.7e-7$ for $f(\mathbf{x}_*)$ (that is, better than those for `fminsearch` and `fninunc`). In general, the performance of `fminhj` is somewhat similar to that of `fminunc`, rather than `fminsearch`. In this particular case the Hooke–Jeeves algorithm (`fminhj`) converged even faster than the gradient-based `fminunc` (six against seven iterations) and employed about the same number of function evaluations (26 versus 27), as shown in Fig.10.34.

To be able to demonstrate the usage of the constrained optimization function `fmincon`, we need to slightly modify the problem. Let us impose the set of following constraints:

$$c_1(\mathbf{x}) = 0.2^2 - (x_1 - 0.6)^2 - (x_2 + 0.9)^2 \le 0$$
$$c_2(\mathbf{x}) = -x_1 - 0.1(x_2 + 0.4)^{-1} \le 0, \quad -1.15 \le x_2 \tag{10.73}$$

Graphically, it means that the solution cannot be in any of the shaded areas as shown on the 2-D projection in Fig. 10.35a. Now, we need to program these constraints. First, we create and save a `confun` function that represents nonlinear constraints as follows:

```
function [c,ceq] = confun(x)
% relative inequality constraints
c(1) = 0.2^2-(x(1)-0.6)^2-(x(2)+0.9)^2;
c(2) = -x(1)-0.1./(x(2)+0.4);
ceq = [ ]; % non-onlinear equality constraints
```

(the last line must appear in this function even though we have no nonlinear constraints). Then, we invoke minimization by using the following set of commands (note that an empty array, `[ ]`, appears in the input arguments list instead of nonexisting constraints everywhere):

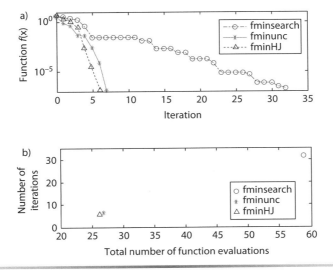

Fig. 10.34 Comparison of a) methods convergence and b) their overall effectiveness.

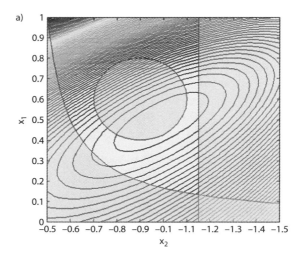

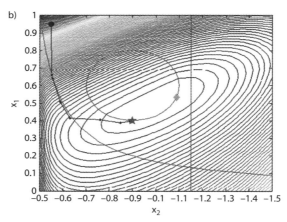

Fig. 10.35 a) Prohibited areas and b) example of `fmincon` performance.

```
>> options=optimset('TolFun',1e-4,'TolX',1e-4,'TolCon',1e-4);
>> x0=[0.95;-0.55];
>> [x,fval,exitflag,output]=fmincon(fun,x0,[],[],[],...
                              [],[0;-1.15],[],@confun,options)
```

The output appear as

```
    x =      0.4000
            -0.9007
    fval =   0.0298
    exitflag = 5
```

```
output = iterations: 9
              funcCount: 30
           lssteplength: 1
               stepsize: 9.3886e-004
              algorithm: 'med.-scale: SQP, Quasi-Newton, line-search'
          firstorderopt: 0.0022
                message: [1x172 char]
```

The fmincon solution follows one of the constraints and rapidly (in 9 iterations and a total of 30 function evaluations) reaches the closest minimum located on another constraint (Fig. 10.35b).

Similar to fminunc, the gradient-based fmincon function allows you to solve feasible nonlinear constrained minimization problems, where the objective and constraint functions are both continuous and have continuous first-order derivatives, fairly fast, the same pitfall though is that fmincon function is not applicable for solving nondifferentiable problems and cannot handle discontinuity. Also, as well as all other functions, fmincon can only find a local minimum. In fact, the solution shown in Fig. 10.35b is a local minimum, whereas the global minimum is located farther to the right (depicted with the rhomb).

Can we compare the performance of unconstrained optimization functions with that of the constrained optimization functions available in MATLAB? In other words, is whether it is possible to employ unconstrained minimization algorithms (fminsearch and fminunc) to solve constrained minimization problems? The answer is yes. You can do it by augmenting an old objective function with weighted (quadratic) penalties for constraints violation. Specifically, for the problem posed as Eqs. (10.61)–(10.64) such augmented function becomes

$$f^*(\mathbf{x}) = f(\mathbf{x}) + \mathbf{w}_{eq} \left\| \mathbf{p}_{eq} \right\|^2 + \mathbf{p}^T diag(\mathbf{w})\mathbf{p} \tag{10.74}$$

where

$$\mathbf{p}_{eq} = \begin{bmatrix} \mathbf{c}_{eq}(\mathbf{x}) \\ \mathbf{A}_{eq}\mathbf{x} - \mathbf{b}_{eq} \end{bmatrix} \text{ and } \mathbf{p} = \max \left(0, \begin{bmatrix} \mathbf{c}(\mathbf{x}) \\ \mathbf{A}\mathbf{x} - \mathbf{b} \\ \mathbf{x}_{lb} - \mathbf{x} \\ \mathbf{x} - \mathbf{x}_{ub} \end{bmatrix} \right) \tag{10.75}$$

(here **max** returns a vector). Having this augmented function, we convert the problem to that of unconstrained minimization, because achieving $\min_{\mathbf{x}} f^*(\mathbf{x})$ implies $\min_{\mathbf{x}} f(\mathbf{x})$ and satisfying all constraints.

As an example, let us develop a new augmented function for Eq. (10.72) with constraints [Eq. (10.73)]

$$f^*(\mathbf{x}) = f(\mathbf{x}) + \max([0, -1.15 - x_2])^2 + \max([0, c_2(\mathbf{x})])^2 + w\max([0, c_1(\mathbf{x})]) \quad (10.76)$$

For the sake of simplicity, we use a single weighting coefficient $w \geq 1$ to weight the most significant (active) inequality constraint. Augmenting the objective function simply means that we are reshaping the problem topology. Figure 10.36 shows an exaggerated landscape with $w=20$. In practice though, there is no need to have such a high weighing coefficient, because it disturbs the original topology too much. Having $w=1$ or $w=2$ may introduce enough information to account for the constraint.

An anonymous function describing the new, augmented objective function $f^*(\mathbf{x})$, can be programmed as

```
>> funA = @(x) exp(x(1))*(4*x(1)^2+2*x(2)^2+4*x(1)*x(2)+2*x(2)+1)...
         + max([0,-1.15-x(2)])^2+max([0,-x(1)-0.1./(x(2)+0.4)])^2 ...
         + w*max([0,0.2^2-(x(1)-0.6)^2-(x(2)+0.9)^2])
```

Now, we can define the initial guess and weighting coefficient

```
>> x0 = [0.95;-0.55]; w = 1;
```

and call either the `fminsearch` function

```
>> x = fminsearch(funA,x0)
```

or the `fminunc` function

```
>> x = fminunc(funA,x0)
```

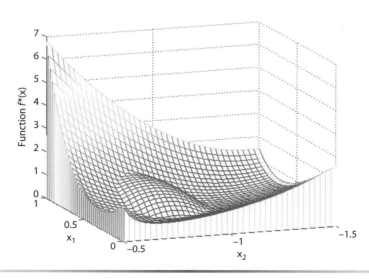

Fig. 10.36 Changing topology of the objective function.

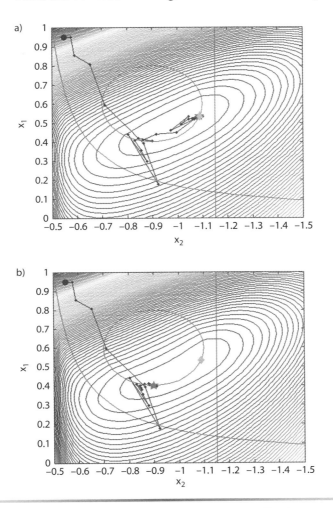

Fig. 10.37 Convergence of the `fminsearch` algorithm with a) `w=1` and b) `w=2`.

The results for `w=1` and `w=2` are presented in Figs. 10.37 (for `fminsearch`) and 10.38 (for `fminunc`). As seen, when using either function with `w=1` the algorithm converges to the correct solution, but when increasing the weighting coefficient to `w=2`, it gets stuck at the local minimum. In terms of performance, it takes 48 iterations (a total of 91 function evaluations) and 43 iterations (a total of 83 function calls) for the `fminsearch` function, and 15 iterations (a total of 183 function calls) and 7 iterations (a total of 111 function calls) for the `fminunc` function to arrive at the solution.

For the sake of comparison, the results of using the `fminhj` function are shown in Fig. 10.39. This algorithm converges to the right solution with `w=1` and $w=2$ (Fig. 10.39a), but gets stuck at the local minimum in the case of $w \geq 3$ (Fig. 10.39b). For the cases shown in Fig. 10.39 it takes 15 iterations (a total of 52 function calls) and 13 iterations (a total of 49 function evaluations), respectively. Figure 10.40 compares all four minimization algorithms discussed earlier.

Compared to fmincon, it takes much more function evaluations for fminsearch and fminunc to find the direction of the next iteration (the first one needs to create a simplex and the second one needs to estimate a gradient). However, fminunc performs more efficiently in terms of using less iterations. The performance of Hooke–Jeeves algorithm

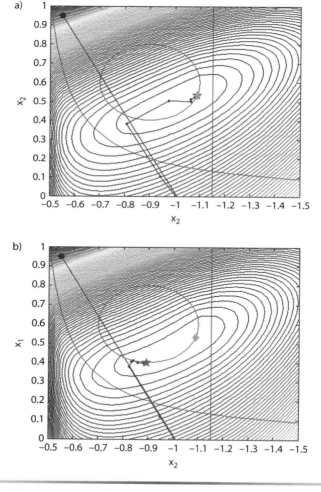

Fig. 10.38 Convergence of the fminunc algorithm with a) w=1 and b) w=2.

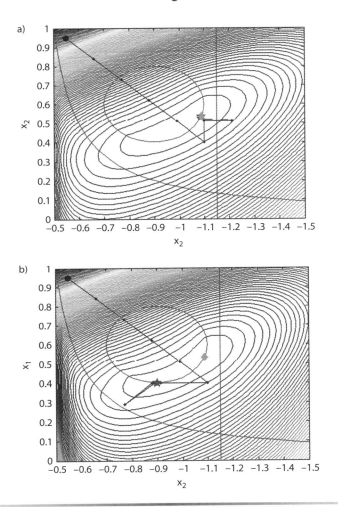

Fig. 10.39 Convergence of the Hooke–Jeeves algorithm with a) w=2 and b) w=3.

happens to be somewhere between that of `fmincon` and `fminunc`. Using the same initial conditions for all the four algorithms, `fmincon` can only find a local minimum, although by varying the weighting coefficient in the augmented objective function [Eq. (10.76)] used by the other three methods, we were able to find a global minimum (in Fig. 10.40, data corresponding to the global minimum solution are marked with a plus sign).

Finding a global minimum is always a difficult problem and that is why in addition to the Optimization Toolbox, Mathworks also offers Global

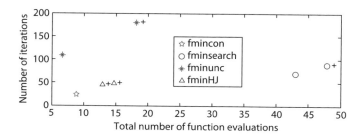

Fig. 10.40 Comparison of solving the constrained optimization problem using different algorithms (a plus sign designates the global as opposed to local solutions).

Optimization Toolbox dedicated to a search of the global minimum. To do that it enables multistart search and also features three additional minimum-finding functions realizing

- A genetic algorithm, heuristic search algorithm that mimics the process of natural evolution
- Another direct search algorithm
- Simulated annealing algorithm

Although the latter one only deals with unconstrained or bound-constrained problems, the first two are applicable for the problems with all types of constraints [Eqs. (10.62)–(10.64)].

Problems

10.1 Try to apply all the functions introduced in this chapter (`roots`, `fzero`, `fminbnd`, `fminsearch`, and `fminunc`) to find a value of the cubic root of 200. Use the `tic-toc` pair to record CPU time and present the results of your findings graphically. When appropriate (for minimization functions) determine how many function evaluations was required.

10.2 Determine the first three (closest to $x=0$) positive roots of the equation

$$4\cos(2x) - e^{0.5x} + 5 = 0$$

Use the `fplot` or `ezplot` function to find out where to look for these roots and then the `lroot` function of Sec. 10.5.1 to bracket the zeros.

10.3   Using the corresponding scripts of Secs. 10.2.2 and 10.3.3, develop two MATLAB functions

```
function x=falseposition(fun,a,b,eps)
```

and

```
function x=secant(fun,a,b,eps)
```

to find a solution of the equation $\cos(x)=2x^3$. Use the interval $[-4;2]$ to apply the first function and points -4 and -3.8 to use the second one. For the same accuracy of `eps=1e-5`, estimate the number of iterations needed to achieve the goal. When developing these two functions, use `fcnchk` and `feval` functions to enable accepting function handles, name strings, and inline objects as `fun`.

Chapter 11

Curve Fitting to Measured Data

- Understanding Least-Squares Regression
- Understanding Piecewise Approximation
- Understanding Fourier Approximation
- Using the MATLAB Functions

`polyfit, spline, pchip, interp1, interp2, interp3, interpn, fft`

Introduction

Fitting experimental data with analytical dependences is one of the most common problems. For instance, you are getting discrete experimental data from the wind tunnel or flight tests and you are required to estimate parameters at the points between these discrete values (interpolation). You may have some complex model you need to reduce to a simpler model and one way to do this is to compute its outputs at a number of discrete settings within the range of interest and then fit these values to some model of lesser complexity (model reduction). Another case is that you know the analytical dependence for your data already and all you have to do is to determine some varied parameters (system identification). You may want to use the data corrected so far to predict the behavior of your system in the future (extrapolation). In all aforementioned cases, you may have to deal with the linear or nonlinear functions of one or several variables. Another application is that you are dealing with systems that oscillate or vibrate, and you want to analyze the spectral content of the output to find natural frequencies and other harmonics (spectrum analysis). All these applications are known as curve fitting. There are two general approaches for curve fitting that are distinguished from each other on the basis of the amount of errors associated with the data. First, if data exhibits a significant degree of error or "noise," the strategy is to derive a single curve that represents the general trend of the data. Because any individual data point may be incorrect, we make no effort to intersect every point. One approach of this nature is known as a least-squares regression, and MATLAB features two functions allowing you to do this: `polyfit` for polynomial curve fitting (regression) and `fft` for discrete Fourier transform (DFT) or fitting with the trigonometric functions. Second, if the data are known to be very precise, the basic approach is to fit a curve or a series of curves that pass directly through each of the points. When an analytical function is established by one way or another, this function can be readily used to obtain estimates of the value without making additional (quite often very costly) experiments. The major functions of MATLAB to address this type of problem are `spline` and `pchip` producing piecewise cubic approximations. In addition to this, MATLAB offers a few more functions for piecewise interpolation of a function of one and more variables, `interp1`, `interp2`, `interp3`, `interpn`. All aforementioned functions are considered in this chapter forestalled by a detailed explanation of the least-squares regression, full-degree polynomial, and piecewise interpolation algorithms behind them. For the sake of completeness, this chapter briefly introduces a special case of parametrically defined splines, Bézier curves, as well.

11.2 Least-Squares Regression

Measured data are typically compared to some "theoretical" value derived from an empirical or physical relationship between two quantities. A function may then be assumed, which correlates to this theoretical relationship, and deviations from the function may be determined. *Regression analysis* is the method used to develop a function, which minimizes this deviation. In what follows, we will start from a simple linear regression, progressing to the higher-order regression and multiple linear regressions, followed by the analysis of regression errors and attempts to employ least-squares regression formulas to a nonlinear relationship. This will allow you to better understand what is programmed in the MATLAB function `polyfit`.

11.2.1 Linear Regression

Suppose, we have a set of data (x_i, y_i), $i=1,2, ..., N$ (denoted with the circles in Fig. 11.1), to which we postulate a linear relationship (dashed lines).

$$y = f(x) = a + bx \qquad (11.1)$$

Two unknown coefficients in Eq. (11.1) are a and $b = \tan(\theta)$, where θ is the slope of regression. We can express the deviation of this function from the actual data points as *residuals*

$$r_i = y_i - f(x_i), \quad i = 1, 2, ..., N \qquad (11.2)$$

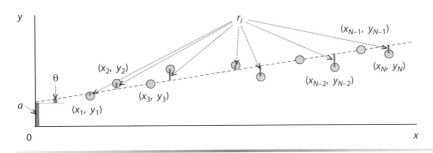

Fig. 11.1   Residuals for the linear regression.

representing distances between the data points and the suggested regression curve along the *y*-axis. The sum of squared residuals

$$S_r = \sum_{i=1}^{N} r_i^2 \tag{11.3}$$

for the linear function [Eq. (11.1)] is then given by

$$S_r(a,b) = \sum_{i=1}^{N} (y_i - a - bx_i)^2 \tag{11.4}$$

The goal is to minimize the value of S_r by varying coefficients *a* and *b*.

Minimizing S_r implies finding such values of *a* and *b*, that all partial derivatives become zero

$$\frac{\partial S_r}{\partial a} = -2\sum_{i=1}^{N} (y_i - a - bx_i) = 0$$

$$\frac{\partial S_r}{\partial b} = -2\sum_{i=1}^{N} x_i (y_i - a - bx_i) = 0 \tag{11.5}$$

Expanding Eq. (11.5) and dropping the "–2" yields

$$\sum_{i=1}^{N} y_i - \sum_{i=1}^{N} a - \sum_{i=1}^{N} bx_i = 0 \qquad Na + b\sum_{i=1}^{N} x_i = \sum_{i=1}^{N} y_i$$

$$\text{or}$$

$$\sum_{i=1}^{N} x_i y_i - \sum_{i=1}^{N} ax_i - \sum_{i=1}^{N} bx_i^2 = 0 \qquad a\sum_{i=1}^{N} x_i + b\sum_{i=1}^{N} x_i^2 = \sum_{i=1}^{N} x_i y_i \tag{11.6}$$

Rewriting Eq. (11.6) in the matrix form gives

$$
\begin{bmatrix} N & \sum_{i=1}^{N} x_i \\ \sum_{i=1}^{N} x_i & \sum_{i=1}^{N} x_i^2 \end{bmatrix} \begin{bmatrix} a \\ b \end{bmatrix} = \begin{bmatrix} \sum_{i=1}^{N} y_i \\ \sum_{i=1}^{N} x_i y_i \end{bmatrix}
\tag{11.7}
$$

Solving for a and b reduces to

$$
\begin{bmatrix} \hat{a} \\ \hat{b} \end{bmatrix} = \begin{bmatrix} N & \sum_{i=1}^{N} x_i \\ \sum_{i=1}^{N} x_i & \sum_{i=1}^{N} x_i^2 \end{bmatrix}^{-1} \begin{bmatrix} \sum_{i=1}^{N} y_i \\ \sum_{i=1}^{N} x_i y_i \end{bmatrix}
$$

$$
= \frac{1}{N\sum_{i=1}^{N} x_i^2 - \left(\sum_{i=1}^{N} x_i\right)^2} \begin{bmatrix} \sum_{i=1}^{N} x_i^2 & -\sum_{i=1}^{N} x_i \\ -\sum_{i=1}^{N} x_i & N \end{bmatrix} \begin{bmatrix} \sum_{i=1}^{N} y_i \\ \sum_{i=1}^{N} x_i y_i \end{bmatrix}
\tag{11.8}
$$

or in the scalar form

$$
\hat{a} = \frac{\sum_{i=1}^{N} x_i^2 \sum_{i=1}^{N} y_i - \sum_{i=1}^{N} x_i \sum_{i=1}^{N} x_i y_i}{N\sum_{i=1}^{N} x_i^2 - \left(\sum_{i=1}^{N} x_i\right)^2} \quad \text{and} \quad \hat{b} = \frac{N\sum_{i=1}^{N} x_i y_i - \sum_{i=1}^{N} x_i \sum_{i=1}^{N} y_i}{N\sum_{i=1}^{N} x_i^2 - \left(\sum_{i=1}^{N} x_i\right)^2}
\tag{11.9}
$$

Coefficients $\hat{a}$ and $\hat{b}$ assure the "best" fit for the given data points, where the "best" means the minimum sum of squared residuals [Eq. (11.4)], and that is why Eq. (11.9) is referred to as a *least-squares solution*.

11.2.2 Regression Analysis with Higher-Order Polynomials

If you do not know for sure that there is a linear relationship between the points in the dataset (x_i, y_i), $i = 1, 2, ..., N$, you may want to try fitting these data with a second- (Fig. 11.2) or even higher-order polynomial. In general, for a Pth-order polynomial $(P < N)$ we may write

$$
y = f(x) = a_0 + a_1 x + a_2 x^2 + \cdots + a_p x^P = \sum_{j=0}^{P} a_j x^j
\tag{11.10}
$$

where $a_j, j = 0, 1, ..., P$ are the constants needed to be determined.

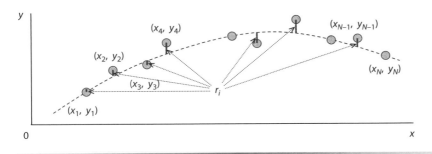

Fig. 11.2   Residuals for the second-order polynomial regression.

Following the least-squares regression approach of the previous section, we start from composing the sum of squared residuals [Eq. (11.3)] for Eq. (11.10)

$$S_r(a_0, a_1, \ldots, a_P) = \sum_{i=1}^{N} \left(y_i - \sum_{j=0}^{P} a_j x^j \right)^2 \tag{11.11}$$

Next, in order to minimize S_r, we need to assure that all partial derivatives of S_r with respect to undetermined coefficients a_j, $j = 0$, 1, ..., P are zero, that is, we have to resolve a system of linear (with respect to these coefficients) algebraic equations

$$\frac{\partial S_r}{\partial a_j} = -2 \sum_{i=1}^{N} x_i^j \left(y_i - \sum_{j=0}^{P} a_j x_i^j \right) = 0 \tag{11.12}$$

Being written in the matrix form this set of equations becomes

$$\begin{bmatrix} N & \sum_{i=1}^{N} x_i & \sum_{i=1}^{N} x_i^2 & \cdots & \sum_{i=1}^{N} x_i^P \\ \sum_{i=1}^{N} x_i & \sum_{i=1}^{N} x_i^2 & \sum_{i=1}^{N} x_i^3 & \cdots & \sum_{i=1}^{N} x_i^{P+1} \\ \vdots & \vdots & \vdots & \ddots & \vdots \\ \sum_{i=1}^{N} x_i^P & \sum_{i=1}^{N} x_i^{P+1} & \sum_{i=1}^{N} x_i^{P+2} & \cdots & \sum_{i=1}^{N} x_i^{2P} \end{bmatrix} \begin{bmatrix} a_0 \\ a_1 \\ \vdots \\ a_P \end{bmatrix} = \begin{bmatrix} \sum_{i=1}^{N} y_i \\ \sum_{i=1}^{N} x_i y_i \\ \vdots \\ \sum_{i=1}^{N} x_i^P y_i \end{bmatrix} \tag{11.13}$$

This system of linear algebraic equations (LAEs) can be solved using any method described in Sec. 9.2.

11.2.3 Multiple Linear Regression

The results of Sec. 11.2.1 can be easily extended to the case when we want to develop a multiple linear regression for the dataset that describes a function depending on not just one, but multiple independent variables. For a function of L independent variables, such linear regression takes the form

$$y = a_0 + \sum_{j=1}^{L} a_j x_j \tag{11.14}$$

(for $L = 2$ this equation represents a plane).

Employing the standard least-squares routine for N data points ($L < N$), we form the sum of squared residuals

$$S_r(a_0, a_1, \ldots, a_L) = \sum_{i=1}^{N} r_i^2 = \sum_{i=1}^{N} \left(y_i - a_0 - \sum_{j=1}^{L} a_j x_{ji} \right)^2 \tag{11.15}$$

followed by equating all its partial derivatives to zero

$$\frac{\partial S_r}{\partial a_0} = -2 \sum_{i=1}^{N} \left(y_i - a_0 - \sum_{j=1}^{L} a_j x_{ji} \right) = 0,$$

$$\frac{\partial S_r}{\partial a_j} = -2 \sum_{i=1}^{N} x_{ji} \left(y_i - a_0 - \sum_{j=1}^{L} a_j x_{ji} \right) = 0, \quad j = 1, 2, \ldots, L \tag{11.16}$$

This results in the following linear algebraic system:

$$\begin{bmatrix} N & \sum_{i=1}^{N} x_{1i} & \sum_{i=1}^{N} x_{2i} & \cdots & \sum_{i=1}^{N} x_{Li} \\ \sum_{i=1}^{N} x_{1i} & \sum_{i=1}^{N} x_{1i}^2 & \sum_{i=1}^{N} x_{1i} x_{2i} & \cdots & \sum_{i=1}^{N} x_{1i} x_{Li} \\ \vdots & \vdots & \vdots & \ddots & \vdots \\ \sum_{i=1}^{N} x_{Li} & \sum_{i=1}^{N} x_{Li} x_{1i} & \sum_{i=1}^{N} x_{Li} x_{2i} & \cdots & \sum_{i=1}^{N} x_{Li}^2 \end{bmatrix} \begin{bmatrix} a_0 \\ a_1 \\ \vdots \\ a_L \end{bmatrix} = \begin{bmatrix} \sum_{i=1}^{N} y_i \\ \sum_{i=1}^{N} x_{1i} y_i \\ \vdots \\ \sum_{i=1}^{N} x_{Li} y_i \end{bmatrix} \tag{11.17}$$

to solve for the vector of unknown coefficients $a_j, j = 0, 1, \ldots, L$.

11.2.4 Relation of Linear Regression Error to that of Statistics

By construction, the regressions obtained in Secs. 11.2.1 and 11.2.2 assure the smallest value of the sum of squared residuals S_r [computed according to Eqs. (11.4), (11.11), and (11.15)] among all other curves of the same class (same-order polynomials). If the distribution of individual residuals happens to be normal, and the spread of the points around the fit curve is of a similar magnitude along the entire range of data, then not only regressions considered earlier are the best in the sense of delivering a minimum of S_r, but also in the sense of the *maximum likelihood* principle. The latter means that the most likely value of the error variance is the variance of residuals. In such a case, a "standard deviation," or *standard error of the estimate* for the regression curve can be determined as

$$s_{y/x} = \sqrt{\frac{S_r}{N-(P+1)}} \qquad (11.18)$$

where N is the number of data points and P the regression polynomial's order (for multiple linear regression $s_{y/x} = \sqrt{S_r/(N-(L+1))}$). The subscript notation "y/x" designates that the error is estimated for a predicted value of y corresponding to a particular value of x.

Note that the denominator of Eq. (11.8) has the $N-(P+1)$ term as opposed to just N for the *standard deviation of the sample* or $N-1$ for the *sample standard deviation*. The reason is that $P+1$ data-derived estimates—polynomial coefficients—were used to compute S_r, decreasing the number of degrees of freedom by exactly this number, $P+1$. Another reasoning is that if the number of data points N is equal to $P+1$, then the Pth-order polynomial passes through all the points with no error (see Sec. 11.3). According to Eq. (11.8), this case corresponds to a distinguishable zero-by-zero division (resulting in NaN in MATLAB).

Just as is the case with the sample standard deviation, the standard error of the estimate [Eq. (11.18)] quantifies the spread of the data. The difference though is that $S_{y/x}$ quantifies the spread *around the regression* curve in contrast to the standard deviation S_y that quantifies the spread *around the mean* value

$$\bar{y} = \frac{1}{N}\sum_{i=1}^{N} y_i \qquad (11.19)$$

so that

$$s_y = \sqrt{\frac{S_t}{N-1}}, \quad S_t = \sum_{i=1}^{N}(y_i - \bar{y})^2 \qquad (11.20)$$

As a result, the aforementioned concepts can be used to quantify the "goodness" of our fit. This is particularly useful for comparison of several regressions. The difference between the two quantities, $S_t - S_r$, quantifies the error reduction due to characterization of data in terms of a regression curve rather than an average value. Because the magnitude of this quantity is scale-dependent, the difference is normalized to S_t to yield

$$R^2 = \frac{S_t - S_r}{S_t} \tag{11.21}$$

where R^2 is called the *coefficient of determination* and R the *correlation coefficient*. For a perfect fit, $S_r = 0$ and $R = R^2 = 1$, signifying that the regression curve explains 100% of data variability. If $S_r = S_t$, then $R = R^2 = 0$ meaning that the regression is not any better than a simple approximation of data with

$$y = f(x) = \bar{y} \tag{11.22}$$

where the mean value $\bar{y}$ is determined by Eq. (11.19).

11.2.5 Application of Linear Regression to Nonlinear Relationship

So far, we only dealt with linear regressions, that is, regressions where unknown parameters entered the approximating equation linearly, like in Eq. (11.14). Generally speaking, if we know that data should be represented by some nonlinear dependence

$$y = f(x, \boldsymbol{\theta}) \tag{11.23}$$

where $\boldsymbol{\theta}$ represents the vector of unknown parameters, then the least-squares regression should be cast as a minimization problem

$$\boldsymbol{\theta} = \arg\min_{\boldsymbol{\theta}} S_r(\boldsymbol{\theta}) = \arg\min_{\boldsymbol{\theta}} \sum_{i=1}^{N} \left[y_i - f(x_i, \boldsymbol{\theta}) \right]^2 \tag{11.24}$$

The reason we cannot use the same approach we used before directly is that taking partial derivatives in this case results in nonlinear equations. As known from Chapter 10, resolving Eq. (11.24) may not be so straightforward, especially in the case of highly nonlinear functions $f(x, \boldsymbol{\theta})$ and a large number of coefficients (the solution may diverge). Besides, there may be multiple minima of S_r and none of minimization methods considered in Chapter 10 guarantees convergence to the global minimum.

Luckily, some nonlinear regression problems can be moved to a linear domain by a suitable transformation of the model formulation, and therefore still be addressed using methods of linear algebra considered in the previous sections of this chapter. For example, consider a *power law* relationship

$$y(x) = ax^b \tag{11.25}$$

with unknown coefficients b and m. By taking the base-10 logarithm of both sides, this equation transfers to

$$\log(y) = \log(a) + b\log(x) \tag{11.26}$$

which produces a linear dependence of the form of Eq. (11.1) in the log–log axes with a slope of $\tan^{-1}(b)$ and a y-intercept of $\log(a)$. Hence, we reduced the original problem [Eq. (11.25)] to

$$y^* = a^* + bx^* \tag{11.27}$$

where $y^*=\log(y)$, $a^*=\log(a)$, and $x^*=\log(x)$. Once resolved, meaning that coefficients $\hat{a}^*$ and $\hat{b}$ of Eq. (11.27) are found via Eq. (11.9), the parameter a of Eq. (11.25) can be computed as

$$\hat{a} = 10^{\hat{a}^*} \tag{11.28}$$

Strictly speaking, by using the above approach, instead of Eq. (11.3), we minimized the following function:

$$S_r^*(a,b) = \sum_{i=1}^{N}\left[\log(y_i) - \log(y^*(x_i))\right]^2$$

$$= \sum_{i=1}^{N}\left(\log\frac{y_i^* + r_i}{y_i^*}\right)^2 = \sum_{i=1}^{N}\left(\log(1+\frac{r_i}{y_i^*})\right)^2 \sim \sum_{i=1}^{N}\left(\frac{r_i}{y_i^*}\right)^2 \tag{11.29}$$

This function represents the sum of weighted squared residuals, paying more attention to the data points with the smaller values of y_i^* [cf. Eq. (11.29) with Eq. (11.3)]. As will be shown in examples of Sec. 11.7, the results of minimization of Eq. (11.29) as opposed to Eq. (11.3) do not look too bad and quite often can be used as is. However if needed, you may use the estimates $\hat{a}$ and $\hat{b}$ obtained via linear least-squares regression in the modified domain [minimization of Eq. (11.29)] as a good initial guess for addressing the minimization problem of Eq. (11.24). This applies to all other examples as follows.

An *exponential law* relationship

$$y(x) = ac^{bx} \tag{11.30}$$

can be handled similarly to that of a power law. Specifically, if we apply the base-c logarithm to both sides of Eq. (11.30), we obtain

$$\log_c(y) = \log_c(a) + bx \tag{11.31}$$

or

$$y^* = a^* + bx \tag{11.32}$$

where $y^* = \log_c(y)$ and $a^* = \log_c(a)$. Upon solving it for $\hat{a}^*$ and $\hat{b}$, we get back to $\hat{a}$ using

$$\hat{a} = c^{\hat{a}^*} \tag{11.33}$$

The only problem with realization of Eqs. (11.31) and (11.32) in MATLAB is that it does not have an arbitrary-base logarithm function. However, using a well-known formula

$$\log_a(b) = \frac{\log_c(b)}{\log_c(a)} \tag{11.34}$$

we can always bring it down to one of the three available logarithm functions, base-2 (`log2`), base-10 (`log10`), or natural logarithm (`log`) (that is, $c = \{2, 10, e\}$).

Another example is a multiple linear regression of the general form

$$y = a_0 \prod_{j=1}^{L} x_j^{a_j} \tag{11.35}$$

Taking the logarithm of both sides yields

$$\log(y) = a_0 + \sum_{j=1}^{L} a_j \log(x_j) \tag{11.36}$$

which now has the form of Eq. (11.14) and can be handled appropriately with $y^* = \log(y)$ and $x_j^* = \log(x_j)$.

Next, consider the not so obvious example of a nonlinear equation (describing Michaelis–Menten model for enzyme kinetics)

$$y = \frac{ax}{b+x} \tag{11.37}$$

In this case, you can bring it to the linear form by simply inverting it, which yields

$$\frac{1}{y} = \frac{1}{a} + \frac{b}{a}\frac{1}{x} \tag{11.38}$$

You can further rewrite it in the form of Eq. (11.1)

$$y^* = a^* + b^* x^* \tag{11.39}$$

with $y^* = y^{-1}$, $a^* = a^{-1}$, $b^* = ba^{-1}$, and $x^* = x^{-1}$. Once solved for $\hat{a}^*$ and $\hat{b}^*$, we get back to $\hat{a}$ and $\hat{b}$ using

$$\hat{a} = \hat{a}^{*-1} \text{ and } \hat{b} = \hat{b}^* \hat{a}^{*-1} \tag{11.40}$$

In this latter case, the actual function being minimized through the linear least-squares regression in the modified domain is

$$S_r^*(a,b) = \sum_{i=1}^{N} \left(\frac{1}{y_i} - \frac{1}{y_i^*} \right)^2 = \sum_{i=1}^{N} \left(\frac{r_i}{(y_i^* + r_i)y_i^*} \right)^2 \approx \sum_{i=1}^{N} \left(\frac{r_i}{y_i^{*2}} \right)^2 \tag{11.41}$$

As in the case of Eq. (11.29), the regression produced in this case pays even more attention to the data points with the smaller values of y_i^*.

Finally, let us slightly change Eq. (11.37) to

$$y = \frac{a\sqrt[3]{x^2}}{b+x} \tag{11.42}$$

This equation resembles the Sutherland formula for dynamic viscosity of gases vs absolute temperature. Suppose you performed some measurements and want to find coefficients $\hat{a}$ and $\hat{b}$, so that your data best fits this formula. By inverting it and multiplying by $\sqrt[3]{x^2}$, you may bring it to the form

$$\frac{\sqrt[3]{x^2}}{y} = \frac{b}{a} + \frac{1}{a}x \qquad (11.43)$$

Using substitutions $y^* = \sqrt[3]{x^2}/y$, $a^* = ba^{-1}$ and $b^* = a^{-1}$, you can further reduce it to the linear form

$$y^* = a^* + b^* x \qquad (11.44)$$

Once solved for $\hat{a}^*$ and $\hat{b}^*$, the original coefficients $\hat{a}$ and $\hat{b}$ can be calculated as

$$\hat{a} = \hat{b}^{*-1} \quad \text{and} \quad \hat{b} = \hat{a}^* \hat{b}^{*-1} \qquad (11.45)$$

Compared to other cases considered in this section, the new variable y^* now contains the independent variable x, which results in the least-squares regression applied to Eq. (11.44) minimizing

$$S_r^*(a, b) = \sum_{i=1}^{N} \left(\frac{\sqrt[3]{x_i^2}}{y_i} - \frac{\sqrt[3]{x_i^2}}{y_i^*} \right)^2 = \sum_{i=1}^{N} \left(\sqrt[3]{x_i^2} \frac{r_i}{(y_i^* + r_i)y_i^*} \right)^2 \approx \sum_{i=1}^{N} \left(\frac{\sqrt[3]{x_i^2}}{y_i^{*2}} r_i \right)^2 \qquad (11.46)$$

with an even more complex scaling coefficient than those in Eqs. (11.29) and (11.41). As mentioned earlier, Sec. 11.7 will present a couple of nonlinear regressions addressed using the linear algebra and function minimization approach.

For completeness, let us finish this section with presenting the *Gauss–Newton iterative method*, which can be employed to find an exact solution to Eq. (11.24) with the good guess obtained as explained in the aforementioned examples.

Suppose, we have a good initial guess on the values of unknown parameters $\boldsymbol{\theta}_0$ and we want to find a shift vector $\Delta\boldsymbol{\theta}$, which would bring it to the exact solution of Eq. (11.24) $\boldsymbol{\theta}_*$

$$\boldsymbol{\theta}_* = \boldsymbol{\theta}_0 + \Delta\boldsymbol{\theta} \qquad (11.47)$$

We can use the Taylor series expansion, similar to that of Eq. (10.30) to linearize the function [Eq. (11.23)] about the approximate solution as

$$f(x, \theta) \approx f(x, \theta_0) + \nabla f_\theta(x, \theta_0)^T \Delta\theta \tag{11.48}$$

where $\nabla f_\theta(x, \theta_0)$ denotes the gradient of $f(x, \theta)$ [see Eq. (10.54)]. Hence, the sum of squared residuals can be represented as

$$S_r(\theta) = \sum_{i=1}^{N} r_i^2 = \sum_{i=1}^{N} [y_i - f(x_i, \theta)]^2$$
$$= \sum_{i=1}^{N} \left[y_i - f(x_i, \theta_0) - \nabla f_\theta(x_i, \theta_0)^T \Delta\theta \right]^2 \tag{11.49}$$

Introducing two notations

$$\Delta\mathbf{y} = \begin{bmatrix} y_1 - f(x_1, \theta_0) \\ y_2 - f(x_2, \theta_0) \\ \vdots \\ y_N - f(x_N, \theta_0) \end{bmatrix} \quad \text{and} \quad \mathbf{J}(\mathbf{x}, \theta_0) = \begin{bmatrix} \nabla f_\theta(x_1, \theta_0)^T \\ \nabla f_\theta(x_2, \theta_0)^T \\ \vdots \\ \nabla f_\theta(x_N, \theta_0)^T \end{bmatrix} \tag{11.50}$$

Eq. (11.49) can be rewritten in the matrix form as

$$S_r(\theta) = (\Delta\mathbf{y} - \mathbf{J}(\mathbf{x}, \theta_0)\Delta\theta)^T (\Delta\mathbf{y} - \mathbf{J}(\mathbf{x}, \theta_0)\Delta\theta) \tag{11.51}$$

The goal is to find the $\Delta\theta$ that minimizes $S_r(\theta)$. In fact, we have dealt with this type of problem already [Eq. (9.81)], so that we can write a solution right away [cf. with Eq. (9.85)]

$$\left(\mathbf{J}(\mathbf{x}, \theta_0)^T \mathbf{J}(\mathbf{x}, \theta_0) \right) \Delta\theta = \mathbf{J}(\mathbf{x}, \theta_0)^T \Delta\mathbf{y} \tag{11.52}$$

The formula can be applied iteratively to find an exact solution to Eq. (11.24) with the desired tolerance.

11.3 Full-Degree Polynomial Approximation

After a functional relation that describes the data is determined, you can use this function to predict other values for the data points that lie within the x-range of the original data. This process is called *interpolation*.

Also, you might need to use this functional relation for *extrapolation*, which is the process of estimating a value of $f(x)$ that lies outside the x-range of the known data points. Because regression analysis minimizes the sum of squared residuals (errors), you can be fairly confident that your prediction is reasonably accurate as long as your original assumptions of the functional relation are valid. However, you will frequently have an occasion to estimate intermediate values between precise data points, meaning that the approximating curve should pass through each of the data points precisely. Because of this requirement, the least-squares regression method cannot be applied (S_r is equal to zero by construction). The following addresses these types of curve fitting problems.

Having $N+1$ data points (x_i, y_i), $i = 1, 2, \ldots, N+1$ (no two x_i are the same), the most straightforward approach we can use is to apply the *full-degree polynomial interpolation* determined by the Nth-order polynomial

$$f_N(x) = a_0 + a_1 x + a_2 x^2 + \cdots + a_N x^N = \sum_{i=0}^{N} a_i x^i \qquad (11.53)$$

This form of the polynomial representation is referred to as the *power* or *monomial-basis form*, and coefficients of Eq. (11.53) assuring it passing through all data points can be found by solving the following system of LAEs:

$$\begin{bmatrix} 1 & x_1 & x_1^2 & \cdots & x_1^N \\ 1 & x_2 & x_2^2 & \cdots & x_2^N \\ \vdots & \vdots & \vdots & \ddots & \vdots \\ 1 & x_{N+1} & x_{N+1}^2 & \cdots & x_{N+1}^N \end{bmatrix} \begin{bmatrix} a_0 \\ a_1 \\ \vdots \\ a_N \end{bmatrix} = \begin{bmatrix} y_1 \\ y_2 \\ \vdots \\ y_{N+1} \end{bmatrix} \qquad (11.54)$$

The matrix on the left-hand side of Eq. (11.54) is called the *Vandermonde* matrix.

For large N, the Vandermonde matrix may be very badly conditioned, and so application of the power form [Eq. (11.53)] is usually limited to problems involving a few well-spaced and well-scaled data points. Fortunately, there are a variety of other mathematical formats rather than that of Eq. (11.53), in which polynomial can be expressed (constructed). Surely, because of the uniqueness of the polynomial that passes through all the data points, all these forms eventually produce exactly the same polynomial.

The most compact of all the forms is the form known as the *Lagrange interpolating polynomial*

$$f_N(x) = \sum_{i=1}^{N+1} L_i(x) y_i \qquad (11.55)$$

In this form, the *Lagrange basis polynomials*

$$L_i(x) = \prod_{\substack{j=1 \\ j\neq i}}^{N+1} \frac{x - x_j}{x_i - x_j} \tag{11.56}$$

are constructed such that at the data points x_i, $i = 1,2, \ldots, N+1$ assume the values of either 1 ($L_i(x_i)=1$) or 0 ($L_i(x_j)=0$, $j \neq i$). For example, the Lagrange form of the linear equation ($N=1$) is

$$f_1(x) = \frac{x - x_2}{x_1 - x_2} y_1 + \frac{x - x_1}{x_2 - x_1} y_2 \tag{11.57}$$

and the second-order polynomial ($N=2$) can be represented as

$$f_2(x) = \frac{x - x_2}{x_1 - x_2} \frac{x - x_3}{x_1 - x_3} y_1 + \frac{x - x_1}{x_2 - x_1} \frac{x - x_3}{x_2 - x_3} y_2 + \frac{x - x_1}{x_3 - x_1} \frac{x - x_2}{x_3 - x_2} y_3 \tag{11.58}$$

For the computational purposes, Eq. (11.55) can be modified as

$$f_N(x) = \prod_{j=1}^{N+1}(x - x_j) \sum_{i=1}^{N+1} \frac{w_i}{x - x_i} y_i \tag{11.59}$$

where the weights w_i can be computed *a priori* as

$$w_i = \prod_{\substack{j=1 \\ j\neq i}}^{N+1} \frac{1}{x_i - x_j} \tag{11.60}$$

This form is known as the first form of the *barycentric interpolation formula*.

Equations (11.57) and (11.58) can also expressed as

$$f_1(x) = y_1 + \frac{y_2 - y_1}{x_2 - x_1}(x - x_1) \tag{11.61}$$

and

$$f_2(x) = y_1 + \frac{y_2 - y_1}{x_2 - x_1}(x - x_1)$$
$$+ \left[\frac{(y_3 - y_2 / x_3 - x_2) - (y_2 - y_1 / x_2 - x_1)}{x_3 - x_1} \right](x - x_1)(x - x_2) \tag{11.62}$$

which constitutes another polynomial form referred to as the *Newton's interpolating polynomial.* Although the second terms in both formulas represent the slope of the line connecting the points x_1 and x_2, the last term in Eq. (11.62) introduces the curvature, proportional to the second-order derivative. In general,

$$f_N(x) = \sum_{i=1}^{N+1} a_i N_i(x) \tag{11.63}$$

with the *Newton's basis polynomials*

$$N_i(x) = \prod_{j=1}^{i-1} (x - x_j) \tag{11.64}$$

and coefficients a_i defined as fractional expressions like those in Eq. (11.62) representing the *finite divided difference* approximations of the corresponding derivatives (Chapter 12). Hence, the Newton's interpolating polynomial is the straightforward differences-based version of the Taylor's polynomial [cf. with Eq. (8.31) that uses instantaneous derivatives].

In practice, Eq. (11.63) can be solved for coefficients a_i using the following set of linear algebraic equations

$$\begin{bmatrix} 1 & 0 & 0 & \cdots & 0 \\ 1 & x_2 - x_1 & 0 & \cdots & 0 \\ \vdots & \vdots & \vdots & \ddots & \vdots \\ 1 & x_{N+1} - x_1 & (x_{N+1} - x_1)(x_{N+1} - x_2) & \cdots & \prod_{j=1}^{N} (x_{N+1} - x_j) \end{bmatrix} \begin{bmatrix} a_1 \\ a_2 \\ \vdots \\ a_{N+1} \end{bmatrix}$$

$$= \begin{bmatrix} y_1 \\ y_2 \\ \vdots \\ y_{N+1} \end{bmatrix} \tag{11.65}$$

Different forms of an interpolation polynomial feature different computational advantages and disadvantages. For example, compared with the Vandermonde matrix of Eq. (11.54), Eq. (11.65) employs a much simpler lower triangular matrix which can be solved faster. Another advantage is that adding more data points (to the right) does not involve recomputation

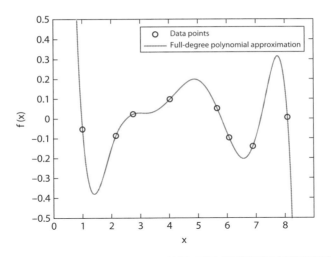

Fig. 11.3 Full-degree polynomial wiggle and extrapolation errors.

of the entire matrix. Some regard the Lagrange interpolating polynomial (specifically its barycentric form) as the general formula of choice. However, using a full-degree polynomial approximation shares the same major pitfalls among all approaches.

First, increasing the number of data points N leads to increasing an oscillatory behavior of approximation (Fig. 11.3). This phenomenon is called *polynomial wiggle*. This is due to the fact that in general the Nth-order polynomial has $N-1$ extremums. Such excessive variations in $f(x)$ between the data points (overshooting its values at the neighboring data points) are especially typical for the first and last subintervals. The situation aggravates when the data points are not equally spaced along the x-axis. As a result, a full-degree polynomial approximation can definitely not be used for extrapolation, when $f(x)$ can easily diverge from the prediction with very large errors.

Second, the higher-order polynomials tend to be ill-conditioned, that is, they tend to be highly sensitive to round-off error (which is also true for the higher-order polynomial regression). This results in coefficients being highly inaccurate, so that at some point a round-off error interferes with the ability to interpolate. Finally, in general, the change of the value of a single point (either x_i or y_i) or adding a few extra points to the dataset leads to a necessity of recomputing all coefficients.

As a result of these disadvantages, the full-degree polynomial interpolation is hardly ever used for data and curve fitting. Its primary application is in the derivation of other numerical methods (Secs. 12 and 13).

The MATLAB `polyfit` function exploiting a monomial form of the interpolating polynomial of the order `n` is concise and straightforward. It first forms the Vandermonde matrix **V**

```
V(:,1) = ones(length(x),1);
for j = 2:n+1
   V(:,j) = x.*V(:,j-1);
end
```

and then uses the backslash operator, \, to solve Eq. (11.54). If necessary, you may modify this function to use other forms of the full-degree polynomial interpolation, but the beauty of this approach is that it also allows you to solve the least-squares problems using the very same M-file function.

Suppose, you have $N > 2$ data points but want to use a linear interpolating polynomial. In this case, Eq. (11.7) becomes

$$\begin{bmatrix} 1 & x_1 \\ 1 & x_2 \\ \vdots & \vdots \\ 1 & x_N \end{bmatrix} \begin{bmatrix} a_0 \\ a_1 \end{bmatrix} = \begin{bmatrix} y_1 \\ y_2 \\ \vdots \\ y_N \end{bmatrix} \tag{11.66}$$

Because in this case, the Vandermonde matrix **V** is not a square matrix, precluding from solving Eq. (11.66) as is, the MATLAB backslash operator premultiplies both sides of this equation by the transpose of matrix **V**[(Eq. (9.85]), which yields

$$\begin{bmatrix} 1 & 1 & \cdots & 1 \\ x_1 & x_2 & \cdots & x_2 \end{bmatrix} \begin{bmatrix} 1 & x_1 \\ 1 & x_2 \\ \vdots & \vdots \\ 1 & x_N \end{bmatrix} \begin{bmatrix} a_0 \\ a_1 \end{bmatrix} = \begin{bmatrix} 1 & 1 & \cdots & 1 \\ x_1 & x_2 & \cdots & x_2 \end{bmatrix} \begin{bmatrix} y_1 \\ y_2 \\ \vdots \\ y_N \end{bmatrix} \tag{11.67}$$

Multiplying matrices on both sides results in Eq. (11.7)! Similarly, you can obtain Eq. (11.13) for the least-squares regression using a Pth-order polynomial $(P < N)$.

11.4 Piecewise Approximation

As seen from the preceding discussion, despite the fact that the full-degree polynomial captures all meanderings suggested by the data points, quite often such high-order functions lead to erroneous results because of round-off errors and overshoot between the data points. An alternative approach is to apply lower-order polynomials to subsets of data points, and then "seamlessly" join these approximations together. Such connecting polynomials are called the *spline functions*.

11.4.1 Linear and Quadratic Splines

To introduce some basic concepts and problems associated with the spline interpolation, first consider a piecewise linear interpolation, in which a linear function is assumed to exist between the each pair of adjacent data points as shown in Fig. 11.4.

The *linear* (first-order) *splines* for a group of $N+1$ ordered data points (x_i, y_i), $i = 1, 2, ..., N+1$ can be defined as a set of the linear functions

$$f(x_i) = y_i + \frac{y_{i+1} - y_i}{x_{i+1} - x_i}(x - x_i), \quad x_i \leq x < x_{i+1}, \quad i = 1, 2, ..., N \quad (11.68)$$

This method of interpolation is fairly accurate when the data points are fairly closely spaced, and any nonlinearities in the function between the data points can be approximated by a linear function. However, even then it has an obvious disadvantage—the first-order piecewise approximation is not smooth. At the point where two lines meet (referred to as a *node*, or *knot*, or *breakpoint*, or *break*), the slope changes abruptly, that is, the first-order derivative is discontinuous. Let us try to overcome this deficiency by using a higher-order polynomial spline.

For each of N intervals, the *quadratic* (second-order) *spline* uses a second-order polynomial)

$$f_i(x) = a_i(x - x_i)^2 + b_i(x - x_i) + y_i, \quad x_i \leq x < x_{i+1}, \quad i = 1, 2, ..., N \quad (11.69)$$

featuring two unknown coefficients (by construction an approximation for each interval passes through the left knot). Consequently for N intervals,

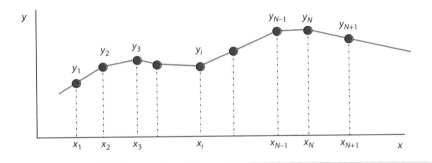

Fig. 11.4 Piecewise linear interpolation.

Table 11.1 Stitching Conditions for the Quadratic Spline

Condition No.	Equation	No. of Equations
1	$a_i h_i^2 + b_i h_i + y_i = y_{i+1}$ for $i = 1, 2, \ldots, N$	N
2	$2a_{i-1} h_{i-1} + b_{i-1} = b_i$ for $i = 2, 3, \ldots, N$	$N-1$

we have $2N$ unknown constants (a_i, b_i, $i = 1,2, \ldots, N$). To find their values, the $2N$ conditions should be imposed. They are as follows:

1. For each interval, the approximation must pass through the right knot, which will assure continuity of a function.
2. At $N-1$ interior nodes, the first-order derivative of two adjacent polynomials must be equal to each other, which assures continuity of the first-order derivative.

These conditions result in equations shown in Table 11.1 (using the notation $h_i = x_{i+1} - x_i$ for the length of the ith subinterval). This table also shows the budget of the number of equations.

As seen, the sum of the number of equations in Table 11.1 is equal to $2N-1$, meaning that we have a freedom of enforcing one more condition. For example, we may require that the second-order derivative at one of the ends is zero resulting in either $a_1 = 0$ or $a_{N+1} = 0$. A visual interpretation of such a condition is that the first (or the last) two points will be connected by a straight line.

Having $2N$ conditions allow you to find $2N$ coefficients a_i, b_i, $i = 1, 2, \ldots, N$ of Eq. (11.69) from the system of LAEs

$$
\begin{bmatrix}
1 & 0 & 0 & 0 & 0 & 0 & \cdots & 0 & 0 & 0 & 0 \\
h_1^2 & h_1 & 0 & 0 & 0 & 0 & \cdots & 0 & 0 & 0 & 0 \\
2h_1 & 1 & 0 & -1 & 0 & 0 & \cdots & 0 & 0 & 0 & 0 \\
0 & 0 & h_2^2 & h_2 & 0 & 0 & \cdots & 0 & 0 & 0 & 0 \\
0 & 0 & 2h_2 & 1 & 0 & -1 & \cdots & 0 & 0 & 0 & 0 \\
\vdots & \vdots & \vdots & \vdots & \vdots & \vdots & \ddots & \vdots & \vdots & \vdots & \vdots \\
0 & 0 & 0 & 0 & 0 & 0 & \cdots & h_{N-1}^2 & h_{N-1} & 0 & 0 \\
0 & 0 & 0 & 0 & 0 & 0 & \cdots & 2h_{N-1} & 1 & 0 & 1 \\
0 & 0 & 0 & 0 & 0 & 0 & \cdots & 0 & 0 & h_N^2 & h_N
\end{bmatrix}
\begin{bmatrix}
a_1 \\ b_1 \\ a_2 \\ b_2 \\ a_3 \\ b_3 \\ \vdots \\ \vdots \\ a_N \\ b_N
\end{bmatrix}
=
\begin{bmatrix}
0 \\ y_2 - y_1 \\ 0 \\ y_3 - y_2 \\ 0 \\ \vdots \\ y_N - y_{N-1} \\ 0 \\ y_{N+1} - y_N
\end{bmatrix}
\quad (11.70)
$$

($a_1 = 0$ was assumed in the first equation). Compared with matrices of Eqs. (11.54) and (11.65), the sparse matrix of Eq. (11.70) is much easier to handle. In fact, you may start from the very first equation solving it for a_1, then find b_1 from the second equation, a_2 from the third equation, b_2 from the fourth equation, an so on.

The two shortcomings of the quadratic splines fit are

- Asymmetry (we have only one extra condition to be imposed on either end)
- Discontinuity of the second-order derivative, which may lead to undesirable high swings between the points

On the positive side, analyzing the linear and quadratic splines leads to a conclusion that to ensure continuity of the kth-order derivative, a spline of at least $(k+1)$th-order must be used. Hence, we should proceed with a cubic spline.

11.4.2 Cubic Splines

The *cubic spline* relies on a third-order polynomial approximation between each pair of knots

$$f_i(x) = a_i(x-x_i)^3 + b_i(x-x_i)^2 + c_i(x-x_i) + y_i, \; x_i \le x < x_{i+1},$$
$$i = 1, 2, ..., N \tag{11.71}$$

For the same $N+1$ data points (x_i, y_i) $i = 1, 2, ..., N+1$, and N intervals, we now have $3N$ unknown constants to evaluate. Hence, in addition to those two types of conditions shown for the quadratic spline, we may now add one more

1. For each interval the approximation must pass through the right knot, which will assure continuity of a function.
2. At $N-1$ interior nodes, the first-order derivative of two adjacent polynomials must be equal to each other, which assures continuity of the first-order derivative.
3. At $N-1$ interior nodes, the second-order derivative of two adjacent polynomials must be equal to each other, which assures continuity of the second-order derivative.

Similar to Table 11.1, Table 11.2 shows the corresponding equations and counts their number in the rightmost column.

At this point, we are two conditions short. Therefore this time, we have a luxury of enforcing two additional conditions, one at each end. These additional conditions may vary as shown in Table 11.3.

Table 11.2 Stitching Conditions for the Cubic Spline

Condition No.	Equation	No. of Equations
1	$a_i h_i^3 + b_i h_i^2 + c_i h_i + y_i = y_{i+1}$ for $i = 1, 2, ..., N$	N
2	$3a_{i-1} h_{i-1}^2 + 2b_{i-1} h_{i-1} + c_{i+1} = c_i$ for $i = 2, 3, ..., N$	$N-1$
3	$6a_{i-1} h_{i-1} + 2b_{i-1} = 2b_i$ for $i = 2, 3, ..., N$	$N-1$

Table 11.3 Endpoint Conditions for a Cubic Spline

	Condition	Name of Spline
A	The second-order derivative at the endpoints is zero	*Natural* or *relaxed* spline (the function becomes a straight line at the end knots)
B	The second-order derivative at the endpoints is *extrapolated* from the adjacent interior knots	*Not-a-knot* or extrapolated spline
C	The first-order derivative at the endpoints is specified	*Clamped, fixed-slope,* or *complete* spline
D	The second-order derivative at the endpoints is specified	*Endpoint curvature-adjusted* spline
E	The third-order derivative at the endpoints is forced to be zero	*Parabolically terminated* spline (the first and the last intervals feature quadratic rather than cubic splines)

Before computers, different drafting tools were used by illustrators and designers to build their drawings by hand. Among them, draftsmen often used long, thin, flexible strips of wood, plastic, or metal called splines, and that is where the term spline came from. The splines were held in place with weights or nails as shown in Fig. 11.5. The elasticity of the spline material would cause the strip to take the smoothest possible shape that minimizes the potential energy required for bending it between the knots. Specifically, Fig. 11.5 shows the natural spline with no bending at the endpoints.

Among five types of splines shown in Table 11.3, the MATLAB `spline` function allows you to create two. They are a not-a-knot end condition spline (by construction requiring no additional inputs from the user beyond a set of data points) and fixed-slope spline (for which the end slopes should be defined explicitly).

Having the aforementioned three conditions described in Table 11.2 complimented with two more for a variety of splines presented in Table 11.3 allows you to compute $3N$ coefficients for approximation [Eq. (11.71)]. We could follow the routine laid out for the quadratic spline and develop a system like [Eq. (11.70)] here as well, but from the practical standpoint it is better to do it slightly differently, starting from explicit notations for the second-order derivative at each knot as m_i. For each interval of a cubic spline, the second-order derivative is linear. Hence, it can be represented in the Lagrange form [Eq. (11.57)] as

$$f_i''(x) = \frac{x - x_{i+1}}{x_i - x_{i+1}} m_i + \frac{x - x_i}{x_{i+1} - x_i} m_{i+1} = \frac{1}{h_i} \left[(x_{i+1} - x) m_i + (x - x_i) m_{i+1} \right],$$

$$i = 1, 2, \ldots, N \tag{11.72}$$

Integrating Eq. (11.72) twice yields

$$f_i(x) = \frac{(x_{i+1} - x)^3 m_i + (x - x_i)^3 m_{i+1}}{6h_i} + p_i(x_{i+1} - x) + q_i(x - x_i) \qquad (11.73)$$

where the last two terms represent two constants of integration (we choose to write them in this form to simplify further analysis). To solve for p_i and q_i, we should rewrite Eq. (11.73) for $x = x_i$ and $x = x_{i+1}$, which yields

$$y_i = \frac{h_i^2 m_i}{6} + p_i h_i \quad \text{and} \quad y_{i+1} = \frac{h_i^2 m_{i+1}}{6} + q_i h_i \qquad (11.74)$$

Hence,

$$p_i = \frac{y_i}{h_i} - \frac{h_i^2 m_i}{6h_i} \quad \text{and} \quad q_i = \frac{y_{i+1}}{h_i} - \frac{h_i^2 m_{i+1}}{6} \qquad (11.75)$$

Substituting these values back to Eq. (11.73) results in an alternative representation for a cubic spline which instead of $3N$ coefficients a_i, b_i, c_i, $i = 1, 2, ..., N$ features $N+1$ unknown second-order derivatives m_i

$$f_i(x) = \frac{(x_{i+1} - x)^3 m_i + (x - x_i)^3 m_{i+1}}{6h_i} + \left(\frac{y_i}{h_i} - \frac{h_i m_i}{6} \right)(x_{i+1} - x)$$

$$+ \left(\frac{y_{i+1}}{h_i} - \frac{h_i m_{i+1}}{6} \right)(x - x_i) \qquad (11.76)$$

Note that for Eqs. (11.72) and (11.74), we have used two of the spline defining conditions already (continuity of the second-order derivative and

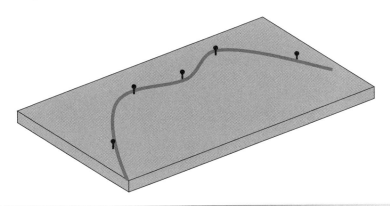

Fig. 11.5 Interpretation of the natural spline.

function itself, respectively). However, we still have $N-1$ conditions assuring continuity of the first-order derivative in the interior nodes plus two more from Table 11.3. These $N+1$ conditions will allow us to find unknown value of m_i at each knot. The only thing is that we will need to express these $N+1$ conditions, so that they become dependent on the second-order derivative.

Let us differentiate Eq. (11.76)

$$f_i'(x) = \frac{-(x_{i+1} - x)^2 m_i + (x - x_i)^2 m_{i+1}}{2h_i} - \left(\frac{y_i}{h_i} - \frac{h_i m_i}{6}\right) + \left(\frac{y_{i+1}}{h_i} - \frac{h_i m_{i+1}}{6}\right) \quad (11.77)$$

This formula defines the first-order derivative within the ith interval. Evaluating it at the left end, that is, at $x = x_i$, yields

$$f_i'(x_i) = \frac{-h_i m_i}{2} - \left(\frac{y_i}{h_i} - \frac{h_i m_i}{6}\right) + \left(\frac{y_{i+1}}{h_i} - \frac{h_i m_{i+1}}{6}\right) \quad (11.78)$$

By using the notation

$$\delta_i = \frac{y_{i+1} - y_i}{h_i} \quad (11.79)$$

for the first divided difference, we can further reduce Eq. (11.78) to

$$f_i'(x_i) = -\frac{h_i m_i}{3} - \frac{h_i m_{i+1}}{6} + \delta_i \quad (11.80)$$

We can write an equation like Eq. (11.77) for the previous interval (by simply replacing i with $i-1$, and $i+1$ with i) and estimate it at the right end, that is, at $x = x_i$ again, which results in

$$f_{i-1}'(x_i) = \frac{h_{i-1} m_i}{3} + \frac{h_{i-1} m_{i-1}}{6} + \delta_{i-1} \quad (11.81)$$

Equating Eqs. (11.80) and (11.81), we get $N-1$ relations involving $N+1$ unknowns

$$h_{i-1} m_{i-1} + 2(h_{i-1} + h_i) m_i + h_i m_{i+1} = 6(\delta_i - \delta_{i-1}), \quad i = 2, 3, \ldots, N \quad (11.82)$$

Table 11.4 specifies two additional relations for all cases listed in Table 11.3. The relations of Table 11.4 supply the expressions for m_1 and m_{N+1} allowing you to exclude them from the first and the last equations of system

Table 11.4 Endpoint Conditions for the Second-Order Derivative of Cubic Spline

	Type of Spline	Expression for the Second-Order Derivative
A	Natural spline	$m_1 = 0$, $m_{N+1} = 0$
B	Not-a-knot spline	$m_1 = m_2 - \dfrac{h_1}{h_2}(m_3 - m_2)$, $m_{N+1} = m_N + \dfrac{h_N}{h_{N-1}}(m_N - m_{N-1})$
C	Clamped spline	$m_1 = \dfrac{3}{h_1}(\delta_1 - y_1') - \dfrac{m_2}{2}$, $m_{N+1} = \dfrac{3}{h_N}(y_{N+1}' - \delta_N) - \dfrac{m_N}{2}$
D	Endpoint curvature-adjusted spline	$m_1 = y_1''$, $m_{N+1} = y_{N+1}''$
E	Parabolically terminated spline	$m_1 = m_2$, $m_{N+1} = m_N$

[Eq. (11.82)], respectively. Therefore, you can finally write a tridiagonal system of $N-1$ linear equations for determining remaining $N-1$ unknown second-order derivatives m_i as

$$
\begin{bmatrix}
\alpha_{1;1} & \alpha_{1;2} & 0 & \cdots & 0 & 0 & 0 \\
h_2 & 2(h_2 + h_3) & h_3 & \cdots & 0 & 0 & 0 \\
\vdots & \vdots & \vdots & \ddots & \vdots & \vdots & \vdots \\
0 & 0 & 0 & \cdots & h_{N-2} & 2(h_{N-2} + h_{N-1}) & h_{N-1} \\
0 & 0 & 0 & \cdots & 0 & \alpha_{N;N-1} & \alpha_{N;N}
\end{bmatrix}
\begin{bmatrix}
m_2 \\ m_3 \\ \vdots \\ m_{N-1} \\ m_N
\end{bmatrix}
$$

$$
=
\begin{bmatrix}
6(\delta_2 - \delta_1) - \Delta_2 \\
6(\delta_3 - \delta_2) \\
\vdots \\
6(\delta_{N-1} - \delta_{N-2}) \\
6(\delta_N - \delta_{N-1}) - \Delta_N
\end{bmatrix}
\tag{11.83}
$$

where six variables for the first and last equations are defined for each type of spline in Table 11.5. After all unknowns m_i, $i = 1, 2, \ldots, N+1$ are determined, coefficients of the original spline [Eq. (11.71)] can be related to them as

$$
a_i = \frac{m_{i+1} - m_i}{6h_i}, \quad b_i = \frac{m_i}{2}, \quad c_i = \delta_i - \frac{h_i(2m_i + m_{i+1})}{6}, \quad i = 1, 2, \ldots, N
\tag{11.84}
$$

Table 11.5 Coefficients of the First and Last Equations for the Trigonal System [Eq. (11.83)]

Number	$\alpha_{1;1}$	$\alpha_{1;2}$	$\alpha_{N;N-1}$	$\alpha_{N;N}$	Δ_2	Δ_N
A	$2(h_1 + h_2)$	h_2	h_{N-1}	$2(h_{N-1} + h_N)$	0	0
B	$3h_1 + 2h_2 + \dfrac{h_1^2}{h_2}$	$h_2 - \dfrac{h_1^2}{h_2}$	$h_{N-1} - \dfrac{h_N^2}{h_{N-1}}$	$2h_{N-1} + 3h_N + \dfrac{h_N^2}{h_{N-1}}$	0	0
C	$\dfrac{3}{2}h_1 + 2h_2$	h_2	h_{N-1}	$2h_{N-1} + \dfrac{3}{2}h_N$	$3(\delta_1 - y_1')$	$3(y_{N+1}' - \delta_N)$
D	$2(h_1 + h_2)$	h_2	h_{N-1}	$2(h_{N-1} + h_N)$	$h_1' y_1''$	$h_N y_{N+1}''$
E	$3h_1 + 2h_2$	h_2	h_{N-1}	$2(h_{N-1} + 3h_N)$	0	0

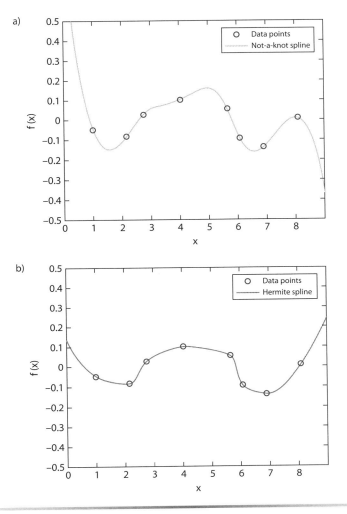

Fig. 11.6 a) Not-a-knot and b) Hermite splines.

An example of not-a-knot spline passing through the same data points as the full-degree polynomial approximation of Fig. 11.3 is shown in Fig. 11.6a. As seen, it exhibits much less overshooting compared to those of Fig. 11.3, but can still return the questionable interpolation values.

11.4.3 Cubic Splines Defined Locally

As shown in the previous section, cubic splines ensure continuity of the first- and second-order derivatives and therefore are frequently used in practice. The discontinuity of the third-order derivative poses no problem and in fact cannot be detected visually anyway. On the contrary, sometimes it is worth sacrificing continuity of even the second-order derivative to preserve a general shape of approximation.

Suppose, we know the data points (x_i, y_i), $i = 1, 2, ..., N+1$, but additionally we have to satisfy the first-order derivatives d_i at all $N+1$ nodes as well. In such a case, using notations of the previous section (the length of subinterval h_i and the first divided difference δ_i), the following interpolating expression for each of N intervals can be derived as

$$f_i(x) = \frac{3h_i s_i^2 - 2s_i^3}{h_i^3} y_{i+1} + \frac{h_i^3 - 3h_i s_i^2 + 2s_i^3}{h_i^3} y_i + \frac{s_i^2 (s_i - h_i)}{h_i^2} d_{i+1}$$

$$+ \frac{s_i (s_i - h_i)^2}{h_i^2} d_i, \quad x_i \le x \le x_{i+1} \tag{11.85}$$

Such interpolations, satisfying conditions on derivatives are known as *Hermite* or *osculatory* interpolants. ("Osculari" means "to kiss" in Latin, which as used in this chapter imply passing through the knots at a certain angle, that is, touching or "kissing" the derivative constraints at the knots.)

The Hermite interpolation [Eq. (11.85)] assumes that the slopes d_i are given, but if not, we need to define them somehow. The approach used in the MATLAB pchip function produces a *shape-preserving* "visually pleasing" interpolation by defining the first-order derivatives at the knots using a set of empirical rules. The key idea in determining the slopes d_i is that the function values in between the nodes do not overshoot those at the endpoints of the current interval as was the case in Fig. 11.3 for the full-degree polynomial approximation or Fig. 11.6a for a regular cubic spline. In doing this, the following intuitive rules apply for all the inner knots:

1. If δ_i and δ_{i-1} have opposite signs, or if either of them is zero (that is, $\delta_i \delta_{i-1} \le 0$), then x_i is considered to be a local minimum or maximum, so that you set

$$d_i = 0 \tag{11.86}$$

2. If δ_i and δ_{i-1} have the same signs, then d_i is computed as a weighted harmonic mean, with the weights determined by the lengths of the two adjacent intervals

$$\frac{w_{1i} + w_{2i}}{d_i} = \frac{w_{1i}}{\delta_{i-1}} + \frac{w_{2i}}{\delta_i} \tag{11.87}$$

where $w_{1i} = 2h_i + h_{i-1}$ and $w_{2i} = h_i + 2h_{i-1}$. For same-length intervals, this gives the harmonic mean of the two slopes

$$\frac{1}{d_i} = \frac{1}{2}\left(\frac{1}{\delta_{i-1}} + \frac{1}{\delta_i}\right) \tag{11.88}$$

(that is, the reciprocal slope of the Hermite interpolant is the average of the reciprocal slopes of the piecewise linear interpolant on either side).

The slopes d_1 and d_{N+1} at either end of the data interval are determined by a slightly different, one-sided (noncentered), three-point formula as

$$d_1 = \frac{(2h_1 + h_2)\delta_1 - h_1\delta_2}{h_1 + h_2} \quad \text{and} \quad d_{N+1} = \frac{(2h_N + h_{N-1})\delta_N - h_N\delta_{N-1}}{h_N + h_{N-1}} \quad (11.89)$$

respectively. If the sign of d_1 differs from the sign of d_2, then it is set to be 0 (the same correction applies for the pair d_{N+1} and d_N). The pchip function also limits the magnitude of the slope d_1 (d_{N+1}) to that of $3\delta_1$ ($3d_N$) in the case the knot 2 (N) was assumed to be a local extremum at the previous steps (while working on the slopes at the inner knots).

An example of Hermite spline is shown in Fig. 11.6b. It passes through the same data points as the full-degree polynomial approximation of Fig. 11.3 and not-a-knot spline approximation of Fig. 11.6a. Compared to those two, there is no overshooting; however, we had to pay with the smoothness of the approximation. Not only does this approach produce a shape-preserving approximation, but it also has a very nice property of being *locally determined*. This means that if one data point was changed or another, additional point introduced, you only have to recompute the coefficients for two adjacent cubic Hermite interpolation splines, not for all of them as would be the case with the regular splines, *determined globally* [via Eq. (11.70) or (11.83)]. Alternatively, this property can be paraphrased as the perturbance introduced at some point for the locally determined splines that do not propagate beyond two adjacent intervals.

Let us make some concluding remarks. There is always a tradeoff between smoothness and a somewhat subjective property of shape preservation or *local monotonicity*. The piecewise linear interpolant [Eq. (11.68)] is at one extreme. It has hardly any smoothness. It is continuous, but there are jumps in its first derivative. On the other hand, it preserves the local monotonicity of data. It never overshoots the data and is increasing, decreasing, or remaining constant within each interval. The full-degree polynomial interpolant is at the other extreme. It is infinitely differentiable. However, quite often it fails to preserve a shape, particularly near the ends of the data interval.

The cubic Hermite shape-preserving and regular spline interpolants are in between these two extremes. The regular spline (the spline function) is smoother than the shape-preserving spline (pchip). The regular spline has two continuous derivatives, whereas a shape-preserving spline has only one. A discontinuous second-order derivative implies discontinuous curvature. The human eye can detect large jumps in curvature in graphs and/or in mechanical parts made by numerically controlled machine tools. However, the cubic Hermite interpolant is guaranteed to preserve a shape, but the regular spline might not.

11.4.4 Interpolation with the Bézier Curves

Just to broaden your horizons, it is worth mentioning that the monomials considered are not the only basis for approximating data. Other basis, for instance Jacobi, Chebyshev, Laguerre, and other polynomials or trigonometric functions, might also be used. To conclude discussion on different types of approximation, let us introduce an example of one more, parametrically defined, polynomial when addition flexibility is required.

The parametrically defined Bézier curves, developed in the late 50s by Paul de Casteljau at Citroen and publicized in the early 60s by Pierre Bézier at Renault (who used them to design automobile bodies) belong to a class of approximating splines. This type of spline forms the basis of the entire Adobe drawing software family and continues to be the primary method of representing curves and surfaces in computer graphics, computer-aided design and computer-aided manufacturing.

Although the data points we dealt with up to this point are treated differently (being referred to as the *control points*), the general idea still preserves. Given the set of points $(x_i, y_i)\, i = 0, 1, ..., N$, the 2-D Bézier curve $y(x)$ is defined parametrically using a dimensionless parameter $\tau \in [0;1]$ as

$$x(\tau) = \sum_{i=0}^{N} x_i B_{i;N}(\tau) \quad \text{and} \quad y(\tau) = \sum_{i=0}^{N} y_i B_{i;N}(\tau) \tag{11.90}$$

with the basis functions $B_{i;N}(\tau)$ being the *Bernstein polynomials*

$$B_{i;N}(\tau) = \frac{N!}{i!(N-i)!}\tau^i (1-\tau)^{N-i} \tag{11.91}$$

These basis functions have certain useful properties to make things work. To illustrate some of these properties, consider a cubic Bézier curve defined by four points. In this case, $N = 3$ and Eq. (11.90) becomes

$$x(\tau) = x_0 B_{0;3}(\tau) + x_1 B_{1;3}(\tau) + x_2 B_{2;3}(\tau) + x_3 B_{3;3}(\tau)$$
$$y(\tau) = y_0 B_{0;3}(\tau) + y_1 B_{1;3}(\tau) + y_2 B_{2;3}(\tau) + y_3 B_{3;3}(\tau) \tag{11.92}$$

According to Eq. (11.91), the Bernstein polynomials of the third-order are

$$B_{0;3}(\tau) = (1-\tau)^3, \ B_{1;3}(\tau) = 3\tau(1-\tau)^2, \ B_{2;3}(\tau) = 3\tau^2(1-\tau), \ B_{3;3}(\tau) = \tau^3 \tag{11.93}$$

Substituting them into Eq. (11.92) yields

$$x(\tau) = x_0(1-\tau)^3 + x_1 3\tau(1-\tau)^2 + x_2 3\tau^2(1-\tau) + x_3\tau^3$$
$$y(\tau) = y_0(1-\tau)^3 + y_1 3\tau(1-\tau)^2 + y_2 3\tau^2(1-\tau) + y_3\tau^3 \tag{11.94}$$

Simplifying Eq. (11.94) results in

$$x(\tau) = (x_3 - 3x_2 - 3x_1 - x_0)\tau^3 + (3x_2 - 6x_1)\tau^2 + 3x_1\tau + x_0$$
$$y(\tau) = (y_3 - 3y_2 - 3y_1 - y_0)\tau^3 + (3y_2 - 6y_1)\tau^2 + 3y_1\tau + y_0 \qquad (11.95)$$

By construction, the Bézier curve always starts at the first data point and ends at the last one. Other points are used to provide guidance on interpolant curvature, so that the curve never passes through them. That is why all together they are called the control points. An example of a four-point Bézier curve $y(x)$ is shown in Fig. 11.7a (with dashed lines). This

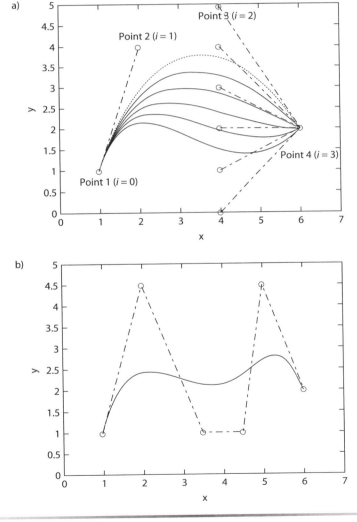

Fig. 11.7 Bézier curve with a) four and b) six control points.

figure also shows how the curve changes its shape when the third point moves up and down. Similarly, you can change the slope at the first point by moving around the second one. Figure 11.7b presents another example with 6 points (4 of which are used to control the curvature of approximation passing through the first and last points).

In the case of Fig. 11.7b, the Bézier curve is defined globally using all the data points as the control points and therefore the approximation does not pass through them. Alternatively, for the curve fitting problem, we could use a sequence of the low-order Bézier curves like the ones shown in Fig. 11.7a sharing common endpoints (*composite Bézier curves*), which are actually the data points. For each interval, we would add one or two control points to vary the shape in between the knots to assure smoothness of the overall approximation. In fact, the idea is quite similar to that used to go from the higher-order full-degree polynomial approximation down to piecewise lower-order (cubic) splines.

11.5 Fourier Approximation

Up to this point, all approximations used either monomials, $1, x, x^2, \ldots,$ x^p, (Secs. 11.2 and 11.3) or Bernstein polynomials (Sec. 11.4) as the basis functions. Let us consider another important class of the basis functions, trigonometric functions: $1, \cos(x), \cos(2x), \ldots, \cos(Px), \sin(x), \sin(2x), \ldots,$ $\sin(Px)$. These functions play a fundamental role in signal processing and modeling of oscillating or vibrating systems. The *Fourier analysis* represents a systematic framework for using trigonometric series for this purpose. In this section, we will use a *Fourier series* to decompose any periodic function or signal into a sum of finite set of simple oscillating functions, sines and cosines (or complex exponentials), casting the problem as the least-squares regression.

Consider the well-known sinusoid (the term *sinusoid* represents any waveform that can be described as a sine or cosine), which can, in general, be expressed as

$$f(t) = a_0 + c_1 \cos(\omega_0 t + \theta) \tag{11.96}$$

Four parameters, the *mean value* a_0, the *amplitude* c_1, the *angular frequency* ω_0 and the *phase shift* θ, serve to characterize the sinusoid. Another way of representing the very same sinusoid [Eq. (11.96)] is

$$f(t) = a_0 + a_1 \cos(\omega_0 t) + b_1 \sin(\omega_0 t) \tag{11.97}$$

This representation is more convenient for our purpose because it is cast in the format of a general linear (with respect to a_0, a_1, and b_1) model with two new parameters, a_1 and b_1, which can be related to those of Eq. (11.96) as

$$a_1 = c_1 \cos\theta \quad \text{and} \quad b_1 = -c_1 \sin\theta \tag{11.98}$$

Having a set of data points, (t_k, y_k), $k = 0, 2, \ldots, N-1$, we can proceed with the standard routine established in Sec. 11.2.1, that is, define the sum of squared residuals for trigonometric approximation [Eq. (11.97)]

$$S_r = \sum_{k=0}^{N-1} r_k^2 = \sum_{k=0}^{N-1} \left[y_k - a_0 - a_1 \cos(\omega_0 t_k) - b_1 \sin(\omega_0 t_k) \right]^2 \tag{11.99}$$

differentiate it with respect to three unknown coefficients, and equate these partial derivatives to zero, which results in the system of linear algebraic equations

$$
\begin{bmatrix}
N & \sum_{k=0}^{N-1} \cos(\omega_0 t_k) & \sum_{k=0}^{N-1} \sin(\omega_0 t_k) \\
\sum_{k=0}^{N-1} \cos(\omega_0 t_k) & \sum_{k=0}^{N-1} \cos^2(\omega_0 t_k) & \sum_{k=0}^{N-1} \cos(\omega_0 t_k)\sin(\omega_0 t_k) \\
\sum_{k=0}^{N-1} \sin(\omega_0 t_k) & \sum_{k=0}^{N-1} \cos(\omega_0 t_k)\sin(\omega_0 t_k) & \sum_{k=0}^{N-1} \sin^2(\omega_0 t_k)
\end{bmatrix}
\begin{bmatrix}
a_0 \\
a_1 \\
b_1
\end{bmatrix}
$$

$$
=
\begin{bmatrix}
\sum_{k=0}^{N-1} y_k \\
\sum_{k=0}^{N-1} y_k \cos(\omega_0 t_k) \\
\sum_{k=0}^{N-1} y_k \sin(\omega_0 t_k)
\end{bmatrix}
\tag{11.100}
$$

However, rather than doing this, let us examine a special case, when N observations are spaced evenly over the range $t \in [t_0; t_{N-1}]$ with $\Delta t = T(N-1)^{-1}$, where $T = t_{N-1} - t_0$. For large N, it is safe to assume the following average values:

$$\sum_{k=0}^{N-1} \cos(\omega_0 t_k) = \sum_{k=0}^{N-1} \sin(\omega_0 t_k) = \sum_{k=0}^{N-1} \cos(\omega_0 t_k)\sin(\omega_0 t_k) = 0,$$

$$\frac{1}{N}\sum_{k=0}^{N-1} \cos^2(\omega_0 t_k) = \frac{1}{N}\sum_{k=0}^{N-1} \sin^2(\omega_0 t_k) = \frac{1}{2} \tag{11.101}$$

Thus, for the evenly spaced data points, Eq. (11.100) may be reduced to

$$
\begin{bmatrix} N & 0 & 0 \\ 0 & 0.5N & 0 \\ 0 & 0 & 0.5N \end{bmatrix}
\begin{bmatrix} a_0 \\ a_1 \\ b_1 \end{bmatrix} =
\begin{bmatrix} \sum_{k=0}^{N-1} y_k \\ \sum_{k=0}^{N-1} y_k \cos(\omega_0 t_k) \\ \sum_{k=0}^{N-1} y_k \sin(\omega_0 t_k) \end{bmatrix}
\tag{11.102}
$$

The inverse of a diagonal matrix is merely another diagonal matrix whose elements are the reciprocals of the original one. Therefore, unknown coefficients can be determined as

$$
a_0 = \frac{1}{N}\sum_{k=0}^{N-1} y_k, \quad a_1 = \frac{2}{N}\sum_{k=0}^{N-1} y_k \cos(\omega_0 t_k), \quad b_1 = \frac{2}{N}\sum_{k=0}^{N-1} y_k \sin(\omega_0 t_k) \tag{11.103}
$$

So far our approximation had only three terms, relying on a single angular frequency ω_0, but in fact we can add more terms with multiple frequencies, so that Eq. (11.97) becomes

$$
f(t) = a_0 + \sum_{j=1}^{P} \left[a_j \cos(j\omega_0 t) + b_j \sin(j\omega_0 t) \right] \tag{11.104}
$$

In this case (for the evenly spaced data), the linear least-squares regression yields

$$
a_0 = \frac{1}{N}\sum_{k=0}^{N-1} y_k, \quad a_1 = \frac{2}{N}\sum_{k=0}^{N-1} y_k \cos(j\omega_0 t_k),
$$

$$
b_1 = \frac{2}{N}\sum_{k=0}^{N-1} y_k \sin(j\omega_0 t_k), \quad j = 1,2,\ldots,P \tag{11.105}
$$

In the case of $N>2p+1$, Eqs. (11.104) and (11.105) can be used to fit data in the least-squares regression sense. Analogous to the full-degree polynomial approximation is the case when $N=2p+1$, so that Eq. (11.104) becomes an interpolant passing through all data points. Increasing N to infinity leads to the continuous Fourier series

$$
f(t) = a_0 + \sum_{j=1}^{\infty} \left[a_j \cos(j\omega_0 t) + b_j \sin(j\omega_0 t) \right] \tag{11.106}
$$

with

$$a_0 = \frac{1}{T}\int_0^T f(t)\,dt, \quad a_j = \frac{2}{T}\int_0^T f(t)\cos(j\omega_0 t)\,dt,$$

$$b_j = \frac{2}{T}\int_0^T f(t)\sin(j\omega_0 t)\,dt, \quad j = 1, 2, \ldots, \infty \qquad (11.107)$$

This basically shows that any arbitrary periodic function $f(t)$ with a period T can be represented by an infinite series of sinusoids of harmonically related frequencies [Eq. (11.106)], where $\omega_0 = 2\pi T^{-1}$ is called the *fundamental frequency* and its constant multiples, $2\omega_0$, $3\omega_0$, etc., are called *harmonics*. The relation

$$P \le \frac{N-1}{2} \qquad (11.108)$$

is basically equivalent to the *Nyquist–Shannon sampling theorem* stating that perfect reconstruction of a signal is possible only when the sampling frequency (number of points N) is greater than twice the maximum frequency of the signal being sampled (P). If the lower sampling rates are used (not enough data points N), the original signal's information may not be completely recoverable from the sampled signal.

Coefficients [Eq. (11.105)] can be combined into a single complex quantity

$$c_j = a_j + ib_j = \frac{2}{N}\sum_{k=0}^{N-1} y_k \left[\cos(j\omega_0 t_k) + i\sin(j\omega_0 t_k)\right]$$

$$= \frac{2}{N}\sum_{k=0}^{N-1} y_k e^{ij\omega_0 t_k}, \quad j = 0, 2, \ldots, P \qquad (11.109)$$

($i = \sqrt{-1}$), which may be computed more efficiently. Generally speaking, in the context of this chapter, the MATLAB `fft` function computes the sums in Eq. (11.109) (with a negative sign of the exponent argument though) using the *fast Fourier transform* (FFT) algorithm. Once c_j are computed you may always go back to

$$a_j = \text{Re}(c_j) \quad (a_0 = \text{Re}(c_0)/2) \quad \text{and} \quad b_j = \text{Im}(c_j) \qquad (11.110)$$

11.6 Using the MATLAB Functions

As mentioned in Sec. 11.3, MATLAB features one major function, `polyfit`, allowing you to perform both least-squares regression and full-degree polynomial interpolation. Its simplest syntax is

```
p=polyfit(x,y,n)
```

where x is the vector of the x-data points (along the abscissa), y the vector containing the y-data points (along the ordinate) and n the power of a polynomial chosen to fit data ($n \le$ `length(x)-1`). This simplest call returns a row vector p of length $n+1$ containing coefficients p_i of a polynomial in the descending order of an argument x (that is, the highest-power coefficient goes first)

$$[p_1 \quad p_2 \quad \cdots \quad p_{n+1}][x^n \quad x^{n-1} \quad \cdots \quad 1]^T = p_1 x^n + p_2 x^{n-1} + \cdots + p_{n+1} \quad (11.111)$$

For example, a set of commands

```
>> x=linspace(1,10,500)'; y=rand(1,500)';
>> p=polyfit(x,y,1)
```

generates two vectors, linearly spaced x and y with 5000 uniformly distributed pseudorandom numbers, and then fits these data with the first-order polynomial. The result is quite obvious

```
>> p=polyfit(x,y,1)
```

generates two vectors, linearly spaced x and y with 5000 uniformly distributed pseudorandom numbers, and then fits these data with the first-order polynomial. The result is quite obvious

```
p = -0.0036 0.5262
```

that is, $y = -0.0036x + 0.5262$, basically meaning that the regression is almost horizontal (has a very small slope) and the y-intercept is very close to the mean value

```
>> mean(y)
ans = 0.5064
```

The `polyfit` function employs the QR factorization [see (Eq. (9.51) of Sec. 9.4] of a matrix defined by Eq. (11.13) or (11.54) (with the vector of coefficients in the reversed order). Hence, the aforementioned linear regression could be obtained using the `qr` function directly:

```
>> [Q,R] = qr([sum(x.^2) sum(x); sum(x) 500],0);
>> p=R(Q'*[sum(x.*y); sum(y)])
```

which returns

```
p = 0.0003
    0.4911
```

Another syntax of `polyfit` features two output parameters

```
[p,s]=polyfit(x,y,n)
```

In this case, along with a vector of polynomial coefficients it returns a structure S that contains three fields providing additional information about the solution. Specifically, these fields are: the upper triangular matrix R, received after the *Cholesky decomposition* (of a symmetric, positive-definite matrix of Eq. (11.13) into the product of a lower triangular matrix and its conjugate transpose), number of degrees of freedom df, and sum of squared residuals normr. You may assess any of structure s fields by issuing an appropriate command, like s.normr. For example, the commands

```
>> x=[1:5]'; y=[2 5 8 20 26]';
>> [p,s]=polyfit(x,y,2)
```

return

```
p = 1.0714    -0.1286    0.8000
s =      R: [3x3 double]
       df: 2
    normr: 3.4393
```

In this return the normr is simply `norm(y-polyval(p,x))`, the degree of freedom df equal to 2 tells you that we could increase the order of the fitting polynomial by two (the order of the full-degree polynomial for five points is four), and the matrix R

```
>> s.R
ans =  -31.2890    -7.1910    -1.7578
             0    -1.8136    -1.3010
             0          0     0.4663
```

is `-chol(V)`, where V is formed as

```
N=3;    % the number of coefficients to be determined
for i=1:N
    for j=1:N
    V(i,j)=sum(x.^(2*N-i-j));
    end
end
```

Using the output parameter s.normr and employing the MATLAB function std to compute a standard deviation, we may compute the coefficient of determination [Eq. (11.21)]. For example, for the 10 data points generated as

```
>> N=10; xd=linspace(0,10,N); yd=xd.^2+5*randn(1,N);
```

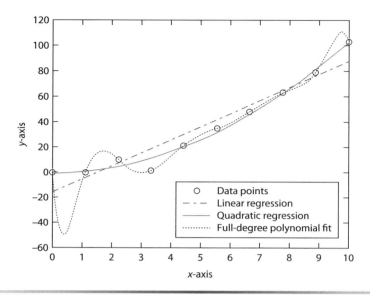

Fig. 11.8 Examples of polynomial regressions.

and the sum of squared residuals around the mean value S_t

```
>> St=std(yd)^2*(N-1);
```

we can explore the goodness of different fitting polynomials as

```
>> [p1,S1]=polyfit(xd,yd,1);    Rsq1=(St-S1.normr^2)/St
>> [p2,S2]=polyfit(xd,yd,2);    Rsq2=(St-S2.normr^2)/St
>> [p9,S9]=polyfit(xd,yd,N-1);  Rsq9=(St-S9.normr^2)/St
```

which returns the values of R^2

```
Rsq1 =      0.9202
Rsq2 =      0.9874
Rsq9 =      1
```

Obviously, in terms of R^2 increasing the order of a polynomial regression makes the fit "better" (R^2 gets closer to 1). However, you should be careful not to ascribe more meaning to it than is warranted, because all it means is that the sum of the squared residuals gets smaller. You should always inspect a result of regression visually. The following script may help you to visualize the results of the regression we just did (Fig. 11.8):

```
plot(xd,yd,'ro'), hold
x=linspace(xd(1),xd(end)); % creating a finer grid
y1=polyval(p1,x);
y2=polyval(p2,x);
```

```
y9=polyval(p9,x);
plot(x,y1,'-.',x,y2,x,y9,':')
legend('Data points','Linear regression','Quadratic regression',...
       'Full-degree polynomial fit',0)
xlabel('x-axis'), ylabel('y-axis')
```

Obviously, despite the fact that for the full-degree polynomial we have $R^2 = 1$, the approximation is really bad.

The most comprehensive syntax of `polyfit` features three output parameters

```
[p,s,mu]=polyfit(x,y,n)
```

where `mu` is a two-element vector. This call finds coefficients of a polynomial in a specially constructed domain with a new argument

$$\widetilde{x} = \frac{x - \mu_1}{\mu_2} \tag{11.112}$$

with

$$\mu_1 = \frac{1}{N}\sum_{i=1}^{N} x_i \quad \text{and} \quad \mu_2 = \sqrt{\frac{1}{N-1}\sum_{i=1}^{N}(x_i - \mu_1)^2} \tag{11.113}$$

being the mean value of vector **x** (`mu(1)=mean(x)`) and standard deviation (`mu(2)=std(x)`). This centering and scaling transformation improves numerical properties of both the polynomial and fitting algorithm. However, you should be careful because this syntax obviously returns the different values of the fitting polynomial. Continuing the aforementioned example, the command

```
>> [pc,s,mu]=polyfit(x,y,2)
```

returns

```
pc =     2.6786    9.9612    11.0571
s =        R: [3x3 double]
          df: 2
       normr: 3.4393
mu =   3.0000
       1.5811
```

Hence, instead of $y = 1.0714x^2 - 0.1286x + 0.8$, we now get

$$y = 2.6786\left(\frac{x - \mu_1}{\mu_2}\right)^2 + 9.9612\frac{x - \mu_1}{\mu_2} + 10.0571 \tag{11.114}$$

To check whether it matches the original polynomial in x, you may use the following commands (involving the Symbolic Math Toolbox functions `syms`, `poly2sym`, `subs`, `expand`, and `simplify`):

```
>> syms x
>> ex = expand(subs(poly2sym(pc),x,(x-mu(1))/mu(2)))
```

to obtain

```
ex = (15*x^2)/14-(9*x)/70+4/5
```

To avoid the vulgar fractions, you may further employ the `vpa` function, for example the command

```
>> vpa(ex,3)
```

transfers the previous result to

```
ans = 1.07*x^2-.129*x+.800
```

Now, let us try to use linear regressions to fit nonlinear data. As discussed in Sec. 11.2.5, we may try employing the following call to fit the power law data [Eq. (11.25)]:

```
coeff=polyfit(log10(x),log10(y),1)
```

The coefficients of the original power law relationship can then be obtained as `a=10^coeff(2)` and `b=coeff(1)`. For the exponential relation [Eq. (11.30)], the `polyfit` function can be used as

```
coeff=polyfit(x,log10(y),1)
```

for base-10 exponentials, and

```
coeff=polyfit(x,log(y),1)
```

for the natural base exponentials. The coefficients a and b of Eq. (11.30) can then be estimated as `b=coeff(1)` and `a=10^coeff(2)` (`a=exp(coeff(2))`).

As mentioned in Sec. 11.2.5, these pseudolinear regressions minimize the sum of squared weighted residuals [like those shown in Eqs. (11.30), (11.41), and (11.46)] and therefore are less accurate than those obtained via minimization [Eq. (11.24)]. As an example, the following script explores the two types of regressions (obtained via linear algebraic least-squares regression and nonlinear minimization) for the nonlinear dependence of Eq. (11.37)

```
%% Defining the function
f= @(a,b,x) a*x./(b+x); at=3; bt=2.55;
%% Producing "experimental" data
x=[1:2:20]'; n=length(x); xp=linspace(x(1),x(end))';
y=f(at,bt,x)+0.2*randn(n,1);
subplot(211)
```

```
plot(x,y,'ok'), hold, xlabel('x-axis'), ylabel('y-axis')
%% Linear regression in the modified domain
subplot(212)
c=polyfit(1./x,1./y,1); % linear LS regression in the modified domain
yp=polyval(c,1./xp);
plot(1./x,1./y,'ok'), hold, plot(1./xp,yp,'r-.')
xlabel('1/x'), ylabel('1/y')
ar=1/c(2); br=c(1)/c(2);
legend('Data points','Algebraic solution',0)
% Converting the results of linear LS regression back to original domain
subplot(211)
plot(xp,f(ar,br,xp),'r-.')
Srr=norm(y-f(ar,br,x)); % sum of squared residuals for regression
text(6,1.0,['S_r^{reg}=' num2str(Srr,3)])
%% Minimization in the original domain
fn=@(a) norm(y-f(a(1),a(2),x));
[c,Sropt]=fminsearch(fn,[ar,br]);
plot(xp,f(c(1),c(2),xp),'b')
text(6,0.4,['S_r^{opt}=' num2str(Sropt,3)])
legend('Data points','Algebraic solution','Minimization',0)
%% Comparing individual errors
figure
title('Residuals for the pseudolinear regression and minimization')
subplot(211), bar(y-f(ar,br,x))
text(5,.25,'... obtained via algebraic solution')
xlabel('Data point'), ylabel('Residual'), ylim(0.3*[-1 1])
subplot(212), bar(y-f(c(1),c(2),x))
text(5,.25,'... obtained via minimization')
xlabel('Data point'), ylabel('Residual'), ylim(0.3*[-1 1])
```

Specifically, this script generates some data points and then follows a routine outlined in Sec. 11.2.5 to convert the physical domain to some other domain, where the linear algebraic regression analysis can be carried (using the `polyfit` function). The bottom portion of Fig. 11.9a shows the linear regression produced in the modified $(x^{-1} - y^{-1})$ domain. The coefficients of this linear approximation are then converted back to the original domain and used to initialize the optimization routine (using the `fminsearch` function). The top portion of Fig. 11.9a features both regressions. As seen, they are slightly different with the one obtained via minimization exhibiting the lesser value of the sum of squared residuals. The individual residuals for both approaches are shown in Fig. 11.9b. As expected, because of inevitable scaling [Eq. (11.41)], the linear regression pays more attention to residuals with a smaller values of y, allowing residuals for the last data points to be larger.

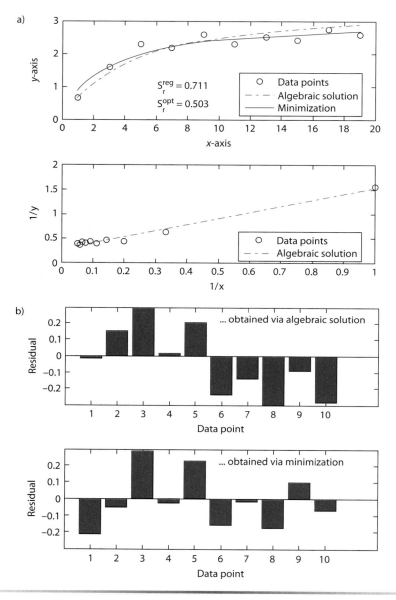

Fig. 11.9 Example of a) nonlinear regression and b) individual residuals.

Now let us consider two major functions for approximating the x–y data points with piecewise cubic splines. They are

`y_int=spline(x,y,x_int)` Computes a not-a-knot or clamped spline interpolation at the set of arguments `x_int` `y_int=pchip(x,y,x_int)` Computes a shape-preserving Hermite interpolating polynomial at the set of arguments `x_int`

If the first two input vectors for the `spline` function have the same length, this function produces a not-a-knot spline. If the vector `y` has two more elements than the vector `x`, the very first and very last elements of the vector `y` are considered to be the end slopes, and therefore the `spline` function produces a clamped spline. The following script resulting in an output shown in Fig. 11.10a presents a self-explanatory example of using the `spline` and `pchip` functions:

```
x=-3:3; y=[-1 -1 -1 0 1 1 1];
t=linspace(x(1),x(end));
p=pchip(x,y,t);              % Hermite spline
s=spline(x,y,t);            % not-a-knot spline
c=spline(x,[-2 y -2],t);    % clamped spline
plot(x,y,'o',t,p,'-',t,s,'-.',t,c,':b')
legend('Data points','Hermite spline','Not-a-knot spline',...
       'Clamped spline with -2 slopes at the endpoints',0)
xlabel('x-axis'), ylabel('y-axis')
```

Figure 11.10b features another example, when for a specific set of data points both the not-a-knot and Hermite piecewise spline interpolations are very close to each other and practically coincide with a full-degree polynomial interpolation (the key is in a special arrangement of x data points).

The simpler syntax for the spline functions is

```
pp = spline(x,y)
```

and

```
pp = pchip(x,y)
```

These calls do not do interpolation, but rather create a structure that contains all information pertaining to a piecewise polynomial defined by the vectors `x` and `y`. This structure can be used later by the `ppval` function and the spline utility `unmkpp`

`yy=ppval(pp,xx)` Returns the value of interpolation `pp` at the points `xx`

`[kn,cf]=unmkpp(pp)` (abbr. for <u>un</u>make <u>p</u>iecewise <u>p</u>olynomial) breaks the piecewise polynomial structure `pp` into a set of knots `kn`, whereas the `cf` matrix contains four coefficients in each of its `length(pp)-1` rows

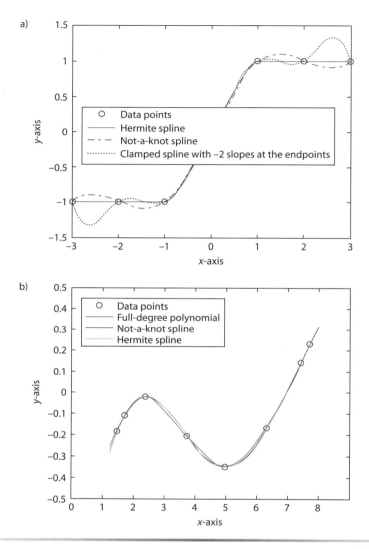

Fig. 11.10 a) Examples of different splines and b) comparison of different types of approximations.

Another utility, pp=mkpp(kn,cf), builds a piecewise polynomial structure pp from its breaks (knots) kn and coefficients cf. For example, a self-explanatory set of commands

```
x=1:5; y=[2 5 6 3 4];
pp = spline(x,y),            [kn,cf]=unmkpp(pp)
xx=linspace(x(1),x(end));    yy=ppval(pp,xx);
plot(x,y,'rs',xx,yy,'.'),    xlabel('x-axis'), ylabel('y-axis')
```

returns

```
pp =    form: 'pp'
      breaks: [1 2 3 4 5]
       coefs: [4x4 double]
      pieces: 4
       order: 4
         dim: 1
kn = 1      2       3      4      5
cf =  -0.7500    1.2500    2.5000    2.0000
      -0.7500   -1.0000    2.7500    5.0000
       1.7500   -3.2500   -1.5000    6.0000
       1.7500    2.0000   -2.7500    3.0000
```

and produces a plot shown in Fig. 11.11a. As seen along with the breaks (kn) and coefficients (`coefs`), the pp structure has some extra fields defining the number of the intervals (`pieces`) and the number of coefficients per each interval (`order`). The last field (`dim`), defines the number of dimensions of a problem based on the size of the array y in the `spline(x,y)` call (in the multidimensional case interpolation is performed for each set of data in y against the same vector x).

MATLAB features one more M-file function to interpolate a function of one variable $y = f(x)$, `interp1`. Its syntax is

```
yy=interp1(x,y,xx,'method')
```

Fig. 11.11   Examples of a) spline produced using the `unmkpp` utility and b) approximations produced using the `interp1` function.

This function returns an interpolated vector yy at the specified values xx using data stored in x and y to produce a piecewise approximation. The default `'method'` is `'linear'`, which produces a piecewise linear interpolation. Other methods include

`'nearest'`	Produces a ladder-type nearest-neighbor interpolation (see Fig. 11.11b)
`'spline'`	Yields a natural cubic spline approximation
`'pchip'` or `'cubic'`	Employs a shape-preserving cubic Hermite approximation

By default, the `interp1` function does not do extrapolation outside the interval spanned by x for the `'linear'` and `'nearest'` methods returning NaN. If you need it, you may request it directly, for example,

```
yy=interp1(x,y,xx,'nearest','extrap')
```

When the Figure window **Tools** were introduced in Sec. 6.11 (Fig. 6.44), one of them was **Basic Fitting**, allowing you to conveniently explore different interpolants. Suppose, you created a plot using the following call

```
>> fplot('sin',[0,2*pi],'.')
```

Choosing the **Basic Fitting** option from the **Tools** menu brings the Basic Fitting tool window shown in Fig. 11.12a (in order to have this expanded view compared to that of Fig. 6.44, you have to click on the **Show next panel** button). Picking a couple of interpolants to compare and checking off several radiobuttons results in the plot shown in Fig.  11.12b. Now that we managed to create all these plots programmatically in the previous examples of this section, you should totally understand all the options for curve fitting that the Basic Fitting tool has to offer and be able to explain the results. (Note that, neither spine generates an equation to be shown as it is the case for the first- through 10th-order interpolants.)

Similar to the one-dimensional case, MATLAB has several functions for multidimensional piecewise approximation as well:

`zz=interp2(x,y,w,xx, yy,'method')`	Interpolates a function of two variables, $w = f(x,y)$, returning an interpolated vector zz at the specified values xx and yy based on data stored in x, y, and w with the method specified in the same manner as for the `interp1` function
`ww=interp3(x,y,z,w,`	Interpolates a function of three

a)

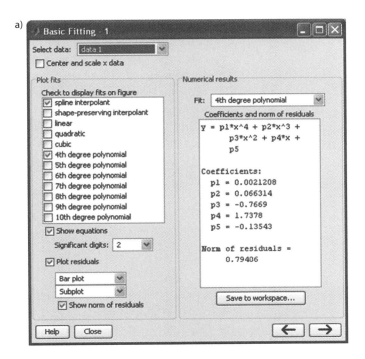

b)

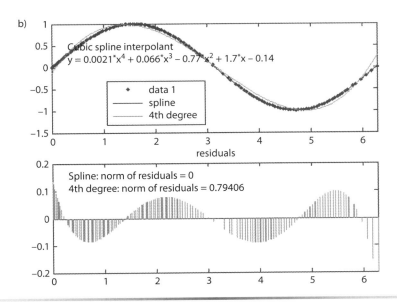

Fig. 11.12 a) Using the Basic Fitting Tool to b) fit a sinusoid with the spline and fourth-order polynomial.

`xx,yy,zz,'method')`	variables $w = f(x,y,z)$, returning an interpolated vector `zz` at the specified values `xx`, `yy`, and `zz` based on data stored in `x`, `y`, `z`, and `w`

`vv=interpn(x1,x2,...,xn,w,`
`  xx1,xx2,...,xxn,`
`  'method')`
	Interpolates a function of *n* variables based on data stored in the multidimensional lookup table $w = f(x_1, x_2, ..., x_n)$

For example, the following set of commands interpolates the MATLAB `peaks` function over a finer grid and generates a plot shown in Fig. 11.13a.

```
[x,y]=meshgrid(-3:0.25:3);
z=peaks(x,y);
[xi,yi]=meshgrid(-3:0.125:3);
zi=interp2(x,y,z,xi,yi);
mesh(x,y,z), hold, mesh(xi,yi,zi+15)
axis([-3 3 -3 3 -5 20])
```

The following script employs a 3-D interpolation to compute a takeoff distance of a Cessna-172 aircraft to clear-off a 50-ft obstacle (Fig. 11.13b), the task all pilots do to better understand airplane performance operating limitations. This distance is computed based on the table data for a variety of ground winds, airport elevation and aircraft weight

```
wind=[0 10 20];
altitude=[0 2500 5000 7500];
weight=[1700 2000 2300];
tkof(:,:,1)=[780 920 1095 1370; 570 680 820 1040; 385 470 575 745];
tkof(:,:,2)=[1095 1325 1625 2155; 820 1007 1250 1685; 580 720 910 1255];
tkof(:,:,3)=[1525 1910 2480 3855; 1170 1485 1955 3110; 850 1100 1480 2425];
We=210; Al=254; Wi=0:20;
w=interp3(altitude,wind,weight,tkof,Al,Wi,We,'spline');
plot(Wi,w), grid
text(2,850,'Aircraft: Cessna-172 Skyhawk','BackgroundColor','w')
text(2,750,[' Gross weight: ' num2str(We) 'lbs'],'BackgroundColor','w')
text(2,650,['Alitude:' num2str(Al), 'ft MSL'],'BackgroundColor','w')
xlabel('Head wind (knots)'), ylabel('Take-off distance (ft)')
title('Take-off distance to clear-off a 50ft obstacle')
```

The next script features another example of interpolation for a 3-D fluid flow data as provided by the MATLAB function `flow`.

```
figure(1), [x,y,z,v]=flow(10);
slice(x,y,z,v,[6 9.5],2,[-2 .2])
xlabel('x-axis'), ylabel('y-axis'), zlabel('z-axis')
figure(2), [xi,yi,zi]=meshgrid(.1:.125:10, -3:.125:3, -3:.125:3);
```

a)

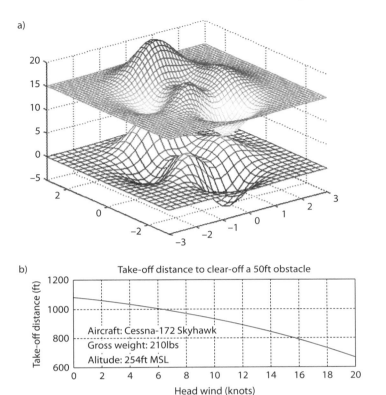

b)

Take-off distance to clear-off a 50ft obstacle

Aircraft: Cessna-172 Skyhawk
Gross weight: 210lbs
Alitude: 254ft MSL

Take-off distance (ft)

Head wind (knots)

Fig. 11.13   Example of a) 2-D interpolation featuring a finer mesh grid and b) 3-D interpolation using the lookup tables.

```
vi=interp3(x,y,z,v,xi,yi,zi); % vi is a 25 by 10 by 25 array
slice(xi,yi,zi,vi,[6 9.5],2,[-2 .2]), shading flat
xlabel('x-axis'), ylabel('y-axis'), zlabel('z-axis')
```

Similar to that of the `peak` function of two variables, the `[x,y,z,w]=flow(n)` call produces a coarse approximation of the flow in a 2n by n by n array w returning knot coordinates x, y, z as well (Fig. 11.14a). The `interp3` function is then employed to interpolate over a finer mesh as shown in Fig. 11.14b.

Now let us show an example of least-squares regression applied to a sinusoid as discussed in Sec. 11.5. The following script, generates the data points based on a sinusoid corrupted with some noise (Fig. 11.15a) and then applies Eq. (11.103) to fit these data with the least-squares regression

```
%% Generating the data points
a0=1; a1=0.5; b1=3; w=1; N=100;
td=linspace(0,4*pi,N);
```

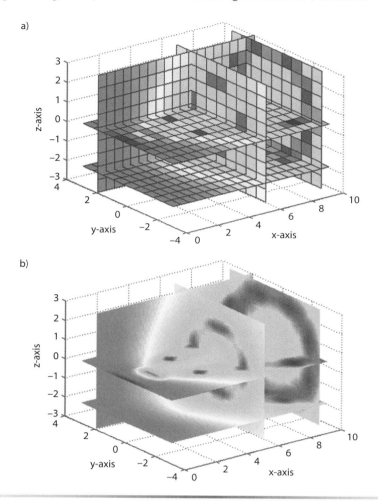

Fig. 11.14 Example of 3-D interpolation converting a) coarse mesh data to b) finer mesh data.

```
yd=a0+a1*cos(w*td)+b1*sin(w*td)+0.3*randn(1,N);
subplot(2,1,1)
plot(td,yd,'r.'), hold
%% Performing a least-squares regression
a0est=sum(yd)/N;
a1est=2*sum(yd.*cos(w*td))/N;
b1est=2*sum(yd.*sin(w*td))/N;
t=linspace(0,4*pi,2*N);
y=a0est+a1est*cos(w*t)+b1est*sin(w*t);
plot(t,y,'--'), xlabel('t'), ylabel('y')
legend('Data points','Regression',0)
```

```
%% Showing individual residuals
subplot(2,1,2)
stem(td,yd-a0est-a1est*cos(w*td)-b1est*sin(w*td))
xlabel('t'), ylabel('y-y^{reg}')
```

Figure 11.15b shows individual residuals.

In the last example of Fig. 11.15a, we knew what frequency to use for a regression, but in the general case this information is not available, and that is when Eq. (11.104) becomes useful. Before we show one practical example consider a general case when you would like to use the continuous Fourier series [Eq. (11.106)] to approximate a rectangular wave function within one period, that is, within $t \in [0;T]$

$$f(t) = \begin{cases} -1 & 0 < t \le \frac{1}{4}T \\ 1 & \frac{1}{4}T < t \le \frac{3}{4}T \\ -1 & \frac{3}{4}T < t \le T \end{cases} \qquad (11.115)$$

Applying Eq. (11.107) for a_0 yields

$$a_0 = \frac{1}{T}\int_0^T f(t)\,dt = \frac{1}{T}\left(-\int_0^{T/4} dt + \int_{T/4}^{3T/4} dt - \int_{3T/4}^T dt\right) = 0 \qquad (11.116)$$

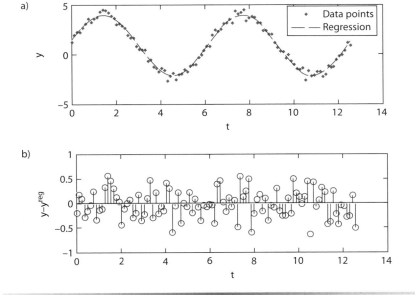

Fig. 11.15 a) Least-squares regression for the noise-corrupted sinusoid and b) individual residuals.

Similarly, you can compute

$$b_j = 0, \; j = 1, 2, \ldots, \infty \quad \text{and} \quad a_j = \begin{cases} -4/(j\pi) & \text{for } j = 1, 5, 9, \ldots \\ 4/(j\pi) & \text{for } j = 3, 7, 11, \ldots \\ 0 & \text{for } j = 2, 4, 6, 8, \ldots \end{cases} \quad (11.117)$$

The following script proves these findings via employing the Symbolic Math Toolbox function `int`:

```
syms T t
for j=0:7
f1=cos(j*2*pi/T*t); f2=sin(j*2*pi/T*t);
a(j+1)=2*(-int(f1,-T/2,-T/4)+int(f1,-T/4,T/4)-int(f1,T/4,T/2))/T;
b(j+1)=2*(-int(f2,-T/2,-T/4)+int(f2,-T/4,T/4)-int(f2,T/4,T/2))/T;
end
[0:7; a; b]
```

returning

```
ans = [ 0,       1,   2,       3,   4,       5,   6,       7]
      [ 0,   -4/pi,   0,   4/3/pi,   0,  -4/5/pi,  0,   4/7/pi]
      [ 0,       0,   0,       0,   0,       0,   0,       0]
```

Therefore, the Fourier series for Eq. (11.115) is

$$f(t) = -\frac{4}{\pi}\cos(\omega_0 t) + \frac{4}{3\pi}\cos(3\omega_0 t) - \frac{4}{5\pi}\cos(5\omega_0 t)$$

$$+ \frac{4}{7\pi}\cos(7\omega_0 t) - \cdots \quad (11.118)$$

where $\omega_0 = 2\pi T^{-1}$. Figure 11.16 shows the results of adding up to eight terms, whereas Fig. 11.17 features an alternative way of presenting information provided by Eq. (11.118) via the *amplitude spectrum*. The following script shows how to employ the `fft` function to produce the same amplitude spectrum as shown in Fig. 11.17a numerically

```
%% Generating the data points
T=2*pi; N=1000; quat=N/4;
t=linspace(0,T,N);
y=-ones(1,N); y(quat:3*quat)=1;
%% Applying the fft function
```

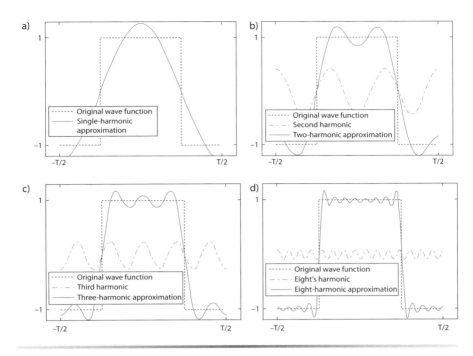

Fig. 11.16 First four terms of the Fourier series to approximate the square of a rectangular wave.

```
d=fft(y);                              % perform discrete Fourier transform
a=2*real(d)/N; b=-2*imag(d)/N;         % evaluate coefficients a and b
c=2*sqrt(d.*conj(d))/N;                % compute the magnitude of coefficients c
%% Plotting amplitude spectrum
subplot(2,1,1)
f=2*pi/T*[0:8]; stem(f,c(1:length(f)))
text(f(2:2:8)'+0.2,c(2:2:8)',num2str(c(2:2:8)','%-5.3f'))
xlabel('Harmonic'), ylabel('Amplitude')
%% Plotting DFT interpolant
subplot(2,1,2)
NUP=ceil((N+1)/2); yf=a(1)/2;
for i=1:NUP
    yf=yf+a(i+1)*cos(2*pi/T*i*t)+b(i+1)*sin(2*pi/T*i*t);
end
plot(t,yf,'r-.'), xlabel('Time, s'), ylabel('Function')
```

Specifically, the command $d=fft(y)$ returns complex quantities of the DFT of vector y. The amplitude of the DFT multiplied by $2N^{-1}$ yields coefficients c_j [Eq. (11.109)]. Figure 11.17b shows the absolute values of these

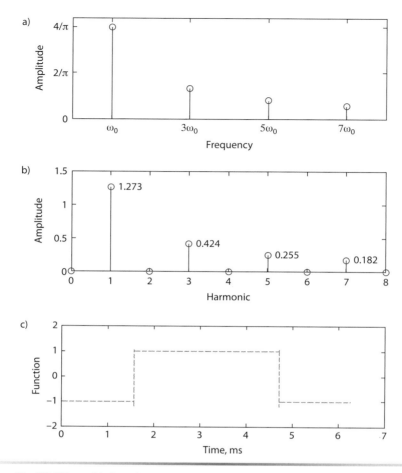

Fig. 11.17 a, b) Amplitude spectrum of the Fourier series (for the first four terms), and c) the resulting DFT approximation of a rectangular wave function.

coefficients matching those of Fig. 11.17a. Note that, we could also plot the values of coefficients a_j explicitly, because the values of b_j are negligibly small (because of the round-off errors, they are not exactly zeros though). The fft call returns the same amount of data as vector y has, but in fact only half of these data happen to be unique, although the second half is just a reflection of the first half. Hence, the line NUP=ceil((N+1)/2) evaluates the number of unique frequencies, followed by computation of the DFT interpolant, shown in Fig. 11.17c. This interpolant passes through all 1000 data points.

Let us conclude this section with another example, when similarly to Fig. 11.15a we want to find a regression to better fit some experimental data. As opposed to the previous formulation, however, we do not know any frequency-related information up front. On the contrary, it is required to find the fundamental frequency of the signal ω_0 and major harmonics.

MATLAB has several examples showing how to generate data formed by two sinusoids corrupted with some noise and then apply the `fft` function to extract two major harmonics, so that the following script goes a little bit further and allows you to change the data and also reconstruct the original signal by automatically finding two harmonics on the power spectrum plot

```
%% Defining the signal
Harmonic1=50;                        % the first harmonic (Hz)
Harmonic2=220;                       % the second harmonic (Hz)
AmpRat=2;                            % ratio of harmonics amplitudes
NoiseLevel=2.5;
%% Creating a noisy signal from two harmonics and
ti=0:0.0005:0.6;                     % zero-mean random noise
t=0:0.001:0.6;                       % considering data sampled at 1000Hz
x=sin(2*pi*Harmonic1*t)+AmpRat*sin(2*pi*Harmonic2*t); % blending 2 sinusoids
y=x+NoiseLevel*randn(size(t));       % adding signal noise
subplot(211)                         % plotting original and corrupted signals
plot(1000*ti(1:100),interp1(t,x,ti(1:100),'spline'),'--g'), hold
plot(1000*t(1:50),y(1:50), '.b'), xlabel('Time, ms'), ylim([-5 15])
%% Extracting signal power spectrum using FFT
Y = fft(y,512);                      % the 512-point fast Fourier transform
Pyy = 2*Y.* conj(Y) / 512; % computing the power at various frequencies
f = 1000*(0:256)/512;        % defining the frequency range (half of DFT return)
subplot(212)
plot(f,Pyy(1:257)), hold % plotting power spectrum
xlabel('Frequency, Hz')
%% Reconstructing the signal from two major harmonics
[peak1,ifreq1]=max(Pyy(1:257));              % finding the first harmonic
Pyy(ifreq1-10:ifreq1+10)=0;                  % excluding the first harmonic
[peak2,ifreq2]=max(Pyy(1:257));              % finding the second harmonic
freq1=(ifreq1-1)*1000/512; freq2=(ifreq2-1)*1000/512;
if freq1 > freq2             % rearranging harmonics in the ascending order
fr=freq1; freq1=freq2; freq2=fr;
pe=peak1; peak1=peak2; peak2=pe;
end
plot([freq1 freq1],[0 peak1],'--r')          % adding both harmonics
plot([freq2 freq2],[0 peak2],'-.m')          % to the power spectrum plot
```

```
legend('Data power spectrum','1^{st}','2^{nd} harmonic',0)
z =                   sin(2*pi*freq1*ti)+...
    sqrt(peak2/peak1)*sin(2*pi*freq2*ti);     % computing reconstructed signal
subplot(211)
plot(1000*ti(1:100),z(1:100),'-.r')           % plotting reconstructed signal
legend('Original uncorrupted signal',...
        'Corrupted signal (data points)','DFT approximation',0)
```

The consequent results of this self-explanatory script accompanied by the extended comments are shown in Fig.11.18.

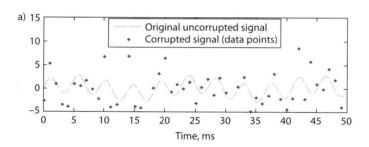

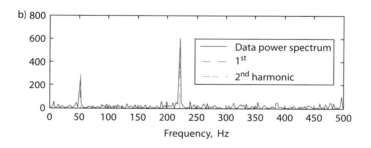

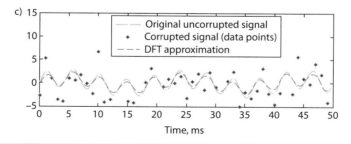

Fig. 11.18   a) Example of a noisy signal, b) frequency content of this signal, and c) DFT reconstruction using two harmonics.

Problems

11.1 For gases, the variation of dynamic viscosity with temperature can be modeled by the Sutherland's formula, which has the form (Eq. 11.42)

$$\mu = \frac{C\sqrt[3]{T^2}}{T+S}$$

where μ is the absolute (dynamic) viscosity, T the absolute temperature, and C and S are empirical constants. The following table gives the absolute viscosity of air at different temperatures

$T\,(°C)$	−20	0	40	100	200	300	400	500	1000
$\mu\,(N\,s/m^2)$ $(\times 10^{-5})$	1.63	1.71	1.87	2.17	2.53	2.98	3.32	3.64	5.04

Using the linear least-squares regression, determine the constants C and S by and present your results as a three-band plot (`subplot(3,1,…)`). The first one should present the dependence of viscosity from the temperature (use the `semilogy` function to plot it). Show the values of C and S along with the coefficient of determination R^2 (use the `num2str` function). The second plot should represent the same data but in coordinates you introduced to perform a linear regression. The third plot should present residuals for your linear regression.

11.2 Compare a locally defined spline, "not-a-knot" and "clamped" globally defined splines applied to approximate the following two "experimental" sequences of points:

a) `x = [0 2 4 5 7.5 10];`   **b)** `x = linspace(0,1.5,10);`
  `y = exp(-x/6).*cos(x);`   `y = humps(x);`

For the "clamped" spline, try to adjust the slope at the endpoints to better fit your data. Obviously, to compute spline coefficients you have to use "experimental" points only, but once you have your splines developed, do not be stingy, increase the number of `x` points to show how smooth your approximations are.

11.3 Repeat the example of retrieving the useful signals from the noise-corrupted data using the Fourier transform (see the very last example of Sec. 11.6). Assume true data containing four sinusoids of the same amplitude of 1 with the frequencies of 40, 70, 170 and 210Hz sampled at 1000Hz. Corrupt it with a double-amplitude normally distributed noise. Plot the signal in the time domain and its retrieved components in the frequency domain, like in Fig. 11.19.

Chapter 12 / Numerical Differentiation

- Two- and Multi-Point Approximations
- Estimating Approximation Errors
- Handling Partial Derivatives
- Using the MATLAB Functions

`diff, gradient, polyder`

12.1 Introduction

Q uite often in engineering practice, there is a need to numerically estimate the first- or even higher-order derivative of a function (a signal) with respect to some argument at the trial points. For instance, you have measurements of a vehicle position along some predetermined direction $x(t)$ and need to know its speed along this direction, that is, you need to estimate the first-order derivative of the coordinate x with respect to time. Another example is that you have the 2-D (planar) coordinates of the vehicle $(x(t), y(t))$ and need to know its heading $\Psi(t)$, that is, you need to estimate the first-order derivative of the coordinate y with respect to the coordinate x. The estimate of the second-order derivative in the latter case might also be needed because it is related to a torque control required to follow a path described by $y(x)$. Of course, we are not talking here about measurement errors or signal noise, but rather we assume that the data are somehow filtered already, so that the aforementioned derivatives make sense. The general problem considered in this chapter is to estimate the first- and higher-order derivatives, f', f'', etc., at n nodes using discrete data (measurements) $(x_i, f(x_i))$, $i = 1, 2, \ldots,$ n. Depending on a specific application, these estimates may use either data from several previous points (in this case, the derivative can be evaluated on-line) or data on multiple points before and after the current one (in this case, the estimates can be computed off-line only, when all data become available). The only MATLAB function that might help in estimating these derivatives is $\texttt{diff(x)}$, which simply calculates the difference between the adjacent elements of the vector $\texttt{x}$. Hence, we will derive high-accuracy approximations (using the Symbolic Math toolbox) that may have a good use of this simple

MATLAB function. The chapter starts from the simplest case, when only the knowledge about two data points is used to evaluate the first-order derivative at each trial point, followed by estimation of local truncation errors associated with this case using the Taylor series expansions. It then proceeds with the development of the more accurate estimates involving several as opposed to just two data points around each trial point, followed by expanding this approach to develop formulas for higher-order derivatives. An alternative way of obtaining the same formulas via using fitting polynomials, followed by assessing the values of partial derivatives, is discussed later. Finally, this chapter presents examples of using the `diff` function to compute the first- and higher-order derivatives, and also introduces two more functions: the `gradient` function to compute a gradient of a multivariable function and the `polyder` function to differentiate a polynomial.

12.2 Two-Point Difference Approximations of the First-Order Derivative

Consider a function $y = f(x)$ shown in Fig. 12.1 (depicted with the dashed line). Of course, this function is not known; instead the multiple data points depicted with circles are available. For the evenly spaced argument, we may present these data points as $x_{i\pm v} = x_i \pm vh$ ($v = 1, 2, ...$) and $y_{i\pm v} = f(x_{i\pm v}) = f(x_i \pm vh)$. The objective is to estimate the first-order derivative at point x_i.

Let us start from the easiest possible estimations of the first-order derivative. These estimations are based on a linear piecewise interpolation as shown in Fig. 12.1 with the solid line. By exploiting the geometry of an appropriate triangle we receive either a *forward* or *backward difference approximation* as follows:

$$f'(x_i) \approx \frac{f(x_{i+1}) - f(x_i)}{h} \tag{12.1}$$

$$f'(x_i) \approx \frac{f(x_i) - f(x_{i-1})}{h} \tag{12.2}$$

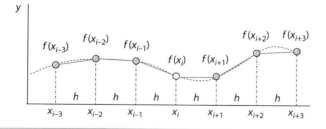

Fig. 12.1   Approximation of the derivative using a two-point difference.

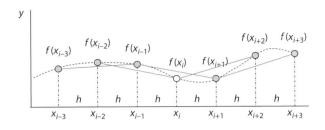

Fig. 12.2 Centered difference approximation of the first-order derivative.

For the particular situation depicted in Fig. 12.1, the forward difference approximation seems to give a better result for the points x_{i-2} and x_{i-1}, whereas the backward difference approximation, for all other points. However in practice, you cannot possibly know which approximation would work better, and that is why it is probably safer to use their combination, that is, Eqs. (12.1) and (12.2) blended together with the weighting coefficients of 0.5, which results in the *centered difference approximation* formula

$$f'(x_i) \approx \frac{f(x_{i+1}) - f(x_{i-1})}{2h} \qquad (12.3)$$

Geometrically, it means connecting the next nearest points as shown in Fig. 12.2.

As seen from Figs. 12.1 and 12.2, not only do the three developed approximations not coincide, but all of them produce somewhat erroneous estimates anyway. It is only in the case of a linear behavior of a function $f(x)$, all three formulas (12.1)–(12.3) would yield the same exact result. Otherwise intuitively, the overall error of approximation (12.3) should be smaller than that of approximations (12.1) and (12.2). None of the discussed formulas, however, provides us with the estimates of the approximation error. To get an estimate, we have to use another formal approach exploiting the Taylor series expansion of the function $f(x)$; refer to Eq. (8.31).

12.3 Errors Estimates from Taylor Series Expansion

Rather than using a geometrical approach or linear interpolation to obtain the formulas of the previous section, let us now use a Taylor series expansion assuming that the grid interval size h is small enough ($h \ll 1$). The Taylor series expansions of $f(x_{i\pm1})$ about the data point x_i yields

$$f(x_{i+1}) = f(x_i) + hf'(x_i) + \frac{h^2}{2!} f''(x_i) + \frac{h^3}{3!} f'''(x_i) + \frac{h^4}{4!} f^{(iv)}(x_i) + \cdots \qquad (12.4)$$

Solving for $f'(x_i)$ results in

$$f'(x_i) = \frac{f(x_{i+1}) - f(x_i)}{h} - \frac{h}{2} f''(x_i) + \text{HOT} \approx \frac{f(x_{i+1}) - f(x_i)}{h} + O(h) \quad (12.5)$$

where the error of the forward difference approximation (the first term in Eq. (12.5) is the same as Eq. (12.1)), denoted as $O(h)$, is proportional to the size of a grid interval h and the second-order derivative of the function $f(x)$ at x_i

$$O(h) \cong -\frac{h}{2} f''(x_i) \quad (12.6)$$

Similarly, for the backward difference approximation (12.2) we start from the Taylor series expansion for $f(x_{i-1})$ about the data point x_i

$$f(x_{i-1}) = f(x_i) - hf'(x_i) + \frac{h^2}{2!} f''(x_i) - \frac{h^3}{3!} f'''(x_i) + \frac{h^4}{4!} f^{(iv)}(x_i) - \cdots \quad (12.7)$$

which yields

$$f'(x_i) = \frac{f(x_i) - f(x_{i-1})}{h} + \frac{h}{2} f''(x_i) + \text{HOT} \approx \frac{f(x_i) - f(x_{i-1})}{h} + O(h) \quad (12.8)$$

Thus, for the backward difference approximation the error is defined by

$$O(h) \cong \frac{h}{2} f''(x_i) \quad (12.9)$$

The centered difference approximation (12.3) can be obtained by subtracting the Taylor series approximation for $f(x_{i-1})$ (Eq. (12.7)) from that of $f(x_{i-1})$ (Eq. (12.4)), yielding

$$f(x_{i+1}) - f(x_{i-1}) = \left(f(x_i) + hf'(x_i) + \frac{h^2}{2!} f''(x_i) + \frac{h^3}{3!} f'''(x_i) + \frac{h^4}{4!} f^{(iv)}(x_i) + \cdots \right)$$
$$- \left(f(x_i) - hf'(x_i) + \frac{h^2}{2!} f''(x_i) - \frac{h^3}{3!} f'''(x_i) + \frac{h^4}{4!} f^{(iv)}(x_i) - \cdots \right)$$

$$f(x_{i+1}) - f(x_{i-1}) = 2hf'(x_i) + \frac{2h^3}{6} f'''(x_i) + \frac{2h^5}{120} f^{(v)}(x_i) + \cdots \quad (12.10)$$

Solving for $f'(x_i)$ in this case yields

$$f'(x_i) = \frac{f(x_{i+1}) - f(x_{i-1})}{2h} - \frac{h^2}{6} f'''(x_i) + \text{HOT} \quad (12.11)$$

Hence, the complete formula for the centered difference approximation (cf. Eq. (12.11) with Eq. (12.3)) should be written as

$$f'(x_i) = \frac{f(x_{i+1}) - f(x_{i-1})}{2h} + O(h^2) \qquad (12.12)$$

with the error defined by

$$O(h^2) \cong -\frac{h^2}{6} f'''(x_i) \qquad (12.13)$$

that is, being proportional to the square of the grid interval size h^2 and the third-order derivative (because of cancellation of the second-order derivative terms in Eq. (12.10)).

To summarize, the two-point forward and backward difference approximations introduced in Eqs. (12.1) and (12.2), respectively, turned out to be the first-order approximations meaning that the error in both cases is proportional to the grid interval size, h (Eqs. (12.6) and (12.9)). These formulas produce an exact solution if the original function $f(x)$ is represented by a first-order polynomial (since in this case, $f''(x) \equiv 0$). As expected, the two-point centered difference approximation (12.3) turned out to be more accurate, namely, the second order, since the error is proportional to the square of the grid interval size, h^2 (Eq. (12.13)). Hence, it produces an exact solution if the original function $f(x)$ is represented by up to a second-order polynomial (since in this case, $f'''(x) \equiv 0$). Other than that the only way to have smaller errors is to decrease the grid interval size h (sampling time in case the argument is time), meaning that experiment measurements need to be taken more often. However, we may not have such luxury. Therefore, we need to develop more accurate formulas.

Before we start developing more accurate formulas in the following section, let us practice with the MATLAB Symbolic Math toolbox in order to obtain Eqs. (12.5), (12.8), and (12.12) (it may seem too easy here, but it will pay off later). For instance, the self-explanatory script

```
%% Defining symbolic variables
syms h f_i f_iplus1 f_prime f_2prime
%% Introducing the Taylor series expansion
Eq = -f_iplus1 + f_i + h*f_prime + h^2*f_2prime/factorial(2);
%% Solving the equation for f_prime
ans = solve(Eq,f_prime);
%% Simplifying and displaying the result for f_2prime
r=simplify(ans); pretty(r)
```

results in the following output, matching Eq. (12.5):

```
f_2prime h      f_i - f_iplus1
- ---------  -  --------------
     2                h
```

Note that the output of the MuPad-based Symbolic Math toolbox is different from that of the Maple-based one. The latter one benefited from using the `collect` function (`collect(ans,h)` in our case) that would always return the predictable result. In the latest versions of MATLAB, you need to play with the `simple` and `simplify` functions to see which one produces a better-looking result.

If we substitute the second and fourth line of the previous script with

```
syms h f_i f_iminus1 f_prime f_2prime
```

and

```
Eq = -f_iminus1 + f_i - h*f_prime + h^2*f_2prime/factorial(2);
```

we get the backward difference approximation formula (12.8)

```
f_2prime h      f_i - f_iminus1
---------- +  ---------------
     2                h
```

Finally, to get the centered difference approximation of Eq. (12.12), we employ the following script:

```
%% Defining symbolic variables
syms h f_i f_iplus1 f_iminus1 f_prime f_2prime f_3prime
%% Introducing two equations
Eq1 = -f_iplus1  +f_i+h*f_prime+h^2*f_2prime/2+h^3*f_3prime/6;
Eq2 = -f_iminus1 +f_i-h*f_prime+h^2*f_2prime/2-h^3*f_3prime/6;
%% Solving two equations for f_prime and f_2prime
ans = solve(Eq1,Eq2,f_prime,f_2prime);
%% Simplifying and displaying the result for f_prime
r=simple(ans.f_prime); pretty(r)
```

yielding

```
                                       2
f_iplus1 - f_iminus1    f_3prime h
-------------------- -  ----------
        2 h                6h
```

Note that in the latter case, we need to keep more terms in both Taylor series expansions. Also, since we used two equations, they had to be resolved with respect to two unknowns rather than a single one as in the previous example. We will use the same script later in this chapter to get the solution for the second-order derivative. In the meantime, we will proceed with the development of more accurate approximations for the first-order derivative.

12.4 More Accurate Multiple-Point Approximations

So far, we only used two points to get an estimate of the first-order derivative of the function. We learned that asymmetric (forward and backward) two-point difference approximations provide the first-order estimates and the two-point centered difference approximation happens to be of the second order. But why not use more points, if available, to increase the accuracy of approximation formulas? In what follows, we will show that the results of the previous section may be generalized further, so that the following statements hold:

1. The first-order derivative approximation involving k points produces an error of the order of h^{k-1} as the worst.
2. Centered approximations are one order more accurate than asymmetric ones, that is, of the order of h^k.

As an example, consider an approximation of the first-order derivative using three points instead of two. Looking at Taylor series expansions for $f(x_{i+1})$ and $f(x_{i+2})$

$$f(x_{i+1}) = f(x_i) + hf'(x_i) + \frac{h^2}{2!}f''(x_i) + \frac{h^3}{3!}f'''(x_i) + \frac{h^4}{4!}f^{(iv)}(x_i) + \cdots$$

$$f(x_{i+2}) = f(x_i) + 2hf'(x_i) + \frac{(2h)^2}{2!}f''(x_i) + \frac{(2h)^3}{3!}f'''(x_i) + \frac{(2h)^4}{4!}f^{(iv)}(x_i) + \cdots$$

$$(12.14)$$

and eliminating the second-order derivative terms by subtracting $f(x_{i+2})$ from $4f(x_{i+1})$ yields

$$4f(x_{i+1}) - f(x_{i+2}) = 4\left(f(x_i) + hf'(x_i) + \frac{h^2}{2!}f''(x_i) + \frac{h^3}{3!}f'''(x_i) + \frac{h^4}{4!}f^{(iv)}(x_i) + \cdots\right)$$

$$- \left(f(x_i) + 2hf'(x_i) + \frac{(2h)^2}{2!}f''(x_i) + \frac{(2h)^3}{3!}f'''(x_i)\right.$$

$$\left. + \frac{(2h)^4}{4!}f^{(iv)}(x_i) + \cdots\right)$$

$$(12.15)$$

$$4f(x_{i+1}) - f(x_{i+2}) = 3f(x_i) + 2hf'(x_i) - \frac{4h^3}{6}f'''(x_i) - \frac{12h^4}{24}f^{(iv)}(x_i) - \cdots$$

Solving for $f'(x_i)$ yields the *three-point forward difference approximation*

$$f'(x_i) = \frac{-f(x_{i+2}) + 4f(x_{i+1}) - 3f(x_i)}{2h} + \frac{h^2}{3} f'''(x_i) + \text{HOT} \quad (12.16)$$

or

$$f'(x_i) = \frac{-f(x_{i+2}) + 4f(x_{i+1}) - 3f(x_i)}{2h} + O(h^2) \quad (12.17)$$

with

$$O(h^2) \cong \frac{h^2}{3} f'''(x_i) \quad (12.18)$$

Hence, the error is proportional to the square of the grid interval size and the third-order derivative just as in the two-point centered difference approximation. The Symbolic Math toolbox can help to produce this result as follows:

```
%% Defining symbolic variables
syms h f_i f_iplus1 f_iplus2 f_prime f_2prime f_3prime
%% Introducing two equations
Eq1 = -f_iplus1+f_i+h*f_prime+h^2*f_2prime/2+h^3*f_3prime/6;
Eq2 = -f_iplus2+f_i+2*h*f_prime+(2*h)^2*f_2prime/2+(2*h)^3*f_3prime/6;
%% Solving two equations for f_prime and f_2prime
ans = solve(Eq1,Eq2,f_prime,f_2prime);
%% Simplifying and displaying the result for f_prime
r=simple(ans.f_prime); pretty(r)
```

yielding

```
          2
f_3prime h     3 f_i - 4 f_iplus1 + f_iplus2
----------- -  -----------------------------
     3                   2 h
```

Similarly, the *three-point backward difference approximation* can be developed using other three points, namely, $f(x_i)$, $f(x_{i-1})$, and $f(x_{i-2})$, yielding

$$f'(x_i) = \frac{3f(x_i) - 4f(x_{i-1}) + f(x_{i-2})}{2h} + O(h^2) \quad (12.19)$$

where

$$O(h^2) \cong \frac{h^2}{3} f'''(x_i) \quad (12.20)$$

If four points were considered for the forward and backward difference approximations, then following the same routine and having three Taylor series expansions, we would eliminate the second-order derivative terms from the first two series expansions and use the third one to eliminate the third-order derivative term. Thus, we would end up with the third-order approximations (with the error proportional to the cube of the grid interval size and the fourth-order derivative). For instance, for the third-order backward difference approximation of the first-order derivative, the script

```
%% Defining symbolic variables
syms h f_i f_im1 f_im2 f_im3 f_pr f_2pr f_3pr f_4pr
%% Introducing three equations
Eq1 = -f_im1+f_i-h*f_pr+h^2*f_2pr/2-h^3*f_3pr/6+h^4*f_4pr/24;
Eq2 = -f_im2+f_i-2*h*f_pr+(2*h)^2*f_2pr/2-(2*h)^3*f_3pr/6 ...
                    +(2*h)^4*f_4pr/24;
Eq3 = -f_im3+f_i-3*h*f_pr+(3*h)^2*f_2pr/2-(3*h)^3*f_3pr/6 ...
                    +(3*h)^4*f_4pr/24;
%% Solving three equations for f_prime, f_2prime and f_3prime
ans = solve(Eq1,Eq2,Eq3,f_pr,f_2pr,f_3pr);
%% Simplifying and displaying the result for f_prime
r=simplify(ans.f_pr); pretty(r)
```

produces

```
          3
f_4pr h        22 f_i - 36 f_im1 + 18 f_im2 - 4 f_im3
-------- + ---------------------------------------------
   4                         12 h
```

In the full correspondence with the second statement formulated in the beginning of this section, if four points, $f(x_{i-2})$, $f(x_{i-1})$, $f(x_{i+1})$, and $f(x_{i+2})$, were used for the *four-point centered difference approximation* even a more accurate, fourth-order accuracy approximation would be derived as follows:

$$f'(x_i) = \frac{-f(x_{i+2}) + 8f(x_{i+1}) - 8f(x_{i-1}) + f(x_{i-2})}{12h} + O(h^4) \qquad (12.21)$$

where

$$O(h^4) \cong \frac{h^4}{30} f^{(v)}(x_i) \qquad (12.22)$$

The MATLAB script employing the Symbolic Math toolbox may look like

```
%% Defining symbolic variables
syms h f_i f_ip1 f_ip2 f_im1 f_im2 f_p f_2p f_3p f_4p f_5p
%% Introducing equations for four points
Eqp1 =-f_ip1+f_i+h*f_p+h^2*f_2p/2+h^3*f_3p/6+h^4*f_4p/24+h^5*f_5p/120;
Eqp2 =-f_ip2+f_i+2*h*f_p+4*h^2*f_2p/2+8*h^3*f_3p/6+16*h^4*f_4p/24 ...
        +32*h^5*f_5p/120;
Eqm1 =-f_im1+f_i-h*f_p+h^2*f_2p/2-h^3*f_3p/6+h^4*f_4p/24-h^5*f_5p/120;
Eqm2 =-f_im2+f_i-2*h*f_p+4*h^2*f_2p/2-8*h^3*f_3p/6+16*h^4*f_4p/24 ...
        -32*h^5*f_5p/120;
%% Resolving four equations for f_p, f_2p, f_3p and f_4p
sol=solve(Eqp1,Eqp2,Eqm1,Eqm2,f_p,f_2p,f_3p,f_4p);
```

```
%% Simplifying and displaying the result for f_p
r=simplify(sol.f_p); pretty(r)
```

resulting in

```
          4
f_5p h       8 f_im1 - f_im2 - 8 f_ip1 + f_ip2
------- -  ------------------------------------
  30                    12 h
```

To summarize, Table 12.1 presents different formulas for the first-order derivatives, which can be easily programmed and used in off-line and on-line (backward difference) applications.

If needed, the higher-accuracy formulas employing more adjacent points can be obtained using the MATLAB scripts similar to the ones shown in this and previous sections.

It should be noted that although the backward difference approximation formulas seem to be less accurate than that of the center difference, they represent a very important class of formulas that can be used on-the-go or on-line, if you wish, because they only rely on the past points.

Table 12.1 Approximations of the First-Order Derivative $f'(x_i)$

Name	Formula	Error Estimate
Two-point forward difference	$\dfrac{f(x_{i+1}) - f(x_i)}{h}$	$-\dfrac{1}{2}hf''(x_i)$
Two-point backward difference	$\dfrac{f(x_i) - f(x_{i-1})}{h}$	$\dfrac{1}{2}hf''(x_i)$
Two-point centered difference	$\dfrac{f(x_{i+1}) - f(x_{i-1})}{2h}$	$-\dfrac{1}{6}h^2 f'''(x_i)$
Three-point forward difference	$\dfrac{-f(x_{i+2}) + 4f(x_{i+1}) - 3f(x_i)}{2h}$	$\dfrac{1}{3}h^2 f'''(x_i)$
Three-point backward difference	$\dfrac{3f(x_i) - 4f(x_{i-1}) + f(x_{i-2})}{2h}$	$\dfrac{1}{3}h^2 f'''(x_i)$
Four-point forward difference	$\dfrac{2f(x_{i+3}) - 9f(x_{i+2}) + 18f(x_{i+1}) - 11f(x_i)}{6h}$	$-\dfrac{1}{4}h^3 f^{(iv)}(x_i)$
Four-point backward difference	$\dfrac{11f(x_i) - 18f(x_{i-1}) + 9f(x_{i-2}) - 2f(x_{i-3})}{6h}$	$\dfrac{1}{4}h^3 f^{(iv)}(x_i)$
Four-point centered difference	$\dfrac{-f(x_{i+2}) + 8f(x_{i+1}) - 8f(x_{i-1}) + f(x_{i-2})}{12h}$	$\dfrac{1}{30}h^4 f^{(v)}(x_i)$

Let us introduce the notation

$$\Delta f_i = f_i - f_{i-1} \tag{12.23}$$

denoting a simple backward difference of f. Applying the difference operator Δ twice will result in

$$\Delta(\Delta f_i) \equiv \Delta^2 f_i = f_i - f_{i-1} - (f_{i-1} - f_{i-2}) = f_i - 2f_{i-1} + f_{i-2} \tag{12.24}$$

and so on. Using these notations, the k-point backward difference approximation of the first-order derivative at the point x_i can be calculated as

$$f'(x_i) \approx \frac{1}{h} \sum_{m=1}^{k} \frac{1}{m} \Delta^m f(x_i) \tag{12.25}$$

The estimate of the local truncation error for approximation is

$$\frac{1}{k} h^{k-1} f^{(k)}(x_i) \tag{12.26}$$

12.5 Second- and Higher-Order Derivatives

From the previous analysis, it is intuitively clear that for a derivative of order k, the smallest number of data points necessary to derive a difference approximation is $k+1$. For example, we saw that a difference approximation for the first-order derivative of the function needs at least two data points. Likewise, the second-order derivative needs not less than three data points.

To derive the difference approximation for the second-order derivative, the basic principle is to eliminate the first-order derivative from the Taylor series expansion and as many higher-order terms (in this case, those greater than two) as possible. For instance, the *forward difference approximation for the second-order derivative* $f''(x_i)$ can be found by considering the same Taylor series expansions as in Eq. (12.14), but treating them slightly differently. Specifically, subtracting $f(x_{i+2})$ from $2f(x_{i+1})$ eliminates the first-order derivative:

$$2f(x_{i+1}) - f(x_{i+2}) = 2\left(f(x_i) + hf'(x_i) + \frac{h^2}{2!} f''(x_i) + \frac{h^3}{3!} f'''(x_i) + \frac{h^4}{4!} f^{(iv)}(x_i) + \cdots \right)$$

$$- \left(f(x_i) + 2hf'(x_i) + \frac{(2h)^2}{2!} f''(x_i) + \frac{(2h)^3}{3!} f'''(x_i) \right.$$

$$\left. + \frac{(2h)^4}{4!} f^{(iv)}(x_i) + \cdots \right) \tag{12.27}$$

$$2f(x_{i+1}) - f(x_{i+2}) = f(x_i) - \frac{2h^2}{2} f''(x_i) - \frac{6h^3}{6} f'''(x_i) - \frac{14h^4}{24} f^{(iv)}(x_i) - \cdots$$

Now, we can solve Eq. (12.27) for $f''(x_i)$, which yields

$$f''(x_i) = \frac{f(x_{i+2}) - 2f(x_{i+1}) + f(x_i)}{h^2} + O(h) \qquad (12.28)$$

where the error for the second-order derivative estimate is proportional to the size of the grid interval h and the third-order derivative:

$$O(h) \cong -hf'''(x_i) \qquad (12.29)$$

In fact, we can employ the very same script found after Eq. (12.18) in the previous section but with the new last line:

```
r=simple(ans.f_2prime,h); pretty(r)
```

which will yield

```
              f_iplus2 + f_i - 2 f_iplus1
-f_3prime h + ---------------------------
                            2
                           h
```

Similarly, the *backward difference approximation for the second-order derivative* can be found using $f(x_i)$, $f(x_{i-1})$, and $f(x_{i-2})$ as

$$f''(x_i) = \frac{f(x_i) - 2f(x_{i-1}) + f(x_{i-2})}{h^2} + O(h) \qquad (12.30)$$

where

$$O(h) \cong hf'''(x_i) \qquad (12.31)$$

The *centered difference approximation for the second-order derivative* is found by taking a Taylor series expansion for $f(x_{i-1})$ (Eq. (12.7)) and $f(x_{i+1})$ (Eq. (12.4)) to yield

$$f(x_{i+1}) = f(x_i) + hf'(x_i) + \frac{h^2}{2!}f''(x_i) + \frac{h^3}{3!}f'''(x_i) + \frac{h^4}{4!}f^{(iv)}(x_i) + \cdots$$
$$\qquad (12.32)$$
$$f(x_{i-1}) = f(x_i) - hf'(x_i) + \frac{h^2}{2!}f''(x_i) - \frac{h^3}{3!}f'''(x_i) + \frac{h^4}{4!}f^{(iv)}(x_i) + \cdots$$

Adding them together (rather than subtracting as in Eq. (12.10)) allows you to get rid of the first-order derivative

$$f(x_{i+1}) + f(x_{i-1}) = \left(f(x_i) + hf'(x_i) + \frac{h^2}{2!} f''(x_i) + \frac{h^3}{3!} f'''(x_i) + \frac{h^4}{4!} f^{(iv)}(x_i) + \cdots \right)$$

$$+ \left(f(x_i) - hf'(x_i) + \frac{h^2}{2!} f''(x_i) - \frac{h^3}{3!} f'''(x_i) + \frac{h^4}{4!} f^{(iv)}(x_i) - \cdots \right)$$

$$f(x_{i+1}) - f(x_{i-1}) = 2f(x_i) + h^2 f''(x_i) - \frac{h^4}{6} f^{(iv)}(x_i) + \cdots$$

(12.33)

Solving for $f''(x_i)$ yields

$$f''(x_i) = \frac{f(x_{i+1}) - 2f(x_i) + f(x_{i-1})}{h^2} + O(h^2) \qquad (12.34)$$

where the error is proportional to the square of the grid interval size, h^2, and the fourth-order derivative of the function

$$O(h^2) \cong -\frac{h^2}{12} f^{(iv)}(x_i) \qquad (12.35)$$

Again, we have developed the MATLAB script that computes this approximation at the end of Sec. 12.3 already. We only need to add one more term to each Taylor series expansion (otherwise, we will not get an error estimate) and modify the last line to output the second-order rather than the first-order derivative

```
%% Defining symbolic variables
syms h f_i f_iplus1 f_iminus1 f_prime f_2prime f_3prime f_4prime
%% Introducing two equations
Eq1 = -f_iplus1+f_i+h*f_prime+h^2*f_2prime/2+h^3*f_3prime/6 ...
    +h^4*f_4prime/24;
Eq2 = -f_iminus1+f_i-h*f_prime+h^2*f_2prime/2-h^3*f_3prime/6 ...
    +h^4*f_4prime/24;
%% Solving two equations for f_prime and f_2prime
ans = solve(Eq1,Eq2,f_prime,f_2prime);
%% Simplifying and displaying the result for f_2prime
r=simple(ans.f_2prime); pretty(r)
```

Running this script results in the following:

```
                                          2
f_iplus1 - 2 f_i + f_iminus1    f_4prime h
---------------------------- - ----------
              2                     12
             h
```

Now, we can summarize that obtaining approximations of higher-order derivatives unfolds as follows:

1. The pth-order derivative, $f_i^{(p)}$, needs at least $p+1$ data points.
2. The difference approximation is derived from the Taylor series expansions about x_i by combining them to eliminate the derivatives of the order less than p.
3. The error estimate is the lowest term truncated.

Let us have another look at the MATLAB script appearing after Eq. (12.20). It exploits three Taylor series expansions (utilizing four points) and, therefore, can be used to obtain an estimate of up to the third-order derivative. In fact, it does it already—look at the second to last command, which solves three equations for three derivatives. Hence, all we have to do is to append two more lines to this script:

```
r=simple(ans.f_2pr); pretty(r)
r=simple(ans.f_3pr); pretty(r)
```

to receive the four-point backward difference approximation for $f''(x_i)$

```
          2
11 f_4pr h      2 f_i - 5 f_im1 + 4 f_im2 - f_im3
----------- + --------------------------------
    12                          2
                               h
```

and $f'''(x_i)$

```
3 f_4pr h     f_i - 3 f_im1 + 3 f_im2 - f_im3
--------- + --------------------------------
    2                        3
    h
```

Similarly, the MATLAB script shown after Eq. (12.22) can be used to obtain approximations of up to the fourth-order derivative. For instance, if we append one more line to this script

```
r=simplify(sol.f_3p); pretty(r)
```

we will get a centered four-point approximation for the third-order derivative

```
         f_im2              f_ip2
f_im1 - ----- - f_ip1 + -----
           2                  2          2
                                      f_5p h
--------------------------------- - ------
                3                       4
               h
```

To get centered four-point approximations for the second- and fourth-order derivative together with the error terms we would need to add one more term to the Taylor series expansion as follows:

```
%% Defining symbolic variables
syms h f_i f_ip1 f_ip2 f_im1 f_im2 f_p f_2p f_3p f_4p f_5p f_6p
%% Introducing equations for four points
Eqp1 =-f_ip1+f_i+h*f_p+h^2*f_2p/2+h^3*f_3p/6+h^4*f_4p/24+h^5*f_5p/120 ...
     +h^5*f_6p/720;
Eqp2 =-f_ip2+f_i+2*h*f_p+4*h^2*f_2p/2+8*h^3*f_3p/6+16*h^4*f_4p/24 ...
     +32*h^5*f_5p/120+64*h^5*f_6p/720;
Eqm1 =-f_im1+f_i-h*f_p+h^2*f_2p/2-h^3*f_3p/6+h^4*f_4p/24-h^5*f_5p/120 ...
     +h^5*f_6p/720;
Eqm2 =-f_im2+f_i-2*h*f_p+4*h^2*f_2p/2-8*h^3*f_3p/6+16*h^4*f_4p/24 ...
     -32*h^5*f_5p/120+64*h^5*f_6p/720;
%% Resolving four equations for f_p, f_2p, f_3p and f_4p
sol=solve(Eqp1,Eqp2,Eqm1,Eqm2,f_p,f_2p,f_3p,f_4p);
%% Simplifying and displaying the result for f_3p and f_4p
r=simplify(sol.f_3p); pretty(r)
r=simplify(sol.f_4p); pretty(r)
```

Running this script yields

```
        f_im2                f_ip2
f_im1 - ----- - f_ip1 +      -----                2
          2                    2        f_5p h
------------------------------------- - ------
                  3                       4
                  h
```

for $f'''(x_i)$ and

```
6 f_i - 4 f_im1 + f_im2 - 4 f_ip1 + f_ip2     f_6p h
----------------------------------------- - ------
                  4                           6
                  h
```

for $f^{(iv)}(x_i)$. Tables 12.2–12.4 summarize our findings.

Analyzing these tables, the observation made in the beginning of Sec. 12.4 can be extended to the case of higher-order derivatives:

1. A pth-order derivative approximation involving k points ($k > p$) produces an error of the order of h^{k-p} at the worst.
2. Centered approximations are one order more accurate that asymmetric ones, that is, of the order of h^{k-p+1}.

Obviously, if there is no control over a grid size h, but a higher accuracy is required, the only way to achieve this goal is to increase a number of points k used to produce an approximation formula.

Finally, if we use the notations of Eqs. (12.23) and (12.24), for the backward difference approximations of the higher-order derivatives we can write

$$f^{(k)}(x_i) \approx \frac{1}{h^k} \Delta^k f(x_i) \tag{12.36}$$

Table 12.2 Approximations of the Second-Order Derivative $f''(x_i)$

Name	Formula	Error Estimate
Three-point forward difference	$\dfrac{f(x_{i+2}) - 2f(x_{i+1}) + f(x_i)}{h^2}$	$-hf'''(x_i)$
Three-point backward difference	$\dfrac{f(x_i) - 2f(x_{i-1}) + f(x_{i-2})}{h^2}$	$hf'''(x_i)$
Three-point centered difference	$\dfrac{f(x_{i+1}) - 2f(x_i) + f(x_{i-1})}{h^2}$	$-\dfrac{1}{12}h^2 f^{(iv)}(x_i)$
Four-point forward difference	$\dfrac{-f(x_{i+3}) + 4f(x_{i+2}) - 5f(x_{i+1}) + 2f(x_i)}{h^2}$	$\dfrac{11}{12}h^2 f^{(iv)}(x_i)$
Four-point backward difference	$\dfrac{2f(x_i) - 5f(x_{i-1}) + 4f(x_{i-2}) - f(x_{i-3})}{h^2}$	$\dfrac{11}{12}h^2 f^{(iv)}(x_i)$
Four-point centered difference	$\dfrac{-f(x_{i+2}) + 16f(x_{i+1}) - 30f(x_i) + 16f(x_{i-1}) - f(x_{i-2})}{12h^2}$	$\dfrac{1}{90}h^3 f^{(vi)}(x_i)$

12.6 Alternative Ways of Obtaining Approximation Formulas

The way of obtaining the approximation formulas presented in the previous sections is the best, because it also allows estimating the errors of approximations. However, there are some alternative ways of obtaining exactly the same approximation formulas. One of them is the *multiple-application scheme*.

As shown in the previous section, any formula provides an estimate of the derivative in the following form:

$$D \approx \hat{D}(h) + \hat{e}(h) \tag{12.37}$$

Here, the derivative approximation $\hat{D}(h)$ and the error approximation $\hat{e}(h)$ both depend on the grid size h. The error approximation also depends on the higher-order derivative at x_i, which is unknown (see the last column in Tables 12.1–12.4). The idea to improve an accuracy of the approximation is to somehow estimate $\hat{e}(h)$ without estimating this higher-order derivative at x_i.

Table 12.3 Approximations of the Third-Order Derivative $f'''(x_i)$

Name	Formula	Error Estimate
Four-point forward difference	$\dfrac{f(x_{i+3}) - 3f(x_{i+2}) + 3f(x_{i+1}) - f(x_i)}{h^3}$	$-\dfrac{3}{2}hf^{(iv)}(x_i)$
Four-point backward difference	$\dfrac{f(x_i) - 3f(x_{i-1}) + 3f(x_{i-2}) - f(x_{i-3})}{h^3}$	$\dfrac{3}{2}hf^{(iv)}(x_i)$
Four-point centered difference	$\dfrac{f(x_{i+2}) - 2f(x_{i+1}) + 2f(x_{i-1}) - f(x_{i-2})}{2h^3}$	$-\dfrac{1}{4}h^2 f^{(v)}(x_i)$

To do so, we apply the same approximation formula twice, with a single step size h and a double-step size, $2h$, which yields

$$D \approx \hat{D}(h) + \hat{e}(h) \approx \hat{D}(2h) + \hat{e}(2h) = \hat{D}(2h) + 2^m \hat{e}(h) \qquad (12.38)$$

where m is the order of approximation error ($k - p$ for noncentered and $k - p + 1$ for centered approximations). From this equation, it immediately follows that

$$\hat{e}(h) \approx \frac{\hat{D}(h) - \hat{D}(2h)}{2^m - 1} \qquad (12.39)$$

Substituting Eq. (12.39) back to Eq. (12.38) yields

$$D \approx \hat{D}(h) + \frac{\hat{D}(h) - \hat{D}(2h)}{2^m - 1} = \frac{2^m \hat{D}(h) - \hat{D}(2h)}{2^m - 1} \qquad (12.40)$$

The error of this later approximation is of the order of the first term in the *HOT* remainder of the exact approximation formula. Specifically, for noncentered approximations it is just the next order, that is, $m + 1$ (for example, see Eqs. (12.15) and (12.26)), for centered approximations because of cancellations it maybe even the next nearest order, that is, $m + 2$ (for example, see Eq. (12.10)). In numerical analysis, this approach is known as *Richardson extrapolation* and is widely used to improve the rate of convergence of numerical algorithms.

Table 12.4 Approximation of the Fourth-Order Derivative $f^{(iv)}(x_i)$

Name	Formula	Error Estimate
Four-point centered difference	$\dfrac{f(x_{i+2}) - 4f(x_{i+1}) + 6f(x_i) - 4f(x_{i-1}) + f(x_{i-2})}{h^4}$	$-\dfrac{1}{6}hf^{(vi)}(x_i)$

For instance, if we apply the suggested double-step scheme to the three-point centered difference approximation for $f''(x_i)$ with $m = 2$ (Table 12.2), we will end up with the following more accurate approximation:

$$D \approx \frac{4}{3}\hat{D}(h) - \frac{1}{3}\hat{D}(2h) \qquad (12.41)$$

Substituting

$$\hat{D}(h) \approx (f(x_{i+1}) - 2f(x_i) + f(x_{i-1}))/h^2$$

and $\qquad (12.42)$

$$\hat{D}(2h) \approx (f(x_{i+2}) - 2f(x_i) + f(x_{i-2}))/4h^2$$

results in

$$D \approx \frac{-f(x_{i+2}) + 16f(x_{i+1}) - 30f(x_i) + 16f(x_{i-1}) - f(x_{i-2})}{12h^2} \qquad (12.43)$$

which happens to be the four-point centered difference shown in the last row of Table 12.2.

Another way of obtaining exactly the same difference approximation formulas is to use k data points to produce a $(k - 1)$th-order fitting polynomial. Then, derivatives can be estimated by differentiating this polynomial and estimating the result at x_i. For instance, the following MATLAB script:

```
%% Defining symbolic variables
syms x_i h f_i f_ip1 f_ip2 f_ip3 x
%% Solving for coefficients of a cubic interpolation
A = [      x_i^3              x_i^2              x_i       1;
        (x_i+h)^3          (x_i+h)^2          x_i+h     1;
        (x_i+2*h)^3        (x_i+2*h)^2        x_i+2*h   1;
        (x_i+3*h)^3        (x_i+3*h)^2        x_i+3*h   1];
b= [f_i;f_ip1;f_ip2;f_ip3];
coef = A\b;
%% Forming the polynomial out of vector of its coefficients
Cpol=[x^3 x^2 x 1]*coef;
%% Computing the first-order derivative at x=x_i
D1 = diff(Cpol,1); D1=subs(D1,x,x_i); pretty(simplify(D1))
%% Computing the second-order derivative at x=x_i
D2 = diff(Cpol,2); D2=subs(D2,x,x_i); pretty(simplify(D2))
%% Computing the third-order derivative at x=x_i
D3 = diff(Cpol,3); D3=subs(D3,x,x_i); pretty(simplify(D3))
```

employs the Symbolic Math Toolbox to

1. Compute coefficients of the third-order polynomial passing through four data points (x_{i+v}, y_{i+v}), $v = 0, 1, 2, 3$.
2. Form the symbolic representation of this polynomial.
3. Find expressions for the first- through third-order derivatives and estimate them at x_i.

Running this script yields three expressions

```
  11 f_i - 18 f_ip1 + 9 f_ip2 - 2 f_ip3
- -------------------------------------
                  6 h


 2 f_i - 5 f_ip1 + 4 f_ip2 - f_ip3
 ---------------------------------
                 2
                h


   f_i - 3 f_ip1 + 3 f_ip2 - f_ip3
- -------------------------------
                 3
                h
```

which are the four-point forward difference approximations of the first-, second-, and third-order derivatives (compare them with that of Tables 12.1–12.3).

12.7 Partial Derivatives

The formulas obtained in the previous sections of this chapter can also be used to estimate partial derivatives in the case of a multivariable function. Consider, for example, the two-variable function $z = f(x, y)$. The difference approximation for the partial derivative

$$f_x = \frac{\partial}{\partial x} f(x, y) \text{ at } x = x_i \text{ and } y = y_i \qquad (12.44)$$

can be derived by fixing y at $y = y_i$ and finding the derivative of f with respect to x. Therefore, the different schemes derived earlier can be used to approximate the partial derivatives as well. For instance, with h_x being the grid interval size along x, the following formulas hold:

$$f_x(x_i, y_i) \approx \frac{f(x_i + h_x, y_i) - f(x_i, y_i)}{h_x} \qquad (12.45)$$

$$f_x(x_i, y_i) \approx \frac{f(x_i, y_i) - f(x_i - h_x, y_i)}{h_x} \qquad (12.46)$$

$$f_x(x_i, y_i) \approx \frac{f(x_i + h_x, y_i) - f(x_i - h_x, y_i)}{2h_x} \tag{12.47}$$

They represent forward, backward, and centered difference approximations of the first-order partial derivative, respectively.

The difference approximations for the second-order partial derivatives can be obtained in a similar manner. For instance, the *centered difference approximation* for *the second-order partial derivatives* of $f(x, y)$ at x_i and y_i, with h_y being the grid interval size along y, are

$$f_{xx} = \frac{\partial^2}{\partial x^2} f(x, y) \approx \frac{f(x_i + h_x, y_i) - 2f(x_i, y_i) + f(x_i - h_x, y_i)}{h_x^2}$$

$$f_{yy} = \frac{\partial^2}{\partial y^2} f(x, y) \approx \frac{f(x_i, y_i + h_y) - 2f(x_i, y_i) + f(x_i, y_i - h_y)}{h_y^2} \tag{12.48}$$

$$f_{xy} = \frac{\partial^2}{\partial x \partial y} f(x, y) \approx \frac{f(x_i + h_x, y_i + h_y) - f(x_i - h_x, y_i + h_y)}{4h_x h_y}$$

$$+ \frac{f(x_i + h_x, y_i - h_y) - f(x_i - h_x, y_i - h_y)}{4h_x h_y}$$

12.8 Using the MATLAB Functions

MATLAB does not have a single function realizing different order approximations discussed in the previous sections of this chapter. The only function that can be used to calculate derivatives (beside the close-related to this issue `gradient` and `polyder` functions to be discussed later in this section) is the `diff` function. Three basic syntaxes for this function are

```
d=diff(x)
d=diff(x,N)
d=diff(x,N,dim)
```

where x is either a vector of length n or matrix with m rows and n columns. If x is a vector then `diff(x)` calculates differences between adjacent elements of x, so that the result d is a vector of length $n - 1$. If x is a matrix, then d is a matrix with $m - 1$ rows and n columns containing the differences between adjacent rows in each column. For example,

```
>> x=[1,3,5,2,4,6];
>> d=diff(x)
```

returns

```
d = 2  2  -3  2  2
```

and

```
>> X=[1,2,3,4,5;5,4,3,2,1;1,3,5,2,4]
>> d=diff(X)
```

yields

```
X =     1    2    3    4    5
        5    4    3    2    1
        1    3    5    2    4
d =     4    2    0   -2   -4
       -4   -1    2    0    3
```

The following call

```
>> diff(x,2)
```

which is equivalent to

```
>> diff(diff(x))
```

returns

```
ans = 0 -5 5 0
```

that is, applies the `diff` function recursively N times (in our case twice because N = 2), resulting in the *N*th difference (Fig. 12.3).

The dimension of calculating the differences in the case where x is a matrix can be changed using a scalar `dim`, for instance, for the above matrix X the call

```
>> d=diff(X,[],2)
```

produces

```
d =     1    1    1    1
       -1   -1   -1   -1
        2    2   -3    2
```

(note that, you have to use a placeholder, [ ], because the order of arguments matters).

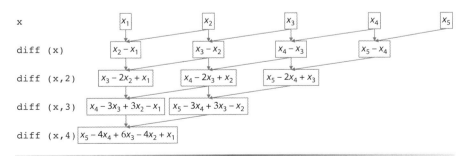

Fig. 12.3 Illustration of recursive application of the `diff` function.

How can we use the `diff` function to calculate derivatives? Assume we have two *n*-element vectors, x and y, containing the arguments and function values. Then, for the first $n-1$ data points we can compute the two-point forward difference approximation of the first-order derivative dy/dx by issuing the following command:

```
diff(y)./diff(x);
```

The very same vector of the dimension of $n-1$ can also be treated as the two-point backward difference approximation of the first-order derivative for the last $n-1$ data points (Table 12.1).

In the case of the evenly spaced argument with the grid interval h, the second-order derivative d^2y/dx^2 can be estimated in accordance with the three-point difference approximation [Eq. (12.30)] as

```
diff(x,2)/h^2;
```

The dimension of the resulting vector is $n-2$. The result can be treated as the three-point backward difference approximation corresponding to the data points 3 through *n*, or the three-point centered difference approximation (Eq. (12.34)) corresponding to the data points 2 through $n-1$, or the three-point forward difference approximation [Eq. (12.28)] corresponding to the data points 1 through $n-2$ (Table 12.2).

Similarly, the result of

```
diff(x,3)/h^3;
```

can be treated as the four-point backward difference approximation of d^3y/dx^3 corresponding to the data points 4 through *n*, or the four-point forward difference approximation corresponding to the data points 1 through $n-3$ (Table 12.3). The command

```
diff(x,4)/h^4;
```

produces the four-point centered difference approximation of d^4y/dx^4 corresponding to the data points 3 through $n-2$ and so on.

The following script shows an example of using the `diff` function to compute derivatives:

```
x=linspace(0,pi,50); h=x(2)-x(1);
y=sin(2*x).*cos(x);
plot(x,y,'o-.'), hold
plot(x(2:end),diff(y)/h,'+m')
plot(x(2:end-1),diff(y,2)/h^2,'pg'), grid
legend('Function','1^{st}-order derivative','2^{nd}-order derivative',0)
```

Running it results in the graph shown in Fig.12.4.

Note that, there is no value for the first-order derivative at the very first data point, and no second-order derivative estimates at the first and last data points (due to shortening the length of the vector every time you apply the `diff` function).

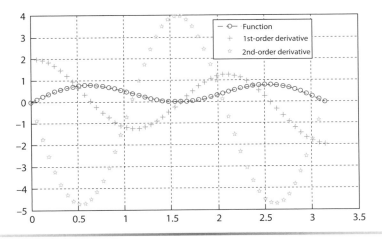

Fig. 12.4 Function $y = \sin(2x)\cos(x)$ and its numerically computed derivatives.

Besides `diff`, there is one more function that allows computing the numerical gradients (differences) of a multivariable function. For a function of N variables, $F(x, y, z,...)$, the gradient is defined as

$$\nabla F = \frac{\partial F}{\partial x}\mathbf{i} + \frac{\partial F}{\partial y}\mathbf{j} + \frac{\partial F}{\partial z}\mathbf{k} + \cdots \tag{12.49}$$

so the MATLAB `gradient` function returns the N components of the N-dimensional numerical gradient. Depending on dimensions of the input array `F`, the syntax for the `gradient` function is one of the following:

```
Fx=gradient(F)
[Fx,Fy]=gradient(F)
[Fx,Fy,Fz,...]=gradient(F)
[...] = gradient(F,h)
[...] = gradient(F,h1,h2,...)
```

If `F` is a vector (meaning that instead of pairs $(x_i, f(x_i))$, $i = 1, 2,..., n$ we are only given a set of $f(x_i)$, $i = 1, 2,..., n$), then `Fx` is a vector corresponding to dF/dx, the differences in the x direction. The two-point forward difference (Eq. (12.1)) is used to estimate the derivative at the first point, the two-point backward difference (Eq. (12.2)) is used to estimate the derivative at the very last point, and the two-point centered difference (Eq. (12.3)) is used otherwise. If the grid size `h` is not defined, it is assumed to be 1. The following example is intended to clarify this (try to calculate the gradient at the first, second, and last point yourself):

```
>> t=[1 3 6 7 3 4 6]
t = 1    3    6    7    3    4    6
```

```
>> gradient(t)
ans = 2.0000 2.5000 2.0000 -1.5000 -1.5000 1.5000 2.0000
>> gradient(t,10)
ans = 0.2000 0.2500 0.2000 -0.1500 -0.1500 0.1500 0.2000
```

If F is a matrix of values of the two-variable function F(x, y), then Fx returns the numerical gradient $\partial F/\partial x$, and Fy returns the numerical gradient $\partial F/\partial y$. Both Fx and Fy are the matrices of partial derivatives computed according to formulas similar to Eqs. (12.45)–(12.47). For the gradient of a function depending on more than two variables, that is, when F is an N-dimensional array, the gradient function returns N N-dimensional arrays.

The last two syntaxes of the gradient function assume either a scalar or vector as the input argument(s). For instance, in the previous example, the default value of 1 and then 10 were used to represent an evenly spaced argument. You could also use a vector input as in the following two examples:

```
>> t=[1 3 6 7 3 4 6]
t = 1    3    6    7    3    4    6
>> gradient(t,t)
ans = 1    1    1    1    1    1    1
>> gradient(t,0:0.1:0.7)
ans = 20.0000 25.0000 20.0000 -15.0000 -15.0000 15.0000 20.0000
```

In this case, the length of the vector h should be the same of that of the vector F.

In the case of a multivariable function, a single spacing value h specifies the spacing between the data points in every direction and N spacing values (h1, h2,...) specify the spacing for each dimension of F individually. Similarly, they can assume the scalar or vector values.

The following script presents an example of computing and displaying the gradient of the function of two variables, $F(x, y) = (x^2 + y^2) \cos(x) \sin(2y)$:

```
px=linspace(-pi/2,pi/2,40); hx=px(2)-px(1); % 1st argument
py=linspace(-pi,-pi/4,40); hy=py(2)-py(1); % 2nd argument
[x,y] = meshgrid(px,py);        % mesh grid of two arguments
F = (x.^2+y.^2).*cos(x).*sin(2.*y); % matrix of function values on the grid
[Fx,Fy] = gradient(F,hx,hy);    % computing the gradient
colormap('Spring')
contourf(px,py,F), hold on       % drawing a filled contour map
quiver(x,y,Fx,Fy), hold off      % adding the gradient
xlabel('x'), ylabel('y')
figure
waterfall(x,y,F), view([-60,30]), axis tight
xlabel('x'), ylabel('y'), zlabel('z')
```

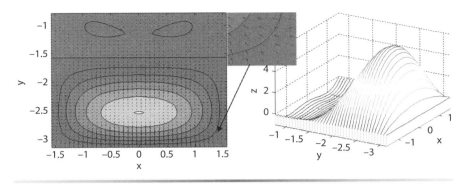

Fig. 12.5 Illustration of computing and showing the 2-D gradient.

First, it creates a mesh grid of arguments and computes the values of the function in each node creating a 40-elements-by-40-elements matrix. Then, it applies the `gradient` function specifically indicating grid size hx and `hy` to compute the 2-D gradient defined as

$$\nabla F = \frac{\partial F}{\partial x}\mathbf{i} + \frac{\partial F}{\partial y}\mathbf{j} \qquad (12.50)$$

This gradient can be thought of as a collection of vectors pointing in the direction of increasing values of F. The two components of these vectors, $\partial F/\partial x$ and $\partial F/\partial y$, are stored in two matrices Fx and Fy. Next, the function itself is being drawn as a filled (colored) contour map, and finally the gradient is added to this contour map as the arrows with components (Fx, Fy) at the points (x, y). For better understanding, the script also produces another 3-D plot of the same function. Running this script results in the two graphs shown in Fig. 12.5 (the inset between them is a zoomed right-bottom corner of the contour plot).

If a function for which we need to compute the derivative is defined analytically rather than a set of data points, you can use the `diff` function from the Symbolic Math toolbox (which is different from the `diff` function for computing numerical differences considered earlier). This function computes the derivative analytically as explained in Sec. 7.4.2. After this, you can compute the exact numerical values of derivatives at any point you want.

In addition to that, MATLAB offers one more function to deal with polynomials. In the special case of finding derivatives of polynomials (defined by their coefficients), the `polyder` function allows you to do it analytically. This function has three syntaxes

b=polyder(p)   Returns a vector containing the
                            coefficients of the derivative of the
                            polynomial defined by the vector of
                            coefficients (in descending powers) p

`b=polyder(p1,p2)`	Returns a vector containing the coefficients of the derivative of the product of the polynomials p1 and p2
`[n,d]=polyder(p1,p2)`	Returns the vectors n and d containing the coefficients of the numerator and denominator of the derivative of the quotient of polynomials p1 and p2

For example, a set of commands

```
p=[1 2 3 4 5];                      % defining a polynomial
disp('Original polynomial')
pretty(poly2sym(p))
b=polyder(p);                       % taking a derivative
disp('Derivative polynomial')
pretty(poly2sym(b))
```

results in the following output

```
Original polynomial
              4     3       2
          x + 2 x + 3 x + 4 x + 5
Derivative polynomial
                3     2
          4 x + 6 x + 6 x + 4
```

The same result can be achieved by employing the Symbolic Math toolbox `diff` function as well. The script

```
p=[1 2 3 4 5];
disp('Original polynomial')
pretty(poly2sym(p))
disp('First derivative')
d1=diff(poly2sym(p)); pretty(d1)
disp('Second derivative')
d2=diff(poly2sym(p),2); pretty(d2)
```

yields

```
Original polynomial
              4     3       2
          x + 2 x + 3 x + 4 x + 5
First derivative
                3     2
          4 x + 6 x + 6 x + 4
```

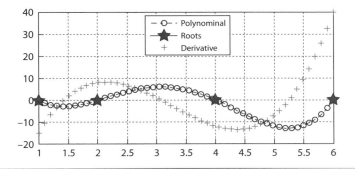

Fig. 12.6 Illustration of computing and showing the gradient.

```
Second derivative

          2
   12 x + 12 x + 6
```

The following script presents another example of first computing the first-order derivative of a polynomial analytically using the `polyder` function, and then using the `polyval` function to evaluate this derivative at a set of arguments:

```
r=[1,2,4,6]; a=poly(r); b=polyder(a);
x=linspace(r(1),r(end),50);
plot(x,polyval(a,x),'o-.'), hold on
plot(r,zeros(length(r),1),'Marker','p','MarkerFaceColor','r',...
                   'MarkerSize',15)
plot(x,polyval(b,x),'+m'), hold off, grid
legend('Polynomial','Roots','Derivative','Location','North')
```

Running this script results in the graph presented in Fig. 12.6. We can visually confirm the validity of the results by noting that the derivative crosses the zero line where the original polynomial has two minimums and one maximum.

Problems

12.1 Calculate the first- and second-order derivatives of $\cos(x)$ at $x = \pi/4$ with $h = \pi/100$ using the centered difference approximation (Eqs. (12.12) and (12.34)) and estimate the errors (for the higher-order derivatives using: a) analytical expressions and b) a multiple-application scheme Eq. (12.39).

12.2   Use the MATLAB Symbolic Math toolbox to obtain the centered six-point finite-difference approximations of the first-, second-, third-, and fourth-order derivatives. Also, find the errors associated with these approximations.

12.3   Use the MATLAB `diff` function to compute the centered difference approximation for the first-order derivative of a function given by

```
x = 0:0.1:10;
f = sin(x);
```

Compare the result with the values of analytically derived derivative ($f' = \cos(x)$) and less accurate estimates based on the forward and backward difference approximations. To this end, first subtract the numerically derived values of the first-order derivative from the true values (computed using an analytical formula). Plot a `bar`-graph showing the values of these discrepancies. Second, plot a graph for the approximation error $O(h^2)$ (Eq. (12.13)) based on analytically derived expression for the third-order derivative ($f''' = -\cos(x)$). At the same plot show the error you would have if the forward or backward difference approximation (Eqs. (12.6) and (12.9)) was used instead of the centered one (scale them if necessary to accommodate within the same axes).

Chapter 13 Numerical Integration

- Trapezoidal and Simpson's Formulas
- Understanding Newton–Cotes Formulas
- Understanding Gauss Quadratures
- Using the MATLAB Functions

Introduction

S uppose that during an experiment (simulation) you measured (recorded) some variable, say acceleration. Now you want to know an integral of this output, that is, velocity. It is possible to solve the problem analytically by using a table of integrals if you have an analytical formula for some variable. Yet, not every analytical expression can be integrated analytically. You may still need to compute the values given by this analytical formula over some interval and then use these "experimental" data values as if you did not have an analytical solution. Anyway, having discrete measurements of some explicit or implicit function, $y = f(x)$, given as n pairs (x_i, y_i), $i = 1, 2, ..., n$ (in our example, x stands for time and y for acceleration) we need to numerically estimate the integral

$$I = \int_{x_1}^{x_n} f(x)\,\mathrm{d}x \tag{13.1}$$

MATLAB offers several functions allowing you to do this, including trapz, accepting the pairs (x_i, y_i), as well as quad, quadgk, and quadl (plus dblquad and trapllequad for double and triple integrals, respectively), that rely on an analytical expression of the integrand. Although the first two functions (trapz and quad) are relatively simple and easy to understand, the latter two (quadgk and quadl) require more detailed explanation implying a broader review of a variety of methods that can be used for numerical integration. That is why this chapter starts from introducing a variety of the methods (Newton—Cotes integration formulas, multiple application scheme, and Gauss quadrature), showing the main ideas behind them, deriving underlying formulas using the Symbolic Math toolbox, and showing how to develop and program these methods yourself. Only after that will we discuss the functions available in MATLAB and see some examples of how to use them effectively.

13.2 Rectangle Method

Recall that integration of a function $f(x)$ represents an area under the curve defined by $f(x)$, but we may not have the curve itself. Instead, we have some discrete points as shown in Fig. 13.1. These points can be spaced unevenly (Fig. 13.1a) or evenly (Fig. 13.1b).

These pairs, (x_i, y_i), constitute some kind of dependence implicitly anyway, so that $y_i = f(x_i)$. Hence, the most obvious approach would be to represent a definite integral of Eq. (13.1), which represents the area under the imaginary curve passing through the data points, as a sum of rectangles with the width $x_i - x_{i-1}$, $i = 2, 3, \ldots, n$, and height of either $y_{i-1} = f(x_{i-1})$ (Fig. 13.2a) or $y_i = f(x_i)$ (Fig. 13.2b)

$$I^{LR} = \sum_{i=2}^{n} f(x_{i-1})(x_i - x_{i-1}) \quad \text{or} \quad I^{RR} = \sum_{i=2}^{n} f(x_i)(x_i - x_{i-1}) \qquad (13.2)$$

These formulas are known as the *left rectangles* (LR) *formula* and the *right rectangles* (RR) *formula*, respectively (Fig. 13.2). There is an obvious

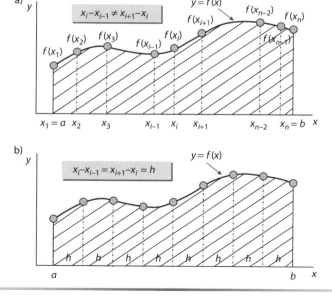

Fig. 13.1   Graphical representation of a definite integral of the function $f(x)$ with a) uneven and b) even sample points distribution.

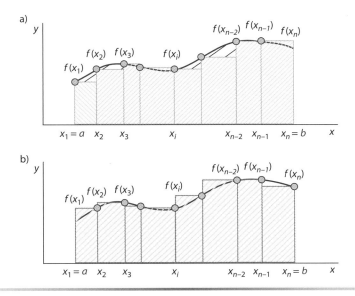

Fig. 13.2 a) Top-left and b) top-right corner approximations.

difference between the imaginary curve and the ladder-type approximation, resulting in an error. The obvious method to mitigate this error would be to have more points, so that the difference $x_i - x_{i-1}$ is very small; however, this is a last resort. Other approaches can be used before this, and the most obvious one would be simply taking an average between I^{LR} and I^{RR}, which results in the trapezoidal rule (discussed later).

13.3 Trapezoidal Rule

The *trapezoidal rule* (also known as the trapezoid or trapezium rule) is a numerical integration technique derived from a linear interpolation of the function $f(x)$ (as opposed to the ladder-type in the previous section) and, therefore, by substituting an exact value of the integral with a sum of individual trapezoids' areas as shown in Fig. 13.3.

Looking at an arbitrary point on the curve x_i, $2 < i \leq n$, the integral to that point can be defined as $I(x_i)$, which is the area under the curve from x_1 to x_i. Similarly, the integral to the next arbitrary point can be defined as $I(x_{i+1})$, which is the area under the curve from x_1 to x_{i+1}. The equation for the single area approximated by a trapezoid ΔI_i^{TR} is given by

$$\Delta I_i^{TR} = I(x_{i+1}) - I(x_i) = \int_{x_i}^{x_{i+1}} f^*(x)\,\mathrm{d}x + e_i = \frac{f(x_i) + f(x_{i+1})}{2}(x_{i+1} - x_i) + e_i$$

$$(13.3)$$

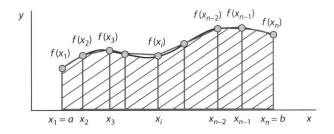

Fig. 13.3 Numerical integration using the trapezoidal rule.

where $f^*(x)$ is a piece-wise linear interpolation of the real (unknown) function $f(x)$, and e_i is the error between the exact solution and the trapezoidal rule solution for a single trapezoid, that is, a local truncation error. As seen, Eq. (13.3) is an average of two formulas of Eq. (13.2).

Applying the previous formula recursively to an approximation of the entire integral, from $x_1 = a$ to $x_n = b$, and assuming that for any $2 < i \leq n$, $x_i - x_{i-1} = h =$ const (we will use this not-restricting-generality assumption to derive compact formulas) we get

$$I^{TR}(x_n) = \int_a^b f^*(x)\,dx = \frac{1}{2}h\left(f(a) + 2\sum_{i=1}^{n-1} f(a+ih) + f(b)\right)$$
$$+ \sum_{i=1}^{n-1} e_i = \frac{1}{2}h\left(f_1 + 2\sum_{i=2}^{n-1} f_i + f_n\right) + E \qquad (13.4)$$

Here, the error E between the exact solution and the trapezoidal rule solution for the entire interval $[a; b]$, global truncation error, is proportional to the value averaged among all $n - 1$ intervals error $\bar{e}$, so that the following holds:

$$E = \sum_{i=1}^{n-1} e_i = \bar{e}(n-1) = \bar{e}\,\frac{b-a}{h} \qquad (13.5)$$

Note that, the order of the global truncation error E (in terms of a grid size h) is always one unit less than that of the local truncation error e_i. Moreover, because of the universality of Eq. (13.5), it applies to any method, and not necessarily to the trapezoidal method alone.

Now let us try to estimate these truncation errors, e_i and E. Assuming $I(x)$ is analytic, a Taylor series approximation can be made about x_i, yielding

$$I(x_{i+1}) = I(x_i) + hI'(x_i) + \frac{h^2}{2!}I''(x_i) + \frac{h^3}{3!}I'''(x_i) + \cdots \qquad (13.6)$$

By making obvious substitutions

$$I'(x_i) = f(x_i), \ I''(x_i) = f'(x_i), \ I'''(x_i) = f''(x_i) \tag{13.7}$$

we can rewrite Eq. (13.6) as

$$I(x_{i+1}) = I(x_i) + hf(x_i) + \frac{h^2}{2!} f'(x_i) + \frac{h^3}{3!} f''(x_i) + \cdots \tag{13.8}$$

Next, using a simple two-point forward difference approximation for $f'(x_i)$ (Eq. (12.5))

$$f'(x_i) \cong \frac{f(x_i + h) - f(x_i)}{h} - \frac{h}{2} f''(x_i) + \cdots \tag{13.9}$$

and substituting it into Eq. (13.8) we obtain

$$I(x_{i+1}) = I(x_i) + hf(x_i) + \frac{h^2}{2!} \left(\frac{f(x_i + h) - f(x_i)}{h} - \frac{h}{2} f''(x_i) + \cdots \right)$$

$$+ \frac{h^3}{3!} f''(x_i) + \cdots \tag{13.10}$$

Collecting the terms and noting that $x_{i+1} = x_i + h$ yields

$$I(x_{i+1}) = I(x_i) + \frac{h}{2} \left(f(x_i) + f(x_{i+1}) \right) - \frac{h^3}{12} f''(x_i) + O(h^4) \tag{13.11}$$

By comparing the terms in Eq. (13.11) with those of Eq. (13.3), we finally find that

$$e_i \cong -\frac{h^3}{12} f''(x_i) \tag{13.12}$$

Using the Mean Value Theorem, we may further write

$$\sum_{i=1}^{n-1} f''(x_i) = (n-1) f''(\overline{x}) = \frac{b-a}{h} f''(\overline{x}) \tag{13.13}$$

where $\overline{x}$ is some representative point within the $[a; b]$ interval ($a \le \overline{x} \le b$), and substitute the result into Eq. (13.12) and further into Eq. (13.5) to obtain an estimate of the error E for the entire interval $[a; b]$ as

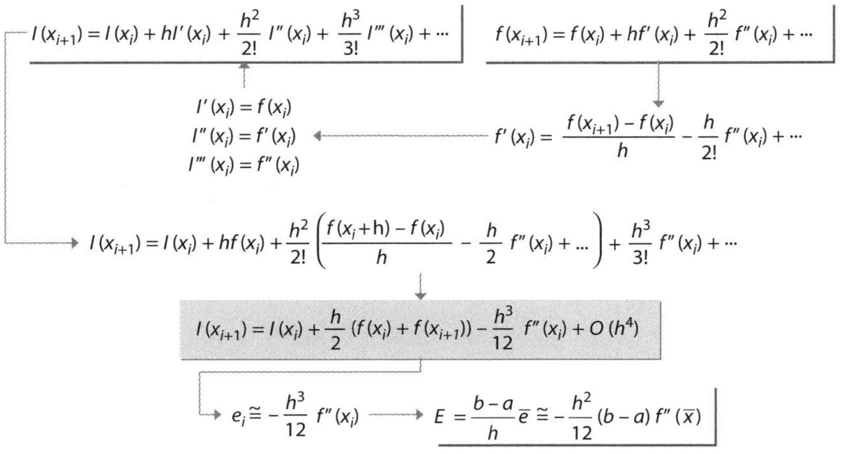

$$I(x_{i+1}) = I(x_i) + hI'(x_i) + \frac{h^2}{2!}I''(x_i) + \frac{h^3}{3!}I'''(x_i) + \cdots \qquad f(x_{i+1}) = f(x_i) + hf'(x_i) + \frac{h^2}{2!}f''(x_i) + \cdots$$

$$I'(x_i) = f(x_i)$$
$$I''(x_i) = f'(x_i) \qquad\qquad f'(x_i) = \frac{f(x_{i+1}) - f(x_i)}{h} - \frac{h}{2!}f''(x_i) + \cdots$$
$$I'''(x_i) = f''(x_i)$$

$$I(x_{i+1}) = I(x_i) + hf(x_i) + \frac{h^2}{2!}\left(\frac{f(x_i + h) - f(x_i)}{h} - \frac{h}{2}f''(x_i) + \cdots\right) + \frac{h^3}{3!}f''(x_i) + \cdots$$

$$I(x_{i+1}) = I(x_i) + \frac{h}{2}(f(x_i) + f(x_{i+1})) - \frac{h^3}{12}f''(x_i) + O(h^4)$$

$$e_i \cong -\frac{h^3}{12}f''(x_i) \qquad\longrightarrow\qquad E = \frac{b-a}{h}\bar{e} \cong -\frac{h^2}{12}(b-a)f''(\bar{x})$$

Fig. 13.4 Formal approach to estimate the truncation error for trapezoidal rule.

$$E \cong -\frac{h^2}{12}(b-a)f''(\bar{x}) \qquad\qquad (13.14)$$

For the sake of clarity, all transformations starting from Eq. (13.6) down to Eq. (13.14) are shown in Fig. 13.4. This helps to better understand the following script that automates the derivation of these formulas using the MATLAB Symbolic Math toolbox:

```
%% Defining symbolic variables
syms I_i I_iplus1 I_prime I_2prime I_3prime h
syms f_i f_iplus1 f_prime f_2prime
%% Introducing Taylor series equation for the integral
Eq = -I_iplus1+I_i+h*I_prime+h^2*I_2prime/factorial(2)+ ...
                    h^3*I_3prime/factorial(3);
%% Making the first set of substitutes
Eq = subs(Eq,{I_prime I_2prime I_3prime},{f_i f_prime f_2prime});
%% Substituting function derivative with forward difference approximation
Eq = subs(Eq,f_prime,(f_iplus1-f_i)/h-h*f_2prime/2);
%% Computing a single area under the curve
ans = solve(Eq,I_iplus1)-I_i;
%% Simplifying and displaying the result
r=collect(ans,h); pretty(r)
```

Running this script results in

```
/   f_2prime \    3   /  f_i     f_iplus1 \
| - ------   |  h  + |  --- +  -------- |  h
\      12    /        \   2        2     /
```

showing resemblance with Eq. (13.11).

To summarize our findings,

1. The trapezoidal rule is based on integration of the linear interpolation of the function.
2. The global truncation error associated with the trapezoidal rule for a fixed interval $[a; b]$ is proportional to h^2 [Eq. (13.14)]. Therefore, the trapezoidal rule is referred to as a second-order method.

For the record, if in Eq. (13.8) we only leave three terms, throwing away the term proportional to the second-order derivative, we will have the formula for the left rectangles along with an estimate of a local truncation error for this method, $e_i \cong (h^2/2) f'(x_i)$, which makes the global error of $E \cong -(h/2)(b - a) f'(x)$.

Hence, by utilizing a linear interpolation of the function as opposed to the ladder-type interpolation for the rectangle method, we improved the accuracy of numerical estimation of a definite integral by one order of h. This accuracy still depends on h, meaning that by increasing the number of points within the same interval (decreasing h), we can decrease the error, but even with the same h the second-order method assures a better result. For example, if $h = 0.1$, then the global error for the rectangles formula will be of the order of 0.1, whereas for the trapezoidal formula it will be of the order of 0.01.

Now it becomes clear how to improve the accuracy even more—we should simply use the higher-order interpolation formulas. And that is exactly what is being done in the following two sections.

13.4 Simpson's 1/3 Rule

So far, we have used a ladder-type and linear interpolation of a function between the data points. Now let us proceed with the second-order interpolation, which results in the *Simpson's 1/3 rule*. Figure 13.5 illustrates our intentions, showing that we will now use double intervals to produce parabolic interpolation passing through every three consecutive points, 1-2-3, 3-4-5, 5-6-7, etc.

A quadratic (second order) polynomial interpolation can be represented as

$$f^*(x) = Kx^2 + Lx + M \tag{13.15}$$

Hence, for each three points (double interval $[x_i; x_{i+2}]$, $i = 2k-1$, $k = 1, 2, \ldots$) we need to determine the coefficients K, L, and M. The requirement that at x_i, $x_{i+1} = x_i + h$, and $x_{i+2} = x_i + 2h$ the interpolation should pass through

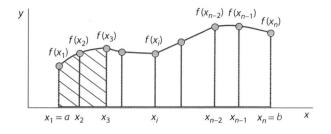

Fig. 13.5 Numerical integration using quadratic interpolation (Simpson's 1/3 rule).

$f_i = f(x_i), f_{i+1} = f(x_{i+1}),$ and $f_{i+2} = f(x_{i+2})$. This results in the following three linear algebraic equations:

$$
\begin{bmatrix}
x_i^2 & x_i & 1 \\
(x_i + h)^2 & x_i + h & 1 \\
(x_i + 2h)^2 & x_i + 2h & 1
\end{bmatrix}
\begin{bmatrix}
K \\
L \\
M
\end{bmatrix}
=
\begin{bmatrix}
f_i \\
f_{i+1} \\
f_{i+2}
\end{bmatrix}
\tag{13.16}
$$

Being resolved, this system yields

$$
K = \frac{f_i - 2f_{i+1} + f_{i+2}}{2h^2}, \quad L = -\frac{3f_i - 4f_{i+1} + f_{i+2}}{2h} \text{ and } M = f_i \tag{13.17}
$$

Now, we have this second-order interpolation, the integral for each double interval may be estimated as

$$
\Delta I_i^{S\frac{1}{3}} = \int_{x_i}^{x_i + 2h} f^*(x)\,dx + e_i = \frac{8}{3} h^3 K + 2h^2 L + 2hM + e_i \tag{13.18}
$$

or substituting the coefficients of the interpolation [Eq. (13.17)] back in it as

$$
\Delta I_i^{S\frac{1}{3}} = \frac{1}{3} h \big(f(x_i) + 4f(x_i + h) + f(x_i + 2h) \big) + e_i \tag{13.19}
$$

This numerical integration technique is credited to the mathematician Thomas Simpson and, therefore, is known as "*Simpson's 1/3 rule*" [because of the 1/3 multiplier in Eq. (13.19)]. The following script employs Symbolic Math toolbox and allows you to arrive at the same Eq. (13.19):

```
%% Defining symbolic variables
syms x_i f_i f_iplus1 f_iplus2 h x
```

```
%% Solving for coefficients of a parabolic interpolation
A = [                 x_i^2              x_i        1;
                 (x_i+h)^2          x_i+h        1;
                 (x_i+2*h)^2        x_i+2*h      1];
b= [f_i; f_iplus1; f_iplus2];
coef = A\b;
%% Computing the integral
dI = int([x^2, x, 1]*coef, x_i, x_i+2*h);
%% Simplifying and displaying the result
[dI,how] = simple(dI); pretty(dI)
```

It returns

```
h (f_i + 4 f_iplus1 + f_iplus2)
-------------------------------
              3
```

To estimate local and global truncation errors, we need to use the Taylor series expansions similarly to how we did it in the previous section. Figure 13.6 shows how to do it step-by-step and utilizes the most accurate centered difference approximation of the second-order derivative from the previous chapter [Eqs. (12.34) and (12.35)]. This results in

$$e_i \cong -\frac{h^5}{90} f^{(iv)}(x_i) \qquad (13.20)$$

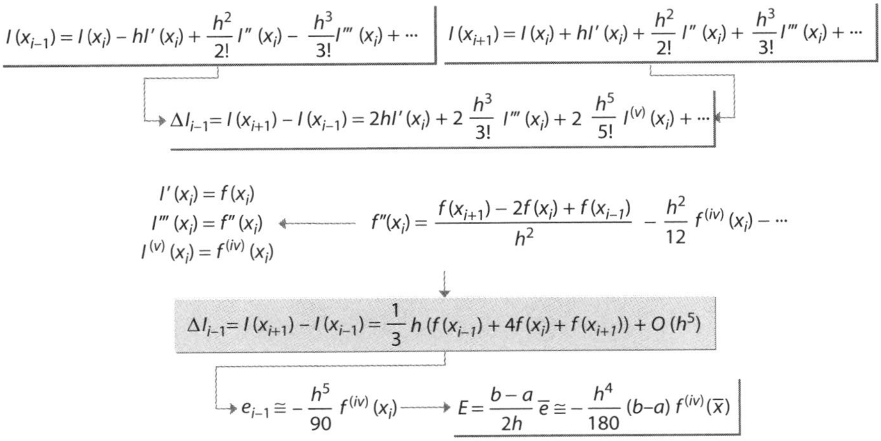

$$I(x_{i-1}) = I(x_i) - hI'(x_i) + \frac{h^2}{2!}I''(x_i) - \frac{h^3}{3!}I'''(x_i) + \cdots \qquad I(x_{i+1}) = I(x_i) + hI'(x_i) + \frac{h^2}{2!}I''(x_i) + \frac{h^3}{3!}I'''(x_i) + \cdots$$

$$\rightarrow \Delta I_{i-1} = I(x_{i+1}) - I(x_{i-1}) = 2hI'(x_i) + 2\frac{h^3}{3!}I'''(x_i) + 2\frac{h^5}{5!}I^{(v)}(x_i) + \cdots$$

$$I'(x_i) = f(x_i)$$
$$I'''(x_i) = f''(x_i) \leftarrow \qquad f''(x_i) = \frac{f(x_{i+1}) - 2f(x_i) + f(x_{i-1})}{h^2} - \frac{h^2}{12} f^{(iv)}(x_i) - \cdots$$
$$I^{(v)}(x_i) = f^{(iv)}(x_i)$$

$$\Delta I_{i-1} = I(x_{i+1}) - I(x_{i-1}) = \frac{1}{3} h \left(f(x_{i-1}) + 4f(x_i) + f(x_{i+1}) \right) + O(h^5)$$

$$\rightarrow e_{i-1} \cong -\frac{h^5}{90} f^{(iv)}(x_i) \longrightarrow E = \frac{b-a}{2h} \bar{e} \cong -\frac{h^4}{180} (b-a) f^{(iv)}(\bar{x})$$

Fig. 13.6 Formal approach to estimate the truncation error for the Simpson's 1/3 rule.

and

$$E \cong -\frac{h^4}{180}(b-a)f^{(iv)}(\overline{x}) \qquad (13.21)$$

where, $a \leq \overline{x} \leq b$ as usual.

The following script allows you to do it automatically rather than by hand:

```
%% Defining symbolic variables (assume middle point to be i-th point)
syms I_i I_prime I_2prime I_3prime I_4prime I_5prime h
syms f_iminus1 f_i f_iplus1 f_prime f_2prime f_3prime f_4prime
%% Tylor series expansions for I_plus1 and I_minus1
Eq1=I_i+h*I_prime+h^2*I_2prime/factorial(2)+h^3*I_3prime/factorial(3)+...
            h^4*I_4prime/factorial(4)+h^5*I_5prime/factorial(5);
Eq2=I_i-h*I_prime+h^2*I_2prime/factorial(2)-h^3*I_3prime/factorial(3)+...
            h^4*I_4prime/factorial(4)-h^5*I_5prime/factorial(5);
Eq3=Eq1-Eq2; % Subtracting I_minus1 from I_plus1
%% Making the first set of substitutes
Eq4=subs(Eq3,{I_prime,I_2prime,I_3prime,I_4prime,I_5prime},...
            {f_i,f_prime,f_2prime,f_3prime,f_4prime});
%% Substituting central diff. approximation for the 2nd-order derivative
Eq5=subs(Eq4,f_2prime,(f_iminus1-2*f_i+f_iplus1)/h^2-1/12*h^2*f_4prime);
%% Simplifying and displaying the result
[r,how]=simple(Eq5); r=collect(r); pretty(r)
```

It returns

```
/    f_4prime  \   5     / 4 f_i    f_iplus1     f_iminus1 \
|  - --------  | h  +    | -----  + --------  +  --------- | h
\      90      /         \   3         3             3     /
```

Applying Eq. (13.20) recursively to all double intervals within $[a; b]$, we get a formula for the *Composite Simpson's 1/3 rule*

$$I^{S\frac{1}{3}} = \int_a^b f(x)\,dx$$

$$= \frac{1}{3}h\left(f(a) + 4\sum_{\substack{i=1 \\ \text{odd}\,i}}^{n-1} f(a+ih) + 2\sum_{\substack{i=2 \\ \text{even}\,i}}^{n-2} f(a+ih) + f(b) \right) + E \qquad (13.22)$$

or

$$I^{S\frac{1}{3}}(x_n) = \frac{1}{3}h(f_1 + 4f_2 + 2f_3 + 4f_4 + 2f_5 + \cdots + 2f_{n-2} + 4f_{n-1} + f_n) + E \qquad (13.23)$$

(Note that Eqs. (13.22) and (13.23) assume odd n, that is, we have an even number of intervals. If we happen to have one extra interval, that is, we have the even number of data points n to start with, then the trapezoidal rule needs to be applied to either the very first or the very last interval, then proceeding with the Simpson's 1/3 rule in a regular manner.)

To summarize, increasing the order of an interpolation from linear to quadratic (by a unit) allowed us to improve the accuracy of integration by two orders of the grid size h, bringing the global truncation error to be of the order of h^4. Hence, the Simpson's 1/3 rule is referred to as a fourth-order method).

Should we expect more if we further increase the order of interpolation? The following section provides you with an answer to this question.

13.5 Simpson's 3/8 Rule

Following the same routine introduced in two previous sections to increase the accuracy of numerical integration, the *Simpson's 3/8 rule* employs a third-order polynomial interpolation technique. As shown in Fig. 13.7, triple intervals—1-2-3-4, 4-5-6-7, etc.—are used instead of double the intervals of Fig. 13.5.

To save space, let us employ the following script to obtain the numerical integration formula in this case

```
%% Defining symbolic variables
syms x_i f_i f_iplus1 f_iplus2 f_iplus3 h x
%% Solving for coefficients of a cubic interpolation
A = [           x_i^3              x_i^2           x_i      1;
              (x_i+h)^3          (x_i+h)^2        x_i+h     1;
            (x_i+2*h)^3        (x_i+2*h)^2      x_i+2*h     1;
            (x_i+3*h)^3        (x_i+3*h)^2      x_i+3*h     1];
b= [f_i; f_iplus1; f_iplus2; f_iplus3];
coef = A\b;
%% Computing the integral
dI = int([x^3, x^2, x, 1]*coef, x_i, x_i+3*h);
%% Simplifying and displaying the result
[dI,how] = simple(dI); pretty(dI)
```

This script solves for the coefficients of the third-order polynomial interpolation between four points, (x_i, f_i), $(x_i + h, f_{i+1})$, $(x_i + 2h, f_{i+2})$, and $(x_i + 3h, f_{i+3})$, and then analytically integrates this interpolation between x_i and $x_i + 3h$ to yield

```
3 h (f_i + 3 f_iplus1 + 3 f_iplus2 + f_iplus3)
-----------------------------------------------
```

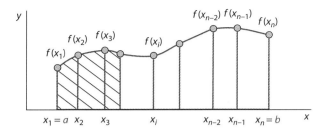

Fig. 13.7 Numerical integration using cubic interpolation (Simpson's 3/8 rule).

Therefore,

$$\Delta I_i^{S\frac{3}{8}} = \int_{x_i}^{x_i+3h} f^*(x)\,\mathrm{d}x + e_i = \frac{3}{8}h\big(f(x_i) + 3f(x_i + h)$$
$$+ 3f(x_i + 2h) + f(x_i + 3h)\big) + e_i \quad (13.24)$$

To estimate a local truncation error, e_i, we need to perform Taylor series expansion computations similar to those shown in Figs. 13.4 and 13.6, which results in

$$e_i \cong -\frac{3h^5}{80} f^{(iv)}(x_i) \quad (13.25)$$

Unexpectedly, this error happens to be of the same order as that of the Simpson's 1/3 rule [Eq. (13.21)]. Hence, increasing the order of interpolation from quadratic to cubic did not lead to significant improvement of the accuracy as it took place when the order on interpolation was increased from linear to quadratic.

The composite Simpson's 3/8 rule is still a fourth-order method with

$$I^{S\frac{3}{8}}(x_n) = \frac{3}{8}h(f_1 + 3f_2 + 3f_3 + 2f_4 + 3f_5 + \cdots + 3f_{n-2} + 3f_{n-1} + f_n) + E \quad (13.26)$$

and

$$E \cong -\frac{3h^4}{160}(b-a)f^{(iv)}(\overline{x}) \quad (13.27)$$

Similar to Eq. (13.23), the number of intervals in Eq. (13.26) is supposed to be divisible by three, that is, $n = 3k+1$, $k = 1, 2, \ldots$. If, however, the number of data points is such that there are one or two extra intervals, the trapezoidal or Simpson's 1/3 rule should be applied to this interval(s), respectively.

13.6 Newton–Cotes Formulas

In the previous sections, it was demonstrated that we can increase the accuracy of numerical integration by employing the higher-order interpolation formulas. These interpolation formulas produce mth-order polynomial equations from $m+1$ data points. Being integrated within the corresponding range (covering these $m+1$ data points), these interpolations result in the *Newton—Cotes integration closed formulas*. The Newton–Cotes formulas are represented as follows:

$$\int_{x_i}^{x_i+mh} f^*(x)\,dx = \alpha h(w_1 f(x_i) + w_2 f(x_{i+1}) + w_3 f(x_{i+2}) + \cdots + w_{m+1} f(x_{i+m})) \quad (13.28)$$

where m is the number of intervals, α the multiplier, and w_i, $i = 1, 2, ...,$ $m+1$ the weighting coefficients. Obviously, the trapezoidal rule and the two Simpson's formulas are the members of the Newton–Cotes integration formulas. Table 13.1 presents them and a few more formulas in the standardized form.

Note that as was the case with the Simpson's 1/3 and Simpson's 3/8 rules, the five- and six-point formulas have the same order error. This general characteristic holds for the higher-point formulas as well and results in the even number of intervals (odd number of points), formulas (for example, Simpson's 1/3 rule and Boole's rule) being the methods of preference.

For the higher number of intervals, some of coefficients w_i may become negative, degrading the accuracy due to round-off errors. Hence, in engineering practice, the higher-order formulas are rarely used, so that it is the Simpson's rules that are used in most applications. That is why MATLAB

Table 13.1   Weighting Coefficients and Errors for the Newton–Cotes Closed Formulas

Number of Intervals m	Multiplier α	Weighting Factors w_i	Order of Error E	Known as
1	1/2	1 1	h^2	Trapezoidal rule
2	1/3	1 4 1	h^4	Simpson's 1/3 rule
3	3/8	1 3 3 1	h^4	Simpson's 3/8 rule
4	2/45	7 32 12 32 7	h^6	Boole's rule
5	5/288	19 75 50 50 75 19	h^6	

offers only two functions, `trapz` and `quad`, realizing the trapezoidal and Simpson's rules, respectively (to be considered in Sec. 13.9).

In addition to that, the Newton—Cotes integration formulas come with the estimates of errors, so if you have a luxury of getting as many data points as you need (for instance, when the function $f(x)$ is given analytically or $f(x_i)$ being the outputs of some script or model), you can always control (improve) the accuracy by adjusting the grid size h appropriately. That allows you to satisfy any error tolerance requirement.

Gauss quadrature (*quadrature* is an old term for the process of measuring areas) is another attractive alternative to the Newton—Cotes integration formulas, and will be considered in Sec. 13.8. Before we proceed with considering this alternative, let us use the multiple-application scheme (introduced in Sec. 12.6), which can be used for either improving the accuracy of Newton—Cotes integration formulas without introducing the new data points, or adjusting the current grid size h, based on the quick estimates of the local truncation error e_i to keep it within the given tolerance.

13.7 Multiple-Application Scheme

Consider the same rule, say trapezoidal or Simpson's 1/3 rule, applied to compute an integral using two different grid sizes, h_1 and h_2 ($h_1 > h_2$). Then we can write

$$I(h_1) + E(h_1) = I(h_2) + E(h_2) \qquad (13.29)$$

Assume that $f''(\overline{x})$ (or $f^{(iv)}(\overline{x})$ for the Simpson's rule) is about the same for both grids. Then, following the notations of Table 13.1, we can write

$$E(h_1) \cong E(h_2)\left(\frac{h_1}{h_2}\right)^{2m} \qquad (13.30)$$

(for trapezoidal rule $m = 1$, for Simpson's 1/3 rule $m = 2$, etc.).

Substituting Eq. (13.30) into Eq. (13.29) and resolving the latter one with respect to $E(h_2)$ yields

$$E(h_2) \cong \frac{I(h_1) - I(h_2)}{1 - (h_1/h_2)^{2m}} \qquad (13.31)$$

[cf. with Eq. (12.39)]. Therefore, without applying formulas for truncation errors, like Eqs. (13.14) or (13.21), we were able to estimate the error of the numerical integral with a thinner grid, $E(h_2)$, so that now we can enjoy a more accurate estimate

$$I^{MA} = I(h_2) + E(h_2) = I(h_2) + \frac{I(h_2) - I(h_1)}{(h_1/h_2)^{2m} - 1} \tag{13.32}$$

[cf. with Eq. (12.40)]. In the special case, when we are using the original grid and the double intervals, that is, when $h_2 = (1/2)h_1 \equiv h$, Richardson extrapolation, given by Eq. (13.32), yields

$$I^{MA} = I(h) + \frac{I(h) - I(2h)}{4^m - 1} = \frac{4^m I(h) - I(2h)}{4^m - 1} \tag{13.33}$$

Specifically, for the trapezoidal and Simpson's 1/3 rule

$$\hat{I}^{TR} = \frac{4}{3} I^{TR}(h) - \frac{1}{3} I^{TR}(2h) \quad \text{and} \quad \hat{I}^{S\frac{1}{3}} = \frac{16}{15} I^{S\frac{1}{3}}(h) - \frac{1}{15} I^{S\frac{1}{3}}(2h) \tag{13.34}$$

where $\hat{I}^{TR}$ and $\hat{I}^{S\frac{1}{3}}$ are more accurate integrals compared with the original ones, I^{TR} and $\hat{I}^{S\frac{1}{3}}$, respectively. Using a formal approach (similar to what was shown in Figs. 13.4 and 13.6, but using more terms in the Taylor expansion series), it can be shown that the error of this multiple-application formula is two orders of h higher than that of the original rule. Therefore, $\hat{I}^{TR}$ features the error of the order of h^4 (compared to h^2 for I^{TR}) and $\hat{I}^{S\frac{1}{3}}$ is accurate within h^6 (compared to h^4 for $\hat{I}^{S\frac{1}{3}}$).

In his online book, Cleve Moler developed an interactive GUI, quadgui, to visualize how the MATLAB quad function, exploiting the Simpson's 1/3 rule, controls the error of numerical integration. Figure 13.8 shows several snapshots of this GUI. As seen, the integration of some function starts with computing a minimum number of data points to enable application of the Simpson's 1/3 rule twice, on two double intervals as shown in Fig. 13.8a. From the other hand, the same integral can be estimated using only three points, first, third, and fifth, that is, on a rougher grid. Since in this specific case, $|I(h) - I(2h)|$ happens to be greater than some specified error tolerance ε, the quad function adds two additional data points halving the original two intervals (Fig. 13.8b).

The entire *step-halving procedure* is then repeated again and since the error tolerance is still not met, the quad function keeps introducing the new data points halving the intervals. Once

$$|I(h) - I(2h)| \le \varepsilon \tag{13.35}$$

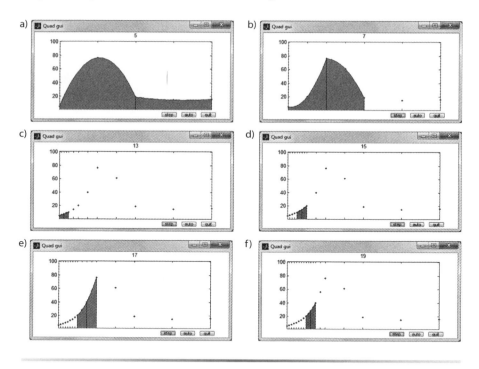

Fig. 13.8 Employing a multiple-application scheme to keep integration error below the tolerance.

(Fig. 13.8c), the value of the integral is computed according to the second of Eq. (13.35). Next, the quad function adds two more data points to halve the following two intervals and proceeds with the multiple-application scheme on them. In our particular case, the error tolerance happens to be met again (Fig. 13.8d). At the very next step (Fig. 13.8e), the difference $|I(h) - I(2h)|$ exceeds ε. Hence, this latter step needs to be repeated on a tinnier grid (Fig. 13.8f) and so on.

13.8 Gauss Quadrature

The Newton—Cotes formulas considered in the previous sections were based on an evenly spaced grid. Consequently, the location of the grid nodes used in these equations was predetermined or fixed. But what if you have a luxury to compute the function values on the set of arbitrary points? Then, by positioning these points wisely we could "balance" local negative and positive errors e_i, so that the accuracy of the overall estimate could be sufficiently improved.

Gauss quadrature is the name for one class of techniques to implement such a strategy. Let us illustrate this technique on the following example. Say, we want to develop a formula to estimate an integral by only using two predetermined points, which are the terminal points of the interval $[a; b]$, using the following approximation:

$$\int_a^b f(x)\,dx \cong c_a f(a) + c_b f(b) \tag{13.36}$$

The two unknown coefficients c_a and c_b of Eq. (13.36) can be determined from the requirement that this formula must yield exact solutions for $f(x)$ being: 1) a constant and 2) a straight line. These two conditions yield two linear algebraic equations

$$c_a + c_b = \int_a^b 1\,dx = b - a \quad c_a a + c_b b = \int_a^b x\,dx = \frac{b^2 - a^2}{2} \tag{13.37}$$

From these equations, it follows that

$$c_a = c_b = \frac{b - a}{2} \tag{13.38}$$

which gives the already known trapezoidal rule formula [Eq. (13.3)].

Now, let us free the location of two points in Eq. (13.36), so that

$$\int_a^b f(x)\,dx \cong c_1 f(x_1) + c_2 f(x_2) \tag{13.39}$$

Here in addition to coefficients c_1 and c_2, we have two more varied parameters x_1 and $x_2, a \le x_1 \le x_2 \le b$. To simplify mathematics and to make the formulation as general as possible, let us map the original domain $[a; b]$ to the new canonical domain $[-1; 1]$, that is, let us assume that $a = -1$ and $b = 1$, so that $-1 \le x_1 \le x_2 \le 1$. The objective is to find an "optimum" location of points x_1 and x_2 and corresponding weighting coefficients c_1 and c_2. For instance, we may request that Eq. (13.39) yields an exact solution for two more functions, that is, 1) constant, 2) linear function, 3) parabolic function, and 4) cubic function, which results in four equations

$$c_1 f(x_1) + c_2 f(x_2) = c_1 + c_2 = \int_{-1}^1 1\,dx = 2$$

$$c_1 f(x_1) + c_2 f(x_2) = c_1 x_1 + c_2 x_2 = \int_{-1}^1 x\,dx = 0$$

$$c_1 f(x_1) + c_2 f(x_2) = c_1 x_1^2 + c_2 x_2^2 = \int_{-1}^1 x^2\,dx = \frac{2}{3} \tag{13.40}$$

$$c_1 f(x_1) + c_2 f(x_2) = c_1 x_1^3 + c_2 x_2^3 = \int_{-1}^1 x^3\,dx = 0$$

Being resolved, these four equations yield

$$c_1 = c_2 = 1 \quad \text{and} \quad x_1 = -x_2 = -3^{-0.5} \approx -0.5773503 \tag{13.41}$$

which can be substituted back into Eq. (13.39) to yield the *two-point Gauss–Legendre formula*

$$\int_{-1}^{1} f(x)\,dx \cong f\left(\frac{-1}{\sqrt{3}}\right) + f\left(\frac{1}{\sqrt{3}}\right) \tag{13.42}$$

This result can be obtained using the Symbolic Math toolbox as follows:

```
%% Defining symbolic variables
syms c1 c2 x1 x2
%% Defining four algebraic equations
Eq1= c1+c2-2;
Eq2= c1*x1+c2*x2;
Eq3= c1*x1^2+c2*x2^2-2/3;
Eq4= c1*x1^3+c2*x2^3;
%% Solving equations
coef=solve(Eq1,Eq2,Eq3,Eq4,c1,c2,x1,x2);
%% Simplifying and displaying the results
Weights = [coef.c1(1) coef.c2(1)];
Location = [coef.x1(1) coef.x2(1)];
pretty(Weights)
pretty(Location)
WeightsD=sym2poly(Weights)
LocationD=sym2poly(Location)
```

This script results in the weighting factors and locations of two points as follows:

```
+-      -+
| 1,  1 |
+-      -+

+-              -+
|  1/2      1/2  |
|  3         3   |
|  ----,  - ---- |
|  3         3   |
+-              -+
WeightsD =   1                1
LocationD =  0.5774         -0.5774
```

even a pretty representation requires some assistance—what is shown is

$$\frac{1}{3}3^{\frac{1}{2}} = \frac{1}{3}\sqrt{3} = \frac{1}{\sqrt{3}}$$

In the more general case, we can use as many points as we want

$$\int_{-1}^{1} f(x)\,dx \cong \sum_{i=1}^{n} c_i f(x_i) \tag{13.43}$$

finding $2n$ varied parameters by providing the exact solution to up to the $(2n-1)$th-order polynomial. For example, with the three points we can obtain exact solutions for up to the fifth-order polynomial. The following script, which is similar to the previous one, allows you to obtain the locations of these three points and their corresponding weights:

```
%% Defining symbolic variables
syms c1 c2 c3 x1 x2 x3 x
%% Defining six algebraic equations
Eq1= sum([c1 c2 c3].*[1 1 1])-2;
Eq2= sum([c1 c2 c3].*[x1 x2 x3]);
Eq3= sum([c1 c2 c3].*[x1^2 x2^2 x3^2])-int(x^2,-1,1);
Eq4= sum([c1 c2 c3].*[x1^3 x2^3 x3^3]);
Eq5= sum([c1 c2 c3].*[x1^4 x2^4 x3^4])-int(x^4,-1,1);
Eq6= sum([c1 c2 c3].*[x1^5 x2^5 x3^5]);
%% Solving equations
coef=solve(Eq1,Eq2,Eq3,Eq4,Eq5,Eq6,c1,c2,c3,x1,x2,x3);
%% Simplifying and displaying the results
Weights =  [coef.c1(1) coef.c2(1) coef.c3(1)];
Location = [coef.x1(1) coef.x2(1) coef.x3(1)];
pretty(Weights), pretty(Location)
WeightsD=sym2poly(Weights)
LocationD=sym2poly(Location)
```

This script returns

```
+-              -+
| 5 5   8  |
| -,  -,  -  |
| 9 9   9  |
+-              -+

+-                            -+
|       1/2        1/2        |
| 15            15            |
| -----,  - -----,  0 |
|    5              5      |
+-                            -+
```

```
WeightsD   =    0.8889    0.5556    0.5556
LocationD = 0-0.7746    0.7746
```

The values for c's and x's for up to and including the seven-point formula are summarized in Table 13.2, whereas Fig. 13.9 shows a graphical representation of the location and weighting coefficients for the two-, three- and six-point schemes. The estimate of an error of the n-point Gauss quadrature rule for the interval $[a;b]$ is as follows:

$$E(\Delta X, n) = \frac{\Delta X^{2n+1}(n!)^4}{(2n+1)\left((2n)!\right)^3} f^{(2n)}(\overline{x}) = \Delta X^{2n+1} E(n), \quad a < \overline{x} < b \qquad (13.44)$$

($\Delta X = b - a$). Comparison of Gauss quadrature rules to Newton–Cotes formulas (Table 13.1) indicates the superiority of the former ones meaning that much higher accuracy can be achieved using less data points (once again, in the case we have a luxury to pick any point we want).

Now, let us go back and show how to map any arbitrary domain $[a;b]$ to the universal domain $[-1; 1]$ that the formulas of Table 13.2 were developed for. Obviously, it can be done by a simple change of original argument x to x^*

$$x = \frac{a+b}{2} + \frac{b-a}{2} x^* \qquad (13.45)$$

(when $x^* = 1$, $x = b$, and when $x^* = -1$, $x = a$). This implies

$$dx = \frac{b-a}{2} dx^* \qquad (13.46)$$

so that

$$\int_a^b f(x)\,dx = \frac{b-a}{2} \int_{-1}^1 f\left(\frac{a+b}{2} + \frac{b-a}{2} x^*\right) dx^*$$

$$\cong \frac{b-a}{2} \sum_{i=1}^n c_i f\left(\frac{a+b}{2} + \frac{b-a}{2} x_i\right) \qquad (13.47)$$

A numerical example will be provided in the following section.

If we attempt to use the multiple-application scheme, described in Sec. 13.7 for the Newton–Cotes formulas, to the Gauss quadrature rules, it will not work. The reason is that when subdividing the original interval $[-1; 1]$, the Gauss evaluation points of the new subintervals will never coincide with the previous evaluation points (except at zero for odd numbers), and thus the

Table 13.2 Arguments, Weighting Factors, and Errors for the Gauss Quadrature Formulas

Number of Points n	Function Argument x_i	Weight Factor c_i	Error $E(n)$
2	$x_1 = -x_2 = -0.577350269$	$c_1 = c_2 = 1$	$2.3 \times 10^{-4}\, f^{(iv)}(\overline{x})$
3	$x_1 = -x_3 = -0.774596669$ $x_2 = 0$	$c_1 = c_3 = 0.5555556$ $c_2 = 0.8888889$	$5.0 \times 10^{-7}\, f^{(vi)}(\overline{x})$
4	$x_1 = -x_4 = -0.861136312$ $x_2 = -x_3 = -0.339981044$	$c_1 = c_4 = 0.3478548$ $c_2 = c_3 = 0.6521452$	$5.6 \times 10^{-10}\, f^{(viii)}(\overline{x})$
5	$x_1 = -x_5 = -0.906179846$ $x_2 = -x_4 = -0.538469931$ $x_3 = 0$	$c_1 = c_5 = 0.2369969$ $c_2 = c_4 = 0.4786287$ $c_3 = 0.5688889$	$3.9 \times 10^{-13}\, f^{(x)}(\overline{x})$
6	$x_1 = -x_6 = -0.932469514$ $x_2 = -x_5 = -0.661209386$ $x_3 = -x_4 = -0.238619186$	$c_1 = c_6 = 0.1713245$ $c_2 = c_5 = 0.3607616$ $c_3 = c_4 = 0.4679139$	$1.9 \times 10^{-16}\, f^{(xii)}(\overline{x})$
7	$x_1 = -x_7 = -0.949107912$ $x_2 = -x_6 = -0.741531186$ $x_3 = -x_5 = -0.405845151$ $x_4 = 0$	$c_1 = c_7 = 0.1294850$ $c_2 = c_6 = 0.2797054$ $c_3 = c_5 = 0.3818301$ $c_4 = 0.4179592$	$6.5 \times 10^{-20}\, f^{(xiv)}(\overline{x})$

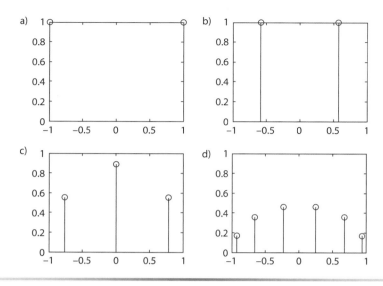

Fig. 13.9 Visualization of the points location and weight factors of a) trapezoidal rule, and b) Gauss two-, c) three- and d) six- point schemes.

integrand will need to be evaluated at every new point. Gauss–Kronrod formulas are extensions of Gauss quadrature formulas constructed by adding $n+1$ points to an original n-point Gauss quadrature rule, so that the function values at these n points can be re-used to compute the higher-order estimates. The difference between the n-point Gauss quadrature and its $(2n+1)$-point Kronrod extension,

$$\left| I_n^G - I_{2n+1}^K \right|$$

is used to estimate the local error and adjust the step (integration interval) in a manner similar to what was shown in Fig. 13.8.

A popular example, realized in the MATLAB function `quadgk`, combines a seven-point Gauss rule presented in the last row of Table 13.2 (Fig. 13.10a) with a 15-point Kronrod rule. This 15-point formula utilizes the seven function values of the seven-point Gauss rule (with the different weights though) and adds eight additional points as shown in Table 13.3 (the seven-point Gauss rule arguments are shown in the boldface). Figure 13.10b presents a portrait of the 15-point Kronrod rule overlaid over the seven-point Gauss rule.

As discussed earlier, the n-point Gauss quadrature rules are constructed to yield an exact result for up to the $(2n-1)$th-order polynomial. Say with only two points, the Gauss quadrature portrait looks as presented in Fig. 13.9b. The trapezoidal rule also uses two points, but is designed using another paradigm, so its portrait given in Fig. 13.9a is different from that of

Fig. 13.9b. Therefore, other approaches in choosing the locations x_i and weights c_i, $i = 1, 2, ..., n$ in Eq. (13.43) are also possible.

It turns out that the Gauss quadrature nodes are the roots of the Legendre polynomials $P_n(x)$

$$P_n(x) = \frac{1}{2^n n!} \frac{d^n}{dx^n} ((x^2 - 1)^n) \tag{13.48}$$

(note, this is an explicit formula compared to the recursive definition of Table 7.6). The Gauss quadrature weights can also be expressed via $P_n(x)$ as

$$c_i = \frac{2}{(1 - x_i^2) P_n'(x_i)^2} \tag{13.49}$$

For example,

$$P_0(x) = 1, \; P_1(x) = x, \; P_2(x) = \frac{1}{2}(3x^2 - 1), \; P_3(x) = \frac{1}{2}(5x^3 - 3x), \text{ etc.} \tag{13.50}$$

and, therefore, the arguments for the two-point and three-point rules (Table 13.2) are

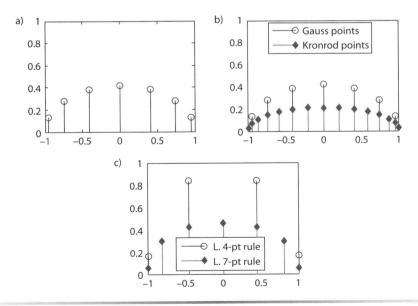

Fig. 13.10 Visualization of the argument location and weight factors for a) seven-point Gauss quadrature rule, b) Gauss–Kronrod 7–15 pair, and c) Lobatto 4–7 pair.

Table 13.3   15-Point Kronrod Rule Points

Function Argument x_i	Weight Factor c_i
$x_1 = -x_{15} = -0.991455372$	$c_1 = c_{15} = 0.0229353$
$x_2 = -x_{14} = \mathbf{-0.949107912}$	$c_2 = c_{14} = 0.0630921$
$x_3 = -x_{13} = -0.864864423$	$c_3 = c_{13} = 0.1047900$
$x_4 = -x_{12} = \mathbf{-0.741531186}$	$c_4 = c_{12} = 0.1406533$
$x_5 = -x_{11} = -0.586087236$	$c_5 = c_{11} = 0.1690047$
$x_6 = -x_{10} = \mathbf{-0.405845151}$	$c_6 = c_{10} = 0.1903506$
$x_7 = -x_9 = -0.207784955$	$c_7 = c_9 = 0.2044330$
$x_8 = 0$	$c_8 = 0.2094821$

$$x_i = \pm\frac{1}{\sqrt{3}} \quad \text{and} \quad x_i = 0; \pm\sqrt{\frac{3}{5}} \tag{13.51}$$

respectively. According to Eq. (13.49), the weight factors for the two-point and three-point rules are

$$c_i = \frac{2}{(1-x_i^2)9x_i^2} = 1 \quad \text{and} \quad c_i = \frac{8}{9(1-x_i^2)(5x_i^2-1)^2} = \frac{8}{9}; \frac{5}{9} \tag{13.52}$$

That is the reason why the Gauss quadrature is often referred to as *Gauss–Legengre quadrature*.

Following the same idea, the nodes x_i in Eq. (13.43) may be chosen to be the roots of other special polynomials: Jacobi, Chebyshev, Laguerre, Hermite (see Table 7.6), and that is why *Gauss–Jacobi, Gauss–Chebyshev, Gauss–Laguerre,* and *Gauss–Hermite* quadratures also exist (for the sake of accuracy, the integrand in this case is represented as $w(x)f(x)$, with $w(x)$ being some weighting function, like $(1-x^2)^{-0.5}$, e^{-x}, or $\exp(-x^2)$, and the limits may differ from that of $[-1; 1]$). These quadratures have some special applications. For instance, in the recent years they are widely used in the *pseudrospectral methods* for optimization of a trajectory of a dynamic system, described by ordinary differential equations.

MATLAB does not have the special functions for these types of quadratures, but it features one more function realizing another adaptive quadrature (like the Gauss–Kronrod 7–15 pair). The *Gauss–Lobatto* or simply Lobatto quadrature is similar to the Gauss quadrature, except that two arguments (integration points) in Eq. (13.43) are fixed and represent the end points of integration interval

$$\int_{-1}^{1} f(x)\,dx \cong I_n^L = c_1 f(-1) + \sum_{i=2}^{n-1} c_i f(x_i) + c_n f(1) \tag{13.53}$$

The remaining nodes x_i, $i = 2, 3, ..., n-1$ are defined as the roots of $P'_{n-1}(x)$ [Eq. (13.48)]. Finding n varied parameters, unknown weighting coefficients c_i, $i = 1, 2, ..., n$, is accomplished by the requirement to provide an exact solution to up to the $(n-1)$th-order polynomial, similar to that of Eq. (13.40).

For example, consider $n = 4$. In this case $P_3(x) = (1/2)(5x^3-3x)$ [Eq. (13.50)], $P'_3(x) = (3/2)(5x^2-1)$, and therefore

$$x_{2;3} = \pm\frac{1}{\sqrt{5}} \tag{13.54}$$

Owing to symmetry, $c_4 = c_1$ and $c_3 = c_2$, for two unknown coefficients c_1 and c_2 we can write

$$2(c_1 f(x_1) + c_2 f(x_2)) = 2(c_1 + c_2) = \int_{-1}^{1} 1\,\mathrm{d}x = 2,\ 2(c_1 f(x_1) + c_2 f(x_2))$$

$$= 2\left(c_1 + c_2\frac{1}{5}\right) = \int_{-1}^{1} x^2\,\mathrm{d}x = \frac{2}{3} \tag{13.55}$$

which results in $c_1 = 1/6$ and $c_2 = 5/6$. Therefore, the four-point Lobatto quadrature rule can be written as

$$I_4^L = \frac{1}{6}(f(-1) + f(1)) + \frac{5}{6}\left(f\left(-\frac{1}{\sqrt{5}}\right) + f\left(\frac{1}{\sqrt{5}}\right)\right) \tag{13.56}$$

Similarly, for say a seven-point Lobatto quadrature we may obtain

$$I_7^L = \frac{11}{210}(f(-1) + f(1)) + \frac{72}{245}\left(f\left(-\sqrt{\frac{2}{3}}\right) + f\left(\sqrt{\frac{2}{3}}\right)\right)$$

$$+ \frac{125}{294}\left(f\left(-\frac{1}{\sqrt{5}}\right) + f\left(\frac{1}{\sqrt{5}}\right)\right) + \frac{16}{35}f(0) \tag{13.57}$$

Notably the seven-point rule uses the four function values of the four-point rule, thus increasing computational effectiveness (Fig. 13.10c). The following script utilizes the MATLAB Symbolic Math toolbox to prove the solution:

```
%% Defining symbolic variables
syms c1 c2 c3 c4 x
```

```
%% Defining four algebraic equations
Eq1= sum([c1 c2 c3 c4].*[2 2 2 1])-2;
Eq2= sum([c1 c2 c3 c4].*[2 2*(2/3) 2/5 0])-int(x^2,-1,1);
Eq3= sum([c1 c2 c3 c4].*[2 2*(2/3)^2 2/5^2 0])-int(x^4,-1,1);
Eq4= sum([c1 c2 c3 c4].*[2 2*(2/3)^3 2/5^3 0])-int(x^6,-1,1);
%% Solving equations
coef=solve(Eq1,Eq2,Eq3,Eq4,c1,c2,c3,c4);
%% Simplifying and displaying the results
Weights = [coef.c1(1) coef.c2(1) coef.c3(1) coef.c4(1)];
pretty(Weights)
```

It returns the weighting coefficients of Eq. (13.57)

```
+-                       -+
|  11    72   125   16   |
|  ---,  ---, ---,  --   |
|  210   245  294   35   |
+-                       -+
```

The `quadl` function uses Eqs. (13.56) and (13.57) as a pair. It estimates I_7^L and the difference

$$\left| I_4^L - I_7^L \right|$$

which serves as an estimate of the local error. Based on this latter estimate, the decision is made on whether to proceed with computations (if the error tolerance is met) or halve the range and apply this pair again (similarly to the idea shown in Fig. 13.8).

13.9 Using the MATLAB Functions

As mentioned in the previous sections, MATLAB features the following functions for numerical integration (as usual more syntax is available via the MATLAB Help system):

`T=trapz(x,y)` — Uses trapezoidal integration to compute the integral of y with respect to x, where y is an array that contains the values of the function at the points contained in x

`Q=quad('fun',a,b,tol)` — Uses a recursive (multiple application) Simpson's 1/3 rule to compute the integral of fun with a as the lower limit and b as the upper limit. The optional parameter tol specifies the desired error tolerance (the default tolerance is 10^{-6}). Alternative syntax, [Q, fcnt]=quad(...), returns

	the number of function evaluations, `fcnt`, as well
`Q=quadgk('fun',a,b)`	Evaluates the integral of `fun` using the Gauss–Kronrod 7- and 15-point quadrature pair. Alternative syntax, `[Q,err]=quadgk(...)`, returns the estimate of the error `err` as well
`Q=quadl('fun',a,b,tol)`	Uses a recursive adaptive Lobatto integration algorithm (four- and seven-point formulas) to compute the integral from `a` to `b` with the error tolerance of `tol`. It may also return the number of function evaluations, if applied as `[q,fcnt]=quadl(...)`
`Q=dblquad('fun',xmin, xmax,ymin,ymax, tol,method)`	Employs either the `quad` (`method` is omitted) or `quadl` (`method` is specified as `@quadl` or `'quadl'`) to evaluate a double integral of `function` with the limits of integration of `xmin` to `xmax` and `ymin` to `ymax`
`Q=triplequad('fun', xmin,xmax,ymin, ymax,zmin,zmax, tol,method)`	Employs either the `quad` (`method` is omitted) or `quadl` (`method` is specified as `@quadl` or `'quadl'`) to evaluate a triple integral of `function` over the 3-D rectangular region

Note that in the aforementioned calls, the integrand `fun` is supposed to be a vectorized expression capable of accepting a vector of parameters rather than a single scalar argument. For example, the call

```
quadl('sin(x)*cos(x)',0,pi/2)
```

will not work and returns an error. To fix the problem, you should either vectorize it explicitly

```
quadl('sin(x).*cos(x)',0,pi/2)
```

or implicitly, using the vectorize function introduced in Sec. 5.8.1

```
quadl(vectorize('sin(x)*cos(x)'),0,pi/2)
```

In the case of the `quad` function, you may also take advantage of its vectorized version, `quadv`, which `vectorizes` the input function automatically

```
>> quadv('sin(x)*cos(x)',0,pi/2)
ans = 0.5000
```

Depending on your specific application, the MATLAB quad function is probably all you will need, especially if you are dealing with nonsmooth integrands, that is, in the case when the function values are the outputs of some script or model and generally take some time to compute. If, however, you are interested in integrating smooth analytical integrands at higher accuracies, you may want to try the quadl function. The quadgk function is more suitable for smooth oscillatory integrands with moderate singularities at the endpoints or infinite intervals.

Let us consider several examples and start from a simple integral

$$\int_0^\pi \sin(x)\,dx \qquad (13.58)$$

First, we can compute the exact value of this definite integral using the Symbolic Math toolbox. To this end, the following commands:

```
>> syms x
>> f=int('sin(x)');
>> fun=inline(char(f));
>> I=fun(pi)-fun(0)
```

return

```
I = 2
```

To be able to use the trapezoidal rule (the trapz function), we need to create the data points (x and y vectors). Let us go with as little as 10 points

```
>> x=linspace(0,pi,9); y=sin(x);
```

Now, applying the following two commands allows you to compute the integral and the relative error

```
>> ITR=trapz(x,y)
>> fprintf('Trapezoidal rule error is %3.2f %% \n',abs(ITR-2)/2*100)
ITR = 1.9797
Trapezoidal rule error is 1.02 %
```

We can also try to use Eq. (13.14) to estimate an absolute error

```
>> Eest=abs((pi/9)^2*pi/12)
```

which returns the upper bound

```
Eest = 0.0319
```

Since the second-order derivative of the function, $|f''(\bar{x})|$, is lesser than one, we can expect the actual error to be even smaller than that estimate. Indeed, it is true

```
>> Eact=2-ITR
Eact = 0.0203
```

Let us also try to implement any of the rectangle formulas of Eq. (13.3) (due to symmetry, both of them return the same result)

```
>> ILR=sum(y(1:end-1).*(diff(x)))
>> fprintf('Rectangle rule error is %3.2f %% \n',abs(ILR-2)/2*100)
ILR = 1.9797
Rectangle rule error is 1.02 %
```

Again, due to the symmetry of the integrand within the specified range of Eq. (13.58), in this particular case the results coincide with that of the trapezoidal formula.

We may also want to explore the quad function. For example,

```
>> IS13=quad('sin',0,pi,1e-1)
>> fprintf('Simpson rule error is %4.4e \n',abs(IS13-2))
```

returns

```
IS13 = 2.0000
Simpson rule error is 6.5035e-006
```

Here, the first command is intended to compute the integral within the specified limits degrading an absolute error tolerance to as low as 0.1, whereas the second command compares the result with the exact value and shows the actual error. As seen in this particular example, the actual error happens to be much smaller than the predefined tolerance.

Then, let us explore Gauss quadrature formulas. To use them (as explained in Sec. 13.8), we need to change the domain. Using Eq. (13.43), we have

$$\int_0^\pi \sin(x)\,dx = \frac{\pi}{2}\int_{-1}^1 \sin\left(\frac{\pi}{2}+\frac{\pi}{2}x^*\right)dx^* \tag{13.59}$$

The following commands compute this integral using the Gauss quadrature formulas for two and three points (Table 13.2), also showing the absolute error:

```
>> x=1/sqrt(3);
>> IG2=pi*(sin((pi+pi*x)/2)+sin((pi-pi*x)/2))/2;
>> x=sqrt(15)/5; c1=5/9; c2=8/9;
>> IG3=pi*(c1*sin((pi+pi*x)/2)+c1*sin((pi-pi*x)/2)+c2*sin(pi/2))/2;
>> disp([IG2, 2-IG2])
>> disp([IG3, 2-IG3])
1.9358    0.0642
2.0014   -0.0014
```

As seen, with only three wisely distributed points we were able to outperform the trapezoidal method with 10 evenly distributed points.

As presented in the beginning of this section, the `quadl` and `quadgk` functions need no domain change and can be used similarly to the `quad` function considered earlier. For the sake of comparison, let us analyze the return of these three functions being called as follows:

```
>> [qG,nG]=quad(@sin,0,pi,1e-1);
>> [qL,nL]=quadl(@sin,0,pi,1e-1);
>> [qK,errK]=quadgk(@sin,0,pi);
>> disp([qG,nG; qL,nL; qK,errK])
 2.0000    13.0000
 2.0000    18.0000
 2.0000     0.0000
```

Surely, the desired tolerance of 0.1 played no effect, so that the accuracy of computations turned out to be much higher than that, even with the minimum number of function evaluations (which happens to be slightly smaller for the Simpson's 1/3 adaptive rule). However, if we request a better accuracy,

```
>> [qG,nG]=quad(@sin,0,pi,1e-6);
>> [qL,nL]=quadl(@sin,0,pi,1e-6);
>> disp([qG,nG; qL,nL])
 2.0000    57.0000
 2.0000    18.0000
```

the number of function evaluations in `quadl` remains the same, whereas `quad` requires more and more resources. The effectiveness of the Gauss quadrature rules compared with Newton–Cotes formulas become even more clear if you require even a better accuracy

```
>> [qG,nG]=quad(@sin,0,pi,1e-10);
>> [qL,nL]=quadl(@sin,0,pi,1e-10);
>> disp([qG,nG; qL,nL])
 2.0000   225.0000
 2.0000    48.0000
```

Let us proceed with a triple integral

$$\int_0^\pi \int_0^1 \int_{-1}^1 (y\sin(x) + z\cos(x))\,\mathrm{d}x\,\mathrm{d}y\,\mathrm{d}z \tag{13.60}$$

The following commands solve the problem using the `triplquad` function

```
>> integrnd=@(x,y,z) y*sin(x)+z*cos(x);
>> I=triplequad(integrnd, 0, pi, 0, 1, -1, 1)
```

returning

```
I = 2.0000
```

which happens to be an exact value.

Finally, let us consider one more example and find the dimensionless volume of the nozzle shown in Fig. 13.11 (similar to that shown in Fig. 6.35c). This nozzle was obtained by a rotation of the curve given by the equation $r(z) = 2 + \cos(z)$, $0.5\pi \le z \le 2\pi$ about the z-axis, and then stretched along this axis by the factor of 5 (the `cylinder` command creates a cylinder 1 unit tall). The volume of this body is given by

$$\int_0^5 \pi r^2 \, dz = \int_0^5 \pi \left(2 + \cos\left(\frac{\pi}{2} + \frac{2\pi - 0.5\pi}{5} z \right) \right)^2 dz \qquad (13.61)$$

The following script computes the volume of the nozzle using trapezoidal, Simpson's 1/3 and 3/8 formulas using six equally spaced intervals, and compares it with a more accurate estimate obtained using the `quad` function:

```
f=@(x) pi*(2+cos(pi/2+x*(2*pi-pi/2)/5)).^2;
z=linspace(0,5,7); s=f(z);
ITR=trapz(z,s);
h=z(2)-z(1);
IS13=h*(s(1)+4*s(2)+2*s(3)+4*s(4)+2*s(5)+4*s(6)+s(7))/3;
IS38=3*h*(s(1)+3*s(2)+3*s(3)+3*s(4)+3*s(5)+3*s(6)+s(7))/8;
I=quad(f,0,5);
fprintf('Method       | Value | Rel. Error\n')
fprintf('Trapezoidal  | %4.2f | %3.2f %%\n',ITR, abs(ITR-I)/I*100)
fprintf('Simpson 1/3  | %4.2f | %3.2f %%\n',IS13,abs(IS13-I)/I*100)
fprintf('Simpson 3/8  | %4.2f | %3.2f %%\n',IS38,abs(IS38-I)/I*100)
fprintf('Exact value  | %4.2f |\n',I)
```

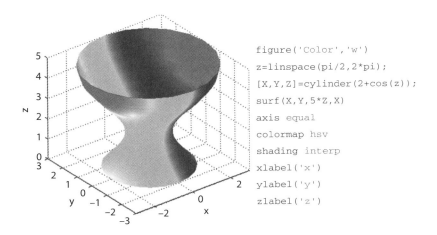

```
figure('Color','w')
z=linspace(pi/2,2*pi);
[X,Y,Z]=cylinder(2+cos(z));
surf(X,Y,5*Z,X)
axis equal
colormap hsv
shading interp
xlabel('x')
ylabel('y')
zlabel('z')
```

Fig. 13.11 Body of revolution about z-axis with corresponding commands.

This results in the following output:

```
Method       | Value | Rel. Error
Trapezoidal  | 58.05 | 1.21 %
Simpson 1/3  | 57.32 | 0.05 %
Simpson 3/8  | 58.92 | 2.73 %
Exact value  | 57.35 |
```

Problems

13.1   Determine the mean value of the function

$$f(x) = -23 + 22.7x - 6.9x^2 + 0.855x^3 - 0.0365x^4$$

between $x = 2$ and $x = 10$. Confirm the result graphically (showing both function and its mean value).

13.2   Use numerical integration formulas (and inline function) to

   a. Evaluate the *x-y* coordinates of the center of gravity (CG) of a plane area defined as

$$2 \le x \le 10, \quad 0 \le y \le f(x)$$

   where $f(x)$ is given in Problem 13.1. The CG coordinates can be evaluated as

$$x_{CG} = \frac{1}{A} \int_a^b x f(x)\, dx$$

   and

$$y_{CG} = \frac{1}{2A} \int_a^b f(x)^2\, dx$$

   where

$$A = \int_a^b f(x)\, dx$$

   b. Evaluate a volume of a 3-D shape defined as

$$-2 \le x \le 2, \quad 0 \le y \le 4, \quad 0 \le z \le \frac{1}{16}x^2 - \frac{1}{4}y^2 + \frac{1}{32}xy^3 + 1.25$$

Visualize the area and CG location in the first case, and the 3-D shape with an equivalent-volume rectangular cuboid (parallelepiped) in the second case.

13.3 Use `trapz`, `quad` and `quadl` commands to compute the value of the following definite integral:

$$\int_{(2\pi)^{-1}}^{2} \sin\left(x^{-1}\right) dx$$

(Keep in mind that command `[q,Nfcn]=quad(...,tol)` returns not only the value of integral `q` but also the number of function evaluations (intervals) required to meet `tol` requirement.)

First, estimate a pseudo exact value of the integral by setting `tol=1e-6`. That will be your reference value. Then, produce graphs 'Nfcn vs. `tol`' and '`q` vs. `tol`' for both `quad` and `quadl` using different tolerance within $[10^{-10}; 10^{-1}]$. Add a point showing the location of the reference value. Label it. Using `trapz` plot a graph 'value of integral vs. number of points'. On the same plot show the value of the integral your curve should converge to while increasing the number of points (take, for example, N=4000, compute the value of the integral and draw a horizontal line). Make a comment on how many function evaluations (intervals) you need to achieve about the same, 10^{-10}-order tolerance.

Chapter 14 / Initial-Value Problem

- Understanding Single-Steps Methods
- Understanding Multi-Steps Methods
- Using MATLAB IVP Solvers
- Using MATLAB BVP and PDE Solvers

```
ode23, ode45, ode113,
ode15s, ode23s, ode23t,
ode23tb, ode15i, dde23,
ddesd, bvp4c, bvp5c,
pdepe, odeexamples
```

14.1 Introduction

Many engineering problems require the determination of a function satisfying an equation containing one or more derivatives of this unknown function with respect to some argument (for example, time). These types of equations are called differential equations. Differential equations can be classified into two main types: *ordinary differential equations* (ODEs), where the function depends on only one variable so that only ordinary derivatives appear in the equation, and *partial differential equations* (PDEs), where the function depends on more than one variable and contains partial derivatives. Section 7.6 introduced the dsolve function allowing you to solve up to six relatively simple ODEs analytically. In Section 9.6, linear ODEs were also solved analytically, using the eigendecomposition of their state matrix. In addition to these capabilities, MATLAB offers a variety of solvers allowing you to solve any ODE or PDE numerically. The primary focus of this chapter is to introduce ODE solvers that are used in the so-called *initial-value problems* (IVPs) that are suitable to model system dynamics. They are ode23, ode45, ode113, ode15s, ode23s, ode23t, ode23tb, and ode15i. The Simulink modeling environment, introduced in Chapter 15, uses the very same and some other fixed-step-size solvers. Hence, to better understand the difference between these solvers and, therefore, use them more efficiently, this chapter provides with a detailed explanation of the algorithms behind them. As usual, the Symbolic Math Toolbox is used to assist in deriving some bulky expressions. The chapter starts from defining the IVP and then proceeds with single and multiple step algorithms, emphasizing their numerical features as implemented in the MATLAB ODE suite. It then specifically addresses handling the so-called *stiff problems* and provides a variety of examples of using the ODE solvers. For completeness, this chapter concludes with introducing some other ODE solvers to address differential equations with delays (dde23 and ddesd), *boundary-value problems* (BVPs) (bvp4c and bvp5c), and PDEs (pdepe).

14.2 Initial-Value Problem Definition

Let us start with a few formal definitions. If y is an unknown function in x, then an equation of the form

$$F(x, y, y', y'', \ldots, y^{(l)}) = 0 \tag{14.1}$$

is called an ODE of order l. The *order of* a differential equation is the order of the highest derivative that appears in the equation. Moreover, Eq. (14.1) features an *implicit* form, whereas in general, an *explicit* form is

$$F(x, y, y', y'', \ldots, y^{(l-1)}) = y^{(l)} \tag{14.2}$$

A differential equation not depending on x is called autonomous. Quite often ODEs use time as argument, so $x \triangleq t$.

A differential equation is said to be *linear* if F can be written as a linear combination of the derivatives of y together with some constant term g, all possibly depending on x

$$a_l(x)\frac{\mathrm{d}^l y}{\mathrm{d}x^l} + a_{n-1}(x)\frac{\mathrm{d}^{l-1} y}{\mathrm{d}x^{l-1}} + \cdots + a_1(x)\frac{\mathrm{d}y}{\mathrm{d}x} + a_0(x)y = g(x) \tag{14.3}$$

Linear differential equations follow a very important theorem called the *superposition principle*. The superposition principle states that if $y_1(x)$, $y_2(x)$, ..., $y_l(x)$ are l linearly independent solutions of an lth-order linear differential equation, then

$$y(x) = c_1 y_1(x) + c_2 y_2(x) + \cdots + c_l y_l(x) \tag{14.4}$$

is also a solution, where $c_1, c_2, \ldots, c_n$ are constant coefficients.

Differential equations that cannot be written in the form of Eq. (14.3) are called *nonlinear* equations. Nonlinear differential equations do not follow the principle of superposition.

Any differential equation of order l can be written as a system of l first-order differential equations. All you have to do is to introduce a family of unknown functions instead of just one, y, as follows:

$$y_k = y^{(k-1)}, \quad k = 1, 2, \ldots, l \tag{14.5}$$

The original ODE, Eq. (14.2), can then be rewritten as the system of l first-order differential equations

$$
\begin{aligned}
y_1' &= y_2 \\
y_2' &= y_3 \\
&\vdots \\
y_l' &= F(x, y_1, y_2, y_3, \ldots, y_l)
\end{aligned}
\tag{14.6}
$$

or concisely in vector notation

$$\mathbf{y}' = \mathbf{F}(x, \mathbf{y}) \tag{14.7}$$

where $\mathbf{y} = [y_1, y_2, \ldots, y_l]^T$ and $\mathbf{F}(x,\mathbf{y}) = [y_2, y_3, \ldots, y_l, F(x, \mathbf{y})]^T$.

Consider the following example. Suppose we have a simple mass–spring–damper system depicted in Fig. 14.1a. The only three forces acting on the cart having a mass m are the damper force $c\dot{y}(t)$, with c being the damping coefficient, the spring force $ky(t)$, with k denoting the spring constant, and some external force $u(t)$. If the cart is moving to the left of its equilibrium state shown in Fig. 14.1a, so that y is positive and growing, then both the damper force and spring force stem this motion, that is, they are negative. Hence, the second Newton's law yields

$$\sum F_y = -c\dot{y} - ky + u = m\ddot{y} \tag{14.8}$$

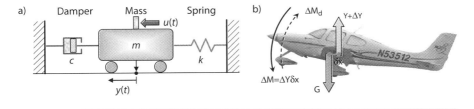

Fig. 14.1   a) Mass–spring–damper system and b) longitudinal motion of an aircraft.

which can be further rewritten as

$$\ddot{y} + \frac{c}{m}\dot{y} + \frac{k}{m}y = \frac{1}{m}u \tag{14.9}$$

According to aforementioned definitions, this equation is a second-order linear ODE and can be represented as a system of two first-order differential equations by introducing two new unknown functions, say,

$$x_1(t) = y(t), \ x_2(t) = \dot{y}(t) \tag{14.10}$$

Hence, Eq. (14.9) becomes

$$\dot{x}_1 = x_2$$
$$\dot{x}_2 = \frac{-k}{m}x_1 - \frac{c}{m}x_2 + \frac{1}{m}u \tag{14.11}$$

or in the matrix form

$$\dot{\mathbf{x}} = \mathbf{A}\mathbf{x} + \mathbf{B}\mathbf{u} \tag{14.12}$$

where

$$\mathbf{x} = \begin{bmatrix} x_1 \\ x_2 \end{bmatrix}, \ \mathbf{A} = \begin{bmatrix} 0 & 1 \\ \dfrac{-k}{m} & \dfrac{-c}{m} \end{bmatrix}, \ \mathbf{B} = \begin{bmatrix} 0 \\ \dfrac{1}{m} \end{bmatrix} \tag{14.13}$$

We can formally express the displacement of the cart via two new unknown functions of Eq. (14.10) in the matrix form as well

$$y = \begin{bmatrix} 1 & 0 \end{bmatrix}\mathbf{x} = \mathbf{C}\mathbf{x} \tag{14.14}$$

The derived second-order linear ODE has a fundamental importance because it is widely used to describe dynamics of mechanical, electrical, fluid, and thermal systems. For example, the model of an aircraft landing gear, its hydrodynamic system, accelerometer, rotational dynamics of the entire aircraft (Fig. 14.1b), etc. can be reduced to Eqs. (14.12) and (14.14) (with the elements of matrices $\mathbf{A}_{2\times2} = [a_{ij}]$ and $\mathbf{B}_{2\times1} = [b_{ij}]$ being some constant coefficients).

As discussed in Sec. 9.5, if Eq. (14.7) happens to be linear with constant coefficients, then you can solve them analytically using the `eig` function. For small l, you can also try the `dsolve` function of the Symbolic Math Toolbox (Sec. 7.6). Otherwise, the numerical methods discussed in this chapter are your only alternative.

A *general solution* of an lth-order ODE is a solution containing l arbitrary variables, corresponding to l constants of integration. A *particular solution* is derived from the general solution by setting the constants to particular values, often chosen to satisfy initial or boundary conditions. Consequently, when solving ODEs numerically we distinguish an IVP with all conditions prescribed at the initial value of the argument x and a BVP with all conditions prescribed at the spatial boundaries. This chapter primarily deals with IVP, however, for completeness, the BVP solvers are also mentioned in Sec. 14.8 among other ODE and PDE solvers of MATLAB.

Formally, the IVP can be formulated as follows. Given

$$\mathbf{y}' = \mathbf{F}(x, \mathbf{y}), \; \mathbf{y}(x_0) = \mathbf{y}_0 \tag{14.15}$$

find the numerical values of components of the vector $\mathbf{y}(x)$—$\mathbf{y}_1, \mathbf{y}_2, \ldots, \mathbf{y}_n, \ldots,$ $\mathbf{y}_f$—corresponding to argument values $x_1, x_2, \ldots, x_n, \ldots, x_f,$ which constitutes a solution over an interval $x \in [x_0; x_f]$. This problem is often referred to as a *Cauchy problem*.

In what follows, we will address problem (14.15) using different numerical methods and assuming that $x_{n+1} = x_n + h$, $n = 0,1,2,\ldots$ with h being the *integration step size*. For simplicity, we will use notations of the first-order ODE, that is, instead of Eq. (14.15) consider

$$y' = f(x, y), \; y(x_0) = y_0 \tag{14.16}$$

Nevertheless, all formulas developed in this case can be easily extended to the case of Eq. (14.15).

Problem (14.16) looks somewhat similar to the numerical integration problem addressed in Chapter 13. The only difference is that there we dealt with integrating an equation $y' = f(x)$, rather than $y' = f(x, y)$, that is, function f did not depend on the unknown variable y. Therefore, neither of the methods developed in Chapter 13 can be applied to the problem (14.6) directly.

For $y' = f(x)$, given analytically, the slope y'_n can be estimated at any argument x_n. In the case of $y' = f(x, y)$, we cannot do this without estimating y_n first. Therefore, we run into the Catch-22 situation. The two groups of methods to deal with it (to estimate y'_n without knowing y_n) rely on using the multiple points that are already available (*multistep methods*) or capability to estimate y'_n based on some trial points within the interval $x \in [x_{n-1}; x_n]$ (*single-step methods*).

14.3 Single-Step Methods

One group of methods for solving the IVP, *Runge–Kutta* (RK) *methods*, exploits an idea of representing the solution of Eq. (14.16) for each argument x_{n+1}, $n = 0, 1, 2, ...$, in the following form:

$$y_{n+1} = y_n + h\sum_{i=1}^{m} a_i k_i \tag{14.17}$$

Formally, a_i are weighting coefficients and k_i are gradients (slopes) evaluated at m points within the nth-interval, $x \in [x_n; x_{n+1}]$, as follows

$$k_1 = f(x_n, y_n)$$
$$k_2 = f(x_n + q_1 h, y_n + p_{11} k_1 h)$$
$$k_3 = f(x_n + q_2 h, y_n + p_{21} k_1 h + p_{22} k_2 h)$$
$$k_4 = f(x_n + q_3 h, y_n + p_{31} k_1 h + p_{32} k_2 h + p_{33} k_3 h) \tag{14.18}$$
$$\vdots$$
$$k_m = f\left(x_n + q_{m-1} h, y_n + h\sum_{j=1}^{m-1} p_{m-1, j} k_j\right)$$

Here q_i, $i = 1, 2, ..., m-1$ ($0 < q_1 \le q_2 \le ... \le q_{m-1} \le 1$) and p_{ij}, $i = 1, 2, ..., m-1$, $j = 1, 2, ..., i$ are coefficients defined using a special procedure discussed in the following section.

To distinguish these m inner points from the solution points (x_{n+1}, y_{n+1}), $n = 0, 1, 2, ...$, Eq. (14.17) is referred to as an *m-stage* formula. Geometrically, the sum

$$\sum_{i=1}^{m} a_i k_i$$

in Eq. (14.17) can be interpreted as a representative slope y' over the interval $x \in [x_n; x_{n+1}]$, based on the slopes estimated at m inner points, x_n and $x_n + q_i h$, $i = 1, 2, ..., m-1$. Hence, weighting coefficients a_i should satisfy the obvious weighting equation

$$\sum_{i=1}^{m} a_i = 1 \tag{14.19}$$

Having m coefficients k_i in the sum of Eq. (14.17), we need to obtain $0.5(m^2 + 3m - 2)$ unknowns, which are the coefficients a_i, $i = 1, 2, ..., m$; q_i, $i = 1, 2, ..., m-1$; and p_{ij}, $i = 1, 2, ..., m-1$, $j = 1, 2, ..., i$. Intuitively, the more intermediate points m we have, the more accurate the integration procedure is. To understand where the values of these unknown coefficients can possibly come from, let us consider the simplest cases when $m = 1$ and $m = 2$. In doing this, we will follow the steps of Runge, who developed integration formulas for $m = 2$ and $m = 3$. Then, we will proceed with the general case of Eqs. (14.17) and (14.18) as formulated by Kutta.

14.3.1 Low-order Formulas

According to Eqs. (14.17) and (14.18) in the case of $m = 1$, we only have to find a single coefficient a_1. The only possible solution comes from Eq. (14.19), so that $a_1 = 1$. Therefore,

$$y_{n+1} = y_n + hf(x_n, y_n) \qquad (14.20)$$

A geometric interpretation of this integration scheme known as the *Euler method* is shown in Fig. 14.2a. The representative slope y' over the interval $x \in [x_n; x_{n+1}]$ is estimated as $k_1 = f(x_n, y_n)$ using a single point (x_n, y_n), the beginning of the interval, and a full step is made along the direction defined by this representative slope.

Equation (14.20) can also be obtained using the Taylor series expansion

$$y_{n+1} = y_n + hf(x_n, y_n) + \frac{h^2}{2!} f'(x_n, y_n) + \frac{h^3}{3!} f''(x_n, y_n) + O(h^4) \quad (14.21)$$

As seen, the first two terms of Eq. (14.21) comprise Eq. (14.20). Moreover, Eq. (14.21) provides us with an estimate of the local truncation error, which happens to be of the order of the very next term, that is, h^2. As seen

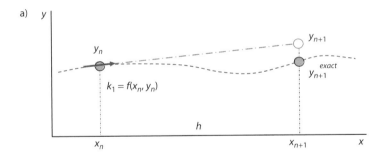

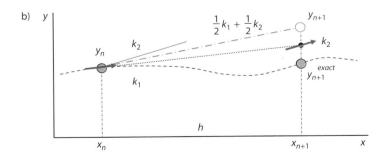

Fig. 14.2 a) Euler and b) modified Euler integration schemes.

in Fig.14.2a, we started an integration step assuming that y_n is an exact solution to Eq. (14.16). Obviously, it is only the case for the very first point given by the initial condition. The global error builds up at each step of integration, so that for the number of steps, inversely proportional to the step size h $((x_f - x_0) h^{-1})$, it is always one order less than a local truncation error. Hence, the Euler's method [Eq. (14.20)] supplies the first-order (of the step size h) formula.

Now, let us consider the case when $m = 2$. According to Eqs. (14.17) and (14.18), it leaves us with the following general formula

$$y_{n+1} = y_n + h(a_1 k_1 + a_2 k_2) \tag{14.22}$$

where two slopes are defined at the origin of the interval x_n and some intermediate point $x_n + q_1 h$

$$\begin{aligned} k_1 &= f(x_n, y_n) \\ k_2 &= f(x_n + q_1 h, y_n + p_{11} k_1 h) \end{aligned} \tag{14.23}$$

The problem is in determining four unknown coefficients: a_1, a_2, q_1, and p_{11}. To address it, let us leave three terms in Eq. (14.21) yielding

$$y_{n+1} = y_n + h f(x_n, y_n) + \frac{h^2}{2!} f'(x_n, y_n) \tag{14.24}$$

(Intuitively, using more terms in Eq. (14.17) should lead to a higher accuracy, so leaving three terms in Eq. (14.21) instead of two makes the local truncation error to be of the order of h^3 and the global error being accrued while integrating Eq. (14.6) within $x \in [x_0; x_f]$ of the order of h^2). Since the function f depends on two arguments, to determine f' in Eq. (14.24) we have to employ a chain-rule differentiation

$$f' = \frac{\partial f}{\partial x} + \frac{\partial f}{\partial y} \frac{dy}{dx} = \frac{\partial f}{\partial x} + \frac{\partial f}{\partial y} f \tag{14.25}$$

Substituting Eq. (14.25) to Eq. (14.24) results in

$$y_{n+1} = y_n + h f(x_n, y_n) + \frac{h^2}{2!} \left(\frac{\partial f}{\partial x} + \frac{\partial f}{\partial y} f \right) \tag{14.26}$$

Now, we have to expand the second equation in Eq. (14.23) as well

$$k_2 = f(x_n + q_1 h, y_n + p_{11} k_1 h) = f(x_n, y_n) + q_1 h \frac{\partial f}{\partial x} + p_{11} k_1 h \frac{\partial f}{\partial y} + O(h^2) \tag{14.27}$$

This result can be substituted along with the first equation in Eq. (14.23) into Eq. (14.22) to yield

$$y_{n+1} = y_n + h(a_2 + a_1) f(x_n, y_n)$$

$$+ h^2 \left(a_2 q_1 \frac{\partial f}{\partial x} + a_2 p_{11} f(x_n, y_n) \frac{\partial f}{\partial y} \right) + O(h^3) \tag{14.28}$$

By comparing the latter equation with Eq. (14.26), we determine that for the two equations to be equivalent, the following *order conditions* must hold

$$a_1 + a_2 = 1, \; a_2 q_1 = \frac{1}{2}, \; a_2 p_{11} = \frac{1}{2} \tag{14.29}$$

(for better understanding, the entire procedure is presented graphically in Fig. 14.3).

The following script employing the Symbolic Math Toolbox:

```
%% Defining symbolic variables
syms h fn a1 a2 k1 k2 q1 p11
syms epx epy dfdx dfdy
%% Making substitutions
Eq1 = h*(a1*k1+a2*k2);
s12 = fn+epx*dfdx+epy*dfdy;
s12 = subs(s12,{epx,epy},{q1*h,p11*h*fn});
Eq1 = subs(Eq1,{k1,k2},{fn,s12});
%% Defining the reference equation
Eq2 = h*fn+h^2/factorial(2)*(dfdx+dfdy*fn);
%% Comparing Eq1 and Eq2 by the orders of h
```

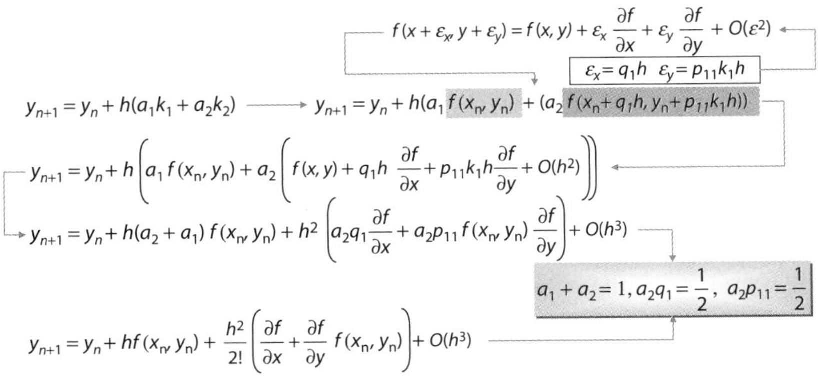

Fig. 14.3 Defining unknown coefficients for the two-stage integration scheme.

```
[Q1,R1] = quorem(Eq1,h^2,h);
[Q2,R2] = quorem(Eq2,h^2,h);
disp([Q1 Q2]);
disp([quorem(R1,h,h) quorem(R2,h,h)]);
```

results in

```
[ a2*(dfdx*q1 + dfdy*fn*p11), dfdx/2 + (dfdy*fn)/2]
[ a1*fn + a2*fn,  fn]
```

allowing us to come up with Eq. (14.29) by comparing the corresponding terms in two expressions in each pair of brackets.

As seen, we ended up having three non-linear equations for four unknowns, which means the existence of multiple solutions (infinite number of solutions to be more precise). For example, allow

$$a_1 = a_2 = \frac{1}{2} \tag{14.30}$$

It gives

$$p_{11} = q_1 = \frac{1}{2a_2} = 1 \tag{14.31}$$

Coefficients (14.30) and (14.31) constitute the valid second-order formula known as *Heun method* or *Modified Euler method*

$$y_{n+1} = y_n + h\left(\frac{1}{2}k_1 + \frac{1}{2}k_2\right) \tag{14.32}$$

with

$$\begin{aligned} k_1 &= f(x_n, y_n) \\ k_2 &= f(x_n + h, y_n + k_1 h) \end{aligned} \tag{14.33}$$

Geometrically, Eqs. (4.32) and (14.33) can be treated as a *prediction–correction method*, where eventually we are averaging slopes at two boundary points (Fig. 14.2b). Since we do not know the terminal point yet, we have to guess or make a prediction

$$y_{n+1}^* = y_n + hf(x_n, y_n) \tag{14.34}$$

Then, we correct this prediction as

$$y_{n+1} = y_n + \frac{h}{2}(f(x_n, y_n) + f(x_n + h, y_{n+1}^*)) \tag{14.35}$$

that is, making a full step in the direction $0.5(k_1 + k_2)$. Equations (14.34) and (14.35) are the same as Eqs. (14.32) and (14.33), but expressed in the predictor–corrector form. Note that, Fig. 14.2 is exaggerated in the sense that the step size h is too big, so that the slope y' varies in between x_n and x_{n+1} a lot. This is the reason that for the second-order method [Eq. (14.32)] (Fig. 14.2b), we ended up with the larger error compared with that of the first-order method [Eq. (14.20)] (Fig. 14.2a). In fact, it may happen in practice as well, because according to Eq. (14.21) the local truncation error is not only proportional to the step size h, but also to the higher-order derivative of the function f as well.

Another way to obtain Eq. (14.35) is to apply the trapezoidal rule to the corresponding integral

$$y_{n+1} = y_n + \int_{x_n}^{x_{n+1}} f(x, y) \, dx \cong y_n + \frac{h}{2}\left(f(x_n, y_n) + f(x_n + h, y_{n+1}^*)\right) \quad (14.36)$$

Since y_{n+1} is not known (when we developed the trapezoidal rule in Sec. 13.3, we had $y' = f(x)$, not $y' = f(x,y)$), we approximate it by using the Euler's method as in Eq. (14.34).

Another geometrically sound solution to Eq. (14.29) is the *Midpoint method*, suggested by Runge, where we allow

$$a_1 = 0 \quad \text{and} \quad a_2 = 1 \quad (14.37)$$

Then, from Eq. (14.39)

$$p_{11} = q_1 = \frac{1}{2a_2} = \frac{1}{2} \quad (14.38)$$

Being substituted into Eqs. (14.22) and (14.23) it yields

$$y_{n+1} = y_n + hk_2 \quad (14.39)$$

with

$$k_1 = f(x_n, y_n)$$
$$k_2 = f\left(x_n + \frac{1}{2}h, y_n + \frac{1}{2}k_1 h\right) \quad (14.40)$$

As seen from Fig. 14.4a, the second slope is estimated at midpoint followed by the full step in this (k_2) direction. Equations (14.39) and (14.40) can also be cast in the predictor–corrector form

$$y_{n+0.5}^* = y_n + \frac{h}{2} f(x_n, y_n) \quad (14.41)$$

$$y_{n+1} = y_n + hf\left(x_n + \frac{1}{2}h, y_{n+0.5}^*\right) \quad (14.42)$$

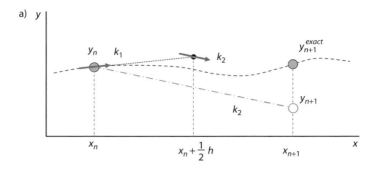

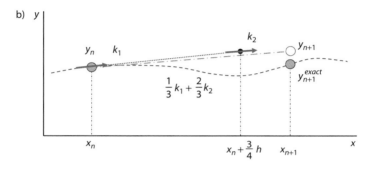

Fig. 14.4 a) Midpoint and b) Ralston integration schemes.

As mentioned, Eqs. (14.32) and (14.33) do not provide a unique solution of Eq. (14.29). From the practical point of view, the second point (along with x_n) can be chosen anywhere within the interval $x \in [x_n + \frac{1}{2}h; x_{n+1}]$ resulting in $a_1 \in [0; 0.5]$. Presumably, among all second-order methods, obeying Eq. (14.29), the coefficients

$$a_1 = \frac{1}{3}, \ a_2 = \frac{2}{3}, \ p_{11} = q_1 = \frac{1}{2a_2} = \frac{3}{4} \tag{14.43}$$

assure the smallest local truncation error. Being substituted into Eqs. (14.22) and (14.23), it results in the *Ralston method*

$$y_{n+1} = y_n + h\left(\frac{1}{3}k_1 + \frac{2}{3}k_2\right) \tag{14.44}$$

with

$$k_1 = f(x_n, y_n)$$
$$k_2 = f\left(x_n + \frac{3}{4}h, y_n + \frac{3}{4}k_1 h\right) \tag{14.45}$$

(Fig. 14.4b) or in the predictor–corrector form

$$y^*_{n+0.75} = y_n + \frac{3}{4} hf(x_n, y_n) \tag{14.46}$$

$$y_{n+1} = y_n + \frac{h}{3}\left(f(x_n, y_n) + 2f\left(x_n + \frac{3}{4}h, y^*_{n+0.75}\right) \right) \tag{14.47}$$

14.3.2 Higher-Order Formulas

Being illustrated for the case of $m = 1$ and $m = 2$, it is now clear how to use even more inner points (stages) within the interval $x \in [x_n; x_{n-1}]$ (the terms in Eqs. (14.17) and (14.18)). Leaving four terms and following the same routine as in Fig. 14.3 leads to a system of six nonlinear algebraic equations, similar to those of Eq. (14.29), with eight unknown coefficients: $a_1, a_2, a_3, q_1, q_2, p_{11}, p_{21}$, and p_{22}. Once again, an infinitive number of solutions exist. As suggested by Kutta to narrow possible variants in addition to conditions [Eq. (14.29)], the following relation can be enforced for all $i = 1, 2, ..., m$

$$q_i = \sum_{j=1}^{i} p_{ij} \tag{14.48}$$

It does not help for the $m=2$ case we just considered, but allows simplifying the derivation of order conditions like Eq. (14.29) for the higher-order methods.

One common version of the *third-order RK method* results in

$$y_{n+1} = y_n + h\left(\frac{1}{6}k_1 + \frac{4}{6}k_2 + \frac{1}{6}k_3 \right) \tag{14.49}$$

where

$$k_1 = f(x_n, y_n)$$
$$k_2 = f\left(x_n + \frac{1}{2}h, y_n + \frac{1}{2}k_1 h \right) \tag{14.50}$$
$$k_3 = f(x_n + h, y_n - k_1 h + 2k_2 h)$$

(that is, the slope estimates are taken at two boundary points and a midpoint). In the predictor–corrector form, it looks as

$$y^*_{n+0.5} = y_n + \frac{h}{2} f(x_n, y_n), \quad y^*_{n+1} = y_n + hf(x_n, y_n) \tag{14.51}$$

$$y_{n+1} = y_n + \frac{h}{6}(f(x_n, y_n) + 4f(x_n + 0.5h, y^*_{n+0.5}) + f(x_n + h, y^*_{n+1})) \tag{14.52}$$

Note that, if we deal with $y' = f(x)$ rather than $y' = f(x,y)$, then this third-order RK method reduces to the Simpson's 1/3 rule, Eq. (13.19). It uses three consecutive points and results in the local truncation error of the order of h^4. This transformation is presented in Fig. 14.5.

The most popular (and accurate) RK methods are fourth-order methods. They are similar to the third-order schemes except that one more intermediate point is used to estimate a representative slope over the interval. The analogue of Eq. (14.29) with an account of Eq. (14.48) is the following set:

$$\sum_{i=1}^{4} a_i = 1 \qquad \sum_{i=2}^{4} a_i q_i = \frac{1}{2} \qquad \sum_{i=2}^{4} a_i q_i^2 = \frac{1}{3} \qquad \sum_{i=3}^{4} a_i \sum_{j=2}^{i-1} p_{ij} q_j = \frac{1}{6}$$

$$\sum_{i=2}^{m} a_i q_i^3 = \frac{1}{4} \qquad \sum_{i=3}^{4} a_i q_i \sum_{j=2}^{i-1} p_{ij} q_j = \frac{1}{8} \qquad \sum_{i=3}^{4} a_i \sum_{j=2}^{i-1} p_{ij} q_j^2 = \frac{1}{12} \qquad a_4 p_{43} p_{32} q_2 = \frac{1}{24}$$

$$(14.53)$$

As seen, we have 8 nonlinear equations for 13 unknown coefficients, again resulting in the infinite number of possibilities.

Among the variety of fourth-order methods, the *classical forth-order RK method* is used the most. It is defined as

$$y_{n+1} = y_n + h \left(\frac{1}{6} k_1 + \frac{2}{6} k_2 + \frac{2}{6} k_3 + \frac{1}{6} k_4 \right) \qquad (14.54)$$

where

$$k_1 = f(x_n, y_n)$$
$$k_2 = f\left(x_n + \frac{1}{2} h, y_n + \frac{1}{2} k_1 h \right)$$
$$k_3 = f\left(x_n + \frac{1}{2} h, y_n + \frac{1}{2} k_2 h \right) \qquad (14.55)$$
$$k_4 = f(x_n + h, y_n + k_3 h)$$

$$y_{n+1} = y_n + h \left(\frac{1}{6} k_1 + \frac{4}{6} k_2 + \frac{1}{6} k_3 \right)$$

$k_1 = f(x_n, y_n)$

$k_2 = f(x_n + \frac{1}{2} h, y_n + \frac{1}{2} k_1 h)$

$k_3 = f(x_n + h, y_n - k_1 h + 2 k_2 h)$

$k_1 = f(x_n)$

$k_2 = f(x_n + \frac{1}{2} h)$

$k_2 = f(x_n + h)$

$y_{n+1} = y_n + \frac{h}{2} \left(\frac{1}{3} f_n + \frac{4}{3} f_{n+0.5} + \frac{1}{3} f_{n+1} \right) \longrightarrow y_{n+2} = y_n + h^* \frac{1}{3} (1 f_n + 4 f_{n+1} + 1 f_{n+2})$

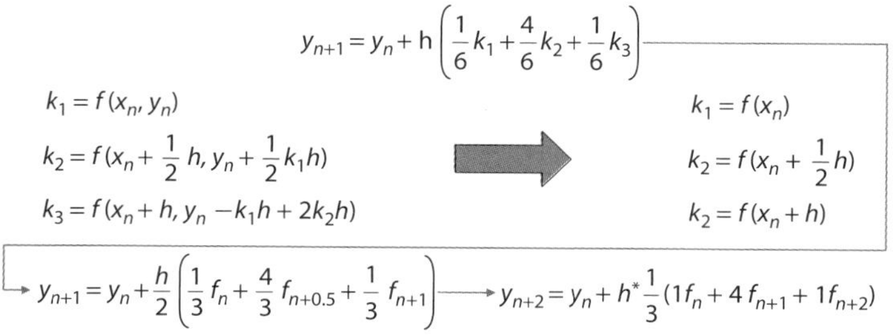

Fig. 14.5 Reduction of the third-order RK method to the Simpson's 1/3 rule for $y' = f(x)$.

$$y_{n+1} = y_n + h\left(\frac{1}{6}k_1 + \frac{2}{6}k_2 + \frac{2}{6}k_3 + \frac{1}{6}k_4\right) \quad\quad y_{n+1} = y_n + h\left(\frac{1}{8}k_1 + \frac{3}{8}k_2 + \frac{3}{8}k_3 + \frac{1}{8}k_4\right)$$

$$k_1 = f(x_n, y_n) \quad\quad\quad\quad\quad\quad\quad\quad\quad k_1 = f(x_n, y_n)$$

$$k_2 = f\left(x_n + \frac{1}{2}h, y_n + \frac{1}{2}k_1 h\right) \quad\quad k_2 = f\left(y_n + \frac{1}{3}k_1 h, x_n + \frac{1}{3}h\right)$$

$$k_3 = f\left(x_n + \frac{1}{2}h, y_n + \frac{1}{2}k_2 h\right) \quad\quad k_3 = f\left(y_n + \frac{1}{3}k_1 h + \frac{1}{3}k_2 h, x_n + \frac{2}{3}h\right)$$

$$k_4 = f(x_n + h, y_n + k_3 h) \quad\quad\quad\quad k_4 = f(y_n + k_1 h - k_2 h + k_3 h, x_n + h)$$

$$\rightarrow y_{n+1} = y_n + h\left(\frac{1}{6}f_n + \frac{2}{6}f_{n+0.5} + \frac{2}{6}f_{n+0.5} + \frac{1}{6}f_{n+1}\right) \quad \rightarrow y_{n+1} = y_n + h\left(\frac{1}{8}f_n + \frac{3}{8}f_{n+1/3} + \frac{3}{8}f_{n+2/3} + \frac{1}{8}f_{n+1}\right)$$

$$\rightarrow y_{n+2} = y_n + h^* \frac{1}{3}(1f_n + 4f_{n+1} + 1f_{n+2}) \quad\quad \rightarrow y_{n+3} = y_n + h^{**}\frac{3}{8}(1f_n + 3f_{n+1} + 3f_{n+2} + 1f_{n+3})$$

Fig. 14.6 Reduction of the fourth-order RK formulas to a) Simpson's 1/3 and b) 3/8 rules for $y' = f(x)$.

or in the predictor–corrector form

$$y^*_{n+0.5} = y_n + \frac{h}{2}f(x_n, y_n), \quad y^{**}_{n+0.5} = y_n + \frac{h}{2}f(x_n + 0.5h, y^*_{n+0.5})$$

$$y^*_{n+1} = y_n + hf(x_n + 0.5h, y^{**}_{n+0.5}) \tag{14.56}$$

$$y_{n+1} = y_n + \frac{h}{6}\left(f(x_n, y_n) + 2f\left(x_n + \frac{1}{2}h, y^*_{n+0.5}\right) \right.$$

$$\left. + 2f\left(x_n + \frac{1}{2}h, y^{**}_{n+0.5}\right) + f(x_n + h, y^*_{n+1}) \right) \tag{14.57}$$

The second, also popular, version of the *fourth-order RK method* is

$$y_{n+1} = y_n + h\left(\frac{1}{8}k_1 + \frac{3}{8}k_2 + \frac{3}{8}k_3 + \frac{1}{8}k_4\right) \tag{14.58}$$

with

$$k_1 = f(x_n, y_n)$$

$$k_2 = f\left(x_n + \frac{1}{3}h, y_n + \frac{1}{3}k_1 h\right)$$

$$k_3 = f\left(x_n + \frac{2}{3}h, y_n + \frac{1}{3}k_1 h + \frac{1}{3}k_2 h\right) \tag{14.59}$$

$$k_4 = f(x_n + h, y_n + k_1 h - k_2 h + k_3 h)$$

As with the third-order RM method [Eq. (14.49)], in the case when $y' = f(x)$ rather than $y' = f(x,y)$, Eqs. (14.54) and (14.58) reduce to the Simpson's 1/3 and 3/8 rule (Fig. 14.6), respectively.

As it was the case with numerical integration (Chapter 13), although the higher-order RK formulas can be developed and are available, their stability properties worsen as m increases. Hence, instead of increasing m, adaptive-step algorithms employing RK formulas with a moderate m were found to be more efficient. These algorithms are discussed in the following sections.

14.4 Adaptive Single-Step Algorithms

As with any numerical method, there are only two basic strategies for increasing the accuracy of the numerical solution to an ODE. They are

1. Reducing step size h
2. Choosing a more accurate scheme

Unfortunately, as mentioned already, the dramatic success obtained by increasing the order from Euler method to the fourth-order RK method does not continue above fourth-order RK methods. The accuracy can be improved further, but the cost in increased number of function evaluations and stability issues render these higher-order formulas less attractive. That leaves us with option 1.

Obviously, decreasing the step size h also leads to the increased number of function evaluations. Although methods presented up to this point have used a fixed h, there is no reason why h cannot be changed from one interval to the next one during a numerical solution. Hence, the key to further increase the efficiency of numerical schemes is to minimize the overall number of steps necessary to achieve a solution with the defined accuracy by adoptively varying h. The only complication is in devising a criterion for automatically choosing the size of h.

To be efficient, an algorithm for adjusting h must relate changes in h to the local accuracy of the solution. The objective is to reduce h when a smaller h is necessary to meet a user-specified error tolerance ε and to increase h when doing so will not cause the error tolerance to be exceeded. For this to work, the ODE integration scheme must somehow measure the errors it makes during the solution, specifically the local truncation error.

One way of estimating the local truncation error is to apply the *step-halving procedure* introduced in Sec. 13.7. It involves taking each step twice, once as a full step and independently as two half steps. The difference in the two results represents an estimate of the local truncation error. If there is no difference between the solution produced by a large and a small value of h, then there is no advantage in decreasing h. If there is a significant difference between the solutions produced by the two h, then we assume the solution obtained with the large h is inaccurate. If y_1 designates the single-step estimate and y_2 designates the estimate using the two half steps, the error Δ can be represented as

$$\Delta = y_2 \left(\tfrac{1}{2}h\right) - y_1(h) \qquad (14.60)$$

similar to that of Eq. (13.35). Along with providing a criterion for the step-size control, Eq. (14.60) can also be used to correct the y_2 estimate (see the Richardson extrapolation formula (13.33)). For fourth-order RK version, the correction is

$$y_2^{MA} = y_2 + \frac{\Delta}{15} \tag{14.61}$$

This estimate is fifth-order accurate.

Obviously, the step-halving procedure requires evaluations of two numerical solutions for each step that is not efficient. Another way to estimate the local truncation error is to simultaneously advance the solution with two different-order schemes using the same value of h and the same estimates of the slope in the intermediate points. If the less accurate scheme results in $\hat{y}_{n+1}$ and the more accurate scheme in y_{n+1}, so that

$$\Delta = |y_{n+1} - \hat{y}_{n+1}| \le \varepsilon \tag{14.62}$$

then there is no need to decrease h. If, however, Eq. (14.62) does not hold, then the step h should be reduced. If Δ happens to be much smaller than ε, the step size h can probably be safely increased. MATLAB adaptive IVP solvers pursue exactly this idea.

The `ode23` solver of MATLAB uses the third-order Ralston formula and second-order Bogaci–Shampine formula to advance the solution and monitor the accuracy. Specifically, formulas for the most and least accurate estimates are as follows:

$$y_{n+1} = y_n + h\left(\frac{2}{9}k_1 + \frac{1}{3}k_2 + \frac{4}{9}k_3\right) \tag{14.63}$$

$$\hat{y}_{n+1} = y_n + h\left(\frac{7}{24}k_1 + \frac{1}{4}k_2 + \frac{1}{3}k_3 + \frac{1}{8}k_4\right) \tag{14.64}$$

where

$$
\begin{aligned}
k_1 &= f(x_n, y_n) \\
k_2 &= f\left(x_n + \frac{1}{2}h, y_n + \frac{1}{2}k_1 h\right) \\
k_3 &= f\left(x_n + \frac{3}{4}h, y_n + \frac{3}{4}k_2 h\right) \\
k_4 &= f(x_n + h, y_{n+1})
\end{aligned}
\tag{14.65}
$$

First, three slopes, k_i, $i = 1, 2, 3$, are computed to produce y_{n+1}, then k_4 and $\hat{y}_{n+1}$ are evaluated. Note that despite the third-order result, y_{n+1} is used to compute one of the slopes for the second-order formula. The latter one's accuracy is still of the order of h^2 (having more terms in (Eq. 14.64) compared with those of Eq. (14.63) does not necessarily mean the superb performance). On the positive side, the choice of k_4 allows minimizing the number of function evaluations since this last slope evaluation is used at the very next interval $[x_{n+1}; x_{n+2}]$ as k_1 (the so-called *first step as last step*, (FSAL) paradigm).

In practical algorithms, the value of $\hat{y}_{n+1}$ is not even being computed. Instead, the local error estimate Δ [Eq. (14.62)] is computed explicitly in terms of the same k_i, $i = 1, 2, \ldots$

$$\Delta = h \left| -\frac{5}{72} k_1 + \frac{1}{12} k_2 + \frac{1}{9} k_3 - \frac{1}{8} k_4 \right| \tag{14.66}$$

This estimate Δ is used to come up with the step size h_{new} for the following step

$$\frac{h_{new}}{h} \sim \left(\frac{\varepsilon}{\Delta} \right)^{1/p} \tag{14.67}$$

with p being the order of the method. For $\texttt{ode23}$ a constant power is, therefore, equal to 1/3.

Similarly, the $\texttt{ode45}$, the default MATLAB solver, uses seven slope evaluations to calculate fourth- and fifth-order accurate solutions using the Dormand–Prince pair

$$y_{n+1} = y_n + h \left(\frac{35}{384} k_1 + \frac{500}{1113} k_3 + \frac{125}{192} k_4 - \frac{2187}{6784} k_5 + \frac{11}{84} k_6 \right) \tag{14.68}$$

$$\hat{y}_{n+1} = y_n + h \left(\frac{5179}{57600} k_1 + \frac{7571}{16695} k_3 + \frac{393}{640} k_4 - \frac{92097}{339200} k_5 + \frac{187}{2100} k_6 + \frac{1}{40} k_7 \right) \tag{14.69}$$

where

$$k_1 = f(x_n, y_n)$$

$$k_2 = f\left(x_n + \frac{1}{5} h, y_n + \frac{1}{5} k_1 h \right)$$

$$k_3 = f\left(x_n + \frac{3}{10} h, y_n + \frac{3}{40} k_1 h + \frac{9}{40} k_2 h \right)$$

$$k_4 = f\left(x_n + \frac{4}{5} h, y_n + \frac{44}{45} k_1 h - \frac{56}{15} k_2 h + \frac{32}{9} k_3 h \right)$$

$$k_5 = f\left(x_n + \frac{8}{9} h, y_n + \frac{19372}{6561} k_1 h - \frac{25360}{2187} k_2 h + \frac{64448}{6561} k_3 h - \frac{212}{729} k_4 h \right)$$

$$k_6 = f\left(x_n + h, y_n + \frac{9017}{3168} k_1 h - \frac{355}{33} k_2 h + \frac{46732}{5247} k_3 h + \frac{49}{176} k_4 h - \frac{5103}{18656} k_5 h \right)$$

$$k_7 = f(x_n + h, y_{n+1})$$

$$\tag{14.70}$$

As in the case of the `ode23` solver, instead of computing the value of $\hat{y}_{n+1}$, the local error estimate is computed as

$$\Delta = h\left|\frac{71}{57600}k_1 - \frac{71}{16695}k_3 + \frac{71}{1920}k_4 - \frac{17253}{339200}k_5 + \frac{22}{525}k_6 - \frac{1}{40}k_7\right| \quad (14.71)$$

The constant power in Eq. (14.67) for the `ode45` solver is equal to 1/5.

A lot of other pairs, sharing the same slope estimates and allowing to monitor the accuracy are also known but only the aforementioned two made it to MATLAB code.

14.5 Multistep Methods

As described in the previous sections, single-step methods use only a single point to compute the following one, that is, the initial point (x_0, y_0) is used to compute (x_1, y_1) and only (x_n, y_n) is needed to compute (x_{n+1}, y_{n+1}). Of course, they use multiple function evaluations to estimate a representative slope within each interval, $x \in [x_n; x_{n+1}]$, but they do not take advantage of the previously obtained points.

On the contrary, multistep methods attempt to gain efficiency by keeping and using the information from several previous steps, specifically $s + 1$ values of the function $f(x_{n-i}, y_{n-i}) = f(x_{n-ih}, y_{n-i}) = f_{n-i}, i = 0,1, ..., s$, obtained at the previous steps, to calculate y_{n+1}.

14.5.1 Adams—Bashforth—Moulton Predictor— Corrector Method

Adams–Bashforth and Adams–Moulton multistep methods rely on the fundamental theorem of calculus

$$y_{n+1} = y_n + \int_{x_n}^{x_{n+1}} f(x, y)\, dx \quad (14.72)$$

where function $f(x, y)$ is substituted with polynomial approximation $f^*(x)$ based on $s + 1$ points. This allows you to take a finite integral in Eq. (14.72) analytically, so instead of Eq. (14.17) have

$$y_{n+1} = y_n + h\sum_{i=0}^{s} a_i f(x_{n-i}, y_{n-i}) \quad (14.73)$$

Specifically, the *Adams—Bashforh—Moulton predictor—corrector method*, incorporated in the MATLAB `ode113` solver, involves computing

Eq. (14.72) twice, so that the difference between two estimates can be used to evaluate a local truncation error. This error is then used to determine whether the step size h is small enough to obtain an accurate value for y_{n+1}, yet large enough so that unnecessary and time-consuming calculations are eliminated. As opposed to the single-step methods, using a combination of a predictor and corrector requires only two function evaluations of $f(x, y)$ per step. Hence, the `ode113` solver may be more efficient than say `ode45`, especially when the function $f(x, y)$ is particularly expensive to evaluate (in terms of CPU time).

Let us consider how this works assuming $s = 3$ (Fig. 14.7), that is, relying on the four points, resulting in the third-order Lagrange polynomial approximation for $f^*(x, y)$ in Eq. (14.72). If the coefficients of this polynomial

$$f^*(x) = Kx^3 + Lx^2 + Mx + N \tag{14.74}$$

are known, substituting Eq. (14.74) into Eq. (14.72) and taking the integral yields the *Adams–Bashforth prediction*

$$y_{n+1}^* = y_n + \int_{x_n}^{x_n+h} f^*(x)\,dx = y_n + \frac{1}{4}h^4 K + \frac{1}{3}h^3 L + \frac{1}{2}h^2 M + hN \tag{14.75}$$

Having four pairs (x_{n-3}, y_{n-3}), (x_{n-2}, y_{n-2}), (x_{n-1}, y_{n-1}), and (x_n, y_n) allows you to solve for unknown coefficients $K, L, M,$ and N by resolving the following system:

$$\begin{bmatrix} x_n^3 & x_n^2 & x_n & 1 \\ (x_n - h)^3 & (x_n - h)^2 & x_n - h & 1 \\ (x_n - 2h)^3 & (x_n - 2h)^2 & x_n - 2h & 1 \\ (x_n - 3h)^3 & (x_n - 3h)^2 & x_n - 3h & 1 \end{bmatrix} \begin{bmatrix} K \\ L \\ M \\ N \end{bmatrix} = \begin{bmatrix} f_n \\ f_{n-1} \\ f_{n-2} \\ f_{n-3} \end{bmatrix} \tag{14.76}$$

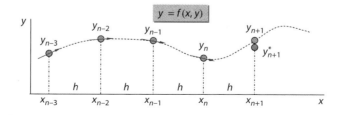

Fig. 14.7 Illustration of the four-point Adams–Bashforth prediction geometry.

As usual, let us employ the Symbolic Math Toolbox to find the solution to Eqs. (14.7) and (14.76). The following script:

```
%% Defining symbolic variables
syms x_n h f_nm3 f_nm2 f_nm1 f_n f_np1 y_n f_nplast x
%% Solving for coefficients of a cubic interpolation
A = [      x_n^3            x_n^2      x_n        1;
         (x_n-h)^3        (x_n-h)^2    x_n-h      1;
        (x_n-2*h)^3      (x_n-2*h)^2   x_n-2*h    1;
        (x_n-3*h)^3      (x_n-3*h)^2   x_n-3*h    1];
b= [f_n;f_nm1;f_nm2;f_nm3];
coef = A\b;
%% Producing a prediction step
In = int([x^3   x^2   x   1]*coef,x_n,x_n+h);
In = simplify(In); pretty(y_n+In)
```

results in

```
        h (55 f_n  -  59 f_nm1 + 37 f_nm2  -  9 f_nm3)
y_n +   ---------------------------------------------
                            24
```

Therefore,

$$y_{n+1}^* = y_n + \frac{h}{24}\left(55 f_n - 59 f_{n-1} + 37 f_{n-2} - 9 f_{n-3}\right) \qquad (14.77)$$

The *Adams–Moulton corrector* is developed similarly, but this time shifting four points to the right, that is, using (x_{n-2}, y_{n-2}), (x_{n-1}, y_{n-1}), (x_n, y_n), and (x_{n+1}, y_{n+1}^*). The following script:

```
%% Defining symbolic variables
syms x_n h f_nm2 f_nm1 f_n f_nplast x y_n
%% Solving for coefficients of a correction cubic interpolation
A = [  (x_n+h)^3      (x_n+h)^2         x_n+h      1;
          x_n^3          x_n^2          x_n        1;
        (x_n-h)^3      (x_n-h)^2        x_n-h      1;
       (x_n-2*h)^3   (x_n-2*h)^2        x_n-2*h    1];
b= [f_nplast; f_n; f_nm1; f_nm2];
coef = A\b;
%% Producing a correction step
In = int([x^3   x^2   x   1]*coef,x_n,x_n+h);
In = simplify(In); pretty(y_n+In)
```

helps to obtain the result

```
        h (19 f_n  -  5 f_nm1 + f_nm2 + 9 f_nplast)
y_n +   ------------------------------------------
                           24
```

Hence,

$$y_{n+1} = y_n + \frac{h}{24}(9f_{n+1}^* + 19f_n - 5f_{n-1} + f_{n-2}) \tag{14.78}$$

where f_{n+1}^* denotes $f(x_n + h, y_{n+1}^*)$.

The alternative way to receive Eqs. (14.77) and (14.78) is to employ Taylor series expansion as we did it in Chapter 13. The benefit is that along a with prediction and correction formulas, we can also receive the estimates of the local truncation error. The following script:

```
%% Defining symbolic variables
syms h y_n f_n f_nm1 f_nm2 f_nm3 f_prime f_2prime f_3prime f_4prime
%% Introducing equation for the forward Taylor series expansion around x_n
y_np1 = y_n+h*f_n+h^2*f_prime/factorial(2)+h^3*f_2prime/factorial(3)...
                +h^4*f_3prime/factorial(4)+h^5*f_4prime/factorial(5);
%% Introducing equations for f_nm3, f_nm2, and f_nm1
Eqm1 = -f_nm1+f_n-h*f_prime+h^2*f_2prime/factorial(2)   ...
              -h^3*f_3prime/factorial(3)+h^4*f_4prime/factorial(4);
Eqm2 = -f_nm2+f_n-2*h*f_prime+(2*h)^2*f_2prime/factorial(2)...
            -(2*h)^3*f_3prime/factorial(3)+(2*h)^4*f_4prime/factorial(4);
Eqm3 = -f_nm3+f_n-3*h*f_prime+(3*h)^2*f_2prime/factorial(2)...
            -(3*h)^3*f_3prime/factorial(3)+(3*h)^4*f_4prime/factorial(4);
%% Resolving three equations for f_prime, f_2prime, and f_3prime
sol=solve(Eqm1,Eqm2,Eqm3,f_prime,f_2prime,f_3prime);
%% Substituting estimates of the derivatives into y_np1 equation
y_np1=subs(y_np1,{f_prime,f_2prime,f_3prime},...
                {sol.f_prime,sol.f_2prime,sol.f_3prime});
%% Simplifying and displaying the result
r=collect(y_np1,h); pretty(r)
```

returns

```
251 f_4prime  5    / 55 f_n   59 f_nm1   37 f_nm2   3 f_nm3 \
------------  h  + |  ------ - -------- + -------- - ------- | h + y_n
   720             \   24        24         24         8    /
```

that is,

$$e_{n+1}^* = y_{n+1}^{exact} - y_{n+1}^* = \frac{251}{720}h^5 y^{(iv)}(\bar{x}^*) = C_4^* h^5 y^{(iv)}(\bar{x}^*) \tag{14.79}$$

Here y_{n+1}^{exact} denotes the exact (unknown) solution at x_{n+1}. Similarly, for the correction stage the script

```
%% Defining symbolic variables
syms h y_n f_n f_nm1 f_nm2 f_nm3 f_prime f_2prime f_3prime f_4prime f_np1
```

```
%% Introducing equation for the forward Taylor series expansion around x_n
y_np1 = y_n+h*f_n+h^2*f_prime/factorial(2)+h^3*f_2prime/factorial(3)...
                +h^4*f_3prime/factorial(4)+h^5*f_4prime/factorial(5);
%% Introducing equations for f_nm2, f_nm1, and f_np1
Eqp1 = -f_np1+f_n+h*f_prime+h^2*f_2prime/factorial(2)...
            +h^3*f_3prime/factorial(3)+h^4*f_4prime/factorial(4);
Eqm1 = -f_nm1+f_n-h*f_prime+h^2*f_2prime/factorial(2)...
            -h^3*f_3prime/factorial(3)+h^4*f_4prime/factorial(4);
Eqm2 = -f_nm2+f_n-2*h*f_prime+(2*h)^2*f_2prime/factorial(2)...
            -(2*h)^3*f_3prime/factorial(3)+(2*h)^4*f_4prime/factorial(4);
%% Resolving three equations for f_prime, f_2prime, and f_3prime
sol=solve(Eqm1,Eqm2,Eqp1,f_prime,f_2prime,f_3prime);
%% Substituting estimates of the derivatives into y_np1 equation
y_np1=subs(y_np1,{f_prime,f_2prime,f_3prime},...
            {sol.f_prime,sol.f_2prime,sol.f_3prime});
%% Simplifying and displaying the result
r=collect(y_np1,h); pretty(r)
```

returns

```
/   19 f_4prime \  5   / 19 f_n     5 f_nm1    f_nm2    3 f_np1 \
| - ----------- | h  + | ------  -  -------  +  -----  +  ------- | h + y_n
\       720     /      \   24         24         24         8    /
```

so that

$$e_{n+1} = y_{n+1}^{\text{exact}} - y_{n+1} = -\frac{19}{720} h^5 y^{(iv)}(\overline{x}) = C_4 h^5 y^{(iv)}(\overline{x}) \qquad (14.80)$$

The local truncation error for both the predictor and corrector happens to be of the same order of h ($O(h^5)$). It is safe to assume that the average fifth-order derivatives $y^{(iv)}(\overline{x}^*)$ and $y^{(iv)}(\overline{x})$ are close enough, so that $y^{(iv)}(\overline{x}^*) \approx y^{(iv)}(\overline{x})$. This enables us to exclude them from Eqs. (14.79) and (14.80) to produce an estimate of the local truncation error based on the two computed values y_{n+1}^* and y_{n+1}

$$e_{n+1} = y_{n+1}^{\text{exact}} - y_{n+1} \approx \frac{C_4}{C_4^* - C_4}(y_{n+1} - y_{n+1}^*) = -\frac{19}{270}(y_{n+1} - y_{n+1}^*) \qquad (14.81)$$

This equation can be used to determine when to change the step size similar to the manner demonstrated in Fig. 13.8. Although elaborate methods are available, the following simple procedure shows how to reduce

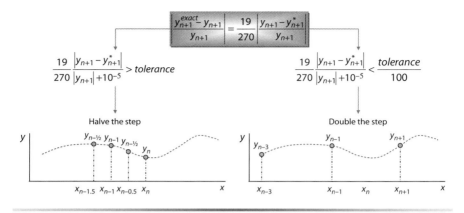

Fig. 14.8 Adaptive application of the Adams–Bashforth–Moulton four-step method.

the step size to $h/2$ or increase it to $2h$. When the predicted and corrected values do not agree to five significant digits (Sec. 8.5.1), that is,

$$\frac{19}{270}\frac{|y_{n+1}-y_{n+1}^*|}{|y_{n+1}|+10^{-5}} > \varepsilon^{rel} \tag{14.82}$$

where ε^{rel} is the desired relative tolerance, then the step size should be reduced to $h/2$ (cf. Eq. (14.82) with Eq. (8.27)). If they agree to seven or more significant digits, that is,

$$\frac{19}{270}\frac{|y_{n+1}-y_{n+1}^*|}{|y_{n+1}|+10^{-5}} < \frac{\varepsilon^{rel}}{100} \tag{14.83}$$

then the step size can be safely increased to $2h$.

Table 14.1 Initialization of the `ode113` Solver

s	Predicting Step	Correcting Step
0	$y_1^* = y_0 + hf_0$	$y_1 = y_0 + hf_1^*$
1	$y_2^* = y_1 + h((3/2)f_1 - (1/2)f_0)$	$y_2 = y_1 + h((1/2)f_2^* + (1/2)f_1)$
2	$y_3^* = y_2 + h((23/12)f_2 - (4/3)f_1 + (5/12)f_0)$	$y_3 = y_2 + h((5/12)f_3^* + (2/3)f_2 - (1/12)f_1)$
3	$y_4^* = y_3 + h((55/24)f_3 - (59/24)f_2 + (37/24)f_1 - (3/8)f_0)$	$y_4 = y_3 + h((3/8)f_4^* + (19/24)f_3 - (5/24)f_2 + (1/24)f_1)$
4	$y_5^* = y_4 + h((1901/720)f_4 - (1387/360)f_3 + (109/30)f_2 - (637/360)f_1 + (251/722)f_0)$	$y_5 = y_4 + h((251/720)f_5^* + (646/720)f_4 - (264/720)f_3 + (106/720)f_2 - (19/720)f_1)$
	$a_{s-k} = ((-1)^k/k!(s-k)!)\int_0^1 \prod_{\substack{i=0\\i\neq k}}^{s}(v+i)\,dv,$ $k = 0,1,\dots,s$	$a_{s-k+1} = ((-1)^k/k!(s-k+1)!)\int_0^1 \prod_{\substack{i=0\\i\neq k}}^{s}(v+i-1)\,dv,\ k = 0,1,\dots,s$

Table 14.2 Coefficients for the Local Truncation Errors Estimates

s	0	1	2	3	4
Predicting step, C^*_{s+1}	1/2	5/12	3/8	251/720	95/288
Correcting step, C_{s+1}	――	-(1/12)	-(1/24)	-(19/720)	-(3/160)

Reducing the step size requires four new starting values. Interpolation [Eq. (14.74)] is used to supply the missing values that bisect the intervals $[x_{n-2}; x_{n-1}]$ and $[x_{n-1}; x_n]$, that is, at $x_{n-1.5}$, and $x_{n-0.5}$. The interpolation formulas needed to obtain the new starting values for the step size $h/2$ are

$$f_{n-0.5} = \frac{-5f_{n-4} + 28f_{n-3} - 70f_{n-2} + 140f_{n-1} + 35f_n}{128}$$

$$f_{n-1.5} = \frac{3f_{n-4} - 20f_{n-3} + 90f_{n-2} + 60f_{n-1} - 5f_n}{128}$$

(14.84)

Increasing the step size is even an easier task. Seven prior points are needed to double the step size. The four new points are obtained by omitting every second one. The entire procedure is presented in Fig. 14.8 (the point (x_{n-5}, y_{n-5}) is not shown).

Some other multistep schemes (*Mine–Simpson method, Hamming method,* etc.) are available, but the self-contained MATLAB `ode113` solver utilizes the Adams–Bashforth–Moulton family of formulas of a variable order (1–13) in a manner similar to that shown earlier for the four-point version. It starts from just one point, —the initial condition y_0, and applies the Euler formula [Eq. (14.20)], being an analog of the left rectangles formula of Eq. (13.2) to produce y_1^*. Then, the corrector uses the backward Euler formula, being the same as the right rectangles formula of Eq. (13.2) to produce y_1 (in fact, this backward step can be skipped). Now that we have two points to play with, the predictor uses a two-point linear approximation in Eq. (14.72) to produce y_2^*, followed by the first-order corrector to come up with y_2 using the analog of a trapezoidal rule of Eq. (13.3), etc. Table 14.1 presents formulas for the first five steps of this procedure. The last row in Table 14.1 shows how to compute coefficients a_1 in Eq. (14.73) for the remaining s, $s = 5, 6, ..., 12$. Once the highest order of 13 is achieved, the solver proceeds with it until the end adjusting the step size to meet the tolerance.

The Milne estimate for the local truncation error is described by Eq. (14.81), where the values of C_i^* and C_i for the first few steps are presented in Table 14.2.

14.5.2 Backward and Numerical Differentiation Formulas

The multistep method considered in the previous section was based on numerical integration, so that the integral in Eq. (14.72) was approximated numerically using Lagrange polynomials. Ironically, another group of multistep methods, developed even before RK methods, rely on an exactly opposite idea and are based on numerical differentiation.

The underlying idea of the backward differentiation method is that if $s + 1$ solutions $y_{n-s}, y_{n-s+1}, ..., y_n$ of Eq. (14.16) [Eq. (14.15)] are known, then the derivative at $(x_{n+1}, y_{n+1}), f_{n+1} = f(x_{n+1}, y_{n+1})$, can be estimated numerically using the backward difference approximation formulas similar to those developed in Chapter 12 for the first-order derivative (Table 12.1). Using notations of this chapter, we can rewrite Eq. (12.25) as

$$hf_{n+1} = \sum_{m=1}^{s} \frac{1}{m} \Delta^m y_{n+1} \tag{14.85}$$

where the difference operator Δ was defined by Eqs. (12.23) and (12.24). The first- to fifth-order *backward differentiation formulas* (BDFs) are shown in Table 14.3.

Unfortunately, these formulas are implicit (the explicit formulas, similar to those of Eq. (14.85), but with f_n on the right-hand side, can be also developed, but they proved to be unstable). Therefore, the solution to Eq. (14.85) should be obtained iteratively, such as, using the simple fixed-point iteration method described in Sec. 10.3.1. The iteration process for solving Eq. (14.85) starts with the initial guess

$$y_{n+1}^{(0)} = y_n + hf_n = y_n + \sum_{m=1}^{s} \frac{1}{m} \Delta^m y_n \tag{14.86}$$

and continues until the tolerance is met

$$\left| y_{n+1}^{(r)} - y_{n+1}^{(r-1)} \right| \le \varepsilon^{rel} \left| y_{n+1}^{(r)} \right| + \varepsilon^{abs} \tag{14.87}$$

Similar to formulas of Table 14.1, the BDF of Table 14.3 constitute a variable-order self-starting algorithm implemented in the ode15s solver. The step size h is typically kept constant unless there is good reason to

Table 14.3 First- to Fifth-Order BDFs

s	Formula
1	$y_{n+1} - y_n = hf_{n+1}$
2	$(3/2)y_{n+1} - 2y_n + (1/2)y_{n-1} = hf_{i+1}$
3	$(11/6)y_{n+1} - 3y_n + (3/2)y_{n-1} - (1/3)y_{n-2} = hf_{n+1}$
4	$(25/12)y_{n+1} - 4y_n + 3y_{n-1} - (4/3)y_{n-2} + (1/4)y_{n-3} = hf_{n+1}$
5	$(137/60)y_{n+1} - 5y_n + 5y_{n-1} - (10/3)y_{n-2} + (5/4)y_{n-3} - (1/5)y_{n-4} = hf_{n+1}$

Table 14.4 The values of parameter κ in the first- through fifth-order NDF

s	1	2	3	4	5
κ	$-(37/200)$	$-(1/9)$	-0.0823	-0.0415	0

change it. To be more precise, the ode15s solver relies on a slightly modified version of the aforementioned approach (and offers BDF as an option). In this approach, to improve convergence of iterative process of finding y_{n+1}, Eq. (14.85) is modified to add an extra term, proportional to the estimate of truncation error

$$hf_{n+1} = \sum_{m=1}^{s} \frac{1}{m} \Delta^m y_{n+1} - \kappa (y_{n+1} - y_{n+1}^{(0)}) \sum_{m=1}^{s} \frac{1}{m} \qquad (14.88)$$

Here, $y_{n+1}^{(0)}$ is defined by Eq. (14.86) and the parameter κ is chosen for each s to reduce the truncation error as much as possible while still assuring the stability of the algorithm. Equation (14.88) is referred to as *numerical differentiation formulas* (NDFs). For example, the first-order NDF reduces to

$$hf_{n+1} = y_{n+1} - y_n - \kappa (y_{n+1} - 2 y_n + y_{n-1}) \qquad (14.89)$$

For the BDFs of Table 14.3, the best value of κ used in NDFs is presented in Table 14.4. As seen, the fifth-order NDF coincides with the fifth-order BDF (since $k = 0$).

The ode15s solver of MATLAB, employing NDFs and BDFs, along with another variable-order solver, ode15i, which only employs NDFs are recommended for the *stiff* problems that cannot be handled effectively by the implicit ode23, ode45, and ode113 solvers. The following section provides more details on the stiffness issue and describes a few more methods to handle it.

14.6 Stiff Problems and Other Issues

For more or less homogeneous functions without singularities (let us refer them as *nonstiff systems*), methods considered in the previous sections work fairly well, so you might ever need nothing more (especially, if you are a beginner). Formally, stiffness means that the solution can change on a time scale that is very short compared with the interval of integration, which causes decreasing the integration step to very small values. Hence, methods that were not designed to handle stiff systems, may work extremely slow (if at all). For these types of problems, the MATLAB ODE suite offers some alternatives briefly described later.

The stiff systems solvers implement more sophisticated algorithms than those in `ode23`, `ode45`, and `ode113`, but the terminology introduced earlier in this chapter still applies. For instance, 23 in the name of these solvers reflect the fact that they use the second- and third-order pairs, similar to what `ode23` does.

Mathematically, the stiffness of the problem can be determined by looking at the ratio of maximum and minimum real parts among all eigenvalues of the Jacobian matrix **J**. That is why all solvers for the stiff problems take this matrix into account.

14.6.1 Incorporation of the Jacobian Matrix

A well-known approach for improving a stability of the algorithm is to use *implicit* rather than explicit formulas. In this case, the general RK solution given by Eq. (14.17) remains the same, but the slope estimates of Eq. (14.18) will have the quantity on the left-hand side appearing inside the function on the right-hand side. For example

$$k_2 = f(x_n + q_1 h, y_n + p_{11} k_1 h + p_{12} k_2 h) \tag{14.90}$$

Instead of applying the iterative process to solve these equations for the slopes (in our specific case, to find k_2), we can try to linearize them, which results in a new class of methods. In our case depending on what function value we linearize it about we may obtain

$$k_2 = f(x_n + q_1 h, y_n + p_{11} k_1 h) + J p_{12} k_2 h \tag{14.91}$$

or even

$$k_2 = f(x_n, y_n) + T q_1 h + J p_{11} k_1 h + J p_{12} k_2 h \tag{14.92}$$

In these formulas, J denotes the partial derivative of the function f with respect to y, and T the partial derivative of the function f with respect to x. (Note that, for Eqs. (14.91) and (14.92) these derivatives are estimated at two different points.) In the case of Eq. (14.15), **J** and **T** denote the Jacobian matrix and a gradient vector

$$\mathbf{J} = \begin{bmatrix} \dfrac{\partial F_1}{\partial y_1} & \dfrac{\partial F_1}{\partial y_2} & \cdots & \dfrac{\partial F_1}{\partial y_l} \\ \dfrac{\partial F_2}{\partial y_1} & \dfrac{\partial F_2}{\partial y_2} & \cdots & \dfrac{\partial F_2}{\partial y_l} \\ \vdots & \vdots & \ddots & \vdots \\ \dfrac{\partial F_l}{\partial y_1} & \dfrac{\partial F_l}{\partial y_2} & \cdots & \dfrac{\partial F_l}{\partial y_l} \end{bmatrix} \text{ and } \mathbf{T} = \begin{bmatrix} \dfrac{\partial F_1}{\partial x} \\ \dfrac{\partial F_2}{\partial x} \\ \vdots \\ \dfrac{\partial F_l}{\partial x} \end{bmatrix} \tag{14.93}$$

By rearranging terms, Eqs. (14.91) and (14.92) become

$$(1 - Jp_{12}h)k_2 = f(x_n + q_1h, y_n + p_{11}k_1h) \tag{14.94}$$

and

$$(1 - Jp_{12}h)k_2 = f(x_n, y_n) + Tq_1h + Jp_{11}k_1h \tag{14.95}$$

respectively. It is exactly the $1 - Jp_{12}h$ term ($1 - hp_{12}J$, in the case of Eq. (14.15)) that makes the algorithm more robust (stable). As earlier, this approach, introduced by Rosenbrock, allows you to have infinite number of the unknown coefficients to choose from. One additional constraint that leads to the computational effectiveness is that the Jacobian matrix is to be computed only once for each interval. Another is that p_{12} is to be a constant value for all slope estimates. Combined with the desire to have an effective FSAL pair like Eqs. (14.63)–(14.65) for the ode23 solver resulted in the ode23s solver with the following underlying equations as suggested by Shampine and Reichelt:

$$y_{n+1} = y_n + hk_2 \tag{14.96}$$

where

$$k_1 = (1 - hdJ)^{-1}(f(x_n, y_n) + hdT)$$

$$k_2 = (1 - hdJ)^{-1}\left(f\left(x_n + \frac{1}{2}h, y_n + \frac{1}{2}hk_1\right) - k_1\right) + k_1$$

$$k_3 = (1 - hdJ)^{-1}\left(f(x_{n+1}, y_{n+1}) - (6 + \sqrt{2})\left(k_2 - f\left(x_n + \frac{1}{2}h, y_n + \frac{1}{2}hk_1\right)\right)\right.$$

$$\left. - 2(k_1 - f(x_n, y_n)) + hdT\right) \tag{14.97}$$

and $d = 1/(2 + \sqrt{2})$. As earlier, instead of estimating the second-order formula the error estimate is computed directly as

$$\Delta = \frac{h}{6}|k_1 - 2k_2 + k_3| \tag{14.98}$$

Although Eqs. (14.96)–(14.98) seem to be much more complicated compared with those of Eqs. (14.63)–(14.65), they preserve the same structure. As opposed to explicit formulas, Eqs. (14.96)–(14.98) feature and benefit from using an additional information about the function $f(x, y)$ contained in J and T.

To be able to evaluate the solution anywhere within the interval $x \in [x_n; x_{n+1}]$ (for example, for the purpose of rescaling) the quadratic polynomial is used

$$y_{n+c} = y_n + h\left(\frac{c(1-c)}{1-2d}k_1 + \frac{c(c-2d)}{1-2d}k_2\right) \tag{14.99}$$

Obviously, at $c=0$ it assumes the value of y_n, and at $c=1$ becomes Eq. (14.96).

14.6.2 Using the Implicit Trapezoidal Rule

MATLAB offers two solvers, ode23t and ode23tb, that use an implicit trapezoidal rule. The ode23tb solver uses it in a specific form

$$y_{n+\gamma} = y_n + \gamma h \frac{1}{2}(f_n + f_{n+\gamma}) \tag{14.100}$$

Equation (14.100) produces the estimate of the solution not at x_{n+1}, but at some point $x_{n+\gamma}$, $0 < \gamma \le 1$. As the second formula, the ode23tb solver uses the second-order BDF. In this case, the parabolic approximation should pass through three points, (x_n, y_n), $(x_{+\gamma}, y_{+\gamma})$, and (x_{n+1}, y_{n+1}). Taking its derivative and estimating it at x_{n+1} produces the estimate of the derivative. Employing the self-explained script

```
%% Defining symbolic variables
syms x_n h gam y_n y_gam y_np1 x
%% Solving for coefficients of a quadratic interpolation
A = [        x_n^2          x_n   1;
       (x_n+gam*h)^2  x_n+gam*h  1;
         (x_n+h)^2       x_n+h   1];
b= [y_n;y_gam;y_np1];
coef = A\b;
%% Forming the polynomial out of the vector of its coefficients
Cpol=[x^2   x   1]*coef;
%% Computing the first-order derivative at x=x_i
D1 = diff(Cpol,1); D1=subs(D1,x,x_n+h); pretty(simplify(D1))
```

yields

```
y_gam - y_n + gam (y_n - y_np1)      y_n - y_np1
-------------------------------  -  -----------
        gam h (gam - 1)                  h
```

that is,

$$(2-\gamma)y_{n+1} - \frac{1}{\gamma}y_{n+\gamma} + \frac{(1-\gamma)^2}{\gamma}y_n = (1-\gamma)hf_{n+1} \tag{14.101}$$

Linearizing Eqs. (14.100) and (14.101) reveals that the corresponding multiplies including the Jacobian matrix (see the previous section) are

$$2\mathbf{I} - \gamma h\mathbf{J} \quad \text{and} \quad (2 - \gamma)\mathbf{I} - (1 - \gamma)h\mathbf{J} \tag{14.102}$$

respectively. They assume the same form if

$$\frac{2}{\gamma} = \frac{2 - \gamma}{1 - \gamma} \tag{14.103}$$

Equation (14.103) results in $\gamma = 2 - \sqrt{2}$. Along with providing the best computational effectiveness (both implicit stages within iterative process for each step use the same multiplier), this choice of γ assures the least truncation error and largest stability region.

This trapezoidal-rule/BDF2 combination (TR-BDF2) can also be viewed as an implicit RK pair of the orders 2 and 3

$$y_{n+1} = y_n + h(wk_1 + wk_2 + dk_3) \tag{14.104}$$

where

$$\begin{aligned}
k_1 &= f(x_n, y_n) \\
k_2 &= f(x_n + \gamma h, y_n + dhk_1 + dhk_2) \\
k_3 &= f(x_{n+1}, y_{n+1})
\end{aligned} \tag{14.105}$$

$d = 0.5\gamma$ and $w = 0.25\sqrt{2}$. As usual, instead of estimating the second-order formula, the error estimate is computed directly as

$$\Delta = h \left| \frac{1 - w}{3} k_1 + \frac{3w + 1}{3} k_2 + \frac{d}{3} k_3 \right| \tag{14.106}$$

Since at each step, the TR-BDF2 produces three solutions and three derivative values at x_n, $x_{n+\gamma}$, and x_{n+1} to interpolate the solution anywhere within the interval $x \in [x_n; x_{n+1}]$, it is natural to employ the piecewise cubic Hermite interpolant (Eq. (11.44))

$$y(x) = (v_3 - 2v_2)\rho^3 + (3v_2 - v_3)\rho^2 + v_1\rho + v_0 \tag{14.107}$$

with the coefficients for $x \in [x_n; x_{n+\gamma}]$ and $x \in [x_{n+\gamma}; x_{n+1}]$ given in Table 14.5.

The iteration procedure for the first stage [Eq. (14.100)] starts with the simplest initial guess

$$f_{n+\gamma}^{(0)} = f_n \tag{14.108}$$

Table 14.5 Coefficients for the `ode23tb` Cubic Hermite Interpolant

Interval	Argument $p(x)$	v_3	v_2	v_1	v_0
$x \in [x_n; x_{n+\gamma}]$	$(x-x_n)/(\gamma h)$	$\gamma h(f_{n+\gamma}-f_n)$	$y_{n+\gamma}-y_n-v_1$	$\gamma h f_n$	y_n
$x \in [x_{n+\gamma}; x_{n+1}]$	$(x-x_{n+\gamma})/((1-\gamma)h)$	$(1-\gamma)h(f_{n+1}-f_{n+\gamma})$	$y_{n+1}-y_{n+\gamma}-v_1$	$(1-\gamma)hf_{n+\gamma}$	$y_{n+\gamma}$

(the desire to maintain a single-step structure prevents using the points from the previous steps) and then proceeds as

$$y_{n+\gamma}^{(r)} = (y_n + dhf_n) + dhf_{n+\gamma}^{(r)}, \quad D^{(r)} = (\mathbf{I} - hd\mathbf{J})^{-1} h(f(x_{n+\gamma}, y_{n+\gamma}^{(r)}) - f_{n+\gamma}^{(r)}) \quad (14.109\text{a})$$

and

$$f_{n+\gamma}^{(r+1)} = \frac{f_{n+\gamma}^{(r)}}{hD^{(r)}} \quad (14.109\text{b})$$

until the accuracy tolerance is met. Once $\hat{y}_{n+\gamma}$ is found, Eq. (14.101) is solved with the initial guess

$$f_{n+1}^{(0)} = (1.5+\sqrt{2})f_n + (2.5+2\sqrt{2})f_{n+\gamma} - (6+4.5\sqrt{2})(\hat{y}_{n+\gamma} - y_n)h^{-1} \quad (14.110)$$

resulting from extrapolating the derivative of the piecewise cubic Hermite spline [Eq. (14.107)] to $x_{n+\gamma}$. (Equation (14.110) is coded in MATLAB as presented at the end of Sec. 7.9.1.) The iterative process for the second stage unfolds as follows:

$$y_{n+\gamma}^{(r)} = (y_n + whf_n + whf_{n+\gamma}) + dhf_{n+1}^{(r)}$$
$$D^{(r)} = (\mathbf{I} - hd\mathbf{J})^{-1} h(f(x_{n+1}, y_{n+1}^{(r)}) - f_{n+1}^{(r)}), \quad \text{and} \quad f_{n+1}^{(r+1)} = \frac{f_{n+1}^{(r)}}{hD^{(r)}} \quad (14.111)$$

One more implicit solver of MATLAB, `ode23t`, uses an implicit trapezoidal rule in its common form

$$y_{n+1} = y_n + h\frac{1}{2}(f_n + f_{n+1}) \quad (14.112)$$

Equation (14.112) is equivalent to Eq. (14.100) with $\gamma = 1$. It relies on using a cubic interpolant to estimate the local truncation error and adjust the step size (to be discussed later).

14.6.3 Controlling the Integration Step Size

As discussed in the previous sections, the local truncationv error is largely defined by the step size h. The decision on whether to keep the current step size or change it is based on this error estimate. Some of methods introduced in this chapter have an embedded capability to estimate these errors (for example, RK pairs). Let us introduce here a general step-size control strategy for other methods that may lack these embedded formulas.

Assume that the local truncation error is proportional to the second derivative, say $e \sim (h^2 / 2)|y''|$ as for the Euler method or BDF1. Suppose we have computed y_{n-1}, y_n and consequently f_{n-1}, f_n corresponding to x_{n-1}, x_n already. Then, the value of y'' can be estimated as

$$\hat{y}'' = \frac{f_n - f_{n-1}}{x_n - x_{n-1}} \tag{14.113}$$

Assuming this estimate stays about the same for the next interval, an attempt to estimate the next step size h can be based on the requirement to meet the error tolerance ε

$$\hat{e} \sim \frac{h^2}{2}\left|\frac{f_n - f_{n-1}}{x_n - x_{n-1}}\right| \le \varepsilon \tag{14.114}$$

It leads to

$$h \le \sqrt{2\varepsilon \left|\frac{x_n - x_{n-1}}{f_n - f_{n-1}}\right|} \tag{14.115}$$

In practice, h is usually selected to be less than that, say about 0.9 of the preceding value. Also we do not want the step size to be either too large, greater than some $h_{\max}$ or too low, so that the computer cannot distinguish between x_n and $x_n + h$. Therefore, the typical size selection strategy would become

$$h \le \max\left(16 * \text{eps} * |x_n|, \min\left(\sqrt{2\varepsilon^{rel} \; |y_n| \left|\frac{x_n - x_{n-1}}{f_n - f_{n-1}}\right|}, h_{\max}\right)\right) \tag{14.116}$$

Here ε^{rel} represents the relative accuracy and eps—the distance between 1 and the next floating point number (Table 8.3).

If the local truncation error is proportional to a higher-order derivative, like $|e_n| \sim (h^2 /12) y_n'''$ for the trapezoidal rule of ode23t, estimating y''' as we did it earlier may not be practical. In this case, several prior points may be used to develop an approximation polynomial, which can be differentiated analytically and estimated at x_n or even at x_{n+1}. For example, the ode23t solver uses y_{n-3}, y_{n-2}, y_{n-1}, and y_n to produce a cubic approximation [Eq. (14.107)].

One more question is how to choose the initial step size, when the only available data is the differential equation itself and the initial conditions. Of course you may start from *a priori* excessively small step size, but it may waste a lot of computing time. An alternative, "smart" approach used in most of the codes is as follows.

First, we need to evaluate the function $\mathbf{F}(x,\mathbf{y})$ at the initial point. Next, we estimate the error as

$$\Delta = \frac{1}{\max\left(|x_0|,|x_f|\right)^{p-1}} + \|\mathbf{F}(x_0,y_0)\|^{p+1} \qquad (14.117)$$

where p is the order of the method and $\|...\|$ denotes the Euclidean norm. Then, the initial step can be chosen as

$$h = \left(\frac{\varepsilon}{\Delta}\right)^{\frac{1}{p+1}} \qquad (14.118)$$

which resembles Eq. (14.67).

Very often, however, the initial conditions are chose to be at an equilibrium point resulting in $\|\mathbf{F}(x_0,y_0)\|$ being very close to zero (because almost all components of $\mathbf{F}(x_0, y_0)$ are zeros). Thus, usually, choosing the initial integration step includes two more computations. First, the Euler method is applied with h followed from Eq. (14.118) to produce $x_1,\mathbf{y}_1$, and second, the estimates (14.117) and (14.118) are repeated for this next point. Finally, the initial integration step size is chosen as the smallest between the two estimates.

14.7 Using MATLAB IVP Solvers

As mentioned in the previous sections, the MATLAB ODE suite offers several solvers allowing handling both stiff and nonstiff problems. These solvers are summarized in Table 14.6. Moreover, some of these solvers are suitable to handle not only the explicit ODEs of the form of Eq. (14.7) but also linearly implicit ODEs of the form

$$\mathbf{M}(x,y)y' = \mathbf{F}(x,y) \qquad (14.119)$$

with $\mathbf{M}(x,\mathbf{y})$ being the *mass matrix*, and even fully implicit ODEs

$$\mathbf{F}(x,y,y') = 0 \qquad (14.120)$$

The most common syntax for MATLAB solvers is one of the following:

```
[T,Y] = solver(odefun,tspan,y0)
[T,Y] = solver(odefun,tspan,y0,options)
[T,Y] = solver(odefun,tspan,y0,options,p1,p2,…)
```

where `solver` is one of ode23, ode45, ode113, ode15s, ode23s,

Table 14.6 MATLAB Solvers for Explicit, Linearly Implicit, and Fully Implicit ODEs

Solver	Order of Accuracy	Description
Solvers for Nonstiff Problems		
ode45	Medium	Explicit single-step RK solver based on the Dormand–Price 4/5 pair [Eqs. (14.68) and (14.69)]. Suitable for nonstiff problems that require moderate accuracy. This is the first solver to try on a new problem.
ode23	Low	Explicit single-step RK solver, based on the Bogacki–Shampine 2/3 pair [Eqs. (14.63) and (14.64)]. Suitable for problems that tolerate lower accuracy or exhibit moderate stiffness.
ode113	Low to high	Multistep Adams–Bashforth–Moulton solver of 1st- to 13th-order (Table 14.1). Suitable for nonstiff problems that require moderate to high accuracy involving problems, where $\mathbf{F}(x, y)$ is expensive to compute. Not suitable for problems where $\mathbf{F}(x, y)$ is not smooth.
Solvers for Stiff Problems		
ode15s	Low to medium	Multistep solver of varying order (up to 5), relying on NDF. Options allow integration with BDF (Table 14.3) and integration with a maximum order less than the default five. It is suitable for stiff problems that require moderate accuracy. This is typically the solver to try if ode45 fails or is too slow.
ode23s	Low	Single-step solver based on the implicit modified second-order Rosenbrock formula [Eqs. (14.96)–(14.98)]. Suitable for stiff problems, where lower accuracy (compared to that of ode15s) is acceptable or where $f(x, y)$ is discontinuous.
ode23t	Low	Implicit single-step solver relying on the trapezoidal rule [Eq. (14.112)] and a cubic interpolant [Eq. (14.117)] to estimate the local truncation error. Suitable for moderately stiff problems. Can be used to solve differential-algebraic equations.
ode23tb	Low	Implicit RK pair solver combining a trapezoidal rule of Eq. (14.100) with a second-order BDF [Eq. (14.101)]. Can be more efficient than ode15s if using crude tolerances.
Solver for Fully Implicit ODEs		
ode15i	Low to medium	A multistep NDF solver of the first to fifth order, especially suitable for solving fully implicit differential equations.

ode23t, ode23tb, or ode15i. In these commands, odefun is the function that evaluates the system of ODEs ($\mathbf{F}(x,\mathbf{y})$), tspan is the vector specifying the interval of integration, y0 is the vector of initial conditions (the solver imposes the vector of initial conditions `y0` at `tspan(1)`, and integrates from `tspan(1)` to `tspan(end))`, `options` is a structure of optional parameters that changes the default integration properties, `p1,p2,...` are additional parameters that can be passed directly to the ODE function as `odefun(t,y,p1,p2,...)`, and to all functions specified in `options`. Each row in the solution array `Y` corresponds to a time returned in the column vector `T`. Additionally,

```
sol = solver(odefun,[t0 tf],y0,...)
```

returns a structure that can be used with the function `deval` to evaluate the solution (and derivative) at any point between `t0` and `tf`.

Before we proceed with examples of using these solvers, let us try to solve an IVP using the simplest formulas developed in Sec. 14.3. Let us start with the problem given by

$$\dot{y} + y = \sin(t), \ y(0) = 5, \ t \in [0;10]\text{s} \tag{14.121}$$

First, we need to rewrite Eq. (14.121) in the form of Eq. (14.16)

$$\dot{y} = f(t, y) = \sin(t) - y \tag{14.122}$$

Now we can implement, for example, the Euler's formula [Eq. (14.20)], which yields

$$y_{n+1} = y_n + h(\sin(t_n) - y_n) \tag{14.123}$$

The following script implements this formula and computes two solutions for two different step sizes as shown in Fig. 14.9a

```
h1=0.01; h2=1;
ff = @(a,b) sin(a)-b;        % defining the function
y(1)=-5; z(1)=-5;
t1=0:h1:10; N1=length(t1); % defining the vector of arguments for h1
t2=0:h2:10; N2=length(t2); % defining the vector of arguments for h2
tic
for i=2:N1, y(i)=y(i-1)+h1*ff(t1(i-1),y(i-1)); end % solution for h1
T1=toc;
tic
for i=2:N2, z(i)=z(i-1)+h2*ff(t2(i-1),z(i-1)); end % solution for h2
T2=toc;
plot(t1,y,t2,z,'--m.')
text(4,-2.0,['T^{h=0.01s}=' num2str(T1*1000,2) 'ms'])
text(4,-2.6,['T^{h=1s} =' num2str(T2*1000,2) 'ms'])
xlabel('Time, s'), ylabel('y'), legend('h=0.01s','h=1s',0)
```

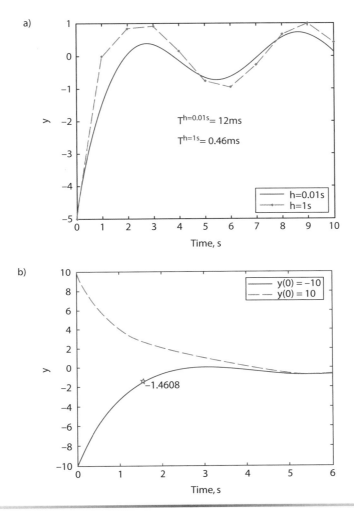

Fig. 14.9 Solving ODE with the Euler method varying a) the step size and b) initial condition.

Note that, we wrapped each cycle with the `tic` and `toc` commands to start a stopwatch timer and measure the elapsed time, respectively. As expected (Fig. 14.9a), the $h = 0.01$s solution demonstrates much better accuracy than that of $h = 1$s, but takes more time to compute. If the size h is increased even further, the system becomes unstable. Figure 14.9b shows that the initial conditions in ODE do matter. Almost the same script as mentioned previously was used in this case with the only difference that both step sizes were set to 0.01s and solutions were initialized with different values of $y(0)$. Also, as an example of how you can use the obtained solution to find the certain values, Fig. 14.9b features a star indicating a solution closest to the point $x = 0.5\pi$. The following piece of code shows how you can do it:

```
delta=t1-pi/2;
[C,ind]=min(abs(delta));      % finding the point closest to pi/2
plot(t1(ind),y(ind),'rp')
text(t1(ind)+0.1,y(ind),num2str(y(ind)))
```

Now, let us consider the classical second-order system introduced in the beginning of this chapter and described by Eq. (14.13). Assuming, $m = 1$ kg, $k = 1$ kg/s², $c = 1$ kg/s, and $u = \sin(t)$, let us rewrite it as

$$\dot{\mathbf{x}} = \begin{bmatrix} 0 & 1 \\ -1 & -1 \end{bmatrix} \mathbf{x} + \begin{bmatrix} 0 \\ 1 \end{bmatrix} \sin(t) \tag{14.124}$$

The following script implements the Euler and Heun methods [Eq. (14.20) and (14.32), respectively] using the step size $h = 0.8$s (Fig. 14.10a).

```
h=0.8;
ff = @(a,b) [b(2); -b(1)-b(2)+sin(a)];
y(:,1)=[-5; 0]; z(:,1)=[-5; 0];
t=0:h:15; N=length(t);
tic
for i=2:N
    y(:,i)=y(:,i-1)+h*ff(t(i-1),y(:,i-1));
end
T1=toc;
tic
for i=2:N
    k1=ff(t(i-1),z(:,i-1));
    k2=ff(t(i),z(:,i-1)+h*k1);
    z(:,i)=z(:,i-1)+h*(k1+k2)/2;
end
T2=toc;
plot(t,y(1,:),'-.m.',t,z(1,:),'.-')
text(8.5,-2,['t_{CPU}^{Euler}=' num2str(T1*1000,2) 'ms'])
text(8.5,-3,['t_{CPU}^{Heun}= ' num2str(T2*1000,2) 'ms'])
xlabel('Time, s'), ylabel('Position, m')
legend('Euler Method','Heun Method',1)
```

As seen, the step size of $h = 0.8$s for the lowest-order possible (Euler method) solution not only produces a huge error as compared with that of the second-order method (Heun method) solution but it also brings it on the verge of stability (with $h = 0.8$s, the Euler method solution diverges).

An almost identical script as mentioned earlier was used to produce another pair of solutions to the same problem, this time with $h = 0.08$s (Fig. 14.10b). Both solutions seem to be more close to the exact solution (which in this case can be obtained analytically), with the Heun method solution being more accurate anyway. Note the difference in computational time in both cases. The Heun method employs twice as much function evaluations as the Euler method, and it clearly shows up for the $h = 0.08$s solution of Fig. 14.10b. For the small number of nodes (Fig. 14.10a), the amount of CPU time spent on function evaluations is compatible to that of other mathematical operations, and that is why the difference in the CPU time for two solutions is smaller.

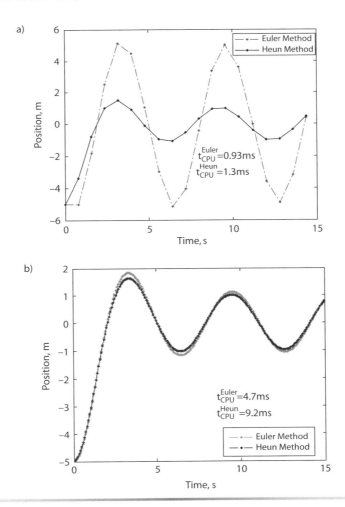

Fig. 14.10 Solving the second-order ODE with the lower-order methods using a) $h = 0.8$s and b) $h = 0.08$s.

By the way, you can also estimate the effectiveness of your code using the **Profiler**, introduced at the end of Sec. 2.2. The tenth icon on the Desktop Toolbar opens a window from which you can run your code, for example, the one introduced earlier with $h = 0.08$s. The return is shown in Fig. 14.11a. Although **Profiler** shows the CPU time consumption for the entire code, not just the fragment you are interested in, you can still find how much time it takes to execute some specific commands. For example, if you click on the second line, the name of the file itself, you may observe more details on CPU time usage (Fig. 14.11b).

a)
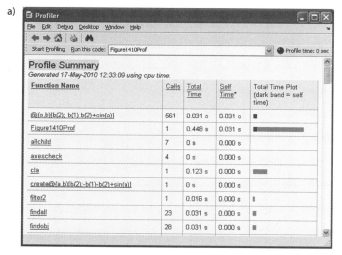

b)
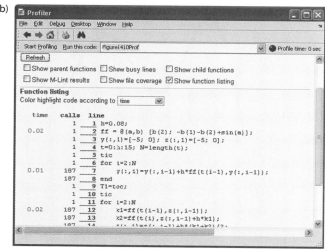

Fig. 14.11 a) Profiler summary and b) more details on CPU usage.

Now let us show how to use MATLAB solvers. For example, the problem described by Eq. (14.122) can be addressed using the `ode23` solver as follows:

```
[t,y]=ode23(@(a,b)[b(2);-b(1)-b(2)+sin(a)],[0 15],[-5 0]);
plot(t,y(:,1),t,y(:,2),'-.'), xlabel('Time, s'), ylabel('y')
legend('Position','Speed',0)
```

This produces the result shown in Fig. 14.12a. Note that we used the same anonymous function as before, but in general the script evaluating ODEs should be stored as an M-file function. For example, such a function for Eq. (14.124) could look like

```
function dy = cart(t,y)
dy(1) = y(2);
dy(2) = -y(1) -y(2)+sin(t);
dy = dy'; % the output must be a column vector
```

We can also use another syntax employing a structure as an input, as shown in the following script, producing Fig. 14.12b featuring both components of the solution ($y_1(t)$ and $y_2(t)$):

```
opt = odeset('RelTol',1e-2);
sol=ode113(@cart,[0 15],[-5 0],opt);
t=linspace(0,15,30);   y = deval(sol,t);
plot(t,y(1,:),'.',t,y(2,:),'-.'), xlabel('Time, s'), ylabel('y')
legend('Position','Speed',0)
xst=pi;                yst = deval(sol,xst,1);
text(xst,yst+0.3,num2str(yst,2))
```

In this case, you can use the `sol` structure with the `deval` function. It allows you to evaluate the ODE solution at any set of points residing within the interval the solution was obtained for. (Another function, `odextend`, allows you to extend the solution beyond this interval, if you need to.) In our particular case, we call the `deval` function twice, first to produce a solution on a vector `t` (note that, solutions returned by the `deval` function are in the row vector format, not the column vector format), and second, to obtain just a single value of position at a certain point (the third argument in this second command, `1`, indicates that we only need an estimate for y_1, being a cart's position).

In the previous script, we also used the command that redefines some of the default `odeset` parameters (the default value of relative tolerance is `1e-3`, which corresponds to 0.1% accuracy). The `odeset` function lets you adjust a lot of other integration parameters of ODE solvers, including properties of a mass matrix $M(x, y)$ in Eq. (14.119) and some specific parameters for `ode15s` and `ode15i` solvers, supplying an analytical Jacobian, which increases the speed and reliability of the solution for stiff problems, etc. Table 14.7 lists these and other properties, which can be obtained using the `odeget` function.

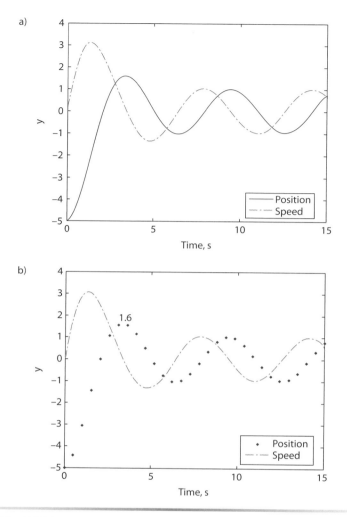

Fig. 14.12 Employing a) ode23 solver and b) ode113 solver.

Note that at every step the solver estimates the local truncation error in each ith component of the solution. This error must be less than or equal to the acceptable error, which is a function of the specified relative tolerance, RelTol, and the specified absolute tolerance vector, AbsTol

```
|e(i)| <= max(RelTol*abs(y(i)),AbsTol(i))
```

By default, MATLAB solvers use this more stringent component-wise error control as opposed to another error control

```
norm(e) <= max(RelTol*norm(y),AbsTol)
```

which becomes available when the NormControl property is on.

Table 14.7 Odeset Parameters Available for Different Solvers

Parameters	ode45	ode23	ode113	ode15s	ode15i	ode23s	ode23t	ode23tb
RelTol, AbsTol, Norm-Control	+	+	+	+	+	+	+	+
OutputFcn, OutputSel, Refine, Stats	+	+	+	+	+	+	+	+
NonNegative	+	+	+	+/−	+/−		+/−	+/−
Events	+	+	+	+	+	+	+	+
MaxStep, InitialStep	+	+	+	+	+	+	+	+
Jacobian, JPattern, Vectorized				+	+	+	+	+
Mass, MStateDependence	+	+	+	+		+/−	+	+
MvPattern, MassSingular				+			+	+/−
InitialSlope				+	+		+	
MaxOrder				+				
BDF				+				

Among other properties of `odeset`, you can specify an output function providing a handle to the `OutputFcn` parameter. For instance, you can use one of the following plotting functions:

`odeplot`	For 1-D plotting (which is a default option when you call the solver with no output arguments and you have not specified an output function)
`odephas2`	For 2-D phase plane plotting
`plottingodephas3`	For 3-D phase plane plotting
`odeprint`	For displaying a solution as it is computed in the Command window

For example, a simple command

```
opt = odeset('OutputFcn',@odeplot);
ode113(@cart,[0 15],[-5 0],opt)
```

which is equivalent to

```
ode113(@cart,[0 15],[-5 0])
```

produces a plot shown in Fig. 14.13a marking all computational nodes. Similarly, the commands

```
opt = odeset('OutputFcn',@ odephas2);
fun=@(a,b) [(b(1)-2*b(2))/3;2*b(1)+b(2)];
ode23t(fun,[-2 2],[1 1],opt)
```

produce a phase plot shown in Fig. 14.13b.

Figure 14.14 demonstrates the solution of the second-order equation, like the one described by Eq. (14.12), with two different state matrices **A** and **B**=0

$$\mathbf{x}' = \mathbf{A}\mathbf{x} \tag{14.125}$$

The two plots in Fig. 14.14 feature

$$\mathbf{A} = \begin{bmatrix} 1 & 1 \\ 0 & 1 \end{bmatrix} \text{(Fig. 14.14a)} \quad \text{and} \quad \mathbf{A} = \begin{bmatrix} -1 & 1 \\ 0 & -1 \end{bmatrix} \text{(Fig. 14.14b)}$$

and are plotted as the phase space (x_2 vs x_1). The dots represent a variety of the initial conditions. The script to produce Fig. 14.14a is presented as follows:

```
hold
for i=-1:0.25:1
  for j=-1:0.1:1
    [t,y]=ode45(@(a,b)[b(1)+b(2);b(2)],[-2 2],[i j]);
    plot(y(1,1),y(1,2),'.r'), plot(y(:,1),y(:,2))
  end
end
xlabel('x'), ylabel('y'), axis([-2 2 -2 2]), axis equal
```

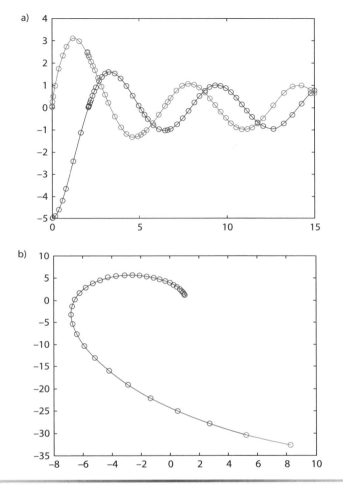

Fig. 14.13 Employing a) `odeplot` and b) `odephas2` functions.

Some minor changes are needed to produce Fig. 14.14b.

Figure 14.15a presents an example of solving a stiff problem

$$y' = \frac{1+x}{x}y, \quad x \in [-1;1] \tag{14.126}$$

using the `ode23s` solver. As seen, this problem features a singularity at $x = 0$. Hence, none of nonstiff solvers would work. The script used to produce this figure provides several initial conditions, so the plot presents a family of solutions with $y_0 \in [-5; 5]$

```
hold
for i=-5:0.5:5
```

a)

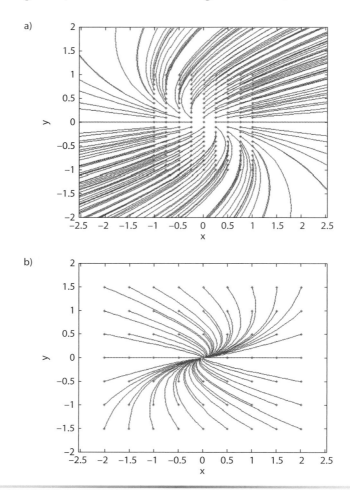

b)

Fig. 14.14 Employing the `ode45` solver for two different state matrices in Eq. (14.125).

```
[x,y]=ode23s(@(x,y)(-1+x)/y,2*[-1 1],i);
plot(x(1),y(1),'.r'), plot(x,y), xlabel('x'), ylabel('y')
end
```

Although for each solution, MATLAB returns a warning saying that it is was not able to meet integration tolerance without reducing the step size below the smallest value allowed, it does return the solution in no time.

Using the `ode15i` solver slightly differs from all other solvers. First, the `odefun` argument has an implicit form of Eq. (14.120). Second, you must supply one extra input argument, `yp0`, like in

```
[T,Y] = ode15i(odefun,tspan,y0,yp0)
```

To compute the vector of initial conditions for a derivative (which may not be given upfront) MATLAB offers the `decic` function, making all initial conditions consistent (by assuring that $\mathbf{F}(x_0, \mathbf{y}_0, \mathbf{y}_0') = \mathbf{0}$)

```
[y0mod,yp0mod] = decic(odefun,t0,y0,fixed_y0,yp0,fixed_yp0)
```

Here you provide the value you do know and the guesses on parameters you do not know. Arguments `fixed_y0` and `fixed_yp0` indicate what is known for sure (1) and what can be varied (0). This function returns the modified (consistent) values of `y0` and `yp0` that can now be passed to the `ode15i`

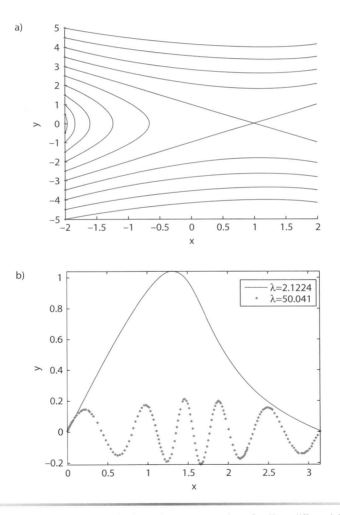

Fig. 14.15 Examples of employing a) `ode23s` solver for the stiff problem and b) `ode15i` solver for the implicit ODE.

solver. Consider the following example. Suppose we deal with the following implicit ODE:

$$((1-0.8\sin^2(x))y')' - (x-\lambda)y = 0, \quad y(0) = 0, \quad y'(0) = 1, \quad x \in [0; \pi] \quad (14.127)$$

First, we need to bring this equation to two first-order equations. Introducing $z_1 = y$ and $z_2 = y'$ we obtain

$$z_1' - z_2 = 0$$
$$-1.6\sin(x)\cos(x)\,z_2 + (1-0.8\sin^2)(x)\,z_2' - (x-\lambda)z_1 = 0 \quad (14.128)$$

Equation (14.128) can be programmed as an M-file function

```
function res = StLi(x,z,zp)
global p
res(1)=zp(1)-z(2);
res(2)=(-1.6*sin(x)*cos(x))*z(2)+(1-0.8*sin(x)^2)*zp(2)-(x-p)*z(1);
res=res';
```

(we used the `global` function to pass in the value of `lambda`).

Second, we need to establish the initial conditions. Although $\mathbf{z}(0) = [0,1]^T$ is given, we still need to find the values of $\mathbf{z}'(0)$. You can do it manually, resolving Eq. (14.128) at x_0 or simply use the `dedic` function as follows:

```
[z0,zp0] = decic(@StLi,0,[0 1],[1 0],[0 0],[0 0]);
```

Here the first vector in the list of arguments is $\mathbf{z}(0)$ and the third one is our guess on $\mathbf{z}'(0)$. Although we know the values of $\mathbf{z}(0)$ for sure, we allowed one of them to be a varied parameter along with both components of $\mathbf{z}'(0)$ (look at the values of the second and forth input vectors).

Now, we can call the ODE solver. To this end, the following script computes two solutions for two values of λ, as shown in Fig. 14.15b

```
global p
p=2.1224; [z0,zp0]=decic(@StLi,0,[0 1],[1 0],[0 0],[0 0]);
sol=ode15i(@StLi,[0 pi],z0,zp0);
plot(sol.x,sol.y(1,:)); hold,
p=50.041; [z0,zp0]=decic(@StLi,0,[0 1],[1 0],[0 0],[0 0]);
sol=ode15i(@StLi,[0 pi],z0,zp0);
plot(sol.x,sol.y(1,:),'.m'), axis tight, xlabel('x'), ylabel('y')
legend('\lambda=2.1224','\lambda=50.041')
```

Let us consider one more practical example showing the importance of monitoring the accuracy of the solution. In the so-called circular restricted three-body problem, two massive bodies move in circular orbits around their common center of mass (known as the *barycenter*), with the third body of a negligibly small mass moving in between. For example, these three bodies may be Sun, Earth, and Moon, or Earth,

Moon, and a spaceship. The normalized equations describing small body dynamics in a rotating reference frame associated with the first two bodies are

$$y_1'' = y_1 + 2y_2' - (1-\mu)\frac{y_1+\mu}{d_1} - \mu\frac{y_1-\mu}{d_2}$$

$$y_2'' = y_2 - 2y_1' - (1-\mu)\frac{y_2}{d_1} - \mu\frac{y_2}{d_2}$$

(14.129)

where

$$d_1(y_1, y_2) = ((y_1 + \mu)^2 + y_2^2)^{1.5}, \quad d_2(y_1, y_2) = ((y_1 - (1-\mu))^2 + y_2^2)^{1.5}$$

and $0 < \mu \le 1$ is the ratio of two larger bodies masses

$$d_1(y_1, y_2) = ((y_1 + \mu)^2 + y_2^2)^{1.5}$$

The fourth-order system of Eq. (14.129) can be represented by the following function

```
function dx = threeBp(t,x)
global mu
dx=zeros(length(x),1); % defining the output as a column vector
d1 = ((x(1)+mu)^2    + x(3)^2)^1.5;
d2 = ((x(1)-(1-mu))^2 + x(3)^2)^1.5;
dx(1) = x(2);
dx(2) = x(1)+2*x(4)-(1-mu)*(x(1)+mu)/d1-mu*(x(1)-(1-mu))/d2;
dx(3) = x(4);
dx(4) = x(3)-2*x(2)-(1-mu)*x(3)/d1    -mu*x(3)/d2;
```

Then we can call this function using the following commands

```
global mu
mu = 0.012277471; % Moon to Earth mass ratio
x0 = [ 0.994, 0, 0, -2.0015851];
tf = 17.065216;
[t,y] = ode45(@threeBp,[0 tf],x0,odeset('RelTol',1e-6));
plot(y(:,1),y(:,3),y(1,1),y(1,3),'.r');
line(-mu,0,'Marker','o','MarkerSize',14,'MarkerFaceColor','b')
text(-mu,0.15,'Earth'), text(1-mu,0.1,'Moon'),axis equal
xlabel('Relative Downrange'), ylabel('Relative Crossrange')
```

These commands result in a famous periodic solution for the Earth–Moon–spaceship problem shown in Fig. 14.16a. However, if you degrade the relative tolerance of the solution (its accuracy) to say 1e-2 instead of 1e-6, you will end up with an erroneous trajectory shown in Fig. 14.16b.

To conclude this section, let us mention that MATLAB provides an interactive GUI allowing you to see more examples of using the ODE solvers introduced in this section as shown in Fig. 14.17.

You can invoke this GUI allowing you to run these examples and preview their source code by typing in

```
>> odeexamples
```

In addition, MATLAB has some more single-step solvers using the variety of methods developed in Sec. 14.3 and programmed as C-code. These solvers are used as the fixed-step solvers in Simulink (to be discussed in Chapter 15).

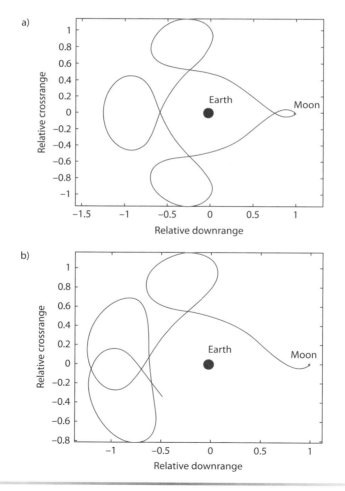

Fig. 14.16 Examples of the three-body problem trajectories.

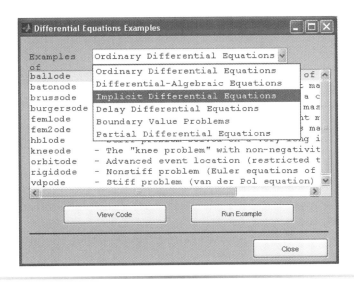

Fig. 14.17   Interactive MATLAB GUI featuring examples of ODE solving.

14.8 Other MATLAB Solvers

For the sake of completeness, let us briefly mention other solvers available in MATLAB to address special-type IVPs, BVPs, and PDEs.

14.8.1 Solvers for IVP with Delays

The dde23 and ddesd solvers allow solving IVPs for delay differential equations (DDEs) with constant and general delays, respectively. The first ones can be described as

$$\dot{\mathbf{y}} = \mathbf{F}(t, \mathbf{y}(t), \mathbf{y}(t-\tau_1), \mathbf{y}(t-\tau_2), \dots, \mathbf{y}(t-\tau_k)) \qquad (14.130)$$

with $\tau_1, \tau_2, \dots, \tau_k$ being the constant delays, and the second ones, with time and state-dependant delays $d_1, d_2, \dots, d_k$, as

$$\dot{\mathbf{y}} = \mathbf{F}(t, \mathbf{y}(t), \mathbf{y}(d_1(t, \mathbf{y})), \mathbf{y}(d_2(t, \mathbf{y})), \dots, \mathbf{y}(d_k(t, \mathbf{y}))) \qquad (14.131)$$

Compared with Eq. (14.130), the difference is that while looking for a solution within the interval $t \in [t_0; t_f]$ the values of $\mathbf{y}$ depend on the values of $\mathbf{y}$ at times prior to t_0. Hence, to be able to address Eqs. (14.130) or (14.131) the history of the states $\mathbf{S}(t)$ prior to t_0 should be supplied.

Generally, the solution $\mathbf{y}(t)$ of an IVP for a system of DDEs has a jump in its first derivative at the initial point t_0 because the first derivative of the history function does not satisfy the DDE there. This discontinuity propagates into the future. The DDEs solvers, dde23 and ddesd, are designed to mitigate this phenomenon, so that the solution becomes smoother as the integration proceeds. Specifically, dde23 tracks low-order discontinuities and integrates the differential equations with the explicit RK 2/3 pair and interpolant used by ode23. The RK formulas are implicit for step sizes longer than the delays. When the solution is smooth enough that steps this big are justified, the implicit formulas are evaluated by a predictor–corrector iteration. The ddesd solver integrates with the classic four-stage, fourth-order explicit RK method [Eq. (14.54)] controlling the magnitude of truncation error.

To use dde23, you must rewrite Eq. (14.128) as an equivalent system of the first-order differential equations, and code it in the M-file function like

```
function dydt = ddex1de(t,y,Z)
ylag1 = Z(:,1);
ylag2 = Z(:,2);
dydt = [ylag1(1)*y(2)
        ylag1(1)-ylag2(2)]; % the output must be a column vector
```

which corresponds to the system

$$\dot{y}_1 = y_1(t-\tau_1)\,y_2(t)$$
$$\dot{y}_2 = y_1(t-\tau_1) - y_2(t-\tau_2) \tag{14.132}$$

Then, you should supply the dde23 solver with the delays organized as a vector lags. You should also code the history function like

```
function S = ddex1hist(t)
S = zeros(3,1);
```

Finally, you can apply the solver as

```
sol = dde23(@ddex1de,lags,@ddex1hist,range);
```

where range is the two-value vector defining the interval of integration. The output structure sol has the following main fields:

sol.x	Contains a mesh selected by the solver to meet the tolerance
sol.y	Represents approximation to $\mathbf{y}(t)$ at the mesh points in sol.x
sol.yp	Provides approximation to $\mathbf{y}(t)$ at the mesh points in sol.x

To evaluate solution at some other mesh, you can use the deval function as described in the previous section.

For the state-dependent delays of Eq. (14.131), you should additionally create a function describing the state-variant delays, like

```
function lags = ddex3delay(t,y)
lags = exp(1 - y(2));
```

Then, you can use the following syntax:

```
sol = ddese(@ddex1de,@lags,@ddex1hist,range);
```

Consider a space colony on the Moon. The Kermack–McKendrick model for the number of people infected with a contagious illness in a closed population over time is given by

$$
\begin{aligned}
\dot{y}_1 &= -\beta y_1(t) y_2(t - \tau_1) \\
\dot{y}_2 &= \beta y_1(t) y_2(t - \tau_1) - \gamma y_2(t - \tau_2) \\
\dot{y}_3 &= \gamma y_2(t - \tau_2)
\end{aligned}
\tag{14.133}
$$

Here t defines the normalized time, y_2 is the number of infected individuals, who can pass on some virus or disease to others, y_1 is the number of susceptible, who have yet to contact the disease and become infectious, y_3 is the number of colony members, who have been infected, but cannot transmit the disease for some reason, for example, they have been isolated from the rest of the colony or developed immunity to the infection, β represents the probability of getting infected, and γ is the isolation effectiveness. This model also assumes the incubation period τ_1 and time τ_2 needed for isolation. For $t \leq 0$, consider the following values: $y_1 = 990$, $y_2 = 10$ and $y_3 = 0$. Equation (14.133) can be coded as

```
function dy = kmckd(t,y,Z)
ylag1 = Z(:,1);
ylag2 = Z(:,2);
dy = zeros(length(y),1);
dy(1)  = - 0.01*y(1)*ylag1(2);
dy(2)  =   0.01*y(1)*ylag1(2) - 0.2*ylag2(2);
dy(3)  =   0.2*ylag2(2);
```

The call

```
>> dde23('kmckd',[.2 .7],[990; 10; 0],[0, 10]);
>> legend('Susceptable','Infected','Izolated')
```

Produces the result shown in Fig. 14.18a.

More examples of how to handle DDEs are provided by the ODE Examples GUI (Fig. 14.18b)

```
>> odeexamples('dde')
```

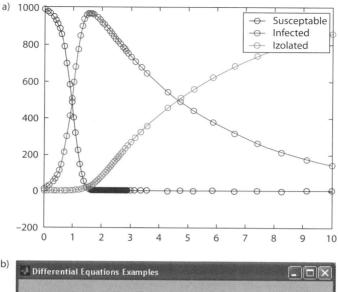

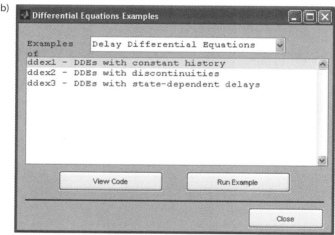

Fig. 14.18 a) Modeling a disease outbreak and b) DDEs examples in MATLAB.

14.8.2 BVP Solver

The `bvp4c` and `bvp5c` solvers allow you to solve BVPs, where unlike an IVP [Eq. (14.15)] the conditions to be satisfied are established at x_0 and x_f

$$\mathbf{y}' = \mathbf{F}(x, \mathbf{y}), \ \mathbf{g}(\mathbf{y}(x_0), \mathbf{y}(x_f)) = \mathbf{0} \qquad (14.134)$$

These solvers can also assist in finding unknown parameters for problems of the form

$$\mathbf{y}' = \mathbf{F}(x, \mathbf{y}, \mathbf{p}), \ \mathbf{g}(\mathbf{y}(x_0), \mathbf{y}(x_f), \mathbf{p}) = \mathbf{0} \qquad (14.135)$$

with **p** being the vector of parameters to be determined, so that Eq. (14.135) has a solution.

As opposed to IVPs, the BVPs may have no solution, or finite number of solutions, or even infinite number of solutions. Because of this, the programs for solving BVPs require users to provide a good guess for the solution desired. The BVPs are much harder to solve than IVPs, so be prepared that the solver may not work even with the good guesses for the solution and unknown parameters.

The syntax of the `bvp4c` function is

```
sol = bvp4c(odefun,bcfun,solinit)
sol = bvp4c(odefun,bcfun,solinit,options)
```

and the initial guess is defined using the `bvpinit` function

```
solinit = bvpinit(x,yinit,params)
```

This and the other input arguments are explained in more detail in Table 14.8.

The `bvp4c` solver is a finite difference code that implements the Simpson's 1/3 rule of Eq. (13.19) (for a single interval, the multiplier becomes 1/6 rather

Table 14.8 Input Arguments for BVP Solvers

Argument	Description

odefun	A function handle that evaluates the right-hand side of differential equations, $\mathbf{F}(x,\mathbf{y})$, or $\mathbf{F}(x,\mathbf{y},\mathbf{p})$, whichever is appropriate
	`dydx = odefun(x,y)`
	`dydx = odefun(x,y,parameters)`
	(note that, y and `dydx` must be the column vectors).

bcfun	A function handle that computes the residual in the boundary conditions, defined by the column vectors `ya` and `yb`
	`res = bcfun(ya,yb)`
	`res = bcfun(ya,yb,parameters)`
	(`res` must be a column vector as well).

solinit	A structure containing the initial guess for a solution and created using the `bvpinit` function. This structure must have the following fields:
	x — Containing the ordered nodes of initial mesh, so that the boundary conditions are imposed at `x0=solinit.x(1)` and `xf=solinit.x(end)`
	y — Containing the initial guess for the solution, so that `solinit.y(:,i)` corresponds to the node `solinit.x(i)`
	parameters — Defining a vector of initial guesses for unknown parameters (if present).
options	An optional argument changing integration options using the `bvpset` function.

than 1/3). To be more specific, the solution $\mathbf{S}(x)$ to Eq. (14.134) is sought as a piecewise global cubic approximation satisfying the boundary conditions

$$\mathbf{g}(\mathbf{S}(x_0), \mathbf{S}(x_f)) = \mathbf{0} \tag{14.136}$$

and differential equations (*collocates*) at both ends and midpoint of each interval $x \in [x_n; x_{n+1}]$

$$\mathbf{S}'(x_n) = \mathbf{F}(x_n, \mathbf{S}(x_n))$$

$$\mathbf{S}'\left(\frac{x_n + x_{n+1}}{2}\right) = \mathbf{F}\left(\frac{x_n + x_{n+1}}{2}, \mathbf{S}\left(\frac{x_n + x_{n+1}}{2}\right)\right) \tag{14.137}$$

$$\mathbf{S}'(x_{n+1}) = \mathbf{F}(x_{n+1}, \mathbf{S}(x_{n+1}))$$

These conditions result in a system of nonlinear algebraic equations for coefficients defining $\mathbf{S}(x)$ and solved iteratively by linearization. Therefore, the BVP solvers rely on the linear equation solvers described in Chapter 9, rather than on IVP solvers.

The piecewise collocation polynomial provides a continuous solution that is fourth-order accurate in the step size h. Mesh selection (the points $x_0 < x_1 < x_2 < \ldots < x_N = x_f$) and error control are based on the residuals at the midpoints, computed using the five-point Lobatto quadrature

$$\mathbf{r} = \mathbf{S}'(x) - \mathbf{F}(x, \mathbf{S}(x)) \tag{14.138}$$

The bvp4c solver can also solve multipoint BVPs, where the values of the function are defined at the points (the points b_i, $i = 0, 1, \ldots, N$ ($x_0 = b_0 < b_1 < \ldots < b_N = x_f$). The only difference in this case is that the bcfun function has a slightly different format.

The bvp5c function is used exactly like bvp4c, but is more efficient for small error tolerances. It implements the four-point Lobatto formula developed in Eq. (13.56). In this particular case for each interval $x \in [x_n; x_{n+1}]$, the four points are established at $x_{n,i}^L = x_n + c_i h_n$, $i = 1, 2, 3, 4$ with $c_{2;3} = (5 \mp \sqrt{5}/10)$, and $c_4 = 1$. As opposed to bvp4c, bvp5c cannot solve multipoint BVPs.

Consider a simple example of a projectile motion in the vertical plan (a shot fired from a cannon). The equations

$$\dot{x} = V \cos(\gamma)$$
$$\dot{y} = V \sin(\gamma)$$
$$\dot{V} = -kV^2 - g \sin(\gamma) \tag{14.139}$$
$$\dot{\gamma} = -g V^{-1} \cos(\gamma)$$

describe the problem with x and y denoting the horizontal and vertical coordinate, V, speed; γ, flight path angle; g, acceleration due to gravity; and k, air resistance (drag). Let us change the argument in these equations to x

$$y' = \tan(\gamma)$$
$$V' = -\frac{kV^2 + g\sin(\gamma)}{V\cos(\gamma)} \tag{14.140}$$
$$\gamma' = -gV^{-2}$$

and develop an M-file function assuming $g = 9.8\text{m/s}^2$ and $k = 0.0002\text{m}^{-1}$

```
function dy = projectile(x,y,param)
dy = zeros(length(y),1);
dy(1) = tan(y(3));
dy(2) = -(9.8*sin(y(3))+0.0002*y(2)^2)/y(2)/cos(y(3));
dy(3) = -9.8/y(2)^2;
```

Now we can use either solver to find a solution for $y(0) = 0\text{m}$, $V(0) = 200\text{m/s}$, and $y(x_f) = 0$ m with a varied parameter $\gamma(0) = var$ (that is the reason the function projectile lists it as an argument even though it is not used in there). These boundary conditions should be saved in a separate M-file function, like

```
function res = BC(ya,yb,param)
res = [  ya(1)
         yb(1)
         ya(2)-200
         ya(3)-param];
```

We can call these two functions in attempt to find a solution as follows:

```
solinit=bvpinit(linspace(0,2000,5),[100 100 0.8],0.8);
sol=bvp4c(@projectile,@BC,solinit);
xbvp = linspace(0,2000); ybvp = deval(sol,xbvp);
plot(xbvp,ybvp(1,:)); hold
text(300,500,['\gamma_0=' num2str(sol.parameters*180/pi,2) '^o'])
axis equal, xlabel('x, km'), ylabel('y, km')
```

Here, the first line defines the range of the argument ($x \in [0;2000]$m) and number of points to estimate the solution at (5), along with the initial guesses on $y(x)$, $V(x)$, and $\gamma(x)$ (constant values of 100 m, 100 m/s, and 0.8, respectively, for all five points) and the varied parameter $\gamma(0)$ (0.8). Note that the solution is quite sensitive to these initial guesses, so they have to be reasonable. Upon finding the solution (the second line), we use the `deval` function to evaluate the numerical solution at 100 equally spaced points (the third line) and plot the result, which is shown in Fig. 14.19a (labeled as the first solution).

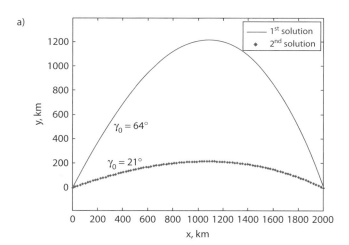

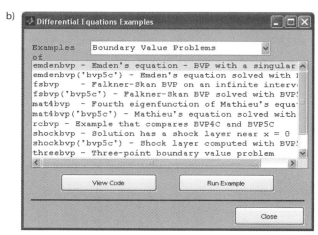

Fig. 14.19 a) Employing the `bvp4c` function and b) BVP examples in MATLAB.

Note that both `bvp4c` and `bvp5c` solvers return only one solution (if at all). If you know that the problem has multiple solutions, you have to play with the initial guesses. In our case, we know that there has to be another solution, so we could run the same script, introducing the only change, decreasing the value of $\gamma(0)$. However, for the educational purpose, in the following script (which continues the first portion presented earlier) we varied the initial guesses on $y(x)$, $V(x)$, and $\gamma(x)$ leaving the guess on $\gamma(0)$ unchanged

```
solinit=bvpinit(linspace(0,2000,5),[1 200 0.2],0.8);
sol=bvp4c(@projectile,@BC,solinit);
ybvp = deval(sol,xbvp);
plot(xbvp,ybvp(1,:),'.r');
```

```
legend('1^{st} solution','2^{st} solution')
text(300,200,['\gamma_0=' num2str(sol.parameters*180/pi,2) '^o'])
```

This demonstrates the sensitivity of the solution to changes in all initial guesses. For instance, if in the previous code you change the initial guess on $V(x)$ to 50, the solver will diverge returning the following error message:

```
??? Error using ==> bvp4c at 203
Unable to solve the collocation equations --
a singular Jacobian encountered
```

Once again the initial guesses must be reasonable.

Other examples of using BVP solvers are accessible via the MATLAB Help system or by calling the interactive GUI

```
>> odeexamples('bvp')
```

which brings the window shown in Fig. 14.19b.

14.8.3 PDE Solver

The pdepe solver solves IVP and BVP for systems of parabolic and elliptic PDEs in one space variable x and time t of the form

$$\mathbf{c}\left(x,t,\mathbf{u},\frac{\partial\mathbf{u}}{\partial x}\right)\frac{\partial\mathbf{u}}{\partial t}=x^{-m}\frac{\partial}{\partial x}\left(x^{m}\mathbf{f}\left(x,t,\mathbf{u},\frac{\partial\mathbf{u}}{\partial x}\right)\right)+\mathbf{s}\left(x,t,\mathbf{u},\frac{\partial\mathbf{u}}{\partial x}\right) \quad (14.141)$$

where $t \in [t_0; f_f]$, $x \in [x_0; x_f]$, m can be 0, 1, or 2, corresponding to slab, cylindrical, or spherical symmetry, $\mathbf{f}(x,t,\mathbf{u}, (\partial\mathbf{u}/\partial x))$ is the flux term, $\mathbf{s}(x,t,\mathbf{u}, (\partial\mathbf{u}/\partial x))$ is the source term, and $\mathbf{c}(x,t,\mathbf{u}, (\partial\mathbf{u}/\partial x))$ is the diagonal matrix with the nonnegative elements. If this matrix is a zero matrix, Eq. (14.141) is referred to as an *elliptic* equation, otherwise a *parabolic* equation (these equations model totally different phenomena, so that the behavior of their solutions is quite different).

For $t = t_0$ and $x \in [x_0; x_f]$, the solution components satisfy the initial conditions

$$\mathbf{u}(x,t_0) = \mathbf{u}_0(x) \quad (14.142)$$

For any t and $x = x_0$ or $x = x_f$, the solution components satisfy a boundary condition

$$\mathbf{p}(x,t,\mathbf{u})+\mathbf{q}(x,t)\mathbf{f}\left(x,t,\mathbf{u},\frac{\partial\mathbf{u}}{\partial x}\right)=\mathbf{0} \quad (14.143)$$

where the elements of $\mathbf{q}(x, t)$ are either zero or never zero.

The pdepe solver is called in any of two forms

```
sol = pdepe(m,pdefun,icfun,bcfun,xmesh,tspan)
sol = pdepe(m,pdefun,icfun,bcfun,xmesh,tspan,options)
```

The pdepe function returns the values of solution on a mesh provided in xmesh. The pdefun, icfun, and bcfun are the handles to the functions describing Eqs. (14.141), (14.142), and (14.143), respectively. Specifically, the pdefun function computes the terms of Eq. (14.141) as follows:

```
[c,f,s] = pdefun(x,t,u,dudx)
```

(c, f, and s should be returned as the column vectors, with c storing the elements of the corresponding diagonal matrix). The icfun evaluates and returns the initial conditions in a column vector u

```
u = icfun(x)
```

The bcfun function evaluates the boundary condition

```
[pl,ql,pr,qr] = bcfun(xl,ul,xr,ur,t)
```

with xl, xr corresponding to the boundaries of the interval, x_0 and x_f, respectively; ul and ur approximating correspondent boundary conditions; and pl, ql, pr, and qr computing $\mathbf{p}(x,t,\mathbf{u})$ and $\mathbf{q}(x,t)$ at x_0 and x_f.

The ODEs resulting from discretization in space (Sec. 12.7) are integrated to obtain approximate solutions at times specified in tspan. For example, the parabolic equation

$$\frac{\partial \mathbf{u}}{\partial t} = a^2 \frac{\partial^2 \mathbf{u}}{\partial x^2} \tag{14.144}$$

would become

$$\frac{\partial \mathbf{u}_i}{\partial t} = \frac{a^2}{\Delta x^2} (\mathbf{u}_{i+1} - 2\mathbf{u}_i + \mathbf{u}_{i-1}) \tag{14.145}$$

(cf. with Eq. (12.46)). The output sol(:,:,i) corresponds to the *i*th component of the solution vector. Hence, the element sol(k,j,i) corresponds to u_i (t_k, x_j) (tspan(k) and xmesh(j)). As with other solvers, you can compute the approximation and its partial derivative at an arbitrary set of points (within the original interval) using the special function, in this case the pdeval function.

As an example, consider a simplest parabolic equation

$$\frac{\partial u}{\partial t} = \frac{\partial^2 u}{\partial x^2} \tag{14.146}$$

with the initial and boundary conditions

$$u(x,0) = 0, \quad u(0,t) = 0, \quad \text{and} \quad u(1,t) = \sin(3\pi t) \qquad (14.147)$$

This equation represents the *heat transfer* equation in one dimension. This specific setup can be treated as heating/cooling (with a sine wave) one end of the thin, homogeneous rod of uniform thickness, which at time $t = 0$ (we are using some normalized time here) has temperature $u = 0$. For convenience, all three functions, required by the pdepe solver and describing Eq. (14.146), initial and boundary conditions of Eq. (14.147) can be placed in a single M-file together with the calling and plotting commands:

```
function rod
x = linspace(0,1,30); t = linspace(0,1,30);
sol = pdepe(0,@heattransfer,@pdeIC,@pdeBC,x,t);
u = sol(:,:,1);
    surf(x,t,u)   , view([40,30])
    xlabel('Distance x'), ylabel('Time t')
    zlabel('Temperature T')
function [c,f,s] = heattransfer(x,t,u,DuDx)
c = 1;
f = DuDx;
s = 0;
    function u0 = pdeIC(x)
    u0 = 0;
function [pl,ql,pr,qr] = pdeBC(xl,ul,xr,ur,t)
pl = ul;
ql = 0;
pr = ur-sin(3*pi*t);
qr = 0;
```

Running this self-explanatory code (saved as *rod.m*) results in the plot shown in Fig. 14.20a.

MATLAB provides several more illustrative examples of how to use the pdepe solver, which can be accessed issuing the command

```
>> odeexamples('pde')
```

(Fig. 14.20b). On top of that the PDEs Toolbox of MATLAB (if installed) offers many more options through the pdetool GUI, representing a self-contained graphical environment for solving PDEs and visualizing obtained solutions.

a)

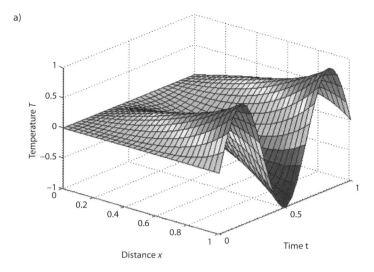

b)

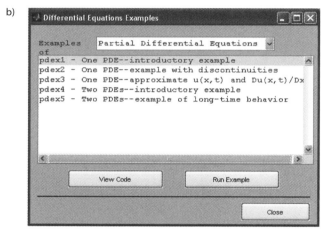

Fig. 14.20 a) Employing the `pdepe` function and b) PDE examples in MATLAB.

Problems

14.1 An airplane uses a parachute and other means of braking as it slows down on the runway after landing. Its acceleration is given by $a = (-0.0045V^2 - 3)\text{m/s}^2$. Consider an airplane with a velocity V of 300 km/h that opens its parachute and starts decelerating at $t = 0$s. Do the following:

   a. By solving the differential equation, determine and plot the velocity as a function of time from $t = 0$s until the airplane stops.

b. Use numerical integration to determine the distance x the airplane travels as a function of time. Make a plot of x vs time.

14.2 Consider the first-order ODE

$$y' = -2x^3 + 12x^2 - 20x + 8.5$$

Solve it from $x = 0$ to $x = 4$ with the initial condition at $x = 0$ being at $y = 1$. Do it using the following methods:

a. Euler's first-order method with the step size of 0.5 (*you have to develop your own script*).
b. Euler's first-order method with the step size of 0.25.
c. Heun's second-order method (predictor–corrector approach) with the step size of 0.5 (*you have to develop your own script*).
d. Midpoint second-order method with the step size of 0.5 (*you have to develop your own script*).
e. RK low-order pair solver with the step size of 0.5 (*use the* ode23 *solver*).
f. RK medium-order pair solver with the step size of 0.5 (*use the* ode45 *solver*).

Present the results of all solutions by pairs on a three-band plot together with a true solution added to each pair. Do not forget the legend. Measure performance using stopwatch timer (tic-toc pair). Compute an average error for every solution and present is as an "Accuracy versus CPU time" plot.

14.3 Solve the following system of ODEs:

$$\dot{x} = 1 - y$$
$$\dot{y} = x^2 - y^2$$

over the interval $t \in [0;5]$. As the initial conditions pick several points within the following two segments: $y(0) = 0$, $x(0) \in [-2;2]$, and $y(0) = 2$, $x(0) \in [-2;2]$. Present your results as a family of curves "y vs x."

Chapter 15　Simulink Basics

- Modeling Continuous Systems
- Modeling Discrete Systems
- Integration of MATLAB and Simulink
- Model Debugging and Profiling

Simulink libraries, `sim`, `simset`, `simget`, `emlBlock`, `set_param`

15.1　Introduction

As a part of the MathWorks' product family, Simulink, built on top of MATLAB, provides a graphical user interface (GUI) that uses a variety of elements called blocks to create a simulation of a dynamic system, that is, a system that can be modeled with differential or difference equations with one independent variable (time). These blocks are as simple as a multiplier, sum, integrator, scope, etc. Simulink uses the same solvers of MATLAB that were considered in Chapter 14, but instead of using them programmatically, it enables high-level interaction with the user, bringing the necessity to know about the syntax of different functions to a minimum. The graphical interface of Simulink enables you to choose a block from a variety of specialized libraries; position, resize, and label it; specify its parameters; interconnect the blocks; combine groups of blocks into a larger subsystem, etc. This leads to a capability of creating the multi-layer models of complex real-world continuous, discrete, and multirate systems. Simulating a dynamic system includes several major steps. First, you create a graphical model of the system to be simulated using the Simulink Model Editor. The model depicts time-dependent mathematical relationships among the system's inputs, states, and outputs. Then, you use Simulink to simulate a behavior of the system over a specified time (independent variable) span with the specified inputs with the parameters you entered into the model to perform the simulation. Last but not the least, Simulink provides tools for debugging, profiling, and analyzing your model. Simulink represents an extensive product, especially if you consider all additional blocksets, blocks libraries, you can extend it with. Although it deserves a separate textbook (and several such textbooks are available), the material of this chapter covers only the basics of Simulink. It starts with introducing the Simulink modeling environment and showing how to build and run the simplest continuous model using three different ways of representing differential equations. Next, it shows the major difference between the two types of systems, continuous and discrete, also supplying you with some simple examples. More sophisticated examples of continuous model discretization and controller design are then followed by a demonstration of how a Simulink model can call MATLAB commands and M-scripts while loading and running, and vice versa, and how a Simulink model can be called from a MATLAB M-script. It also shows the ways of creating the blocks from the symbolic expressions developed a priori and available in the MATLAB workspace. The chapter ends with examples of real-world systems to provide you with a guidance for further exploration.

15.2 Simulink Development Environment

Let us start exploring the Simulink development environment by clicking the eighth button of the MATLAB Desktop toolbar, as shown in Fig. 15.1. As a result, you should see a Simulink Library Browser window popping up as shown in Fig. 15.2.

You may also bring up this window by typing

```
>> simulink
```

in the MATLAB's Command window. The Simulink Library Browser presents a collection of different blocks called blocksets that are used to build up a model. Clicking on any icon of the right pan (Fig. 15.2) opens a specific library.

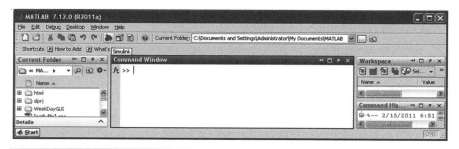

Fig. 15.1   Simulink button on the MATLAB toolbar.

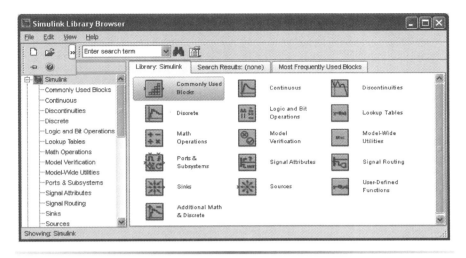

Fig. 15.2 Simulink Library Browser.

For example, Fig. 15.3 shows the content of the Sources library, Fig. 15.4 the Sinks library, and Fig. 15.5 the Math Operations library. Because you always need to feed your model with some input signal, perform some operations on it and display the results; these three libraries contain the most often used blocks. The last three toolbar instruments of the Simulink Library Browser toolbar allow you to search for a specific block in case you forgot where to find it.

You will use the Simulink Library Browser very often to grab different blocks from it, and so it is not a bad idea to keep it open. For this purpose, the **Stay on top** button allows you to keep Simulink Browser on top of all open windows all the time. Another way to set it up this way is to check off this option in the **View** menu.

In the Simulink Library Browser window, click on the **Create new model** button (<Ctrl>+<N>) which brings up a blank window you will use to create your model (Fig. 15.6). The first nine buttons of the Simulink Model Editor toolbar are pretty much standard buttons, similar to those of the MATLAB M-file Editor (Fig. 2.5). The **Go to parent system** button allows you to return to the parent system in case you have a multilevel model with subsystems (to be discussed in Sec. 15.3.2). **Undo** and **Redo** buttons allow you to undo (<Ctrl>+<Z>) and redo (<Ctrl>+<Y>) the latest operation, followed by **Start** and **Stop** simulation buttons, intended to start your simulation (<Ctrl>+<T>) and stop it (prematurely).

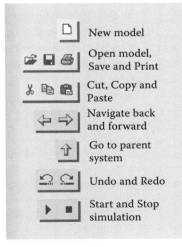

Band-Limited White Noise	Chirp Signal	Clock	Constant	Counter Free-Running
Counter Limited	Digital Clock	Enumerated Constant	From File	From Workspace
Ground	In1	Pulse Generator	Ramp	Random Number
Repeating Sequence	Repeating Sequence Interpol...	Repeating Sequence Stair	Signal Builder	Signal Generator
Sine Wave	Step	Uniform Random Number		

Fig. 15.3 Sources library blocks.

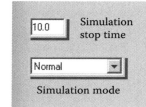

Time of simulation, t_f, defining the span of an independent variable, appears in the next window. By default, it is 10. You may run your simulation in different modes with the Normal speed mode being the default one. This mode allows the greatest flexibility in making model adjustments and displaying the results when you run it, but if for any reason you want to speed it up, other modes are also available. Specifically, the Accelerator and Rapid Accelerator modes use portions of the Real-Time Workshop (Table 1.20) to replace the normal interpreted code with compiled (executable) code. Accelerator mode creates and uses the standalone executable from your model, whereas Rapid Accelerator mode creates and uses the standalone executable from your model and a solver, and therefore runs the fastest. Neither of these two modes, however, support the debugger or profiler and they work only with those models for which C code is available for all of the blocks in the model (for instance, it does not work with any model containing an algebraic loop).

Among the remaining group of buttons, there are buttons allowing you to refresh model blocks, update diagram and bring up a Library Browser window in case you closed it or it is obscured by other windows. The **Toggle Model Browser** button adds a blocks browser in a separate pan on the left of the Model Editor, two other buttons bring up the Model Explorer and Debug windows. The tools behind the latter three buttons will be briefly introduced in Sec. 15.8. The remaining three buttons, are to be used by a proficient user and are not considered in this text.

When you save your model, a default extension .mdl is added automatically. If you have your model created and saved already, to open it in the

Fig. 15.4 Sinks library blocks.

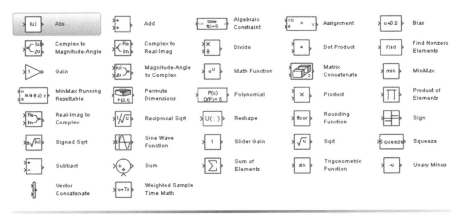

Fig. 15.5 Math Operations library blocks.

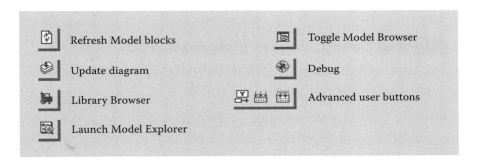

Simulink Model Editor, you can use the **Open a model** button of the Library Browser (Fig. 15.2) or Simulink Model Editor (Fig. 15.6). Alternatively, you may call it up by simply typing its name in the MATLAB Command Window or using the `open('ModelName')` command.

As you can see by the names of Simulink libraries in Fig. 15.2, Simulink allows you to create two different types of models, continuous (analog) and discrete. The following two sections consider some simple examples, allowing you to familiarize yourself with the both types.

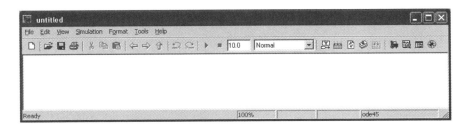

Fig. 15.6 New (empty) Simulink model window.

15.3 Modeling Continuous Dynamics

To start with, let us open the Continuous library blocks, as presented in Fig. 15.7.

The three major blocks of this library, Integrator, Transfer Fcn, and State-Space, allow you to simulate a continuous (analog) model of a dynamic system described by the differential equations in three different ways:

- Using primitive linear blocks, representing each mathematical operation separately
- Taking advantage of using a transfer function block (for linear systems)
- Employing a state-space block (for linear systems)

Depending on the problem, you may choose any of them as presented in the following.

15.3.1 Configuring an Integrator Block

As an example, consider Eq. (14.9). This second-order differential equation is typically represented slightly differently

$$\ddot{x} + 2\zeta\omega_n\dot{x} + \omega_n^2 x = 0 \tag{15.1}$$

Here, ζ is called the damping ratio of the second-order dynamic system and ω_n the undamped natural frequency (we changed y to x because some Simulink blocks use x notation rather than y; it does not really matter). Obviously, Eq. (14.9) can be rewritten in the form of Eq. (15.1) with

$$\omega_n = \sqrt{\frac{k}{m}}, \; \zeta = \frac{c}{2\sqrt{km}} \tag{15.2}$$

The numerical values of ζ and ω_n define the behavior of the dynamic system (transient response).

For example, for an aircraft (Fig. 14.1b), the response to disturbances in the longitudinal channel can be described by two sets of Eq. (15.1) defining *short-period motion* (with x being the angle of attack) and long-period or

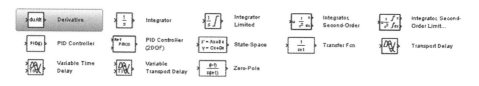

Fig. 15.7 Continuous library blocks.

phugoid motion (with x representing the pitch angle). The typical values (the order of magnitude) for the short-period motion are

$$\omega_n = 5s^{-1}, \; \zeta = 0.5 \text{ (that is, } 2\zeta\omega_n = 5s^{-1} \text{ and } \omega_n^2 = 25s^{-2}) \tag{15.3}$$

whereas for the phugoid motion,

$$\omega_n = 0.1s^{-1}, \; \zeta = 0.05 \text{ (that is, } 2\zeta\omega_n = 0.01s^{-1} \text{ and } \omega_n^2 = 0.01s^{-2}) \tag{15.4}$$

Let us model Eq. (15.1) for the phugoid motion of an aircraft.

Because Eq. (15.1) describes a second-order system, to model its behavior we will need two integrators. Hence, we start developing a new model by opening a blank (new) window of the Simulink Model Editor and dragging two Integrator blocks from the Continuous library (Fig. 15.7), as shown in Fig. 15.8.

Then, let us label both Integrator blocks as shown in Fig. 15.9 and connect them. To change the default name of a block, you simply click on this name and type in a new name. To connect the output of the Rate integrator to the input of the Angle integrator, drag from the output port of the first block to the input port of the second one. Alternatively, you can do it by selecting the source block, holding down the <Ctrl> key and left-clicking the destination block.

To be able to use this model later, save it as *SimulinkC1.mdl* (the name of the model then shows up as the name of the window as seen in Fig. 15.9a).

The output of the Rate integrator is $\dot{x}$; therefore, its input must be $\ddot{x}$. Let us rewrite Eq. (15.1) to have an explicit expression for the angular acceleration $\ddot{x}$ as the function of x and $\dot{x}$

$$\ddot{x} = -2\zeta\omega_n\dot{x} - \omega_n^2 x \tag{15.5}$$

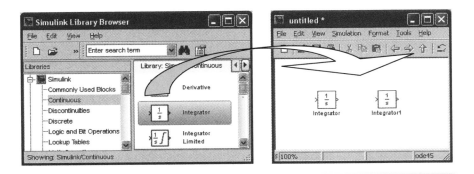

Fig. 15.8 Dragging an Integrator block of the Continuous library to the new Simulink model.

a)

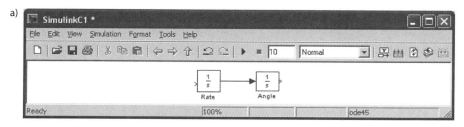

b)

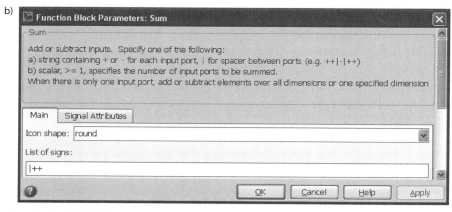

Fig. 15.9   a) Labeling and connecting two blocks and b) the Sum's block Parameters window.

Substituting parameters of the phugoid motion Eq. (15.4) yields

$$\ddot{x} = -0.01\dot{x} - 0.01x \qquad (15.6)$$

Let us also assume some disturbance or initial condition (IC) of

$$x(0) = 0 \text{ and } \dot{x}(0) = 0.01\,\text{s}^{-1} \qquad (15.7)$$

Now, you may proceed with developing the rest of the model, as shown in Fig. 15.10a.

First, you add a Sum block from the Math Operations library (Fig. 15.5) to compute $\ddot{x}$. All inputs of this block are drawn from 12 o'clock down to 6 o'clock counterclockwise. By double-clicking on it, which brings up its Parameters window shown in Fig. 15.9b, you can see that it is the **List of signs** line "| + +" that defines two positive inputs and their orientation (the vertical line | serves as a spacer or virtual input, making 12 o'clock input vacant). Let us change the **List of signs** line to "| − −" [because Eq. (15.5) features two negative signs]. (Figure 15.9b also shows that you have an option of choosing a rectangular Sum block, in which case all inputs appear on one side of the block.)

Second, from the same library add two Gain blocks. For them to appear as shown in Fig. 15.10a you need to right-click on each of them and choose **Format** and then the **Flip Block** option from a drop-down menu shown in Fig. 15.10b. These menus are the same for any block and as seen you can introduce some other changes beyond flipping a block as well. It may also be useful to use the following two shortcuts:

<Ctrl>+<I>   To flip a selected block
<Ctrl>+<R>   To rotate it

To enter the numerical values for both gains, double-click on each block with the left button, which opens the Gain block Parameters window presented in the Fig. 15.11a. The gain values of Eq. (15.6) are all you have to enter, but for future use, note that when operating with matrices, the gain K is represented by a matrix as well, and you should be careful of how you want multiplication to occur and choose the proper option from the drop-down menu.

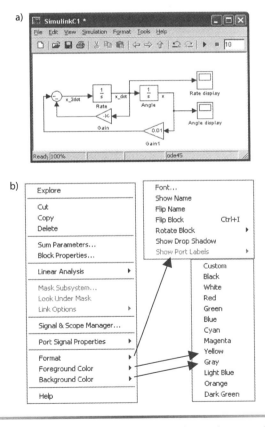

Fig. 15.10   a) Simulink model the second-order system and b) available options of the right-click drop-down menu.

a)
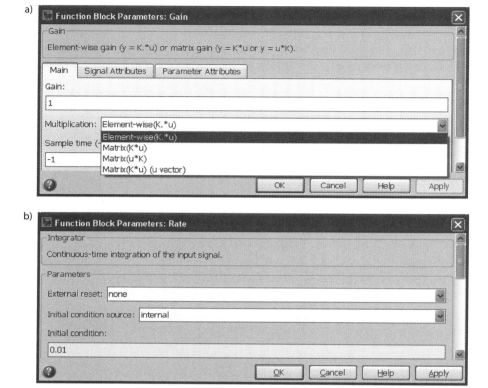

b)

Fig. 15.11 a) Gain and b) Integrator block Parameter windows.

Note that because the numerical value of the gains are too long, the Gain block appears with a symbol -K- upon it (Fig. 15.10a). To see the value you have to resize (enlarge) the block by selecting it (clicking on it), placing a mouse pointer over one of its corners, and stretching it, what was done for the block Gain1 (for the sake of comparison, the Gain block in Fig. 15.10a is left unchanged). Connect these two new blocks and the summation block with integrators. Note that to connect the Gain block with -K- on it you should start dragging a line from its input port and drop it (release mouse button) on the line connecting two integrators, which creates a branch point.

Third, from the Sinks library Eq. (15.4) add two Scope blocks, to display angular rate and angle itself, and rename them as shown in Fig. 15.10a. (Note that you cannot use the same name for two blocks—if you do it inadvertently Simulink will alter one of their names.) Connect these two blocks with the corresponding lines. To name the connecting lines, double-click on each of them and enter the name. If needed, you may move the line and block labels by dragging them to the desired location.

Finally, double-click on each Integrator block and set the **Initial condition** field in the Rate integrator dialog box to 0.01 (as shown in Fig. 15.11b) and the **Initial condition** field in the Angle integrator dialog box to 0. These are the ICs defined by Eq. (15.7). In the latest versions of Simulink, when you alter any parameter in the Parameters window from its default value, the corresponding field becomes darker, like in Fig. 15.11b. It stays this way until you click the **Apply** or **OK** button (in the latter case, the Parameters window will be closed with the values you entered applied).

Note that the **Initial condition source** in the Integrator Parameters window (Fig. 15.11b) is by default set to *internal* (that is why every time you need to change the IC, you have to open the Block Parameters window of the Integrator block and introduce a new IC). Simulink also allows you to do it in a more elegant way by setting the **Initial condition source** (in the Rate integrator dialog box) to *external* (Fig. 15.12a). This results in an extra input to the Rate integrator as shown in Fig. 15.12b. Now, you may add a Constant block from the Sources library (Fig. 15.3), change its value to

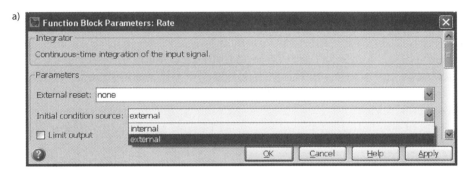

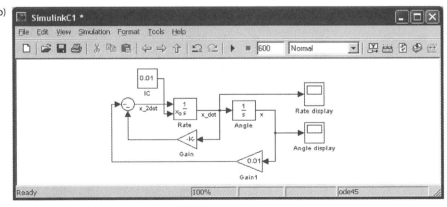

Fig. 15.12 a) Integrator block Parameters window and b) modified model with IC set externally.

Configuration Parameters: SimulinkC1/Configuration (Active)

Select:
- Solver
- Data Import/Export
- Optimization
- Diagnostics
  - Sample Time
  - Data Validity
  - Type Conversion
  - Connectivity
  - Compatibility

Simulation time

Start time: 0.0   Stop time: 600

Solver options

Type: Variable-step   Solver: ode45 (Dormand-Prince)

Max step size: auto   Relative tolerance: 1e-3

Min step size: auto   Absolute tolerance: auto

Initial step size: auto   Shape preservation: Disable all

OK   Cancel   Help   Apply

Fig. 15.13   Configuration Parameters window for variable-step solvers.

0.01 and connect its output with the just-created second input of the Rate integrator block (Fig. 15.12b).

Now, you are almost ready to run your model. Before you do so, change the default simulation time in the corresponding window of the Model Editor toolbar to 600 s (seconds). Alternatively, you can do the same by opening the **Simulation** menu (on the model's window menu bar), choosing **Configuration Parameters** option (<Ctrl>+<E>), which brings the window shown in Fig. 15.13, and changing **Stop time** to 600.

Push **Start Simulation** button on the model's toolbar, or choose the **Start** option (<Ctrl>+<T>) from the **Simulation** menu to start the model. Note that, as opposed to the M-File Editor, if you made changes to your model and attempt to run it without saving, you will not be prompted to save your model first. Double-click on each Scope display to observe the results of simulation as shown in Fig. 15.14. You may need to click on the **Autoscale** button to autoscale the plots.

Other buttons on the Scope toolbar are as follows. The first two buttons allow you to print your plot and set additional parameters. Among other parameters you may set up additional axes, so that you have several inputs to a single Scope block as will be shown later. The three following buttons

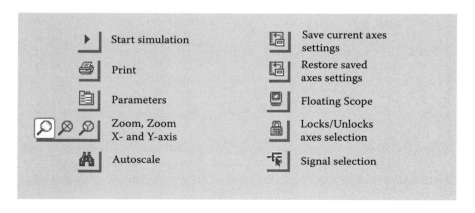

▶ Start simulation		Save current axes settings
Print		Restore saved axes settings
Parameters		Floating Scope
Zoom, Zoom X- and Y-axis		Locks/Unlocks axes selection
Autoscale		Signal selection

enable zooming the entire plot, zooming in x and y directions. The two buttons following the binoculars (autoscaling) button allow you to save and restore current axes settings, so that when you run your model next time you could have some predetermined setup. The next button-enables switching to another type of scope, the so-called Floating Scope, which is not explicitly connected to any other block. If you switch to this type of scope, you will have to run your model again, after which you could observe any signal within your model by choosing it using the **Signal Selector** window called by pushing the **Signal selection** button. The last but one button on the Scope toolbar, locks/unlocks axes selection in the Floating-Scope mode.

If you click on the **Expand** button, shown to the right of the solver name in Fig. 15.13, the drop-down list of all available variable-step solvers will show up as presented in Fig. 15.14c. As seen, this list is identical to that given in Table 14.6, that is, Simulink uses the same solvers presented and discussed in Chapter 14. The default solver is `ode45`. The name of the current solver shows up at the Status bar at the bottom of the Model Editor (Figs. 15.8, 15.9, 15.10a, and 15.12b).

Along with variable-step solvers, Simulink features fixed-step solvers, and so if you change the solver **Type** (Fig. 15.13) to a fixed step, then the **Configuration Parameters** window changes to that of Fig. 15.15. The list of available solvers in this case is shown in Fig. 15.16a. These simple

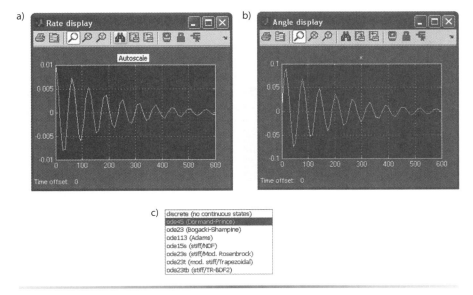

Fig. 15.14 Time histories of a) pitch rate and b) pitch with the default relative tolerance of `1e-3`, and c) list of variable-step solvers.

Fig. 15.15 Configuration Parameters window for fixed-step solvers.

solvers, discussed in Sec. 14.3, belong to the Real-Time Workshop and are programmed as C code as opposed to the M-functions for adaptive variable-step solvers (you may find their source codes in the ... \MATLAB\ R2011a\rtw folder).

Before we proceed with the next model, let us make a comment. Quite often, when you run your model, you do not have enough smoothness of your results (have a closer look at both plots in Figs. 15.14a and 15.14b). It happens because you are using some predetermined (not small enough) step for the default integration method (see Chapter 14 for details). If this happens to be the case, to achieve a better result (for the sake of speed of integration), you may manually change certain parameters of the solver. For example, changing **Relative tolerance** in the **Configuration Parameters** window of Fig. 15.13 to, say, $1e-8$ (as opposed to the default $1e-3$) produces

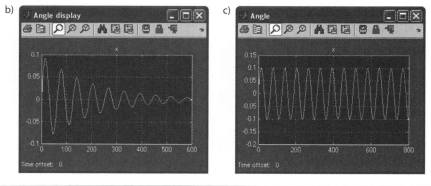

Fig. 15.16 a) List of fixed-step solvers, b) angle profile with the relative tolerance of $1e-8$, and c) erroneous result obtained with a fixed 1-second integration step.

much smoother result, as shown in Fig. 15.16b (do not forget that after changing any parameter in the **Configuration Parameters** window, you have to rerun your model).

However, if your primary concern is the speed of integration, so that you are willing to degrade relative tolerance, be aware that you may end up with a completely erroneous result. In our example, changing the solver type to a fixed step (Fig. 15.15), choosing the `ode1` (Euler method) solver (the last but one from the list of Fig. 15.16a), and running simulation with a fixed-step of `1`, leads to an erroneous result shown in Fig. 15.16c, indicating that the system is on the verge of stability (for larger step sizes it becomes unstable). This result certainly differs from what we had earlier (15.14b and 15.16b). In fact, establishing such a big value for the step size is equivalent to introducing an artificial delay, changing Eq. (15.1) we deal with.

15.3.2 Improving Model's Readability

It is always worth spending some extra time to make your model more readable and easy to use (not to forget what it is supposed to do). That is why in the previous section we suggested labeling of all the blocks and even signal lines between them. Table 15.1 provides with a few tricks of handling signal lines and their labels, to avoid crossing blocks and introduce multiple labels for the same line if needed.

You may also want to double-click on the background (any empty space within your model) to type in some comments. Another way to add comments is to use a Model Info block from the Model-Wide Utilities library of the Simulink Library Browser (presented in Sec. 15.5) to add extended semi-automatic comments to your model or employ the DocBlock block from the same library capable of carrying an ASCII text.

By right-clicking on any block and choosing the **Foreground Color** or **Background Color** option from the drop-down menu of Fig. 15.10b, you may choose the text and block colors for the selected block, respectively. Say, you want to have all blocks that require your input to be green, and all integrators to be magenta, etc. You may also want to add a shadow to specific (or all) blocks. You can do so by selecting these blocks and choosing the **Show Drop Shadow** option from the **Format** submenu (Fig. 15.10b).

Other enhancements may include adding symbolic notations. Click somewhere by the Gain block and type in `2\zeta\omega_n`; click by the Gain1 block and type in `\omega^2_n`. These strings use the TeX language introduced in Sec. 6.5. To enable using the TeX commands, you should select the text it applies to and check off the **Enable TeX Commands** option of the **Format** menu of the Model Editor menu bar. An alternative way of setting annotation properties is right-clicking on it (which brings

Table 15.1 Effective Handling of the Simulink Signal Lines and Labels

Operation	Method
Branch from a signal line	Right mouse click on the desired location of a branch point and drag with the right mouse button pushed; alternatively do the same with the left button holding down the <Ctrl> key
Copy a signal line label	Right mouse click on the label and drag it to a desired location with the right mouse button pushed; alternatively do the same with the left button holding down the <Ctrl> key
Draw a signal line in segments	Drag from the output port to the first bend and release a mouse button, drag from the first bend to second bend, and so on
Delete all occurrences of a signal line label	Select the label and delete all characters
Delete one occurrence of a signal line label that is repeated somewhere else	Holding down the <Shift> key, click on the label (select it) and press <Delete>
Split a signal line	Select a line, holding down the <Shift> key and clicking the left mouse button, drag the newly created vertex to the desired location

the menu shown in Fig. 15.17a) and choosing the **Annotation properties** option, resulting in a dialog box of Fig. 15.17b. All aforementioned improvements are incorporated in the modified version of the `Simulink1` model (Fig. 15.12b), as shown in Fig. 15.18a.

Shown in Fig. 15.19a is essentially the same Simulink model as in Fig. 15.12b, but with a few changes. First, the parameters of the model are changed to accommodate the short-period motion of Eq. (15.3) with $t_f = 3$. Second, the

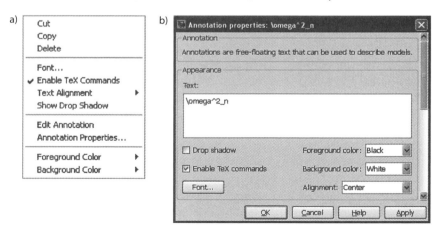

Fig. 15.17 a) Annotation drop-down menu and b) properties window.

a)

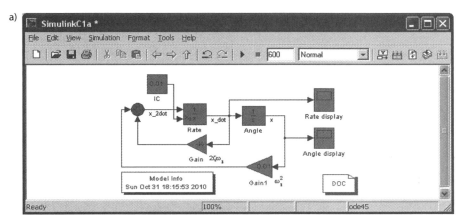

b)

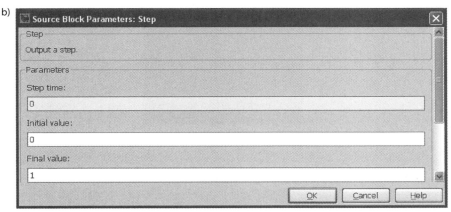

Fig. 15.18   a) Example of a model with enhanced readability and b) Step block Parameters window.

ICs for both Integrator blocks are set to zero (internally). To perturb the system, the Step input block (from the Sources library of Fig. 15.3) is employed, and so instead of the original Eq. (15.1), we now use

$$\ddot{x} + 2\zeta\omega_n\dot{x} + \omega_n^2 x = \overline{f}(t) \tag{15.8}$$

where $\overline{f}(t)$ represents some external forcing input (for example, a *step function* $1(t)$). The parameter window for the Step block is shown in Fig. 15.18b. Note that the **Step time** attribute is changed to 0 as opposed to the default value of 1. To normalize the output (to make the steady-state value of an angle to be equal to 1), the Gain2 block with the value of $\omega_n^2 = 25$ is added to the model.

Third, the Mux block (from the Signal Routing library, presented in Sec. 15.4) is added to blend the two output signals together, so that they

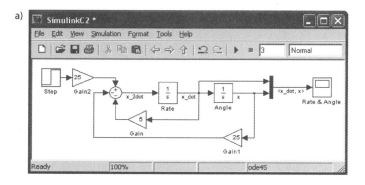

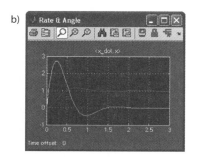

Fig. 15.19   b) Step response of a) the damped second-order system.

could be presented on a single scope as shown in Fig. 15.19b. Note that by clicking on the line connecting the Mux block to the Scope block and entering the symbol <, you force the propagation of the Mux's input names farther downstream. To see the result of this propagation after you entered the symbol < (which immediately changes to < >), you have to either update the diagram by pushing the **Update diagram** button or run your simulation.

Combining two signals within the same scope raises an issue of distinguishing them. Luckily, Simulink uses a certain default pattern of colors. It is yellow, magenta, cyan, red, green, blue, etc., that is, the first signal always appears in yellow, the second in magenta and so on. The way to remember it is to use some kind of memorable abbreviation, say "You Must Color RGB." The MathWorks' developers themselves prefer using the following trick. They add two dummy blocks (the Constant block from the Sources library and the Scope block from the Sinks library) as shown in Fig. 15.20a, so that they always have a color palette to look at handy (Fig. 15.20b).

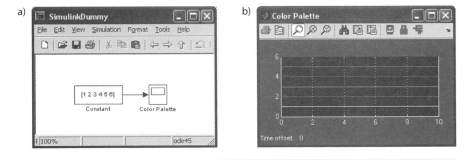

Fig. 15.20 Developers trick allowing distinguishing multiple outputs colors.

Alternatively, you may always use multiple axes on a single scope to present each output individually. To this end, Fig. 15.21a provides with an example of changing the **Number of axes** attribute of the Scope block Parameters window (called by clicking the **Parameters** button) from the default 1 to 2. Using the model of Fig. 15.19a without the Mux block and connecting two outputs to two inputs now to the Scope block results in Fig. 15.21b.

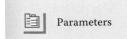

Now that you have started developing quite complex (bulky) models, it is time to introduce one more trick, allowing you to improve the model's readability. If you have a complex model with a lot of blocks, you might want to consider creating *subsystems*. For example, let us select almost all blocks of the model of Fig. 15.19a excluding the input Step block and output Scope block as shown in Fig. 15.22a. To select several blocks, left-click at the location of one of the corner blocks and then, continuing to depress the mouse button, drag the bounding box to enclose the desired area. If this does not work (the

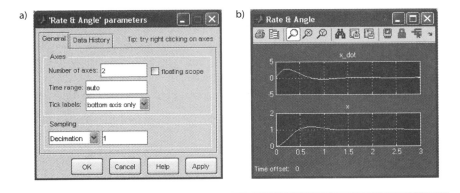

Fig. 15.21 Example of using a multiple-input scope.

blocks happen to have several unwanted blocks between them), hold down the <Shift> key and click on each object you want to select/deselect.

With all necessary blocks selected, click the right button of the mouse, which brings the menu shown in Fig. 15.22b. Choosing the **Create Subsystem** option results in the model shown in Fig. 15.22c, where all the selected blocks became a part of new subsystem, renamed from the default "Subsystem" to "2nd-order systems." This subsystem is shown in Fig. 15.23, featuring the In and Out blocks from the Sources and Sinks libraries (Figs.15.3, and 15.4) added automatically during the creation of this subsystem. Also, you can see that the **Go to parent system** button, is now acti-vated, and allows you to go one level up (in our case, to the *root level*).

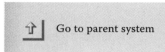

One more option to make your subsystem block more universal, so that it has its own Block Parameters window, as all other Simulink blocks do, is to mask it. This advanced feature is discussed in Appendix E (Sec. E.3).

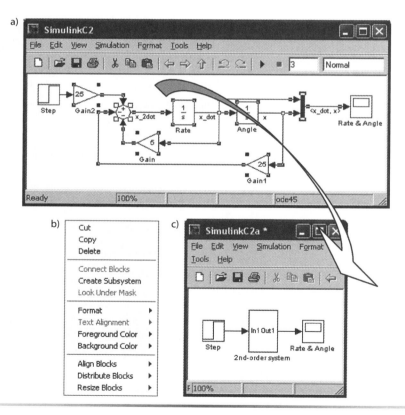

Fig. 15.22 Creating a subsystem.

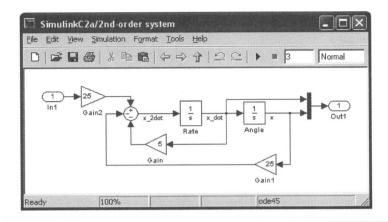

Fig. 15.23 Content of a subsystem.

15.3.3 Using the Tranfer Function and State-Space Blocks

Consider the same second-order system of Eq. (15.8) with the normalized input that is, having $\omega_n^2 f(t)$ on the right-hand side of equation. Assuming zero ICs (that is, $x(0) = 0$ and $\dot{x}(0) = 0$) and taking the *Laplace transform* yields

$$s^2 X(s) + 2\zeta\omega_n X(s) + \omega_n^2 X(s) = \omega_n^2 F(s) \tag{15.9}$$

The ratio of the Laplace transform of the output (angle), $X(s)$, to the Laplace transform of the input, $F(s)$, forms the standard *transfer function* $G(s)$ of the second-order system

$$G(s) = \frac{X(s)}{F(s)} = \frac{\omega_n^2}{s^2 + 2\zeta\omega_n s + \omega_n^2} \tag{15.10}$$

That is the form of a dynamic system that the Transfer Fcn block of the Continuous library (Fig. 15.7) deals with.

Let us replace a Subsystem block of Fig. 5.22c with the Transfer Fcn block as shown in Fig. 15.24a. The Parameters window for this block is presented in Fig. 15.24b. It shows how the numerator and the denominator of the transfer function Eq. (15.10) are introduced as a regular MATLAB-type representation of polynomials by their coefficients. The only disadvantage is that the transfer function Eq. (15.10) links the input to the angle only. The input-to-rate transfer function needs to be developed separately. The response of the system presented in Fig. 15.24a is shown in Fig. 15.24c.

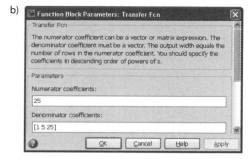

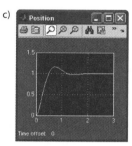

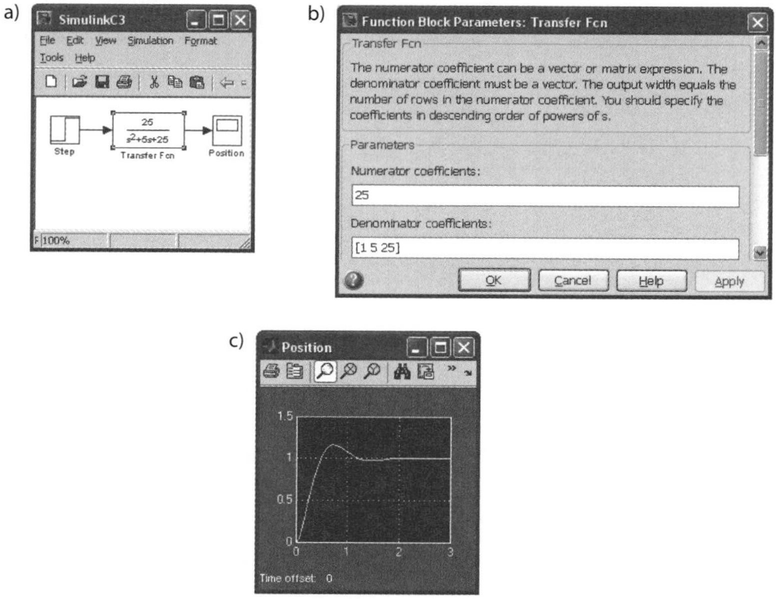

Fig. 15.24   a) Second-order system modeled with a Transfer Fcn block, b) parameters of this block, and c) response of the system.

Next, let us now convert Eq. (15.8) to the state-space form similar to that of Eqs. (1.12), and (14.14). Introducing two new variables $x_1 = x$ and $x_2 = \dot{x}$ results in

$$\dot{x}_1 = x_2$$
$$\dot{x}_2 = -2\zeta\omega_n x_2 - \omega_n^2 x_1 + \overline{f}(t) \tag{15.11}$$

These two first-order equations can be further expressed in the canonical state-space representation form

$$\dot{\mathbf{x}} = \mathbf{A}\mathbf{x} + \mathbf{B}\overline{f}(t)$$
$$y = \mathbf{C}\mathbf{x} + \mathbf{D}\overline{f}(t) \tag{15.12}$$

where

$$\mathbf{x} = \begin{bmatrix} x_1 \\ x_2 \end{bmatrix}$$

is the state vector,

$$\mathbf{A} = \begin{bmatrix} 0 & 1 \\ -2\zeta\omega_n & -\omega_n^2 \end{bmatrix}$$

is the system matrix, and

$$\mathbf{B} = \begin{bmatrix} 0 \\ 1 \end{bmatrix}$$

is the input matrix. Equation (15.12) also defines the *output vector y* through the *output matrix* **C** and *feedthrough* (or *feed forward*) *matrix* **D**. For instance, if we are only concerned about the time history of the angle, we may set **C** = [**1** 0] and **D** = [0] (meaning that there is no direct transmittance).

Let us model system Eq. (15.12) using the State-Space block from the Continuous library (Fig. 15.7) using the combination of the Constant block and Step input block with the **Step time** set to 0.01 (Fig. 15.17b) to generate an impulse input

$$\bar{f}(t) = 100\big(1(0) - 1(0.01)\big) \tag{15.13}$$

The upper part of the model presented in Fig. 15.25a shows a Simulink model for Eq. (15.12) with the parameters of the short-period motion defined in matrices A, B, C, and D in the Parameters dialog box of the State-Space block (Fig. 15.25b). The bottom portion of this Simulink model uses one more State-Space block featuring

$$\mathbf{C} = \begin{bmatrix} 1 & 0 \\ 0 & 1 \end{bmatrix}$$

[introduced in the State-Space block dialog box as C=eye(2)] and

$$\mathbf{D} = \begin{bmatrix} 0 \\ 0 \end{bmatrix}$$

enabling observation of both the angle and rate on the same display. The model also includes the Sine Wave and Manual Switch blocks taken from Sources and Signal Routing libraries, respectively, and allowing you to use the two different input signals for the second State-Space block.

15.4 Modeling Discrete Systems

So far, we dealt with analog- or continuous-type systems. Simulink, however, provides full support for discrete-type systems as well. You can create discrete systems in the same way that you created analog systems, but using the Discrete rather than Continuous library of the Simulink Library Browser. The content of this library is presented in Fig. 15.26 and the following two sections introduce some major blocks implemented for very simple systems to avoid getting into much detail of defining discrete systems per se.

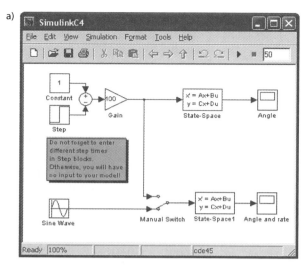

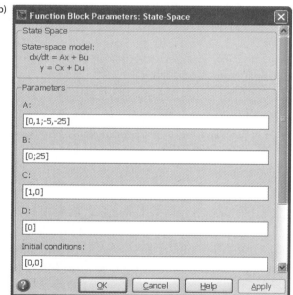

Fig. 15.25 a) State-space model of the second-order system and b) Parameters window of a State-Space block.

15.4.1 Using the Discrete Library Blocks to Model Simple Systems

Figure 15.27 shows the Parameters windows of three major (most often used) blocks of the Discrete library of Fig. 15.26. The Memory block (Fig. 15.27a) simply outputs its input from the previous time step.

The Unit Delay block (Fig. 15.27b) delays its input by the specified sample period. The Discrete-Time Integrator (Fig. 15.27c) performs discrete-time integration or accumulation of an input signal. As seen from Figs. 15.27a–5.27c, all three blocks require ICs (more parameters not appearing in Figs. 15.27a–5.27c can also be set for each individual block beyond the IC).

In addition to these three blocks, the Zero-Order Hold block samples and holds its input for the specified sample period. It is pretty similar to the Unit Delay block, but does not require the IC. The only parameter of this block is the time between samples specified with the **Sample time** parameter (similar to that appearing at the bottom of Fig. 15.27b). A setting of -1 means the **Sample time** is inherited. For the record, during the compilation phase of a simulation, Simulink determines the sample time of the block from its **Sample time** parameter (if it has it), sample-time inheritance or block type (continuous blocks have a continuous sample time). It is this compiled sample time that determines the sample rate of a block during simulation.

Consider several examples of using the aforementioned four blocks in the simple models. The first model shown in Fig. 15.28a simply introduces three discrete blocks. By now you should be able to develop such a model with no problem. The three Discrete library blocks are taken as is, with no changes in their Parameters windows. Instead of the Step block, the Ramp block of the Sources library of Fig. 15.3 is used to provide an input signal. The Sum block (Fig. 15.9b) is set to be rectangular. The **Number of axes** in the Scope block of Fig. 15.21a is set to 3.

When you run this model and then double-click on the Scope block (and apply autoscale), the results of the simulation will show up as presented in Fig. 15.28b. At the very first step, all signals are zero. At $t = 1$, the Ramp block outputs 1 and the Memory block outputs its old value, 0, making the output of the Sum block equal to 1. The output of the Zero-Order Hold jumps to 1 to be held until the next cycle. The Discrete-Time Integrator, using the default Euler forward integration formula of Eq. (14.20), produces 0 ($y = 0 + 1.0$). At $t = 2$, the Ramp block outputs 2 and the Memory block

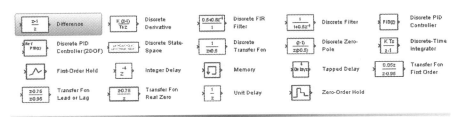

Fig. 15.26 Discrete library blocks.

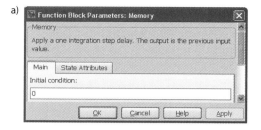

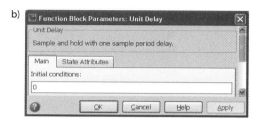

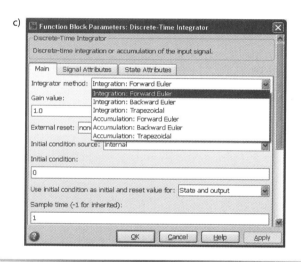

Fig. 15.27 Block Parameters window for a) Memory, b) Unit Delay and c) Discrete-Time Integrator blocks.

outputs 1, making the output of the Sum block equal to 1. The output of the Zero-Order Hold jumps to 2 to be held until the next cycle. The Discrete-Time Integrator produces 1 ($y = 0 + 1.1$). At $t = 3$, the Ramp block outputs 3, the Memory block 2, the Sum block 1, the Zero-Order Hold 3, the Discrete-Time Integrator 3 ($y = 1 + 1.2$), and so on.

Next, consider the model of Fig. 15.29. Two Unit Delay blocks are used here to generate a Fibonacci number sequence introduced in Sec. 10.2.3

[Eq. (10.15)]. The IC for the most right Unit Delay block is set to 0 (the very first number in the Fibonacci sequence) and the left block to 1 (the second number). The Divide block from the Math Operations library (Fig. 15.5) is used to model Eq. (10.16), which is supposed to converge to the golden ratio of Eq. (10.14), presented by the exact value in the Constant block. Figure 15.29a shows the result of running this model for five steps, so that the three displays at the bottom of the model (taken from the Sinks library of Fig. 15.4) show the sixth, seventh, and eighth Fibonacci numbers (from right to left), whereas the Scope in Fig. 15.29b proves that this sequence of numbers does converge to the golden ratio.

Finally, consider a loan amortization model. As well known, at the end of each month, the loan balance b_k is the sum of the balance at the beginning of the month b_{k-1} plus the interest for the month minus the end-of-the-month payment p_k. The interest rate is usually set as an annual interest rate r (in %). Hence, the loan balance at the end of month k can be computed using the following iterative formula:

$$b_k = \left(1 + \frac{1}{12}\frac{r}{100}\right) b_{k-1} - p_k \tag{15.14}$$

Assume that upon completing this textbook you became a proficient programmer and knowledgeable engineer and that this allows you to think about setting aside \$3000 a month to make payments for you own airplane. Would it be a two-seat general aviation aircraft worth \$100,000 (Fig. 15.30a) or perhaps a \$3,000,000 turbojet (Fig. 15.30b)?

a)
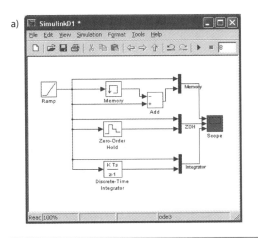

b)
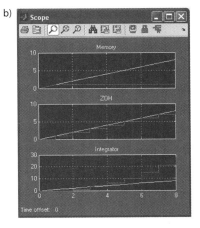

Fig. 15.28 Exploring Discrete library blocks.

a)
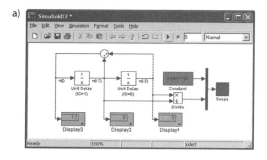

b)
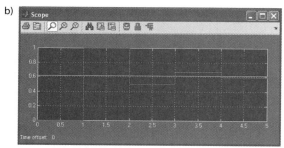

Fig. 15.29 a) Generating Fibonacci numbers and b) finding the golden ratio.

The model presented in Fig. 15.31 allows you to compute when you could pay your loan off. The Unit Delay block with the IC of 103 (thousands of dollars) to offset the $3000 (virtual) payment occurring at month 0 is used to model the right part of Eq. (15.14). The Fcn block (from the User-Defined Functions library of Fig. 15.2 presented in Sec. 15.5), displaying a mathematical expression as typed in it, computes the interest. The Discrete-Time Integrator block (with the default Euler forward method of integration) is used to compute the total amount of payments. The Compare To Zero block from the Logic and Bit Operations library (see Sec. 15.5) outputs 1 when the input value becomes negative (0 otherwise) and the Stop Simulation block from the Sinks library stops simulation when the input signal becomes 1. The Clock block of the Sources library

a)

b)

Fig. 15.30 a) Two-seat Skycatcher and b) five-seat Citation Mustang.

a)

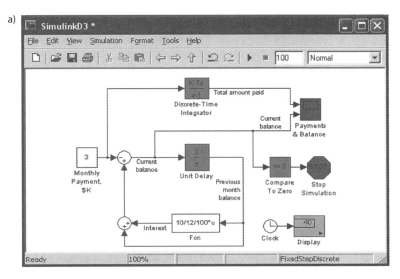

b)

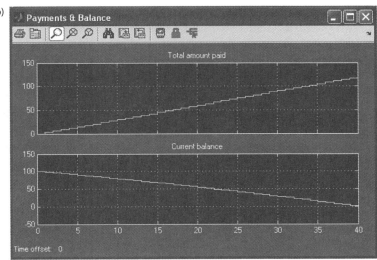

Fig. 15.31   Aircraft loan amortization model.

outputs the current simulation time. The number of steps is set to 100, meaning that you are counting on paying off your loan within eight to nine years (100 months).

When you run your model, the balance becomes negative (meaning that you paid your loan off) sooner than in 100 months. To be exact, the simulation stops at the 40th month triggered by the Stop Simulation block (negative balance means that the actual last payment should be less than $3000). The results of simulation are shown in Fig. 15.31b. The bottom line is, you can probably afford this buy. What about turbojet? Well, we leave this question as the last problem set for this textbook. To answer it you will need to demonstrate your programming skills.

15.4.2 Continuous Model Discretetization and Controller Design

The type of examples considered in the previous section is probably more applicable to address problems in finances, logistics, maintenance, etc. Systems dynamics and controls deal with slightly different models. If you open the **Tools** menu of the Model Editor and choose the **Control Design** option (Fig. 15.32a), you will see that Simulink offers some specialized tools in this area as well.

Open the *Simulink1C.mdl* model shown in Fig. 15.12b and save it as *Simulink1Cd.mdl*. Choose the **Model Discretizer** option from the **Control Design** menu of Fig. 15.32a. The Model Discretizer window will show up as shown in Fig. 15.32b. This tool analyzes your model, automatically identifies all continuous blocks conflicting with the discretized version (two integrator in our case), marks them with a red foreground, and lists them in the **Contains continuous blocks** pan, suggesting to discretize them (replace with the corresponding blocks from the Discrete library).

Pushing the **Discretize current selection** button on the right bottom of the Model Discretizer window (or the corresponding button on the Discretizer toolbar) converts the selected blocks (or all of them) to their discrete version. Specifically, the *Simulink1Cd.mdl* model becomes as shown in Fig. 15.33a

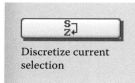

Discretize current selection

with two converted blocks highlighted with a blue foreground. Running this model with $t_f = 25$ results in a discrete (ladder-type) response of Fig. 15.33b as compared to the smooth output of the original model presented in Fig. 15.16b. To be more precise, the output of *Simulink1Cd.mdl* is a discrete version of that displayed in Fig. 15.16c.

In practice, you may not need to discretize all the blocks. For example, you want to design a *controller* for some analogous (continuous) system with the goal of improving its performance or making it follow some reference signal. You first create a model of your system verifying that it matches the real one, that is, that the outputs of your model (called *plant*) are about the same as the real system. The next step is to develop a *feedback* controller that would take the outputs of your system, compare them to the desired outputs, and calculate necessary control actions. Before you transfer the controller, developed and working properly for your

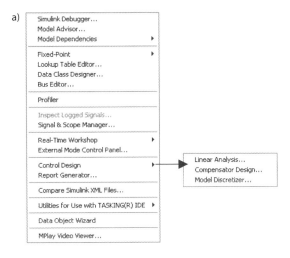

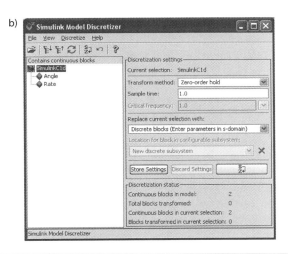

Fig. 15.32 a) Tools menu of the Simulink Model Editor and b) Model Discretizer window.

continuous model, onto the real system, you would need to assure that the real-life constraints on the controller will not degrade the performance of the systems. These real-life constraints may be based on the fact that the measurements of the output parameters of your system can only be taken with a certain rate, say 2 Hz or 100 Hz, and that the control actions are also discrete rather than continuous. In other words, you would want to discretize the controller portion of your model leaving the model itself

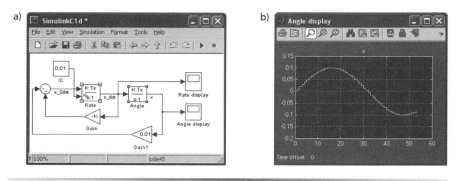

Fig. 15.33 Discrete version of a) second-order system and b) its output.

continuous. Simulink allows you to do this. It also allows you to create multi-rate simulations. Although this subject is obviously beyond the scope of this textbook, the model of Fig. 15.34 illustrates what we are talking about.

The second-order plant of Fig. 15.34a

$$F(s) = \frac{1}{s^2 + 2s + 1} \tag{15.15}$$

is modeled with the Transfer Fcn block. The plant itself (see the top portion of Fig. 15.34a) produces a kind of a sluggish response to the step input shown in the top portion of Fig. 15.34b. We would like to improve it, and for this purpose, we designed and tuned a continuous proportional-integral-derivative (PID) controller, as shown in the middle portion of the model of Fig. 15.34a. (By the way, as seen from Fig. 15.7, the newest versions of Simulink feature a single block, PID Controller, which allows you to design and tune all the gains of such a controller). The improved response of this controller is illustrated in the middle portion of Fig. 15.34b. Although overshooting, it converges to the reference signal (step input) much quicker than the original system.

Is this controller ready to be implemented on a real system? Not quite. The bottom portion of the model in Fig. 15.34a represents the case when the very same controller was discretized, while the plant remained continuous. As seen from the bottom portion of Fig. 15.34b, the discretized version of this controller (with the sample time of 0.05s) results in divergence, that is, inability to follow the reference signal. Hence, the controller must be redesigned.

a)
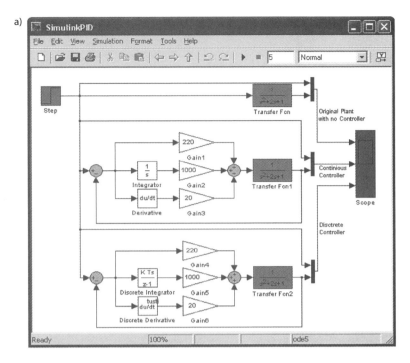

b)
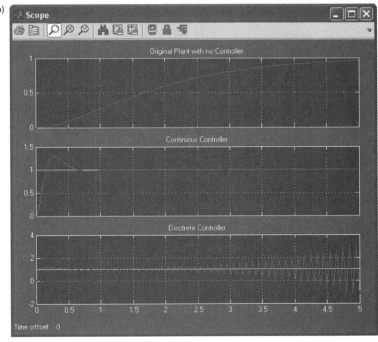

Fig. 15.34 Example of a controller design.

15.5 Other Libraries of Simulink

For completeness and your convenience, Figs. 15.35–15.44 show the content of the remaining libraries of Simulink, complementing those shown in Figs. 15.3–15.5, 15.7, and 15.26.

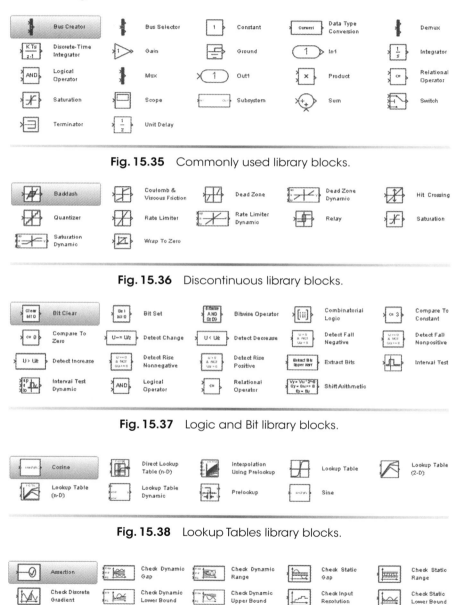

Fig. 15.35 Commonly used library blocks.

Fig. 15.36 Discontinuous library blocks.

Fig. 15.37 Logic and Bit library blocks.

Fig. 15.38 Lookup Tables library blocks.

Fig. 15.39 Model Verification library blocks.

Fig. 15.40 Model-Wide Utilities library blocks.

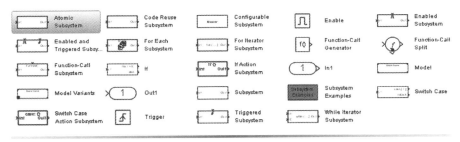

Fig. 15.41 Port & Subsystems library blocks.

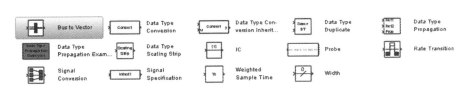

Fig. 15.42 Signal Attributes library blocks.

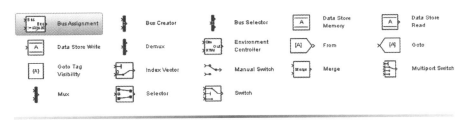

Fig. 15.43 Signal Routing library blocks.

Fig. 15.44 User-Defined Functions library blocks.

15.6 Integration of MATLAB and Simulink

15.6.1 Calling the MATLAB Commands from Simulink

Now that you know both MATLAB and Simulink, you may want to use them together. This will make your models more flexible and usefull. In fact, you have used MATLAB syntax to enter inputs into the Parameters dialog boxes already.

First of all, the User-Defined Functions library (Fig. 15.44) features several blocks, where you would call MATLAB functions by their namesizuse mathematical expressions or even employ M-file functions. The MATLAB Fcn block (Fig. 15.45a) allows you to pass the input values to any MATLAB function or your own function for evaluation. The constraints

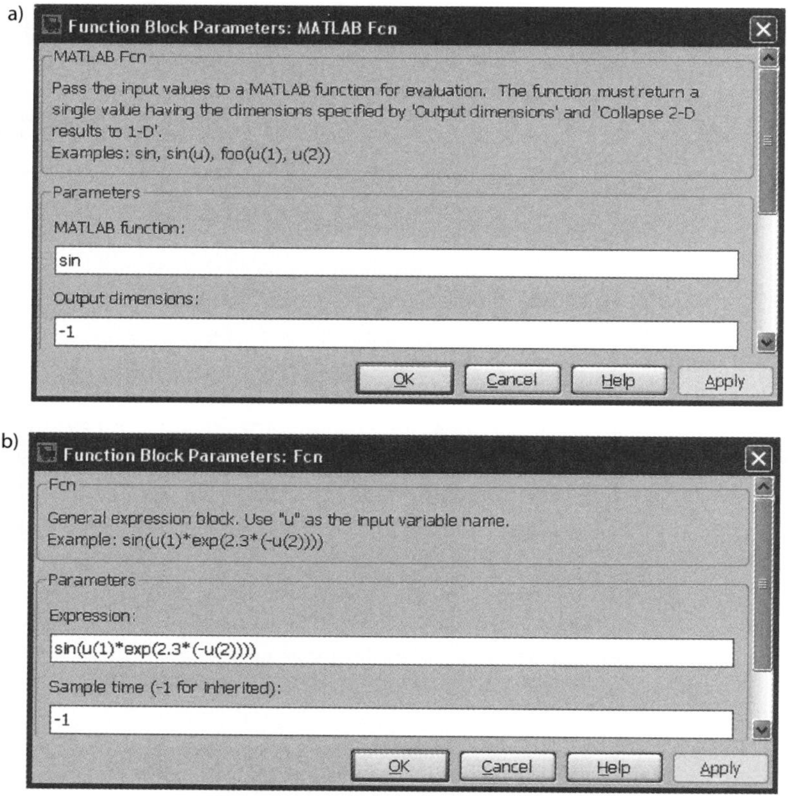

a)

b)

Fig. 15.45 Block Parameters window for a) MATLAB Fcn block and b) Fcn block.

Fig. 15.46 Blank Embedded MATLAB Editor window.

are that this function should return a single value and that the list of its input arguments were composed of the elements of the input vector to the MATLAB Fcn block, that is, if some M-file function `fun` requires three input arguments, it needs to be called as `fun(u(1),u(2),u(3))`. (Some other options for the MATLAB Fcn block that are not shown in Fig. 15.45a are also available.) Simple mathematical expressions, including using function evaluations, can be introduced in the Fcn block (Fig. 15.45b). An expression that appears in the **Expression** window by default could be modeled using simple blocks as well, but this block simplifies your model by allowing you to dedicate more space to more important parts of your model that cannot be modeled as simple expressions.

The Embedded MATLAB Function block from the User-Defined Functions library allows you to open the MATLAB Editor/Debugger and have the MATLAB function created while you are still working with the Simulink model (Fig. 15.46). The simplest default command indicates that you should use u as a single input and y as a single output (they both may be arrays though).

Second, you may use any workspace variable that you have in the current directory (where you started your Simulink model from) as a parameter in any block. For instance, you may modify the model developed in Sec. 15.3.2, so that the subsystem of Fig. 15.23 looks like the one shown in Fig. 15.47a. Specifically, instead of the constant values, the Gain blocks now contain `2*z*wn` (Gain) and `wn^2` (Gain1 and Gain2). In addition to this, the initial value for the Angle integrator is defined as `x0`. Now, if before running your Simulink model you issue the following command:

```
>> wn=5; z=0.5; x0=0;
```

these variables will be defined in the MATLAB current workspace and therefore visible to your model. When you run it, the values from the workspace will be used to compute the Gain values. You may want to create a GUI, dedicated to defining all parameters for your model interactively, and then running it.

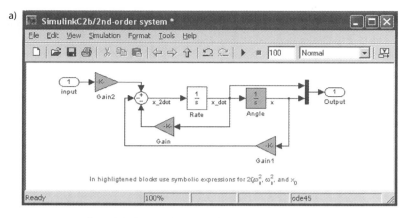

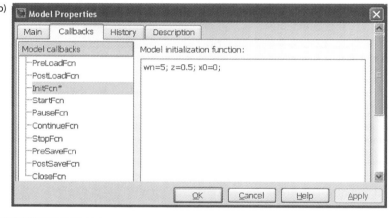

Fig. 15.47 Example of a) using variables' names rather than numerical values and b) defining these variables during model initialization using model callbacks.

Another way of defining these variables, used by your Simulink model, is accessing the Callbacks dialog window via choosing **File => Model Properties => Callbacks** (Fig. 15.47b). As seen, Simulink enables multiple callback functions being executed at different stages of working with your model. Figure 15.47b shows an example where the aforementioned MATLAB call defining three variables, occurs every time before you initialize (run) your model (after being executed, these variables appear in the current MATLAB workspace). The asterisk by InitFcn simply means that you have code in there whereas other callbacks have nothing. This way, your Simulink MDL-file constitutes a complete self-sufficient mdel, you would pass to the end user. Instead of messing with your model, the user will only need to change the values of parameters in the InitFcn callback. This way, if you type a text describing your model in the

Description folder of the Model Properties window (Fig. 15.47b), this text will appear in the MATLAB Commnd window in response to the help call

```
>> help SimulinkC2b
```

Finally, you may use the special blocks of Simulink Sinks library (Fig. 15.4), To File and To Workspace, to store the results of simulation in a file and MATLAB workspace, respectively. To read data from these sources into your simulation, you may use the From File and From Workspace blocks from the Sources library (Fig. 15.3). The dialog boxes for these four blocks are shown in Figs. 15.48 and 15.49. The only comment is that when you store data to the current workspace thinking of looking through it afterwards, you may want to choose the **Array** option, as shown in Fig. 15.49a, as opposed to the **Structure** option.

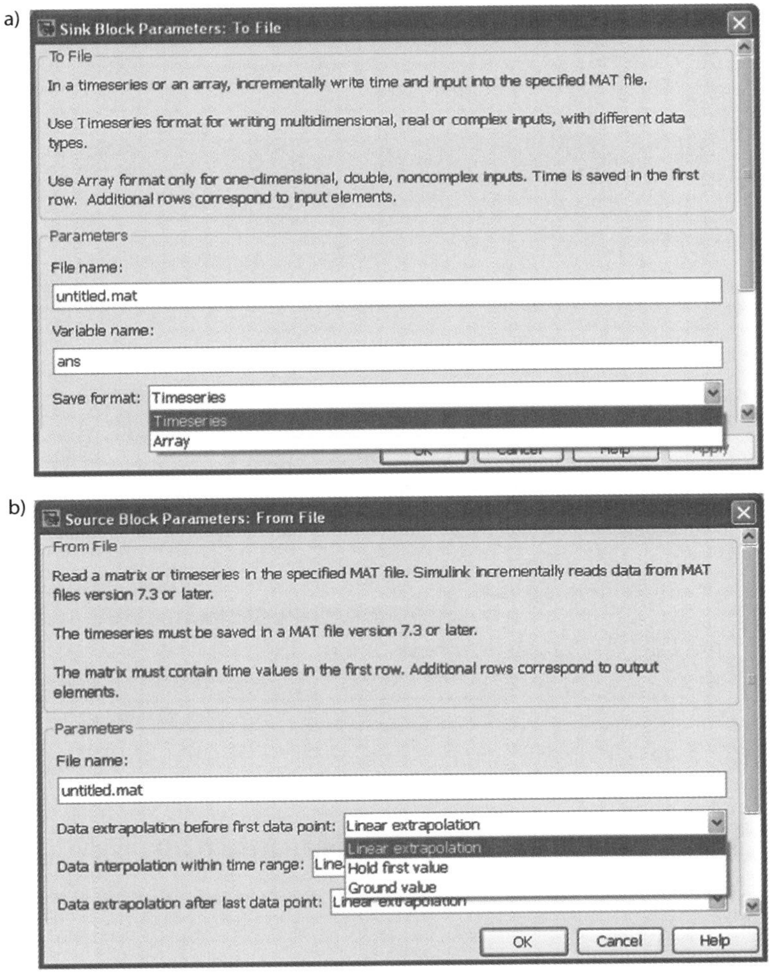

Fig. 15.48 Block Parameters windows for a) To File and b) From File blocks.

a)
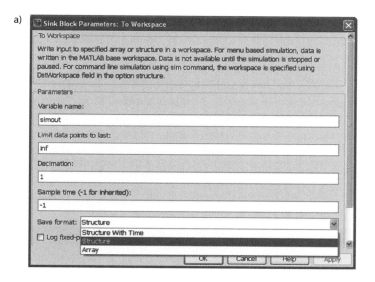

b)
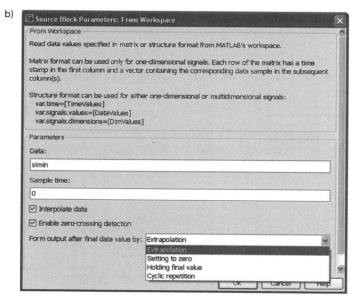

Fig. 15.49 Block Parameters windows for a) To Workspace and b) From Workspace blocks.

15.6.2 Calling the Simulink Models from MATLAB

MATLAB allows you to run Simulink models without even starting Simulink. For instance, you may run one of the models developed in the previous sections right from the MATLAB Command window

```
[t,x]=sim('SimulinkC1',400);
[ax,h1,h2]=plotyy(t,x(:,1)*180/pi,t,x(:,2)*180/pi);
```

```
xlabel('Time, s')
set(get(ax(1),'Ylabel'),'String','Pitch, ^o')
set(get(ax(2),'Ylabel'),'String','Pitch rate, ^o/s')
set(h2,'LineStyle','.')
```

As a result, you will see the plot shown in Fig. 15.50, which combines the two plots depicted in Fig. 15.14a and 15.14b. (Note that, to produce a multiple *y*-axis plot and change its properties, we used the script preceding Fig. 6.13.)

It is the Simulink function `sim`, used in the very first line of the aforementioned call, which allows you to run a Simulink model *SimulinkC1.mdl* and return its results without opening Simulink. This function outputs a time vector `t`, as well as angle and rate data in a two-column vector `x`. By default, the `sim` function runs the Simulink model with all settings you had used in the model when creating (saving) it. However, you may alter some of them. For instance, in the aforementioned code, the second argument in the `sim` function overrides the **Stop time** for simulation. (As you may recall, in the Simulink model itself we set it to be `600`.)

In general, when you want to run a Simulink model from the MATLAB Command window or from MATLAB script (function), you issue a set of two commands specifying simulation options using the `simset` function and running simulation with these specified options using the `sim` function. The general syntax for the `simset` function is

```
options=simset('name1',value1,'name2',value2,...)
```

which creates a Simulink options structure where the named properties `'name1'`,`'name2'`, ... have the specified values `'value 1'`,`'value`

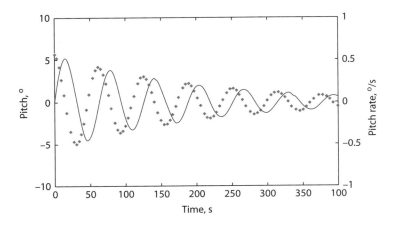

Fig. 15.50 Plotting the results of the Simulink model run.

2', ... The list of properties you may alter can be obtained using the `simget` function. For example,

```
>> simget('SimulinkC1')
```

returns

```
ans =                              AbsTol: 'auto'
                                    Debug: 'off'
                               Decimation: 1
                             DstWorkspace: 'current'
                           FinalStateName: ''
                                FixedStep: 1
                             InitialState: []
                             InitialStep: 'auto'
                                 MaxOrder: 5
               ConsecutiveZCsStepRelTol: 2.8422e-013
                        MaxConsecutiveZCs: 800
                               SaveFormat: 'Array'
                            MaxDataPoints: 1000
                                  MaxStep: 'auto'
                                  MinStep: 'auto'
                   MaxConsecutiveMinStep: 1
                             OutputPoints: 'all'
                          OutputVariables: 'ty'
                                   Refine: 1
                                   RelTol: 1.0000e-008
                                   Solver: 'ode45'
                             SrcWorkspace: 'base'
                                    Trace: ''
                                ZeroCross: 'on'
                            SignalLogging: 'on'
                        SignalLoggingName: 'sigsOut'
                        ExtrapolationOrder: 4
                   NumberNewtonIterations: 1
                                  TimeOut: []
       ConcurrencyResolvingToFileSuffix: []
               ReturnWorkspaceOutputs: []
          RapidAcceleratorUpToDateCheck: []
          RapidAcceleratorParameterSets: []
```

Most of these properties are the default properties taken from the **Configuration Parameters** dialog box (Fig. 15.13 or 15.15).

If you alter some of properties of a Simulink options structure, you pass them to the `sim` function as

```
[t,x]=sim('ModelName',timespan,options)
```

If you do not need to alter the `timespan`, but still want to pass a new option structure, you should replace `timespan` argument with the placeholder `[]`. The `timespan` can be either a scalar, defining the **Stop time** `TFinal` or a two-element vector `[TStart TFinal]`, defining both the **Start time** and **Stop time** or even a longer vector `[TStart OutputTimes TFinal]`, defining some additional time points of interest `OutputTimes` to be returned in `t`. In the latter case, `t` will contain more time points, but will definitely include those defined in a sequence `OutputTimes`. You may also choose to run your model as it is, that is,

```
[t,x]=sim('ModelName')
```

15.6.3 Generating Embedded MATLAB Function Blocks from Symbolic Expressions

In addition to the aforementioned capabilities of MATLAB–Simulink interaction, MATLAB also allows you to create Embedded MATLAB Function blocks from symbolic expressions created using the Symbolic Math Toolbox functions. For example, the following commands

```
new_system('SimulinkC6')
open_system('SimulinkC6')
```

create and open a new (empty) Simulink model. Now, you can develop some symbolic expressions and convert them to Simulink blocks using the `emlBlock` function. Consider a couple of commands creating a 10-term Taylor series expansion of $f = \sin(x)$ around $x = a$ and solving the second-order differential equation with the ICs specified as $y(0) = a$ and $y'(0) = b$

```
syms x a b
f=sin(x)
y=dsolve('D2y=cos(2*x)-y','y(0)=a','Dy(0)=b','x');
emlBlock('SimulinkC6/Taylor',taylor(f,10,a))
emlBlock('SimulinkC6/DiffEq',y)
```

The two `emlBlock` calls create two Embedded MATLAB Function blocks, as shown in Fig. 15.51a. Figure 15.51b shows the content of the DiffEq block with vectorized optimized code in it.

You may add other blocks to these two to build your model in a usual fashion. For example, Fig. 15.52a features the two Constant, Clock, Display, and Scope blocks added to the *Simulink6* model, which produces the output shown in Fig. 15.52b.

When creating the blocks be careful with the names, so that you do not override the definition of an existing block.

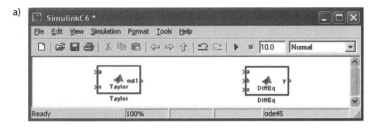

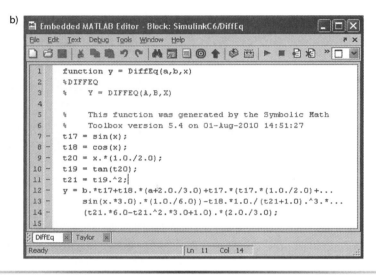

Fig. 15.51 a) Creating blocks from MATLAB and b) an example of automatically generated function.

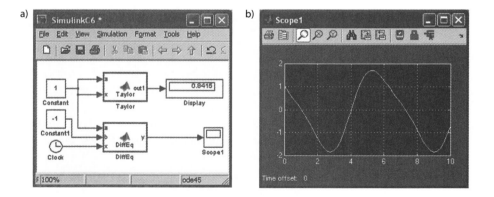

Fig. 15.52 a) Utilizing two blocks created in MATLAB and b) a Scope block output.

15.7 Practical Examples

This section presents an example of a more or less complex model, built in Simulink, with the goal of setting some trends to follow. Figure 15.53 shows a root level of the model of the integrated INS and GPS. As seen, the larger pieces of this model are combined into subsystems, so that the root level has only 10 superblocks (subsystems) and just a few simple blocks. Another point to make is that to avoid numerous lines connecting different blocks between themselves, the Goto and From blocks from the Sinks and Sources libraries are employed.

For example, the Constants block in Fig. 15.53 seems to be not connected to anything. In fact, this block, featuring some constants (scalars and matrices) used in multiple blocks throughout the model, passes its values to these blocks using the Goto blocks as presented in Fig. 15.54a. When you connect the Goto blocks with other blocks, you define the tag of each specific signal (Fig. 15.55a). In another block, where this signal is used, for example, in the block shown in Fig. 15.54b, you introduce the From block and then find the tag you need from a drop-down list, which shows up when you click on the **Expand** button (to the left of the **Update Tags** button) (Fig. 15.55b).

In the models where you use the same constant values in several places, it is worth defining them somewhere once and then passing to other blocks rather than creating the Constant blocks all over the place anyway. In the case of the INS/GPS model of Fig. 15.53, where you have to use several

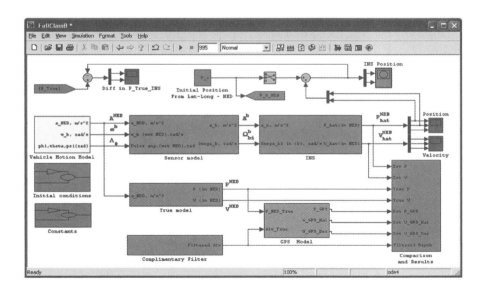

Fig. 15.53 Model of the INS/GPS system.

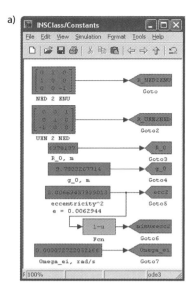

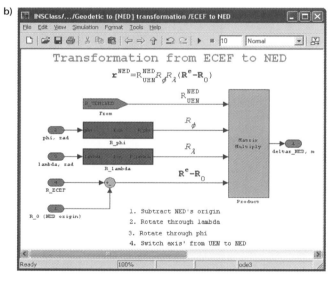

Fig. 15.54 Using a) Goto and b) From blocks to avoid connecting lines.

values with a lot of significant digits, like Earth's radius, gravity, eccentricity, and rotational velocity, a single typo, causing mismatch of the values in different blocks, would prevent your system from working properly, so that you may spend days and weeks before figuring it out (if at all).

As seen, the subsystem of Fig. 15.54b uses TeX language to display symbolic expressions by its blocks. This helps in understanding and debugging the model. Figure 15.56 demonstrates an example of the Embedded MATLAB Function block, R_phi, used in subsystem of Fig. 15.54b. This block performs a rotation from one coordinate frame to another, using the rotation matrix

$$R_\phi = \begin{bmatrix} \cos(\phi) & 0 & \sin(\phi) \\ 0 & 1 & 0 \\ -\sin(\phi) & 0 & \cos(\phi) \end{bmatrix} \tag{15.16}$$

Figure 15.57 gives one more example showing a model of the GPS constellation. It shows how the signals from multiple subsystems are combined together (to minimize the number of signal lines between subsystems) and also demonstrates an example of how the very same subsystem, in this case performing some coordinate-frame transformation, can be duplicated, so that you do not need to develop this transformation for every satellite from scratch.

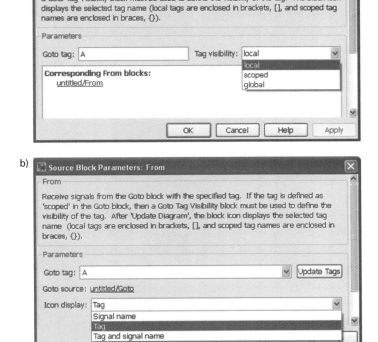

Fig. 15.55 Block Parameters windows for a) Goto and b) From blocks.

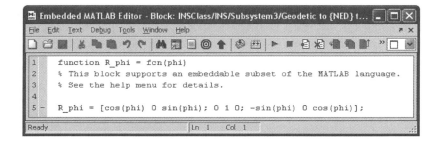

Fig. 15.56 Example of the Embedded MATLAB Function block.

When you develop a complex model containing many blocks, you may want to align and distribute them evenly. With several blocks selected, the **Format** menu of the Simulink Model Editor repeats some of the options available for a single block (Fig. 15.10b), but in addition to them offers tools allowing you to further enhance the readability of your model (Fig. 15.58a).

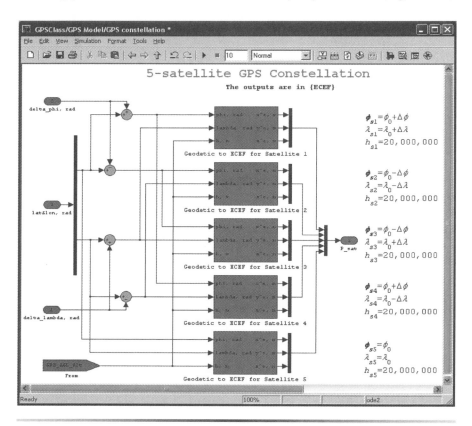

Fig. 15.57 Model of a five-satellite GPS constellation.

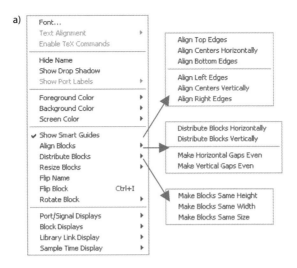

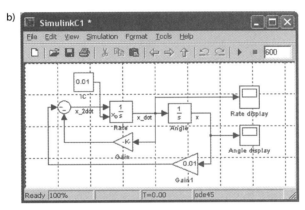

Fig. 15.58   a) Format menu and its model editing options, and b) visualizing the grid lines.

When aligning the blocks, Simulink uses an invisible five-pixel grid to simplify this process. When you move a block to a new location, the block snaps to the nearest line on the grid. To see the grid for the opened model, say *SimulinkC1.mdl*, you should issue the following command at the MATLAB command prompt:

```
>>set_param('SimulinkC1','showgrid','on')
```

If needed, you may also adjust the size of the grid. For example,

```
>>set_param('SimulinkC1','gridspacing',50)
```

sets a visible grid spacing to 50 pixels, as shown in Fig. 15.58b. By default, the width of the visible grid is 20 pixels.

15.8 Model Reviewing and Profiling

As mentioned in Sec. 15.2, Simulink offers some additional tools, allowing you to review your model and analyze it. Without going deeply into it, let us briefly present these tools here.

Clicking the **Toggle Model Browser** button, supplements the Model Editor window with the **Model Browser** pan. For example, Fig. 15.59 shows the content of this pan for the multilevel model of

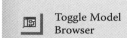

Toggle Model Browser

Fig. 15.57. Browsing through the model allows you to conveniently access the content of its subsystems without opening a new Model Editor window for each of them (as it would take place by simply double-clicking on each module). As seen, the GPS Constellation subsystem of Fig. 15.57 shows up in the Browser pan, whereas another subsystem, Disturbed measurements is chosen to be viewed. This subsystem computes actual ranges from an observer to each of the five satellites, as modeled in Fig. 15.57, and adds some bias and measurement noises.

It should be mentioned that using the Band-Limited White Noise block from the Sources library of Fig. 15.3 has one feature. One of the parameters that needs to be defined is called the **Seed**. The way the random numbers (noise) are generated is that they are represented by a long sequence (say 10^{13}) of randomly generated numbers, stored in computer memory. When you introduce one of the blocks relying on these numbers, you have to specify where to start your sequence from (see Fig. 15.60 featuring the default seeds

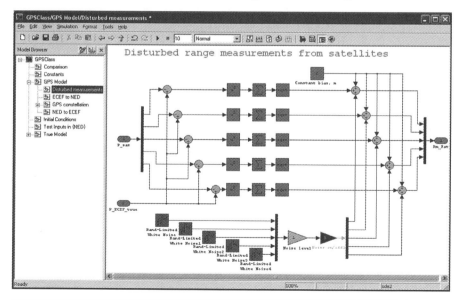

Fig. 15.59   Model Browser pan added to the Model Editor window.

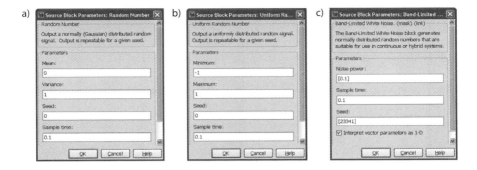

Fig. 15.60 Parameters dialog boxes for a) the Random Number, b) Uniform Random Number, and c) Band-Limited White Noise blocks of the Sources library.

for three blocks of this nature). If your model uses only one block relying on random numbers, the starting point does not matter, but when you use two or more blocks, you should define different seed values for each of them (even if they differ just by one). To make the point, Fig. 15.61a presents a simple model, where the same two blocks feed the *x*- and *y*-axes of the XY Graph block of the Sinks library of Fig. 15.4. If the seeds for both the blocks are the same, they output exactly the same numbers, which results in the linear dependence between two sequences (Fig. 15.61b). Changing one of the seeds by one unit, changes the picture completely, now featuring two independent sequences of random numbers (Fig. 15.61c).

The **Launch Model Explorer** button opens the Model Explorer window (see example in Fig. 15.62). This tool also shows the model hierarchy in the left pan, but allows you to find any specific block and access its parameters in the right pan. For example, the Explorer shown in Fig. 15.62 features the 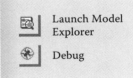 Parameters window of the Omega_ei constant block of a subsystem, presented in Fig. 15.54a, enabling modification of any of its parameters. Pushing the **Debug** button brings up the Simulink Debugger window, as shown in Fig. 15.63. Here, you can set the break/display points in your model and observe what happens when you run it. In this way, you can pinpoint problems in your model to specific blocks, parameters, or interconnections.

Finally, if you check off the **Profiler** option of the **Tools** menu (Fig. 15.32a), every time you run your model, Simulink Profiler Report will show up upon conclusion of your simulation. The Simulink profiler captures data while your model runs and identifies the parts of your model requiring the most time to simulate. You can then use this information to decide where to focus your model optimization efforts (if at all). Among others, it shows the total time required to simulate the model (Total recorded time) and

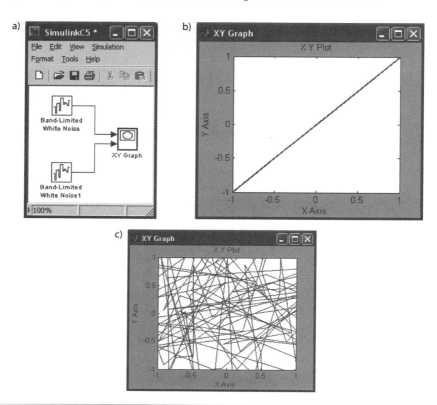

Fig. 15.61   a) Model employing two random number generators, with b) same two seeds and c) different seeds.

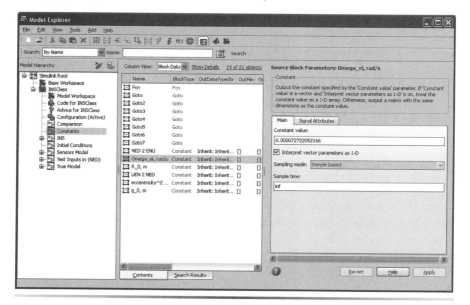

Fig. 15.62   Model Explorer window.

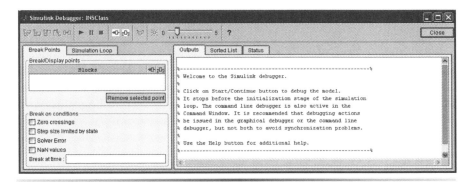

Fig. 15.63 Simulink Debugger window.

total number of invocations of block-level functions (Number of Block Methods). The function list features

- Total time spent executing all invocations of each particular function (Time)
- Number of times each function was invoked (Calls)
- Average time per each invocation, including the time spent in functions invoked by this function (Time/call)
- Average time required to execute this function, excluding time spent in functions called by this function (Self time)

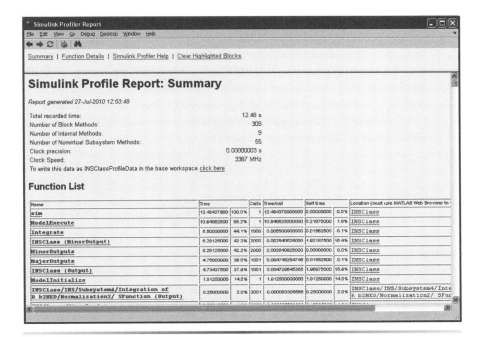

Fig. 15.64 Simulink Profiler Report window.

- Location of each function, that is, the block or model it belongs to (Location)

Specifically, from Fig. 15.64, you may learn that integration (indeed) eats most of the time (44%), whereas, for example, normalization of the rotation matrices accounts for 2% of the time. Clicking on the location link in the rightmost column of the report opens the corresponding subsystem with the block of interest highlighted with a red foreground.

Problems

15.1 Use the *SimulinkC2a.mdl* model created in Sec. 15.3.2 as a base to develop a more sophisticated model shown on the left:

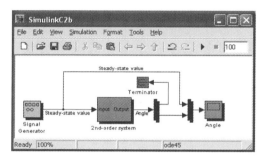 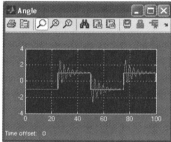

Use a Signal Generator block from the Sources library of Fig. 15.3 to feed your system with the square-form wave with an amplitude of 1 and frequency of 0.02 Hz. Use **File => Model Properties => Callbacks** = > **InitFcn** to enter the numerical values for $\omega_n = 2$, $\zeta = 0.075$, and $x_0 = -1$ (similar to that of Fig. 15.46b). Check the correctness of your model by comparing its output with what is shown in the figure to the right of the model given above.

15.2 Consider the following two-body mass–spring–damper system (which may describe, for instance, an aircraft gear), excited by some external input $u = \varepsilon \sin(\omega t)$ (runway surface)

Assume parameters of the system to be $m_1 = 1$, $m_2 = 1000$, $k_1 = 10$, $k_2 = 2$, and $c = 5$. Do the following:

a. Develop a Simulink model for this mechanical system
b. Check the output of a homogeneous part of the system for the case when $x(0) = 0$, $y(0) = 0$, $\dot{x}(0) = 1$, and $\dot{y}(0) = 0$
c. Simulate the behavior of the system with zero ICs as a response to input u (pick the values of ε and ω that seem reasonable yourself).

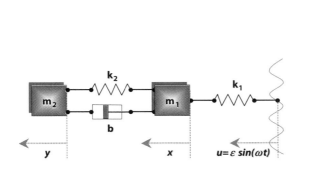

15.3 Using the loan amortization model of Sec. 15.4.1, find how many months it would take to pay off a 3 million dollar loan with a 10% annual interest rate making $3000 monthly payments. Then, try to adjust the monthly payment to be able to pay it off in 40 months, as it was the case in the example considered in the aforementioned section.

This appendix contains solutions to the problem sets posted at the end of each chapter (except the first one). These solutions represent a result of publishing the script containing solution code (and accompanied by extensive comments) from the MATLAB Editor window. Therefore, the output of the script follows the script itself. Each solution set starts from the following three lines clearing the workspace and screen

```
close all        % Deletes all open figures (if any)
clear all        % Clears all variables from the current workspace
clc              % Clears the Command window and homes the pointer
```

Chapter 2 Problem Set Solutions

Problem 2.1

```
fprintf('\nProblem 2.1 Output:\n');
C_L=2;
AirDensity=1;
S=4;
W=100;
g=9.81;
V=sqrt(2*W*g/C_L/AirDensity/S)         % Output in m/s
V=V*3.6                                % Output in km/h
```

Problem 2.1 Output:

V = 15.6605
V = 56.3777

Problem 2.2

```
fprintf('\nProblem 2.2 Output:\n');
% In order to save space you can have several commands per line
```

```
C_L=0.3;  C_D=0.03;  c_T=0.0003;  AirDensity=1;
S=20;     W0=2300;   Wf=2100;     g=9.81;
R1=2/c_T*sqrt(2*C_L*g/AirDensity/S/C_D^2)*...
   (sqrt(W0)-sqrt(Wf))/1000                      % Output in km
R2=1/c_T*sqrt(2*W0*C_L*g/...
AirDensity/S/C_D^2)*log(W0/Wf)/1000              % Output in km
Dif=(R2-R1)/R1*100                               % Diffeference in %
```

Problem 2.2 Output:

```
R1 = 257.0890
R2 = 262.9803
Dif = 2.2915
```

Problem 2.3

```
fprintf('\nProblem 2.3 Output:\n');
% This was your first m-script written in MATLAB.
% When you run it you will be prompted to input the date
% (in the Command window), for example
%         >> Enter the year (YYYY): 1957
%         >> Enter the month:       10
%         >> Enter the day:          4
% The program reads the input data back to you
%         >> You entered  10/ 04/1957
% and then computes and outputs the weekday:
%         >> Your weekday is: Friday
%
% The following input menus are commented for publishing
%{
Year =input('\nEnter the year (YYYY): ');
Month=input('Enter the month:        ');
Day  =input('Enter the day:          ');
%}
Year =2010;  Month=8;   Day  =1;
    fprintf('\nYou entered %3.0f/%2.0f/%4.0f', Month,Day,Year)
d=Day;
    m=Month-2;
    if m <= 0
    m=m+12; Year=Year-1; end % month's number in Ancient Rome
Y=mod(Year,100);            % year's number in the century
c=(Year-Y)/100;             % number of centuries
% here goes your formula
```

```
WeekDay = mod((d+fix((13*m-1)/5)+Y+fix(Y/4)+fix(c/4)-2*c+777),7);
Week={'Sunday' 'Monday' 'Tuesday' 'Wednesday' 'Thursday' 'Friday'...
      'Saturday'};
fprintf('\n\nYour weekday is: %s\n', Week{WeekDay+1});   % result
```

Problem 2.3 Output:

You entered 8/1/2010

Your weekday is: Sunday

Chapter 3 Problem Set Solutions

Problem 3.1

```
fprintf('\nProblem 3.1 Output:\n');

format short g        % shortens the output (try 'format rat')
% Creating matrix B1
B1 = [linspace(1,19,7);linspace(72,36,7);linspace(0,0.75,7)];
B1 = [B1; horzcat(.3:.1:.6,1.2:.2:1.6); 9:-1:3]
% Creating matrix B2
B2=B1;
B2(2,[1:2:7])=zeros(1,4);
B2(3,1:4)=ones(1,4);
B2(4:5,1:3)=zeros(2,3);
B2([1,3,4],5:7)=magic(3)     % shows the final result
% Finding indices of zero elements
[r,c]=find(~B2);
[r,c]                        % shows the final result
% Showing nonzero elements
spy(B2)
format                      % returns format to default
```

Problem 3.1 Output:

```
B1 =    1         4         7        10       13       16       19

             72        66        60        54       48       42       36

              0     0.125      0.25     0.375      0.5     0.625      0.75

            0.3       0.4       0.5       0.6      1.2       1.4       1.6

              9         8         7         6        5         4         3

B2 =  1     4     7    10     8     1     6

       0    66     0    54     0    42     0

       1     1     1     1     3     5     7

       0     0     0    0.6     4     9     2

       0     0     0     6     5     4     3

ans =    2         1
```

4	1
5	1
4	2
5	2
2	3
4	3
5	3
2	5
2	7

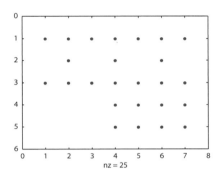

Problem 3.2

```
fprintf('\nProblem 3.2 Output:\n');

g = 9.8;

t = linspace(0,10,11)';

h = 1000-0.5*g*t.^2;

format short g

[t,h]

format

% For a better looking output you may use

fprintf('At t = %02.0f sec the altitude is %6.1f meters\n',[t,h]');
```

Problem 3.2 Output:

```
ans =    0       1000

         1       995.1

         2       980.4
```

3	955.9
4	921.6
5	877.5
6	823.6
7	759.9
8	686.4
9	603.1
10	510

```
At t = 00 sec the altitude is 1000.0 meters
At t = 01 sec the altitude is  995.1 meters
At t = 02 sec the altitude is  980.4 meters
At t = 03 sec the altitude is  955.9 meters
At t = 04 sec the altitude is  921.6 meters
At t = 05 sec the altitude is  877.5 meters
At t = 06 sec the altitude is  823.6 meters
At t = 07 sec the altitude is  759.9 meters
At t = 08 sec the altitude is  686.4 meters
At t = 09 sec the altitude is  603.1 meters
At t = 10 sec the altitude is  510.0 meters
```

Problem 3.3

```
fprintf('\nProblem 3.3 Output:\n');
% Part A: Using vectors to handle polynomial operations
a=ceil(6*rand(1,7));
b=ceil(10*rand(1,3));
[q,r]=deconv(a,b);          % finding the quotient and reminder
s=poly2sym(q); pretty(s) % symbolic polynomial representation
% Part B: Using vectors to handle text strings
n1=ceil(26*rand(1,26));   % randomly distributed integers within [1;26]
c1=char(n1+64)            % randomly distributed letters A-Z
n2=findstr(c1,'A')        % finding letter A
c2=reshape(c1,2,13)       % reshaping
find(~(c2(1,:)-c2(2,:))) % looking for a column with the same letter
```

Problem 3.3 Output:

$$x^4 - 2/5\, x^3 - 7/25\, x^2 + \frac{193}{125}\, x - \frac{207}{625}$$

c1 = POXHTTJOBBNUYDOMAIEUINEPGR

n2 =   17

c2 = PXTJBNYOAEIEG

    OHTOBUDMIUNPR

ans =   3   5

Chapter 4 Problem Set Solutions

Problem 4.1

```
fprintf('\nProblem 4.1 Output:\n');
Record.Date      = '7-Nov-45';
Record.Country   = 'Great Britain';
Record.Vehicle   = 'Gloster Meteor F Mk4';
Record.Speed     = 606;
% Comment: You may also use a one-line instruction like:
Record(2:9) = struct('Date',{'19-Jun-47','7-Sep-53','3-Oct-53',...
'10-Mar-56','12-Dec-57','31-Oct-59','15-Dec-59','7-Jul-62'},...
'Country',{'USA','Great Britain','USA','Great Britain','USA',...
'USSR','USA','USSR'},'Vehicle',{'XP-80R','Mk3','XF4D-1','FDT',...
'F-101A','E-66','F-106A','E-166'},'Speed',{624,728,753,...
1132,1208,1484,1526,1666})
% The 'Record' now is a 1-by-9 structure (to see that, a
% semicolon after the above command is omitted).
% You can look it up in the Array Editor as well.
% Part A
% Speed in m/s rounded to the nearest integer (for compactness)
SpeedSI = round([Record.Speed]*0.44704)
% Part B
Date=ones(9,1);
    for j=1:9
    Date(j) = datenum(Record(j).Date,'dd-mmm-yyyy'); % Serial date
    end
[x,i]=max(diff(Date));
fprintf(['It was %s holding the speed record the longest,'...
        'for %4.0f days\n'], Record(i).Country,x);
[x,i]=min(diff(Date));
fprintf(['%s held the record the shortest, it was broken ',...
        'in %2.0f days by %s\n'],Record(i).Country,x,...
        Record(i+1).Country);
```

Problem 4.1 Output:

```
Record = 1x9 struct array with fields:
            Date
            Country
            Vehicle
            Speed
```

SpeedSI = Columns 1-8

   271   279   325   337   506   540   663   682

Column 9

   745

It was the USA holding the speed record the longest for 2272 days
Great Britain held the record the shortest, it was broken in
26 days by the USA

Problem 4.2

```
fprintf('\nProblem 4.2 Output:\n');
record(1,1,:) = {Record.Date};
record(1,2,:) = {Record.Country};
record(2,1,:) = {Record.Vehicle};
record(2,2,:) = {Record.Speed};
cellplot(record)
% Part A
% Speed in km/h rounded to the nearest integer (for compactness)
SpeedSI = round([record{2,2,:}]'*1.61)
% Part B
Date=ones(9,1);
    for j=1:9
    Date(j) = datenum([record{1,1,j}],'dd-mmm-yyyy'); % Serial date
    end
T=diff(Date);
    for j=1:8
fprintf('%s held the record for %2.0f days beaten by %s\n',...
        record{1,2,j},T(j),record{1,2,j+1})
    end
```

Problem 4.2 Output:

SpeedSI = 976
         1005
         1172
         1212
         1823
         1945
         2389
         2457
         2682

Great Britain held the record for 589 days beaten by the USA

```
The USA held the record for 2272 days beaten by Great Britain
Great Britain held the record for 26 days beaten by the USA
The USA held the record for 889 days beaten by Great Britain
Great Britain held the record for 642 days beaten by the USA
The USA held the record for 688 days beaten by USSR
USSR held the record for 45 days beaten by the USA
The USA held the record for 935 days beaten by USSR
```

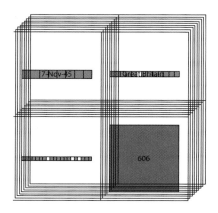

Problem 4.3

```
fprintf('\nProblem 4.3 Output:\n');
% We start from reading headers to a cell array ColHdr using
% the 'textread' function (2 means that the specified format, %s, has to be
% applied twice):
ColHdr = textread('Temp_Conversion.txt','%s',2);
% Note that the following command would do the job too:
% [ColHdr(1),ColHdr(2)] = textread('Temp_Conversion.txt','%s %s',1);
% Now, we proceed with reading numeric data into two vectors
% (the parameter 'headerlines' tells to ignore the specified number
% of lines at the beginning of the file):
[DegreesF,DegreesC]=textread('Temp_Conversion.txt',...
                            '%f %f','headerlines',1);
% Note that the following two lines, employing the 'dlmread'
% function would do the same ('\t' specifies a tab delimiter,
% and 1,0 specify the second row and the first column of the
% upper left corner of the data to read):
% Data = dlmread('Temp_Conversion.txt','\t',1,0);
% DegreesF=Data(:,1); DegreesC=Data(:,2);
% Outputting the data in the predetermined format
fprintf(['\nImporting data using textread/dlmread and'...
         'displaying formatted table:\n']);
```

```
fprintf('|  %s  |  %s  |\n',ColHdr{:});
fprintf('|  %-6g  |  %-6.2f  |\n',[DegreesF,DegreesC]');
```

Problem 4.3 Output:

Importing data using textread/dlmread and displaying formatted table:

TempF		TempC	
70		21.11	
65		18.33	
60		15.56	
55		12.78	
50		10.00	
45		7.22	
40		4.44	
35		1.67	
30		-1.11	
25		-3.89	
20		-6.67	
15		-9.44	
10		-12.22	
5		-15.00	
0		-17.78	
-5		-20.56	
-10		-23.33	
-15		-26.11	
-20		-28.89	
-25		-31.67	
-30		-34.44	

Chapter 5 Problem Set Solutions

Problem 5.1

```
warning off
fprintf('\nProblem 5.1 Output:\n');
TCan = [ 62 50 46 44 43 41 43 45 41 36 41 42 48 51 60 68 71 61 56 ...
          55 66 72 65 70 69 59 56 54 59 67 61];
TEdw = [ 39 40 37 37 37 40 42 47 43 45 44 45 55 46 41 44 48 49 45 ...
          40 38 41 41 39 42 42 47 46 43 43 41];

% Part A
TCanavg = mean(TCan);
TEdwavg = mean(TEdw);
fprintf('\nPart a)\n')
fprintf(['Average daily temperature at Cape Canaveral in January'...
         'of 2010 was %5.2f degrees F.\n'],TCanavg);
fprintf(['Average daily temperature at Edwards AFB in January of'...
         '2010 was %5.2f degrees F.\n'],TEdwavg);

% Part B
DAYSbaCan = length(find( TCan<TCanavg));
DAYSbaEdw = length(find(TEdw<TEdwavg));
fprintf('\nPart b)\n')
fprintf(['The temperature was below average %2.0f days at'...
         'Cape Canaveral\n'], DAYSbaCan);
fprintf(['The temperature was below average %2.0f days at'...
         'Edwards AFB\n'], DAYSbaEdw);

% Part C
EdwgTCan = find(TEdw>TCan);
fprintf('\nPart c)\n')
fprintf(['The temperature at Edwards AFB was greater than the'...
         'temperature at Cape Canaveral \n %2.0f days,'...
         'specifically on \n'],length(EdwgTCan));
% To print the results you would probably use something as simple as
%          fprintf('January %2g\n',EdwgTCan);
% However, to save some space we suggest doing it in a more elegant
% way by employing the 'mod' function as follows:
lenEdwgTCan=length(EdwgTCan);
tail=mod(lenEdwgTCan,3); index=lenEdwgTCan-tail;
   if     index > 1
   fprintf('January %2g  January %2g  January  %2g\n',...
           EdwgTCan(1:index));
```

```
    end
      if     tail == 2
      fprintf('January %2g  January %2g\n',EdwgTCan (index+1:end));
      elseif tail == 1
      fprintf('January %2g\n',EdwgTCan(index+1:end));
      end
% Part D
T3 = find(abs(TEdw-TCan) <= 3);
fprintf('\nPart d)\n')
fprintf(['The temperature at Edwards AFB was within 3 degrees'...
        'of that of Cape Canaveral \n %2.0f days, specifically'...
        'on\n'],length(T3));
fprintf('January %2.0f\n',T3);
% Part E
T3E = find(abs(TEdw-TCan) <= 3 & TEdw > TCan);
fprintf('\nPart e)\n')
fprintf(['Among those %2.0f days of Part D it was hotter at'...
        'Edwards AFB on' '\n'],length(T3));
fprintf('January %2g   January %2g   January %2g January %2g\n',T3E);
```

Problem 5.1 Output:

Part a)
Average daily temperature at Cape Canaveral in January 2010
was 54.90 degrees F.
Average daily temperature at Edwards AFB in January 2010
was 42.81 degrees F.

Part b)
The temperature was below average 14 days at Cape Canaveral
The temperature was below average 16 days at Edwards AFB

Part c)
The temperature at Edwards AFB was greater than the temperature
at Cape Canaveral 6 days, specifically on
January 8 January 9 January 10
January 11 January 12 January 13

Part d)
The temperature at Edwards AFB was within 3 degrees of that of
Cape Canaveral 6 days, specifically on
January 6
January 7
January 8
January 9

January 11

January 12

Part e)

Among those 6 days of Part D it was hotter at Edwards AFB on

January 8   January 9   January 11   January 12

Problem 5.2

```
fprintf('\nProblem 5.2 Output:\n');
% You have several ways of defining your function. First, you can
%           construct an inline function object:
cossin1 = inline('cos(x).*sin(10*x)');
%     Second, you can create an anonymous function (in this case you
%     are actually creating a handle to a function):
cossin2 = @(x) cos(x).*sin(10*x);
%     Third, you may write a primary function to be stored in a
%     separate M-file. In this case the following two lines should be
%     stored as cossin.m:
%{
function y=cossin(x)
y=cos(x).*sin(10*x);
%}
% Now you can proceed with the remaining tasks as follows:
% Part A
fprintf('\nPart a)\n')
% We will use the fplot(fun,lim) call here. Remembering that
% fun must be
%         - the name of an M-file function,
%         - a string with variable x (that may be passed to eval),
%         - an inline function (which is essentially the same as above), or
%         - a function handle for an M-file or anonymous function,
% we conclude, that ANY of the following syntax should work:
fplot('cossin',[0,0.5])       % Passing the name of an M-file function
fplot('cos(x)....
    *sin(10*x)',[0,0.5])      % Passing the string with variable x
fplot(cossin1,[0,0.5])        % Passing the the inline function
fplot(@cossin,[0,0.5])        % Passing the handle for an M-file function
fplot(cossin2,[0,0.5])        % Passing the handle for an anonymous function
% Part B
fprintf('\nPart b)\n')
x0=2;                         % Defining the initial guess for 'fzero'
% In a similar manner ANY of the following syntax will find a zero
```

```
% closest to x0
zero=fzero('cossin',x0);
zero=fzero('cos(x).*sin(10*x)',x0);
zero=fzero(cossin1,x0);
zero=fzero(@cossin,x0);
zero=fzero(cossin2,x0);
fprintf(['The zero of the function (in the vicinity of x=2) is '...
        'x=%1.4f\n'],zero)
% Part C
fprintf('\nPart c)\n')
% Again, ANY of the following syntax will do the job:
[xmin,ymi]=fminbnd('cossin',0,pi);
[xmin,ymi]=fminbnd('cos(x).*sin(10*x)',0,pi);
[xmin,ymi]=fminbnd(cossin1,0,pi);
[xmin,ymi]=fminbnd(@cossin,0,pi);
[xmin,ymi]=fminbnd(cossin2,0,pi);
fprintf(['The minimum of the function (within [0;pi] range) is '...
        'y=%1.4f at x=%1.4f\n'],ymi,xmin)
% Part D
fprintf('\nPart d)\n')
% Well, we do not have 'fmaxbnd' function, therefore the only
% syntax that will work now (without changing the function) is:
[xmax,yma]=fminbnd('-cos(x).*sin(10*x)', 0.5,1.0); yma=-yma;
fprintf(['The maximum of the function (within [0.5;1.0] range) is '...
        'y=%1.4f at x=%1.4f\n'],yma,xmax)
% It is always a good idea to visualize your results. To this end, the
% following commands show all our findings on a single plot:
fplot('cossin',[0,pi])              % Plots the function
grid, hold                          % Adds a grid and holds
plot(zero,0,'pr',...                % Puts a red star at zero
    [xmin xmin],[0 ymi],'m--',...   % Draws a vertical line at a minimum
    [xmax xmax],[0 yma] ,'g-.')     % Draws a vertical line at a maximum
xlabel('x-axis')
ylabel('y-axis')                    % Adds axes names
legend(char(cossin1),'local zero','minimum','maximum')   % Adds a legend
```

Problem 5.2 Output:

```
Part a)

Part b)
The zero of the function (in the vicinity of x=2) is x=1.8850

Part c)
The minimum of the function (within [0;pi] range) is y=−0.4624 at
```

```
x=1.0810
```

```
Part d)
```

```
The maximum of the function (within [0.5;1.0] range) is y=0.7106 at
x=0.7756
```

```
Current plot held
```

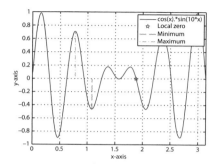

Problem 5.3

```
% The function may look as follows (we commented it to be able to
% publish to the HTML document)
```

```
%{
```

```
function [Density,Pressure,Temperature]=STatmosA(alt)
% Calculation of the 1976 standard atmosphere up to 86 km
% www.grc.nasa.gov/WWW/K-12/airplane/atmos.html
% The model has three zones with separate curve fits for the troposphere,
% the lower stratosphere, and the upper stratosphere. The units are
%   alt      - ft                    (*0.3048    to produce m)
%   Density  - slugs per cubic feet  (*515.378733 to produce kg/m^3)
%   Pressure - pounds per square feet (*47.88025898 to produce Pa)
%   Temperature  - Fahrenheit degrees
% Run ezplot('STatmosA(x)',0,130000) to plot density vs altitude
```

```
% The troposphere runs from the surface of the Earth to 36,152 feet
if alt<36152
Temperature = 59 - .00356 * alt;
Pressure    = 2116 * [(Temperature + 459.7)/ 518.6]^5.256;
elseif alt<82345
% The lower stratosphere runs from 36,152 feet to 82,345 feet
Temperature = -70;
Pressure    = 473.1 * exp(1.73 - .000048 * alt);
else
% The upper stratosphere model is used for altitudes above 82,345 feet
Temperature = -205.05 + .00164 * alt;
Pressure    = 51.97 * [(Temperature + 459.7)/ 389.98]^-11.388;
```

```
end
% In each zone the density is derived from the equation of state
Density = Pressure / [1718 * (Temperature + 459.7)];
return
%}

% Here comes another function based on the tabular data
%{
function [Density,Pressure,Temperature]=STatmosT(alt)
% Calculation of the 1976 standard atmosphere up to 86 km
% Code source:  http://www.pdas.com/programs/atmos.f90
% Run ezplot('STatmosT(x)',0,86000) to plot density vs altitude
%
alt=alt/1000;     % Convert altitude from m to km
% --- Initialize values for 1976 atmosphere
REARTH=6369.0;      % Earth radius (km), depends on Latitude
GMR=34.163195;    % Gas Constant
htab=[0.0, 11.0, 20.0, 32.0, 47.0,...
    51.0, 71.0, 84.852];              % Geometric altitude
ttab=[288.15, 216.65, 216.65, 228.65,...
    270.65, 270.65, 214.65, 186.946]; % Temperature
8.5666784E-3,1.0945601E-3,...
ptab=[1.0, 2.233611E-1, 5.403295E-2,
    6.6063531E-4, 3.9046834E-5, 3.68501E-6]; % Relative pressure
gtab=[-6.5, 0.0, 1.0, 2.8, 0.0,...
    -2.8, -2.0, 0.0];                 % Temperature gradient
P0=101325.0; Ro0=1.225;

% --- Convert geometric to geopotential altitude
    if alt>250
        alt = 100
    end
    h=alt*REARTH/(alt+REARTH);
% --- Binary search for altitude interval
    i = 1;
    j = 8;
while j > i+1
    k=fix((i+j)/2);
    if h<htab(k);
      j = k;
    else
      i = k;
    end
end
end
```

```
% --- Calculate local temperature
    tgrad  = gtab(i);
    tbase  = ttab(i);
    deltah = h - htab(i);
    tlocal = tbase + tgrad*deltah;
    theta  = tlocal/ttab(1);
% --- Calculate local pressure
if (tgrad == 0.0)
  delta = ptab(i)*exp (-GMR*deltah/tbase);      % Isothermal layers
else
  delta = ptab(i)*(tbase/tlocal)^(GMR/tgrad);   % Non-isothermal layers
end
% --- Calculate local density
sigma = delta/theta;
% --- Current atmosphere parameters corresponding to Altitude alt
Temperature=tlocal; Density=Ro0*sigma; Pressure=P0*delta;
return
%}

ezplot('STatmosA(x)',0,130000)
xlabel('Altitude, ft'), ylabel('Density, slug/ft^3')
ezplot('STatmosT(x)',0,40000)
xlabel('Altitude, m'), ylabel('Density, kg/m^3')
```

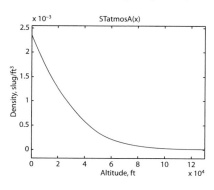

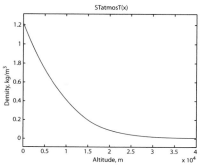

Chapter 6 Problem Set Solutions

Problem 6.1

```matlab
fprintf('\nProblem 6.1 Output:\n');
figure('color','w')
xlabel('x-axis'), ylabel('y-axis')
% We just produced emply (default) axes. The x- and y range is [0;1].
% Let us try to use the text function, that uses the data range
text(0.5,0.5,'Text')
% Obviosly, this command placed the text, so that its left bottom
% corner resides at x=0.5; y=0.5. What about the annotation command?
% Suppose we want to draw an arrow from the point (0;0) to
% the point (0.2;0.2)
annotation('arrow',[0 0.2],[0 0.2],'Color','b');
% It does not work! The arrow is drawn using the normalized figure units.
% Let us record some data about current axis...
apn=get(gca,'Position');
xlp=get(gca,'Xlim');
ylp=get(gca,'Ylim');
% ... an pass it together with some points defined in the data
% coordinates
x=[0.2;0.8]; y=[0.2;0.8];
% to the conv2norm function that performs the transformation from the data
% coordinats to the normalized figure unit (this function appears below)
[xn,yn]=conv2norm(x,y,apn,xlp,ylp);
% Now that the data point x and y are converted, let us go back to the
% annotation function and draw a box consisting of arrows
annotation('arrow',[xn(1) xn(2)],[yn(1) yn(1)]);
annotation('doublearrow',[xn(2) xn(2)],[yn(1) yn(2)],'LineWidth',3);
annotation('arrow',[xn(2) xn(1)],[yn(2) yn(2)],'HeadStyle','rose');
annotation('doublearrow',[xn(1) xn(1)],[yn(2) yn(1)]);

% The function conv2norm (saved as a separate M-file) may look like:
%{
function [xn,yn]=conv2norm(x,y,PlotPos,Xrange,Yrange)
% This function computes normalized figure coordinates of the data point(s)
% (x,y) give n in vectors x and y using the 4-element PlotPos vector
% (defining the normalized coordinates of the left) bottom corner of the
% current axes, their width and high) and also the range of axis in
% x and y directions (Xrange and Yrange, respectively)

for i=1:length(x)
% Step 1: Subtract the first element of Xrange and Yrange
pop=[x(i) y(i)]-[Xrange(1) Yrange(1)];
```

```
% Step 2: Find the relative position of the data point within the range
rel=pop./[Xrange(2)- Xrange(1) Yrange(2)- Yrange(1)];
% Step 3: Scale it to the normal coordinates
relS=[PlotPos(3) PlotPos(4)].*rel;
% Step 4: Add the coordinates of the left bottom corner
X(:,i)=[PlotPos(1) PlotPos(2)] + relS;
end
xn=X(1,:); yn=X(2,:);
%}
```

Problem 6.1 Output:

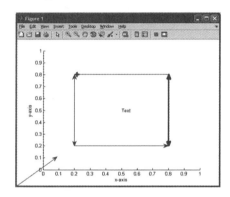

Problem 6.2

```
fprintf('\nProblem 6.2 Output:\n');
scale=1;
G= 27.7/scale; b=0.286*scale; v=.334; C=-G*b/(2*pi*(1-v));
X=linspace(-5*scale, 5*scale, 100);
Y=linspace(-5*scale, -1*scale, 100);
[x,y]=meshgrid(X,Y);
D=(x.^2+y.^2).^-2;
sigxx=C*(D.*y).* (3*(x.^2) + (y.^2));
sigyy=C*(D.*y).*   ((x.^2) - (y.^2));
tauxy=C*(D.*x).*   ((x.^2) - (y.^2));
figure('NumberTitle','off','Name','2a','Color','w')
mesh(x,y,sigxx);
xlabel('x, nm'); ylabel('y, nm'); zlabel('\sigma_{xx}, 10^{9}Pa');
figure('NumberTitle','off','Name','2b','Color','w')
mesh(x,y,sigyy);
xlabel('x, nm'); ylabel('y, nm'); zlabel('\sigma_{yy}, 10^{9}Pa');
figure('NumberTitle','off','Name','2c','Color','w')
mesh(x,y,tauxy);
xlabel('x, nm'); ylabel('y, nm'); zlabel('\tau_{xy}, 10^{9}Pa');
```

Problem 6.2 Output:

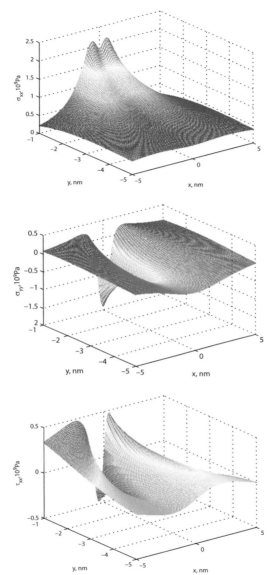

Problem 6.3

```
fprintf('\nProblem 6.3 Output:\n\n');
% First, we need to create a membrane. As opposed to the the square
% meshgrids we dealt with so far let us see how it can be done in
% a more elegant manner by using the circular meshgrid:
r = [0:0.05:1]';              % Defining a radius vector
```

```
phi = 0:pi/20:2*pi;              % Defining a phi angle vector
x = r*cos(phi);                  % Computing x-coordinates of a meshgrid
y = r*sin(phi);                  % Computing y-coordinates of a meshgrid
z = besselj...
(1,3.8316*r)*cos(phi);           % Computing z-coordinates on a meshgrid
% Comment: Note how we managed to create matrices out of the column
%          vector r and row vector phi.
% Now we can plot the membrane
figure('Color','w')
mesh(x,y,z)
xlabel('x-axis'),ylabel('y-axis'),zlabel('z-axis')
axis tight
title('The alive membrane')
% Before we create a move we must issue the following command to allow
% redrawing the membrane while it breathes:
set(gca,'nextplot','replacechildren');
% Record the movie using a smooth (sinusoidal) transition of the
% membrane state (the scale premultiplied by the scale factor
% that changes smoothly from 0 to 1 and then back to 0)
%
for j = 1:20
    mesh(x,y,sin(2*pi*j/20)*z,z);
    F(j) = getframe(gcf);
end
pause      % Pause is used here to separate the stage of movie recording
           % from playing it a couple of times
movie(F,2)
% One of the frames is is shown below:
```

Problem 6.3 Output:

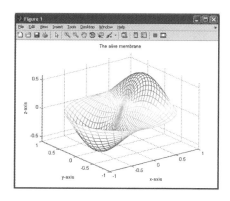

Chapter 7 Problem Set Solutions

Problem 7.1

```
fprintf('\nProblem 7.1 Output:\n\n');
clear all
syms x
Expr=(1-x)^(x^-2);
pretty(Expr)
Lr=limit(Expr,x,0,'right');
Ll=limit(Expr,x,0,'left');
fprintf('lim((1-x))^x^(-2)) as x->0+ is equal to %i\n', eval(Lr))
fprintf('lim((1-x))^x^(-2)) as x->0- is equal to %i\n', eval(Ll))
% It is always a good idea to check your results somehow.
% Let us plot the function to see if we computed it right.
figure('Color','w')
ezplot(Expr), hold
plot(0,eval(Lr),'<r')
```

Problem 7.1 Output:

```
        1
        --
         2
        x
  (1 - x)
lim((1-x))^x^(-2)) as x->0+ is equal to 0
lim((1-x))^x^(-2)) as x->0- is equal to Inf
Current plot held
```

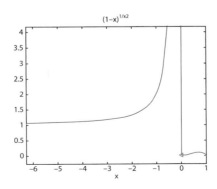

Problem 7.2

```
fprintf('\nProblem 7.2 Output:\n\n');
syms r theta a
% Defining the equation for Lemniscate in polar coordinates
f=a^2*cos(2*theta)-r^2;
% Resolving the equation for Lemniscate in Cartesian coordinates
f=subs(f, {theta,r}, {'atan(y/x)','(x^2+y^2)^(1/2)'});
f=expand(f);
disp('The equation of Lemniscate in Cartesian coordinates is')
pretty(f);
figure('Color','w')
style={'.' '--' '-.' '-'};
color='rgbm';
hold
% Plotting a family of the 2-D curves
for i=1:4
   fp=subs(f,{a},i); h_line=ezplot(fp);
   set(h_line,'Color',color(i),'LineStyle',style{i});
   text(-i+0.2,0,['a=' num2str(i,'%i')])
end
% Adding a third dimention (optionaL)
   for i=0.1:0.1:4
   fp=subs(f,{a},i);  h_line=ezplot(fp);
     xx=length(get(h_line(1),'XData'));
   set(h_line,'ZData',i*ones(xx,1));
   end
title(char(f)), axis tight
zlabel('Parameter a'), view([-10 15])
```

Problem 7.2 Output:

The equation of Lemniscate in Cartesian coordinates is

$$\frac{a^2}{y^2} - y^2 - x^2 - \frac{a^2 y^2}{x^2 + y^2}$$

$$\frac{y}{x^2} + 1$$

Current plot held

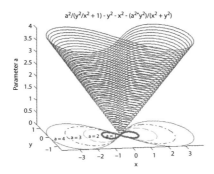

$$a^2/(y^2/x^2 + 1) - y^2 - x^2 - (a^2 y^2)/(x^2 + y^2)$$

Problem 7.3

```
fprintf('\nProblem 7.3 Output:\n\n');
[x,y]=dsolve('Dx=y','Dy=-2*y-5*x','x(0)=0','y(0)=1');
fx=fcnchk(char(x),'vectorized');
fy=inline(vectorize(char(y)));
figure('Color','w')
fplot(fx,[0, 5]),  hold
fplot(fy,[0, 5],'-.g')
for i=1:3
xz(i)=fzero(fx,2*(i-1));
yz(i)=fzero(fy,2*(i-1));
end
plot(xz,fx(xz),'rp')
plot(yz,fy(yz),'md')
xlabel('Time, s'), ylabel('x(t) and y(t)')
text(1,0.2,['x(t)=' char(x)])
text(0.2,0.7,['y(t)=' char(y)])
```

Problem 7.3 Output:

Current plot held

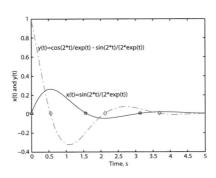

Chapter 8 Problem Set Solutions

Problem 8.1

```
fprintf('\nProblem 8.1 Output:\n\n');
%{
NumSys=[8,3,2];
k = menu('Choose the base of the number system','octal (base-8)',...
         'ternal (base-3)','binary (base-2)');
base=NumSys(k);      % Defining the base via the menu function
% Using abs and round as a safeguard from entering a negative or real number
x=input('Input nonnegative integer number to be converted: ');
x=abs(round(x));
%}
% When publishing neither the menu nor input function can be used,
% so let's define some two mumbers manually:
base=3;              % Base
x=7649734;           % Number
i=1;
    m(1)=mod(x,base);
    n=x;
    while n>0
        i=i+1;
        n=(n-m(i-1))/base;
        if n==0, break, end              % To avoid a leading zero
        m(i)=mod(n,base);
    end
disp('Here you are')
disp(num2str(m(end:-1:1)))      % num2str is optional
disp('... and here is a correct result')
disp(dec2base(x, base))         % Comparing with the correct solution
```

Problem 8.1 Output:

Here you are
1 1 2 1 0 1 1 2 2 1 1 0 1 1 1
... and here is a correct result
112101122110111

Problem 8.2

```
fprintf('\nProblem 8.2 Output:\n\n');
syms x
```

```
disp(taylor(sin(x),10)) % Shows the Maclauren series expansion
clear x       % Clears the symbolic x to reuse it as a numeric variable
x=1;
s=0;
range=1:5;
fprintf(['# of terms  Value   eps_t, %%  eps_a, %% N(eps_t)'...
          'N(eps_a)\n'])
for i=range              % Note how the loop range was determined
   ns=(-1)^(i-1)*x^(2*i-1)/factorial(2*i-1);
   s=s+ns; S(i)=s;
epst(i)=abs(s-sin(x))/sin(x)*100;
epsa(i)=abs(ns)/s*100;     % Note that the first value makes no sence
nsa(i)=fix(2-log10(2*epsa(i)));
nst(i)=fix(2-log10(2*epst(i)));      % The same here
end
fprintf('%6g  %9.4f %9.4f %9.2f  %6g %8g\n', [range',...
         S',epst',epsa',nst',nsa']')
figure('Color','w')
subplot(211)                  % Plotting the results
plot([1 5],sin(x)*[1 1],'-.r',range,S,'o')
xlabel('Number of terms in the series')
ylabel('Approximate of sin(1)')
legend('True value','Estimate')
subplot(212)
semilogy(range,epst,'-.r',range(2:end),epsa(2:end),'o')
xlabel('Number of terms in the series'), ylabel('Error, %')
legend('True error','Error''s estimate')
```

Problem 8.2 Output:

$x^9/362880 - x^7/5040 + x^5/120 - x^3/6 + x$

No. of terms	Value	eps_t, %	eps_a, %	N(eps_t)	N(eps_a)
1	1.0000	18.8395	100.00	0	-0
2	0.8333	0.9671	20.00	1	0
3	0.8417	0.0233	0.99	3	1
4	0.8415	0.0003	0.02	5	3
5	0.8415	0.0000	0.00	7	5

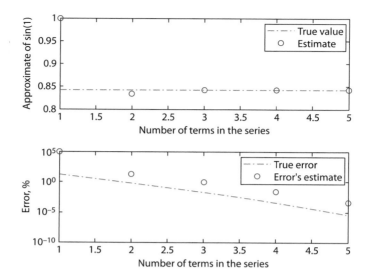

Problem 8.3

```
fprintf('\nProblem 8.3 Output:\n\n');
x=10;                  % Defines the size of an integration interval
It=x^4/4;              % Defines the exact value
    for j=1:7          % Computes an approximate value
    N=10^j;            % Defines the number of intervals
    h=x/N;             % Defines the step size
    I=0;
    tic
        for i=1:N
            I=I+(i-0.5)^3;
        end
    I=I*h^4;           % Extimates the integral
    t=toc;
    T(j)=t;            % Stores the value of T_CPU
    Er(j)=abs(It-I);   % Computes and stores the error
    H(j)=h;            % Stores the value of the istep size h(n)
    figure('Color','w')
    end
loglog(H,T,'-.ob',H,Er,'-.gs',H,Er+0.01*T,'r')
xlabel('Log Step Size'), ylabel('Log Value')
legend('t_{CPU}, s','Error \epsilon','Penalty P',4)
% The plot resembles that of Fig.8.18, indicating that h~0.01
% corresponds to the point of diminishing returns
```

Problem 8.3 Output:

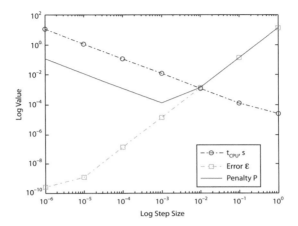

Chapter 9 Problem Set Solutions

Problem 9.1

```
fprintf('\nProblem 9.1 Output:\n\n');
% Here is an example of the Gauss-Jordan elimination procedure
A=[0   0   3 5;...
   3   7  -4 5;...
   5  -6   7 8;...
   0   4   0 5];
b=rand(4,1);
disp('Augmented matrix')
Ab=horzcat(A,b)       % Creating an augmented matrix
% Getting the solution using /
xG=A\b;
% Getting the solution using inv
xI=inv(A)*b;
% Getting the solution from the reduced row echelon form
Abc=rref(Ab); Nc=length(A); xR=Abc(:,Nc+1);
% Getting the solution using the linsolve function
xl=linsolve(A,b);
% Getting the solution using the LU decomposition
[l,u]=lu(A); c=l\b; xL=u\c;
% Getting the solution using the QR decomposition
[q,r]=qr(A); y=q'*b; xQ=r\y;
disp('Side-by-side comparison between six solutions')
disp([xG xI xR xl xL xQ])
disp('Verification of the solution')
  if A*xG==b
    fprintf('Whether Ax=b?    Yes. It is true.\n')
  else
    fprintf(['Whether Ax=b?    No. Somehow Ax~=b. Check the'...
                           'solution.\n'])
  end
% Comment: The problem you run into checking the result is that
%          solutions are only good up to the round-off error. Hence,
fprintf('\nWhether A*x=b?\n')
A*xG==b
% do not returns ones. The following shows how to avoid this type
% of problem by using different ways of comparing approximate
% quantities as opposed to the straight in comparison as in A*x==b.
% We start by visualizing the result (expecting to have 1's, not 0's)
figure('Color','w')
```

```
subplot(2,2,1), spy(A*xG==b)
title('\bfA*x==b')
% Let us have a close look
fprintf('\nWhether A*x-b?\n')
A*xG-b
% reveals the reason, - roundoff errors (as you may recall
% eps=2.2204e-016, i.e. has about the same order.
subplot(2,2,2)
%spy(A*xG-b),
imagesc(A*xG-b)
title('\bfA*x-b')
colorbar
% You may want to try something else, for instance
fprintf('\nWhether A*x=b now?\n')
single(A*xG)==single(b)
% which returns all 1's
subplot(2,2,3), spy(ans)
title('\bfsingle(A*x)==single(b)')
% Another way of proving the equality would be issuing A*x./b or
% A*x.\b command:
fprintf('\nWhether A*x./b returns all 1''s?\n')
A*xG./b
subplot(2,2,4), spy(ans)
title('\bfA*x./b')
```

Problem 9.1 Output:

Augmented matrix

Ab = Columns 1-5

0	0	3.0000	5.0000	0.8147
3.0000	7.0000	-4.0000	5.0000	0.9058
5.0000	-6.0000	7.0000	8.0000	0.1270
0	4.0000	0	5.0000	0.9134

Side-by-side comparison between six solutions

Columns 1-6

-0.8716	-0.8716	-0.8716	-0.8716	-0.8716	-0.8716
-1.0610	-1.0610	-1.0610	-1.0610	-1.0610	-1.0610
-1.4475	-1.4475	-1.4475	-1.4475	-1.4475	-1.4475
1.0315	1.0315	1.0315	1.0315	1.0315	1.0315

Verification of the solution

Whether Ax=b?   No. Somehow Ax~=b. Check the solution.

Whether A*x=b?

```
ans = 0
        0
        0
        0
```

Whether A*x-b?

```
ans = 1.0e-014*
       -0.0333
        0.2109
        0.0111
        0.0111
```

Whether A*x=b now?

```
ans = 1
       1
       1
       1
```

Whether A*x./b returns all 1's?

```
ans = 1.0000
       1.0000
       1.0000
       1.0000
```

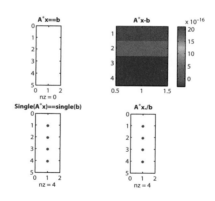

Problem 9.2

```
fprintf('\nProblem 9.2 Output:\n\n');

% Let us first compare two methods for the small system
A=magic(5);
b=[1;5;3;5;9];
tic
xG=A\b;
```

```
tG=toc;
    tic
    xC=cramer(A,b);
    tC=toc;
disp('For 5 equations:')
disp('Side-by-side comparison between two solutions')
disp([xG xC])
disp('Error between them')
% Maximum error (will not work for permutated solutions)
disp(max(xG-xC))
disp('CPU time spent solving a system of LAEs')
disp([tG tC])

% Now let us test them on a large system
A=rand(300);
b=[1:300]';
tic
xG=A\b;
tG=toc;
    tic
    xC=cramer(A,b);
    tC=toc;
disp('For 300 equations:')
%disp('Side-by-side comparison between two solutions')
%disp([xG xC])
disp('Error between two solutions')
disp(max(xG-xC))    % Maximum error
disp('CPU time spent solving a system of LAEs')
disp([tG tC])

%The cramer function (to be stored as a separate M-file) may look
%as follows:
%{
function x=cramer(A,b)
% This function solves a system of LAEs defined via the square matrix A
% and vector b using the Cramer's rule.
N=length(b);
b=reshape(b,N,1); % Reshapes b to the column vector
detA=det(A);
x=zeros(N,1);
for i=1:N
    x(i)=det([A(:,1:i-1) b A(:,i+1:N)])/detA;
end
%}
```

```
Problem 9.2 Output:
```

```
For five equations:
Side-by-side comparison between two solutions
        0.1810      0.1810
       -0.0805     -0.0805
        0.3374      0.3374
       -0.1113     -0.1113
        0.0272      0.0272

Error between them
   5.5511e-017

CPU time spent solving a system of LAEs
        0.0002      0.0050

For 300 equations:
Error between two solutions
   1.5802e-010

CPU time spent solving a system of LAEs
        0.0499      2.8784
```

Problem 9.3

The system of LAEs can be written in the following form:

$$f_1 + f_2 = 100$$

$$f_1 + f_2 + f_3 + f_4 = 600$$

$$f_3 + f_4 + f_5 + f_6 = 800$$

$$f_5 + f_6 + f_7 + f_8 = 800$$

$$f_1 - f_2 = 0$$

$$f_7 - f_8 = 100$$

$$f_6 = 200$$

Therefore,

```
fprintf('\nProblem 9.3 Output:\n\n');

A=[1  1 0 0 0 0 0  0;   % Total cargo handled in MRY
   1  1 1 1 0 0 0  0;   % Total cargo handled in SFO
```

```
     0   0 1 1 1 1 0  0;    % Total cargo handled in ORD
     0   0 0 0 1 1 1  1;    % Total cargo handled in IAD
     1  -1 0 0 0 0 0  0;    % Cargo left in MRY
     0   0 0 0 0 0 1 -1;    % Cargo left in IAD
     0   0 0 0 0 1 0  0];   % Cargo that left IAD on its way to ORD
b=[100; 600; 800; 800; 0; 100; 200];
disp('Solution of the original system of LAEs')
f=A\b                      % Attempting to find a solution
disp('Whether this solution is unique?')
rankA=rank (A)             % Checking a rank of the matrix A
disp(rref([A b]))          % Analyzing a solution
% Obviously, the above solution is not unique. As of now the system has
% infinite number of solutions. The particular solution obtained
% above can be combined with the solution of homogeneous system to
% produce a general solution. As explained in Section 9.7.1
% the solution of homogeneous system
% can be found for instance using the null function
xh=null(A);
disp(['All possible solutions of the problem can be described'...
      'parametrically as'])
for i=1:8
  if i ~= 4
  fprintf('   |f(%i)|   |%3.0f|        |%7.2f|\n',i,f(i),xh(i))
  else
  fprintf('   |f(%i)| = |%3.0f| + p * |%7.2f|\n',i,f(i),xh(i))
  end
end
% As seen only two cargo trafics are affected, therefore let us
% define cargo trafic f_4 explicitly, which adds another equation
disp('Adding another equation')
A(8,:)=[0 0 0 1 0 0 0 0]; % Cargo that left ORD on its way to SFO
b(8)=200;
% Check whether the matrix A has a full rank
rankA=rank(A)
% Yes, it does! Therefore the unique solution can be found as follows:
disp('The unique solution')
f=A\b
% which can be confirmed with the reduced row eschelon form
disp(rref([A b]))
```

Problem 9.3 Output:

Solution of the original system of LAEs

```
f = 50.0000
    50.0000
   500.0000
         0
   100.0000
   200.0000
   300.0000
   200.0000
```

Whether this solution is unique?

```
rankA = 7

  Columns 1-9
```

1	0	0	0	0	0	0	0	50
0	1	0	0	0	0	0	0	50
0	0	1	1	0	0	0	0	500
0	0	0	0	1	0	0	0	100
0	0	0	0	0	1	0	0	200
0	0	0	0	0	0	1	0	300
0	0	0	0	0	0	0	1	200

All possible solutions of the problem can be described parametrically as

```
|f(1)|     | 50|          |  0.00|
|f(2)|     | 50|          |  0.00|
|f(3)|     |500|          | -0.71|
|f(4)| = |  0| + p *  |  0.71|
|f(5)|     |100|          |  0.00|
|f(6)|     |200|          |  0.00|
|f(7)|     |300|          | -0.00|
|f(8)|     |200|          | -0.00|
```

Adding another equation

```
rankA = 8
```

The unique solution

```
f = 50
    50
   300
   200
   100
   200
   300
   200
```

Columns 1-9

1	0	0	0	0	0	0	0	50
0	1	0	0	0	0	0	0	50
0	0	1	0	0	0	0	0	300
0	0	0	1	0	0	0	0	200
0	0	0	0	1	0	0	0	100
0	0	0	0	0	1	0	0	200
0	0	0	0	0	0	1	0	300
0	0	0	0	0	0	0	1	200

Chapter 10 Problem Set Solutions

Problem 10.1

```
fprintf('\nProblem 10.1 Output:\n\n');
figure('Color','w')
    tic
    disp(200^(1/3))
    t(1)=toc;
tic
disp(roots([1 0 0 -200]))
t(2)=toc;
    tic
    disp(fzero('x^3-200',7))
    t(3)=toc;
tic
disp(fminbnd('(x^3-200)^2',3,10))
t(4)=toc;
    tic
    disp(fminsearch('(x^3-200)^2',7))
    t(5)=toc;
tic
disp(fminunc('(x^3-200)^2',7))
t(6)=toc;
    tic
    disp(nthroot(200,3))
    t(7)=toc;
tic
disp(power(200,1/3))
t(8)=toc;
figure('Color','w')
subplot(211), ezplot('x^3-200')
subplot(212), plot(t,'p')
xlabel('Method'), ylabel('t_{CPU}')
set(gca,'XTickLabel',{'200^(1/3)';'roots';'fzero';'fminbnd';'fminsearch';...
'fminunc';'nthroot';'power'})

[x,fval,exitflag,output]=fminbnd('(x^3-200)^2',3,10);
disp(output.algorithm), disp(output.funcCount)
[x,fval,exitflag,output]=fminsearch('(x^3-200)^2',7);
disp(output.algorithm), disp(output.funcCount)
[x,fval,exitflag,output]=fminunc('(x^3-200)^2',7);
disp(output.algorithm), disp(output.funcCount)
```

```
[x,fval,exitflag,output]=fminbnd('(x^3-200)^2',3,10);
disp(output.algorithm), disp(output.funcCount)
```

Problem 10.1 Output:

```
   5.8480
  -2.9240 + 5.0645i
  -2.9240 - 5.0645i
   5.8480
   5.8480
   5.8480
   5.8481
```

Local minimum found.

Optimization completed because the size of the gradient is less than the default value of the function tolerance.

```
   5.8480
   5.8480
   5.8480
```

golden section search, parabolic interpolation
```
      13
```

Nelder-Mead simplex direct search
```
      32
```

Local minimum found.

Optimization completed because the size of the gradient is less than the default value of the function tolerance.

medium-scale: Quasi-Newton line search
```
      12
```

golden section search, parabolic interpolation
```
      13
```

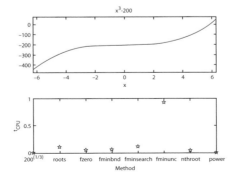

Problem 10.2

```
fprintf('\nProblem 10.2 Output:\n\n');
% ezplot('4*cos(2*x)-exp(0.5*x)+5');
interval=[-0.5 4.5];
% Bracketing all roots within the interval that was determined by
% looking at the function plot
brackets = lroot('4*cos(2*x)-exp(0.5*x)+5',interval(1),interval(2));
% Creating a plot of the given function
ezplot('4*cos(2*x)-exp(0.5*x)+5',interval); hold
xlabel('x'); ylabel('4cos(2x)-e^{0.5x}+5');
plot(interval,[0 0],'--g');    % Adding a zero line
title('Finding Function''s Zeros');
% Finding the zeros in each subrange defined by the lroot function
lb=length(brackets);
for i=1:lb;
root(i) = fzero('4*cos(2*x)-exp(0.5*x)+5',...
[brackets(i,1) brackets(i,2)]);
end
% Displaying the roots of the given equation & plot them on the
% graph of the function
fprintf('\nThe roots of the equation are:\n')
for p=1:lb;
    fprintf('Root %1.0g: %4.4f\n',p,root(:,p));
    plot(root(:,p),0,'*','MarkerEdgeColor','r','MarkerFaceColor','r');
    text(root(:,p)-0.2,1.2,...
    num2str(root(:,p), '%4.4f'),'BackgroundColor','w')
end
% Comment: The lroot function (stored in a separate M-file)
%          bracketing all roots was presented in Section 10.5.1
```

Problem 10.2 Output:

Current plot held

The roots of the equation are:

Root 1: 1.2374

Root 2: 2.0665

Root 3: 3.7375

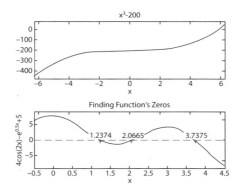

Problem 10.3

```
fprintf('\nProblem 10.3 Output:\n\n');
% Defining the function and accuracy
func = 'cos(x)-2*x^3';
tol   = 1e-5;
% Defining the interval where to look for the root
% ezplot(func);
interval=[0 1];
% Plotting the function
figure('Color','w')
ezplot(func,interval);   hold
plot(interval,[0 0],'--g');        % Adding a zero line
title('Root Finding Using Different Methods');
xlabel('x'), ylabel('cos(x)-2x^3')
% Part A
a = -4;
b = 2;
Xfp = falseposition(func,a,b,tol);
plot(Xfp,0,'or');
fprintf(['\nThe root found using the falsepostion method: x='...
         '%6.6f\n'], Xfp);
% Part B
a = -4;
b = -3.8;
Xsec = secant(func,a,b,tol);
plot(Xsec, 0, '*r');
fprintf('The root found using the secant method is x= %6.6f\n',Xsec);
% The user-developed functions falseposition and secant (stored
% separately) may look like follows:
%{
```

```
function x = falseposition(func,a,c,eps)
f=fcnchk(func);
% Evaluating the initial points
fa = feval(f,a);
fc = feval(f,c);
% Checking the initial points to ensure they are on opposite sides of
% the x-axis and if so, computing an intermediate point
     if fa*fc<=0
     b = c - fc*(a-c)/(fa - fc);
     fb = feval(f, b);
     midpt = (a+c)/2;
     fm = feval(f,midpt);
% Checking what interval to choose
   if abs(b-a)<eps
   x = b;
   else
     if fa*fb<=0
       if fa*fm>0
       a = midpt;
       end
     x = falseposition(f,a,b,eps);
     else
       if fc*fm>0
       c = midpt;
       end
     x = falseposition(f,b,c,eps);
     end
   end
  else
  error('Error: Check the graph and redefine the interval!');
end
%*******************
function x = secant(func,x1,x2,eps)
f=fcnchk(func);
% Evaluating the initial estimates
fx1=feval(f,x1);
fx2=feval(f,x2);
% Using the initial estimates to calculate the next estimate
x3=x2-fx2*(x2-x1)/(fx2-fx1);
fx3=feval(f,x3);
% Checking for convergence
  if abs(x3-x2)<=eps
  x = x3;
```

```
  else
  x = secant(f,x2,x3,eps);       % Recursive call of the function
  end
%}
```

Problem 10.3 Output:

Current plot held

The root found using the falsepostion method: x= 0.721406

The root found using the secant method is   x= 0.721406

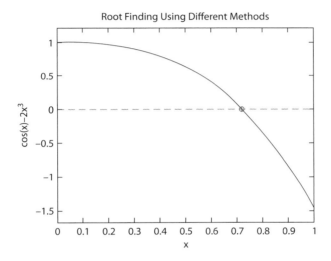

Chapter 11 Problem Set Solutions

Problem 11.1

The equation we are dealing with is

$$\mu = \frac{CT^{2/3}}{T+C}$$

We need to rewrite it so that we could use the `polyfit` function. To do so, let us divide both sides of the original equation by

$$T^{2/3}$$

and then invert both sides. That leads to

$$\frac{T^{2/3}}{\mu} = \frac{T}{C} + \frac{S}{C}$$

Now, we can use the `polyfit` function as shown below.

```
fprintf('\nProblem 11.1 Output:\n\n');
% Defining the data points
TC = [-20 0 40 100 200 300 400 500 1000]; TK=TC+273.1;
mu = [1.63 1.71 1.87 2.17 2.53 2.98 3.32 3.64 5.04];
TKm=linspace(TK(1),TK(length(TK)),100);
% Plotting the table data
figure('Color','w')
subplot(3,1,1)
semilogy(TC,mu,'mo')
axis([-200 1200 1.5 5.5]), hold
title('Viscosity of air vs. temperature')
ylabel('\mu, N s/m^2 x 10^5')
% Finding the coefficients of a linear regression
P=polyfit(TK,TK.^(2/3)./mu,1);
c=1/P(1);
s=P(2)/P(1);
M = c*TK.^(2/3)./(TK+s);
% Drawing a Sutherland's Curve
mu2=c*(TKm.^(2/3))./(TKm+s);
semilogy(TKm-273.15,mu2,'b--')
legend ('Data points','Sutherland''s curve',0)
```

```
% Plotting the linear regression
subplot(3,1,2)
plot(TK-273.15,TK.^(2/3)./mu,'mo')
hold
plot(TKm-273.15,P(1)*TKm+P(2),'g')
axis([-200 1200 22 25])
text(220,24.5,['C=' num2str(c) '   S=' num2str(s)])
title ('Linear regression')
ylabel ('T^{2/3}/\mu')
% Computing and plotting the errors of the linear regression
subplot(3,1,3)
bar(TK-273.15,TK.^(2/3)./mu-TK.^(2/3)./M,'g')
axis([-200 1200 -1 1])
E = mu-M; ybar = mean(mu);      % Computing errors
St = sum((ybar-mu).^2); Sr=sum(E.^2);
R2 = (St-Sr)/St;                % Computing the correlation coefficient
text(220,0.5,['R^2=' num2str(R2)])
xlabel('Temperature, ^oC'), ylabel('Regression errors')
title('Errors at the table data points')
```

Problem 11.1 Output:

Current plot held
Current plot held

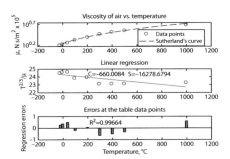

Problem 11.2

```
fprintf('\nProblem 11.2 Output:\n');
figure
set(gcf,'Color','white')            % Setting figure's background color to white
% Part a)
x=[0 2 4 5 7.5 10];
y=exp(-x/6).*cos(x);
```

```
xx=linspace(x(1),x(end),100);  % Defining xx for approximations
NAK=spline(x,y,xx);            % Computing not-a-knot spline
CLAMP=spline(x,[-1 y 0],xx);   % Computing clamped spline
% Comment: Note how we ajusted the derivative at the initial point
% to be -1.
subplot(2,1,1)
plot(x,y,'pr')                 % Plotting data points: red pentagram
hold on
plot(xx,NAK,'g')               % Adding not-a-knot spline: green line
plot(xx,CLAMP,'b-.')           % Adding clamped spline: blue dashed line
grid on
xlabel('x-axis'), ylabel('y-axis')
title(['Spline approximations for the data points belonging to '...
        'y(x)=e^{(-x/6)}*cos(x)'])
legend('Experimental data','Not-a-knot spline','Clamped spline',0)
clear CLAMP NAK h len step x xx y % Clearing all variables from Part a)
% Part b)
x=linspace(0,1.5,10);
y=humps(x);
len=length(x); h=mean(x);
step=h/(10*len);               % Dividing by 10 for smoothness
xx=[x(1)-step:step:x(len)+step]; % Defining xx for approximations
NAK=spline(x,y,xx);            % Computing not-a-knot spline
CLAMP=spline(x,[100 y 0],xx);  % Computing clamped spline
% Comment: Note how we ajusted the derivative at the initial point
% to be 100.
subplot(2,1,2)
plot(x,y,'pr')                 % Plotting data points: red pentagram
hold on
plot(xx,NAK,'g')               % Adding not-a-knot spline: green line
handle=plot(xx,CLAMP,'b-.');   % Adding clamped spline: blue dashed line
grid on
xlim([x(1) x(end)]);
xlabel('x-axis'), ylabel('y-axis')
title(['Spline approximations for the data points belonging to '...
        'y(x)=humps(x)'])
legend('Experimental data','Not-a-knot spline','Clamped spline',0)
% This completes the script and the result will look like:

% For advanced users:
% You might notice that we used a handle when we plotted a
%         clamped spline for the second case. Actually, there is no need
%         for that. The reason we did it is to show you how easily you
%         could add some interactivity to your plot To this end, by
```

```
%        issuing the two following self-explanatory commands we are
%        adding two sliders:
h_slider1=uicontrol('Style','Slider','Units','Normalized',...
'Position',[0.160 0.121 0.2 0.05],'Value',0.5,'Callback',...
'sliders_Callback');
h_slider2=uicontrol('Style','Slider','Units','Normalized',...
'Position',[0.675 0.121 0.2 0.05],'Value',0.5,'Callback',...
'sliders_Callback');
%     Now, every time you move one of the sliders, a separately stored
%       set of commands, changing the derivatives and redrawing
%       the spline will be executed.
% The content of this sliders_Callback.m file is as follows:
%{
% Computing clamp spline with new initial and final gradients and
% redrawing it
yi=4000*(get(h_slider1,'Value')-0.5);
yf=2000*(get(h_slider2,'Value')-0.5);
CLAMP=spline(x,[yi y yf],xx);
set(handle,'Ydata',CLAMP)
%}
```

Problem 11.2 Output:

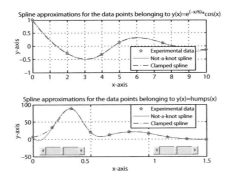

Problem 11.3

```
fprintf('\nProblem 11.3 Output:\n\n');
% Inputting data
NHarm=4;                          % Number of harmonics
Harmonic(1)=40;                   % First harmonic (Hz)
Harmonic(2)=70;                   % Second harmonic (Hz)
Harmonic(3)=170;                  % Second harmonic (Hz)
Harmonic(4)=210;                  % Second harmonic (Hz)
NoiseLevel=2;                     % Noise-to-signal ratio
```

```matlab
% Creating a noisy signal from four harmonics and zero-mean random noise
t = 0:0.001:0.6;                    % Considering data sampled at 1000 Hz
lt=size(t);
ti = 0:0.0005:0.6;
lti=size(ti);
x = zeros(lt);
for i=1:NHarm
x = x+sin(2*pi*Harmonic(i)*t);   % Composing the signal
end
y = x + NoiseLevel*randn(lt);    % Introducing signal noise
figure('Color','w')
subplot(1,2,1)
plot(1000*ti(1:100),interp1(t,x,ti(1:100),...
     'spline'),'--g')               % Showing the original signal
hold
plot(1000*t(1:50),...
        y(1:50),'.b')               % Showing the corrupted signal
title('Noisy signal')
xlabel('Time, \itMilliseconds'), ylabel ('Signal')
axis tight
% Extract signal power spectrum using FFT
Y = fft(y,512);                     % The 512-point fast Fourier transform
Pyy = 2*Y.*conj(Y)512;              % Computing the power
f = 1000*(0:256)/512;
% Plotting a power spectrum of the corrupt signal and finding
% major harmonics
subplot(1,2,2)
plot(f,Pyy(1:257))                          % Plotting power spectrum
hold
title('Power spectrum and major harmonics')
xlabel('Frequency, \itHz'), ylabel('Power spectrum')
xlim([0 500])
% Finding harmonics
for i=1:NHarm
[peak(i),ifreq(i)]=max(Pyy(1:257));  % Finding the i-th harmonic
Pyy(ifreq(i)-10:ifreq(i)+10)=0;         % Excluding the i-th harmonic
freq(i)=(ifreq(i)-1)*1000/512;          % Finding the i-th frequency
end
[freq,Ind] = sort(freq);                % Rearranging harmonics in the
                                        % ascending order
peak=peak(Ind);
z=zeros(lti);
```

```
% Marking harmonics and reconstructing the original signal
for i=NHarm:-1:1
plot([freq(i) freq(i)],[0 peak(i)],'--r') % Marking harmonics on the plot
text(freq(i)-20,peak(i)-20,['Harmonic at ' num2str(freq(i),'%.0f')...
    '\itHz'], 'BackgroundColor','w');
z=z+sqrt(peak(i)/peak(1))*sin(2*pi*freq(i)*ti);
% Computing reconstructed signal
end
% Showing reconstructed signal on the first subplot
subplot(1,2,1)
plot(1000*ti(1:100),z(1:100),'-.r')
legend('Incorrupt signal','Corrupt signal','FFT approximation',0)
% Comment: Every time you run this script you are going to get a
% different result because you are adding a random poise).
% Therefore, you figure may be different from what is shown below.
```

Problem 11.3 Output:

Current plot held
Current plot held

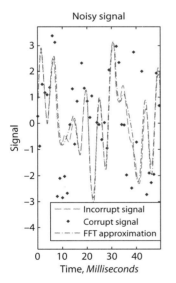

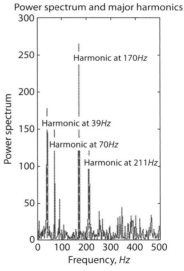

Chapter 12 Problem Set Solutions

Problem 12.1

```
fprintf('\nProblem 12.1 Output:\n\n');
% The three points needed in calculations
x(2)=pi/4; h=pi/100;
x(1)=x(2)-h;
x(3)=x(2)+h;
f=cos(x);
% Analytically determined derivatives
f_pa=-sin(x(2));
f_2pa=-cos(x(2));
f_3pa=sin(x(2));
f_4pa=cos(x(2));
% The two-point centered difference approximation of the
% first-order derivative
d1=(f(3)-f(1))/2/h;
e1=-h^2*f_3pa/6;
% The two-point centered difference approximation of the
% second-order derivative
d2=diff(f,2)/h^2;
e2=-h^2*f_4pa/12;
% Displaying the results
fprintf('For the two-point centered-difference approximation of\n')
fprintf('the 1st-order derivative of cos(x) at x=pi/4 with h=pi/100:\n');
fprintf('Exact value:      %15.10f\n',f_pa);
fprintf('Estimate:         %15.10f\n',d1);
fprintf('Actual error:     %15.10f or %6.3f%%\n',f_pa-d1,...
                           (f_pa-d1)/f_pa*100);
fprintf('Error estimate: %15.10f or %6.3f%%\n',e1,e1/f_pa*100);
fprintf('\nFor the two-point centered-difference approximation of\n')
fprintf(['the 2nd-order derivative of cos(x) at x=pi/4 with'...
         'h=pi/100:\n']);
fprintf('Exact value:      %15.10f\n',f_2pa);
fprintf('Estimate is:      %15.10f\n',d2);
fprintf('Actual error:     %15.10f or %6.4f%%\n',f_2pa-d2,...
                           (f_pa-d2)/f_pa*100);
fprintf('Error estimate: %15.10f or %6.4f%%\n', e2,e2/f_2pa*100);
% Estimating higher-order derivatives using the
% multiple-application scheme (see Eq.(12.34))
fprintf(['\nUsing multiple-application scheme yields the'...
         'following estimates:\n']);
```

```
d1d=(cos(x(2)+2*h)-cos(x(2)-2*h))/2/2/h;
e1m=(d1-d1d)/(2^2-1);
fprintf('For the 1st-order derivative: %15.10f or %6.3f%%\n',...
        e1m,e1m/f_pa*100);
d2d=(cos(x(2)+2*h)-2*f(2)+cos(x(2)-2*h))/(2*h)^2;
e2m=(d2-d2d)/(2^2-1);
fprintf('For the 2nd-order derivative: %15.10f or %6.4f%%\n',...
        e2m,e2m/f_2pa*100);
fprintf('\nThey happen to be very accurate!\n');
```

Problem 12.1 Output:

For the two-point centered-difference approximation of the first-order
derivative of cos(x) at x=π/4 with h=π/100:

Exact value:	*-0.7071067812*
Estimate:	*-0.7069904725*
Actual error:	*-0.0001163087 or 0.016%*
Error estimate:	*-0.0001163144 or 0.016%*

For the two-point centered-difference approximation of the second-
order derivative of cos(x) at x=π/4 with h=π/100:

Exact value:	*-0.7071067812*
Estimate is:	*-0.7070486259*
Actual error:	*-0.0000581553 or 0.0082%*
Error estimate:	*-0.0000581572 or 0.0082%*

Using multiple-application scheme yields the following estimates:
For the first-order derivative: -0.0001162857 or 0.016%
For the second-order derivative: -0.0000581476 or 0.0082%

They happen to be very accurate!

Problem 12.2

```
fprintf('\nProblem 12.2 Output:\n\n');
% Defining symbolic variables
syms h f_i f_ip1 f_ip2 f_ip3 f_im1 f_im2 f_im3
syms f_p f_2p f_3p f_4p f_5p f_6p f_7p f_8p
% Introducing Taylor series expansion for six points keeping eight terms
Eqp3 =-f_ip3+f_i+3*h*f_p+(3*h)^2*f_2p/2+(3*h) ^3*f_3p/factorial(3)+...
        (3*h)^4*f_4p/factorial(4)+(3*h)^5*f_5p/factorial(5)+...
        (3*h)^6*f_6p/factorial(6)+(3*h)^7*f_7p/factorial(7);
Eqp2 =-f_ip2+f_i+2*h*f_p+(2*h)^2*f_2p/2+(2*h)...
        ^3*f_3p/factorial(3)+(2*h)^4*f_4p/factorial(4)+(2*h)^5*f_5p/...
        factorial(5)+(2*h)^6*f_6p/factorial(6)+(2*h)^7*f_7p/...
        factorial(7);
```

```
Eqp1 =-f_ip1+f_i+h*f_p+h^2*f_2p/2+h^3*f_3p/factorial(3)+...
        h^4*f_4p/24+h^5*f_5p/ factorial(5)+h^6*f_6p/factorial(6)+...
        h^7*f_7p/factorial(7);
Eqm1 =-f_im1+f_i-h*f_p+h^2*f_2p/2-h^3*f_3p/factorial(3)+...
        h^4*f_4p/24-h^5*f_5p/factorial(5)+h^6*f_6p/factorial(6)-...
        h^7*f_7p/factorial(7);
Eqm2 =-f_im2+f_i-2*h*f_p+(2*h)^2*f_2p/2-(2*h)^3*f_3p/factorial(3)+...
        (2*h)^4*f_4p/factorial(4)-(2*h)^5*f_5p/factorial(5)+...
        (2*h)^6*f_6p/factorial(6)-(2*h)^7*f_7p/factorial(7);
Eqm3 =-f_im3+f_i-3*h*f_p+(3*h)^2*f_2p/2-(3*h)^3*f_3p/factorial(3)+...
        (3*h)^4*f_4p/factorial(4)-(3*h)^5*f_5p/factorial(5)+...
        (3*h)^6*f_6p/factorial(6)-(3*h)^7*f_7p/factorial(7);
% Resolving six equations for f_p, f_2p, f_3p, f_4p, f_5p and f_6p
sol=solve(Eqp3,Eqp2,Eqp1,Eqm1,Eqm2,Eqm3,f_p, f_2p,f_3p,f_4p,f_5p, f_6p);
% Simplifying and displaying the results
disp('1st-order derivative'), r=collect(sol.f_p,h); pretty(r)
disp('2nd-order derivative'), r=collect(sol.f_2p,h); pretty(r)
disp('3rd-order derivative'), r=collect(sol.f_3p,h); pretty(r)
disp('4th-order derivative'), r=collect(sol.f_4p,h); pretty(r)
disp('5th-order derivative'), r=collect(sol.f_5p,h); pretty(r)
disp('6th-order derivative'), r=collect(sol.f_6p,h); pretty(r)
% Reviewing the results and making the necessary changes
disp('')
disp(['As seen because of cancellations for even-order'...
      'derivatives (2, 4 and 6)'])
disp(['more terms in the Taylor series expansion are needed'...
      'to obtain the errors'])
disp('')
Eqp3 = Eqp3 + (3*h)^8*f_8p/factorial(8);
Eqp2 = Eqp2 + (2*h)^8*f_8p/factorial(8);
Eqp1 = Eqp1 +     h^8*f_8p/factorial(8);
Eqm1 = Eqm1 +     h^8*f_8p/factorial(8);
Eqm2 = Eqm2 + (2*h)^8*f_8p/factorial(8);
Eqm3 = Eqm3 + (3*h)^8*f_8p/factorial(8);
sol=solve(Eqp3,Eqp2,Eqp1,Eqm1,Eqm2,Eqm3,f_p,...
          f_2p,f_3p,f_4p,f_5p,f_6p);
disp('1st-order derivative with the error estimate')
pretty(collect(sol.f_p,h))
disp('2nd-order derivative with the error estimate')
pretty(collect(sol.f_2p,h))
disp('3rd-order derivative with the error estimate')
pretty(collect(sol.f_3p,h))
disp('4th-order derivative with the error estimate')
```

```
pretty(collect(sol.f_4p,h))
disp('5th-order derivative with the error estimate')
pretty(collect(sol.f_5p,h))
disp('6th-order derivative with the error estimate')
pretty(collect(sol.f_6p,h))
```

Problem 12.2 Output:

First-order derivative

$$
(((-3) \ f\_7p) \ h^7 + 63 \ f\_im2 \ - \ 315 \ f\_im1 \ -
$$

$$
7 \ f\_im3 \ + \ 315 \ f\_ip1 \ - \ 63 \ f\_ip2 \ + \ 7 \ f\_ip3) \ /
$$

$$
(420 \ h)
$$

Second-order derivative

$$
(270 \ f\_im1 \ - \ 490 \ f\_i \ - \ 27 \ f\_im2 \ + \ 2 \ f\_im3 \ +
$$

$$
270 \ f\_ip1 \ - \ 27 \ f\_ip2 \ + \ 2 \ f\_ip3) \ / \ (180 \ h^2)
$$

Third-order derivative

$$
((7 \ f\_7p) \ h^7 + 195 \ f\_im1 \ - \ 120 \ f\_im2 \ + \ 15 \ f\_im3 \ -
$$

$$
195 \ f\_ip1 \ + \ 120 \ f\_ip2 \ - \ 15 \ f\_ip3) \ / \ (120 \ h^3)
$$

Fourth-order derivative

$$
(56 \ f\_i \ - \ 39 \ f\_im1 \ + \ 12 \ f\_im2 \ - \ f\_im3 \ -
$$

$$
39 \ f\_ip1 \ + \ 12 \ f\_ip2 \ - \ f\_ip3) \ / \ (6 \ h^4)
$$

Fifth-order derivative

$$
(((-2) \ f\_7p) \ h^7 + 12 \ f\_im2 \ - \ 15 \ f\_im1 \ - \ 3 \ f\_im3 \ +
$$

$$
15 \ f\_ip1 \ - \ 12 \ f\_ip2 \ + \ 3 \ f\_ip3) \ / \ (6 \ h^5)
$$

Sixth-order derivative

$$
(15 \ f\_im1 \ - \ 20 \ f\_i \ - \ 6 \ f\_im2 \ + \ f\_im3 \ + \ 15 \ f\_ip1 \ -
$$

$$
6 \ f\_ip2 \ + \ f\_ip3) \ / \ (h^6)
$$

As seen because of cancelations for even-order derivatives (2, 4, and 6) more terms in the Taylor series expansion are needed to obtain the errors First-order derivative with the error estimate

```
                     7
(((-3) f_7p) h   + 63 f_im2 - 315 f_im1 -

    7 f_im3 + 315 f_ip1 - 63 f_ip2 + 7 f_ip3) /

    (420 h)
```
Second-order derivative with the error estimate

```
                       8
(((-9) f_8p) h   + 7560 f_im1 - 13720 f_i -

    756 f_im2 + 56 f_im3 + 7560 f_ip1 -

                                  2
    756 f_ip2 + 56 f_ip3) / (5040 h )
```
Third-order derivative with the error estimate

```
               7
((7 f_7p) h   + 195 f_im1 - 120 f_im2 + 15 f_im3 -

                                     3
    195 f_ip1 + 120 f_ip2 - 15 f_ip3) / (120 h )
```
Fourth-order derivative with the error estimate

```
                 8
((7 f_8p) h   + 2240 f_i - 1560 f_im1 +

    480 f_im2 - 40 f_im3 - 1560 f_ip1 +

                            4
    480 f_ip2 - 40 f_ip3) / (240 h )
```
Fifth-order derivative with the error estimate

```
                    7
(((-2) f_7p) h   + 12 f_im2 - 15 f_im1 - 3 f_im3 +

                                 5
    15 f_ip1 - 12 f_ip2 + 3 f_ip3) / (6 h )
```
Sixth-order derivative with the error estimate

```
               8
((-f_8p) h   + 60 f_im1 - 80 f_i - 24 f_im2 +

    4 f_im3 + 60 f_ip1 - 24 f_ip2 + 4 f_ip3) /

         6
    (4 h )
```

Problem 12.3

```
fprintf('\nProblem 12.3 Output:\n\n');
% The following presents an example of how you could address the
```

```
% problem of computing the central difference for the
% first-order derivative of y=sin(x)
% Setting domain, range and interval
h=0.1; a=0; b=10; x=[a:h:b]; n=length(x);
% Computing function and its derivative
y   = sin(x);        % Function itself (y=sin(x))
dy  = cos(x);        % First derivative
% Computing differences between the points
fwd=diff(y);
% Computing central differences, e.g. (y3-y2)+(y2-y1)=y3-y1
cent=fwd(2:end)+fwd(1:end-1);
% Computing central difference approximation for the first-order
% derivative
cdapp=cent/2/h;
% Plotting the results
figure('Color','w')
subplot(2,1,1)
hold on
grid on
plot(x,y,'b-.')                % Plotting sinusoid itself
plot(x,dy,'k')                 % Plotting analytical derivative
plot(x(2:end-1),cdapp,'mo')    % Plotting approximation
title(['Sinusoid, its analytic and numeric (centered-difference)'...
       'derivatives'])
legend('sin(x)','d{sin(x)}/dx',...
       'centered-diff approx of the derivative',0)
xlabel('x-axis'),ylabel('y-axis')
% Computing error between cent.-diff approximation and analytic
% derivative
disc=dy(2:end-1)-cdapp;
% Plotting the discrepancies
subplot(2,1,2)
bar(x(2:end-1),disc,'m'), hold
title('Approximation errors')
xlabel('x-axis'), ylabel('Approximation Error')
% Now, we estimate the errors based on known analytical expressions
% First, let us compute second and third derivatives of the function
% (y=sin(x)) that will be used in analytic formulas later on:
d2y = -sin(x);         % Second-order derivative
d3y = -cos(x);         % Third-order derivative
% Let us use those formulas to get error estimates for the centered-, forward-
% and backward-difference approximations of the first-order derivative
centO=-(h^2/6)*d3y;
```

```
fwdO=-(h/2)*d2y;
backO=(h/2)*d2y;
% Finally, let us add them to the latest plot (To accommodate them
% all at the same plot keep in mind that the last two
% of the above error estimates are much bigger than the first one,
% therefore you might want to scale the last
% two, i.e. devide them by say factor of 20):
plot(x,centO,'b+')        % Plotting the centered-diff. app. error
plot(x,fwdO/20,'r-')      % Plotting scaled forward-diff. app. error
plot(x,backO/20,'g--')    % Plotting scaled backward-diff. app. error
legend_handle=legend('Experimentally determined',...
                'Analyt. for the cent-diff. error approx.',...
                'Scaled (by 20) fwd-diff. error approx.',...
                'Scaled (by 20) bwd-diff. error approx',0);
set(legend_handle,'FontSize',8);
text(2,-0.0032,'Centered: O(h^2)=-h^2/6*f^3(x_i)')
text(0.2,0.003,'Forward: O(h)=-h/2*f^2(x_i)')
text(3.9,0.003,'Backward: O(h)=h/2*f^2(x_i)')
```

Problem 12.3 Output:

Current plot held

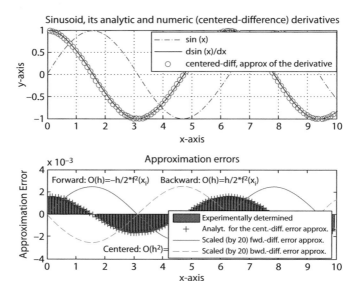

Chapter 13 Problem Set Solutions

Problem 13.1

```
fprintf('\nProblem 13.1 Output:\n\n');
% Comment: Finding the mean value for a function means integrating it
%          and then dividing by the range
% Defining the range
a=2; b=10;
% Defining the function
% Comment: The expression is bulky so it is better to define it as
%          inline function once and for all)
f=inline('-46+45.4*x-13.8*x^2+1.71*x^3-0.0729*x^4');
% Vectorizing inline function, i.e. adding . (we want to preserve the
% nice look of the function for the further use in the figure title)
fv=vectorize(f);
% Presenting all results on a single graph
figure('Color','white')
% Plotting a function
fplot(fv,[a b],'r')                % Using the vectorized inline function
hold
meanval=(quad(fv,a,b)/(b-a));
% Plotting line at mean value
plot([a b], [meanval meanval],'g--')
% Printing mean value on the graph
text(((b+a)/2.1),meanval,num2str(meanval),'BackgroundColor','w')
legend('Function',['Mean value within [',num2str(a,'%2.2g'),';',...
       num2str(b,'%2.2g'),']'],0)
title(['Mean value of f(x)=' char(f)])  % Using the inline function
xlabel('x-axis'); ylabel('y-axis')
% Note how we handled the range. We never referred to it explicitly.
%      Only by names, a and b. That allows us to change them in the
%      beginning of the script and everything else will stay the same.

fprintf('\nProblem 13.2 Output:\n\n');
% Part a
a=2; b=10;
% Following formulas for x_CG, y_CG and A introduce the following three
% vectorized inline functions
funct='(-23+22.7*x-6.9*x^2+0.855*x^3-0.0365*x^4)';
f  = inline(funct);                f  = vectorize(f);
fx = inline(['(' funct ')^2/2']); fx = vectorize(fx);
```

```
fy = inline(['(' funct ')*x']);    fy = vectorize(fy);
% Now the CG coordinates can be computed as
xcg=quad(fy,a,b)/quad(f,a,b);
ycg=quad(fx,a,b)/quad(f,a,b);
plot(xcg,ycg,'o')
% Visualizing the results
figure('Color','w')
xd=linspace(a,b); yd=f(xd);
area(xd,yd,'FaceColor','g'), axis equal, hold
plot(xcg,ycg,'+k',xcg,ycg,'ok','MarkerSize',13)
text(xcg+0.3,ycg,'Center of Gravity')
text(xcg+0.6,ycg-0.3,['(' num2str(xcg,3) ',' num2str(ycg,3) ')'])
title(['CG of an area under f(x)=' char(f)])
xlabel('x-axis'); ylabel('y-axis')

% Part b
% First, let us define an inline function
fundbl=inline('x^2/16-y^2/4+x*y^3/32+1.25'); fundbl=vectorize(fundbl);
% Now, the volume can be computed using a double integral
dblintegral=dblquad(fundbl,0,4,-2,2)
% An equivalent-volume rectangular cuboid will have a hight
zE=dblintegral/4/4;
% To visualize the results let us create a grid ...
[X,Y]=meshgrid(linspace(0,4,40),linspace(-2,2,40));
% ... and compute z values of the surface and rectangular cuboid
Z=fundbl(X,Y);
ZE=zE*ones(40);
% Finally, we can visualize everything
figure('Color','w'), meshz(X,Y,Z)
xlabel('x'),ylabel('y'),zlabel('z'), hold
surf(X,Y,ZE)
h1=text(0,1.5,3,...
['Volume of the 3-D shape is ' num2str(dblintegral,3) ' units']);
set(h1,'FontWeight','bold','BackgroundColor','y');
h2=text(0.3,1.5,2.7,...
['(Equivalent hight is ' num2str(zE,3) ' units)']);
set(h2,'FontWeight','bold','BackgroundColor','y');
```

Problem 13.1 Output:

Current plot held

Problem 13.2 Output:

Current plot held

dblintegral =   20

Current plot held

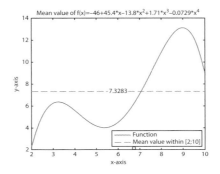

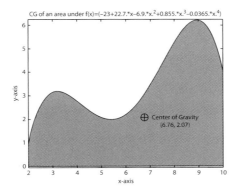

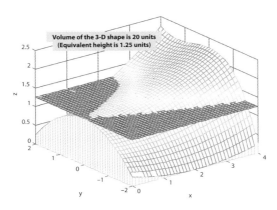

Problem 13.3

```
fprintf('\nProblem 13.3 Output:\n\n');
clear all              % Clearing the workspace after symbolic operations
% Defining the integral
f=inline('sin(1./x)'); % Defining the integrand
a=1/2/pi; b=2;         % Defining the range
% Computing a definite integral using quad and quadl functions
t=logspace(-10,0,50);
for i=1:50
  [q(i), N1fcn(i)]=quad(f,a,b,t(i));
  [ql(i),N2fcn(i)]=quadl(f,a,b,t(i));
end
% Plotting the results
figure('Color','w')
subplot(3,1,1)
semilogx(t,N1fcn,'b-',t,N2fcn,'g-.')
title(['Number of function evaluations vs. tolerance: '...
       'Quad and Quadl approximations'])
xlabel('Tolerance (Accuracy)')
ylabel('# of function evaluations')
legend('Quad function','Quadl function',1)
subplot(3,1,2)
% Computing and showing a reasonably accurate estimate at tol=1e-6
toll=1e-6;                          % Setting a tolerance
[quad_app,iter]=quad(f,a,b,toll);   % Using the quad function
semilogx(t,q,'b-',t,ql,'g-.',toll,quad_app,'rp')
title('Integral value vs. tolerance: Quad and Quadl approximations')
xlabel('Tolerance (Accuracy)')
ylabel('Integral value'), ylim([1.1 1.16])
legend('Quad','Quadl','Quad (tol=1e^{-6})',3)
text(toll,quad_app*1.005,num2str(quad_app,'%10.6f'));
text(toll,quad_app*0.995,['(with' num2str(iter,'%-2.0f')...
' function evaluations)']);
% Comment: Obviously, as the tolerance required decreases in accuracy,
%          the number of function evaluations (or intervals) required to
%          achieve the designated accuracy should decrease.
% Computing a definite integral using trapz function
  for i=1:110
  numpts(i)=i+1;                % Setting the number of points
  x=linspace(a,b,numpts(i));
  value(i)=trapz(x,f(x));       % Computing the integral
  clear x
  end
```

```
subplot(3,1,3)
plot(numpts,value,'b-')
title('Integral value vs. number of points: Trapz approximation')
xlabel('Number of points')
ylabel('Integral value')
hold
% Computing and plotting a pseudoexact value with a high number of
% points
N=4000;
x=linspace(a,b,N);
v=trapz(x,f(x));
plot([numpts(1) numpts(end)],[v v],'g-.')
legend('Trapz approximation',['Pseudoexact value for N='...
num2str(N,'%-6g')],4)
text(numpts(end)/2,v*1.1,num2str(v,'%10.6f'));
% Comment: Obviously, as the number of trapezoids used to accumulate the
%    integration increases, the calculated value of the integral should
%    converge to the (pseudo)exact value. Comparing the result with
%    N=4000 with the previous one, which employed the quad(l) function
%    with the tolerance of 1e-6 resulting in about 60 function
%    computations, we can clearly see that the trapz function is not
%    even nearly as efficient as quad(l).
```

Problem 13.3 Output:

Current plot held

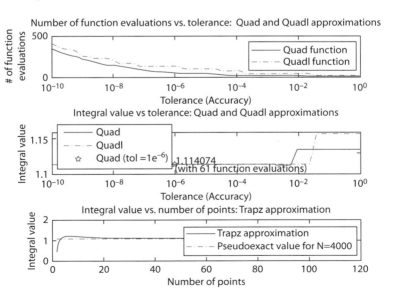

Chapter 14 Problem Set Solutions

Problem 14.1

```
fprintf('\nProblem 14.1 Output:\n\n');
% Defining initial conditions and constants
v0=300;                        % Initial velocity in km/h
v0=v0*1000/3600;               % Initial velocity in m/s
vel(1)=v0;
dist(1)=0;
time(1)=0;                     % Start braking at time t=0
dt=0.1;                        % Time interval
% Solving for the velocity and distance time histories (numerical
% integration)
n=2;
  while vel(n-1) > 0
  a = -0.0045*vel(n-1)^2-3;
  vel(n)  = vel(n-1) + a*dt;
  dist(n) = dist(n-1) + (vel(n-1)+vel(n))*dt/2;
  time(n) = time(n-1) + dt;
  n=n+1;
  end
lt=length(time);
% Plotting the data
figure('Color','w')
% Velocity vs. time
subplot(2,1,1)
plot(time,vel), hold on, grid on
plot(time(1),vel(1),'ro')
plot(time(end),vel(end),'r*')
title(['Velocity vs time: parachute assisted, post-landing'...
       'runway braking'])
text(0.5,vel(1),'\bfVelocity = 300km/h when braking started')
text(time(end)-3,30,'\bf{Airplane fully stopped}')
text(time(end)-2.8,15,['\bfTime to stop: ' num2str (time(end),...
'%6.2f') '\its'])
ylim([0 100])
ylabel('Velocity, m/s')
% Distance vs. time
subplot (2,1,2)
plot(time,dist), hold on, grid on
plot(time(end),dist(end),'r*')
```

```
title(['Distance traveled vs time: parachute assisted, '...
        'post-landing runway braking'])
text(time(end)/3,dist(end)/2,['\bfTotal distance traveled '...
      'while braking: ' num2str(dist(end),'%6.2f') '\itm'])
xlabel('Time, seconds'),ylabel('Distance, m')
```

Problem 14.1 Output:

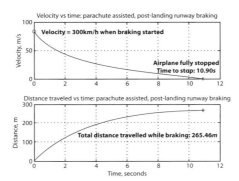

Problem 14.2

```
fprintf('\nProblem 14.2 Output:\n\n');
% Defining the differential equation and initial conditions
strDy='-2*x^3+12*x^2-20*x+8.5';      % Returns the string
Dy=inline(strDy,'x','y');% Returns the inline function of 2 arguments
% Comment: The syntax of this inline function is dictated by the
%          ODE23 and ODE45 solvers you will be using.
%          Otherwise, inline(strDy) would suffice.
xi=0; xf=4; yi=1;          % Defines the initial conditions
% Finding a true solution (you might do if by hand but here is how
% you would do it using the MATLAB Symbolic Math Toolbox int function)
TrueS=int(strDy)+yi;       % Returns the indefinite integral of strDy
TrueS=char(TrueS);         % Converts TrueS from symbolic to char
TrueS=inline(TrueS);       % Converts TrueS inline function
TrueS=vectorize(TrueS);    % Vectorizes TrueS for the use with vectors
% Defining an integral error as a mean discrepancy
error = @ (TrVal,Approx,Npoints) sum(abs(TrVal-Approx)./TrVal)/...
                                 length(TrVal);
% Employing Euler's first-order method with the step size of h=0.5
y(1)=yi; x(1)=xi; h=0.5; x=xi:h:xf; n=1;
t0=cputime;
```

```
  while x(n)<xf,
  y(n+1)=y(n)+h*Dy(x(n),0);
  yTrue(n+1)=TrueS(x(n));
  n=n+1;
  end
t(1)=cputime-t0;
e(1)=error(yTrue(2:end),y(2:end));
% Plotting the true and the first Euler's solution on the upper plot
figure('Color','w')
subplot(3,1,1);
fplot(TrueS,[xi,xf],'r-'), hold on
plot(x,y,'^g');
title(strcat('Solutions for y''=', strDy))
clear x y yTrue
% Employing Euler's first-order method with the step size of h=0.25
y(1)=yi; x(1)=xi; h=0.25; x=xi:h:xf; n=1;
t0=cputime;
  while x(n)<xf,
  y(n+1)=y(n)+h*Dy(x(n),0);
  yTrue(n+1)=TrueS(x(n));
  n=n+1;
  end
t(2)=cputime-t0;
e(2)=error(yTrue(2:end),y(2:end));
% Adding the second Euler's solution to the upper plot
plot(x,y,'sb')
legend({'True solution', 'Euler 0.5', 'Euler 0.25'},'Position',...
       [0.67 0.64 0.24 0.14])
ylim([0 8]);
clear x y yTrue
% Employing Heuns's second-order method with the step size of h=0.5
y(1)=yi; x(1)=xi; h=0.5; x=xi:h:xf; n=1;
t0=cputime;
  while x(n)<xf
  k1=Dy(x(n),0); k2=Dy(x(n)+h,0);
  y(n+1)=y(n)+h*(k1/2+k2/2);
  yTrue(n+1)=TrueS(x(n));
  n=n+1;
  end
t(3)=cputime-t0;
e(3)=error(yTrue(2:end),y(2:end));
subplot(3,1,2);
% Plotting the true and the Heun's solution on the middle plot
subplot(3,1,2)
```

```
fplot(TrueS, [xi, xf], 'r-'), hold on
plot(x,y,'^g')
clear x y yTrue
% Employing Midpoint second-order method with the step size of h=0.5
y(1)=yi; x(1)=xi; h=0.5; x=xi:h:xf; n=1;
t0=cputime;
  while x(n)<xf
  k1=Dy(x(n),0); k2=Dy(x(n)+h/2,0);
  y(n+1)=y(n)+h*k2;
  yTrue(n+1)=TrueS(x(n));
  n=n+1;
  end
t(4)=cputime-t0;
e(4)=error(yTrue(2:end),y(2:end));
% Adding the Midpoint method solution to the middle plot
plot(x,y,'sb');
legend({'True solution', 'Heun', 'Midpoint'}, 'Position',...
      [0.38 0.51 0.24 0.14])
ylim([0 8]);
clear x y yTrue
% Employing Runge-Kutta third-order method with the step size of h=0.5
options=odeset('InitialStep',h,'MaxStep',h);
t0=cputime;
[x,y]=ode23(Dy, [xi, xf], yi, options);
t(5)=cputime-t0;
yTrue=TrueS(x);
e(5)=error(yTrue(2:end),y(2:end));
% Plotting the true and the Runge-Kutta solution on the bottom plot
subplot(3,1,3);
fplot(TrueS, [xi, xf], 'r-'), hold on
plot(x,y,'^g');
clear x y yTrue
% Employing Runge-Kutta fourth-order method with the step size of
% h=0.5
t0=cputime;
[x,y]=ode45(Dy, [xi, xf], yi, options);
yTrue=TrueS(x);
t(6)=cputime-t0;
e(6)=error(yTrue(2:end),y(2:end));
% Adding the Runge-Kutta fourth-order method solution to the bottom
% plot
plot(x,y,'sb')
legend({'True solution','Runge-Kutta 3^{rd}',...
        'Runge-Kutta 4^{th}'},'Position',[0.38 0.20 0.27 0.16])
```

```
ylim([0 8])
xlabel('x-axis')
% Plotting Error/Accuracy vs. CPU time data
figure('Color','w')
t=t*1000;       % Converts the CPU time to milliseconds
e=e*100;        % Converts average error to percent
markers='^Vspho';
hold
for i=1:6
plot(t(i), e(i), ['r' markers(i)], 'MarkerFace Color','r')
end
legend ('Euler     0.5','Euler      0.25','Heun 0.5','Midpoint 0.5',...
        'ODE23   0.5','ODE45   0.5',0)
title('Accuracy vs. CPU time')
xlabel('CPU time, \itms\rm'), ylabel('Average error, %')
```

Problem 14.2 Output:

Current plot held

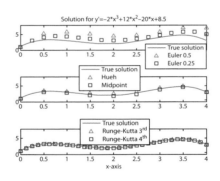

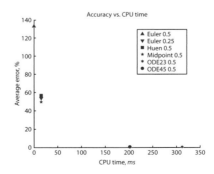

Problem 14.3

```
fprintf('\nProblem 14.3 Output:\n\n');
% Defining differential equations and the initial conditions
strDy='[1-y(2); y(1)^2-y(2)^2]';
Dy=inline(strDy, 't', 'y');
ti=0; tf=5; y0=[0,2]; xi=-2; xf=2;
% Solving and plotting the solutions starting from different
% initial points
figure('Color','w')
n=20;
fmt='rb';
x0=linspace(xi, xf, n);
axis([-2.5 2.5 -0.3 2.5]); hold on;
title(sprintf('Family of solutions for y''=%s', strDy));
xlabel('x-axis'), ylabel('y-axis')
   for i=1:2
     for j=1:n
       [T,Y]=ode45(Dy,[ti,tf],[x0(j),y0(i)]);
       plot(Y(:,1),Y(:,2),fmt(i));
       text(Y(1,1),Y(1,2),'_o','Color','g');
     end
   end
text(-1.4,2.2,['Initial points (0) belong to x_0= [-2;2],'...
               'y_0=0|2'],'Color','g');
```

Problem 14.3 Output:

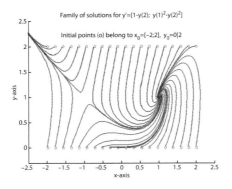

Chapter 15 Problem Set Solutions

Problem 15.1

```
% The following command opens a Simulink model SimulinkP151.mdl

open('SimulinkP151.mdl')
% The top layer of the model looks like follows:
```

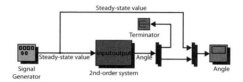

```
% You can use the model of Simulink2a.mdl as a base, define the names of
% variables in the Gain blocks rather than their numerical values) and
% introduce their numerical values in the Model Properties window
% (Files - Model Properties - Callbacks - InitFcn):
```

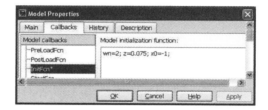

```
% You should also define the proper input signal in the Signal Generator
% Block (leftmost block):
```

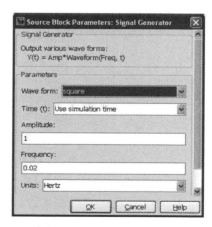

```
% and specify the Stop time (Simulation - Configuration Parameters).
```

% Now, when you run your model, double-click on the display (rightmost
% block) when the simulation stops, and click the binoculars button
% (to autoscale the output), to see the result

Problem 15.2

% The following command opens a Simulink model SimulinkP152.mdl

open('SimulinkP152.mdl')
% The model is pretty similar to that of shown in Fig.15.12b with
% interconnections between the two masses. Running the model
% results in a plot indicating that while the mass 1 basically
% follows the disturbance, the movement of the mass 2 appears to
% be much smoother.

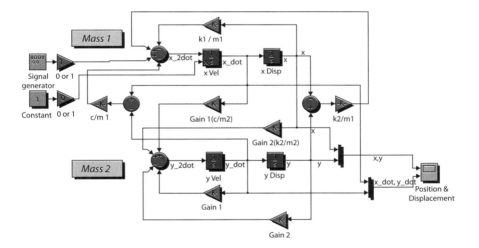

Problem 15.3

```
% The following command opens a Simulink model SimulinkP153.mdl
open('SimulinkP153.mdl')
% The model of Fig.15.31 with the monthly payment of $3K will not
% work, because the interest payment for the $3M loan exceeds $25K.
% So in the model above we assume a payment of  $30K to start with.
% It is defined as a variable payment in the constant block. The
% Initial conditions in the Unit Delay block is defined as loan+
% payment. Both loan and payment are specified in the Model
% Properties Callbacks as
% loan=3000;
% payment=30;
```

```
% We also increased the Stop time to 1000. When you run the model,
% you will see that you'd need 18.5 years (222 months) to pay off
% your jet aircraft loan. Playing with the monthly payment, you may
% find that in order to pay it off in 40 months, you would need to
% contribute $90K a month.
```

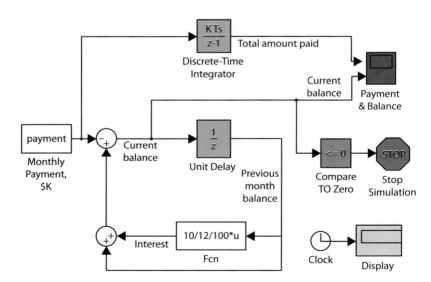

Appendix B Hooke–Jeeves Minimization Function

B.1 Introduction

In this appendix, you will find a listing of the `fminhj` function written by the author. It realizes the Hooke–Jeeves pattern search algorithm mentioned in Sec. 10.8. As shown in the examples of Sec. 10.9.2, this algorithm may out perform that of the `fminsearch` function of MATLAB and that is why the author has been using it in a variety of applications, including real-time optimization, for more than two decades. However, this function is listed here not to promote the method, but rather to show that after completing this book you should have all the knowledge to analyze and understand almost every line of any MATLAB function and also develop your own, professional looking functions, like `fminhj`.

B.2 Description

The `fminhj` function is intended to work with MATLAB and that is why its inputs and outputs are similar to those used by, say, the `fminsearch` function:

- `x=fminhj(fun,x0)` Starts at `x0` and attempts to find a local minimum `x` of the function `fun`
- `x=fminhj(fun,x0,options)` Minimizes with the default optimization parameters replaced by values in a structure, `options`, created with the `optimset` function of MATLAB
- `[x,fval]=fminhj(...)` Returns the value of the objective function, described in `fun`, evaluated at `x`
- `[x,fval,exitflag]=fminhj(...)` Returns an `exitflag` that describes the exit condition of `fminhj`. Possible values of `exitflag` and the corresponding exit conditions are
- 1   Indicates that `fminhj` converged to a solution `x`
- 0   Notifies that the maximum number of function evaluations was reached
- `[x,fval,exitflag,output]=fminhj(...)` Returns a structure `output` with the number of iterations stored in `output`.

iterations, the number of function evaluations in output. funcCount, the algorithm name in output.algorithm, and the exit message in output.message.

Specifically, the fminhj function uses the following fields of MATLAB's optimset structure: Display, TolX, TolFun, and MaxFunEvals. Particularly, the Display field may be one of the following:

'none' or 'off'	To display neither termination message nor intermediate steps
'final'	To display the termination message
'iter'	To display the termination message and all intermediate steps
'notify'	To supplement the previous option with a graphical output

B.3 M-file Function Listing

```
function[x,fval,exitflag,output]=fminhj(fun,x,options)
% FMINHJ Multidimensional unconstrained nonlinear minimization
   (Hooke-Jeeves).
% X = FMINHJ(FUN,X0) starts at X0 and attempts to find a local minimizer
% X of the function FUN. FUN is a function handle. FUN accepts input X
% and returns a scalar function value F evaluated at X. X0 can be a scalar
% or vector.
%
% X = FMINHJ(FUN,X0,OPTIONS) minimizes with the default optimization
% parameters replaced by values in the structure OPTIONS, created
% with the OPTIMSET function. FMINHJ uses the following options:
% Display, TolX, TolFun and MaxFunEvals. Display is one of the
   following:
%
%       'none','off' - no detailed output
%       'final' - display termination message
%       'iter' - display termination message and all intermediate steps
%       'notify' - show the graphical output as well
%
% [X,FVAL]= FMINHJ(...) returns the value of the objective function,
% described in FUN, evaluated at X.
%
% [X,FVAL,EXITFLAG] = FMINHJ(...) returns an EXITFLAG that describes
```

```
%  the exit condition of FMINHJ. Possible values of EXITFLAG and the
%  corresponding exit conditions are
%
%  1    FMINHJ converged to a solution X
%  0    Maximum number of function evaluations is reached
%
%  [X,FVAL,EXITFLAG,OUTPUT] = FMINHJ(...) returns a structure
%  OUTPUT with the number of iterations taken in OUTPUT.iterations, the
%  number of function evaluations in OUTPUT.funcCount, the algorithm name
%  in OUTPUT.algorithm, and the exit message in OUTPUT.message.
%
%    Examples:
%      FUN can be anonymous function or specified using @:
%         fminhj('x(1)^2+3*x(2)^2',[1,5]')
%         [X,FVAL,EXITFLAG]=fminhj(@cos,3)
%
%  FMINHJ uses the Hooke-Jeeves pattern search (direct search) algorithm.
%
%  Reference: Hooke, R., and Jeeves, T.A., "'Direct Search' Solution of
%  Numerical and Statistical Problems," Journal of the Assoc. Comput. Mach.,
%  Vol.8, No.2, 1961, pp.212-229.
%
%  Copyright 1987-2010 Oleg Yakimenko, Revision: 2010/07/13

defaultopt = struct('Display','final','MaxFunEvals',...
    '200*numberOf Variables','TolX',1e-4,'TolFun',1e-4,'OutputFcn',[]);
%  If just 'defaults' passed in, return the default options in X
if nargin==1 && nargout <= 1 && isequal(funfcn,'defaults')
    x = defaultopt;
    return
end


if nargin < 2,
    error('MATLAB:fminhj:NotEnoughInputs', ...
    'FMINHJ requires at least two input arguments');
end

if nargin<3, options = []; end
%  Check for non-double inputs
if ~isa(x,'double')
    error('MATLAB:fminhj:NonDoubleInput',...
    'FMINHJ only accepts inputsof data type double.')
end
```

```matlab
n = numel(x); numberOfVariables = n; % number of varied parameters

printtype = optimget(options,'Display',defaultopt,'fast');
tolx = optimget(options,'TolX',defaultopt,'fast');
tolf = optimget(options,'TolFun',defaultopt,'fast');
maxfun = optimget(options,'MaxFunEvals',defaultopt,'fast');

fun = fcnchk(fun);          % convert to function handle if needed

switch printtype
    case {'none','off'}     % supress any output
            prnt = 0;
    case 'final'            % display final values
            prnt = 1;
    case 'iter'             % printout all intermediate steps
            prnt = 2;
    case 'notify'           % add a plot
            prnt = 3;
    otherwise
            prnt = 1;
end

% Define the initial step based on the largest absolute value among all
% elements of the vector of initial guesses
scale = abs(x(n)); if scale == 0, scale=n; end
k = scale/5.;               % initial step size (20% of the scale)
                            % (final step size is determined by x-tolerance)

%% Check the original (basic) point
indexbp = 0;                % set the basic point (BP) search index
indexps = 0;                % set the pattern search (PS) index

    y = x;                  % set the latest basic point
    p = x;                  % set the suggested pattern search point
    b = x;                  % set the previous pattern search point

fnew = feval(fun,x);        % call minimization function
indexbp=indexbp+1;          % increment the basic-point search index

fold = fnew;
ps = 0;                     % set the pattern search flag
bp = 0;                     % set the basic-point search flag

index=indexbp+indexps;
if prnt >= 2,
varpar(index)=x(1);
bpflag(index)=bp;
```

```
perindex(index)=fnew;
step(index)=k; end

if prnt >= 2
header = ' f(x) eval.   x f(x)    Step    Procedure';
disp(header)
end

status = 'initial function evaluation';

%% Keep looking for the minimum ...
% ... while the step size k is greater than the x-tolerance and the value
%    of the objective function is greater than the function tolerance
  while (k > tolx) & (abs(fnew) > tolf)

index=indexbp+indexps;
  if prnt >= 2
  disp(sprintf('%6.0f %10.3g %10.3g %10.3g %-s',...
               index,x(1),fnew,k,status));
  end

%% Continue the pattern search ...
%  ... if the objective function decreased compared to the previous
%      'pattern' trial, continue the pattern search, i.e. make the same
%      move in the same direction
if (fold - fnew > tolf) & (ps == 1)
status = 'pattern search';

   p = 2.*y-b;              % compute the suggested PS point as b+2*(y-b)
% !!! constrained optimization code !!!
b = y;              % reassign the latest BP to the previous PS point
x = p;              % assign the suggested PS point to the trial point
y = x;

z=feval(fun,x);              % check the latest trial point
indexps=indexps+1;           % increment the PS index

index=indexbp+indexps;
if prnt >= 2,
varpar(index)=x(1);
bpflag(index)=bp;
perindex(index)=fnew;
step(index)=k; end

fold=fnew;
fnew=z;
```

```
%% Switch from searching around the basic point to the pattern search ...
%  ... if the objective function decrease was achieved during a search
%      around the basic point
elseif (fold - fnew > tolf) & (ps == 0)
status = 'switching from BP to PS';

bp = 0;                           % lower the BP flag
ps = 1;                           % rise the PS flag

%% Stop PS, make one backward step & perform a new basic point search ...
%  ... if the last pattern step failed
elseif (fold - fnew <= tolf) & (ps == 1)
status = 'stepping back to start a new BP search';

p = b;              % set everything to be equal to the previous PS point
y = b;
x = b;

fnew=feval(fun,x);
indexps=indexps+1;                % increment the PS index

index=indexbp+indexps;
if prnt >= 2,
varpar(index)=x(1);
bpflag(index)=bp;
perindex(index)=fnew;
step(index)=k; end

fold=fnew;
ps=0;                             % lower the PS flag

%% Proceed with the search around the basic point
elseif (fold - fnew <= tolf) & (ps == 0)
status = 'BP search';

% if a search around the basic point failed, decrease the step size and
% re-examine a vicinity of the current basic point
    if bp == 1
    k=k/10.;                      % decrease the step size
    end

% explore the basic point by making two steps (forward and backward) in
% each direction
            for j = 1:n
            x(j) = y(j) + k;
% !!! constrained optimization code !!!
```

```
                    z=feval(fun,x);
                    indexbp=indexbp+1; % increment the BP index
        index=indexbp+indexps;
        if prnt >= 2
        varpar(index)=x(1);
        bpflag(index)=1;
        perindex(index)=fnew;
        step(index)=k; end
                        if z < fnew
                        y(j) = x(j);
                        else
                        x(j) = y(j) - k;
% !!! constrained optimization code !!!
                    z=feval(fun,x);
                    indexbp=indexbp+1;  % increment the BP index
        index=indexbp+indexps;
        if prnt >= 2
        varpar(index)=x(1);
        bpflag(index)=1;
        perindex(index)=fnew;
        step(index)=k; end
                    if z < fnew
                    y(j) = x(j);
                    else
                    x(j) = y(j);
                    end
                end
        fnew=min(z,fnew);
        end

bp = 1;                       % rise the BP flag

end                           % if end
if index >= maxfun,           % break the loop if
break, end                    % index>=maxfun
end                           % while end

%% Display the results
index=indexbp+indexps;
if prnt >= 2
   disp(sprintf('%6.0f %10.3g %10.3g %10.3g %-s',...
                index,x(1),fnew,k,status));
end
```

```
fval=fnew;

if prnt >= 3
figure
subplot(2,1,1)
semilogy(0:index-1,perindex.*bpflag,'.k'), hold
semilogy(0:index-1,tolf*ones(length(perindex)),'--r')
legend('Search around BP','TolFun',1)
semilogy(0:index-1,perindex)
xlabel('Function evaluation'), ylabel('Function value')
subplot(2,1,2)
semilogy(0:index-1,step.*bpflag,'.k'), hold
semilogy(0:index-1,tolx*ones(length(step)),'--r')
legend('Search around BP','TolX',3)
semilogy(0:index-1,step)
xlabel('Function evaluation'), ylabel('Step size')
end

output.iterations = indexps-1; % Initial f(x) eveluation is 0th iteration
output.funcCount  = index;
output.algorithm  = 'Hooke-Jeeves pattern search';

if index >= maxfun
        msg=...
        sprintf(['Exiting:\n' '   While ' status ', the maximum' ...
        'number\n   of function evaluations has been exceeded\n' ...
        '   - increase MaxFunEvals option.\n   Current function' ...
        'value %f'], fval);
        if prnt > 0
             disp(msg)
        end
        exitflag = 0;
else
        msg=...
        sprintf(['Optimization terminated:\n''   While 'status ...
        ', the current x met\n    the termination criteria ' ...
        'OPTIONS.TolX of %e and F(X)\n   satisfied the convergence' ...
        'criteria OPTIONS.TolFun of %e'], tolx, tolf);
        if prnt >0
             disp(msg)
        end
        exitflag = 1;
end
output.message = msg;
return
```

Comment: This function was originally developed to handle constrained minimization problems, so that the penalty function is minimized before the search for the minimum of the objective function starts. For the purpose of this book, however, these portions of code were eliminated and substituted by the line

```
% !!! constrained optimization code !!!
```

Example

Consider a couple of examples of using the `fminhj` function. The call

```
>> [x,fval,exitflag,output]=fminhj(@cos,5,optimset('display','off'))
```

returns the minimum of .the $f = \cos(x)$ function closest to $x = 5$ ($x_* = \pi$)

```
x = 3.1411
fval = -1.0000
exitflag = 1
output = iterations: 8
         funcCount: 21
         algorithm: 'Hooke-Jeeves pattern search'
           message: [1x199 char]
```

Another call degrades default tolerance and requests for graphical output

```
>> options=optimset('Display','notify','TolFun',1e-3,'TolX',1e-3);
>> X=fminhj('(x(1)-1.5)^2+3*(x(2)+2.22)^2',[1;5],options)
```

It returns all iterations as shown below and also presents them graphically (Fig. B.1)

f(x) eval.	x	f(x)	Step	Procedure
1	1	157	1	initial function evaluation
5	1	116	1	BP search
5	1	116	1	switching from BP to PS
6	1	82	1	pattern search
7	1	53.7	1	pattern search
8	1	31.4	1	pattern search
9	1	15	1	pattern search
10	1	4.72	1	pattern search
11	1	0.395	1	pattern search
12	1	2.08	1	pattern search
13	1	0.395	1	stepping back to start a new BP search

17	1	0.395	1	BP search
20	1.1	0.203	0.1	BP search
20	1.1	0.203	0.1	switching from BP to PS
21	1.2	0.0912	0.1	pattern search
22	1.3	0.0592	0.1	pattern search
23	1.4	0.107	0.1	pattern search
24	1.3	0.0592	0.1	stepping back to start a new BP search
26	1.4	0.0112	0.1	BP search
26	1.4	0.0112	0.1	switching from BP to PS
27	1.5	0.0432	0.1	pattern search
28	1.4	0.0112	0.1	stepping back to start a new BP search
31	1.5	0.0012	0.1	BP search
31	1.5	0.0012	0.1	switching from BP to PS
32	1.6	0.0112	0.1	pattern search
33	1.5	0.0012	0.1	stepping back to start a new BP search
37	1.5	0.0012	0.1	BP search
41	1.5	0.0003	0.01	BP search

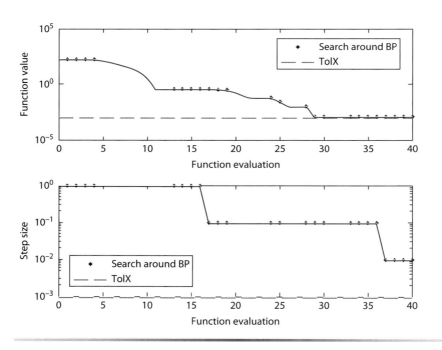

Fig. B.1   Example of `fminhj` graphical output.

```
Optimization terminated:
    While BP search, the current x met
    the termination criteria OPTIONS.TolX of 1.000000e-003 and F(X)
    satisfied the convergence criteria OPTIONS.TolFun of 1.000000e-003
X = 1.5000
   -2.2100
```

Appendix C

GUI Development Environment (GUIDE)

C.1 **Introduction**

For the sake of completeness, this appendix introduces the graphical user interface development environment (GUIDE) of MATLAB, providing a set of tools for creating problem-specific GUIs. These tools greatly simplify the process of designing and buildinzg GUIs. You can use the GUIDE tools not only to lay out the GUI but also to program it. Using the GUIDE Layout Editor, you can lay out a GUI easily by clicking and dragging GUI components—such as panels, buttons, text fields, sliders, and menus—into the layout area. GUIDE stores the GUI layout in a *fig*-file and automatically generates an M-file that controls how this GUI operates. The M-file initializes the GUI and contains a framework for the most commonly used callbacks for each component—the commands that execute when a user clicks a GUI component. Using the M-file Editor, you can then add code to the callbacks to perform the functions you want. This appendix presents a general idea behind the user interface control (uicontrol) graphics objects followed by a GUIDE overview. An example of the development of a simple GUI is shown in the following section. Knowing the general idea, two simple GUIs of MATLAB are examined from the standpoint of a GUI programming. Finally, a couple of practical GUIs are introduced as an example of a complex interface involving numerous uicontrol objects.

C.2 **Uicontrol Graphics Objects**

Let us start from a simple example. Suppose you need to run through a sequence of images and do some image processing. You want to be able to start processing, pause it if you found something you are interested in, go back and forth between the images and restart bulk processing again. MATLAB allows you to add such control capabilities to the figure objects quite easily. For example,

Fig. C.1 features five control buttons that were created using the following commands:

```
hstart = uicontrol('Style','pushbutton','String','Start','Position',...
                   [470 230 50 24],'Value',1);
hpause = uicontrol('Style','pushbutton','String','Pause','Position',...
                   [470 200 50 24],'Value',1);
hnext  = uicontrol('Style','pushbutton','String','Next','Position',...
                   [470 170 50 24],'Value',1);
hback  = uicontrol('Style','pushbutton','String','Back','Position',...
                   [470 140 50 24],'Value',1);
hstop  = uicontrol('Style','pushbutton','String','Stop','Position',...
                   [470 110 50 24],'Value',1);
```

The `uicontrol` function used in these calls creates uicontrol objects, which you use to implement GUIs. Its syntax is as simple as

```
handle=uicontrol('PropertyName',PropertyValue,…)
```

and in the aforementioned example we defined a style, text string, position x-y coordinates of the left-bottom corner of a button, its width, and height, all in pixels (by default), and a value (where `1` stands for button being not pushed).

Beyond push buttons, MATLAB supports other styles of uicontrol objects, introduced in Sec. C.3 along with the `uipanel` function, assisting in grouping different components and the `uibuttongroup` function, also grouping radio and toggle buttons with exclusive selection behavior.

As you may recall from Chapter 6 (Sec. 6.4), now that you have the handles to the buttons created with the aforementioned calls, you can use them to access and change any of their properties. In the specific example of Fig. C.1, somewhere in the M-file code the `get` function is used to capture the state of the button. When pushed, its `'Value'` attribute changes from `1` to `0`, which signals that some predefined action should take place. The following shows an example of how the states of these five

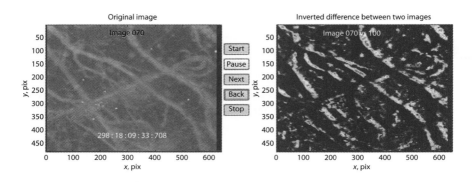

Fig. C.1 Example of adding control buttons to a regular figure.

buttons are captured inside the `while` loop and what actions in terms of the number of the image that needs to be processed are taken

```
%% Waiting for a Start button to be pushed (for a minute)
for i=1:1000
valstart = get(hstart,'Value');
if valstart==0, break, end
pause(0.06)
end
set(hstart,'Value',1);

%% Processing images
while ImNum <= ImNumMax
...
%% Checking status of push buttons
valstart = get(hstart,'Value');
valpause = get(hpause,'Value');
valnext  = get(hnext, 'Value');
valback  = get(hback, 'Value');
valstop  = get(hstop, 'Value');
   if valstop==0, break
   elseif valnext==0,
           ImNum=ImNum+1;
           set(hnext, 'Value',1);
           set(hpause,'Value',0);
           set(hstart,'Value',1);
   elseif valback==0
           ImNum=ImNum-1;
           set(hback, 'Value',1);
           set(hpause,'Value',0);
           set(hstart,'Value',1);
   elseif valpause==0 & valstart==1
           set(hnext, 'Value',1);
           set(hback, 'Value',1);
           set(hstart,'Value',1);
   else
           ImNum=ImNum+1;
           set(hpause,'Value',1);
           set(hstart,'Value',1);
   end
pause(0.05)
end
```

Using the `uicontrol` function allows you to enhance your interactive capabilities even beyond those presented in Sec. 4.9. The GUIDE tool elaborates on them and brings the development of interactive GUIs to the new level, when all you have to do is to create a uicontrol object whereas the code skeleton beyond it is generated automatically. After this, you simply proceed with programming the constructions like the ones considered earlier to specify what actions need to be taken when a specific uicontrol is activated.

For completeness, the following code shows an example of how the images shown in Fig.C.1 could be produced (this code would appear after the `while` command instead of an ellipsis):

```
A=imread(['Frame' num2str(ImNum,'%03.0f\n')],'jpg');
B=imread(['Frame' num2str(ImNum-30,'%03.0f\n')],'jpg');
subplot(1,2,2)
image(A)
xlim([1 640]),ylim([1 480])
axis tight, axis equal
subplot(1,2,2)
image(A(:,:,2)-B(:,:,2)) % use G channel (RGB)
xlim([1 640]),ylim([1 480])
axis tight, axis equal
```

As mentioned in Sec. 4.7.1 (Table 4.3) the `imread` function having the general syntax

```
A=imread(filename,format)
```

reads a variety of different-type grayscale or color images into the data array A. This can be a *M* by *N* array for a grayscale image, or *M* by *N* by 3 array for a truecolor (256^3-color) image (as explained in Sec. 6.9), or an *M* by *N* by 4 array for the *tiff* files, which use a cyan-magenta-yellow-black palette as opposed to the RGB palette. The class of A may also vary depending on the bits-per-sample of the image data (say "logical" for a 1-bit grayscale image or "unit8" for 24-bit color). In the above example the `imread` function reads 24-bit *jpeg* images into "unit8" 480 by 640 by 3 arrays.

The `image(A)` call displays matrix A as an image by interpreting each of *M* by *N* elements as RGB values or an index into the figure's color map (depending on the dimension and class of matrix A). In the above case the first call addresses a 3-D matrix A directly as RGB values, and the second call interprets the 2-D matrix as a matrix of indices.

C.3 GUIDE Overview

As mentioned in Chapter 2, the ninth icon on the MATLAB toolbar brings up the GUIDE tool (Fig. C.2a). Alternatively, you can simply type `guide` in the command window.

 GUIDE

a)
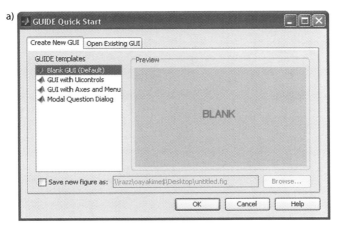

b)
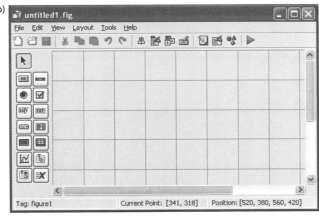

Fig. C.2 a) GUIDE Quick Start window and b) Layout
Editor with a blank GUI template.

To start creating a new GUI, you should choose the **Blank GUI** option, which opens a Layout Editor window with a blank GUI template allowing you to work on your GUI layout (Fig. C.2b). You create a GUI by dragging buttons from the Component Palette on the left of the Layout Editor to the layout area (or clicking a button once and then clicking in the layout area once) followed by aligning them and adjusting their properties to meet your preferences. A brief description of a variety of different `uicontrol` `objects` and three more user interface tools, `uitable`, `uibutton-` `group`, and `uipanel`, is given in following box. In terms of hierarchy of the graphics objects (Fig. 6.8), an uicontrol object is a child of a figure, uipanel, or uibuttongroup, and therefore does not require axes to exist when placed in a figure window, uipanel, or uibuttongroup.

	Push button, generating an action when pressed
	Radio button, providing a user with a number of mutually exclusive choices
	Editable text field, enabling users to enter or modify the text values
	Pop-up (drop-down) menu, displaying a list of choices to choose from
	Toggle button, executing callbacks when clicked on and indicating their state (`on` or `off`)
	Axes, enabling your GUI to display graphics
	Button group, creating container object to exclusively manage radio buttons and toggle buttons (`uibuttongroup` function)
	Slider bar, accepting numeric input within a specific range by enabling a user to move a sliding bar
	Check box, providing a user with a number of independent choices
	Static text, displaying lines of text
	List box, displaying a list of items and enabling users to select one or more of them
	Table, allowing you to display and work with table data within your GUI (`uitable` function)
	Opaque panel, providing a visual enclosure for regions of a figure window with related controls (`uipanel` function)
	ActiveX Controls, enabling your GUI to work with Microsoft ActiveX controls add-ons

At the bottom of the Layout Editor, you may see the tag of the selected component, location of the mouse pointer, and the position of the selected component (x-, y- coordinates of the left-bottom corner, width and height). Several unique buttons showing up in the Layout Editor toolbar are introduced in following box. These functions are also accessible via the **Tools** menu of the Layout Editor (Fig. C.3a). Specifically, the **Align Objects** option (Fig. C.3b) allows you to arrange selected graphics objects within your GUI vertically and horizontally, whereas the Toolbar Editor (Fig. C.3c) is used to create a toolbar for your GUI, similar to that of a standard figure window, like Fig. 6.43. As usual, you may adjust editable properties of GUIDE and Layout Editor via its Preferences window (Fig. C.4).

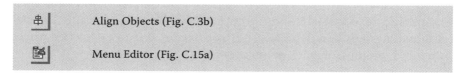

	Align Objects (Fig. C.3b)
	Menu Editor (Fig. C.15a)

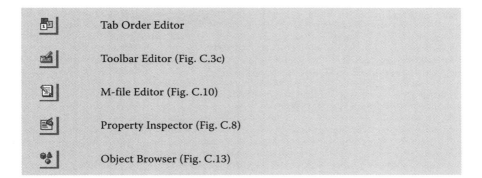

	Tab Order Editor
	Toolbar Editor (Fig. C.3c)
	M-file Editor (Fig. C.10)
	Property Inspector (Fig. C.8)
	Object Browser (Fig. C.13)

The GUIs developed using the GUIDE tool are saved as FIG-files. When you save your GUI (or try to run an unsaved GUI by pushing the **Run** button), the M-file is automatically generated. It bears the same name as the FIG-file and contains the functions needed to open and run the GUI you have developed. Once you are done with editing your GUI, you can call it by typing its name in the command window of MATLAB. The following M-file corresponds to the blank GUI shown in Fig. C.2b and contains a primary function `untitled1` along with two subfunctions, `untitled1_Open-ingFcn` and `untitled1_OutputFcn`.

Run

```
function varargout = untitled1(varargin)
% UNTITLED1 M-file for untitled1.fig
% UNTITLED1, by itself, creates a new UNTITLED1 or raises the existing
% singleton*.
% H = UNTITLED1 returns the handle to a new UNTITLED1 or the handle to
% the existing singleton*.
% UNTITLED1('CALLBACK',hObject,eventData,handles,...) calls the local
% function named CALLBACK in UNTITLED1.M with the given input arguments.
% UNTITLED1('Property','Value',...) creates a new UNTITLED1 or raises the
% existing singleton*.  Starting from the left, property value pairs are
% applied to the GUI before untitled1_OpeningFcn gets called.  An
% unrecognized property name or invalid value makes property application
% stop. All inputs are passed to untitled1_OpeningFcn via varargin.
% *See GUI Options on GUIDE's Tools menu.  Choose "GUI allows only one
% instance to run (singleton)".
%
% See also: GUIDE, GUIDATA, GUIHANDLES

% Edit the above text to modify the response to help untitled1
% Last Modified by GUIDE v2.5 16-Jul-2010 00:31:54

% Begin initialization code - DO NOT EDIT
```

a)

b)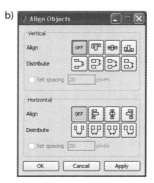

c)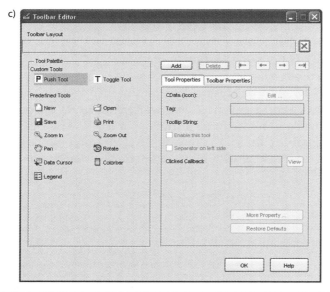

Fig. C.3   a) Tools menu of the Layout Editor, b) Align Objects window, and c) Toolbar Editor window.

```
gui_Singleton = 1;
gui_State = struct('gui_Name',mfilename,'gui_Singleton',gui_Singleton,...
     'gui_OpeningFcn',@untitled1_OpeningFcn,'gui_OutputFcn',...
     @untitled1_OutputFcn,'gui_Lay outFcn',[],'gui_Callback',[]);
if nargin && ischar(varargin{1})
   gui_State.gui_Callback = str2func(varargin{1});
end

if nargout
   [varargout{1:nargout}] = gui_mainfcn(gui_State, varargin{:});
else
   gui_mainfcn(gui_State, varargin{:});
end
```

```
% End initialization code - DO NOT EDIT

% --- Executes just before untitled1 is made visible.
function untitled1_OpeningFcn(hObject, eventdata, handles, varargin)
% This function has no output args, see OutputFcn.
% hObject  handle to figure
% eventdata        reserved - to be defined in a future version of MATLAB
% handles  structure with handles and user data (see GUIDATA)
% varargin command line arguments to untitled1 (see VARARGIN)

% Choose default command line output for untitled1
handles.output = hObject;
% Update handles structure
guidata(hObject, handles);
% UIWAIT makes untitled1 wait for user response (see UIRESUME)
% uiwait(handles.figure1);

% --- Outputs from this function are returned to the command line.
function varargout = untitled1_OutputFcn(hObject, eventdata, handles)
% varargout        cell array for returning output args (see VARARGOUT);
% hObject  handle to figure
% eventdata        reserved - to be defined in a future version of MATLAB
% handles  structure with handles and user data (see GUIDATA)
% Get default command line output from handles structure
varargout{1} = handles.output;
```

These three functions are accompanied with the extended comments, explaining the inputs and outputs, and are needed for opening the GUI window itself. They usually remain unchanged for any GUI (sometimes, a few lines of code are added at the end of the ..._OpeningFcn function to initialize the GUI, that is, set its default values). Adding each new ui-control that assumes some actions adds at least one more function to this

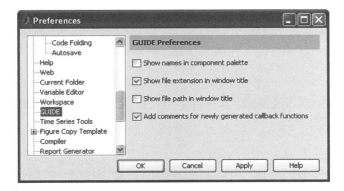

Fig. C.4 GUIDE preferences window.

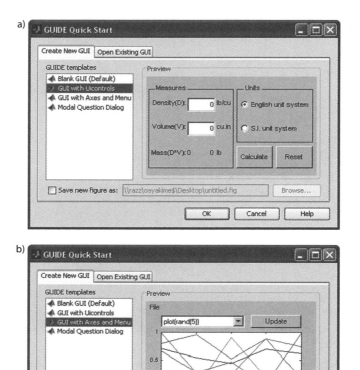

Fig. C.5 Examples of simple GUIs provided with MATLAB.

M-file as explained in Sec. C.4 Information about properties of uicontrol objects is passed between the functions using a single structure, `handles`, containing handles to the uicontrol objects handles and user data.

Figure C.5 shows examples of two simple GUIs as provided by Mathworks. The first one (Fig. C.5a) may be used to learn how to handle arithmetic calculations, the second one (Fig. C.5b) demonstrates how to use a pop-up menu to change a graphical output. We will return to these examples in Sec. C.4, but now let us proceed with developing a simple GUI from scratch.

C.4 Laying Out and Programming a Simple GUI

Consider a simple example, when we would like to supplement the very first code you developed for the Problem 2.3 of Chapter 2 (day-of-the-week calculator) with a nice graphical interface. We start from opening a blank GUI as shown in Fig. C.2b and saving it as *WeekDayGUI.fig*. Immediately after pushing the **Save** button, the M-file *WeekDayGUI.m* is automatically generated for you as well. It is exactly the same as presented in the previous section except that `untitled1` is replaced with `WeekDayGUI` everywhere throughout the code. If you try to change the name of the GUI, its code will immediately be renamed too.

Let us use the buttons of the Component Palette and create several objects including 10 Static texts, 3 Pop-up menus, and 2 Push buttons as shown in Fig. C.6.

Next, let us change some properties of these graphics objects and align them using the **Align Objects** button, so that your GUI looks as shown in Fig. C.7. To adjust the properties of a graphics object, you simply double click on it, which brings up a Property Inspector window, pretty much the same for all uicontrols (Fig. C.8). (In fact, among other properties, you can change the **Style** option of your uicontrol object as well—clicking the **Expand** button at the far right of the **Style** option line brings up the menu of Fig. C.9a.)

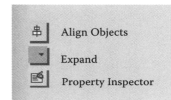

The changes resulted in Fig. C.7 are that: a) we tagged all objects with meaningful names (like `close`, `month`, and `day3` in Figs. C.8a, C.8b, C.8c, respectively), so that we can refer to these objects in our code, changed the String attributes (`Close GUI` and `Wednesday` in Figs. C.8a and C.8c, respectively), and changed the background color for weekdays to white (for pop-up menus, white is a default color anyway). For three pop-up menus instead of assigning a single **String**, we filled their menu lists with years, names of months, and days (1 through 31). You can bring the menu list up by pushing the **String text** button. As an example, the **String** list of Fig. C.8b is presented in Fig. C.9b (that is why January shows up as a String attribute).

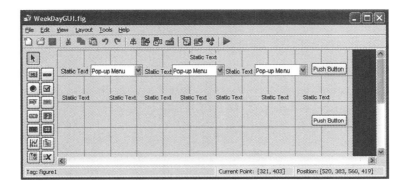

Fig. C.6   Original layout of a sample GUI.

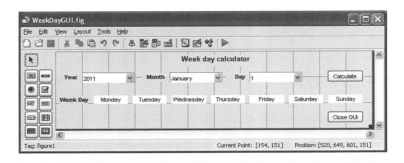

Fig. C.7   Final layout of the WeekDayGUI.

When you save this GUI, it automatically updates the *WeekDayGUI.m* M-file. Now we will need to complete the programming portion of GUI development, hence we need to open this M-file in the Editor (for example, by clicking the **M-file Editor** button of the GUIDE Layout Editor toolbar). Figure C.10 shows the content of the *WeekDayGUI.m* file after we added and modified all aforementioned uicontrol objects

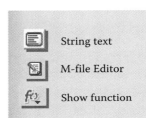

(the list of the functions in *WeekDayGUI.m* shows up in alphabetical order when you click the **Show Function** button of the Editor toolbar).

As mentioned in the previous section, each new object that assumes some actions adds new functions to those three of the blank GUI. As seen from Fig. C.10, the Static text uicontrol assumes no actions, and therefore, adds no functions. Each of the two push buttons adds a Callback function. Each of the three pop-up menus adds two functions, ..._CreateFcn and ..._Callback. These two functions for the Year pop-up menu are shown as follows:

```
% --- Executes during object creation, after setting all properties
function year_CreateFcn(hObject, eventdata, handles)
% hObject  handle to year (see GCBO)
% eventdata      reserved - to be defined in a future version of MATLAB
% handles  empty - handles not created until after all CreateFcns called

% Hint: popupmenu controls usually have a white background on Windows.
% See ISPC and COMPUTER.
if ispc && isequal(get(hObject,'BackgroundColor'),...
    get(0,'defaultUicontrolBackgroundColor'))
    set(hObject,'BackgroundColor','white');
end

% --- Executes on selection change in year.
function year_Callback(hObject, eventdata, handles)
```

```
% hObject  handle to year (see GCBO)
% eventdata  reserved - to be defined in a future version of MATLAB
% handles  structure with handles and user data (see GUIDATA)

% Hints:
% contents = get(hObject,'String') returns year contents as cell array
%           contents{get(hObject,'Value')} returns selected item from year
```

The functions for the pop-up menus are good as they are and need no additional tweaking. The idea is that you would use these three pop-up menus to choose the calendar date. This date will then be used to compute a weekday by pushing the Calculate push button, and this is where all programming should occur. Before we proceed, let us show the programming that needs to be done in another function, responsible for

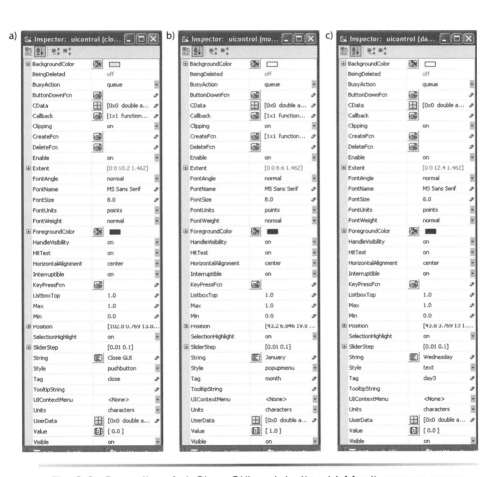

Fig. C.8   Properties of a) Close GUI push button, b) Month pop-up menu, and c) Wednesday static text.

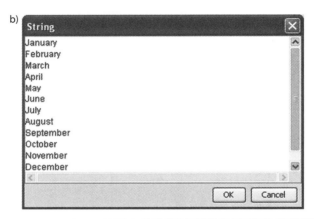

Fig. C.9   a) Style property of uicontrol and b) String content for the Month pop-up menu.

closing the GUI. Your **Close** button callback function may look as simple as

```
% --- Executes on button press in close.
function close_Callback(hObject, eventdata, handles)
% hObject    handle to close (see GCBO)
% eventdata  reserved - to be defined in a future version of MATLAB
% handles    structure with handles and user data (see GUIDATA)
close
```

and the only programming we did here was adding the `close` command at the very end. Now back to the **Calculate** callback function. Suppose that to display the calculated weekday we want to simply change the background color of the corresponding Static text uicontrol, say to green. We want to be able to calculate the weekday for several different dates while the GUI is running, so the first action would be bringing the background color for all weekdays back to white (we could have more sophisticated code, remembering what weekday was colored green at the previous calculation and then only reverting a background color for this specific weekday, but for simplicity, let us apply white color to all weekday static texts, even though six of them are white anyway). The complete `calculate_Callback` function for Problem 2.3 of Chapter 2 may then look as follows:

```
% --- Executes on button press in calculate
function calculate_Callback(hObject, eventdata, handles)
% hObject   handle to calculate (see GCBO)
% eventdata       reserved - to be defined in a future version of MATLAB
% handles  structure with handles and user data (see GUIDATA)
%% Changing the background color to default
set(handles.day1,'BackgroundColor','white')
set(handles.day2,'BackgroundColor','white')
set(handles.day3,'BackgroundColor','white')
set(handles.day4,'BackgroundColor','white')
set(handles.day5,'BackgroundColor','white')
set(handles.day6,'BackgroundColor','white')
set(handles.day7,'BackgroundColor','white')
%% Preliminary computations
Year=2012-get(handles.year,'value');
m=get(handles.month,'value')-2;
d=get(handles.day,'value');
if m <= 0        % Making corrections for January and February
m=m+12;          % Correcting the month's number to that of in Ancient Rome
Year=Year-1;     % Correcting the year to that of in Ancient Rome
end
Y=mod(Year,100); % Computing the year's number in the century
c=(Year-Y)/100;  % Computing the number of centuries
```

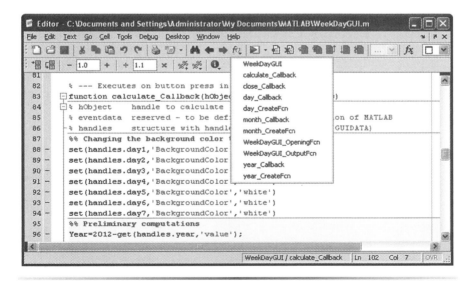

Fig. C.10   Browsing through GUI functions in the Editor.

```
%% Now goes the formula
WeekDay = mod((d+fix((13*m-1)/5)+Y+fix(Y/4)+fix(c/4)-2*c+777),7);
%% The output
switch WeekDay
   case 1
set(handles.day1,'BackgroundColor','green')
   case 2
set(handles.day2,'BackgroundColor','green')
   case 3
set(handles.day3,'BackgroundColor','green')
   case 4
set(handles.day4,'BackgroundColor','green')
   case 5
set(handles.day5,'BackgroundColor','green')
   case 6
set(handles.day6,'BackgroundColor','green')
   case 7
set(handles.day7,'BackgroundColor','green')
end
```

Note that the only difference with the code you developed in Chapter 2 is that in the beginning of this function we are using the `handles` structure to access and change the background color of uicontrols tagged `day1`, `day2`, ..., `day7` (Fig. C.8c), and at the end, when the weekday is determined, we are employing the `switch` statement to color the static text of the corresponding weekday green (using the `handles` structure again).

If you still have your Layout Editor open, you may run this GUI by pushing the **Run** button. You choose the date and click the Calculate push button to see the result (Fig. C.11). You can also run this GUI from the Editor (**Save and run** button) or from the Command window by typing in `Week-DayGUI` and hitting <Enter>.

Fig. C.11   WeekDay GUI in action.

When you know the basics of GUIDE, you may proceed with exploring its other features on your own. For your convenience, let us make a few remarks. First, it should be mentioned that the GUIDE Layout Editor allows you to automatically create some other functions, beyond those two (`CreateFcn` and `Callback`) we have just considered. If you right-click anywhere in the layout area, aside from uicontrol objects you will see the menu shown in Fig. C.12a. Choosing the **View Callbacks** option brings up another menu listing other possible action functions (Fig. C.12b). Right-clicking on the uibutton results in the menu of Fig. C.12c. Choosing the **View Callbacks** option in this case brings up another menu (Fig. C.12d). Other user interfaces, `uitable`, `uibuttongroup`, and `uipanel`, have their own action functions list.

Second, the Object Browser represents a very convenient way of listing all uicontrols within the open GUI providing you with their tags and text strings (names). In our case, this browser appears as shown in Fig. C.13a. Clicking on any uicontrol brings up its Property Inspector window.

Object Browser

ActiveX Control

Finally, when you become proficient, you may want to use some ActiveX add-ons. Clicking on the **ActiveX**

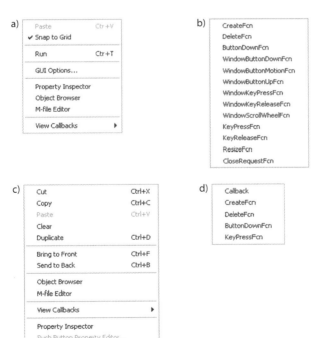

Fig. C.12 Menus in response to right-clicking a) anywhere within blank layout area, c) on any uicontrol graphics object, and the corresponding View Callbacks options (b and d).

Control button brings up the window that lists all available add-ons (their specific list depends on what programs are installed on your computer). Obviously, MATLAB does not necessarily support all of them, but you may find some useful controls like the ones shown in Fig. C.14.

C.5 Examples of Simple and Applied GUIs

Now that you know the basics of how to lay out a GUI and what programming needs to be done behind it, let us return to the two simple GUIs shown in Fig. C.5. The Object Browser for the first one (Fig. C.5a) is shown in Fig. C.13b.

If you look into the code for this GUI, you will find that compared to the blank GUI template code presented in Section C.2, one extra line is added to the end of the ..._OpeningFcn function

```
initialize_gui(hObject, handles, false);
```

This calls for a user-written function that initializes the GUI as follows:

```
function initialize_gui(fig_handle, handles, isreset)
% If the metricdata field is present and the reset flag is false, it means
```

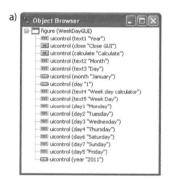

Fig. C.13   a) Object Browser for the WeekDay GUI, along with MATLAB GUIs with b) uicontrols, and c) axes.

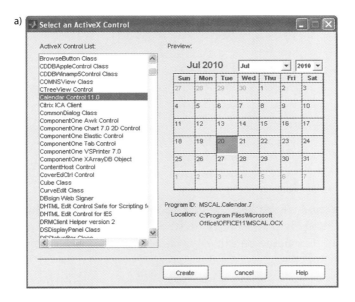

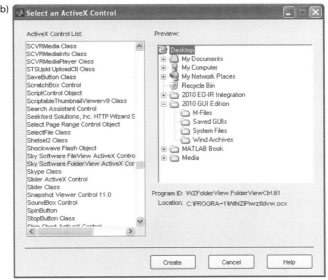

Fig. C.14 Examples of the ActiveX controls: a) Calendar and b) Folder View.

```
% we are just re-initializing a GUI by calling it from the cmd line
% while it is up. So, bail out as we dont want to reset the data.
if isfield(handles, 'metricdata') && ~isreset
    return;
end
```

```
handles.metricdata.density = 0;
handles.metricdata.volume  = 0;
set(handles.density, 'String', handles.metricdata.density);
set(handles.volume,  'String', handles.metricdata.volume);
set(handles.mass, 'String', 0);
set(handles.unitgroup, 'SelectedObject', handles.english);
set(handles.text4, 'String', 'lb/cu.in');
set(handles.text5, 'String', 'cu.in');
set(handles.text6, 'String', 'lb');
% Update handles structure
guidata(handles.figure1, handles);
```

This function simply sets the GUI to work with the English units. If the units are changed to the metric system, the function behind the panel unitgroup, managing an exclusive choice between two systems of units, handles it as follows:

```
% --- Executes when selected object changed in unitgroup
function unitgroup_SelectionChangeFcn(hObject, eventdata, handles)
if (hObject == handles.english)
   set(handles.text4, 'String', 'lb/cu.in');
   set(handles.text5, 'String', 'cu.in');
   set(handles.text6, 'String', 'lb');
else
   set(handles.text4, 'String', 'kg/cu.m');
   set(handles.text5, 'String', 'cu.m');
   set(handles.text6, 'String', 'kg');
end
```

The callback functions behind the two input windows are made foolproof, so that if a user attempts to enter non-numerical data, he gets the warning message

```
function density_Callback(hObject, eventdata, handles)
density = str2double(get(hObject, 'String'));
if isnan(density)
   set(hObject, 'String', 0);
   errordlg('Input must be a number','Error');
end
% Save the new density value
handles.metricdata.density = density;
guidata(hObject,handles)

function volume_Callback(hObject, eventdata, handles)
volume = str2double(get(hObject, 'String'));
if isnan(volume)
```

```
    set(hObject, 'String', 0);
    errordlg('Input must be a number','Error');
end
% Save the new volume value
handles.metricdata.volume = volume;
guidata(hObject,handles)
```

Finally, two push buttons are responsible for a simple calculation and resetting the GUI

```
% --- Executes on button press in calculate
function calculate_Callback(hObject, eventdata, handles)
mass = handles.metricdata.density * handles.metricdata.volume;
set(handles.mass, 'String', mass);

% --- Executes on button press in reset
function reset_Callback(hObject, eventdata, handles)
initialize_gui(gcbf, handles, true);
```

The Object Browser for the second GUI (Fig. C.5b) is shown in Fig. C.13c. Compared with the previous GUIs considered earlier, it features the **Tools** menu, which was created using the Menu Editor. Pushing the corresponding button for menu Editor brings its window as shown in Fig. C.15a. In the 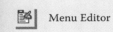 Menu Bar area, you may type any new items and sub items that will appear while the GUI is running as shown in Fig. C.15b. Obviously, each new entry automatically generates the corresponding function header. In this case, programming included writing the following code for the three subitems of the **File** menu:

```
function OpenMenuItem_Callback(hObject, eventdata, handles)
file = uigetfile('*.fig');
if ~isequal(file, 0)
    open(file);
end

function PrintMenuItem_Callback(hObject, eventdata, handles)
printdlg(handles.figure1)

function CloseMenuItem_Callback(hObject, eventdata, handles)
selection = questdlg(['Close' get(handles.figure1,'Name') '?'],...
            ['Close' get(handles.figure1,'Name') '...'],...
            'Yes','No','Yes');
if strcmp(selection,'No')
    return;
end
delete(handles.figure1)
```

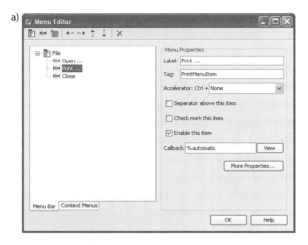

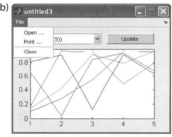

Fig. C.15 Example of using a) Menu Editor and b) corresponding menu items.

Compared to the case considered in Sec. C.3, the pop-up menu ..._CreateFcn function is used to define a list of possible options via the setting menu's string attribute

```
% --- Executes during object creation, after setting all properties
function popupmenu1_CreateFcn(hObject, eventdata, handles)
if ispc && isequal(get(hObject,'BackgroundColor'),...
                  get(0,'defaultUicontrolBackgroundColor'))
   set(hObject,'BackgroundColor','white');
end
set(hObject,'String',{'plot(rand(5))','plot(sin(1:0.01:25))',...
                  'bar(1:.5:10)','plot(membrane)','surf(peaks)'});
```

Once the choice is made, the push-button Callback function produces a plot using the switch statement

```
% --- Executes on button press in pushbutton1
function pushbutton1_Callback(hObject, eventdata, handles)
axes(handles.axes1);
```

```
cla;
popup_sel_index = get(handles.popupmenu1,'Value');
switch popup_sel_index
   case 1
          plot(rand(5));
   case 2
          plot(sin(1:0.01:25.99));
   case 3
          bar(1:.5:10);
   case 4
          plot(membrane);
   case 5
          surf(peaks);
end
```

Note that this latter function starts from taking the axes that have been created already and clears them to draw a new plot. This very first plot is created during the `OpeningFcn` function call, which compared to the case of a blank GUI of Sec. C.2 contains the line

```
plot(rand(5));
```

It is this plot that shows up when you call this GUI for the first time (Fig. C.15b).

MATLAB offers examples of more sophisticated GUIs explaining how to achieve certain functionality. They include

- GUI with multiple axes
- GUI for animating a 3-D view
- GUI to interactively explore data in a table
- GUI to set Simulink model parameters
- GUI to read an address book

For more details on these and other GUIs, you should refer to the MATLAB Help Navigator and choose MATLAB => Examples => Creating GUIs. To encourage you to use the GUI development environment, let us show a couple of practical examples.

Figure C.16 shows an example of a two-page GUI developed as an interface for addressing the so-called pose-estimation problem (when the 3-D position and attitude of an object needs to be estimated using synchronized video streams from multiple cameras). First (Fig. C.16a), a user chooses the drop zone to operate within, he then establishes multiple cameras around the intended point of impact, and a folder containing the test data (by the way, one of the ActiveX controls, `mwxpccontrolsx3x2.PathChooserCtrl`, is implemented here for this purpose). Then (Fig. C.16a), the user jumps to another page where he/she has a choice of

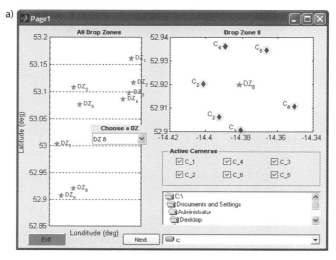

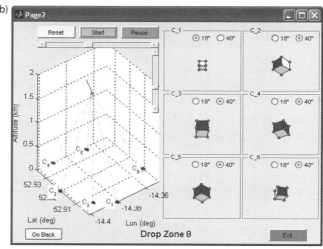

Fig. C.16 Two-page GUI to assist in solving a pose-estimation problem.

choosing the focal length of each camera to emulate what it would see while some aerospace system descends down to the ground.

Figure C.17 shows another GUI to address a safety issue when deploying multiple multistage cargo aerial delivery systems. As seen, this GUI enables a lot of interactive actions, features graphical interface, and allows you to do measurements on the underlying map.

In conclusion, Fig. C.18 presents the main aspects of the process of designing application-specific GUIs as suggested by the Mathworks. It is natural and quite easy to follow.

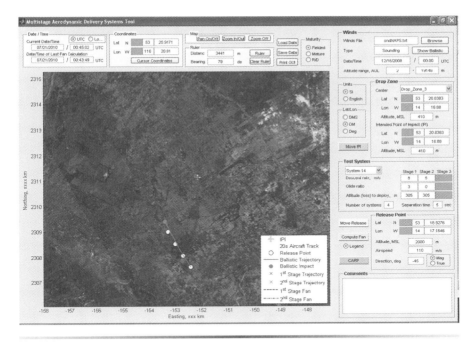

Fig. C.17   Safety fans GUI.

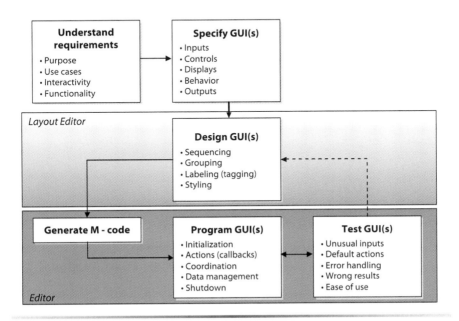

Fig. C.18   Designing professional GUIs.

Appendix D | Creating Standalone Applications

Introduction

One of the most frequently asked questions is whether it is worth spending so much time on developing complex user-friendly GUIs and making programs look professional, that is, foolproof, knowing that they could only be used by a developer, on a computer that has the particular version of MATLAB with all necessary toolboxes installed. The answer is yes, it is totally worth it, because once you developed your program, you have the capability to convert it into either an executable file or shared library that can be used on another computer, which has an older version of MATLAB, lacks some toolboxes, or even does not have MATLAB at all. It is the MATLAB Compiler, one of the MATLAB's toolboxes, which allows you to do this. Although covering this toolbox in detail is well beyond the scope of this textbook, the following simply shows how easy it is to make your MATLAB software available to other non-MATLAB users. Specifically, this appendix starts from introducing the MATLAB Compiler, followed by presenting the GUI for the MATLAB Compiler that assists in organizing all your files into one project. It then shows how to build and package these files for deployment. Finally, it overviews what needs to be done on the deployment computer in order to be able to run your standalone application.

D.2 **Overview of the MATLAB Compiler**

The MATLAB Compiler allows you to run your MATLAB applications outside the MATLAB environment. You can build a standalone application or interface to use your code as a shared library if you need to integrate it into C or C++. If you need to integrate it into another development environments (Java, .NET, Excel) MATLAB offers separate builder products (listed in Table 1.11) that need to be bought and installed separately. In either case, the MATLAB code becomes encrypted, so that the end user cannot view or modify it.

Once the MATLAB Compiler is installed, you should run the

```
>> mbuild -setup
```

command to select C or C++ compiler to use with the MATLAB Compiler. You will be advised to choose from any compiler available on your machine

```
Please choose your compiler for building standalone MATLAB applications:
Would you like mbuild to locate installed compilers [y]/n?
```

If you have none, in response to the previous question you will see

```
Select a compiler:
[1] Lcc-win32 C 2.4.1 in C:\PROGRA~1\MATLAB\R2011a\sys\lcc
[0] None
```

The Lcc-win32 C 2.4.1 compiler is the one included with MATLAB. Alternatively, if you have C or C++ installed on your computer, you may see some other options to choose from, such as `Microsoft Visual C++ 2008`. Once you make your choice, MATLAB updates a couple of files

```
Trying to update options file:
C:\...\Administrator\Application Data\MathWorks\MATLAB\R2011a\compopts.bat
From template:
C:\PROGRA~1\MATLAB\R2011a\bin\win32\mbuildopts\lcccompp.bat
Done . . .
```

and you are ready to go.

The `mcc` function invokes the MATLAB Compiler

```
mcc [-options] mfile1 [mfile2 … mfileN]
```

It prepares MATLAB file(s) for deployment outside of the MATLAB environment, generates wrapper files in C or C++, optionally builds standalone binary files, and writes any resulting files into the current folder (by default). Among `options,` the major ones are

`mcc -m test.m`	To generate a standalone application from the M-file *test.m*
`mcc -l test.m`	To build a shared library from the M-file *test.m*

Other options include

`-f project`	To generate an application from the project file *project.prj* (see following section)
`-a file`	To allow you to add a file or folder to the encrypted archive (to be explained later)
`-o outputname`	To specify the name of the final executable file
`-d outputfolder`	To specify an output folder

We will not discuss other syntax options here, because we will use a graphical interface to build a standalone application rather than programming the

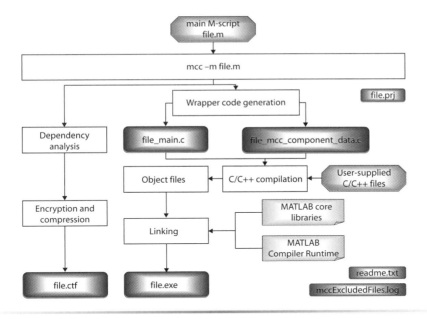

Fig. D.1 Creating a standalone executable application from *file.m* code.

process in the command line. However, as an example, let us make a standalone application from the WeekDayGUI file developed in Appendix C. All you have to do is issue the following command:

```
>> mcc -m WeekDayGUI
```

As a result, several files will be added to the current folder (Fig. D.1). Among theses files, the *WeekDayGUI.prj* file keeps compilation settings, the Component Technology File (CTF) *WeekDayGUI.ctf* contains encrypted and compressed MATLAB functions and data that define your application, two wrapper C-files provide an interface to the compiled M-code, and finally the *WeekDayGUI.exe* file is your standalone executable application. Two more files, *readme.txt* and *mccExcludedFiles.log*, are auxiliary: the first one features a detailed checklist to successfully deploy your application on another computer, and the second one contains the list of various toolbox functions that are not automatically included in the CTF file, so that if you intend to use any of them you have to add them explicitly using the -a option of the compiler.

As seen from Fig. D.1, the MATLAB Compiler uses the MATLAB Compiler Runtime (MCR), a standalone set of shared libraries that enables the execution of MATLAB files on computers without an installed version of MATLAB. Although the MCR provides most of MATLAB's functionality, it differs from MATLAB by not using its desktop graphical interface and having MATLAB files encrypted for portability and integrity.

You may now run the *WeekDayGUI.exe* file on your computer, not relying on MATLAB anymore. Simply double click on the name of this file, which will bring up the DOS window and then the GUI window of Fig. C.11 (if your code requires some inputs, you may use the DOS prompt). If you want to deploy your file on another computer, using the same operating system, but having no MATLAB installed on it at all, you will need to do one extra step, involving the MCR installation, as discussed in Sec. D.3.

D.3 Using the Deployment Tool

Instead of using the `mcc` function explicitly, MATLAB offers a user-friendly graphical interface of the Deployment Tool. The call

```
>> deploytool
```

opens the window shown in Fig. D.2a. The Deployment Tool allows you to specify the target (executable file or shared library), designate your main MATLAB script, add any supporting files that will not be found automatically through dependency checking, save compilation, and packaging preferences.

You start from choosing the goal of your project (standalone application in our case) and naming the project file that will keep information about all

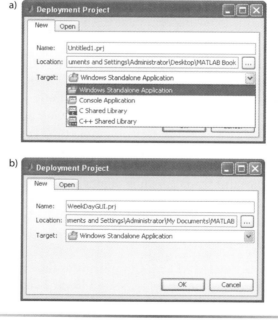

Fig. D.2 a) Opening the Deployment Tool dialog box and b) naming the project.

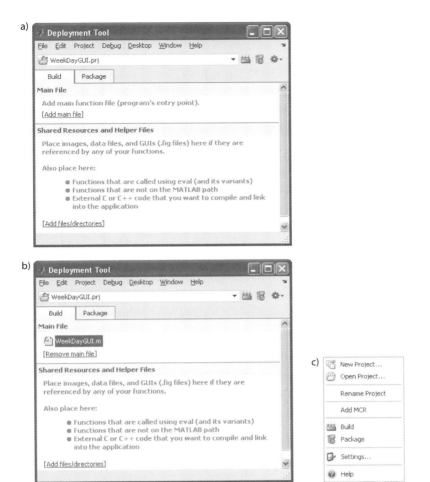

Fig. D.3 a) Building the package dialog box, b) choosing the main file of the project, and c) accessing the Build option via the File menu.

settings for your standalone application in case you would want to modify it in the future (for example, as shown in Fig. D.2b). Clicking **OK** brings another dialog box allowing you to choose the file to compile from (Fig. D.3a).

Let us click on the **Add main file** link and navigate to the *WeekDayGUI.m* file (the result is shown in Fig. D.3b). Now you are ready to proceed with building your standalone application. You may simply click on the **Build** button on the Deployment Tool toolbar, or go to the **File** menu and choose the **Build** option from there (Fig. D.3c). The status bar shows up (Fig. D.4a), and in a while the compilation is reported to be completed (Fig. D.4b). (Behind the scene, the Deployment Tool runs the `mcc` function with all necessary options.)

As discussed in Sec. D.1, when you build your application, the MATLAB Compiler determines which MATLAB functions are necessary to support your application, encrypts, and wraps them with C code (Fig. D.1). As a result of building your application using the Deployment Tool, the *WeekDayGUI.prj* file and the *WeekDayGUI* folder are created. The latter one contains two sub-folders: *src,* which holds intermediate output files (shown in Fig. D.1) and *distrib*, holding two files. These two files are the auxiliary *readme.txt* file and your standalone application *WeekDayGUI.exe*. The names of these two files show up in the package folder of the Deployment Tool (Fig. D.5a).

In the case of more sophisticated applications, before building your executable file you may need to manually add some other files (such as data files and images) or directories that are referenced by your code (Fig. D.1). You can do it by following the link **Add files/directories** (Fig. D.2a). For example, if your GUI uses one of the ActiveX controls, GUIDE creates the separate files for them. You must add them to your CTF archive explicitly, otherwise your application will not work on another computer that has no such add-ons on it (this CTF archive is then embedded in the executable file). You can add files or directories to your deployment package (Fig. B.5a) as well (by pushing **Add files/directories** link and navigating to them).

Eventually, you will need to run the Packaging tool, to bundle your standalone application (or shared library) with additional files you selected. You will be asked where to save the package at, 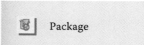 after which a self-extracting executable *file_pkg.exe* is created (on Windows computers, otherwise the results are saved in a *zip* archive).

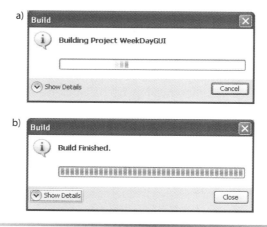

Fig. D.4   Building the project.

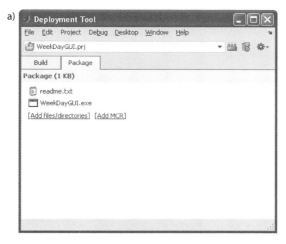

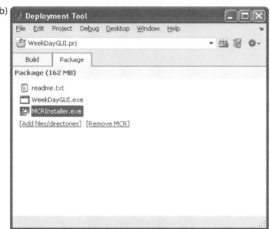

Fig. D.5 a) Standalone application package and b) enhanced package, including the MCR Installer.

D.4 Running Applications

Let us summarize what you should have done up to this point. You copied the MATLAB files needed for your application into the current folder. You run the main file in MATLAB to ensure that everything works properly. You created a standalone application (or shared library) encapsulating your MATLAB code in a C or C++ class. You tried to run an executable on your computer to ensure that it still works properly (if it is not, you have to fix whatever errors you may have). You run the Package tool to bundle your application with additional files you may need. As a result, you have everything you need in the *distrib* folder plus

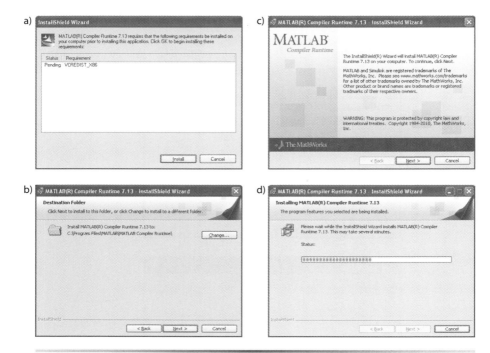

Fig. D.6   MCR installation.

a self-extracting executable package file. Now, you may share this self-extracting executable package file with end users.

To run your executable or shared library on another machine, the end user must first install the MCR (you have to have Administrator privileges to do this). You must run your applications with the version of the MCR associated with the version of the MATLAB Compiler with which it was created. That is why when packaging your application you have an option to include the self-extracting MCR library installer, *MCRInstaller.exe* (Fig. D.5b). The MCR supports the full MATLAB language and lets you include functions from MATLAB toolboxes. If you have it installed on that new computer, you will not need to include it to your deployment package again in the future (as long as you are using the same version when creating your standalone applications). If needed to be shared separately from the deployment package, the MCR installer can be found in *...MATLAB\toolbox\compiler\deploy\win32* directory.

If MCR is not included in the package, running a self-extracting executable package file results in extracting the two files shown in Fig. D.5a. If MCR is included, a self-extracting executable package extracts all three files shown in Fig. D.5b and initiates MCR installation. The program asks to choose a language of installation and suggests to install the MCR (Fig. D.6a).

Upon confirmation (Fig. D.6b), it suggests to choose a folder for installation (Fig. D.6c) and then proceeds installing it (Fig. D.6d). On Windows, you do not need to do anything else because the installer automatically modifies both the system registry and path. If you install the MCR on a computer that already has MATLAB on it, you must adjust the library path according to your needs. It means that to run deployed components using the MCR, *mcr_root\ver\runtime\win32* must appear in your system path before *matlabroot\runtime\win32* and vise versa.

The MCR installer supports the installation of multiple versions of the MCR on a target computer. This allows applications compiled with different versions of the MCR to execute side by side on the same machine. However, if you do not want multiple MCR versions on the target computer, you can remove the unwanted ones. On Windows, run Add or Remove Programs from the Control Panel to remove any of the previous versions.

Now that you have MCR installed on the end-user computer, you may run your standalone application by simply clicking on its name in a usual fashion.

Appendix E Aerospace Systems Modeling Tools

E.1 Introduction

As mentioned in Sec. 1.1.3 of Chapter 1, one of the MATLAB toolboxes and one Simulink blockset are intended to enhance Mathworks' product with scripts and blocks dedicated to modeling and simulation of aerospace vehicles. Although these tools are not part of the basic version of MATLAB/Simulink (that is, need to be acquired separately), we think it is appropriate to mention these tools here and illustrate that once you have completed this book, you are quite ready to proceed towards creating your own professionally looking functions and models. To this end, this appendix introduces one advanced topic, Simulink blocks masking, which was not covered in Chapter 15.

E.2 Aerospace Toolbox

Excluding the specific functions intended to create and manipulate graphic objects for animation within Flight Gear Simulator, the remaining functions, used to model aerospace systems in the MATLAB technical computing environment, can be subdivided into several major categories as presented in Table E.1.

To be more specific, these functions are as follows:

Axes Transformations

angle2dcm	Creates direction cosine matrix from rotation angles
angle2quat	Converts rotation angles to quaternion
dcm2alphabeta	Converts direction cosine matrix to angle of attack and sideslip angle
dcm2angle	Creates rotation angles from direction cosine matrix
dcm2latlon	Converts direction cosine matrix to geodetic latitude and longitude
dcm2quat	Converts direction cosine matrix to quaternion

Table E.1   Aerospace Toolbox Subgroups

Subgroup	Brief Description
Axes transformations	Transforms axes of coordinate systems to different types
Environment	Simulates various aspects of aircraft environment, such as atmosphere conditions, gravity, magnetic fields, and wind
File reading	Reads standard aerodynamic file formats into the MATLAB interface
Flight parameters	Computes various flight parameters, including ideal airspeed correction, Mach number, and dynamic pressure
Gas dynamics	Provides various gas dynamics tables
Quaternion math	Assures common mathematical and matrix operations on a quaternion
Time	Enables time calculations, including Julian dates, decimal year, and leap year
Unit conversion	Converts common measurement units from one system to another

`dcmbody2wind`	Creates a direction cosine matrix from the angle of attack and sideslip angle
`dcmecef2ned`	Creates a direction cosine matrix from the geodetic latitude and longitude
`ecef2lla`	Converts Earth-centered Earth-fixed (ECEF) coordinates to geodetic coordinates
`flat2lla`	Estimates array of geodetic latitude, longitude, and altitude coordinates from flat Earth position
`geoc2geod`	Converts geocentric latitude to geodetic latitude
`geod2geoc`	Converts geodetic latitude to geocentric latitude
`igrf11magm`	Calculates Earth magnetic field using eleventh generation of International Geomagnetic Reference Field
`lla2ecef`	Converts geodetic coordinates to ECEF coordinates
`lla2flat`	Estimates flat Earth position from geodetic latitude, longitude, and altitude
`quat2angle`	Converts quaternion to rotation angles
`quat2dcm`	Converts quaternion to direction cosine matrix

Environment

`atmoscira`	Uses COSPAR International Reference Atmosphere 1986 model
`atmoscoesa`	Uses 1976 COESA model
`atmosisa`	Uses International Standard Atmosphere model
`atmoslapse`	Uses Lapse Rate Atmosphere model
`atmosnonstd`	Uses climatic data from MIL-STD-210 or MIL-HDBK-310

`atmosnrlmsise00`	Implements mathematical representation of the 2001 U.S. Naval Research Laboratory Mass Spectrometer and Incoherent Scatter Radar Exosphere
`atmospalt`	Calculates pressure altitude based on ambient pressure
`geoidegm96`	Calculates geoid height as determined from EGM96 Geopotential Model
`geoidheight`	Calculates geoid height
`gravitycentrifugal`	Implements centrifugal effect of planetary gravity
`gravityspharicalharmonic`	Implements spherical harmonic representation of planetary gravity
`gravitywgs84`	Implements 1984 World Geodetic System (WGS84) representation of Earth's gravity
`gravityzonal`	Implements zonal harmonic representation of planetary gravity
`wrldmagm`	Uses World Magnetic Model

File Reading

`datcomimport`	Brings DATCOM file into MATLAB environment

Flight Parameters

`airspeed`	Computes airspeed from velocity
`alphabeta`	Computes incidence and sideslip angles
`dpressure`	Computes dynamic pressure using velocity and density
`geocradius`	Estimates radius of ellipsoid planet at geocentric latitude
`machnumber`	Computes Mach number using velocity and speed of sound
`rrdelta`	Computes relative pressure ratio
`rrsigma`	Computes relative density ratio
`rrtheta`	Computes relative temperature ratio

Gas Dynamics

`flowfanno`	Computes Fanno line flow relations
`flowisentropic`	Calculates isentropic flow ratios

flownormalshock	Produces normal shock relations
flowprandtlmeyer	Calculates Prandtl-Meyer functions for expansion waves
flowrayleigh	Computes Rayleigh line flow relations

Quaternion Math

quatconj	Calculates conjugate of quaternion
quatdivide	Divides quaternion by another quaternion
quatinv	Calculates inverse of quaternion
quatmod	Calculates modulus of quaternion
quatmultiply	Calculates product of two quaternions
quatnorm	Calculates norm of quaternion
quatnormalize	Normalizes quaternion
quatrotate	Rotates vector by quaternion

Time

decyear	Calculates decimal year
juliandate	Calculates Julian date
leapyear	Determines leap year
mjuliandate	Calculates modified Julian date

Unit Conversion

convacc	Converts from acceleration units to desired acceleration units
convang	Converts from angle units to desired angle units
convangacc	Converts from angular acceleration units to desired angular acceleration units
convangvel	Converts from angular velocity units to desired angular velocity units
convdensity	Converts from density units to desired density units
convforce	Converts from force units to desired force units
convlength	Converts from length units to desired length units

convmass	Converts from mass units to desired mass units
convpres	Converts from pressure units to desired pressure units
convtemp	Converts from temperature units to desired temperature units

For example, the `airspeed` function from the Flight Parameters group is as simple as

```
function as = airspeed(vel)
%   AIRSPEED Compute airspeed from velocity.
%     AS=AIRSPEED(V) computes M airspeeds, AS, from an M-by-3 array
%                   of velocities, V.
%     ...
if ~isnumeric(vel)
    error('aero:airspeed:notnumeric',...
        'Velocity input was not a numeric value.');
end
if (size(vel,2)==3)
    as = sqrt(vel(:,1).^2 + vel(:,2).^2 + vel(:,3).^2);
else
    error('aero:airspeed:wrongdim','Velocity array is not M-by-3.');
end
```

All it does is compute the norm of a three-element vector, in this case velocity vector. This operation could be programmed in a single line, however this function contains several foolproof elements, which makes it more robust. It warns you when you try to pass a non-numeric value (the `isnumeric` function was presented in Table 5.2). It also notifies you in the case when a velocity vector has fewer or greater than three elements (three columns, to be more precise). The `error` function simply displays a message supplemented by an optional message identifier, enabling you to better identify the source of an error, and aborts the function.

As seen, the key mathematical expression is vectorized, which allows you to process an M by 3 array of velocities as opposed to just one 1 by 3 vector. Following the header line, there is an extended comment section (only part of it is shown above for compactness), explaining the usage of this function and showing up in response to the help call

```
>> help airspeed
```

To find a location of this Aerospace Toolbox function and open it in the Editor, you may use the following call:

```
>> open(which('airspeed'))
```

Consider one more example, the `atmoslapse` function of the Environment group, which computes parameters of the atmosphere using a lapse rate for ambient temperature

```
function [T,a,P,rho]=atmoslapse(h,g,gamma,R,L,hts,htp,rho0,P0,T0)
%   ATMOSLAPSE Use Lapse Rate Atmosphere Model.
%   [T,A,P,RHO]=ATMOSLAPSE(H,G,GAMMA,R,L,HTS,HTP,RHO0,P0,T0) implements
%   the mathematical representation of the lapse rate atmospheric
%   equations for ambient temperature, pressure, density, and speed of
%   sound for the input geopotential altitude.
%   ...
%   Limitation:
%   Below the geopotential altitude of 0 km and above the geopotential
%   altitude of the tropopause, temperature and pressure values are held.
%   Density and speed of sound are calculated using a perfect gas
%   relationship.
%   Example:
%   Calculate the atmosphere at 1000 meters with the International
%   Standard Atmosphere input values:
%       [T,a,P,rho]=atmoslapse(1000,9.80665,1.4,287.0531,0.0065,...
%                              11000,20000,1.225,101325,288.15);
%   ...
error(nargchk(10,10,nargin,'struct'));
  if ~(isscalar(g)     && isscalar(gamma) && isscalar(R)    &&...
       isscalar(L)     && isscalar(hts)   && isscalar(htp)  &&...
       isscalar(rho0)  && isscalar(P0)    && isscalar(T0))
        error('aero:atmoslapse:nonscalar',...
            'All inputs other than altitude must be scalars.');
  end
if ~isnumeric(h)
    error('aero:atmoslapse:notnumeric',...
          'Height input was not a numeric value.');
end
 for i = length(h):-1:1
   if (h(i)>htp)
         error('aero:atmoslapse:gthtp',...
               'Height cannot exceed the tropopause altitude.');
   end
   if (h(i)<0)
        error('aero:atmoslapse:ltzero','Height cannot be below zero.');
   end
   if (h(i)>hts)
        T(i) = T0 - L*hts;
        expon(i) = exp(g/(R*T(i))*(hts - h(i)));
   else
        T(i) = T0 - L*h(i);
        expon(i) = 1.0;
```

```
   end
end
a = sqrt(T*gamma*R);
theta = T/T0;
P = P0*theta.^(g/(L*R)).*expon;
rho = rho0*theta.^((g/(L*R))-1.0).*expon;
```

(By the way, you may compare the output of this function with that of yours, developed in Problem 5.3.)

As was the case for the `airspeed` function, the `atmoslapse` function has an extended help comments section, and the actual computations are carried out at the four last lines of code. The remaining lines represent the multiple guards from possible erroneous inputs. The `nargchk` function, used in the beginning of the function, has a general syntax

```
nargchk(low,high,N,'struct')
```

and returns an appropriate error message structure if `N` is not between `low` and `high`. When used with the `error` function as above, the `error` function accepts this error message structure and passes it as an input to `error`. Note the usage of short-circuit logical AND, discussed in Sec. 5.2. If any of the `isscalar` returns is `false`, there is no need to continue with the remaining ones.

From the standpoint of programming, there is no problem using `nargchk` here, however, it would be good if the developer would allow multiple syntax. For example, it would be natural to allow a single input argument, making other parameters assume some default values, or two input parameters, in which case the second input could be a structure having nine fields and carrying nine required scalars.

E.3   Aerospace Blockset

The Aerospace Blockset features a dozen of libraries as shown in Fig. E.1. Some of them have as little as just one block. Others group their blocks into sublibraries. All available blocks are presented in Figs. E.2–E.12.

If you drag any block into your model and double click on it, the Block Parameters window will pop-up as was the case with any basis Simulink blocks (Figs. 15.9b, 15.11, 15.12a, etc.). For example, the Three-axis Accelerometer block of the Guidance, Navigation and Control (GNC) library, shown in Fig. E.11, features two tabbed panes, where the user can define different parameters (Fig. E.14a,b). (By the way, note how the Noise seeds are defined by default in the Noise pane of Fig. E.14b, they differ one from another by 1—this issue was discussed in Fig. 15.61.)

The Block Parameters window can also be opened by right clicking on the block and choosing the **Mask Parameters** option from the drop-down menu (Fig. E.14c). Note that instead of "block" Simulink refers to "mask."

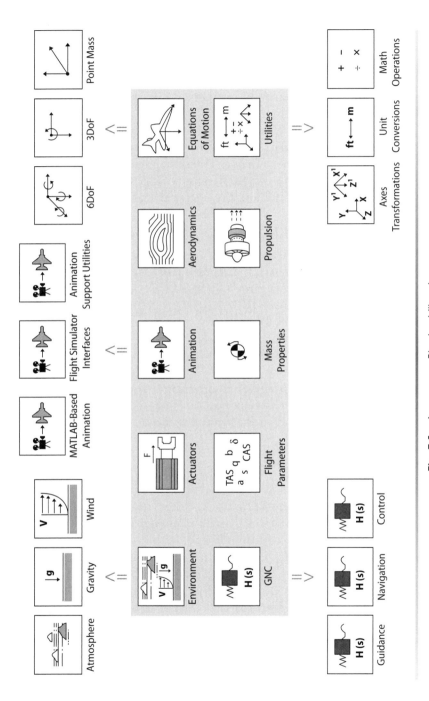

Fig. E.1 Aerospace Blockset libraries.

Fig. E.2 Actuators library blocks.

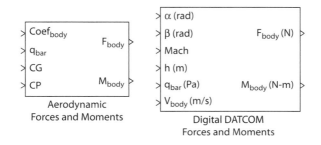

Fig. E.3 Aerodynamics library blocks.

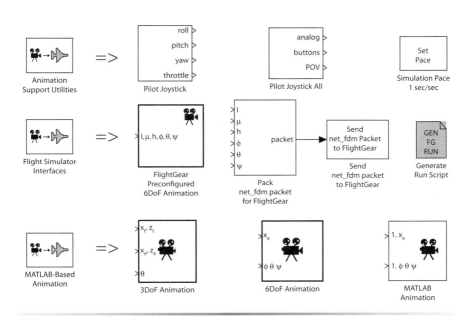

Fig. E.4 Animation libraries blocks.

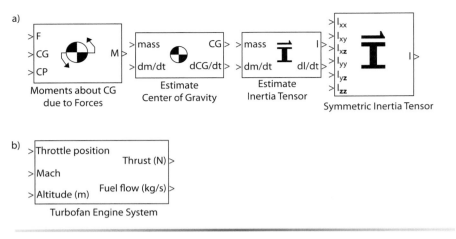

Fig. E.5 a) Mass properties and b) propulsion libraries blocks.

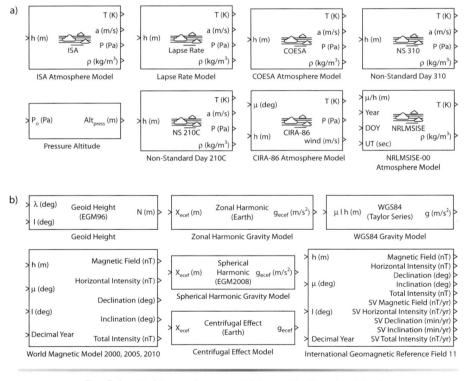

Fig. E.6 a) Atmosphere and b) gravity libraries blocks.

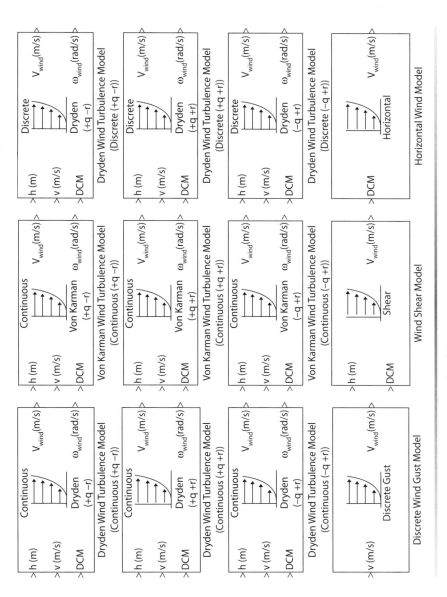

Fig. E.7 Winds library blocks.

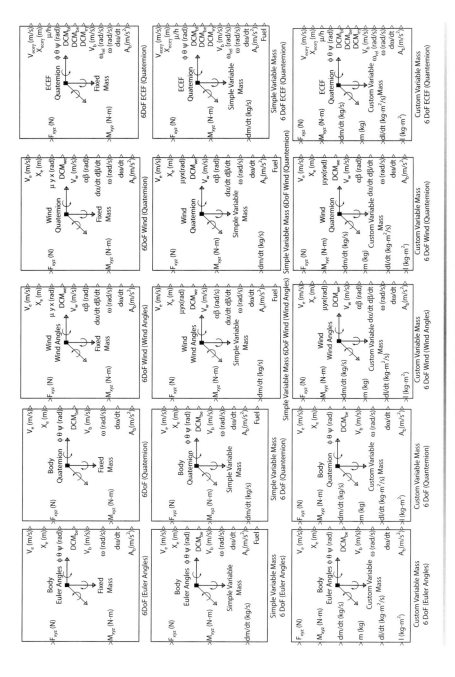

Fig. E.8 Six DoF library blocks.

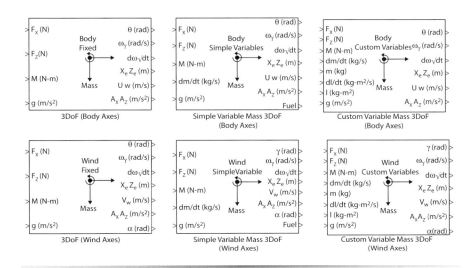

Fig. E.9 Three DoF library blocks.

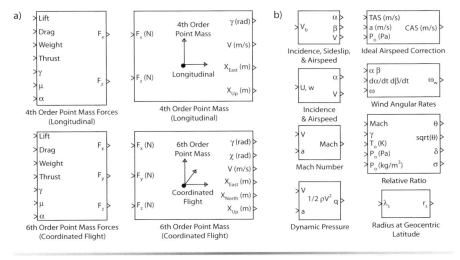

Fig. E.10 a) Point mass library and b) flight parameters libraries blocks.

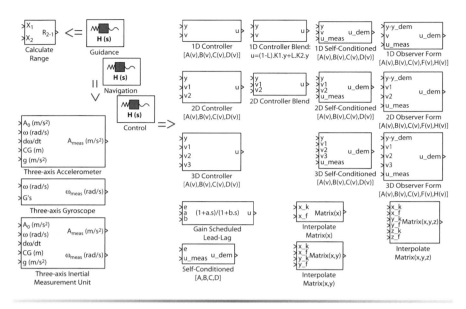

Fig. E.11 GNC library blocks.

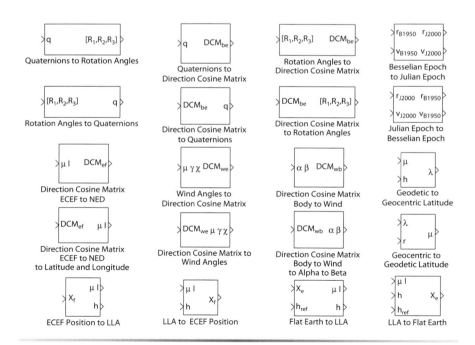

Fig. E.12 Axes transformations library blocks.

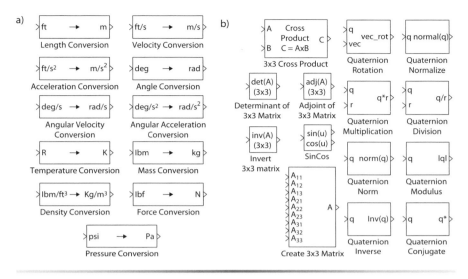

Fig. E.13 Unit conversions a) and b) Math operations libraries blocks.

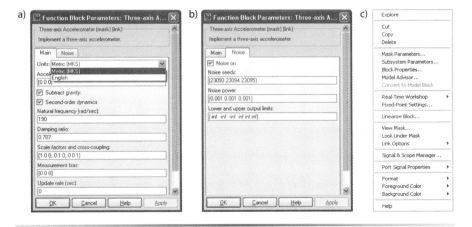

Fig. E.14 Example of the tabbed panes of the masked block (a and b) and the options menu c).

The reason for this is that every block of the Aerospace Blockset represents a subsystem like the one we created in Sec. 15.3.2 (Figs. 15.22 and 15.23). The only difference is that the developers made one additional step—they masked the content of the subsystem.

E.4 Block Masking

There are several reasons for masking a block. Masking

- Prevents unintended modification by hiding its content behind a mask
- Enables creating a single block parameter dialog with its own description, parameter prompts, and help text
- Makes it possible to replace a subsystem's standard icon with a customized icon that depicts its purpose
- Creates a custom block that can be placed in a library

To see what resides under the mask, you should choose the **Look Under Mask** option of the drop-down menu of Fig. E.14c, which in the case of the Three-axis Accelerometer block, opens a model as shown in Fig. E.15.

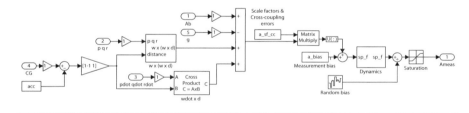

Fig. E.15   Three-axis accelerometer.

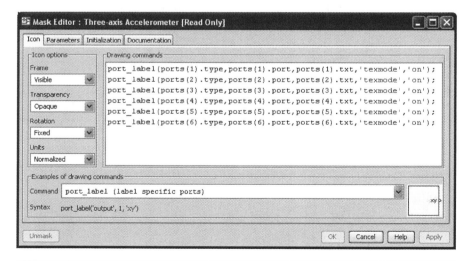

Fig. E.16   Icon pane of the Mask Editor.

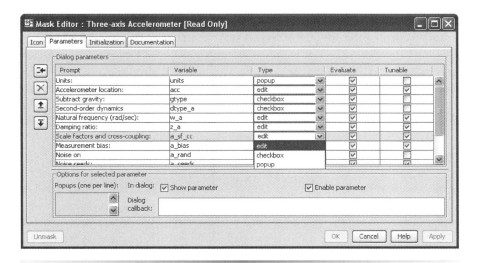

Fig. E.17 Parameters pane of the Mask Editor.

As seen, this model relies on parameters defined in two pans of the Block Properties window (Figs. E.14a,b). Three Constant blocks refer to variables `acc`, `a_sf_cc`, and `a_bias`. The Dynamics subsystem block uses `w_a`, `z_a`, and `dtype_a`. The Saturation block uses a parameter `a_sat`. The noise power and seeds in the Random bias block are defined as `a_pow` and `a_seed`. To see where/how the values for these parameters are defined, you should choose the **View Mask** option from the drop-down menu of Fig. E.14c, which brings up the Mask Editor window, shown in Fig. E.16. Choosing the second tab, Parameters, opens a window (Fig. E.17) defining the names of the fields in Block Parameters panes (Fig. E.14a, b), their types (editable window, check box, or pop-up menu), and the names of parameters. The values entered in two panes (Fig. E.14a, b) are stored in the mask workspace (each mask has its own workspace), and that is where the model reads them from. A block parameter can refer to variables defined in the mask workspaces of the subsystem or nested subsystems that contain the block or the base (MATLAB) workspace. If a variable is defined on more than one level in the model hierarchy, the most local definition is used.

The name of the block, its description and the prompt, appearing when the Help button is pushed, are defined in the Documentation pane of the Mask Editor (Fig. E.18a). The Initialization pane of the Mask Editor is used to accommodate MATLAB code initializing the model when it loads and starts. This code can refer only to variables in its local workspace. For example, the key initialization command shown in Fig. E.18b is the first one. It calls the *airoicon.m* function that defines the `ports` structure, used to represent the names on inputs and outputs for a specific block. You can find this function and look at it by issuing the following call

```
>> open(which('airoicon'))
```

Specifically, for the Three-axis Accelerometer block shown in Fig. E.11, the drawing commands of Fig. E.16 rely on the `txt` field of the `port` structure to label all six ports using TeX language (to produce Greek letter w, superscripts, and subscripts). Examples of appropriate syntax for a variety of allowable drawing functions show up at the bottom of the Icon pane (Fig. E.16). Other options (showing up in the drop-down menu) are shown in Table E.2. The drawing functions have access to all the variables in the mask workspace, so your icon can be dynamic and reflect the current values of block parameters (see an example for the `text` command in Table E.2).

The Icon options on the Icon pane (Fig. E.16) allow you to set up Icon properties that control its appearance. Making the Frame **Invisible** (as opposed to default **Visible**) hides the rectangle that encloses the block. Choosing **Transparent** (as opposed to default **Opaque**) allows you to see underneath the icon. The **Rotates** option (as opposed to default **Fixed**) of the rotation property makes the icon rotation and flipping when the block is rotated or flipped. The Units property options are **Autoscale**, **Pixels**, and **Normalized** (default).

a)
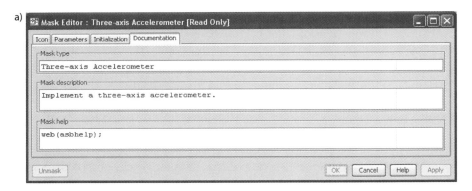

b)
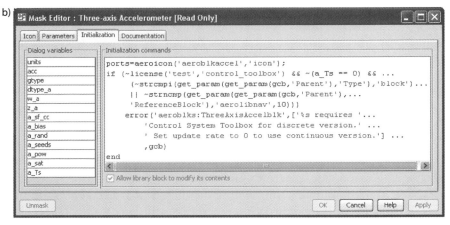

Fig. E.18 a) Documentation and b) Initialization panes of the Mask Editor.

Table E.2 Mask Editor Drawing Functions

Function	Usage	Example
`port_label`	Labels a specific port	`port_label('output',...` `1,'Output 1')`
`disp`	Displays text centered on mask icon	`disp('MK611-8 Engine')`
`text`	Shows text at a specific location	`text(5,10,...` `'F-22 Model')`
`fprintf`	Prints (variable) formatted text	`fprintf('Damping'...` `'Ratio is %5.3f',.7)`
`plot`	Plots 2-D curves	`plot([5 0 5 10 5],...` `[0 5 2 5 0],'r')`
`patch`	Draws filled shapes (see example in Fig. 6.37b)	`patch([5 0 5 10 5],...` `[0 5 10 5 0],'y')`
`image`	Shows a picture on the block	`image(imread...` `('Cessna172.jpg'))`
`dpoly`	Displays a transfer function based on numerator and denominator polynomials (see example in the Transfer Fcn block of Fig.15.24a)	`dpoly([0 0 1],[1 2 1])`
`droots`	Displays a transfer function based on zero-pole-gain representation (allowing you to change the default independent variable name)	`droots([],[-2 -3],...` `5,'z')`
`color`	Changes color of the subsequent mask icon drawing command	`color('b');port_label...` `('output',1,'X_b')`

You may access parameters of the current block using the `get_param` function. For instance, if you have the Three-axis Accelerometer block of your model selected and issue the following command (in the Command window of MATLAB)

```
>> get_param(gcb,'DialogParameters')
```

you get a cell array containing the names of the dialog parameters of the specified block

```
ans =

    units: [1x1 struct]
      acc: [1x1 struct]
    gtype: [1x1 struct]
```

```
 dtype_a: [1x1 struct]
     w_a: [1x1 struct]
     z_a: [1x1 struct]
 a_sf_cc: [1x1 struct]
  a_bias: [1x1 struct]
  a_rand: [1x1 struct]
 a_seeds: [1x1 struct]
   a_pow: [1x1 struct]
   a_sat: [1x1 struct]
    a_Ts: [1x1 struct]
```

(you will not be able to access the values of variables from the base MAT-LAB workspace). The `gcb` returns the full pathname of a current block in the current system (the `get(gcbh)` call will query selected block properties via its handle, `gcbh`.)

The `get_param` function is helpful in initialization code (see code in Fig. E.18a) and may also be used in the Callback window of the Parameters pane (Fig. E.17). For example, having the following set of commands entered in this window:

```
if str2num(get_param(gcb,'z_a'))<0
    error('Damping Ratio must be within [0;1]')
end
```

will bring the error message if you attempt to change the damping ratio to some negative number (that is, when you push **Apply** and **OK** buttons on the Block Parameters dialog box of Fig. E.14a).

Now that you know how the Mask Editor works, you can go back to a subsystem created in Sec. 15.3.2, right click on its block and choose the **Mask Subsystem** option (alternatively, you can choose this option from the **Edit** menu or use <Ctrl>+<M>). This brings an empty Mask Editor window you can work on to convert the second-order system you created back there (Fig. 15.47a) into a professional-looking masked block (you will not need to define variables in InitFcn callback of Fig. 15.47b then).

After you created a mask, you can always turn it off by pushing the **Unmask** button of the Mask Editor or even from the MATLAB Command window (having your block selected)

```
>> set_param(gcb,'Mask','off')
```

Note that, for the masked blocks of Simulink blocksets (including the Aerospace Blockset), this latter command is the only option.

BIBLIOGRAPHY

Abramowitz, M., and Stegun, I.A., (eds) *Handbook of Mathematical Functions with Formulas, Graphs, and Mathematical Tables*, 10th edn., Dover Publications, New York, 1972.

Ashino, R., Nagase, M., and Vaillancourt, R., "Behind and Beyond the MATLAB ODE Suite," CRM-2651 paper, *Centre de recherches mathématiques*, Université de Montréal, Québec, 2000.

Bank, R.E., Coughran, W.C., Fichtner, W.J., Grosse, E.H., Rose, D.J., and Smith, R.K., "Transient Simulation of Silicon Devices and Circuits," *IEEE Transactions on Computer-Aided Design of Integrated Circuits and Systems*, Vol. 4, No. 4, 1985, pp. 436–451.

Bogacki, P., and Shampine, L.F., "A 3(2) Pair of Runge-Kutta Formulas," Applied Mathematics Letters, Vol. 2, No. 4, 1989, pp. 321–325, doi: 10.1016/0893-9659(89)90079-7.

Chapman, S.J., *MATLAB Programming for Engineers*, 3rd edn., Thomson, Scarborough, UK 2005.

Chapra, S.C., *Applied Numerical Methods with MATLAB for Engineers and Scientists*, McGraw-Hill, New York, 2005.

Chapra, S.C., and Canale, R.P., *Numerical Methods for Engineers*, 5th edn., McGraw-Hill, New York, 2006.

Colgren, R., *Basic MATLAB, Simulink and Stateflow*, American Institute of Aeronautics and Astronautics, Reston, VA, 2006.

Dabney, J.B., and Harman, T.L., *Mastering Simulink*, Prentice Hall, Upper Saddle River, NJ, 2004.

Dharmaraja, S., Wang, Y., and Strang, G., "Optimal Stability for Trapezoidal-Backward Difference Split-Steps," *IMA Journal of Numerical Analysis*, Vol. 30, No. 1, 2010, pp. 141–148, doi: 10.1093/imanum/drp022.

Dormand, J. R., and Prince, P.J., "A Family of Embedded Runge-Kutta Formulae," *Journal of Computational and Applied Mathematics*, Vol. 6, No. 1, 1980, pp. 9–26, doi: 10.1016/0771-050X(80)90013-3.

Fausett, L.V., *Numerical Methods: Algorithms and Applications*, Prentice Hall, Upper Saddle River, NJ, 2003.

Fletcher, R., *Practical Methods of Optimization*, 2nd edn., Wiley, New York, 2000.

Forsythe, G., Malcolm, M., and Moler, C., *Computer Methods for Mathematical Computations*, Prentice-Hall, New Jersey, 1977.

Fröberg C.-E., *Numerical Mathematics: Theory and Computer Applications*, Addison Wesley, Reading, MA, 1985.

Gander, W., and Gautschi, W., "Adaptive Quadrature—Revisited," *BIT Numerical Mathematics*, Vol. 40, 2000, pp. 84–101.

Gilat, A., *MATLAB: An Introduction with Applications*, 2nd edn., Wiley, New York, 2004

Gill, P.E., Murray, W., and Wright, M.H., *Practical Optimization*, Academic Press, London, UK, 1981.

Goldberg, D.E., *Genetic Algorithms in Search, Optimization, and Machine Learning*, Addison Wesley, Upper Saddle River, NJ, 1989.

Hahn, B.D., *Essential MATLAB for Scientists and Engineers*, Elsevier, New York, 2005.

Hairer, E., and Wanner, G., *Solving Ordinary Differential Equations II. Stiff and Differential-Algebraic Problems*, 2nd edn., Springer-Verlag, Berlin, New York, 2002.

Hairer, E., Nørsett, S.P., and Wanner, G., Solving Ordinary Differential Equations I: Nonstiff Problems, 2nd edn., Springer-Verlag, Berlin, New York, 2000.

Hanselman, D., and Littlefield, B.R., *Mastering MATLAB 7*, Prentice Hall, Upper Saddle River, NJ, 2005.

Herniter, M.E., *Programming in MATLAB*, 2nd edition, McGraw-Hill, New York, 2003.

Higham, D.J., and Higham, N.J., *MATLAB Guide*, 2nd edn., SIAM, Philadelphioa, 2005.

Hooke, R., and Jeeves, T.A., "Direct Search Solution of Numerical and Statistical Problems," *Journal of the Association for Computing Machinery*, Vol. 8, No. 2, 1961, pp. 212–229.

Hosea, M.E., and Shampine, L.F., "Analysis and Implementation of TR-BDF2," *Applied Numerical Mathematics*, Vol. 20, No. 1-2, 1996, pp. 21–37.

IEEE Standard for Binary Floating-Point Arithmetic (IEEE 754-1985).

IEEE Standard for Floating-Point Arithmetic (IEEE 754-2008).

Kahaner, D., Moler, C., and Nash, S., *Numerical Methods and Software*, Prentice-Hall, Englewood, NJ, 1988.

Kierzenka, J., and Shampine, L.F., "A BVP Solver based on Residual Control and the MATLAB PSE," *ACM Transactions on Mathematical Software*, Vol. 27, No. 3, 2001, pp. 299–316.

Kierzenka, J., and Shampine, L.F., "A BVP Solver that Controls Residual and Error," *Journal of Numerical Analysis, Industrial and Applied Mathematics*, Vol. 3, No. 1-2, 2008, pp. 27–41.

Krylov, V.I., *Approximate Calculation of Integrals*, Macmillan, New York, 1962 (Translated from Russian).

Magrab, E.B, Azarm, S., et al. *An Engineer's Guide to MATLAB*, 2nd edn., Prentice Hall, Upper Saddle River, NJ, 2005.

Mathews, J.H., and Fink, K.K., *Numerical Methods Using MATLAB*, 4th edn., Prentice Hall, Upper Saddle River, NJ, 2004.

Michels, H.H., "Abscissas and Weight Coefficients for Lobatto Quadrature," *Mathematics of Computation*, Vol. 17, No. 83, 1963, pp. 237–244.

Moler, C., *Experiments with MATLAB*, The MathWorks, 2009.

Moler, C., *Numerical Computing with MATLAB, Revised Reprint*, SIAM, Philadelphia, 2008.

Moler, C., "The Growth of MATLAB® and The MathWorks over Two Decades," *The Mathworks News and Notes*, January, 2006.

Nelder, J.A., and Mead, R., "A Simplex Method for Function Minimization," *Computer Journal*, Vol. 7, 1965, pp. 308–313.

Netlib Repository at The University of Tennessee at Knoxville and Oak Ridge National Laboratory, www.netlib.org.

Nocedal, J., and Wright, S.J., *Numerical Optimization*, Springer Series in Operations Research and Financial Engineering, Springer Verlag, Berlin, Germany, 1999.

Palm, W.J., III, *Introduction to MATLAB 7 for Engineers*, McGraw-Hill, New York, 2005.

Pratap, R., *Getting Started with MATLAB 7*, Oxford University Press, New York, 2006.

Ralston, A., A First Course in Numerical Analysis, McGraw-Hill, New York, 1965.

Recktenwald, G.W., *Introduction to Numerical Methods and MATLAB: Implementations and Applications*, Prentice Hall, Upper Saddle River, NJ, 2000.

Scarborough, J.B., *Numerical Mathematical Analysis*, 6th edn., Johns Hopkins University Press, Baltimore, MD, 1966.

Shampine, L.F., "Some Practical Runge-Kutta Formulas," *Mathematics of Computation*, Vol. 46, No. 173, 1986, pp. 135–150, doi: 10.2307/2008219.

Shampine, L.F., "Vectorized Adaptive Quadrature in MATLAB," *Journal of Computational and Applied Mathematics*, Vol. 211, No. 2, 2008, pp. 131–140.

Shampine, L.F., and Gordon, M.K., *Computer Solution of Ordinary Differential Equations: the Initial Value Problem*, W.H.Freeman & Co. Ltd, San Francisco, 1975.

Shampine, L.F., and Hosea, M.E., "Solving ODEs and DDEs with Residual Control," *Applied Numerical Mathematics*, Vol. 52, 2005, pp. 113–127.

Shampine, L.F., and Reichelt, M.W., "The MATLAB ODE Suite," *SIAM Journal on Scientific Computing*," Vol. 18, No. 1, 1997, pp. 1–22, doi: 10.1137/S1064827594276424.

Shampine, L.F., and Thompson, S., "Solving DDEs in MATLAB," *Applied Numerical Mathematics*, Vol. 37, 2001, pp. 441–458.

Shampine, L.F., Gladwell, I., and Thompson, S., *Solving ODEs with MATLAB*, Cambridge University Press, Cambridge, 2003.

Shampine, L.F., Reichelt, M.W., and Kierzenka, J., "Solving Boundary Value Problems for Ordinary Differential Equations in MATLAB with bvp4c," 2000, www.mathworks.com/bvp_tutorial.

The MathWorks, www.mathworks.com.

The Wikipedia, www.wikipeida.org.

Wilkinson, J.H., *Rounding Errors in Algebraic Processes*, Dover Publications, New York, 1994.

Wolfram MathWorld, mathworld.wolfram.com.

Ying, W., Henriquez, C.S., and Rose, D.J., "Composite Backward Differentiation Formula: an Extension of the TR-BDF2 Scheme," 2009, www.math.sjtu.edu.cn/faculty/wying/mypapers/CBDFs.pdf.

MATLAB Functions Index

This index lists the most important icons used by MATLAB, MuPAD, Simulink, and GUIDE.

INDEX

SUPPORTING MATERIALS

To access the supporting materials that accompany this work, visit http://www.aiaa.org/books and click on the "Supporting Materials" link. Select this work from the list provided and enter the password

computing

when prompted.

Many of the topics introduced in this book are discussed in more details in other AIAA publications. For a complete listing of titles in the AIAA Education Series, as well as other AIAA publications, please visit www.aiaa.org.

AIAA is committed to devoting resources to the education of both practicing and future aerospace professionals. In 1996, the AIAA Foundation was founded. Its programs enhance scientific literacy and advance the arts and sciences of aerospace. For more information, please visit www.aiaafoundation.org.

a)

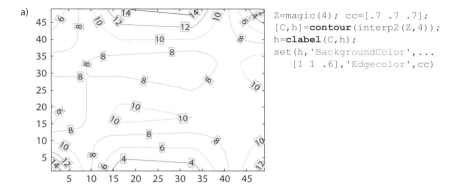

```
Z=magic(4); cc=[.7 .7 .7];
[C,h]=contour(interp2(Z,4));
h=clabel(C,h);
set(h,'BackgroundColor',...
    [1 1 .6],'Edgecolor',cc)
```

b)

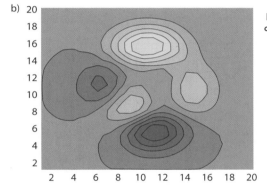

```
[C,h]=contourf(peaks(20),10);
colormap autumn
```

c)

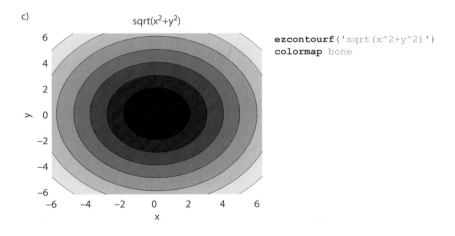

```
ezcontourf('sqrt(x^2+y^2)')
colormap bone
```

Fig. 6.20 Examples of contour plots.

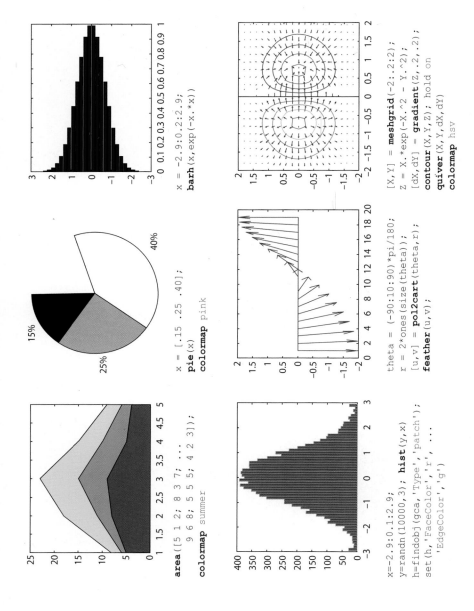

Fig. 6.21 Examples of area plots, histograms, and direction-type plots.

Fig. 6.30 a) Triangular mesh and b) surface plots.

a)

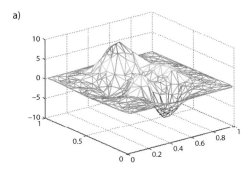

b)

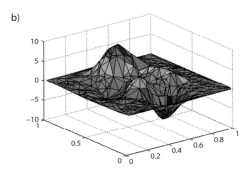

Fig. 6.31 a) Examples of cropping the surface and b) using `meshc` to visualize a matrix.

a)

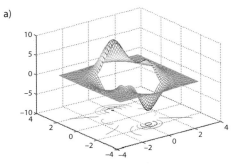

```
[x,y]=meshgrid(-3:.125:3);
z=peaks(x,y);
z(15:35,15:35)=NaN;
meshc(x,y,z)
```

b)

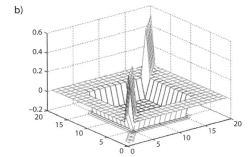

```
w=0.5*eye(20);
w(5:15,5:15)=-0.2*ones(11);
meshc(w)
```

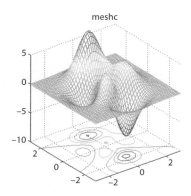

meshc

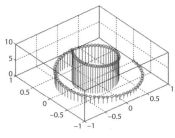

```
t=0:.1:10; y=exp(-(.1+i)*t);
stem3(real(y),imag(y),t)
hold, view(-39.5,62)
plot3(real(y),imag(y),t,'r')
```

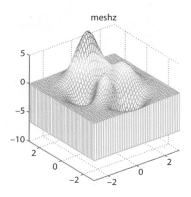

meshz

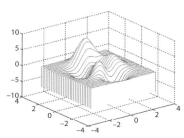

```
[x,y,z] = peaks(30);
waterfall(x,y,z)
colormap winter
```

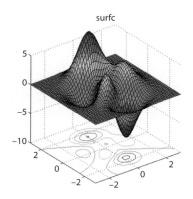

surfc

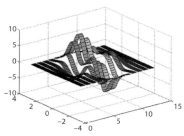

```
x=-3:0.5:3; y=-3:0.1:3;
[A,B]=meshgrid(x,y);
Z=peaks(A,B);
ribbon(B,Z)
colormap jet
```

Fig. 6.28 Illustration of a variety of the surface plotting functions.

Fig. 6.32 3-D analogous of the stem function and special surface visualizing functions.

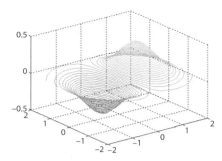

```
[X,Y]=meshgrid([-2:.25:2]);
Z=X.*exp(-X.^2-Y.^2);
contour3(X,Y,Z,50)
colormap spring
```

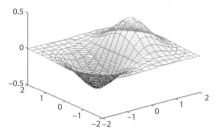

```
[X,Y]=meshgrid([-2:.25:2]);
Z=X.*exp(-X.^2-Y.^2);
contour3(X,Y,Z,50)
surface(X,Y,Z,...
        'EdgeColor','m',
        'FaceColor','none')
grid off, colormap gray
```

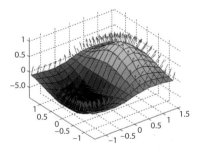

```
[X,Y]=meshgrid(-1.4:.2:1.4);
Z=2*X.*exp(-X.^2-Y.^2);
[U,V,W] = surfnorm(X,Y,Z);
quiver3(X,Y,Z,U,V,W,1);
hold, surf(X,Y,Z);
colormap copper
axis equal
```

Fig. 6.33   Examples of the 3-D
contour plots and surface with
normals.

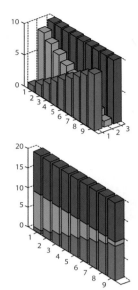

```
x=[1:9;9:-1:1]';
x(:,3)=x(:,1)+x(:,2);
subplot(1,2,1)
bar3(x,'detached')
subplot(1,2,2)
bar3(x,'stacked')
colormap(eye(3))
```

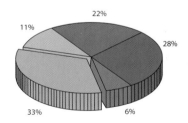

```
x=[1 3 0.5 2.5 2];
det=[0 1 0 0 0];
pie3(x,det), colormap cool
```

Fig. 6.34   Examples of the 3-D bars
and pie.

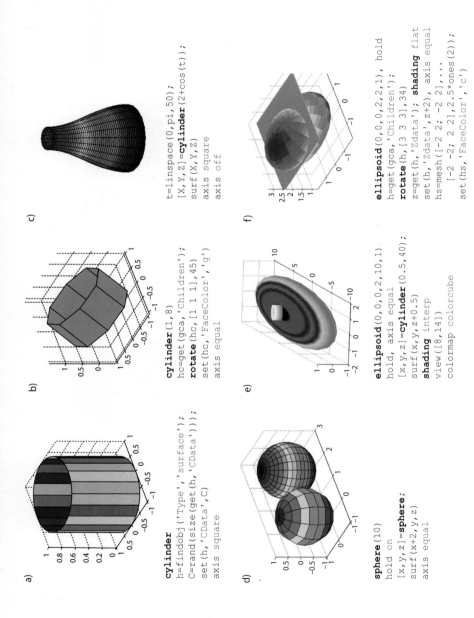

a)

```
cylinder
h=findobj('Type','surface');
C=rand(size(get(h,'CData')));
set(h,'CData',C)
axis square
```

b)

```
cylinder(1,8)
hc=get(gca,'Children');
rotate(hc,[1 1 1],45)
set(hc,'FaceColor','g')
axis equal
```

c)

```
t=linspace(0,pi,50);
[X,Y,Z]=cylinder(2+cos(t));
surf(X,Y,Z)
axis square
axis off
```

d)

```
sphere(10)
hold on
[x,y,z]=sphere;
surf(x+2,y,z)
axis equal
```

e)

```
ellipsoid(0,0,0,2,10,1)
hold, axis equal
[x,y,z]=cylinder(0.5,40);
surf(x,y,z+0.5)
shading interp
view([8,14])
colormap colorcube
```

f)

```
ellipsoid(0,0,0,2,2,1), hold
h=get(gca,'Children');
rotate(h,[3 3 3],34)
z=get(h,'Zdata'); shading flat
set(h,'Zdata',z+2), axis equal
hs=mesh([-2 2; -2 2],....
    [-2 -2; 2 2],2.5*ones(2));
set(hs,'FaceColor','c')
```

Fig. 6.35   Examples of visualizing 3-D shapes.

a)

```
x = [0 1; 1 0; 0 1];
y = [1 0; 1 1; 0 1];
z = [1 1; 1 1; 0 1];
c = [ 1;    0;    1];
fill3(x,y,z,c)
xlabel('x'), ylabel('y')
```

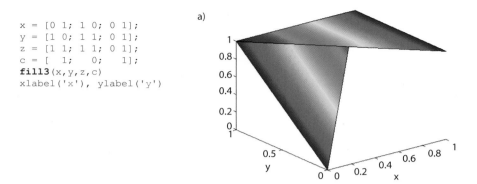

b)

```
t=linspace(0,2*pi,11);
x=cos(t); y=sin(t);
x(2:2:10)=.4*cos(t(2:2:10));
y(2:2:10)=.4*sin(t(2:2:10));
patch(x,y,'b')
axis square
```

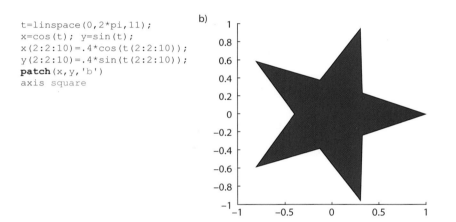

c)

```
x=[0 0 0 0;0 1 1 0;0 1 1 0]';
y=[0 1 1 0;0 0 0 0;1 1 1 1]';
z=[0 0 1 1;0 0 1 1;0 0 1 1]';
patch(x,y,z,-z)
axis square, view([35,35])
zlabel('z'), ylabel('y')
```

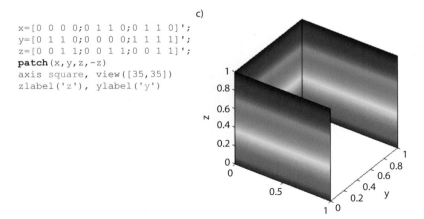

Fig. 6.37 Examples of 3-D polygons filling and patch graphics objects.

```
[x,y,z]=sphere;
X=[x(:)*0.5 x(:)*0.75 x(:)];
Y=[y(:)*0.5 y(:)*0.75 y(:)];
Z=[z(:)*0.5 z(:)*0.75 z(:)];
S=repmat([1 .75 .5]*10,prod(size(x)),1);
C=repmat([1 2 3],prod(size(x)),1);
scatter3(X(:),Y(:),Z(:),S(:),C(:),'filled')
view(-60,60)
```

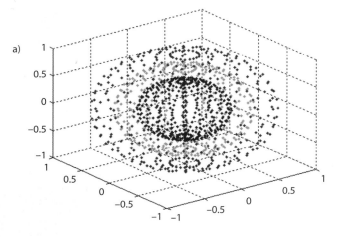

```
[x,y,z]=meshgrid(-2:.2:2,-2:.25:2,-2:.16:2);
v=x.*exp(-x.^2-y.^2-z.^2);
xsl=[-1.2,.8,2]; zsl=-1;
slice(x,y,z,v,xsl,0,zsl)
colormap hot
```

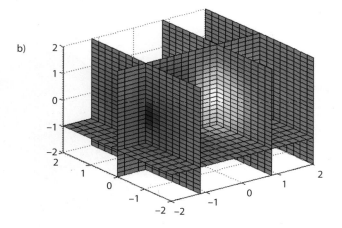

Fig. 6.39 Examples of 3-D scatter plot and volume visualization.

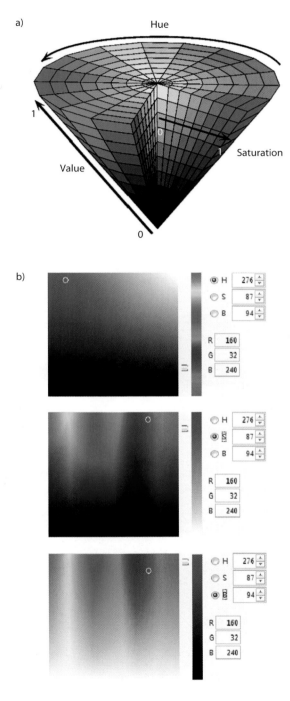

Fig. 6.40 HSV color model.

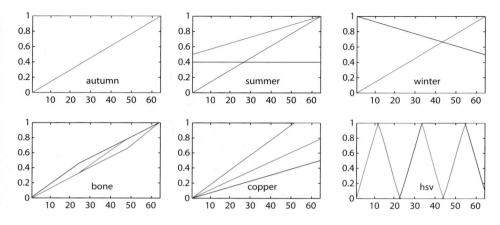

Fig. 6.41 Examples of preset color maps.

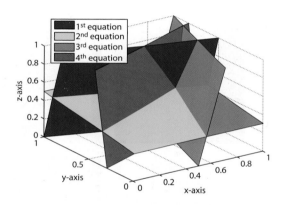

Fig. 9.10b Special case of linearly dependent equations in 3-D.

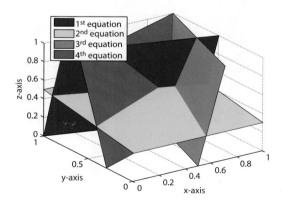

Fig. 9.11b General case of linearly independent equations in 3-D.

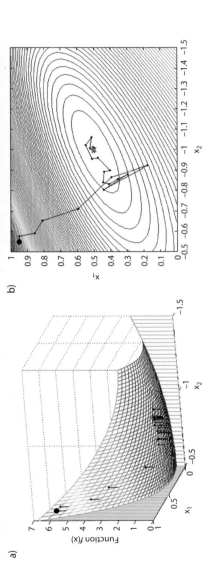

Fig. 10.32 Example of `fminsearch` performance.

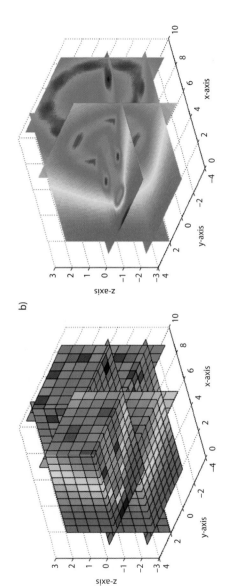

Fig. 11.14 Example of 3-D interpolation converting a) coarse mesh data to b) the finer mesh data.

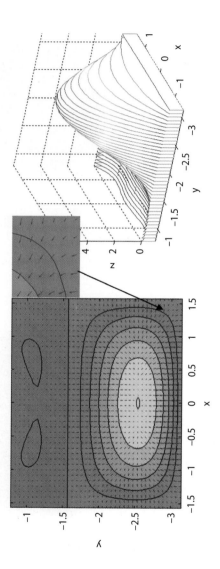

Fig. 12.5 Illustration of computing and showing the 2-D gradient.

```
figure('Color','w')
z=linspace(pi/2,2*pi);
[X,Y,Z]=cylinder(2+cos(z));
surf(X,Y,5*Z,X)
axis equal
colormap hsv
shading interp
xlabel('x')
ylabel('y')
zlabel('z')
```

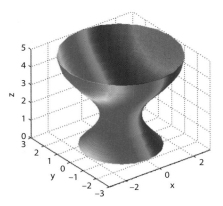

Fig. 13.11 Body of revolution about z-axis with corresponding commands.

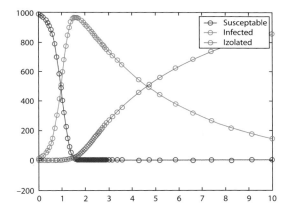

Fig. 14.18a Modeling a disease outbreak.

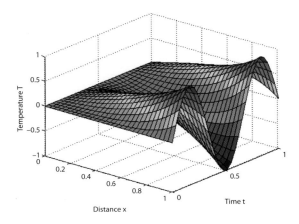

Fig. 14.20a Employing the pdepe function.

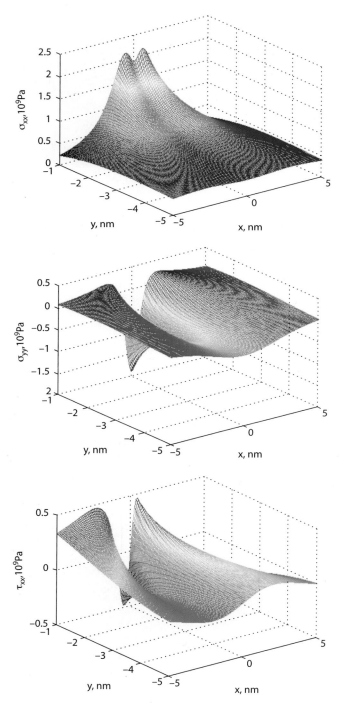

Problem 6.2

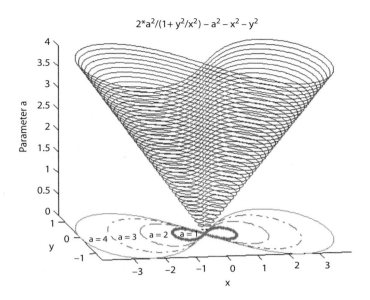

$$2*a^2/(1+y^2/x^2) - a^2 - x^2 - y^2$$

Problem 7.2

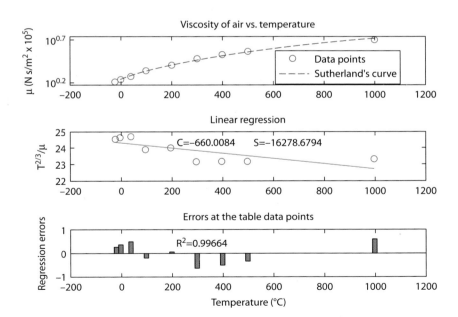

Problem 11.1

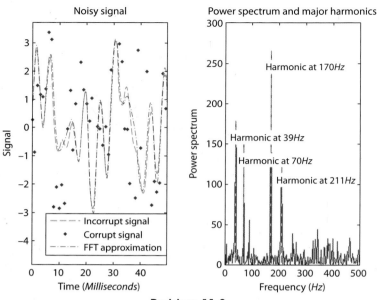

Problem 11.3

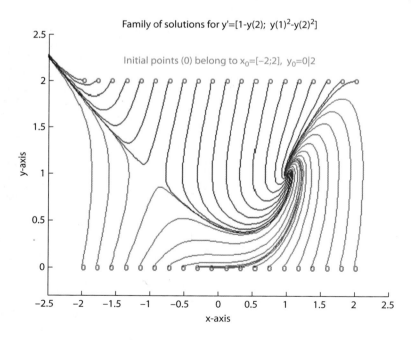

Problem 14.3